Quantum Principles & Particles

Textbook Series in Physical Sciences

This textbook series offers pedagogical resources for the physical sciences. It publishes high-quality, high-impact texts to improve understanding of fundamental and cutting edge topics, as well as to facilitate instruction. The authors are encouraged to incorporate numerous problems and worked examples, as well as making available solutions manuals for undergraduate and graduate level course adoptions. The format makes these texts useful as professional self-study and refresher guides as well. Subject areas covered in this series include condensed matter physics, quantum sciences, atomic, molecular, and plasma physics, energy science, nanoscience, spectroscopy, mathematical physics, geophysics, environmental physics, and so on, in terms of both theory and experiment.

Understanding Nanomaterials
Malkiat S. Johal

Concise Optics
Concepts, Examples, and Problems
Ajawad I. Haija, M. Z. Numan, W. Larry Freeman

A Mathematica Primer for Physicists
Jim Napolitano

Understanding Nanomaterials, Second Edition
Malkiat S. Johal, Lewis E. Johnson

Physics for Technology, Second Edition
With Applications in Industrial Control Electronics
Daniel H. Nichols

Time-Resolved Spectroscopy
An Experimental Perspective
Thomas Weinacht; Brett J. Pearson

No-Frills Physics
A Concise Study Guide for Algebra-Based Physics
Matthew D. McCluskey

Quantum Principles and Particles, Second Edition
Walter Wilcox

For more information about this series, please visit:
[www.crcpress.com/Textbook-Series-in-Physical-Sciences/book-series/TPHYSCI]

Quantum Principles & Particles

Walter M. Wilcox

CRC Press is an imprint of the
Taylor & Francis Group, an **informa** business

CRC Press
Taylor & Francis Group
6000 Broken Sound Parkway NW, Suite 300
Boca Raton, FL 33487-2742

Printed in the United States of America on acid-free paper

International Standard Book Number-13: 978-1-138-09041-5 (Hardback)
International Standard Book Number-13: 978-1-138-09037-8 (Paperback)

Visit the Taylor & Francis Web site at
http://www.taylorandfrancis.com

and the CRC Press Web site at
http://www.crcpress.com

For Diana and Christopher

Contents

Preface to the Second Edition

Quantum Principles and Particles has been thoroughly revised for the purposes of completeness, correctness, and clarity. I have added a much-needed chapter on time-dependent processes in quantum mechanics in Section II as well as an appendix on the exciting subject of quantum computing (Appendix G). The instructor should now have ample material to design a complete two-semester undergraduate quantum course from these contents. The majority of the figures in the first edition were done by myself using a simple drawing window on an old Mac computer. The new figures, prepared by Judy Waller, result in a more pleasing and polished book. I have also added updates on new developments in particle physics in Chapter 12, especially reflecting the exciting discovery of the Higgs boson at the Large Hadron Collider (LHC). The new edition also incorporates typo and text corrections as well as new problems. I hope that I have continued to balance simplification and rigor, and that intrepid physics students will find much that will encourage them within these pages.

–Walter Wilcox
Waco, Texas

Preface to the First Edition

A physical understanding is a completely unmathematical, imprecise, and inexact thing, but absolutely necessary for a physicist.

R. Feynman

Quantum Principles and Particles (QPP) is an introductory, undergraduate, two-semester quantum mechanics textbook that emphasizes basic quantum principles deduced from fundamental experiments. QPP introduces the primary principles of the microscopic world via an analysis of the simplest possible quantum mechanical system, spin 1/2, using hypothetical quantum devices I call "process diagrams" (PDs). PDs can be thought of as spin filter/modulation/transition devices that help students visualize the states and operators, and which make transparent the means of computing the amplitudes for quantum mechanical processes. QPP takes the student from the beginning small steps of quantum mechanics, with confident steady strides through the principles and applications, to a final victorious leap to overview the particles and forces of the standard model of particle physics.

The emphasis on particle physics in this text makes it attractive to those students who are on the fast track to high-energy particle physics, superstrings, and cosmology. It also interfaces well with an introductory undergraduate course on nuclear and particle physics. On the other hand, for those programs that do not have such a separate course, it could be used to at least provide a glimpse of the subject matter. It also can be used as a single-semester course on quantum mechanics, since almost all the essentials are contained in the first six chapters. Helpful mathematical background for such an introductory course would include aspects of linear algebra, complex numbers, and partial differential equations. The physics background should include some statistical and classical mechanics and electromagnetism, but not necessarily in a comprehensive sense.

The spin analysis approach in QPP is a teaching viewpoint pioneered by Julian Schwinger. (Feynman also used a similar "state vector" approach in Vol. III of *The Feynman Lectures on Physics*.) The first six chapters present the essential principles in the development of quantum mechanics, starting with the PD spin state analysis in Chapter 1. After drilling down to the basic theory, QPP switches to an experimentally oriented wave mechanics approach to build up the quantum description from experiment in Chapter 2. The Schrödinger equation is deduced and reconfirmed by connecting to the discrete spin lessons from Chapter 1. The formalism is developed to the point that the solution of some simple one-dimensional problems, and their lessons, can be considered in Chapter 3. Chapter 4 is devoted to formulating additional mathematical background, generalizing the ideas introduced in the first chapter. Chapter 5 shows how to develop approximate perturbative, semiclassical, or variational solutions to the Schrödinger equation when exact solutions are not available. The three-dimensional generalization is carried out and studied in Chapter 6, and the radial Schrödinger equation is formulated. The emphasis after this point is on developing a consistent picture of particle descriptions and interactions, with a further emphasis on applications. Various potentials are invoked to connect to atomic, nuclear, and quark models in Chapter 7. The mathematics necessary for adding angular momenta is developed and demonstrated in the hydrogen atom in Chapter 8. The spin/statistics connection for particles is explained and many-body particle theory is discussed in Chapter 8, leading to the penultimate consideration of particle scattering in Chapter 10. Proton–proton scattering serves as a segue to Chapter 11, which connects to the standard model of particles and interactions. I like to think of the first six chapters being characterized as "quantum principles" and the last five as "quantum particles."* There are also six appendices touching on points that deserve special mention, but which I felt are not essential to the flow of ideas in the main chapters.†

The principles in the text are extended and often illustrated in the end-of-chapter problems, which are designed to build on the ideas in the chapter, test fundamental understanding, and stimulate original thought. The problems are essential teaching tools organically connected to and flowing from the text. They sometimes extend a treatment or mathematical aspect, and are often connected to several sections in the chapter; in such a case they are associated with the last section of the text that would be necessary to completely answer the question. Often, the final answers to problems are disclosed at the end in an encouraging hint. The emphasis, as in the title, is on principles, so these problems are not as time consuming as in many texts. I unabashedly emphasize the quantum melody over the calculational mechanics. The more difficult and time-consuming problems are identified by an asterisk, and are often to be regarded as a helpful, but not essential, enhancement to the text.

The universe of quantum texts is already quite wide and expanding, and exists at the highest possible level of intellectual and publishing quality. In addition to historical and personal motivations discussed below, I have felt compelled to attempt to contribute to these endeavors for two major reasons:

- First, I really believe that spin interactions are the best possible way to introduce the ideas of quantum mechanics, and that a visually oriented quantum text designed along these lines could be very helpful. Spin 1/2 measurements form the simplest quantum mechanical system. The deduction of the quantum principles from idealized Stern–Gerlach experiments emphasizes the experimental nature of physics, and the spin 1/2 starting point shows the radical departure from classical thinking that is necessary in the quantum world. Even the word "spin" conveys a classical concept that is only applicable to the quantum version as an analogy. The student is jolted at the beginning, preparing her or him to think in a radically new way. The process diagrams are my way of presenting this material visually, which is the way I best understand things and is my attempt at Feynman's "physical understanding." I hope it will help others also.
- Second, my emphasis on physics principles is not just a titular device for this book, but expresses my belief that one need not and should not confront the beginning student with complicated harmonic variations until the melody is appreciated. When I have endeavored to balance simplification and rigor in the text or problems, I aim to error, if necessary, on the side of simplification. In this respect, as expressed above, I believe the problems should have a simple point to make, and are not enhanced by adding unnecessary complexity, length, or complications. They should be doable and encouraging.

On a historical note, this book evolved from my desire to develop high-quality lecture notes for my various classes into a convenient form for the students. The point was to save, improve, and refine the raw ore of teaching, and, eventually, to provide a web-published coherent accumulation point for further additions, revisions, and improvements. In addition, I was and am still

* Yes, there are some principles in the particles and vice versa. Chapter 9 on spin and statistics is such a culprit, but this placement allows me to discuss some more advanced topics than would otherwise be possible, as well as to connect smoothly to the identical particle scattering considerations of Chapter 10.

† Alas, this text, like all others, cannot possibly be comprehensive. There are many important subjects left out, including spin density formalism, time-dependent perturbation theory, adiabatic considerations, quantum computing, and many more. However, when these things are glancingly touched upon, helpful references are mentioned.

motivated to make the web-published texts available as an economic alternative for the many budding physics students throughout the world who cannot afford to buy polished end-product texts. Thus the Baylor Physics Department's Open Text Project Web site (http://homepages.baylor.edu/open_text) has arisen and developed since 2003. Unfortunately, the magnitude of the work involved and the pressure of other projects never allowed a polishing of the teaching ore, although accumulation has never stopped. This, thankfully, has been rectified for at least one of these accumulation points, and hopefully others will follow. Concomitant and consistent with these efforts, a limited special offer has been negotiated with the publisher for sale at a substantial discount to students or programs with demonstrable need, as long as the supply lasts. I would like to acknowledge Taylor & Francis for this significant contribution.

The most casual reader will find *QPP* peppered with references to J.J. Sakurai and J.S. Schwinger. Besides their numerous and original physics contributions, both Sakurai and Schwinger were magnificent and inspirational teachers. I was a beginning graduate student at UCLA in Sakurai's introductory quantum classes of 1976/1977, and was Schwinger's PhD student from my second year on. I was also the teaching assistant for Schwinger's three-quarter undergraduate quantum mechanics class beginning in January 1979. (This class differed markedly from the mid-1980s quantum classes he taught later.) My appreciation of their pedagogical efforts has grown with the years, as I have confronted the challenges of explaining physics anew. Both have inspired me, but I have attempted a different synthesis here based upon a visual introduction, an emphasis on basic principles, and a commitment to including particle physics. Since the inspirations are many, I have endeavored to be as accurate as possible with respect to these authors and others in my attributions throughout the text. I am also very appreciative of an insightful Howard Georgi *Physics Today* article on weak quark flavor mixing, which I adapted and turned into my Appendix E.

In the important matter of units, in equations where electromagnetic quantities are required, I use the Gaussian system. These are more convenient than SI units for atomic and subatomic applications, and their use interfaces well with more advanced texts such as Sakurai's or Merzbacher's. For comparison of formulas with other texts and unit conversions to SI, one can consult J.D. Jackson's *Classical Electrodynamics*, 3rd ed., Appendix, Section 4, and Tables 3 and 4.

This book would not have come about without the strong encouragement, understanding, allowances, and support of the Baylor University administration, as well as my fellow faculty friends and colleagues throughout the university. A Baylor summer sabbatical was also key. My Open Text Project site lists some helpful contributing typists and formatters, for the work has been long. I am also indebted to the understanding and loving support I have received from my wife, Diana, as this project has drawn to a close.

What has evolved through intension and selection is a text that incorporates the best explanations or examples of the various subject matters I have encountered or been fortunate enough to think up myself, presented in as simple a manner as possible. I hope you enjoy reading this as much as I have enjoyed writing it!

—Walter Wilcox
Waco, Texas

Quantum Principles

Perspective and Principles*

Synopsis: The failure of classical mechanics to describe the microscopic world is made clear and a possible electron model is shown to be unacceptable. Spin 1/2 measurements are analyzed and new quantities called wave functions are identified. Measurement symbols (or operators) are introduced and hypothetical quantum devices called Process Diagrams are used to orient and guide the reader through the emerging formalism. Spin operators are defined, the bra–ket formalism is set up, and the wave functions are shown to correspond to quantum transition amplitudes. The formalism is demonstrated to be equivalent to matrix algebra. Expectation values, corresponding to appropriately averaged experimental outcomes, are defined within the formalism.

1.1 PRELUDE TO QUANTUM MECHANICS

In order to set the stage, let us begin our study of the microscopic laws of nature with some experimental indications that the laws of classical mechanics, as applied to such systems, are inadequate. Although it seems paradoxical, we turn first to a consideration of matter in bulk to learn about microscopic behavior. Let us consider the simplest type of bulk matter: gases. Indeed, let us consider first the simplest sort of gas: monotonic. It was known in the nineteenth century that for these gases one had the relationship:

$$\text{Internal energy} \propto \text{Absolute temperature} \times \text{Number of molecules}$$

This can be systematized as

$$\bar{E} = \left(\frac{1}{2}kT\right) \times N_{\mathrm{a}} \times 3, \tag{1.1}$$

where

$\bar{E}$ = average internal energy (per mole),
$k = 1.38 \times 10^{-16}\,\mathrm{erg}/(\mathrm{deg} \cdot \mathrm{Kelvin})$ (Boltzmann's constant),

* This chapter is based on Schwinger's introductory quantum mechanics notes for his class at UCLA in 1979, but contains additions, deletions, and diagrammatic interpretations. For those who would like to follow Schwinger's original line of argument, please see my original handwritten notes, posted on the Baylor University Open Text Project website.

and

$N_a = 6.022 \times 10^{23}$ (Avogadro's number).

(Note that a "mole" is simply the number of molecules in *amu* (atomic mass units) grams of the material. There are always $N_a = 6.022 \times 10^{23}$ molecules in a mole. One can simply view N_a as just a conversion factor from *amus* to grams.) The factor of 3 comes from the classical *equipartition theorem*. This law basically says that the average value of each independent quadratic term in the energy of a gas molecule is 1/2 kT. This comes from using *Maxwell–Boltzmann statistics* for a system in thermal equilibrium. Let us use Maxwell–Boltzmann statistics to calculate the average value of a single independent quadratic energy term:

$$\bar{E}_i = \frac{\int_{-\infty}^{\infty} e^{-\beta E(q_1, q_2, \ldots, p_1, p_2, \ldots)} E_i \, dq_1 dq_2 \ldots dp_1 dp_2 \ldots}{\int_{-\infty}^{\infty} e^{-\beta E(q_1, q_2, \ldots, p_1, p_2, \ldots)} dq_1 dq_2 \ldots dp_1 dp_2 \ldots}, \tag{1.2}$$

where $\beta = 1/kT$. The factor $e^{-\beta E} dq_1 \ldots dp_1 \ldots$ is proportional to the probability that the system has an energy E with position coordinates taking on values between q_1 and $q_1 + dq_1$, q_2 and $q_2 + dq_2$, and similarly for the momentum coordinates. Just like all probabilistic considerations, our probabilities need to add to one; the denominator factor in Equation 1.2 ensures this. Notice that because of the large number of particles involved, Maxwell–Boltzmann statistics does not attempt to predict the motions of individual gas particles, but simply assigns a *probability* for a certain configuration to exist. Notice also that the Boltzmann factor, $e^{-\beta E}$, discourages exponentially the probability that the system is in an $E \gg kT$ state. Now let us say that

$$E_i = ap_i^2 \quad \text{or} \quad bq_i^2, \tag{1.3}$$

where a and b are just constants, representing a typical kinetic or potential energy term in the total internal energy, E. Then we have

$$\bar{E}_i = \frac{\int_{-\infty}^{\infty} e^{-\beta a p_i^2} a p_i^2 \, dp_i}{\int_{-\infty}^{\infty} e^{-\beta a p_i^2} a p_i^2} = \frac{-a \frac{\partial}{\partial \beta'} \left(\int_{-\infty}^{\infty} e^{-\beta' p_i^2} dp_i \right)}{\int_{-\infty}^{\infty} e^{-\beta' p_i^2} dp_i}, \tag{1.4}$$

where $\beta' = \beta a$. Introduce the dimensionless variable $x = (\beta')^{1/2} p_i$. Then

$$\int_{-\infty}^{\infty} dp_i \, e^{-\beta' p_i^2} = (\beta')^{-1/2} \int_{-\infty}^{\infty} dx \, e^{-x^2}, \tag{1.5}$$

and therefore

$$\bar{E}_i = -\frac{a}{(\beta')^{-1/2}} \frac{\partial (\beta')^{-1/2}}{\partial \beta'} = \frac{a}{2\beta'} = \frac{1}{2\beta} = \frac{1}{2} kT. \tag{1.6}$$

If we accept the validity of the equipartition theorem, we have that

$\bar{E} = \bar{E}_i \times$ (Total number of quadratic terms in the energy of a mole of gas), so that (remember that $\vec{p}^2 = p_x^2 + p_y^2 + p_z^2$)

$$\bar{E} = \frac{3}{2} kN_a T. \tag{1.7}$$

Let us define the "molar specific heat at constant volume" (also called "molar heat capacity at constant volume"), c_V:

$$c_V \equiv \left(\frac{\partial \bar{E}}{\partial T} \right)_V \tag{1.8}$$

(The subscript V reminds us to keep the variable representing volume of a constant during this differentiation.) In our case, for a simple monotonic gas, we get

$$c_V = \frac{3}{2} kN_a \equiv \frac{3}{2} R \left(= 12.5 \frac{\text{joules}}{\text{mole} \cdot \text{deg}} \right). \tag{1.9}$$

How does this simple result stack up against experiments carried out at room temperature?

Monotonic gas	**c_V (experiment)**
He	12.5
Ar	12.5

A success!

Well, what about diatomic molecules? To get our theoretical prediction, based on the equipartition theorem, all we need to do is just count the quadratic terms in the energy for a single molecule. If we say that the energy of such a molecule is a function of only the relative coordinate, r, separating the two atoms, then from classical mechanics we have,*

$$E = \frac{\vec{P}^2}{2M} + \frac{\vec{L}^2}{2\mu r^2} + \frac{{p_r}^2}{2\mu} + U(r), \tag{1.10}$$

$$\begin{array}{l} \Rightarrow \text{Quadratic degs. of freedom} = 3 \ + \ 2 \ + \ 2, \\ \qquad\qquad\qquad\qquad\qquad\quad \downarrow \quad \downarrow \quad \downarrow \\ \qquad\qquad\qquad\qquad \text{Translation Rotation Radial vibration} \end{array} \tag{1.11}$$

(μ = reduced mass and $\vec{L}$ = center of mass angular momentum) if we say that $U(r) \sim r^2$. Thus, for a diatomic molecule we would expect

$$\bar{E} = \left(\frac{1}{2} kT \right) N_a \times (3+2+2) = \frac{7}{2} kN_a T. \tag{1.12}$$

$$\Rightarrow c_V = \frac{7}{2} R \left(= 29.1 \frac{\text{joules}}{\text{mole} \cdot \text{deg}} \right) \tag{1.13}$$

* If the diatomic atoms are not considered *point* particles, one of these degrees of freedom would increase by one. Can you understand which one and why?

How does this result stack up against experiments at room temperature?

Diatomic gas	c_V (experiment)
N_2	20.6
O_2	21.1

Something is wrong. We seem to be "missing" some degrees of freedom. Notice that

$$\frac{5}{2}R = 20.8\frac{\text{joule}}{\text{mole}\cdot\text{deg}}$$

seems to be a better approximation to the experimental situation than does our (7/2)R prediction. Later considerations have shown that the vibrational degrees of freedom are the "missing" ones. Historically, this was the first experimental indication of a failure in classical physics applied to atoms and was already known in the 1870s.

Another application of these ideas is to solids. Let us treat the atoms of a solid as point masses "locked in place" to a first approximation. Then we have for the energy, E, of a single atom

$$E = \frac{\vec{p}^2}{2M} + ax^2 + by^2 + cz^2, \tag{1.14}$$

where x, y, and z measure the displacement of the ideal atom from its equilibrium position. There are now six quadratic degrees of freedom, which means that

$$\bar{E} = 3kN_aT, \tag{1.15}$$

$$\Rightarrow c_V = 3R\left(=25\frac{\text{joule}}{\text{mole}\cdot\text{deg}}\right). \tag{1.16}$$

This law, known before the preceding theoretical explanation, is called the law of Dulong–Petit. What happens in experiments, again at room temperature?

Solid	$c_p \approx c_V$ (experiment)
Copper	24.5 (23.3)
Silver	25.5
Carbon (diamond)	6.1

(These data have been taken from F. Reif, *Fundamentals of Statistical and Thermal Physics* [Waveland Pr. Inc., 2008]. For solids and liquids we have $c_V \approx c_p$, c_p being the molar specific heat at constant pressure, which is easier to measure than c_V. An estimate of the value of c_V for Copper from Reif is given in parentheses.) Although copper and silver seem to obey the Dulong–Petit rule, diamond obviously does not. What is even harder to understand is that, for example, the c_V for diamond is strongly temperature dependent at room temperature. This is not accounted for by the classical physics behind the Dulong–Petit prediction of the universal value, 3R.

Although copper and silver look rather satisfactory from the point of view of the aforementioned law, there is still a paradox associated with them according to classical mechanics. If N_a atoms each give up m valence electrons to conduct electricity, and if the electrons are freely mobile, the heat capacity of a conductor should be

$$c_V = \underset{\text{"Atomic" piece}}{\underset{\downarrow}{3kN_a}} + \underset{\text{"Electronic" piece}}{\underset{\downarrow}{\frac{3}{2}mkN_a}} \tag{1.17}$$

Thus, in these materials the electronic component of specific heat seems not to be present or is greatly suppressed. Classical mechanics is silent as to the cause.

Another place where experimental results have pointed to a breakdown in the application of classical mechanics to atomic systems was in a classic experiment done by H. Geiger (of counter fame) and E. Marsden in the 1909–1911 time frame. They scattered α particles (helium nuclei) off gold foil and found that a larger number of α particles were backscattered by the atoms from the foil than could be accounted for by then-popular atomic models. This led Rutherford to hypothesize that most of the mass of the atom is in a central core or "nucleus." Electrons were supposed to orbit the nucleus like planets around the sun in order to give atoms their known physical sizes. For example, the hydrogen atom was supposed to have a single electron in orbit around a positively charged nucleus. Although Rutherford's conclusions came via classical reasoning (it turns out that the classical scattering cross section derived by Rutherford is essentially unmodified by the new mechanics we will study here), he could not account for the *stability* of his proposed model by classical arguments since his orbiting electrons would quickly radiate away their energy produced by their accelerated motion.

All of these experimental shortcomings—the "missing" vibrational degrees of freedom in diatomic molecules, the failure of the law of Dulong and Petit for certain solids, the missing or suppressed electronic component of c_V, and the instability of Rutherford's atomic model—pointed to a breakdown in classical mechanics. Thus, the time was ripe for a new, more general mechanics to arise.

1.2 STERN–GERLACH EXPERIMENT

We will begin our study of *quantum* mechanics with another experimental finding, which was at variance with classical ideas.

Consider the following simple, static, neutral charge distribution in an external electric field:

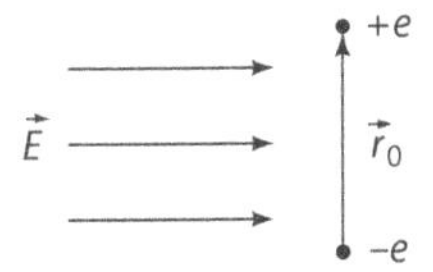

$\vec{E}$ → → → $+e$ ↑ $\vec{r}_0$ $-e$

Clearly, this system is more stable in the orientation

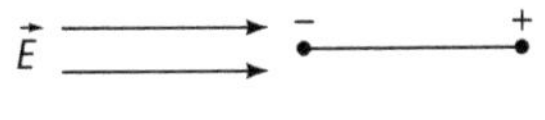

$\vec{E}$ → → $-$ ——— $+$

than

This system is called an electric dipole, and there is an energy associated with its orientation. We know that

$$\text{Energy} = \text{Charge} \times \text{Potential}$$

so that (e is a positive charge)

$$U = e\phi(+) - e\phi(-) \tag{1.18}$$

Now we may expand

$$\phi(+) \cong \phi + \frac{\vec{r}_0}{2} \cdot \vec{\nabla}\phi, \tag{1.19}$$

$$\phi(-) \cong \phi - \frac{\vec{r}_0}{2} \cdot \vec{\nabla}\phi, \tag{1.20}$$

where ϕ represents the potential of the external field at the midpoint of the dipole. Then

$$\phi(+) - \phi(-) = \vec{r}_0 \cdot \vec{\nabla}\phi. \tag{1.21}$$

But by the definition of the electric field

$$\vec{E} = -\vec{\nabla}\phi, \tag{1.22}$$

so that

$$U = -e\vec{r}_0 \cdot \vec{E}. \tag{1.23}$$

Define $\vec{d} \equiv e\vec{r}_0$, the "electric dipole moment." Then Equation 1.23 becomes

$$U = -\vec{d} \cdot \vec{E}. \tag{1.24}$$

Equation 1.24 is consistent with the picture that $\vec{r}_0$ prefers to point along $\vec{E}$ since this minimizes the potential energy.

We also know that

$$\text{Force} = \text{Charge} \times \text{Electric field},$$

so

$$\vec{F} = e\vec{E}(+) - e\vec{E}(-) = e(\vec{r}_0 \cdot \vec{\nabla})\vec{E} = (\vec{d} \cdot \vec{\nabla})\vec{E}. \tag{1.25}$$

Since $\vec{E} = -\vec{\nabla}\phi$, then we may also write this as

$$\vec{F} = -(\vec{d} \cdot \vec{\nabla})\vec{\nabla}\phi = -\vec{\nabla}(\vec{d} \cdot \vec{\nabla}\phi) = \vec{\nabla}(\vec{d} \cdot \vec{E}). \tag{1.26}$$

This makes sense since we expect that $\vec{F} = -\vec{\nabla}U$ and $U = -\vec{d} \cdot \vec{E}$. Notice that if $\vec{E}$ is uniform, there is no net force on the system.

There is also a torque on the system since

$$\text{Torque} = \text{Lever arm} \times \text{Force}.$$

Therefore,

$$\vec{\tau} = \frac{\vec{r}_0}{2} \times \left(e\vec{E}(+)\right) + \left(-\frac{\vec{r}_0}{2}\right) \times \left(-e\vec{E}(-)\right), \tag{1.27}$$

so,

$$\vec{\tau} = \vec{d} \times \vec{E}, \tag{1.28}$$

where $\vec{E}$ is the value of the electric field at the center of the dipole.

In the following, we will really be interested in magnetic properties of individual particles. Rather than deriving similar formulas in the magnetic case (which is trickier), we will simply depend on an electromagnetic analogy to get the formulas we need. The analogy,

Electric	**Magnetic**
$\vec{E}$	$\vec{H}$
$\vec{d}$	$\vec{\mu}$

where $\vec{\mu}$ is the "magnetic moment" then leads to ($\vec{H} = \vec{B}$ for us)

$$U = -\vec{\mu} \cdot \vec{H}, \tag{1.29}$$

$$\vec{F} = (\vec{\mu} \cdot \vec{\nabla})\vec{H} = \vec{\nabla}(\vec{\mu} \cdot \vec{H}), \text{ and} \tag{1.30}$$

$$\vec{\tau} = \vec{\mu} \times \vec{H}. \tag{1.31}$$

These formulas will help us understand the behavior of magnetic dipoles subjected to external magnetic fields. Remember, in order to produce a force on a magnetic dipole, we must first construct an *inhomogeneous* magnetic field. Consider, therefore, the schematic experimental arrangement shown in Figure 1.1.

Looking face-on to the magnets, we would see something like Figure 1.2. The magnetic field lines near the pole faces are highly nonuniform. The field looks something like Figure 1.3.

If we take a z-axis centered on the beam and direct it upward as in Figure 1.3, a nonuniform magnetic field with

$$\frac{\partial H_z}{\partial z} < 0$$

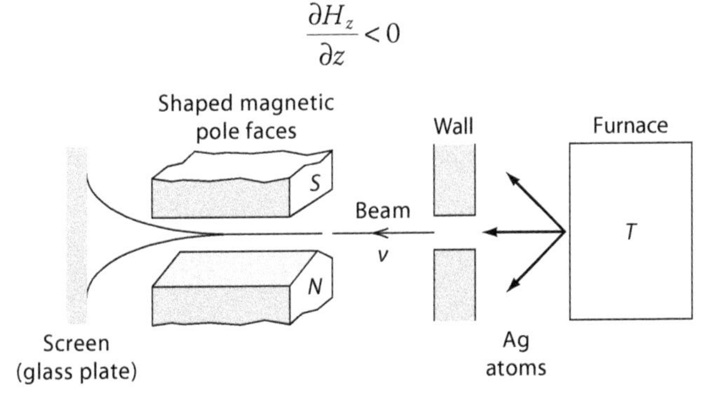

FIGURE 1.1 Schematic representation of the experimental arrangement used by O. Stern and W. Gerlach in 1922.

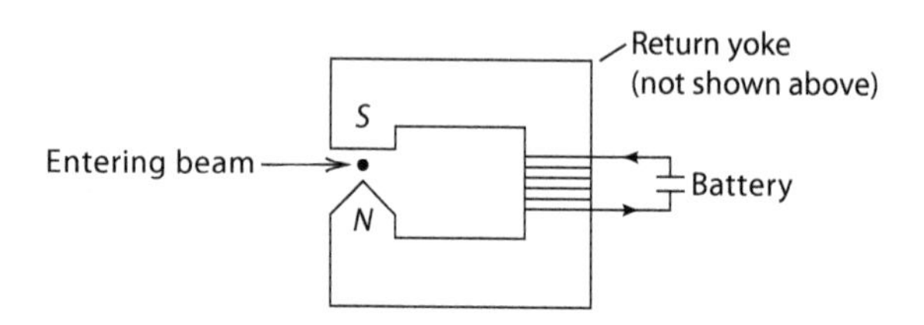

FIGURE 1.2 Schematic face-on view of the electromagnet in Figure 1.1.

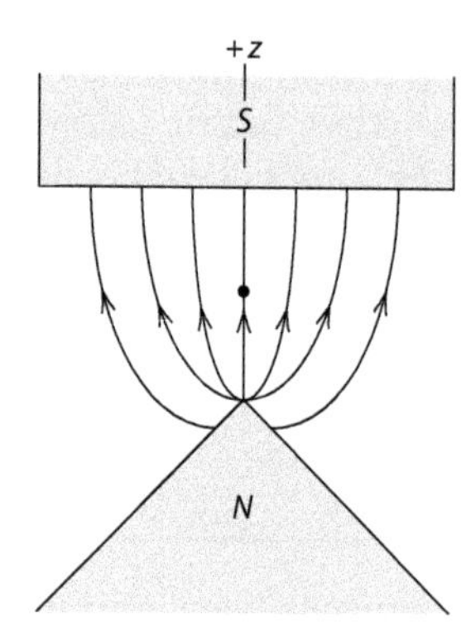

FIGURE 1.3 Close-up view of field lines of the magnet in Figure 1.2.

will be produced. This type of experimental setup was first used by Otto Stern and Walther Gerlach in an experiment on Ag (silver) atoms in 1922. The explanation for their experimental results had to wait until 1925 when Samuel Goudsmit and George Uhlenbeck, on the basis of some atomic spectrum considerations, deduced the physical property responsible.

From Equation 1.30, the force on an Ag atom at a single instant in time is approximately ($H_z \to H$)

$$F_z \approx \frac{\partial}{\partial z}\mu_z H = \mu_z \frac{\partial H}{\partial z}. \tag{1.32}$$

One can imagine measuring the force on a given atom by its deflection in the magnetic field:

$$\frac{\mathrm{d}p_z}{\mathrm{d}t} \approx \mu_z \frac{\partial H}{\partial z}. \tag{1.33}$$

Let us assume that the quantity $\partial H/\partial z$ is approximately a constant in time, fixed by the experimental apparatus. Then, we have a situation that looks something like Figure 1.4.

The change in the z-component of the momentum of an Ag atom is then

$$\Delta p_z \approx \mu_z \frac{\partial H}{\partial z} t. \tag{1.34}$$

But

$$t \approx \frac{L}{v}, \tag{1.35}$$

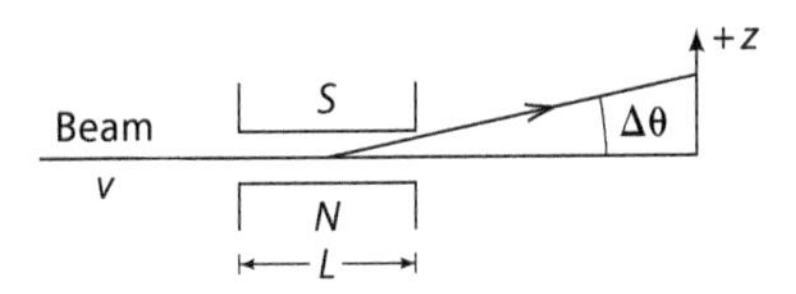

FIGURE 1.4 Schematic representation of the classical deflection angle in the Stern–Gerlach experiment.

where v is the velocity of the atoms, so that

$$\Delta p_z \approx \mu_z \frac{\partial H}{\partial z}\frac{L}{v}. \tag{1.36}$$

The small angular deflection caused by the magnetic field is then

$$\Delta\theta \approx \frac{|\Delta p_z|}{|\vec{p}|} \approx \left|\mu_z \frac{\partial H}{\partial z}\right| \frac{L}{mv^2}. \tag{1.37}$$

Let us get some numerical feeling for this situation. The particular values we will take in the following are:

$m = 1.79 \times 10^{-22}$ gm (Ag atom mass)

$T = 10^3$ K (furnace temperature)

$\left|\frac{\partial H}{\partial z}\right| = \frac{10^3 \text{gauss}}{10^{-1}\text{cm}} = 10^4 \frac{\text{gauss}}{\text{cm}}$ (field gradient)

$L = 10$ cm (magnet length)

$|\mu_z| \approx 10^{-20} \frac{\text{erg}}{\text{gauss}}$ (Ag z-component magnetic moment).

Using these values, we can estimate the angular deviation $\Delta\theta$ as follows. From the equipartition theorem, we expect the mean energy of an Ag atom leaving the furnace to be

$$\frac{1}{2}m\overline{v^2} = \frac{3}{2}kT, \tag{1.38}$$

which gives

$$m\overline{v^2} = 4.14 \times 10^{-13}\text{erg}.$$

Then from Equation 1.37

$$\Delta\theta \approx \frac{10^{-20} \cdot 10^4 \cdot 10}{4.14 \times 10^{-13}} = 2.4 \times 10^{-3} \text{ radians},$$

or about 0.14°. Naively, we would always expect to be able to back off far enough from the magnets to see this deflection.

Classically, what would we expect to see on the glass screen as a result of the beam of Ag atoms passing through the magnetic field? Since the atoms will emerge from the furnace with randomly oriented μ_zs, and since, from Equation 1.37, we expect the deflection of a given particle to be proportional to μ_z, the classical expectation was to see a single continuous line of atoms on the screen (see Problem 1.2.3). However, our idealized experiment will actually yield only two spots. In a real experiment, the "spots" would be smeared because of the spread in particle velocities from the furnace and the nonuniformity of the magnetic field. (We will discuss another source of smearing in Section 1.4.) Originally, this unexpected two-value-only result was referred to as "space quantization." However, this is misleading since the thing that is quantized here is certainly not space.

1.3 IDEALIZED STERN–GERLACH RESULTS

Now let us catalog some experimental results from other setups of Stern–Gerlach apparatuses. In the following, we will ignore the experimental details of this experiment and will be considering *idealized* Stern–Gerlach-like experiments.

(a) First, rotate the magnet, as in Figure 1.5. We would see that the beam is now split along the new z-axis.

FIGURE 1.5 A rotated magnet in idealized Stern–Gerlach experiment.

Let us now add a second magnet to the system at various orientations relative to the first. Let θ represent the angular orientation of magnet 1 with respect to magnet 2. For three specific orientations, Figure 1.6 (b), (c) and (d) shows the experimental results for the intensity of the outgoing beam.

In fact, for an arbitrary orientation θ, the intensity of the "up" orientation is $\cos^2(\theta/2)$ and that of the "down" is $\sin^2(\theta/2)$. Here "result" means whether the final beam emerges in an up or down orientation relative to the *second* magnet.

As mentioned earlier, one would measure the intensity of the outcoming beams to reach these conclusions. However, let us accept the fact that our description of what is occurring must be based on probabilities. Instead of the *intensity* of a *beam* of particles, let us talk about *intrinsic probabilities* associated with *individual, independent particles.*

Let us define the following:

$p(\overset{z}{\pm},\overset{\bar{z}}{\pm})$: Probability that a particle deflected in the $\pm\bar{z}$ direction from the first S–G gives a particle deflected in the $\pm z$ direction relative to the second S–G. (S–G represents Stern–Gerlach experiment.)

The axes are related like this:

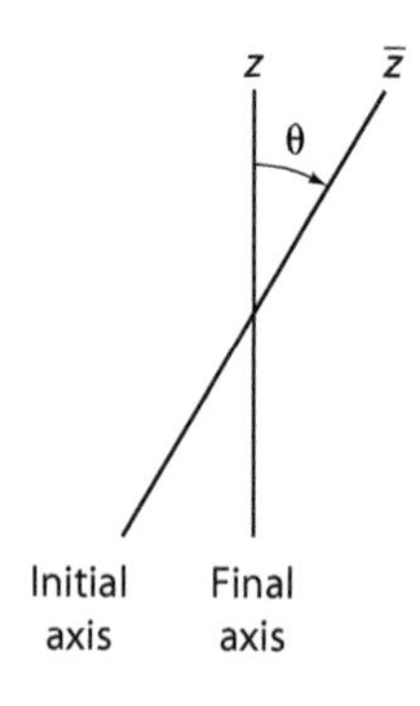

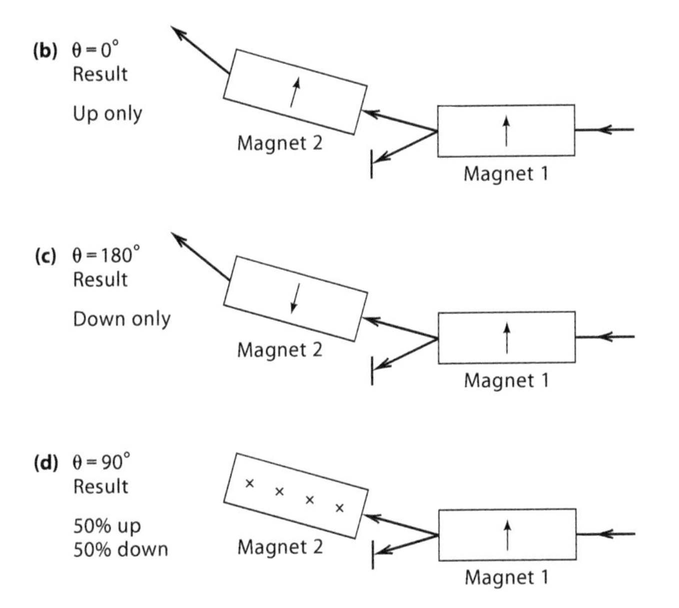

FIGURE 1.6 Three further experimental outcomes of Stern–Gerlach experiments using a second magnet.

There are four probabilities here:

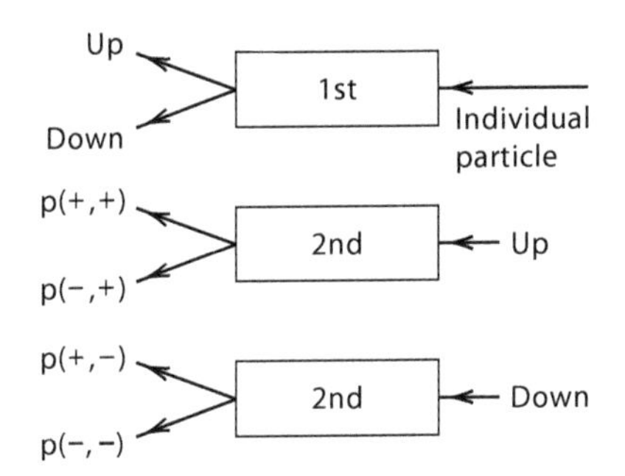

From the preceding illustrations we identify $p(+,+) = \cos^2(\theta/2)$ and $p(-,+) = \sin^2(\theta/2)$.

We must have our probabilities adding to one. Therefore, we must have

$$p(+,+) + p(+,-) = 1$$

$$\Rightarrow \cos^2\frac{\theta}{2} + p(+,-) = 1$$

$$\Rightarrow p(+,-) = \sin^2\frac{\theta}{2}.$$

Also

$$p(-,+) + p(-,-) = 1$$

$$\Rightarrow p(-,-) = \cos^2\frac{\theta}{2}.$$

More abstractly, we have ($a' = +\,-$, $a'' = +\,-$ independently)

$$\sum_{a'} \mathrm{p}(a'', a') = 1, \tag{1.39}$$

and

$$\sum_{a''} \mathrm{p}(a'', a') = 1. \tag{1.40}$$

Notice that

$$\mathrm{p}(a', a'') = \mathrm{p}(a'', a'), \tag{1.41}$$

and that using Equation 1.41, 1.40 follows from Equation 1.39 or vice versa. Thus Equation 1.41 may be viewed as a way of ensuring probability conservation. Therefore, only one probability is independent, p(+,+) say; the rest follow from Equations 1.39 and 1.41 (or Equations 1.40 and 1.41).

From Equation 1.32, we realize that the upward deflected beam is associated with $\mu_z < 0$, while the downward beam must have $\mu_z > 0$. We now ask the question: Given the selection of the up beam along the initial $\bar{z}$-axis ($\mu_{\bar{z}} < 0$), what is the mean value expected for μ_z measured along the final z-axis? The situation looks like Figure 1.7.

Classically, the answer to this question is given by just picking out the projection of $\vec{\mu}$ along the z-axis. Thus, the classical answer is $-\mu \cos\theta$. What do we get from our new, probability point of view? From this point of view, our mean value of μ_z is the weighted average of the two probabilities for finding an upward deflected beam from the second S–G ($\mu_z < 0$) and a downward beam from the second S–G ($\mu_z > 0$). Therefore, we have

$$\langle \mu_z \rangle_+ \equiv \text{average value of } \vec{\mu} \text{ along the } z \text{ given an initial selection of the upward deflected beam along } \bar{z}.$$

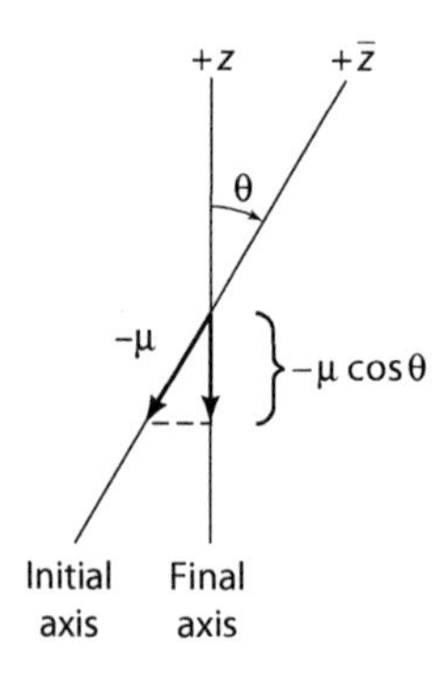

FIGURE 1.7 Relationship of rotated z-axes for a two-magnet Stern–Gerlach experiment and an associated magnetic moment, μ, directed along the $-\bar{z}$ axis.

$$\begin{aligned}\langle\mu_z\rangle_+ &= (-\mu)p(+,+) + (+\mu)p(-,+)\\ &= -\mu\left(\cos^2\frac{\theta}{2} - \sin^2\frac{\theta}{2}\right)\\ &= -\mu\cos\theta.\end{aligned}$$

We thus get the same result as expected classically, although the way we have reached our conclusion is not classical at all.

1.4 CLASSICAL MODEL ATTEMPTS

Let us try to build a classical model of the basic S–G experiment. Magnetic moments are classically produced by the motion of charged particles. (There are no magnetic *monopoles*, at least so far.) A reasonable connection is thus

$$\vec{\mu} = \gamma\vec{S}, \tag{1.42}$$

where $\vec{S}$ is a type of angular momentum associated with the Ag atom. The symbol γ is just a proportionality constant, usually called the "gyromagnetic ratio." What $\vec{S}$ represents is not clear yet. Equation 1.42 is patterned after a classical result. If one has a current loop in a plane,

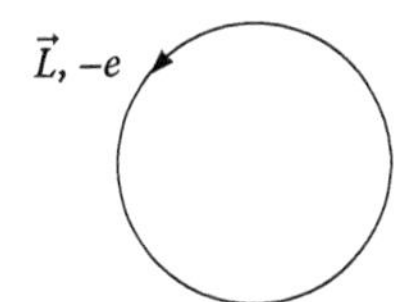

where the elementary charge carriers have a charge $-e$ like an electron ($e > 0$ here), the magnetic moment produced by these moving charges is

$$\vec{\mu} = -\frac{e}{2mc}\vec{L} \tag{1.43}$$

where m refers to the charge carrier's mass and $\vec{L}$ is the angular momentum.* If Equation 1.42 holds for the Ag atom, because the beam is seen to split into two discrete components, we can associate discrete values of S_z with the two spots observed. This behavior of Ag atoms in a magnetic field is due to its internal structure: one unpaired electron outside a closed shell of electrons (which possess no net magnetic dipole moment). Thus, the property of the Ag atoms we are studying is really due to a property of the electron. This property, called "spin," sounds very classical but is far from being a classically behaving angular momentum. Since the magnetic moment we are measuring in the S–G experiment really refers to a property of electrons, it is natural to expect that the gyromagnetic ratio in Equation 1.42 be not too different from the classical one in Equation 1.43, which also refers to the electron. In fact, the actual gyromagnetic ratio is approximately a factor of two larger than that in Equation 1.43:

$$\gamma \approx -\frac{e}{mc}. \tag{1.44}$$

* J.D. Jackson, *Classical Electrodynamics*, 3rd ed. (John Wiley & Sons, 1999), p. 187.

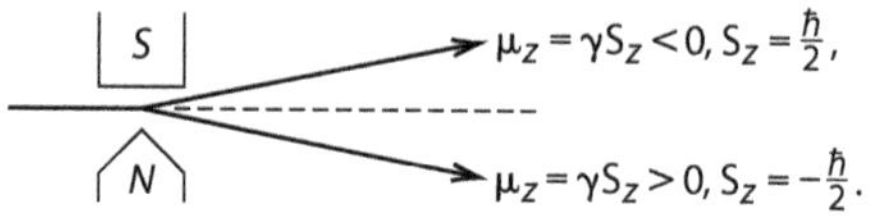

S

N

$\mu_z = \gamma S_z < 0,\ S_z = \frac{\hbar}{2},$

$\mu_z = \gamma S_z > 0,\ S_z = -\frac{\hbar}{2}.$

FIGURE 1.8 The electron spin values associated with up and down deflected beams.

Given this value of γ and the experimental determination of the deflection angle $\Delta\theta$ in Equation 1.37, one can in principle deduce the allowed values of the electron spin along the z-axis. The results are presented in Figure 1.8.

The quantity h ($\hbar \equiv h/(2\pi)$) is known as *Planck's constant*. The results of many experiments show that the z-component of the electron's spin is *quantized*, that is, limited to two discrete values, which are given by $\hbar/2$ and $-\hbar/2$.

Let us continue to develop our classical model. From Equation 1.31 we have $\vec{\tau} = \vec{\mu} \times \vec{H}$. From Equation 1.42 we have $\vec{\mu} = \gamma \vec{S}$, so

$$\vec{\tau} = \gamma \vec{S} \times \vec{H}. \tag{1.45}$$

Newton's laws relate $\vec{\tau}$ to the rate of change of angular momentum:

$$\vec{\tau} = \frac{d\vec{S}}{dt}. \tag{1.46}$$

Putting Equation 1.45 and Equation 1.46 together gives

$$\frac{d\vec{S}}{dt} = \gamma \vec{S} \times \vec{H}. \tag{1.47}$$

Let us take $\vec{H} = H\hat{e}_z$, where H is a constant. Then we have

$$\frac{dS_z}{dt} = 0; \quad \frac{dS_x}{dt} = \gamma H S_y; \quad \frac{dS_y}{dt} = -\gamma H S_x, \tag{1.48}$$

Then, for example,

$$\frac{d^2 S_x}{dt^2} = \gamma H \frac{dS_y}{dt} = -(\gamma H)^2 S_x. \tag{1.49}$$

This is a differential equation of the form $\ddot{x} + \omega^2 x = 0$, where the angular frequency is given by

$$\omega = |\gamma H|. \tag{1.50}$$

The picture that emerges is that of a precessing $\vec{S}$ vector, as shown in Figure 1.9.

Notice that neither $|\vec{S}|$ nor S_z changes in time. Since the time to pass through the magnet poles is given in Equation 1.35 as $t \approx L/v$, the *total* precession angle for an Ag atom is

$$\phi = \omega t \approx \gamma H \frac{L}{v}. \tag{1.51}$$

Again, let us get some feeling for order of magnitude here. Using our previous result for v (below Equation 1.38), we get

$$\phi = \frac{1.76\times10^7 \cdot 10^3 \cdot 10}{4\cdot 81\times10^4} = 3.7\times10^6 \text{ radians!}$$

This is equal to 5.8×10^5 complete revolutions.

To see how far we can push this classical description of spin, we would like to try to "catch" an atom while it is in the act of rotating. Classically, one should in principle be able to accomplish this by, say, decreasing the value of H and L in Equation 1.51. Then, the deflection angle (Equation 1.37) will become smaller, but one can always move the screen far enough away to see such a deflection. However, nature makes it impossible to accomplish this goal. To see why, let us examine the experimental arrangement in more detail.

When calculating the deflection angle, $\Delta\theta$, we have assumed we know exactly where the atom is in the magnetic field. In fact, we do not know exactly where an individual atom is since the wall the beam had to pass through actually has a finite width, as suggested in Figure 1.10.

δz represents the finite width of the slit. In our idealized experiment, up to this point, we have been imagining two separate operations to be done on the beam: first, collimation by the wall; second, the measurement done on the beam by the magnets. Let us idealize our experiment even further. Imagine that the action of the thin wall and the beginning of the effect of the magnets on the beam both take place at the same time or at least approximately simultaneously. Then δz represents an uncertainty in

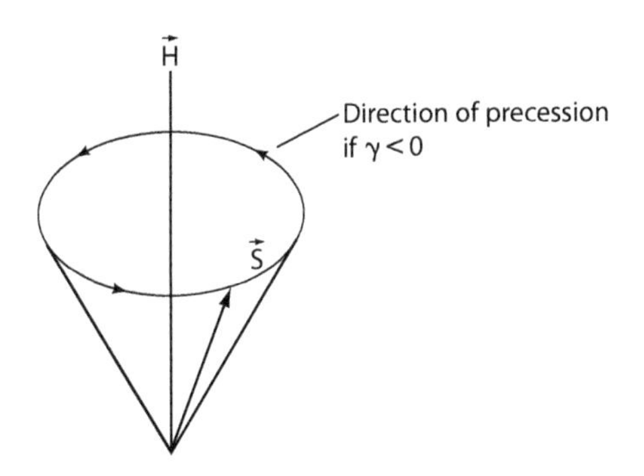

FIGURE 1.9 A precessing spin vector, $\vec{S}$, in a constant magnetic field.

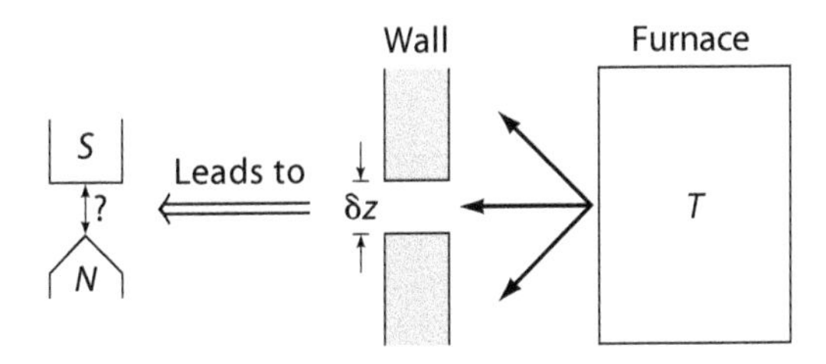

FIGURE 1.10 Illustrating how wall slit width implies an uncertainty in particle position during the Stern–Gerlach experiment.

the position of the Ag atoms as they begin their traverse through the magnetic field. Because of the gradient in H, this will cause an uncertainty in the value of the field acting on the atoms,

$$\delta H = \frac{\delta H}{\delta z}\delta z.$$

This then implies an uncertainty in the precession angle

$$\delta\phi = \gamma\delta H\frac{L}{v} = \gamma\frac{\partial H}{\partial z}\frac{L}{v}\delta z. \tag{1.52}$$

Along with the uncertainty in position, δz, there is also an uncertainty in the z-component of momentum, δp_z, of the Ag particles after they have emerged from the slit, as illustrated in Figure 1.11. This spread in momentum values will, in fact, wash out our magnetically split beam if it is too large. To ensure that the experiment works, that is, that the beam is split so we can tell which way an individual atom is rotating, we need that

$$(\Delta p_z)_+ - (\Delta p_z)_- > \delta p_z, \tag{1.53}$$

where the $(\Delta p_z)_\pm$ represents the up (+) or down (–) "kick" given to the atoms by the field. From Equation 1.36 we know that (remember, $\mu_z = \gamma S_z$, with $S_z = \hbar/2$ or $-\hbar/2$)

$$(\Delta p_z)_+ = \gamma\frac{\hbar}{2}\frac{\partial H}{\partial z}\frac{L}{v}, \tag{1.54}$$

$$(\Delta p_z)_- = -\gamma\frac{\hbar}{2}\frac{\partial H}{\partial z}\frac{L}{v}. \tag{1.55}$$

From Equation 1.52 we then have that

$$(\Delta p_z)_+ - (\Delta p_z)_- = \frac{\hbar\delta\phi}{\delta z}. \tag{1.56}$$

Equation 1.53 says that for the experiment to work, we must have

$$\frac{\hbar\delta\phi}{\delta z} > \delta p_z, \quad \text{or} \quad \hbar\delta\phi > \delta p_z\delta z. \tag{1.57}$$

If nature is such that (as we will learn in Chapter 2)

$$\delta p_z\delta z \gtrsim \hbar, \tag{1.58}$$

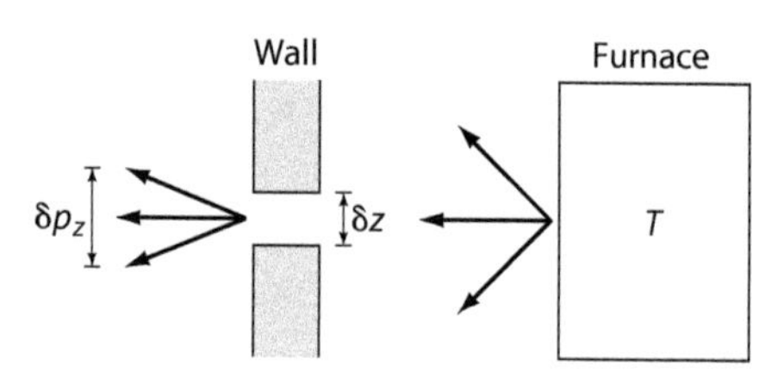

FIGURE 1.11 Illustrating the uncertainty in particle z-component momentum values arising during the collimation step.

then we must conclude that

$$\delta\phi \gtrsim 1 \tag{1.59}$$

cannot be avoided. Relations such as Equation 1.58 or 1.59 are called "uncertainty relations" and are an intrinsic part of quantum theory. Equation 1.58 is Heisenberg's famous momentum/position uncertainty relation, which will be motivated and discussed extensively in the upcoming chapters. Given this input, Equation 1.59 says that the classical picture of a rotating spin angular momentum, whose precession angle should be arbitrarily localizable, is untenable. Ag atoms are *not* behaving as just scaled-down classical tops; we cannot "catch" an Ag in the act of rotating.

As said before, the term "spin" when applied to a particle like an electron sounds classical but it is not. It is, in fact, impossible to construct a classical picture of an object with the given mass, charge, and angular momentum of an electron. Let us try to. Consider a spherical electron model as illustrated in Figure 1.12.

The "electron" consists of an infinitely thin, spherical shell of charge, spinning at the rate that gives it an angular momentum along z of $\hbar/2$. The moment of inertia of this system is

$$I = \rho_s \int_s (a^2 - z^2)\mathrm{d}s, \tag{1.60}$$

where we are doing a surface integral, and

$$\mathrm{d}s = a^2 \sin\theta \, \mathrm{d}\theta \mathrm{d}\phi,$$

$$z = a\cos\theta,$$

$$\rho_s = \frac{m}{4\pi a^2}.$$

Doing the integral gives $\left(\int_0^\pi \sin^3\theta \mathrm{d}\theta = 4/3\right)$

$$I = \frac{2}{3} m a^2. \tag{1.61}$$

Classically, we have (for a principle axis)

$$L = I\omega \tag{1.62}$$

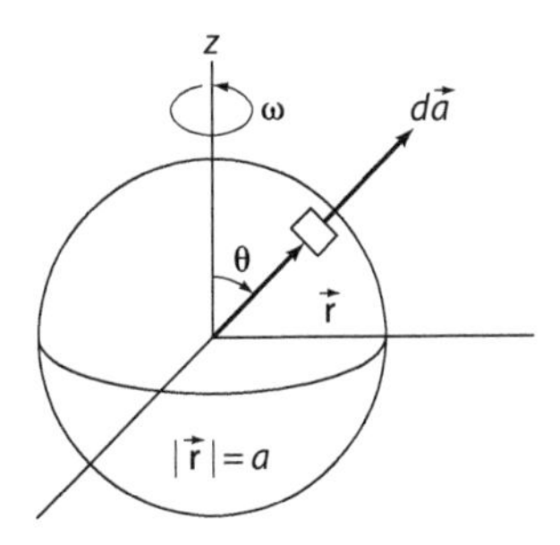

FIGURE 1.12 A classical electron model, considered as a spinning spherical shell of charge.

Setting $L = \hbar/2$, we find that the classical electron's angular velocity must be

$$\omega = \frac{3\hbar}{4ma^2}. \tag{1.63}$$

Now it takes energy to assemble this positive shell of charge because of the electrostatic forces of repulsion. This energy must be in the order of

$$E \sim \frac{e^2}{a}. \tag{1.64}$$

We know from special relativity that mass and energy are equivalent ($E = mc^2$). Thus Equation 1.64 gives a mass for the electron, which, if we hypothesize supplies the entire observed electron mass, implies a radius

$$a \sim \frac{e^2}{mc^2}. \tag{1.65}$$

(This is called the "classical electron radius.") But now notice that the velocity of the electron's surface at its "equator" is given by

$$\omega a \sim \frac{3\hbar}{4ma} = \left(\frac{3}{4\alpha}\right)c \sim 103c!$$

The surface is moving much faster than the speed of light, which is impossible by special relativity. The combination

$$\alpha = \frac{e^2}{\hbar c},$$

is called the *fine structure constant* and has the approximate value $\alpha \approx 1/137$. The impossible surface speed of the electron is not the only thing wrong with this model; we are still stuck with the classical result (Equation 1.43) for the magnetic moment produced by this spinning charge distribution, which gives the wrong magnetic moment.

Our conclusion is that electron spin, called "spin 1/2" since $S_z = \pm\,\hbar/2$ only, is a completely nonclassical concept. Its behavior (as in the S–G setup) and its origin (as shown earlier) are not accounted for by classical ideas.

1.5 WAVE FUNCTIONS FOR TWO-PHYSICAL-OUTCOMES CASES

Now let us go back to the S–G experiment again and look at it from a more general coordinate system. Our experimental results from the two-magnet S–G setup are as follows:

$$p(+,+) = \cos^2\frac{\theta}{2}, \quad p(-,+) = \sin^2\frac{\theta}{2}. \tag{1.66}$$

The "+" or "−" are labeling whether the particles are deflected "up" ($\mu_z < 0$) or "down" ($\mu_z > 0$), respectively. We now know that the upward deflected particles have $S_z = +\hbar/2$ and the downward ones have $S_z = -\hbar/2$. Instead of regarding the signs in Equation 1.66 as labels of being *deflected* up or down, let us regard them instead as labeling the value of the selected S_z value in units of $\hbar/2$. (We often call $S_z = +\hbar/2$ spin "up" and $S_z = -\hbar/2$ spin "down"). The results (Equation 1.66) can also be written as

$$p(+,+) = \frac{1+\cos\theta}{2}, \quad p(-,+) = \frac{1-\cos\theta}{2}. \tag{1.67}$$

Remember, θ is the relative rotation angle of magnet 1 with respect to magnet 2. Picking our z-axis arbitrarily relative to the two S–G apparatuses leads to the situation in Figure 1.13.

Assuming that the θ's in Equation 1.67 retain their meaning as the relative rotational angle, the probabilities in this new picture should be written as

$$p(+,+) = \frac{1+\cos\Theta}{2}, \quad p(-,+) = \frac{1-\cos\Theta}{2}. \tag{1.68}$$

An identity relating Θ to the other angles in Figure 1.13 is

$$\hat{e}_1 \cdot \hat{e}_2 = \cos\Theta$$

$$= \sin\theta\cos\phi\sin\theta'\cos\phi' + \sin\theta\sin\phi\sin\theta'\sin\phi' + \cos\theta\cos\theta', \tag{1.69}$$

$$\Rightarrow \cos\Theta = \cos\theta\cos\theta' + \sin\theta\sin\theta'\cos(\phi - \phi'). \tag{1.70}$$

Let us write $\cos\Theta$ as

$$\cos\Theta = \overbrace{\left(\cos^2\frac{\theta}{2} - \sin^2\frac{\theta}{2}\right)}^{(\cos\theta)}\overbrace{\left(\cos^2\frac{\theta'}{2} - \sin^2\frac{\theta'}{2}\right)}^{(\cos\theta')}$$

$$+ \overbrace{2\sin\frac{\theta}{2}\cos\frac{\theta}{2}}^{(\sin\theta)} \cdot \overbrace{2\sin\frac{\theta'}{2}\cos\frac{\theta'}{2}}^{(\sin\theta')}\cos(\phi - \phi')$$

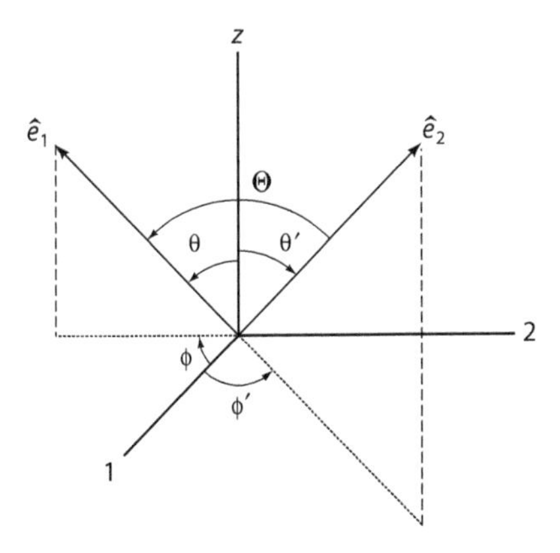

FIGURE 1.13 Two-magnet measurement axes, labeled as $\hat{e}_1$ and $\hat{e}_2$, relative to a third arbitrary z-axis.

or

$$\cos\Theta = \cos^2\frac{\theta}{2}\cos^2\frac{\theta'}{2} + \sin^2\frac{\theta}{2}\sin^2\frac{\theta'}{2} - \sin^2\frac{\theta}{2}\cos^2\frac{\theta'}{2} - \cos^2\frac{\theta}{2}\sin^2\frac{\theta'}{2}$$
$$+ 2\sin\frac{\theta}{2}\cos\frac{\theta}{2}\cdot 2\sin\frac{\theta'}{2}\cos\frac{\theta'}{2}\cos(\phi-\phi').$$

Using $1 = (\cos^2(\theta/2)+\sin^2(\theta/2))(\cos^2(\theta'/2)+\sin^2(\theta'/2))$, we get

$$\begin{aligned} p(+,+) = \frac{1+\cos\Theta}{2} &= \cos^2\frac{\theta}{2}\cos^2\frac{\theta'}{2} + \sin^2\frac{\theta}{2}\sin^2\frac{\theta'}{2} \\ &+ 2\cos\frac{\theta}{2}\cos\frac{\theta'}{2}\sin\frac{\theta}{2}\sin\frac{\theta'}{2}\cos(\phi-\phi'). \end{aligned} \tag{1.71}$$

Now suppose that $(\phi - \phi') = 0$. Then we can write $p(+,+)$ in a matrix formulation as

$$p(+,+) = \left| \underbrace{\begin{pmatrix} \cos\frac{\theta'}{2} & \sin\frac{\theta'}{2} \end{pmatrix}}_{\text{"row matrix"}} \cdot \underbrace{\begin{pmatrix} \cos\frac{\theta}{2} \\ \sin\frac{\theta}{2} \end{pmatrix}}_{\text{"column matrix"}} \right|^2. \tag{1.72}$$

(The absolute value signs are not needed here, yet.) This interesting structure can be repeated when $(\phi - \phi') \neq 0$. Consider the quantity:

$$Q = \left| \begin{pmatrix} e^{-i\phi'/2}\cos\frac{\theta'}{2} & e^{i\phi'/2}\sin\frac{\theta'}{2} \end{pmatrix}^* \cdot \begin{pmatrix} e^{-i\phi/2}\cos\frac{\theta}{2} \\ e^{i\phi/2}\sin\frac{\theta}{2} \end{pmatrix} \right|^2. \tag{1.73}$$

Working backward, we can express this as

$$Q = \left| \cos\frac{\theta}{2}\cos\frac{\theta'}{2} + \sin\frac{\theta}{2}\sin\frac{\theta'}{2}e^{i(\phi-\phi')} \right|^2. \tag{1.74}$$

Now

$$\begin{aligned} |a+b|^2 &= (a+b)^*(a+b) \\ &= |a|^2 + |b|^2 + \underbrace{a^*b + b^*a}_{2\,\mathrm{Re}(a^*b)}, \end{aligned} \tag{1.75}$$

$$\begin{aligned} \Rightarrow Q &= \cos^2\frac{\theta}{2}\cos^2\frac{\theta'}{2} + \sin^2\frac{\theta}{2}\sin^2\frac{\theta'}{2} \\ &+ 2\cos\frac{\theta}{2}\cos\frac{\theta'}{2}\sin\frac{\theta}{2}\sin\frac{\theta'}{2}\cos(\phi-\phi'). \end{aligned} \tag{1.76}$$

This is just the form we want! In general then

$$p(+,+) = \left| \left(e^{-i\phi'/2}\cos\frac{\theta'}{2} \quad e^{i\phi'/2}\sin\frac{\theta'}{2} \right)^* \cdot \begin{pmatrix} e^{-i\phi/2}\cos\frac{\theta}{2} \\ e^{i\phi/2}\sin\frac{\theta}{2} \end{pmatrix} \right|^2. \tag{1.77}$$

This factored sort of form would not have been possible without using complex numbers. This is actually a general lesson about quantum mechanics: complex numbers are a necessity.

Now let us do the same thing for p(–,+):

$$p(-,+) = \frac{1-\cos\Theta}{2} = \cos^2\frac{\theta}{2}\sin^2\frac{\theta'}{2} + \sin^2\frac{\theta}{2}\cos^2\frac{\theta'}{2}$$
$$-2\cos\frac{\theta}{2}\sin\frac{\theta'}{2}\sin\frac{\theta}{2}\cos\frac{\theta'}{2}\cos(\phi-\phi'), \tag{1.78}$$
$$\rightarrow p(-,+) = \left| -\sin\frac{\theta'}{2}\cos\frac{\theta}{2} + \cos\frac{\theta'}{2}\sin\frac{\theta}{2}e^{i(\phi-\phi')} \right|^2.$$

With a little hindsight, this can be seen to be equivalent to

$$p(-,+) = \left| \left(-e^{-i\phi'/2}\sin\frac{\theta'}{2} \quad e^{i\phi'/2}\cos\frac{\theta'}{2} \right)^* \cdot \begin{pmatrix} e^{-i\phi/2}\cos\frac{\theta}{2} \\ e^{i\phi/2}\sin\frac{\theta}{2} \end{pmatrix} \right|^2. \tag{1.79}$$

Let us define the column matrices

$$\psi_+(\theta,\phi) \equiv \begin{pmatrix} e^{-i\phi/2}\cos\frac{\theta}{2} \\ e^{i\phi/2}\sin\frac{\theta}{2} \end{pmatrix}, \tag{1.80}$$

$$\psi_-(\theta,\phi) \equiv \begin{pmatrix} -e^{-i\phi/2}\sin\frac{\theta}{2} \\ e^{i\phi/2}\cos\frac{\theta}{2} \end{pmatrix}. \tag{1.81}$$

Then we may write (the explicit matrix indices are not shown):

$$p(+,+) = |\psi_+(\theta',\phi')^\dagger\ \psi_+(\theta,\phi)|^2, \tag{1.82}$$

$$p(-,+) = |\psi_-(\theta',\phi')^\dagger\ \psi_+(\theta,\phi)|^2, \tag{1.83}$$

where the superscript "†" means "complex conjugation + transpose." (The transpose of a column matrix is a row matrix.)

In general, one may show that ($a',a'' = +-$ independently)

$$p(a'',a') = |\psi_{a''}(\theta',\phi')^\dagger\ \psi_{a'}(\theta,\phi)|^2. \tag{1.84}$$

To make sure we have not made a mistake, set $\theta' = 0$ in the above expressions. We should recover our old results, since this means the z-axis is now taken along the $\hat{e}_2$ direction (i.e., along the direction of the field in the final S–G apparatus). From Equation 1.77 we get

$$p(+,+) = \left| \begin{pmatrix} e^{-i\phi'/2} & 0 \end{pmatrix}^* \cdot \begin{pmatrix} e^{-i\phi/2} \cos\frac{\theta}{2} \\ e^{i\phi/2} \sin\frac{\theta}{2} \end{pmatrix} \right|^2$$

$$= \left| e^{i(\phi'-\phi)/2} \cos\frac{\theta}{2} \right|^2 = \cos^2\frac{\theta}{2}.$$

From Equation 1.79 we get

$$p(-,+) = \left| \begin{pmatrix} 0 & e^{i\phi'/2} \end{pmatrix}^* \cdot \begin{pmatrix} e^{-i\phi/2} \cos\frac{\theta}{2} \\ e^{i\phi/2} \sin\frac{\theta}{2} \end{pmatrix} \right|^2$$

$$= \left| e^{-i(\phi'-\phi)/2} \sin\frac{\theta}{2} \right|^2 = \sin^2\frac{\theta}{2}.$$

No mistakes.

The $\psi_{a'}(\theta,\phi)$ are called "wave functions." To find an interpretation for such objects, as well as to learn about other aspects of quantum mechanical systems, we will now try to generalize our S–G type of measurements.

1.6 MEASUREMENT SYMBOLS AND COMPLETENESS

Before, in the S–G case, we were measuring S_z (or μ_z). The physical outcomes were $S_z = \hbar/2$ or $S_z = -\hbar/2$. The whole measurement can be idealized as follows:

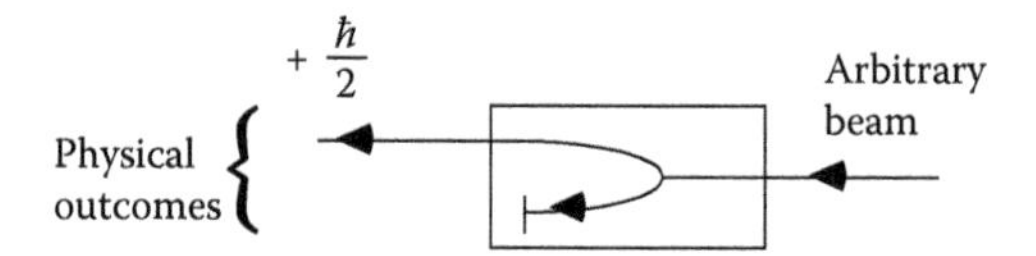

The line entering the box is indicative of a beam of particles entering an S–G apparatus. The *separation* of the beam suggests the effect of the magnets on the atoms. In addition, a *selection* is being performed whereby only particles with a given physical attribute ($S_z = \hbar/2$, say) are permitted to exit.

Let us generalize the above as follows:

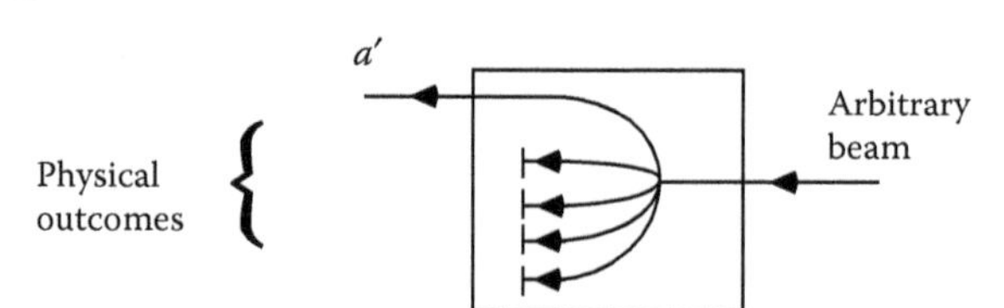

Just as the earlier S–G apparatus selects the outcome of $\hbar/2$ of the physical property S_z, we are imagining this setup to select an outcome a' of some more general physical property A. We will adopt a symbol that represents the process and call it a "measurement symbol" or an "operator." The measurement symbol for our case is

$$|a'|.$$

We will let a', a'', … be typical outcomes of such measurements; we will sometimes explicitly label specific outcomes as a_1, a_2, …. For right now think of the outcomes a', a'', … as dimensionless numbers, to keep things simple.

What sorts of manipulations are appropriate to these measurement symbols? Consider the S–G type of process:

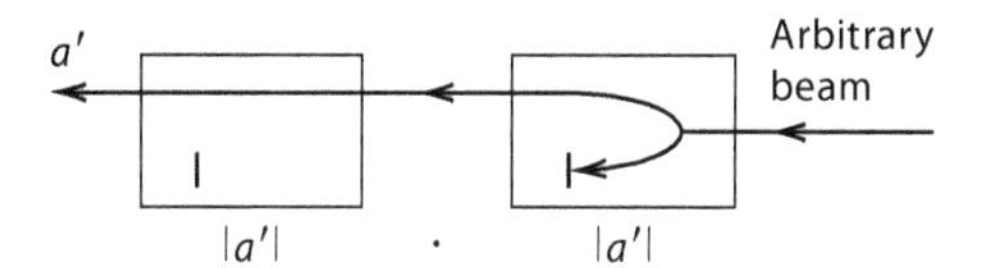

This is clearly the same as just

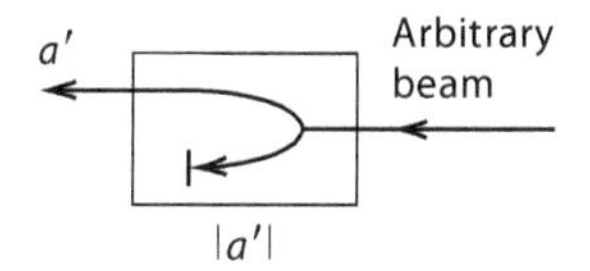

This suggests the following rule:

$$|a'||a'|=|a'|. \tag{1.85}$$

On the other hand, consider ($a' \neq a''$):

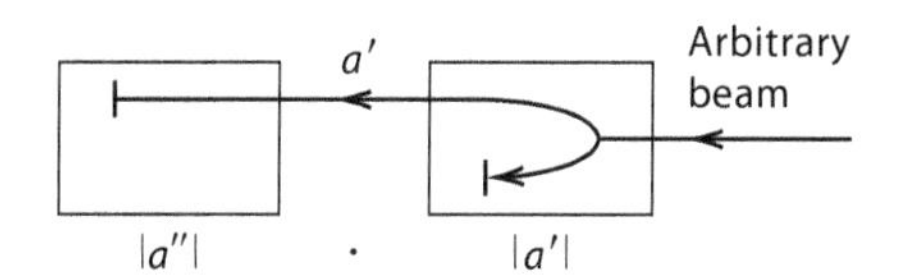

This is equivalent to an apparatus that blocks everything:

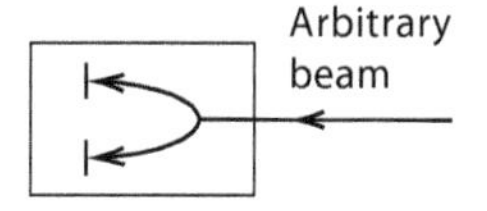

We will call the above a "null measurement" and associate it with the usual null symbol, 0. Therefore, we adopt

$$|a''||a'|=0, \quad a' \neq a''. \tag{1.86}$$

Notice

$$|a'||a''|=|a''||a'|, \tag{1.87}$$

which says that selection experiments are commutative. This defines multiplication in this context. What about addition? Let us start at the opposite end to the null measurement in a system with four physical outcomes, say

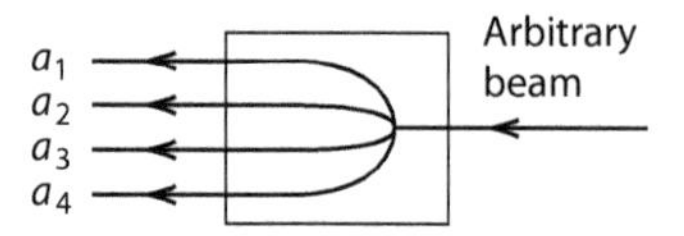

Such a measurement apparatus can perform a separation but no selection. Our symbol for this will be the usual identity character: 1. Clearly, we have

$$1 \cdot 1 = 1. \tag{1.88}$$

Now start blocking out physical outcomes one by one:

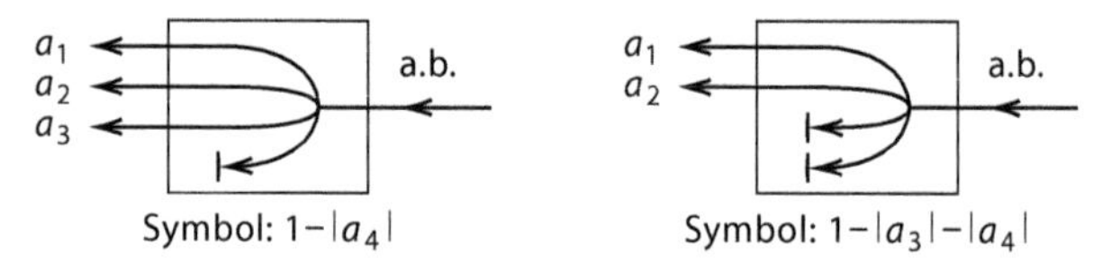

(a.b. means "arbitrary beam.") Now, block all of the outcomes:

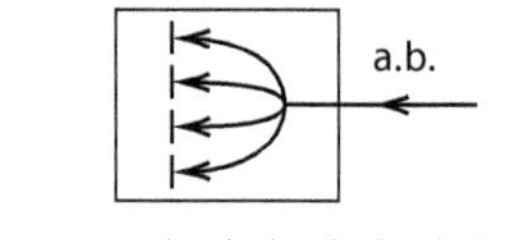

This is obviously just the null measurement again. The two characterizations must be the same:

$$1 - \sum_i |a_i| = 0. \tag{1.89}$$

We require that

$$0 + |a'| = |a'|, \tag{1.90}$$

so that Equation 1.89 can be written as

$$\sum_i |a_i| = 1. \tag{1.91}$$

Equation 1.91 will be called "completeness." It is no exaggeration to say it is the foundation stone of all of quantum mechanics.

1.7 PROCESS DIAGRAMS AND OPERATOR PROPERTIES

We can write down other mathematical results suggested by the above type of diagrams. Consider:

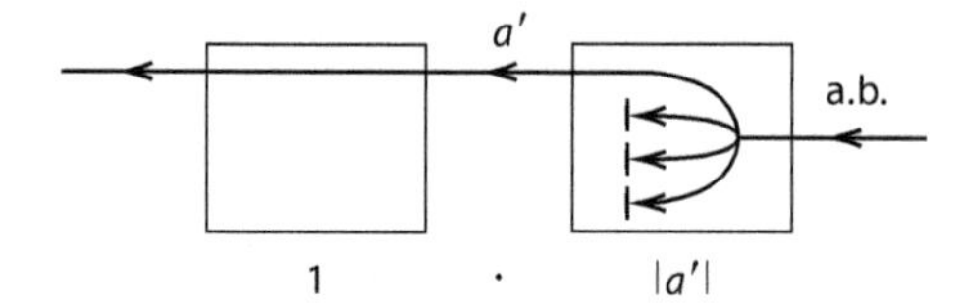

This is clearly equivalent to the opposite order:

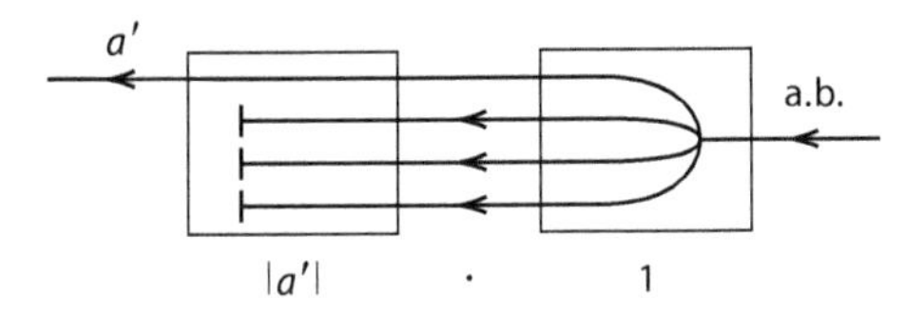

In fact, both are the same as just

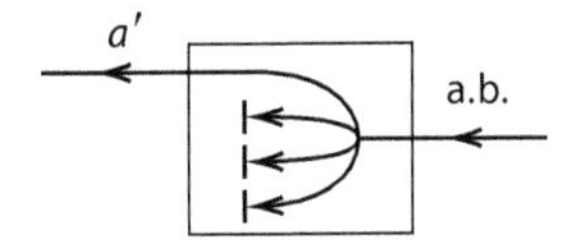

Mathematically, these diagrams tell us that

$$1\cdot|a'|=|a'|\cdot 1=|a'|. \tag{1.92}$$

One can also show

$$1\cdot 0=0;\quad 0\cdot 1=0. \tag{1.93}$$

We have to tie in this discussion with real numbers eventually. All experiments have results and all results are numbers. There has to be more to an experiment than just accepting or rejecting physical attributes. There is also the possibility of *modulating* a signal. If C represents, in general, a complex number, we adopt the simple rules that

$$\left.\begin{aligned} &C\cdot 0=0\\ &C|a'|=|a'|C\\ &C1=1C \end{aligned}\right\}. \tag{1.94}$$

These properties ensure that no distinction between 1,0 (measurement symbols) and 1,0 (numbers) is necessary. For now, let us also regard the numbers C as being dimensionless. We will suggest a modulating device as follows:

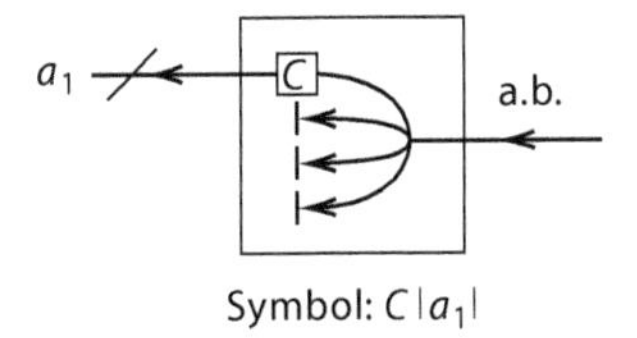

Symbol: $C|a_1|$

The amplitude of the a_1 beam has been modified by a factor $|C|$, and its phase has been changed by $\tan^{-1}(\mathrm{Im}(C)/\mathrm{Re}(C))$, just like for an electronic circuit. A slash through an emerging beam will sometimes be used to denote its modified character, and we can also, if we wish, write the modulating factor in the little box thus – $\boxed{C}$. Using

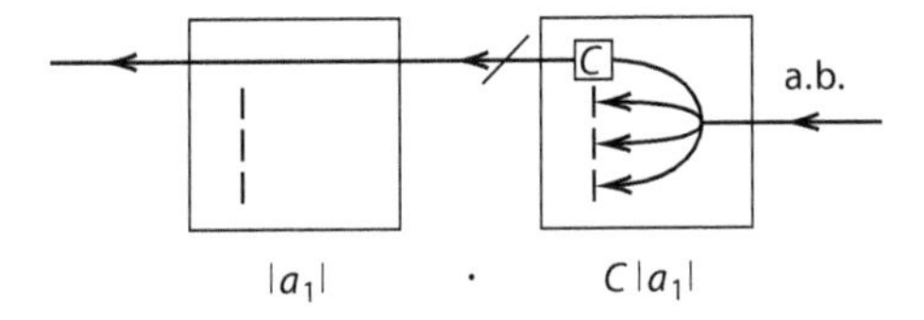

we see that, for example,

$$|a_1|(C|a_1|) = C|a_1|, \tag{1.95}$$

which also follows mathematically from Equations 1.94 and 1.85. We adopt the rule that our beam always travels from right to left and will write down our measurement symbols in the same order as they appear in the diagrams. An example of a more general modulated measurement is as follows:

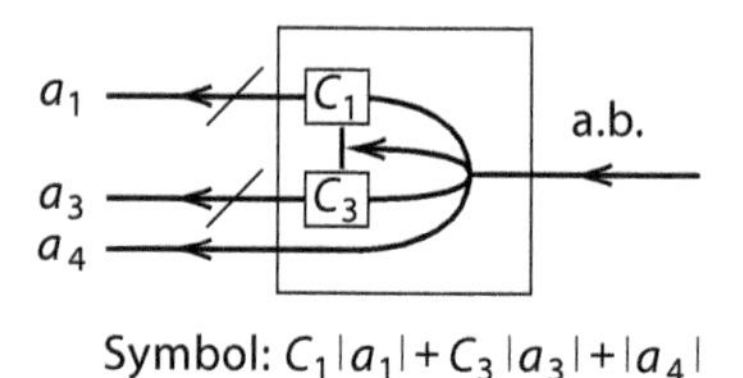

Symbol: $C_1|a_1| + C_3|a_3| + |a_4|$

Now that we know how to associate numbers with measurement symbols, we may write Equations 1.85 and 1.86 together as

$$|a'||a''| = \delta_{a'a''}|a'|, \tag{1.96}$$

where $\delta_{a'a''}$ is the Kronecker delta symbol:

$$\delta_{a'a''} = \begin{cases} 1, & a' = a'' \\ 0, & a' \neq a'' \end{cases}. \tag{1.97}$$

In addition, one can show that the distributive law is operative here:

$$(|a'| + |a''|)|a'''| = |a'||a'''| + |a''||a'''|. \tag{1.98}$$

Let us now define a very special sort of modulated operator. If we choose

$$C_i = a_i, \tag{1.99}$$

that is, the amplification factors are chosen as the values of the physical outcomes (which are real), then we have for this measurement:

$$A \equiv \sum_i a_i |a_i|. \tag{1.100}$$

We have been thinking of the a_i as dimensionless, but we may want to associate physical dimensions with the property A, just as we associate physical dimensions with spin S_z. We can always supply dimensions by multiplying both sides of Equation 1.100 by a single dimensionful constant:

$$A' = CA = \sum_i Ca_i |c a_i| = \sum_i \bar{a}_i |\bar{a}_i|. \tag{1.101}$$

$\uparrow$

related as $Ca_i = \bar{a}_i$

This mathematical act is somewhat mysterious from the point of view of our diagrams, since it cannot be represented in such a manner. However, every experiment has a readout in units of some kind. Let us assume the preceding conversion to physical units represents the machinery's read-out of the result in some appropriate units. For now, we will continue to use dimensionless physical outcomes a_i; we can always supply a dimensionful constant later.

Let us deduce some properties of the aforementioned A. First, notice that

$$A\,|a'| = \left(\sum_i a_i\,|a_i|\right)|a'| = (a_1\,|a_1| + a_2\,|a_2| + \cdots)\,|a'|$$

$$= a'\,|a'||a'| = a'\,|a'|,$$

so

$$A|\,a'| = a'|\,a'|. \tag{1.102}$$

Also

$$|\,a'|\,A = a'\,|\,a'|. \tag{1.103}$$

A has the important property of singling out the value of the physical outcome a' when it acts in concert with the selection $|a'|$. Pictorially, Equation 1.102 is saying

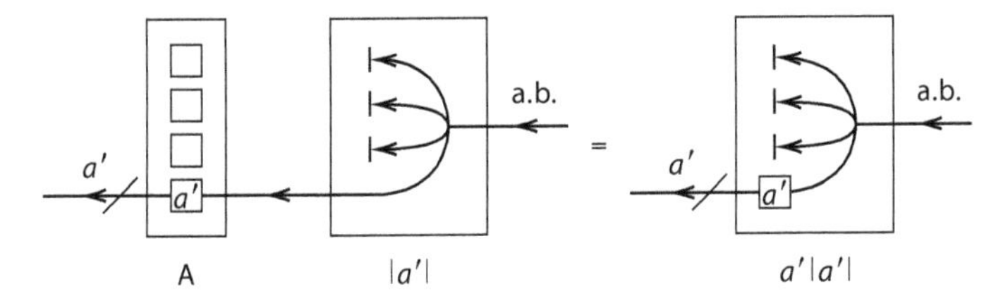

Equation 1.103 can be seen as

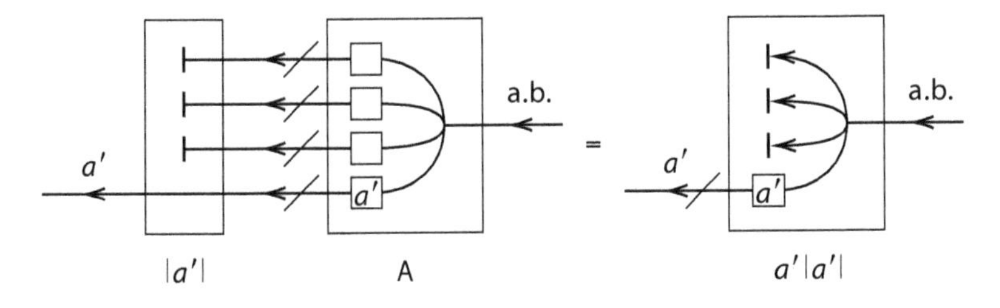

The order of these operations or measurements is not important yet.

It is time to say a little bit more about what the diagrams I have been drawing represent. Although we have used the S–G experimental apparatus to model these idealized measurements, the preceding manipulations on the incoming "beam" do not actually represent physical operations carried out in real space. Instead, they represent operations carried out on individual particle characteristics in a mathematical "space" or arena where the concepts "amplitude" and "phase" makes sense. This mathematical space has been given the name of "Hilbert space" after the great German mathematician

David Hilbert. Although the preceding manipulations do not represent real-space experimental setups, there is still a correspondence between what happens in a real experiment (involving spin, say) and in our Hilbert space idealizations; this connection will be stated shortly. I will call these ideal manipulations on arbitrary beams ("arbitrary" in the sense of containing nonzero amplitudes for all physical outcomes, a') "Process Diagrams."

Some other properties of A are now detailed. Notice that

$$A^2 = A \cdot A = A \sum_i a_i \,|a_i| = \sum_i a_i A \,|a_i|.$$

But $A|a_i| = a_i|a_i|$, so

$$A^2 = \sum_i a_i^2 \,|a_i|. \tag{1.104}$$

This can be pictured as

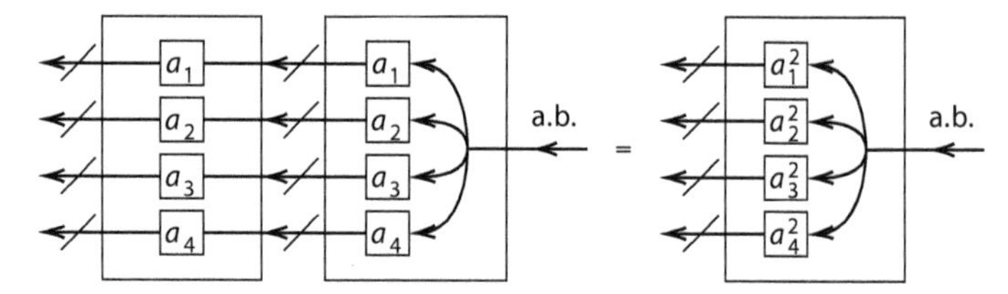

where the amplitude factors a_1, a_2, ... associated with each one of the physical outcomes have been written explicitly.

The generalization of the aforementioned rule for A^2 is

$$f(A) = \sum_i f(a_i) \,|\, a_i \,|, \tag{1.105}$$

for some $f(A)$ power series in A. Let us take some examples to understand Equation 1.105 better. First, which $f(A)$ results from the choice of $f(a_i) = 1$ for all i?

$$f(A) = \sum_i |\, a_i \,| = 1.$$

Next, which $f(A)$ results from $f(a_i) = 0$ for all i?

$$f(A) = \sum_i 0 |\, a_i | = 0.$$

Which $f(A)$ results from the following choice?

$$f(a_j) = 1,$$

$$f(a_i) = 0, \quad \text{all } a_i \neq a_j?$$

This also is easy:

$$f(A) = |\, a_j \,|.$$

However, what is this in terms of A? (This is not so easy.) Now

$$A|a_j| = a_j|a_j| \quad \Rightarrow (A - a_j)\,|a_j| = 0,$$

where I have suppressed the unit symbol 1 in writing the second form. Now consider the statement

$$\left[\prod_{i=1}^{n}(A - a_i)\right]|a_j| = 0. \tag{1.106}$$

To see that this is true, write this out more explicitly:

$$\begin{aligned}&(A - a_1)(A - a_2)\ldots(A - a_j)\ldots(A - a_n)\,|a_j| \\ &= (a_j - a_1)(a_j - a_2)\ldots(a_j - a_j)\ldots(a_j - a_n)\,|a_j| = 0.\end{aligned} \tag{1.107}$$

Since this is true for *any* $|a_j|$, we must have

$$\prod_{i}(A - a_i) = 0. \tag{1.108}$$

So this represents a new way of writing the null measurement. (Can you think of a Process Diagram to represent the left-hand side of Equation 1.108?) Now comparing Equation 1.108, written in the form

$$\prod_{i}(A - a_i) = (A - a_j)\prod_{i \neq j}(A - a_i) = 0, \tag{1.109}$$

with the statement $(A - a_j)\,|a_j| = 0$ leads to the conclusion that

$$|a_j| = C\prod_{i \neq j}(A - a_i), \tag{1.110}$$

where C is some unknown constant. I will just supply this constant:

$$|a_j| = \prod_{i \neq j}\left(\frac{A - a_i}{a_j - a_i}\right). \tag{1.111}$$

It is easy to show that Equation 1.111 works correctly. First, let us show that $|a_j|\,|a_j| = |a_j|$:

$$\begin{aligned}\left[\prod_{i \neq j}\left(\frac{A - a_i}{a_j - a_i}\right)\right]|a_j| &= \left(\frac{A - a_1}{a_j - a_1}\right)\ldots\left(\frac{A - a_n}{a_j - a_n}\right)|a_j| \\ &= \left(\frac{a_j - a_1}{a_j - a_1}\right)\ldots\left(\frac{a_j - a_n}{a_j - a_n}\right)|a_j| \\ &= 1 \ldots 1\,|a_j| = |a_j|.\end{aligned}$$

This tells us we have chosen the constant C correctly. Next, let us check that $|a_j||a_k| = 0(j \neq k)$:

$$\left[\prod_{i \neq j}\left(\frac{A - a_i}{a_j - a_i}\right)\right]|a_k| = \left(\frac{A - a_1}{a_j - a_1}\right)\cdots\left(\frac{A - a_n}{a_j - a_n}\right)|a_k|$$

$$= \left(\frac{a_k - a_1}{a_j - a_1}\right)\cdots\left(\frac{a_k - a_k}{a_j - a_k}\right)\cdots\left(\frac{a_k - a_n}{a_j - a_n}\right)|a_k|$$

$$= 0.$$

Let us study the two physical-outcome case in detail. Let $a_1 = 1, a_2 = -1$. Then

$$\prod_i (A - a_i) = 0 \Rightarrow (A-1)(A+1) = 0 \Rightarrow A^2 = 1. \tag{1.112}$$

This is the algebraic equation satisfied by the physical property A. Also (let $|1| \equiv |+|$, $|-1| \equiv |-|$)

$$|+| = \prod_{i \neq 1}\left(\frac{A - a_i}{a_1 - a_i}\right) = \frac{A+1}{2}, \tag{1.113}$$

$$|-| = \prod_{i \neq 2}\left(\frac{A - a_i}{a_2 - a_i}\right) = \frac{1-A}{2}. \tag{1.114}$$

In addition, we have completeness:

$$\sum_i |a_i| = |+| + |-| = \frac{A+1}{2} + \frac{1-A}{2} = \frac{1}{2} + \frac{1}{2} = 1.$$

We can explicitly check some properties here:

$$|-||+| = \left(\frac{1-A}{2}\right)\left(\frac{A+1}{2}\right) = \frac{1-A^2}{4} = 0$$

$$|+||+| = \left(\frac{A+1}{2}\right)^2 = \frac{1+2A+A^2}{4} = \frac{2+2A}{4} = |+|$$

$$|-||-| = \left(\frac{1-A}{2}\right)^2 = \frac{1-2A+A^2}{4} = \frac{2-2A}{4} = |-|.$$

There is another operation we can imagine performing on our "arbitrary beam" of particles that has a quantum mechanical basis. Besides the selection and amplification operations, one can also imagine the following Process Diagram:

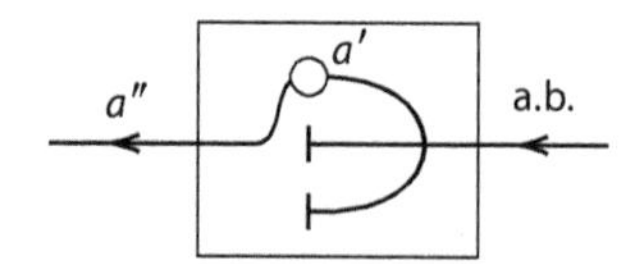

That is, we are imagining an experiment that performs a *transition*. In the preceding illustration, the beam with physical outcome a' is transformed into a beam with physical outcome a'', keeping the a' beam's amplitude and phase information. In the S–G case, the preceding illustration would represent an apparatus that turned spin $S_z = \hbar/2$ into $S_z = -\hbar/2$, say. The symbol we will adopt for this is (the 1 is implicit)

$$|a''\ a'|.$$

↑ ↑

exiting property entering property

This interpretation is the reason the Process Diagram beam is chosen to travel from right to left.* The connection to our earlier measurement symbol is clear:

$$|a'\ a'| = |a'|. \tag{1.115}$$

Consider the following:

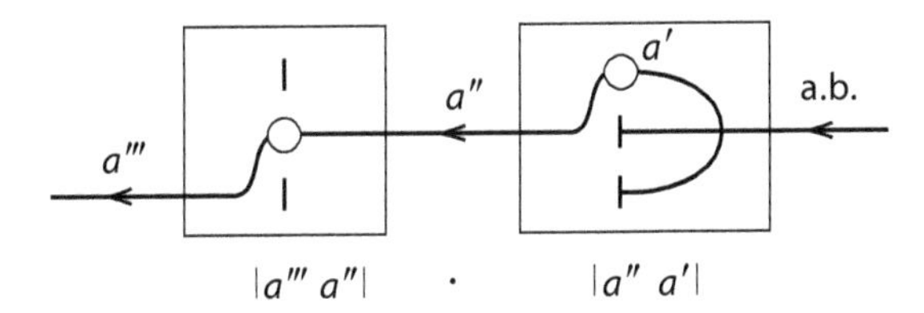

This is the same as

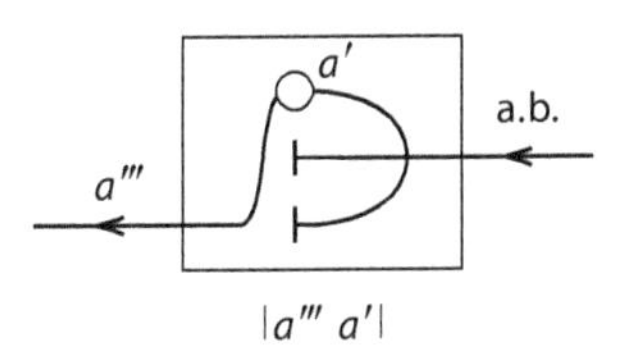

So we adopt the rule that

$$|a'''\ a''||a''\ a'| = |a'''\ a'|. \tag{1.116}$$

It is also clear that

$$|a''''\ a'''||a''\ a'| = 0, \quad a''' \neq a''. \tag{1.117}$$

The algebraic properties of our two types of measurement symbols are summarized in

$$|a'||a''| = \delta_{a'\,a''}|a'|, \tag{1.118}$$

$$|a''''\ a'''||a''a'| = \delta_{a'''\,a''}|a''''a'|. \tag{1.119}$$

* Such time ordering is conventionally seen in quantum mechanical transition matrix-elements, as we will see in Chapters 9 and 11.

As pointed out before, the $|a'|$-type measurements are reversible (i.e., the order of the operations does not matter). However, these new types of measurements are not reversible. An example is as follows:

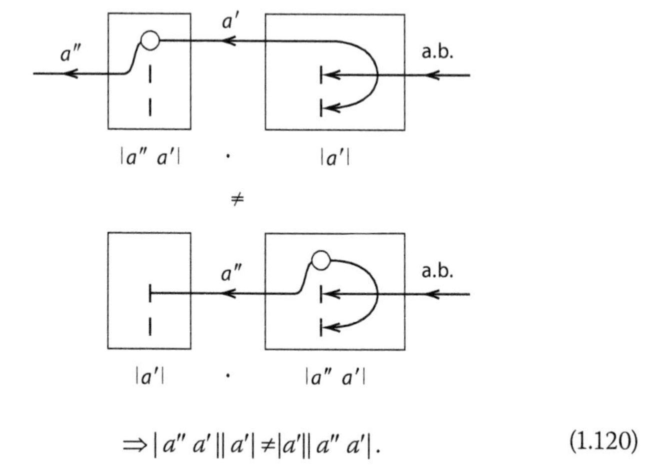

$$\Rightarrow |a''\ a'||a'| \neq |a'||a''\ a'|. \tag{1.120}$$

Equation 1.120 follows mathematically from Equation 1.119 and the fact that $|a'| = |a'\ a'|$. Another example is

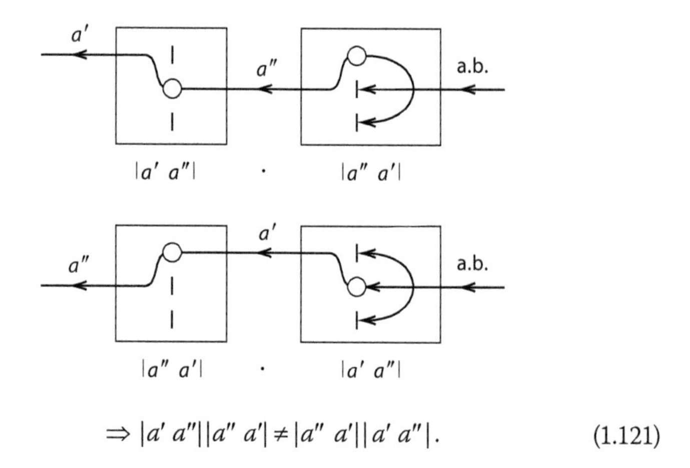

$$\Rightarrow |a'\ a''||a''\ a'| \neq |a''\ a'||a'\ a''|. \tag{1.121}$$

The algebra of the $|a'\ a''|$-type symbols is *noncommutative*. Each side of Equations 1.120 and 1.121 can be reduced further using Equation 1.119. We can always label the O symbols in our transition-type Process Diagrams with the actual transition this device performs to remove any ambiguity. We will do this occasionally in the following.

Let us concentrate on the two-physical-outcomes case, since this is the simplest situation. The symbol associated with

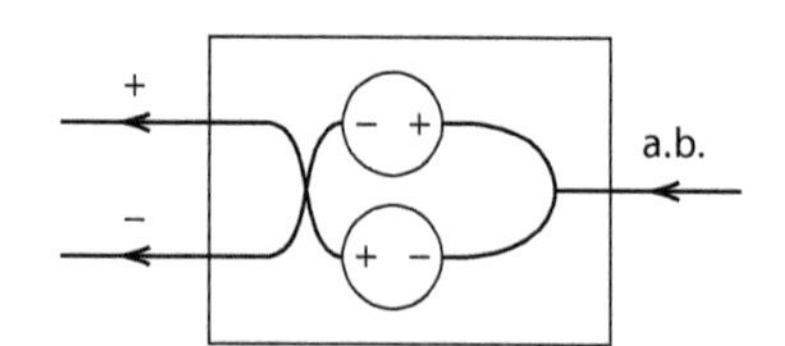

is

$$|-+|+|+-|.$$

This suggests that the operation

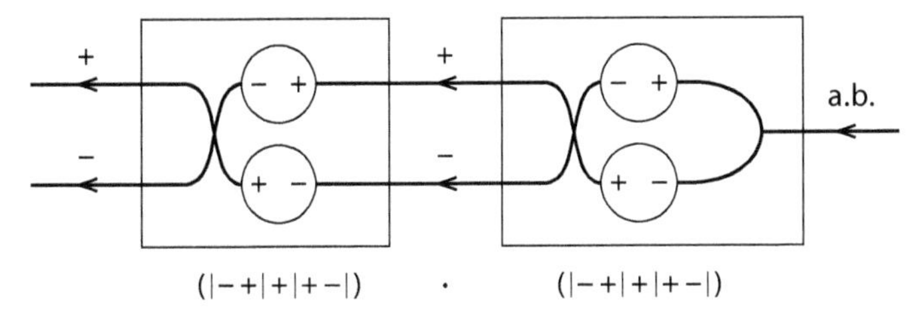

simply reconstitutes the original beam. Let us confirm this mathematically:

$$\begin{aligned}(|-+|+|+-|)\cdot(|-+|+|+-|)\\ =|-+||-+|+|-+||+-|+|+-||-+|+|+-||+-|\\ =|-|+|+|=1.\end{aligned}$$

Here are some more examples and the equations that go along with them.

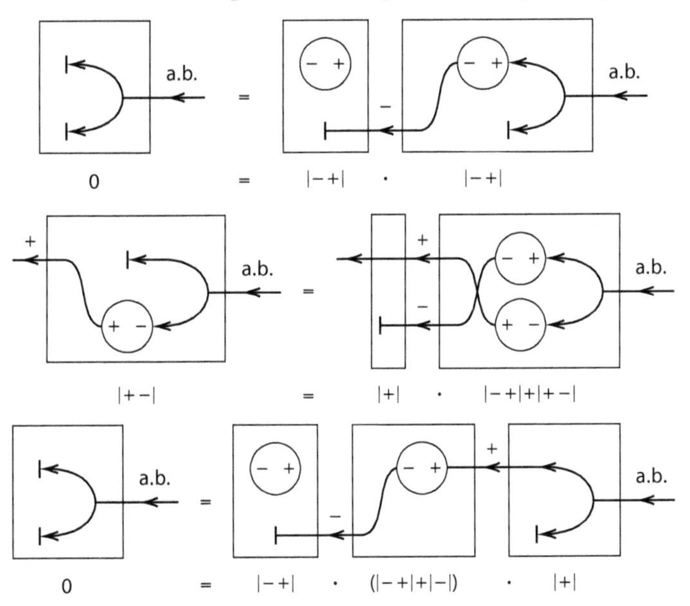

Beams can sometimes combine, as in

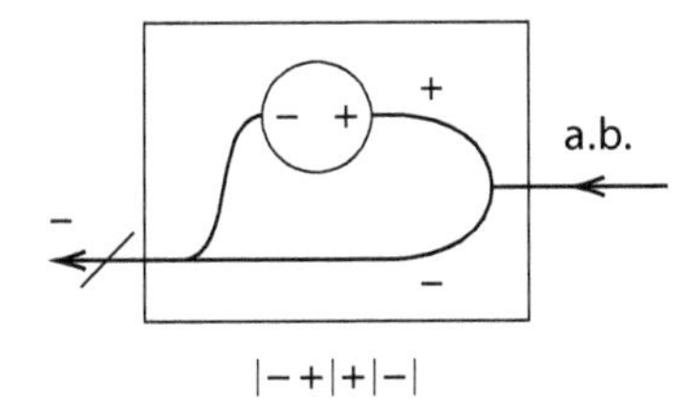

Of course, the "–" beam's amplitude has in general been modified.

1.8 OPERATOR REFORMULATION

Let us continue to investigate the two-physical-outcomes case. There are four independent measurement symbols:

1. $|+\,+| = |+|$
2. $|-\,-| = |-|$
3. $|+\,-|$
4. $|-\,+|$.

We will make a different, more convenient choice of the four independent quantities. We choose the unit symbol

$$1 = |+| + |-|, \tag{1.122}$$

as one of them. Another independent choice is

$$\sigma_3 = |+| - |-|, \tag{1.123}$$

which we can write as

$$\sigma_3 = \sum_{\sigma_3'} |\sigma_3'|\sigma_3', \tag{1.124}$$

where $\sigma_3' = \pm 1$ is just a renaming of the special modulated operator A we investigated earlier. Let us confirm that Equation 1.112 holds:

$$\sigma_3^2 = (|+| - |-|)(|+| - |-|) = |+| + |-| = 1.$$

We will take the other two independent combinations to be

$$\sigma_1 = |-+| + |+-|, \tag{1.125}$$

$$\sigma_2 = i|-+| - i|+-|. \tag{1.126}$$

In our Process Diagram language, σ_1 performs a double transition (see some of the examples already drawn), while σ_2 performs the same transition but additionally modifies the phases of the signals. It is easy to see that, in addition to $\sigma_3^2 = 1$, we have

$$\sigma_1^2 = 1, \; \sigma_2^2 = 1. \tag{1.127}$$

Some other mathematical properties of these new combinations are as follows:

$$\begin{aligned} \sigma_1\sigma_2 &= (|-+| + |+-|)(i|-+| - i|+-|) \\ &= -i|-| + i|+| = i\sigma_3, \end{aligned} \tag{1.128}$$

$$\begin{aligned} \sigma_2\sigma_1 &= (i|-+| - i|+-|)(|-+| + |+-|) \\ &= i|-| - i|+| = -i\sigma_3. \end{aligned} \tag{1.129}$$

Therefore, we see that

$$\sigma_1\sigma_2 = -\sigma_2\sigma_1. \tag{1.130}$$

One can also show the following:

$$\sigma_2\sigma_3 = i\sigma_1, \quad \sigma_3\sigma_2 = -i\sigma_1, \tag{1.131}$$

$$\sigma_3\sigma_1 = i\sigma_2, \quad \sigma_1\sigma_3 = -i\sigma_2. \tag{1.132}$$

Summarizing the properties of the σ's, we have

$$\sigma_k^2 = 1 \quad (k = 1,2,3), \tag{1.133}$$

$$\sigma_k\sigma_\ell = -\sigma_\ell\sigma_k \quad (k \neq \ell), \tag{1.134}$$

$$\sigma_k\sigma_\ell = i\sigma_m \ (k,\ell,m \text{ in cyclic order}) \tag{1.135}$$

Cyclic order: (1,2,3), (2,3,1) or (3,1,2).

Some mathematical phraseology we will use is as follows:

If $AB = BA$, then A and B *commute*;

If $AB = -BA$, then A and B *anticommute.*

Now our two-physical-outcomes case has just been modeled after spin 1/2. Electron spin, being a type of angular momentum, should involve three components; we hypothesize that these spin components are represented by the σ_i, which are seen to satisfy Equation 1.112, the basic operator equation for the two-physical-outcomes case. However, to supply the correct *physical units*, we write ($i = 1,2,3$ denote the x,y,z axes)

$$S_i = \frac{\hbar}{2}\sigma_i, \quad i = 1,2,3. \tag{1.136}$$

The multiplication by $\hbar/2$ is the "somewhat mysterious" part of the measurement that cannot be represented by a diagram. Equation 1.136 is the crucial connection that allows us to tie our developing formalism to the real world.

We may summarize our operator algebra so far as follows. Define the idea of the "commutator":

$$[A,B] \equiv AB - BA. \tag{1.137}$$

Using Equations 1.135 and 1.136, we may then write

$$[S_i, S_j] = i\hbar S_k, \tag{1.138}$$

where i, j, and k are in cyclic order. The entire content of the algebra of σ_i can be summarized as

$$\sigma_i\sigma_j = 1\delta_{ij} + i\sum_k \varepsilon_{ijk}\sigma_k, \tag{1.139}$$

where the permutation symbol ε_{ijk} has been introduced. It is defined as

$$\varepsilon_{ijk} \equiv \begin{cases} 0, & \text{if any index is equal to any other index} \\ +1, & \text{if } i,j,k \text{ form an even permutation of } 1,2,3 \\ -1, & \text{if } i,j,k \text{ form an odd permutation of } 1,2,3. \end{cases} \tag{1.140}$$

Using the permutation symbol, one may now write Equation 1.138 compactly as

$$\left[S_i, S_j\right] = i\hbar \sum_k \varepsilon_{ijk} S_k, \tag{1.141}$$

which automatically enforces the cyclic rule.

1.9 OPERATOR ROTATION

It is important to realize that the aforementioned quantities S_i are *operators*, not numbers. Nevertheless, if they represent an angular momentum they must *behave* like a vector under rotations in real space. Let us confirm our identification of the σ_i as angular momentum components under the following passive rotation ("passive" means the axis, not the vector, is rotated) about the third axis shown in Figure 1.14. A vector $\vec{v}$ should now have new components $\bar{v}_i$ given by

$$\left.\begin{aligned} \bar{v}_3 &= v_3 \\ \bar{v}_1 &= v_1 \cos\phi + v_2 \sin\phi \\ \bar{v}_2 &= -v_1 \sin\phi + v_2 \cos\phi \end{aligned}\right\}. \tag{1.142}$$

v_1, v_2, v_3 are the components along the old (unbarred) axes. We will choose the angle ϕ in Figure 1.14 to be positive by convention. In analogy to Equation 1.142, we require that the spin components in the new coordinates be

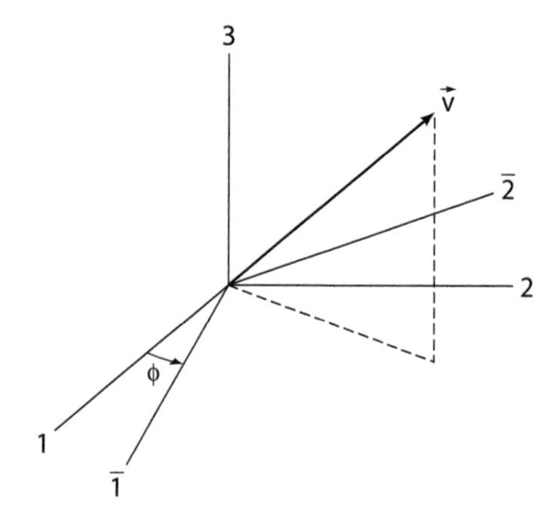

FIGURE 1.14 Showing a passive rotation by angle ϕ about the 3-axis.

$$\left.\begin{aligned}\bar{\sigma}_3 &= \sigma_3\\ \bar{\sigma}_1 &= \sigma_1 \cos\phi + \sigma_2 \sin\phi\\ \bar{\sigma}_2 &= -\sigma_1 \sin\phi + \sigma_2 \cos\phi\end{aligned}\right\}. \tag{1.143}$$

Now, is it still true that $\bar{\sigma}_i^2 = 1$?

$$\bar{\sigma}_3^2 = \sigma_3^2 = 1.$$

$$(\bar{\sigma}_1)^2 = \sigma_1^2\cos^2\phi + \sigma_2^2\sin^2\phi + \overbrace{(\sigma_1\sigma_2 + \sigma_2\sigma_1)}^{=0} \cos\phi\sin\phi$$
$$= \cos^2\phi + \sin^2\phi = 1.$$

$$(\bar{\sigma}_2)^2 = \sigma_1^2\sin^2\phi + \sigma_2^2\cos^2\phi - \overbrace{(\sigma_1\sigma_2 + \sigma_2\sigma_1)}^{=0}\cos\phi\sin\phi$$
$$= \sin^2\phi + \cos^2\phi = 1.$$

What about the cyclic property? One can show that

$$\left.\begin{aligned}\bar{\sigma}_1\bar{\sigma}_2 &= i\bar{\sigma}_3, & \bar{\sigma}_2\bar{\sigma}_1 &= -i\bar{\sigma}_3\\ \bar{\sigma}_2\bar{\sigma}_3 &= i\bar{\sigma}_1, & \bar{\sigma}_3\bar{\sigma}_2 &= -i\bar{\sigma}_1\\ \bar{\sigma}_3\bar{\sigma}_1 &= i\bar{\sigma}_2, & \bar{\sigma}_1\bar{\sigma}_3 &= -i\bar{\sigma}_2\end{aligned}\right\}. \tag{1.144}$$

I would like to show now that the properties $\bar{\sigma}_i^2 = 1$ and the cyclic properties (Equation 1.144) hold for a more general rotation. To start off, let us rewrite σ_1 as

$$\begin{aligned}\bar{\sigma}_1 &= \sigma_1 \cos\phi + \sigma_2 \sin\phi = \sigma_1 \cos\phi - i\sigma_3\sigma_1 \sin\phi\\ \Rightarrow \bar{\sigma}_1 &= (\cos\phi - i\sigma_3 \sin\phi)\sigma_1,\end{aligned} \tag{1.145}$$

or

$$\bar{\sigma}_1 = \sigma_1(\cos\phi + i\sigma_3 \sin\phi). \tag{1.146}$$

One can show in a problem that

$$e^{i\lambda\sigma_3} = \cos\lambda + i\sigma_3\sin\lambda, \tag{1.147}$$

where λ is just a real number. In fact, more generally, one can show that

$$e^{i\lambda(\hat{n}\cdot\vec{\sigma})} = \cos\lambda + i(\hat{n}\cdot\vec{\sigma}) \sin\lambda, \tag{1.148}$$

where $\hat{n}\cdot\vec{\sigma} = n_1\sigma_1 + n_2\sigma_2 + n_3\sigma_3$ and $\hat{n}$ is a unit spatial vector, $\hat{n}^2 = 1$. Using Equation 1.147 or 1.148 with $\hat{n} = \hat{e}_3$, we can write Equations 1.145 and 1.146 as

$$\bar{\sigma}_1 = e^{-i\phi\sigma_3}\sigma_1 \tag{1.149}$$

or

$$\bar{\sigma}_1 = \sigma_1 \, e^{i\phi\sigma_3}. \tag{1.150}$$

There are many ways of seeing the equality of Equations 1.149 and 1.150. Since σ_1 and σ_3 anticommute, we have, for example,

$$\sigma_3\sigma_1 = \sigma_1(-\sigma_3)$$

and

$$\sigma_3^2\sigma_1 = \sigma_1(-\sigma_3)^2.$$

Generalizing from these examples, it is easy to see that

$$f(\sigma_3)\sigma_1 = \sigma_1 f(-\sigma_3),$$

for $f(\sigma_3)$ any power series function of σ_3.

We now wish to prove that

$$e^{i\lambda\sigma_3} \, e^{i\lambda'\sigma_3} = e^{i(\lambda+\lambda')\sigma_3}. \tag{1.151}$$

We can rewrite

$$e^{i\lambda\sigma_3} = \cos\lambda + i\sigma_3\sin\lambda,$$

$$\Rightarrow e^{i\lambda\sigma_3} = \cos\lambda(|+|+|-|) + i(|+|-|-|)\sin\lambda,$$

$$\Rightarrow e^{i\lambda\sigma_3} = |+|e^{i\lambda} + |-|e^{-i\lambda}. \tag{1.152}$$

Using Equation 1.152, we now have

$$e^{i\lambda\sigma_3} \, e^{i\lambda'\sigma_3} = (|+|e^{i\lambda} + |-|e^{-i\lambda})(|+|e^{i\lambda'} + |-|e^{-i\lambda'})$$

$$= |+|e^{i(\lambda+\lambda')} + |-|e^{-i(\lambda+\lambda')} = e^{i(\lambda+\lambda')\sigma_3},$$

which proves Equation 1.151. Therefore, we can write from Equation 1.149, for example,

$$\bar{\sigma}_1 = e^{-i\phi\,\sigma_3}\sigma_1 = e^{-i\phi/2\,\sigma_3}\overbrace{e^{-i\phi/2\sigma_3}\sigma_1}^{=\sigma_1 e^{i\phi/2\sigma_3}} = e^{-i\phi/2\sigma_3}\sigma_1 \, e^{i\phi/2\sigma_3}. \tag{1.153}$$

Equation 1.150 leads to the same conclusion. Now, if we call $U = e^{i\phi/2\sigma_3}$, then $U^{-1} = e^{-i\phi/2\sigma_3}$, since

$$e^{-i\phi/2\sigma_3} e^{i\phi/2\sigma_3} = e^{i(\phi/2-\phi/2)\sigma_3} = e^0 = 1$$

by Equation 1.151. Then, we may write

$$\bar{\sigma}_1 = U^{-1}\sigma_1 U. \tag{1.154}$$

$\bar{\sigma}_2$ and $\bar{\sigma}_3$ can also be written as in Equation 1.150:

$$\begin{aligned} U^{-1}\sigma_2 U &= e^{-i\phi/2\sigma_3}\sigma_2 e^{i\phi/2\sigma_3} = \sigma_2 e^{i\phi\sigma_3} \\ &= \sigma_2(\cos\phi + i\sigma_3 \sin\phi) = \sigma_2\cos\phi - \sigma_1 \sin\phi \\ &= \bar{\sigma}_2, \\ U^{-1}\sigma_3 U &= \bar{\sigma}_3. \quad \text{(trivial)} \end{aligned}$$

Thus, we have $(S_i = (\hbar/2)\,\sigma_i)$

$$\bar{S}_i = U^{-1}S_i U. \tag{1.155}$$

These forms shed a new light on why the algebraic properties of the σ_i are preserved under a rotation. Taking a particular example

$$\sigma_1\sigma_2 = i\sigma_3,$$

we can write this as

$$U^{-1}\sigma_1 U U^{-1}\sigma_2 U = iU^{-1}\sigma_3 U$$

or therefore

$$\bar{\sigma}_1\bar{\sigma}_2 = i\bar{\sigma}_3,$$

and the algebra has been preserved.

Let us take another example of a rotation (rotation in the 1,3 plane). Again, the angle θ shown in Figure 1.15 is positive by our conventions. The components of the new $\bar{\sigma}_i$ operators are clearly

$$\left.\begin{aligned} \bar{\sigma}_1 &= \sigma_1\cos\theta - \sigma_3\sin\theta \\ \bar{\sigma}_2 &= \sigma_2 \\ \bar{\sigma}_3 &= \sigma_3\cos\theta + \sigma_1\sin\theta \end{aligned}\right\}. \tag{1.156}$$

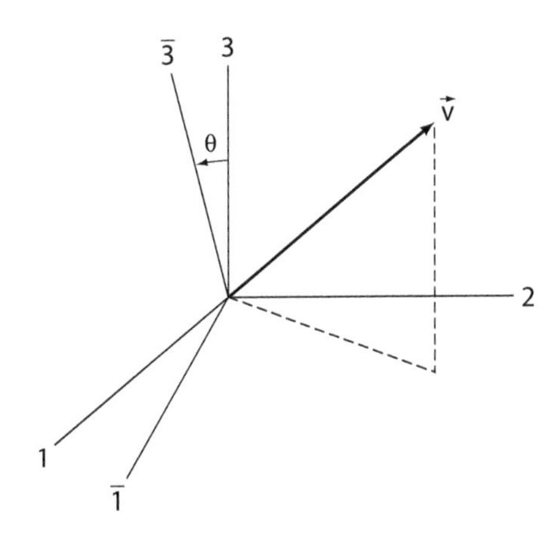

FIGURE 1.15 Showing another passive rotation by angle θ about the 2-axis.

Now let us try to produce the same $\bar{\sigma}_i$ with the following guess for U:

$$U = e^{i\theta/2\sigma_2}. \tag{1.157}$$

We find

$$\begin{aligned} U^{-1}\sigma_1 U &= e^{-i\theta/2\sigma_2}\sigma_1 e^{i\theta/2\sigma_2} = \sigma_1 e^{i\theta\sigma_2} \\ &= \sigma_1(\cos\theta + i\sigma_2 \sin\theta) = \sigma_1\cos\theta - \sigma_3\sin\theta \\ &= \bar{\sigma}_1. \end{aligned}$$

Likewise,

$$U^{-1}\sigma_2 U = \bar{\sigma}_2 \text{ (trivial)}$$

$$\begin{aligned} U^{-1}\sigma_3 U &= e^{-i\theta/2\,\sigma_2}\sigma_3 e^{i\theta/2\,\sigma_2} = \sigma_3 e^{i\theta\sigma_2} \\ &= \sigma_3(\cos\theta + i\sigma_2 \sin\theta) = \sigma_3\cos\theta + \sigma_1\sin\theta \\ &= \bar{\sigma}_3. \end{aligned}$$

Summing up, we have found that

$U = e^{i\phi/2\sigma_3}$ describes a rotation by ϕ about the 3-axis.
$U = e^{i\theta/2\sigma_2}$ describes a rotation by θ about the 2-axis.

Using these operators, we can now describe the general rotation, as seen in Figure 1.16.

We can generate the new (barred) axes from the old ones by performing the following two steps:

1. Rotate by θ about the 2-axis.
2. Rotate by ϕ about the 3-axis.

Of course, this specification of how to get the orientation of the new axes from the old is not unique. What is the U that describes this transformation?

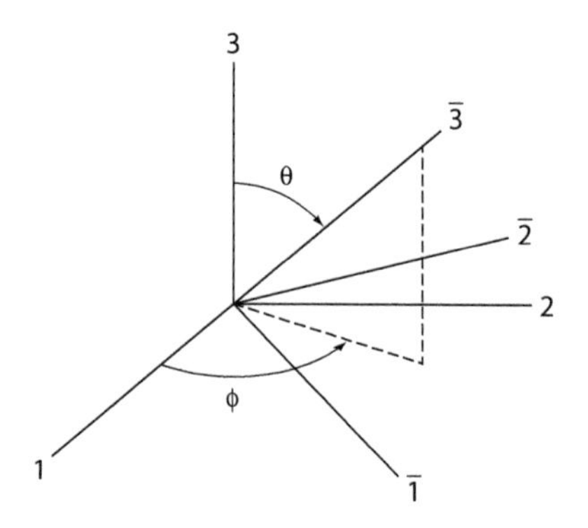

FIGURE 1.16 Showing a compound rotation of axes by rotating first by angle θ about the 2-axis and then angle ϕ about the 3-axis.

At the end of the first step, we have for the spin components S_i,

$$e^{-i\theta/2\,\sigma_2}S_i e^{i\theta/2\,\sigma_2}.$$

At the end of the second step then

$$\bar{S}_i = e^{-i\phi/2\,\sigma_3}(e^{-i\theta/2\,\sigma_2}S_i e^{i\theta/2\,\sigma_2})\,e^{i\phi/2\,\sigma_3}.$$

By comparison with the usual form $\bar{S}_i = U^{-1}S_iU$, we then find that

$$\left.\begin{aligned} U &= e^{i\theta/2\,\sigma_2}\,e^{i\phi/2\,\sigma_3} \\ U^{-1} &= e^{-i\phi/2\,\sigma_3}\,e^{-i\theta/2\,\sigma_2} \end{aligned}\right\}, \tag{1.158}$$

describes this more general rotation. Notice in Equation 1.158 that the individual exponential factors appear not only with extra minus signs, but also show up in the opposite order. This is really not so mysterious since if

$$U = U_1U_2$$

then

$$\begin{aligned} &(U_1U_2)(U_1U_2)^{-1} = 1 \\ &\Rightarrow U_2(U_1U_2)^{-1} = U_1^{-1} \\ &\Rightarrow U^{-1} = (U_1U_2)^{-1} = U_2^{-1}U_1^{-1} \end{aligned}$$

assuming U_1 and U_2 both possess inverses. Because of the noncommutative algebra, this is not the same as $U_1^{-1}U_2^{-1}$.

Using the U-transformation, we have alternate but equivalent sets of measurement symbols in the two frames of reference, which are connected by a rotation. In the "old" description we have the symbols

$$|a'a''|.$$

In the "new" description, we have the symbols

$$|\overline{a'a''}|.$$

The connection is

$$|\overline{a'a''}| = U^{-1}\,|a'a''|\,U. \tag{1.159}$$

Notice that "completeness" is preserved:

$$\sum_{a'}|\overline{a'a'}| = U^{-1}\left(\sum_{a'}|a'a'|\right)U = U^{-1}U = 1.$$

The new symbols will allow a more complete description of our original S–G two-magnet setup. In particular, they will allow us to describe the situation where the two magnets have a relative rotational angle. For

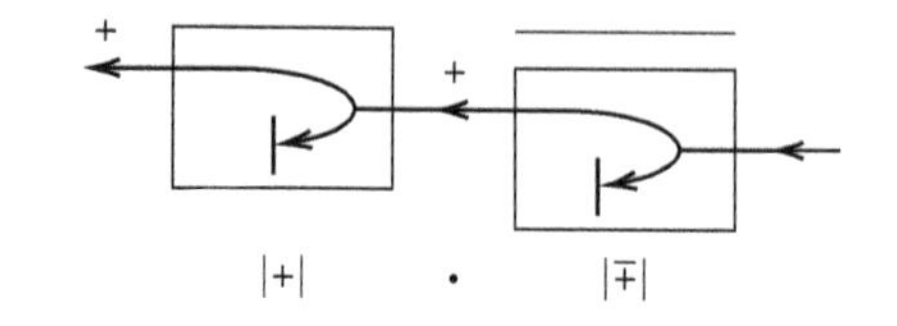

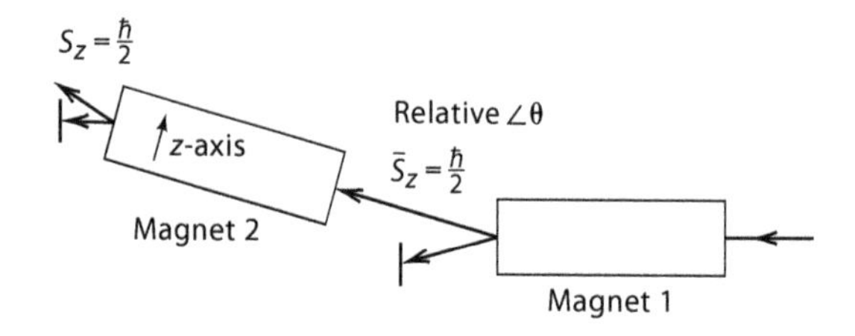

FIGURE 1.17 An idealized two-magnet Stern–Gerlach experiment selecting "up" along $\bar{z}$, then "up" along z.

example, we will model the real-space setup, as shown in Figure 1.17, where magnet 1 is rotated an angle θ with respect to magnet 2, with the Process Diagram:

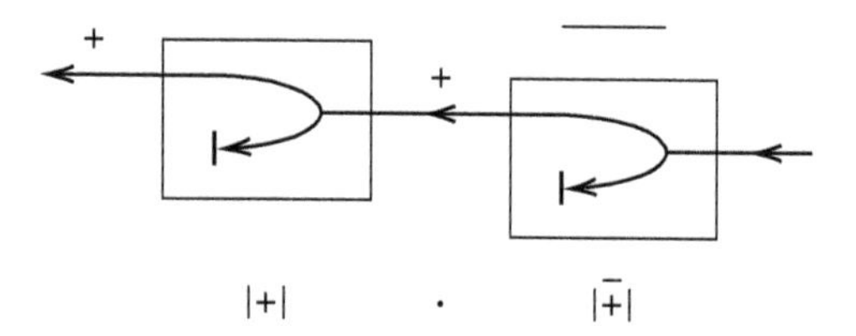

The first measurement symbol, $|\bar{+}|$, selects spin "up" along the $\bar{z}$-axis. The resultant beam is then operated on by the second symbol, $|+|$, in which a further splitting of the beam, suggestive of what happens in magnet 2 of the real-space setup in Figure 1.17, takes place. The first selection of spin up along $\bar{z}$ is suggested in the Process Diagram by barring the first selection apparatus.

1.10 BRA–KET NOTATION/BASIS STATES

We still need to know how to calculate *probabilities* based on the aforementioned ideas. First, we have a vital realization to make. Consider the transition-type measurement device:

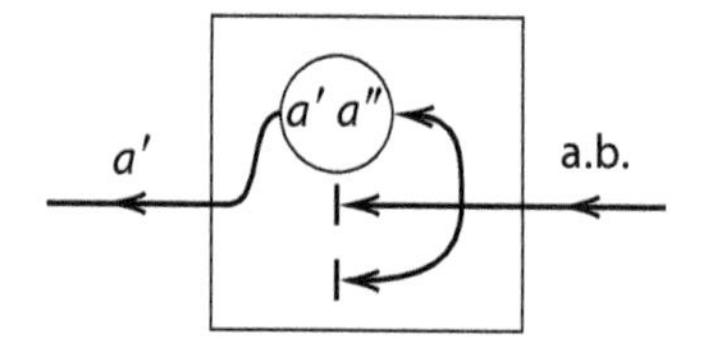

The realization that we need to make at this point is that this can be viewed as a two-step process. This is symbolized by a division of the measurement symbol for the preceding illustration into two parts—destruction and creation:

$$\underset{\text{Creation part}}{\underset{\uparrow}{|a'}}\ \underset{\text{Destruction part}}{\underset{\uparrow}{a''|}}$$

Viewing these as independent acts we will then write (even when $a' = a''$)

$$|a'a''| = |a'\rangle\langle a''|. \tag{1.160}$$

We will call

$$\left.\begin{array}{l} |a'\rangle: a \text{ "bra"} \\ \langle a''|: a \text{ "ket"} \end{array}\right\} \text{together they make a "bra–ket."}$$

"Completeness" now appears as

$$\sum_i |a_i\rangle\langle a_i| = 1. \tag{1.161}$$

The algebraic condition Equation 1.96 $\left(|a'||a''| = \delta_{a'a''}|a'|\right)$ now becomes

$$|a'\rangle\langle a'|a''\rangle\langle a''| = \delta_{a'a''}\,|a'\rangle\langle a'|. \tag{1.162}$$

Notice, in writing Equation 1.162 we have shortened $\langle a'||a''\rangle$ into $\langle a'|a''\rangle$. Now because Equation 1.162 is true for *any* $|a'\rangle\langle a''|$ combination, we must have that

$$\langle a'|a''\rangle = \delta_{a'a''}. \tag{1.163}$$

Equation 1.163 is called "orthonormality." Thus we learn that the bra–ket combinations $\langle a'|a''\rangle$ (called the "inner product") are just ordinary numbers. The connection Equation 1.159 now says

$$|\overline{a'}\rangle\langle\overline{a''}| = U^{-1}|a'\rangle\langle a''|U. \tag{1.164}$$

We separate the independent pieces:

$$|\overline{a'}\rangle = U^{-1}|a'\rangle, \tag{1.165}$$

$$\langle\overline{a''}| = \langle a''|U. \tag{1.166}$$

Using Equations 1.165 and 1.166, we easily see that

$$\langle\overline{a'}|\overline{a''}\rangle = \langle a'|UU^{-1}|a''\rangle = \langle a'|a''\rangle = \delta_{a'a''}. \tag{1.167}$$

That is, orthonormality is true in the rotated system also. Without making a distinction between bras and kets, I will often refer to a particular member $|a'\rangle$ or $\langle a'|$ as a *state*. Also in the same sense, I will call all possible states of a system in a particular description a *basis*. The collections $\{|a'\rangle\}$ and $\{|\bar{a}'\rangle\}$ represent different bases, connected however by a U-transformation.

1.11 TRANSITION AMPLITUDES

We are now ready to model our S–G probabilities. We model the general situation shown in Figure 1.18 (we are selecting either spin up or down from the magnets) with

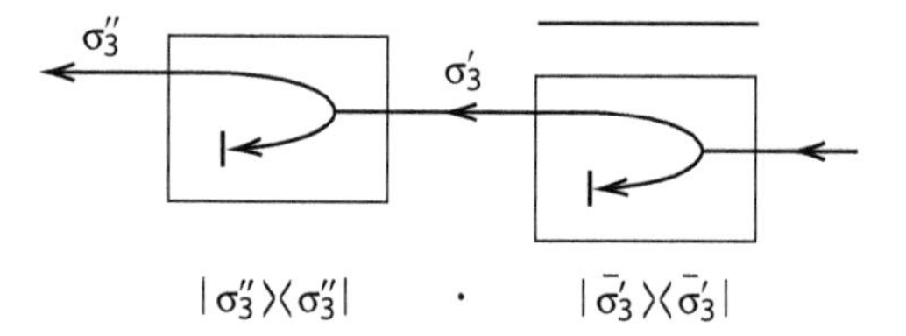

We have performed the separation (Equation 1.160) on the measurement symbols. We can now see that this last diagram is equivalent to (the physical outcomes are given by $S_3' = (\hbar/2)\sigma_3'$, $\sigma_3' = \pm 1$):

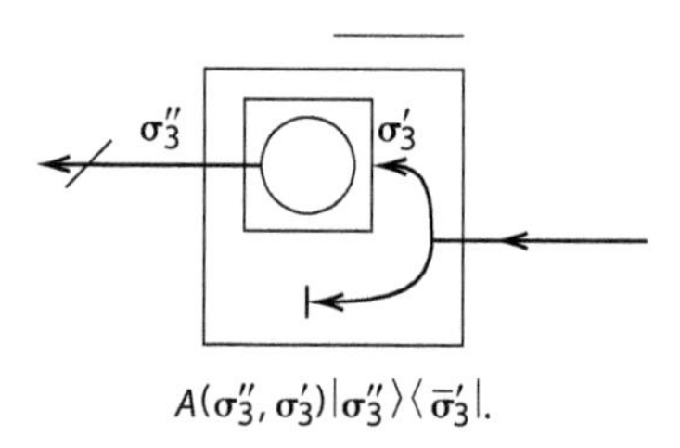

What we have drawn here is a hybrid sort of object that destroys particles with physical property σ_3' in the barred basis, transforming them into particles with physical property σ_3'' in the unbarred basis. In addition, a modulation on the selected beam has been performed. This happens, in general, because the action of the second magnet in the beam produces a further selection and thus a loss of some particles. From the equivalence of the last two Process Diagrams, we find that

$$|\sigma_3''\rangle\langle\sigma_3''||\overline{\sigma_3'}\rangle\langle\overline{\sigma_3'}| = A(\sigma_3''', \sigma_3')|\sigma_3''\rangle\langle\overline{\sigma_3'}|,$$

so

$$A(\sigma_3'', \sigma_3') = \langle\sigma_3''|\overline{\sigma_3'}\rangle. \tag{1.168}$$

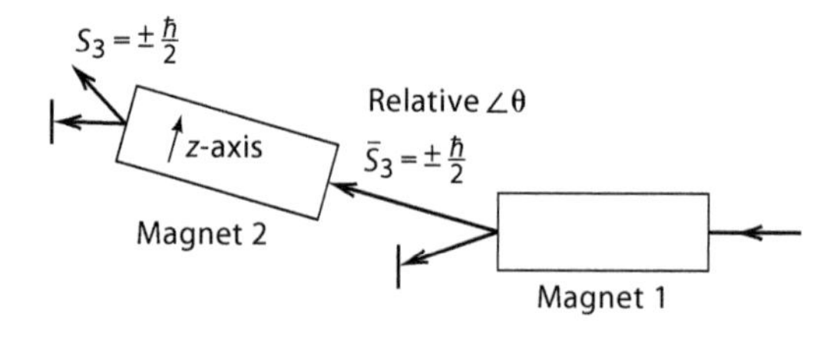

FIGURE 1.18 An idealized, general two-magnet Stern–Gerlach experiment selection.

When the relative angle $\theta = 0$, it is easy to figure out the value of $A(\sigma''_3, \sigma'_3)$. Then the barred and unbarred bases coincide and we have

$$A(\sigma''_3, \sigma'_3)_{\theta=0} = \delta_{\sigma'_3\sigma''_3}, \tag{1.169}$$

that is, either 0 or 1. This suggests in the general case that the modulus (magnitude) of $A(\sigma''_3, \sigma'_3)$ lies between these two values. However, as we will see explicitly later, $A(\sigma''_3, \sigma'_3)$ is in general a complex number. Therefore, it is not directly interpretable as a probability. Now $A(\sigma''_3, \sigma'_3)$ is just a *transition amplitude* for finding a particle with physical outcome σ''_3 in one basis given a particle with the property σ'_3 in a different, physically rotated, basis. If we say that probabilities are like *intensities*, not amplitudes, one might guess from an analogy with the relation between intensity and amplitude of waves that

$$\mathrm{p}(\sigma''_3, \sigma'_3) = (A(\sigma''_3, \sigma'_3))^2,$$

where $\mathrm{p}(\sigma''_3, \sigma'_3)$ is the associated probability for this transition. However, because $A(\sigma''_3, \sigma'_3)$ is not real, this relation does not yield a real number in general. The next simplest guess is that

$$\mathrm{p}(\sigma''_3, \sigma'_3) = |A(\sigma''_3, \sigma'_3)|^2, \tag{1.170}$$

where $|\ldots|$ denotes an absolute value. We will now explicitly calculate $A(\sigma''_3, \sigma'_3)$ using the mathematical machinery we have developed to show that Equation 1.170 gives the correct observed probabilities given in Equation 1.66.

Let us recall the geometrical situation, illustrated in Figure 1.19. This situation was studied before, where we found that

$$U = e^{i\theta/2\sigma_2}\, e^{i\phi/2\sigma_3} \tag{1.171}$$

describes the transformation between the barred and unbarred frames. Although an angle ϕ in Figure 1.19 is necessary to describe the general orientation of the 3 and $\overline{3}$ axes, we expect from our experimentally observed results, Equation 1.66, that our answers for the various probabilities will actually be independent of ϕ. Let us see if this is so.

Using Equation 1.165 in 1.168 gives

$$A(\sigma''_3, \sigma'_3) = \langle \sigma''_3 | U^{-1} | \sigma'_3 \rangle. \tag{1.172}$$

[Equation 1.172 allows a more elegant interpretation of $A(\sigma''_3, \sigma'_3)$ in terms of Process Diagrams, but one that is more unlike the real S–G setup. Instead of the earlier measurement involving the $|\sigma''_3\rangle\langle\sigma''_3|$ and $|\overline{\sigma'_3\rangle\langle\sigma'_3}|$ symbols, we can also deduce $A(\sigma''_3, \sigma'_3)$ from the diagrammatic equation

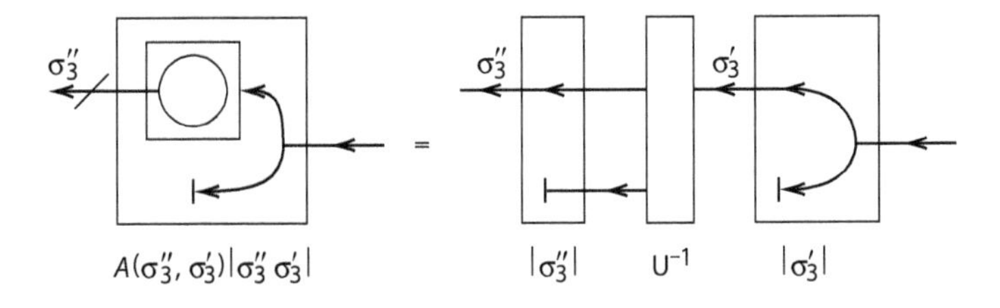

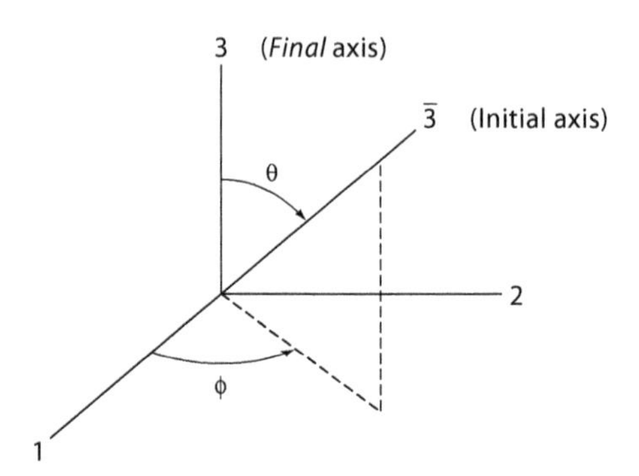

FIGURE 1.19 Relation of the original $\bar{3}$-axis to the final set for the experiment of Figure 1.18.

That is, instead of rotating the magnets, we can think of the U^{-1} above rotating the beam.]

To begin our evaluations, let us choose $\sigma_3' = \sigma_3'' = +$. Then

$$A(+,+) = \langle + | e^{-i\phi/2\,\sigma_3}\, e^{-i\theta/2\,\sigma_2} | + \rangle. \tag{1.173}$$

Some reminders:

$$\sigma_1 = |-\rangle\langle +| + |+\rangle\langle -|,$$
$$\sigma_2 = i\left(|-\rangle\langle +| - |+\rangle\langle -|\right),$$
$$\sigma_3 = |+\rangle\langle +| - |-\rangle\langle -|.$$

Therefore from orthonormality,

$$\sigma_1 |+\rangle = |-\rangle,\ \sigma_1 |-\rangle = |+\rangle,$$
$$\sigma_2 |+\rangle = i|-\rangle,\ \sigma_2 |-\rangle = -i|+\rangle,$$
$$\sigma_3 |+\rangle = |+\rangle,\ \sigma_3 |-\rangle = -|-\rangle.$$

Equation 1.173 now becomes

$$\begin{aligned} A(+,+) &= \langle + | e^{-i\phi/2} \left(\cos\frac{\theta}{2} - i\sigma_2 \sin\frac{\theta}{2} \right) | + \rangle \\ &= e^{-i\phi/2} \langle + | \cdot \left(|+\rangle \cos\frac{\theta}{2} + |-\rangle \sin\frac{\theta}{2} \right) \\ &= e^{-i\phi/2} \cos\frac{\theta}{2}. \end{aligned} \tag{1.174}$$

Thus, we get the correct result:

$$p(+,+) = |A(+,+)|^2 = \cos^2\frac{\theta}{2}.$$

Likewise, you will show in a problem that

$$A(+,-) = -e^{-i\phi/2}\sin\frac{\theta}{2}, \tag{1.175}$$

$$A(-,+) = e^{i\phi/2}\sin\frac{\theta}{2}, \tag{1.176}$$

$$A(-,-) = e^{i\phi/2}\cos\frac{\theta}{2}, \tag{1.177}$$

which also give correct probabilities.

We have now found the explicit connection between the Process Diagrams and the real-space experimental setup:

> **Drawing the Process Diagram of some experimental measurement identifies the transition amplitude of that process; the corresponding probability is the absolute square of the amplitude.**

1.12 THREE-MAGNET SETUP EXAMPLE—COHERENCE

We know that this works now for the two-magnet S–G setup; let us test this out on a three-magnet experiment. It will yield a valuable lesson. Consider the idealized setup (set $\phi = \phi' = 0$ for simplicity) in Figure 1.20. The Process Diagram is

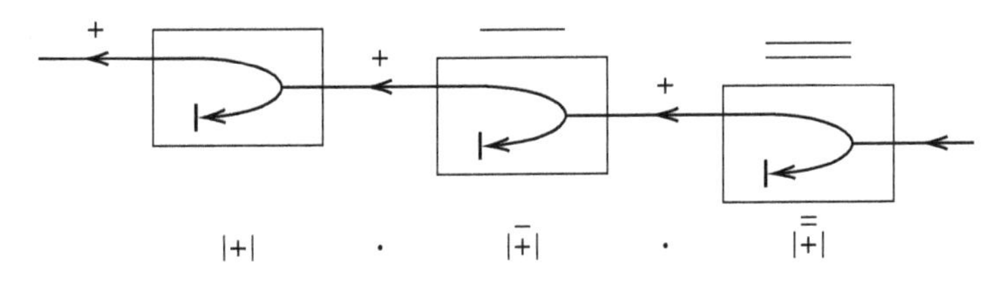

The transition amplitude is identified from

$$|+|\cdot|\bar{+}|\cdot|\bar{\bar{+}}| = |+\rangle\langle+|\bar{+}\rangle\langle\bar{+}|\bar{\bar{+}}\rangle\langle\bar{\bar{+}}|$$

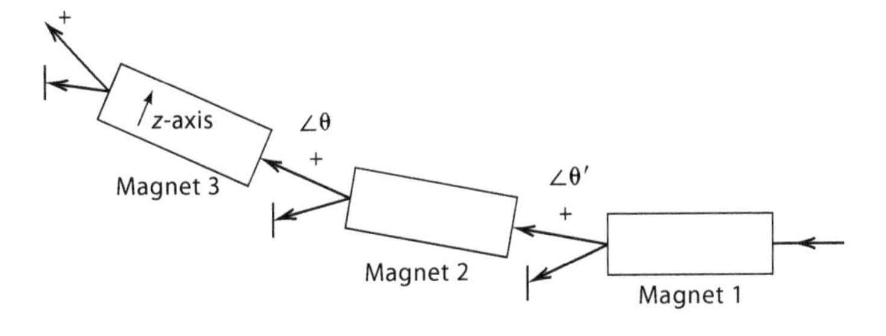

FIGURE 1.20 An idealized three-magnet Stern–Gerlach experiment.

as

$$A = \langle + | \bar{+} \rangle \langle \bar{+} | \bar{\bar{+}} \rangle.$$

The probability is then

$$p_1 = |A|^2 = |\langle + | \bar{+} \rangle|^2 |\langle \bar{+} | \bar{\bar{+}} \rangle|^2 = \cos^2 \frac{\theta}{2} \cos^2 \frac{\theta'}{2},$$

where we have used Equation 1.174, with a proper understanding of the role of the angles θ and θ'. What would we get if we had chosen spin "down" from the second magnet? The appropriate diagram is as follows:

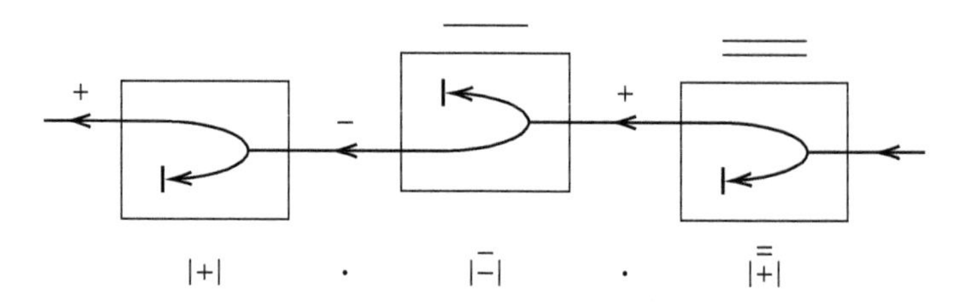

The transition amplitude from

$$|+| \cdot |\bar{-}| \cdot |\bar{\bar{+}}|$$

is

$$A = \langle + | \bar{-} \rangle \langle \bar{-} | \bar{\bar{+}} \rangle.$$

The probability is then

$$p_2 = |A|^2 = |\langle + | \bar{-} \rangle|^2 |\langle \bar{-} | \bar{\bar{+}} \rangle|^2 = \sin^2 \frac{\theta}{2} \sin^2 \frac{\theta'}{2}.$$

Given *both* choices of the intermediate magnet, the probability that final "up" is selected, given the selection "up" from magnet 1, is

$$p = p_1 + p_2 = \cos^2 \frac{\theta}{2} \cos^2 \frac{\theta'}{2} + \sin^2 \frac{\theta}{2} \sin^2 \frac{\theta'}{2}. \tag{1.178}$$

Now let us remove the second magnet entirely, without changing the orientation of magnets 1 and 3. We know the probability for

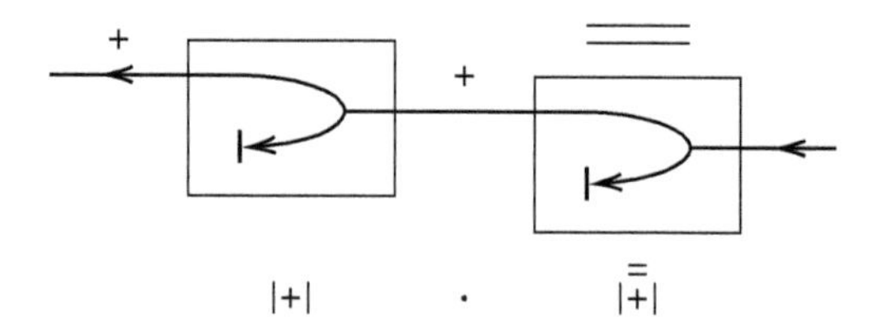

is just

$$p' = |\langle + | \bar{\bar{+}} \rangle|^2 = \cos^2 \left(\frac{\theta + \theta'}{2} \right). \tag{1.179}$$

We get a useful alternate view of the latter probability by inserting completeness (in the single-barred basis),

$$1 = |+|+|\bar{-}| = |\bar{+}\rangle\langle\bar{+}| + |\bar{-}\rangle\langle\bar{-}|,$$

as follows:

$$p' = \left|\langle\bar{+}|1|\bar{\bar{+}}\rangle\right|^2 = \left|\langle+|\bar{+}\rangle\langle\bar{+}|\bar{\bar{+}}\rangle + \langle+|\bar{-}\rangle\langle\bar{-}|\bar{\bar{+}}\rangle\right|^2.$$

Using Equations 1.174 through 1.176 here yields

$$p' = \left|\cos\frac{\theta}{2}\cos\frac{\theta'}{2} - \sin\frac{\theta}{2}\sin\frac{\theta'}{2}\right|^2. \tag{1.180}$$

Mathematically, this is the same as Equation 1.179, of course. Comparing Equation 1.180 to 1.178 we see a strong similarity. Equation 1.178 represents the sum of two *probabilities*, whereas Equation 1.180 arises from the sum of two *amplitudes*. We say that the two amplitudes that make up Equation 1.178 add *incoherently*, while the amplitudes in Equation 1.180 add *coherently*. Using this terminology, we state an important quantum mechanical principle that is being seen here as an example. That is

$$\begin{pmatrix}\textbf{Distinguishable}\\ \textbf{Indistinguishable}\end{pmatrix} \textbf{ processes add } \begin{pmatrix}\textbf{Incoherently}\\ \textbf{Coherently}\end{pmatrix}$$

Thus, there is a difference in the final outcome whether the second magnets are present or not. The act of observing whether an individual particle has spin "up" or "down" in an intermediate stage has altered the experimental outcome. Although the preceding principle has been stated in the context of the behavior of spin, it is actually a completely general quantum mechanics rule. A paraphrase of it could be: The fundamental quantum mechanical objects are amplitudes, not probabilities.

1.13 HERMITIAN CONJUGATION

We can now recover and get an interpretation of some of our previous results. Let us consider the situation of the two-magnet S–G setup described from an arbitrary orientation, as illustrated in Figure 1.21. We already saw the figure in Section 1.5; $\hat{e}_1$ represents the initial $\bar{z}$-axis and $\hat{e}_2$ represents the final $\bar{\bar{z}}$-axis. The spin basis in the single-barred frame we take to be $\{|\overline{\sigma_3'}\rangle\}$ and in the double-barred frame we use $\{|\overline{\overline{\sigma_3'}}\rangle\}$. Then, from the Process Diagram we identify the transition probability,

$$p(\sigma_3'', \sigma_3') = \left|\langle\overline{\overline{\sigma_3''}}|\overline{\sigma_3'}\rangle\right|^2, \tag{1.181}$$

where σ_3', σ_3'' = ±1 independently as usual. We are referring both S–Gs to a third, independent axis. This is accomplished by writing

$$p(\sigma_3'',\sigma_3') = \left|\langle \bar{\bar{\sigma}}_3'' \,|\, 1 \,|\, \bar{\sigma}_3' \rangle\right|^2, \tag{1.182}$$

with

$$\sum_{\sigma_3} |\sigma_3\rangle\langle\sigma_3| = 1, \tag{1.183}$$

where the basis $\{|\sigma_3\rangle\}$ refers to the unbarred frame. Therefore,

$$p(\sigma_3'',\sigma_3') = \left|\sum_{\sigma_3} \langle \bar{\bar{\sigma}}_3'' \,|\, \sigma_3' \rangle\langle \sigma_3 \,|\, \bar{\sigma}_3' \rangle\right|^2. \tag{1.184}$$

We have already calculated the following:

$$\langle + | \bar{+} \rangle = e^{-i\phi/2}\cos\frac{\theta}{2},$$

$$\langle + | \bar{-} \rangle = -e^{-i\phi/2}\sin\frac{\theta}{2},$$

$$\langle - | \bar{+} \rangle = e^{i\phi/2}\sin\frac{\theta}{2},$$

$$\langle - | \bar{-} \rangle = e^{i\phi/2}\cos\frac{\theta}{2}.$$

The new objects we need to calculate are as follows:

$$\langle \bar{\bar{\sigma}}_3' \,|\, \sigma_3 \rangle = \langle \sigma_3'' \,|\, U \,|\, \sigma_3 \rangle, \tag{1.185}$$

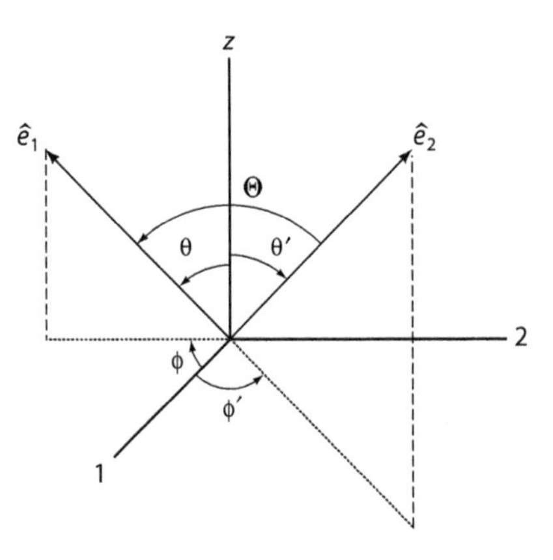

FIGURE 1.21 Same as Figure 1.13. See the text for the interpretation of the axes in terms of a two-magnet experiment.

with the same U as in Section 1.11 with $\theta,\phi \rightarrow \theta',\phi'$. Using this form in the same way as before, we find

$$\langle \bar{\bar{+}} | + \rangle = e^{i\phi'/2} \cos\frac{\theta'}{2}, \tag{1.186}$$

$$\langle \bar{\bar{+}} | - \rangle = e^{-i\phi'/2} \sin\frac{\theta'}{2}, \tag{1.187}$$

$$\langle \bar{\bar{-}} | + \rangle = -e^{i\phi'/2} \sin\frac{\theta'}{2}, \tag{1.188}$$

$$\langle \bar{\bar{-}} | - \rangle = e^{-i\phi'/2} \cos\frac{\theta'}{2}. \tag{1.189}$$

Let us examine the structure of p(+, +):

$$\begin{aligned}
p(+,+) &= \left| \langle \bar{\bar{+}} | + \rangle \langle + | \bar{+} \rangle + \langle \bar{\bar{+}} | - \rangle \langle - | \bar{+} \rangle \right|^2 \\
&= \left| \left(e^{i\phi'/2} \cos\frac{\theta'}{2} \right) \left(e^{-i\phi/2} \cos\frac{\theta}{2} \right) + \left(e^{-i\phi'/2} \sin\frac{\theta'}{2} \right) \left(e^{i\phi/2} \sin\frac{\theta}{2} \right) \right|^2 \\
&= \left| \begin{pmatrix} e^{-i\phi'/2} \cos\frac{\theta'}{2} & e^{i\phi'/2} \sin\frac{\theta'}{2} \end{pmatrix}^* \cdot \begin{pmatrix} e^{-i\phi/2} \cos\frac{\theta}{2} \\ e^{i\phi/2} \sin\frac{\theta}{2} \end{pmatrix} \right|^2 \\
&= \left| \psi_+(\theta',\phi')^\dagger \, \psi_+(\theta,\phi) \right|^2,
\end{aligned} \tag{1.190}$$

where we have used the previous matrix definition of $\psi_+(\theta,\phi)$, Equation 1.80. Likewise, we may write

$$\begin{aligned}
p(-,+) &= \left| \langle \bar{\bar{-}} | + \rangle \langle + | \bar{+} \rangle + \langle \bar{\bar{-}} | - \rangle \langle - | \bar{+} \rangle \right|^2 \\
&= \left| \left(-e^{i\phi'/2} \sin\frac{\theta'}{2} \right) \left(e^{-i\phi/2} \cos\frac{\theta}{2} \right) + \left(e^{-i\phi'/2} \cos\frac{\theta'}{2} \right) \left(e^{i\phi/2} \sin\frac{\theta}{2} \right) \right|^2 \\
&= \left| \begin{pmatrix} -e^{-i\phi'/2} \sin\frac{\theta'}{2} & e^{i\phi'/2} \cos\frac{\theta'}{2} \end{pmatrix}^* \cdot \begin{pmatrix} e^{i\phi/2} \cos\frac{\theta}{2} \\ e^{i\phi/2} \sin\frac{\theta}{2} \end{pmatrix} \right|^2 \\
&= \left| \psi_-(\theta',\phi')^\dagger \, \psi_+(\theta,\phi) \right|^2,
\end{aligned} \tag{1.191}$$

where we have now used the definition of the matrix $\psi_-(\theta,\phi)$, Equation 1.81. Equations 1.190 and 1.191 are exactly the same as the forms for p(+,+) and p(−,+) found in Equations 1.82 and 1.83.

We make several realizations from our recovery of the previous results. First, we notice that when $\theta' = \theta$ and $\phi' = \phi$ (then, the distinction between the $\{|\bar{\bar{\sigma}}_3'\rangle\}$ and $\{|\bar{\sigma}_3'\rangle\}$ states disappears), we have from Equations 1.174 through 1.177 (listed again above Equation 1.185) and Equations 1.186 through 1.189 that

$$\langle \bar{\sigma}_3' | \sigma_3 \rangle^* = \langle \sigma_3 | \bar{\sigma}_3' \rangle, \tag{1.192}$$

for any value of σ_3, σ_3'. Notice that Equation 1.192 is consistent with Equation 1.41 and so conserves probabilities. Equation 1.192 suggests that the act of complex conjugation, or actually some *extension* of its usual meaning, interchanges bras and kets. Let us define an operation "†" that does this. (We have already introduced this symbol in Equations 1.82 and 1.83 for example, and I will comment on this situation later.) In general for any bra or ket,

$$(C\langle a'|)^\dagger \equiv C^*|a'\rangle, \tag{1.193}$$

$$(C|a'\rangle)^\dagger \equiv C^*\langle a'|, \tag{1.194}$$

where C is any complex number. (C^* is the complex conjugate of C.) For any operator,

$$(|a''\rangle\langle a'|)^\dagger \equiv |a'\rangle\langle a''|. \tag{1.195}$$

In addition, we define

$$\langle a'|a''\rangle^\dagger \equiv \langle a''|a'\rangle. \tag{1.196}$$

In Equation 1.196 we are requiring that the "†" operation relates one inner product to another in a one-to-one fashion. Thus, "†" reduces to complex conjugation when acting on a number, but when acting on an operator just gives another operator. What's more, we assume this operation is *distributive* over addition.

Let us test out the effect of "†," which we will call *Hermitian conjugation* (or "Hermitian adjoint" or just "adjoint") on the operators $\sigma_{1,2,3}$:

$$\sigma_3^\dagger = [|+\rangle\langle+| - |-\rangle\langle-|]^\dagger = \sigma_3, \tag{1.197}$$

$$\sigma_2^\dagger = [i|-\rangle\langle+| - i|+\rangle\langle-|]^\dagger = \sigma_2, \tag{1.198}$$

$$\sigma_1^\dagger = [|-\rangle\langle+| + |+\rangle\langle-|]^\dagger = \sigma_1. \tag{1.199}$$

Operators that satisfy $A^\dagger = A$ are said to be *Hermitian* or *self-conjugate.* Such operators are of fundamental importance in quantum mechanics; we will discuss the reason for this later in this chapter. Notice now that we have two operations designated as "†." The first time we used this symbol, it was in a matrix context, and it meant "complex conjugation + transpose." In the present context of Dirac bra–ket notation it now means "complex conjugation + bra–ket interchange." Mathematically, it would be better to introduce a distinction between these two uses of the same symbol. In practice, however, physicists let the context tell them which usage is appropriate. We will follow this point of view here. That there is a connection between these sets of mathematical operations will be evident in a moment.

Let us be crystal clear about what we have done in introducing "†" in this context. Because the quantity $\langle\bar{\sigma}_3'\,|\,\sigma_3\rangle$ is just a (complex) number, we have the trivial statement that

$$\langle\bar{\sigma}_3'\,|\,\sigma_3\rangle^{\dagger} = \langle\bar{\sigma}_3'\,|\,\sigma_3\rangle^{*}. \tag{1.200}$$

On the other hand, using Equation 1.196 gives

$$\langle\bar{\sigma}_3'|\,\sigma_3\rangle^{\dagger} = \langle\sigma_3\,|\,\bar{\sigma}_3'\rangle. \tag{1.201}$$

Comparing Equations 1.200 and 1.201 gives Equation 1.192. Thus, the existence of an operation that interchanges bras and kets, but reduces to ordinary complex conjugation when applied to complex numbers, renders Equation 1.192 an identity. We will assume the existence of such an operation.

1.14 UNITARY OPERATORS

The existence of the Hermitian conjugation operation has another important consequence. From the left-hand side of Equation 1.201 we must have

$$\langle\bar{\sigma}_3'|\,\sigma_3\rangle^{\dagger} = \langle\sigma_3'|\,U\,|\,\sigma_3\rangle^{\dagger} = \langle\sigma_3\,|\,U^{\dagger}\,|\,\sigma_3'\rangle, \tag{1.202}$$

whereas the equality stated there demands this is the same as

$$\langle\sigma_3\,|\,\bar{\sigma}_3'\,\rangle = \langle\sigma_3\,|\,U^{-1}\,|\,\sigma_3'\rangle. \tag{1.203}$$

Comparing the right-hand sides of Equations 1.202 and 1.203, for any states $\langle\sigma_3\,|, |\,\sigma_3'\rangle$, demands that

$$U^{-1} = U^{\dagger}. \tag{1.204}$$

Such an operator as in Equation 1.204 is called *unitary*. We now ask the question: Are the U-transformation operators we have written down so far unitary? Let us concentrate on the transformation Equation 1.171. First, we derive an important rule for Hermitian conjugation. We have by definition

$$(\langle a'\,|\,U_1U_2)^{\dagger} = (U_1U_2)^{\dagger}\,|\,a'\rangle. \tag{1.205}$$

However, we may also write

$$(\langle a'|U_1U_2)^\dagger = (\langle\bar{a}'|U_2)^\dagger = U_2^\dagger|\bar{a}'\rangle, \tag{1.206}$$

where

$$|\bar{a}'\rangle = (\langle\bar{a}'|)^\dagger = (\langle a'|U_1)^\dagger = U_1^\dagger|a'\rangle. \tag{1.207}$$

Substituting Equation 1.207 in 1.206 gives

$$(\langle a'|U_1U_2)^\dagger = U_2^\dagger U_1^\dagger|a'\rangle. \tag{1.208}$$

Then comparing Equations 1.205 and 1.208 for any $|a'\rangle$ tells us that

$$(U_1U_2)^\dagger = U_2^\dagger U_1^\dagger. \tag{1.209}$$

Although the rule Equation 1.209 has been stated for unitary operators, it is in fact true of *any* product of operators.

Now applying Equation 1.209 to Equation 1.171 gives

$$U^\dagger = (e^{i\theta/2\,\sigma_2} e^{i\phi/2\,\sigma_3})^\dagger = (e^{i\phi/2\,\sigma_3})^\dagger (e^{i\theta/2\,\sigma_2})^\dagger. \tag{1.210}$$

Remembering that (Equation 1.148)

$$e^{i\lambda(\hat{n}\cdot\vec{\sigma})} = \cos\lambda + i(\hat{n}\cdot\vec{\sigma})\sin\lambda,$$

and from Equations 1.197 through 1.199, we then have

$$U^\dagger = (e^{i\phi/2\,\sigma_3})^\dagger (e^{i\theta/2\,\sigma_2})^\dagger = e^{-i\phi/2\,\sigma_3}\, e^{-i\theta/2\,\sigma_2}. \tag{1.211}$$

The last form on the right is identically U^{-1}, so we have shown that this U is unitary. We will have a lot more to say about unitary operators as we go along.

1.15 A VERY SPECIAL OPERATOR

I will now introduce the spin operator combinations

$$S_+ \equiv S_1 + iS_2, \tag{1.212}$$

$$S_- \equiv S_1 - iS_2. \tag{1.213}$$

Note that $S_+ = S_-^\dagger$. Using the commutator Equation 1.141 from Section 1.8, one may show that (also note that $[A+B,C] = [A,C] + [B,C]$):

$$[S_+, S_-] = 2\hbar S_3, \tag{1.214}$$

$$[S_+, S_3] = -\hbar S_+, \tag{1.215}$$

$$[S_-, S_3] = \hbar S_-. \tag{1.216}$$

(One more note: Equation 1.216 follows from Equation 1.215 by Hermitian conjugation.) The operators S_+ and S_- are called "raising" and "lowering" operators, respectively. Although these are written down for the spin 1/2 case, we will see that Equations 1.214 through 1.216 have very general validity. The Process Diagram illustration of the scaled quantity $S_+ / \hbar = |+-|$ for spin 1/2 is seen to correspond to

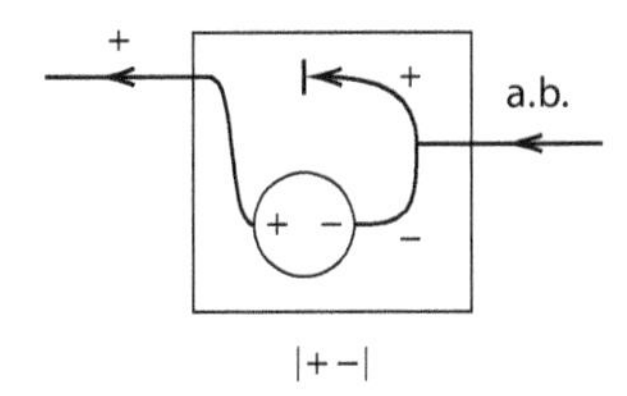

which is not really a pure "raising" operation because of the blocking of the "+" beam, but does promote "−" to "+".

The three-physical-outcomes case ($a_1 = 1, a_2 = 0, a_3 = -1$) can be used to represent the next highest spin value, "spin one," if one assigns S_3 as the three-outcome modulated operator with $S_3' = (\hbar, 0, -\hbar)$ ($|++| \equiv |+|$, etc.):

$$S_3 = \sum_{S_3'} S_3' |S_3' / \hbar| = \hbar(|+| - |-|). \tag{1.217}$$

One can now show that (not unique)

$$S_+ = \sqrt{2}\,\hbar(|+0| + |0-|) \tag{1.218}$$

satisfies the commutation relations Equations 1.214 and 1.215. (Can you draw the implied Process Diagram for $S_+ / \hbar$?) Of course, this result also gives concrete representations for S_1 and S_2 from inverting Equations 1.212 and 1.213.

We will return to higher spin considerations in Chapter 8, where we will display the general form for S_+ for any spin value.

1.16 MATRIX REPRESENTATIONS

Generalizing Equations 1.190 and 1.191, we have

$$\mathrm{p}(\sigma_3'', \sigma_3') = |\,\psi_{\sigma_3''}(\theta', \phi')^{\dagger}\, \psi_{\sigma_3'}(\theta, \phi)\,|^2, \tag{1.219}$$

which is identical to Equation 1.84 if we change the names a', a'' to σ_3', σ_3''. We called the $\psi_{\sigma_3}(\theta, \phi)$ "wave functions." Another realization we make comes from comparing Equation 1.219 with the earlier expression Equation 1.184, written in the form

$$\mathrm{p}(\sigma_3'', \sigma_3') = \left| \sum_{\sigma_3} \langle \sigma_3 | \bar{\bar{\sigma}}_3'' \rangle^* \langle \sigma_3 | \bar{\bar{\sigma}}_3' \rangle \right|^2, \tag{1.220}$$

which now reveals explicitly that

$$\left[\psi_{\sigma_3'}(\theta,\phi)\right]_{\sigma_3} = \langle\sigma_3|\bar{\sigma}_3'\rangle, \tag{1.221}$$

where the σ_3 on the left-hand side of this equation is being viewed as the row index of $\psi_{\sigma_3'}(\theta,\phi)$. Thus, the components of $\psi_{\sigma_3'}(\theta,\phi)$ are revealed as just the *transition amplitudes* in Equation 1.168. This gives us a concrete interpretation of the wave function, for which we use Figure 1.22.

> $\left[\psi_{\sigma_3'}(\theta,\phi)\right]_{\sigma_3}$ **is the transition amplitude for spin** $\frac{1}{2}$ **selected along the** $\bar{3}$**-axis** ($\sigma_3' = \pm 1$) **having component** $\sigma_3 = \pm 1$ **along the 3-axis.**

Explicitly again

$$\psi_+(\theta,\phi) = \begin{pmatrix} \langle +|\bar{+}\rangle \\ \langle -|\bar{+}\rangle \end{pmatrix} = \begin{pmatrix} e^{-i\phi/2}\cos\frac{\theta}{2} \\ e^{i\phi/2}\sin\frac{\theta}{2} \end{pmatrix}, \tag{1.222}$$

$$\psi_-(\theta,\phi) = \begin{pmatrix} \langle +|\bar{-}\rangle \\ \langle -|\bar{-}\rangle \end{pmatrix} = \begin{pmatrix} -e^{-i\phi/2}\sin\frac{\theta}{2} \\ e^{i\phi/2}\cos\frac{\theta}{2} \end{pmatrix}. \tag{1.223}$$

The fact that the preceding wave functions can be displayed as column matrices suggests that the rest of the algebra we have introduced involving the spin operators $\sigma_{1,2,3}$ can also be viewed as matrix manipulations. Indeed this is so, and we will discuss later how this can be done for spin 1/2, deferring a more general discussion until we come to Chapter 4. The basic idea is as follows. Instead of using the spin 1/2 operators $\sigma_{1,2,3}$, we may instead use the *matrix representations* of them. Given a spin basis $\{|\sigma_3'\rangle\}$ and an operator A, one may produce a two-indexed object $A_{\sigma_3'\sigma_3''}$ as in

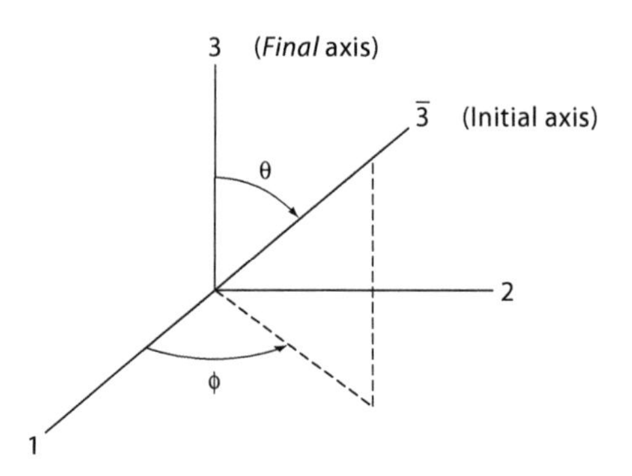

FIGURE 1.22 Same as Figure 1.19. Used for interpretation of the wave function.

$$A_{\sigma_3'\sigma_3''} = \langle \sigma_3' \,|\, A \,|\, \sigma_3'' \rangle. \tag{1.224}$$

If we now interpret σ_3' as the row index and σ_3'' as the column index, we may regard this object as a matrix. Explicitly, for $\sigma_{1,2,3}$ we find

$$\sigma_1 = |-\rangle\langle +| + |+\rangle\langle -| \Rightarrow \sigma_1 = \begin{array}{c|cc} \sigma_3'/\sigma_3'' & + & - \\ \hline + & 0 & 1 \\ - & 1 & 0 \end{array}$$

$$\sigma_2 = i|-\rangle\langle +| - i|+\rangle\langle -| \Rightarrow \sigma_2 = \begin{array}{c|cc} \sigma_3'/\sigma_3'' & + & - \\ \hline + & 0 & -i \\ - & i & 0 \end{array}$$

$$\sigma_3 = |+\rangle\langle +| - |-\rangle\langle -| \Rightarrow \sigma_3 = \begin{array}{c|cc} \sigma_3'/\sigma_3'' & + & - \\ \hline + & 1 & 0 \\ - & 0 & -1 \end{array}.$$

Notice that I have not attempted to assign a different symbol to the σ's when they are regarded as matrices, although to be mathematically precise, we probably should. This again is the common usage; the context will tell us what is meant. These $\sigma_{1,2,3}$ matrices are called the *Pauli matrices.*

Using our matrix language now, let us verify some of the prior algebraic statements involving the σ's. First, let us notice what Equations 1.222 and 1.223 become, when $\theta = \phi = 0$,

$$\psi_+(0,0) = \begin{pmatrix} 1 \\ 0 \end{pmatrix}, \quad \psi_-(0,0) = \begin{pmatrix} 0 \\ 1 \end{pmatrix}. \tag{1.225}$$

We usually call $\begin{pmatrix} 1 \\ 0 \end{pmatrix}$ "spin up" and $\begin{pmatrix} 0 \\ 1 \end{pmatrix}$ "spin down." Then in terms of this matrix language, let us write a few examples of how the matrix algebra works.

Operator statement	**Matrix realization**
$\langle +\|\sigma_1 = \langle -\|$	$(1\ \ 0)\begin{pmatrix} 0 & 1 \\ 1 & 0 \end{pmatrix} = (0\ \ 1)$
$\langle -\|\sigma_1 = \langle +\|$	$(0\ \ 1)\begin{pmatrix} 0 & 1 \\ 1 & 0 \end{pmatrix} = (1\ \ 0)$
$\langle +\|\sigma_2 = -i\langle -\|$	$(1\ \ 0)\begin{pmatrix} 0 & -i \\ i & 0 \end{pmatrix} = (0\ \ -i)$

$\langle -|\sigma_2 = i\langle +|$ $\quad (0\ 1)\begin{pmatrix} 0 & -i \\ i & 0 \end{pmatrix} = (i\ 0)$

$\sigma_1|+\rangle = |-\rangle$ $\quad \begin{pmatrix} 0 & 1 \\ 1 & 0 \end{pmatrix}\begin{pmatrix} 1 \\ 0 \end{pmatrix} = \begin{pmatrix} 0 \\ 1 \end{pmatrix}$

$\sigma_1|-\rangle = |+\rangle$ $\quad \begin{pmatrix} 0 & 1 \\ 1 & 0 \end{pmatrix}\begin{pmatrix} 0 \\ 1 \end{pmatrix} = \begin{pmatrix} 1 \\ 0 \end{pmatrix}$

$\sigma_2|+\rangle = i|-\rangle$ $\quad \begin{pmatrix} 0 & -i \\ i & 0 \end{pmatrix}\begin{pmatrix} 1 \\ 0 \end{pmatrix} = \begin{pmatrix} 0 \\ i \end{pmatrix}$

$\sigma_2|-\rangle = -i|+\rangle$ $\quad \begin{pmatrix} 0 & -i \\ i & 0 \end{pmatrix}\begin{pmatrix} 0 \\ 1 \end{pmatrix} = \begin{pmatrix} -i \\ 0 \end{pmatrix}.$

One may also easily check that the algebraic properties of the σ_i stated in Equation 1.139 also hold for the Pauli matrices. Although there is an ambiguity in our notation now, when we write σ_3, say, we mean the matrix representation or the more abstract operator quantity, let us be clear about the difference of the two in our minds. The more fundamental of the two is the Hilbert space operator quantity. The matrix representation is just a realization of the basic operator quantity in a particular basis. By changing to a rotated basis $\{|\bar{\sigma}_3'\rangle\}$ we could in fact change the representation, although the basic underlying operator structure has not changed. The relationship between operator and representation is summed up very picturesquely in a footnote in Chapter 1 of Sakurai and Napolitano's *Modern Quantum Mechanics**: "The operator is different from a representation of the operator just as the actor is different from a poster of the actor."

The fact that there are two languages of spin, operator and matrix, explains why the same symbol "†" was introduced in two apparently different contexts. In a matrix context, if you remember, it meant "complex conjugation + transpose." In an operator context it meant "complex conjugation + bra–ket interchange." For example, we can now verify the statements in Equations 1.197 through 1.199 using the first meaning of "†":

$$\sigma_3^\dagger = \begin{pmatrix} 1 & 0 \\ 0 & -1 \end{pmatrix}^\dagger = \begin{pmatrix} 1 & 0 \\ 0 & -1 \end{pmatrix} = \sigma_3,$$

$$\sigma_2^\dagger = \begin{pmatrix} 0 & -i \\ i & 0 \end{pmatrix}^\dagger = \begin{pmatrix} 0 & -i \\ i & 0 \end{pmatrix} = \sigma_2,$$

$$\sigma_1^\dagger = \begin{pmatrix} 0 & 1 \\ 1 & 0 \end{pmatrix}^\dagger = \begin{pmatrix} 0 & 1 \\ 1 & 0 \end{pmatrix} = \sigma_1.$$

* J.J. Sakurai and J. Napolitano, *Modern Quantum Mechanics*, 2nd ed. (Addison-Wesley, 2011).

It should now be clear why I did not introduce a distinction between the two senses of the adjoint symbol. It is because one is carrying out the same basic mathematical operation but in different contexts.

1.17 MATRIX WAVE FUNCTION RECOVERY

While we are on the subject of the matrix language of spin, let us see if we can recover the $\psi_{\sigma_3'}(\theta,\phi)$ wave functions, which we have already derived from operator methods, using instead matrix manipulations.

From Equation 1.102, we know that

$$A|a'| = a'|a'|.$$

Since $|a'| = |a'\rangle\langle a'|$, multiplying on the right by $|a'\rangle$ gives

$$A|a'\rangle = a'|a'\rangle. \tag{1.226}$$

We know that σ_3 plays the role of "A" in the two-physical-outcomes case, which we know now as spin 1/2. Therefore

$$\sigma_3|\sigma_3'\rangle = \sigma_3'|\sigma_3'\rangle. \tag{1.227}$$

The matrix realization of this statement is

$$\sigma_3\,\psi_{\sigma_3'}(0,0) = \sigma_3'\,\psi_{\sigma_3'}(0,0), \tag{1.228}$$

where σ_3 is now a matrix. Likewise, in a rotated basis, the analog of Equation 1.227 is

$$\bar{\sigma}_3|\overline{\sigma_3'}\rangle = \sigma_3'|\overline{\sigma_3'}\rangle. \tag{1.229}$$

There are of course only two values allowed for σ_3' ($\sigma_3' = \pm 1$ as usual), but $\bar{\sigma}_3$ is the rotated version of σ_3. We already know how to compute $\bar{\sigma}_3$. It is given by

$$\bar{\sigma}_3 = U^{-1}\sigma_3 U, \tag{1.230}$$

with

$$U = e^{i\theta/2\sigma_2}\, e^{i\phi/2\sigma_3}. \tag{1.231}$$

(You will carry out the evaluation of $\bar{\sigma}_3$ in a problem.) Another, easier way of deriving this quantity is simply to project out the component along the new $\bar{3}$-axis. With the help of the unit vector

$$\hat{n} = (\sin\theta\cos\phi,\ \sin\theta\sin\phi, \cos\theta), \tag{1.232}$$

pointing in the direction given by the spherical coordinate angles θ and ϕ, we find $\bar{\sigma}_3$ as

$$\bar{\sigma}_3 = \vec{\sigma}\cdot\hat{n} = \sigma_1\,\sin\theta\cos\phi + \sigma_2\,\sin\theta\sin\phi + \sigma_3\,\cos\theta. \tag{1.233}$$

We therefore have

$$\bar{\sigma}_3 = \sin\theta\cos\phi\begin{pmatrix} 0 & 1 \\ 1 & 0 \end{pmatrix} + \sin\theta\sin\phi\begin{pmatrix} 0 & -i \\ i & 0 \end{pmatrix} + \cos\theta\begin{pmatrix} 1 & 0 \\ 0 & -1 \end{pmatrix},$$

$$\Rightarrow \bar{\sigma}_3 = \begin{pmatrix} \cos\theta & e^{-i\phi}\sin\theta \\ e^{i\phi}\sin\theta & -\cos\theta \end{pmatrix}. \tag{1.234}$$

The matrix realization of Equation 1.229 is then

$$\begin{pmatrix} \cos\theta & e^{-i\phi}\sin\theta \\ e^{i\phi}\sin\theta & -\cos\theta \end{pmatrix}\begin{pmatrix} \psi_{\sigma_3'}(+) \\ \psi_{\sigma_3'}(-) \end{pmatrix} = \sigma_3'\begin{pmatrix} \psi_{\sigma_3'}(+) \\ \psi_{\sigma_3'}(-) \end{pmatrix}, \tag{1.235}$$

where we are labeling the upper and lower matrix components of $\psi_{\sigma_3'}(\theta,\phi)$ as $\psi_{\sigma_3'}(+)$ and $\psi_{\sigma_3'}(-)$. Equation 1.221 represents two equations in two unknowns. Written out explicitly, we have

$$\cos\theta\,\psi_{\sigma_3'}(+) + e^{-i\phi}\sin\theta\,\psi_{\sigma_3'}(-) = \sigma_3'\,\psi_{\sigma_3'}(+), \tag{1.236}$$

$$e^{i\phi}\sin\theta\,\psi_{\sigma_3'}(+) - \cos\theta\,\psi_{\sigma_3'}(-) = \sigma_3'\,\psi_{\sigma_3'}(-), \tag{1.237}$$

or

$$(\cos\theta - \sigma_3')\psi_{\sigma_3'}(+) + e^{-i\phi}\sin\theta\,\psi_{\sigma_3'}(-) = 0, \tag{1.238}$$

$$e^{i\phi}\sin\theta\,\psi_{\sigma_3'}(+) - (\cos\theta + \sigma_3')\,\psi_{\sigma_3'}(-) = 0, \tag{1.239}$$

We know from the theory of linear equations that for the preceding two (homogeneous) equations to have a nontrivial solution, we must have that the determinant of the coefficients is zero. Therefore:

$$\begin{aligned} &-(\cos\theta - \sigma_3')(\cos\theta + \sigma_3') - \sin^2\theta = 0 \\ &\Rightarrow \sigma_3' - \cos^2\theta - \sin^2\theta = 0 \\ &\Rightarrow \sigma_3'^2 = 1. \end{aligned} \tag{1.240}$$

We have recovered the known result that $\sigma_3' = \pm 1$ only. Let us look at the $\sigma_3' = 1$ case now. Equation 1.238 gives

$$\begin{aligned} &-2\sin^2\frac{\theta}{2}\psi_+(+) + 2\sin\frac{\theta}{2}\cos\frac{\theta}{2}e^{-i\phi}\psi_+(-) = 0 \\ &\Rightarrow \frac{\psi_+(+)}{\psi_+(-)} = e^{-i\phi}\frac{\cos\theta/2}{\sin\theta/2}. \end{aligned} \tag{1.241}$$

All that is determined is the *ratio* $\psi_+(+)/\psi_+(-)$. (Equation 1.239 gives the same result.) Therefore, a solution is

$$\psi_+(+) = Ae^{-i\phi/2}\cos\frac{\theta}{2}, \tag{1.242}$$

$$\psi_+(-) = Ae^{i\phi/2}\sin\frac{\theta}{2}, \tag{1.243}$$

where A is a common factor. It is not completely arbitrary because these transition amplitudes give probabilities that must add to one:

$$|\psi_+(+)|^2 + |\psi_+(-)|^2 = 1. \tag{1.244}$$

A statement like Equation 1.244 is called *normalization*. Equation 1.244 shows that we must have $|A|^2 = 1$. However, A is still not completely determined because we can still write $A = e^{i\alpha}$ with α being some undetermined phase. Such an overall phase has no effect on probabilities however, since probabilities are the absolute square of amplitudes. The operator analog of the normalization condition Equation 1.244 is just the statement

$$\langle \mp | \mp \rangle = 1. \tag{1.245}$$

(Compare Equation 1.245 with Equation 1.167.) Question: Can you derive Equation 1.244 from Equation 1.245? [Hint: use completeness.]

So, outside of an arbitrary phase, we have recovered the result (Equation 1.222) for $\psi_+(\theta,\phi)$ which we previously derived using operator techniques. Similarly, we can recover the result (Equation 1.223) for $\psi_-(\theta,\phi)$ if we choose to look at either Equation 1.238 or 1.239 for $\sigma_3' = -1$. In mathematics, an equation of the form $A|a'\rangle = a'|a'\rangle$ is called an *eigenvalue* equation, the state $|a'\rangle$ is called an *eigenvector* (or *eigenket*), and the number a' is called the *eigenvalue*. For spin 1/2 the eigenvectors are just $|+\rangle$ and $|-\rangle$ and the eigenvalues of the operator S_z are just $\pm\hbar/2$. The above is just a numerical treatment of the more abstract Hilbert space statements in a rotated frame.

1.18 EXPECTATION VALUES

I would like to return now to our spin measurements that we began this chapter with to see how they now look in our emerging formalism. We found that the average value of an electron's magnetic moment along the z-axis, given an initial selection along the $+\bar{z}$-axis was

$$\langle \mu_z \rangle_+ = -\mu(p(+,+) - p(-,+)). \tag{1.246}$$

We saw that $\langle \mu_z \rangle_+$ was obtained as the weighted sum of physical outcomes. In our new language of operators (remember $\mu_z = \gamma S_3$), consider the quantity (we have inserted completeness twice):

$$\langle \mp | \mu_z | \mp \rangle = \gamma \sum_{\sigma_3', \sigma_3'' = \pm 1} \langle \mp | \sigma_3' \rangle \langle \sigma_3' | S_3 | \sigma_3'' \rangle \langle \sigma_3'' | \mp \rangle. \tag{1.247}$$

But

$$\langle \sigma_3' | S_3 | \sigma_3'' \rangle = \langle \sigma_3' | \frac{\hbar}{2} \sigma_3 | \sigma_3'' \rangle = \frac{\hbar}{2} \delta_{\sigma_3' \sigma_3''} \sigma_3', \tag{1.248}$$

so

$$\begin{aligned} \langle \mp | \mu_z | \mp \rangle &= \frac{\gamma\hbar}{2} \sum_{\sigma_3'} \sigma_3' \langle \mp | \sigma_3' \rangle \langle \sigma_3' | \mp \rangle \\ &= \frac{\gamma\hbar}{2} \sum_{\sigma_3'} \sigma_3' | \langle \sigma_3' | \mp \rangle |^2 . \end{aligned} \tag{1.249}$$

This gives

$$\langle\mp|\mu_z|\mp\rangle=\frac{\gamma\hbar}{2}(p(+,+)-p(-,+)). \tag{1.250}$$

Since $\mu=-\gamma\hbar/2$ for the electron, we learn that

$$\langle\mu_z\rangle_+ = \langle\mp|\mu_z|\mp\rangle. \tag{1.251}$$

This is in a spin context. In more generality, we say that the *expectation value* of the operator A in the state $|\psi\rangle$ is

$$\langle A\rangle_\psi \equiv\langle\psi|A|\psi\rangle. \tag{1.252}$$

If $A=A^\dagger$, then expectation values are real:

$$\begin{aligned}\langle A\rangle_\psi^* &=\langle\psi|A|\psi\rangle^* =\langle\psi|A|\psi\rangle^\dagger\\ &=\langle\psi|A^\dagger|\psi\rangle=\langle A\rangle_\psi.\end{aligned} \tag{1.253}$$

This is the reason, mentioned in Section 1.13, that Hermiticity of operators is so fundamental. I will have more to say about expectation values, in a *continuum* eigenvalue context, in Chapter 2, Section 2.16.

1.19 WRAP-UP

I have three final points to make before closing this (already too long) chapter. First, you may have noticed that we have never asked any questions about the results of an S–G measurement of the original beam of atoms coming directly from the hot oven. That is because we have not developed the necessary formalism to describe such a situation. The formalism we have developed assumes that any mixture of spin up and down that we encounter is a *coherent* mixture, that is, the beam of particles is a mixture of the *amplitudes* of spin up and down. The furnace beam, however, is a completely *incoherent* mixture of spin up and down, that is, the *probabilities* (or intensities) of the components are adding. There is a formalism for dealing with the more general case of incoherent mixtures, but we will not discuss such situations here. In fact, our Process Diagrams are incapable of representing an incoherent beam of particles. The "arbitrary beam" entering from the right in such diagrams is really not arbitrary at all but only represents the most general coherent mixture. Nevertheless, we have seen the usefulness of these diagrams in modeling situations where the behavior of coherent beams is concerned. If you are interested, you will find a useful discussion of these matters in the Sakurai and Napolitano book,* starting in Section 3.4. Such considerations become important in any realistic experimental situation where one must deal with beams that are only partially coherent.

* Sakurai and Napolitano, *Modern Quantum Mechanics.*

An important point of connection concerns the mathematics used for spin. The three spin 1/2 matrices, $\vec{\sigma}$, are the "generators" of a structure called the *SU*(2) Lie group. "*SU*" denotes *s*pecial (denoting determinant one) *u*nitary matrices. These matrices can be used to form the most general *SU*(2) matrix, which can be written in the form of Equation 1.148. We will see this group, and others, appear again in the Chapter 12 discussion regarding the group structure of gauge theories.

The last point I wish to make concerns the Process Diagrams themselves. They are only meant to be a stepping-stone in our efforts to learn the principles of quantum mechanics. These principles have a structure of their own and are independent of our diagrams, which are an attempt to simply make *manifest* some of these principles. The basic purpose of these diagrams is to illustrate the existence of quantum mechanical states with certain measurable physical characteristics and to make transparent the means of computing associated amplitudes. We will get a more systematic view of these principles and properties from our study of Hilbert space in Chapter 4.

You should now be in possession of an introductory intuitive and mathematical understanding of the simplest of all quantum mechanical systems, spin 1/2. We will now go on to see how the principles of quantum mechanics apply to other particle and system characteristics.

PROBLEMS

1.1.1 The instantaneous power radiated from a nonrelativistic, accelerated charge is

$$\mathrm{P} = \frac{2}{3}\frac{e^2}{c^3}|\vec{a}|^2,$$

where $\vec{a}$ is the acceleration, and e and c are the particle's charge and the speed of light, respectively (cgs units). According to this classical result, a hydrogen atom should collapse in a very short amount of time because of energy loss. *Estimate* the time period for the hydrogen atom's electron to radiate away its kinetic energy, $E = e^2/2a_0$. Take as a crude model of the atom an electron moving in a circular orbit of radius $a_0 = \hbar^2/me^2$, with a velocity $v^2 = e^2/ma_0$. How many electron orbits does this represent? [These numerical values for a_0 and v come from the simple Bohr model of the hydrogen atom, which we will study in the next chapter.]

1.2.1 I used the estimate $|\mu_z| \approx 10^{-20}$ erg/gauss for the magnitude of the Ag z-component magnetic moment. Reach a more accurate estimate of this value by looking up and plugging in values in

$$|\mu_z| = \frac{e}{mc}|S_z|,$$

assuming $|S_z| = \hbar/2$. Should one use the Ag atom or the electron mass for m?

1.2.2 Think of a gas in thermal equilibrium at a temperature T, each atom acting as a tiny magnetic dipole, $\vec{\mu}$. Let's not worry what the source of this field is at present. (Classically, magnetic dipole fields arise from current loops.) Imagine putting such atoms in an external magnetic field, $\vec{H}$. Of course, without $\vec{H}$, we would expect there to be no preferred direction and so no *net* gas magnetic moment. Let us use Maxwell–Boltzmann statistics to describe the gas:

$$\left.\begin{array}{l}\text{The probability that an atom}\\ \text{has a magnetic moment}\\ \text{between } \vec{\mu} \text{ and } \vec{\mu}+d\vec{\mu}.\end{array}\right\} = \frac{d\Omega e^{-\beta(-\vec{\mu}\cdot\vec{H})}}{\int d\Omega e^{-\beta(-\vec{\mu}\cdot\vec{H})}}$$

Therefore

$$\langle\vec{\mu}\rangle_T = \frac{\int d\Omega\, e^{\beta\,\vec{\mu}\cdot\vec{H}}\,\vec{\mu}}{\int d\Omega\, e^{\beta\,\vec{\mu}\cdot\vec{H}}}.$$

(a) Consider the situation where $|\vec{\mu}\cdot\vec{H}| << kT$.

$$\langle\vec{\mu}\rangle_T = \frac{\int d\Omega\vec{\mu}\left(1+\left(\vec{\mu}\cdot\vec{H}/kT\right)\right)}{\int d\Omega} = \frac{0+(1/kT)\int d\Omega\,\vec{\mu}\left(\vec{\mu}\cdot\vec{H}\right)}{4\pi}.$$

Since $\vec{H}$ is the only direction in the problem, we hypothesize that

$$\int d\Omega\vec{\mu}\left(\vec{\mu}\cdot\vec{H}\right) = C\vec{H},$$

where C is just some constant. By taking $\vec{H}$ along the z-direction and using $\mu_z = |\vec{\mu}|\cos\theta$, find the value of C and show that for weak magnetic fields

$$\langle\vec{\mu}\rangle_T \approx \frac{\vec{\mu}^2}{3kT}\vec{H}.$$

This result is called the "Curie law" and was established by Pierre Curie. We see that the collection of gas molecules has a small net magnetic moment pointed *along* $\vec{H}$. Such behavior is called *paramagnetism.*

(b) Calculate $\langle\vec{\mu}\rangle_T$ without the weak-field approximation. Take the z-axis along $\vec{H}$, so that $\vec{\mu}\cdot\vec{H} = \mu H\cos\theta$. Then $\langle\mu_x\rangle = \langle\mu_y\rangle = 0$, but

$$\langle\mu_z\rangle = \frac{\mu\int d\Omega\,\cos\theta\; e^{\beta\mu H\cos\theta}}{\int d\Omega\, e^{\beta\mu H\cos\theta}},$$

where $d\Omega$ is the element of solid angle. Plot (qualitatively) your result for $\langle \mu_z \rangle$ as a function of H. Find the limits,

$$\lim_{H\to 0}\langle \mu_z \rangle_T = ?,$$

$$\lim_{H\to \infty}\langle \mu_z \rangle_T = ?.$$

and give a physical interpretation of these two limits.

[Hint: The top integral can be done by considering the derivative of the denominator with respect to β. The answer to (b) is

$$\langle \mu_z \rangle_T = \mu\left(\coth(\mu\beta H) - \frac{1}{\mu\beta H}\right).]$$

1.2.3 The expected *classical* result for the single magnet Stern–Gerlach experiment was to find a single continuous line of atoms on the screen positioned beyond the magnet. We now want to find an expression for the expected classical number of atoms, $N(\theta)$, as a function of deflection angle, θ, as illustrated in Figure 1.23. Do this in two parts:

(a) Assuming that the magnetic moments, $\vec{\mu}$, of the particles in the arbitrary beam are randomly oriented, and taking the $+z$-axis along the magnet's magnetic field, show that the number of atoms with $\mu_z = \mu\cos\theta'$ relative to those with $\mu_z = \mu$ (i.e., at $\theta' = \pi/2$) is given by

$$\frac{N(\theta')}{N(\theta' = \pi/2)} = |\sin\theta'|.$$

[Hint: Consider thin strips of area dA and dA' in Figure 1.24 and compare the relative number of atoms included.]

(b) Now, establish a relationship between the angles θ and θ' in Figures 1.23 and 1.24. Use your result in part (a) to show that

$$\begin{cases} N(\theta) = N(\theta = 0)\left(1 - \dfrac{\theta^2}{\theta_{max}^2}\right)^{1/2}, \\ \theta_{max} = \left|\mu \dfrac{\partial H}{\partial z}\dfrac{L}{mv^2}\right|. \end{cases}$$

θ_{max} is the classical maximum deflection angle.

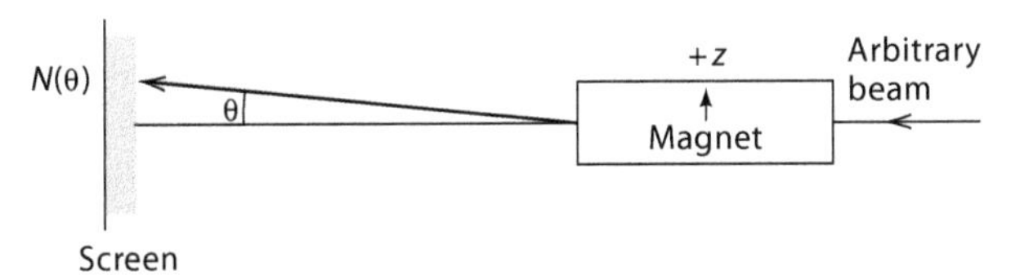

FIGURE 1.23 Idealized setup for measuring the classical outcome of a spin measurement.

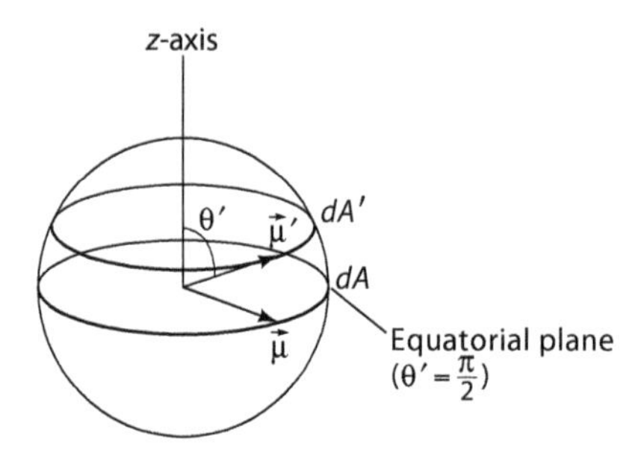

FIGURE 1.24 Illustrating the small areas dA and dA' on a sphere.

1.3.1 Define

$\langle \mu_z \rangle_- \equiv$ average value of $\vec{\mu}$ along the z-axis, given an initial selection of the *downward* deflected beam along z'.

Find $\langle \mu_z \rangle_-$. Does it agree with what you expect?

1.4.1 (a) The energy of a charge, q, moving with velocity $\vec{v}$ in an external electromagnetic field is (classically)

$$E = q\,\Phi - \frac{q}{c}\vec{v}\cdot\vec{A},$$

where Φ is the scalar potential and $\vec{A}$ is the vector potential of the electromagnetic field. The relationship between $\vec{A}$ and $\vec{B}$ is

$$\vec{B} = \vec{\nabla}\times\vec{A}.$$

For a constant $\vec{B}$ field in space, verify that

$$\vec{A} = \frac{1}{2}\vec{B}\times\vec{r},$$

is a possible form of the vector potential. [Hint: The vector identity,

$$\vec{\nabla}\times(\vec{a}\times\vec{b}) = \vec{a}(\vec{\nabla}\cdot\vec{b}) - \vec{b}(\vec{\nabla}\cdot\vec{a}) + (\vec{b}\cdot\vec{\nabla})\vec{a} - (\vec{a}\cdot\vec{\nabla})\vec{b}$$

may be of use.]

(b) From the $-(q/c)\vec{v}\cdot\vec{A}$ term in E, identify a form for $\vec{\mu}$ from Equation 1.29.

(c) A moving charge will execute circular (or helical) motion in a constant $\vec{B}$ field. Using (b), show that $\vec{\mu} = (q/2mc)\vec{L}$, where $\vec{L}$ is the angular momentum of the particle. Assuming the plane of the orbit is perpendicular to $\vec{B}$, and given that $\vec{F} = q(\vec{E} + (\vec{v}/c)\times\vec{B})$ is the force, does $\vec{\mu}$ point along $\vec{B}$ or opposite to $\vec{B}$? (Extra: Can you think of a famous law of electromagnetism that explains this direction?)

1.4.2 A well-known formula of classical electromagnetism states that

$$|\vec{\mu}| = \frac{\mathcal{I}}{c} A$$

where $\vec{\mu}$ is the magnetic moment and $\mathcal{I}(= dq/dt$, where "q" is the charge) is the current in a circuit of area A. Evaluate $|\vec{\mu}|$ for the classical spinning electron model (thin spherical shell of radius a model), and show that

$$|\vec{\mu}| = -\frac{1}{3}\frac{e\omega a^2}{c} = \left(-\frac{e}{2mc}\right)L,$$

where $-e$ is the electron charge and $L = I\omega$, where I is the moment of inertia, is the magnitude of the electronic angular momentum. [This gives the classical gyromagnetic ratio seen in Equation 1.43.]

1.4.3 Replace the thin-shelled spherical electron model with a solid spherical ball of charge, throughout which the charge is uniformly distributed.

(a) Find the moment of inertia of this object (replacing Equation 1.61).

(b) Show that the relation between $|\vec{\mu}|$ and L for this new model is still given by

$$|\vec{\mu}| = -\frac{e}{2mc}L,$$

[Use the technique in Problem 1.4.2 to compute $|\vec{\mu}|$.]

1.5.1 (a) Show that Equation 1.79 of the text is equivalent to the preceding Equation 1.78.

(b) Evaluate p(–,–) and p(+,–) from Equation 1.84 in the case $\theta' = 0$. Are the results as expected?

1.6.1 Illustrate, using Process Diagrams, the product of measurement symbols,

$$(|a'| + |a''|)|a'''|,$$

in the cases where

(a) $a' = a'' \neq a'''$

(b) $a' \neq a'' = a'''$

(c) $a' = a'' = a'''$.

(Consider the case with, say, four possible physical outcomes, just to be concrete.)

1.7.1 In the two-physical-outcomes case ($a_1 = 1$, $a_2 = -1$) evaluate

$$e^{i\lambda A}$$

as a function of A. [Hint: Expand $e^{i\lambda A}$ in a power series and use Equation 1.112, then sum the resulting infinite series.]

1.7.2 In the three-physical-outcomes case ($a_1 = 1$, $a_2 = 0$, $a_3 = -1$) find
(a) The algebraic equation satisfied by $A = \sum_i a_i | a_i |$.
(b) $|+|$, $|0|$, $|-|$ as functions of A.
(c) $e^{i\lambda A}$ as a function of A using part (a).
[Answer: $e^{i\lambda A} = A^2 \cos\lambda + iA\sin\lambda + (1 - A^2)$.]
(d) Give a Process Diagram interpretation to the equation found in part (a). (Supplement your diagram with an explanation in words, if you think this is necessary.)

1.8.1 Draw Process Diagrams representing the following equations:
(a) $(|+-|)^2 = 0$

(b) $\sigma_1^3 = \sigma_1$

(c) $\sigma_1\,\sigma_3 = -i\,\sigma_2$

1.8.2 Using

$$\sigma_1 = |-+| + |+-|,\ \sigma_2 = i(|-+| - |+-|),$$
$$\sigma_3 = |+| - |-|,$$

show
(a) $\sigma_2\,\sigma_3 = i\sigma_1$
(b) $\sigma_3\,\sigma_1 = i\sigma_2$

1.9.1 Using

$$\bar{\sigma}_3 = \sigma_3$$
$$\bar{\sigma}_2 = -\sigma_1\ \sin\phi + \sigma_2\ \cos\phi$$
$$\bar{\sigma}_1 = \sigma_1\ \cos\phi + \sigma_2\ \sin\phi$$

and

$$\sigma_i^2 = 1$$
$$\sigma_i\,\sigma_j = i\sigma_k\ \ (i, j, k \text{ cyclic})$$
$$\sigma_i\,\sigma_j = -\sigma_j\,\sigma_i\ \ (i \neq j),$$

show
(a) $\bar{\sigma}_1\bar{\sigma}_2 = i\bar{\sigma}_3$
(b) $\bar{\sigma}_2\bar{\sigma}_3 = i\bar{\sigma}_1$
(c) $\bar{\sigma}_3\bar{\sigma}_1 = i\bar{\sigma}_2$

1.9.2 Show that

$$e^{i\phi(\hat{n}\cdot\vec{\sigma})} = \cos\phi + i(\hat{n}\cdot\vec{\sigma})\sin\phi,$$

where $\hat{n}$ is an arbitrary unit vector ($\hat{n}^2 = 1$). [Hint: First show that $(\hat{n}\cdot\vec{\sigma})^2 = 1$, then expand the exponential in a power series, remembering Problem 1.7.1.]

1.9.3 Show that

$$\frac{1}{2}\vec{\sigma}\times\frac{1}{2}\vec{\sigma} = \frac{i}{2}\vec{\sigma}.$$

Remember that $(\vec{A}\times\vec{B})_i = \sum_{j,k}\varepsilon_{ijk} A_j B_k$. [Hint: It may be useful to recall that

$$\sum_{j,k}\varepsilon_{ijk}\varepsilon_{\ell jk} = 2\delta_{i\ell}.]$$

1.9.4 Using the "commutator,"

$$[A,B] \equiv AB - BA,$$

finish the equality:

$$[\vec{\sigma}\cdot\vec{a}, \vec{\sigma}\cdot\vec{b}] = ?$$

[$\vec{a}$ and $\vec{b}$ are vectors whose components are numbers, whereas $\vec{\sigma} = \sigma_1\hat{i} + \sigma_2\hat{j} + \sigma_3\hat{k}$, where the σ's are operators.]

1.9.5 Finish the equation:

$$\sum_{k=1}^{3}\sigma_k\vec{\sigma}\sigma_k = ?$$

1.9.6 Verify that

$$\bar{\sigma}_3 = \sigma_1 \sin\theta\cos\phi + \sigma_2 \sin\theta \sin\phi + \sigma_3 \cos\theta,$$

is produced by

$$\bar{\sigma}_3 = U^{-1}\sigma_3 U, \quad U = e^{i\theta/2\sigma_2}\, e^{i\phi/2\sigma_3}$$

[Hints: Simply use the algebraic properties of the σ's, write the exponentials in their factorial forms (e.g., $e^{i\phi/2\sigma_3} = \cos(\phi/2) + i\sigma_3 \sin(\phi/2)$) and do the algebra.]

1.11.1 Verify the results stated in Equations 1.175 through 1.177.

1.14.1 For any polynomial or infinite series in A, $f(A)$, show that

$$U^{-1} f(A) U = f(\bar{A}),$$

where $U^{-1} = U^{\dagger}$ and $\bar{A} \equiv U^{-1}AU$.

1.15.1 (a) Verify the Equations 1.214 through 1.216 from the commutation relations, Equation 1.141.

(b) Show that

$$S_+ = \sqrt{2}\hbar\ (|+\rangle\langle 0| + |0\rangle\langle -|),$$

satisfies the commutation relations in Equations 1.214 and 1.215 for spin 1.

1.16.1 (a) Find matrix representations for S_3 as well as the S_1 and S_2 operators for spin 1 from the S_+ found in Problem 1.15.1. [Answer (for +,0,– row/column ordering):

$$S_1 = \frac{\hbar}{\sqrt{2}}\begin{pmatrix} 0 & 1 & 0 \\ 1 & 0 & 1 \\ 0 & 1 & 0 \end{pmatrix},$$

$$S_2 = \frac{i\hbar}{\sqrt{2}}\begin{pmatrix} 0 & -1 & 0 \\ 1 & 0 & -1 \\ 0 & 1 & 0 \end{pmatrix},$$

$$S_3 = \hbar\begin{pmatrix} 1 & 0 & 0 \\ 0 & 0 & 0 \\ 0 & 0 & -1 \end{pmatrix}.]$$

(b) Show that $\vec{S}^2 = S_1^2 + S_2^2 + S_3^2 = 2\hbar^2 .1$.

1.16.2 A beam of atoms with spin one ($S_3' = -\hbar, 0, \hbar$) passes through two Stern–Gerlachs, which have a relative polar angle θ between their *z*-axes (Figure 1.25). Given that the probability p(+,+), associated with the selection $S_3'' = +\hbar$ along the second magnet's 3-axis, is given by

$$p(+,+) = |<+|e^{i\theta S_2/\hbar}|+>|^2,$$

show that

$$p(+,+) = \cos^4\left(\frac{\theta}{2}\right).$$

[Hint: Remember the result of the three-physical-outcomes case, Problem 1.7.2:

$$e^{i\lambda A} = A^2\cos\lambda + iA\sin\lambda + (1 - A^2).$$

Expand the exponential and use the representation for S_2 from Problem 1.16.1.]

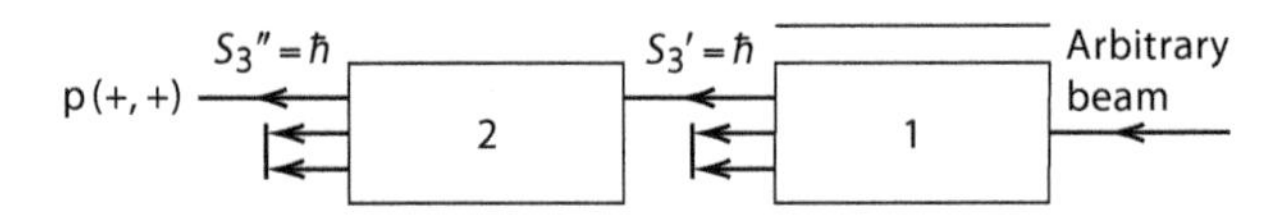

FIGURE 1.25 Illustrating an idealized Stern–Gerlach selection experiment for spin "up" along two different axes for a spin 1 system.

***1.16.3** (a) Given the new basis in the spin 1 (three-physical-outcomes) case,

$$|\bar{+}\rangle \equiv \frac{1}{\sqrt{2}}(|+\rangle + i|0\rangle),$$

$$|\bar{0}\rangle \equiv -|-\rangle, \text{ and}$$

$$|\bar{-}\rangle \equiv -\frac{1}{\sqrt{2}}(|+\rangle - i|0\rangle),$$

find the unitary operator, U, such that

$$|\overline{a_j}\rangle = U|a_j\rangle,$$

where we will associate the (matrix) index $j=1$ with the + ($\bar{+}$) state, $j=2$ with the 0 ($\bar{0}$) state and $j=3$ with the – state ($\bar{-}$).

(b) Construct the spin 1 matrices in this representation (for +, 0, – row/column ordering):

$$\bar{S}_1 = i\hbar\begin{pmatrix} 0 & 0 & 0 \\ 0 & 0 & -1 \\ 0 & 1 & 0 \end{pmatrix},$$

$$\bar{S}_2 = i\hbar\begin{pmatrix} 0 & 0 & 1 \\ 0 & 0 & 0 \\ -1 & 0 & 0 \end{pmatrix},$$

$$\bar{S}_3 = i\hbar\begin{pmatrix} 0 & -1 & 0 \\ 1 & 0 & 0 \\ 0 & 0 & 0 \end{pmatrix}.$$

[Hint: I wrote $\bar{S}_+ = \sqrt{2}\,\hbar(|\overline{+0}| + |\overline{0-}|)$, $\bar{S}_3 = \hbar\,(|\overline{++}| - |\overline{--}|)$, and used the transformation on states from (a).]

1.16.4 Given the most general rotation possible,

$$U = e^{i\psi/2\,\sigma_3}\, e^{i\theta/2\,\sigma_2}\, e^{i\phi/2\,\sigma_3},$$

find components of the spin 1/2 wave function:

$$\left[\psi_{\sigma_3'}\left(\psi,\theta,\phi\right)\right]_{\sigma_3} = ?$$

[The change in ψ is very trivial compared to our earlier form for $[\psi_{\sigma_3'}(\theta,\phi)]_{\sigma_3}$.]

1.17.1 Derive the normalization condition,

$$|\psi_{+}(+)|^2 + |\psi_{+}(-)|^2 = 1,$$

where $\psi_{\sigma_3'}(\sigma_3) = \langle \sigma_3 | \bar{\sigma}_3' \rangle$ (see Equation 1.221) from the orthonormality condition,

$$\langle \bar{+} | \bar{+} \rangle = 1.$$

(As usual, the bar denotes a rotated state. This is the spin one-half case.)

Particle Motion in One Dimension

Synopsis: After drilling down to the basic theory, the process of building up the quantum description from experiment is continued. The reader is introduced to the Schrödinger equation and related concepts. The emphasis is on what experiment tells us about particle properties, and various experiments as well as the Bohr model are used to motivate wave–particle dualism ideas. It is argued that the Fourier transform is the mathematical concept needed to build in this dualism. The Schrödinger equation is deduced from an analogy with the wave equation, and connection is made to the ideas of probability. Finally, the entire formalism is recovered and illuminated by reconnecting to the more formal spin basis ideas of the first chapter.

2.1 PHOTOELECTRIC EFFECT

We started the last chapter with experimental indications of a breakdown in classical mechanics. We start this chapter by citing two famous experiments that helped to begin to construct a new picture.

It had been known since the work of H. Hertz in 1887 that electromagnetic waves incident on a metal surface can eject electrons from that material. This was in the context of an experiment where he noticed that a spark could jump a gap between two metallic electrodes more easily when the gap was illuminated. In particular, it was established before Einstein's explanation of the effect in 1905 that shining light on metallic surfaces leads to ejected electrons. Einstein's formulas were not verified until 1916 by R.A. Millikan. Einstein received the Nobel Prize in 1921 for this work.

Consider the experimental arrangement shown in Figure 2.1 for measuring this *photoelectric effect*. We can shine light of varying intensities and frequencies on the metal surface indicated; a stream of electrons will then be emitted, some of which will strike the other metal plate inside the vacuum tube, thus constituting a current flow. The classical picture of the behavior of the electrons under such circumstances did not explain the known facts. In particular, it was known that there existed a *threshold frequency*, dependent on the type of metal being studied, below which no photoelectrons were emitted. Once this threshold frequency has been exceeded, one may then adjust the magnitude of the stopping voltage to cancel the photocurrent. (Let us call this voltage V_{max}.) In his explanation of the photoelectric effect, Einstein

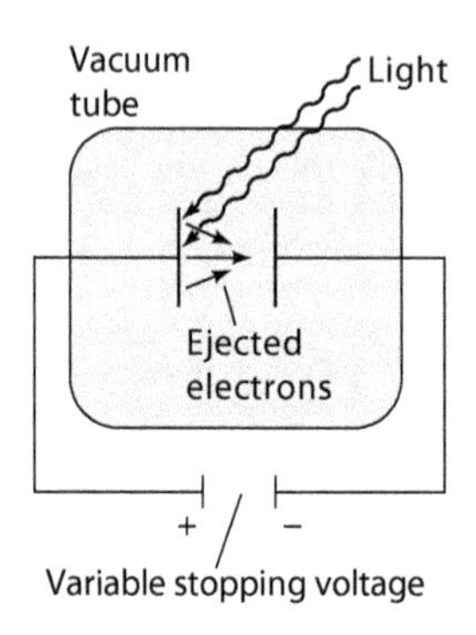

FIGURE 2.1 Schematic experimental setup for the photoelectric effect.

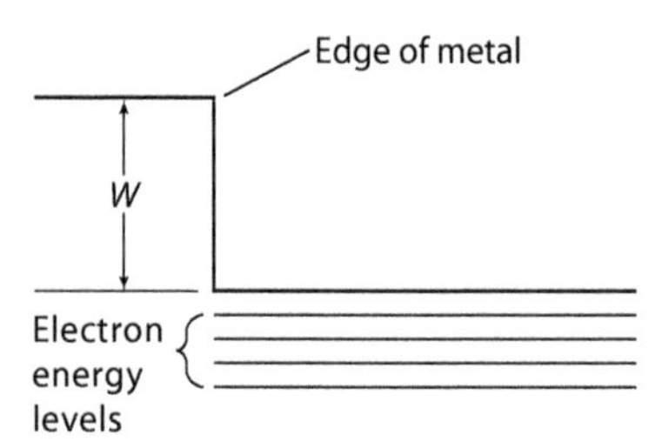

FIGURE 2.2 Schematic illustration of the work function, *W*, seen by electrons at the edge of a metal.

assumed that the incoming beam of light could be viewed as a stream of particles traveling at the speed of light, each of which carried an energy

$$E = h\nu, \tag{2.1}$$

where h is Planck's constant. Planck had originally postulated a relation like Equation 2.1 to hold for the atoms in the wall of a hot furnace, the so-called *black-body problem*.* Einstein then supposed that, based on this particle picture of light, the maximum energy of an electron ejected from the metal's surface could be written as

$$T_{max} = h\nu - W \tag{2.2}$$

Equation 2.2 is based on energy conservation. The phenomenological picture Einstein was led to was one in which the energy transferred to an electron by the photon, $h\nu$, is used to overcome the *minimum* energy necessary to remove an electron from that metal, W, which is called the *work function*. This is illustrated in Figure 2.2.

The work function, W, is not accounted for in Einstein's theory but is assumed to be a constant characteristic of a particular metal. A clear implication of Equation 2.2 is that there exists a frequency ν_0, given by

$$\nu_0 = \frac{W}{h}, \tag{2.3}$$

* We will deal with black body radiation again when we come to discuss Bose–Einstein statistics in Chapter 9. There we will see that the black body radiation law is simply a consequence of the particle nature of light, independent of assumptions about the atoms in the wall of the furnace.

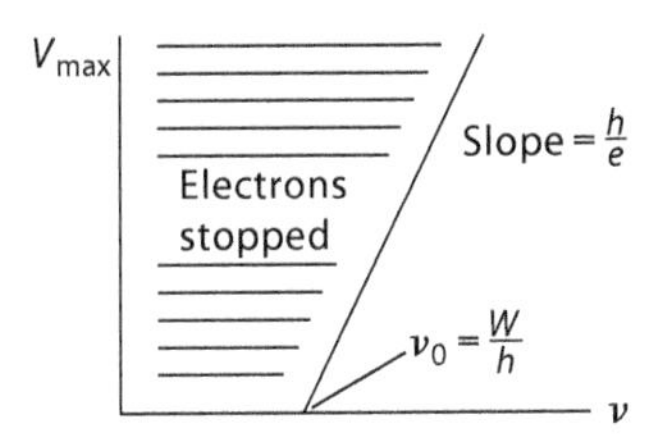

FIGURE 2.3 Relationship between V_{max} and frequency, ν, for the photoelectric effect.

below which we would expect to see no ejected electrons since the energy available to any single electron will not be large enough to overcome that material's work function. Now we know that V_{max}, the maximum voltage that allows a photocurrent to flow, is related to T_{max} by

$$eV_{max} = T_{max}, \tag{2.4}$$

where e is the magnitude of the charge on the electron. Plugging Equation 2.4 into Equation 2.2 then gives

$$V_{max} = \frac{h}{e}\nu - \frac{W}{e}. \tag{2.5}$$

Therefore, another implication of Einstein's theory is that if we plot V_{max} as a function of photon frequency, we should get experimental results as in Figure 2.3.

The first implication explains the threshold frequency behavior observed in metals. The second implication involving V_{max} was the part that was verified later by Millikan. Einstein's theory also agreed nicely with another experimental observation; namely, that the photocurrent was directly proportional to the light intensity. This is because light intensity is proportional to the number of light quanta, as is the number of electrons emitted from the surface.

2.2 COMPTON EFFECT

Another famous experiment that has added to our understanding of a new mechanics was done by Arthur Compton in 1923 and has since been known as the *Compton effect*. Other investigators had actually performed versions of this experiment before Compton but had not reached his conclusions. A schematic representation of his setup is shown in Figure 2.4, which involves an X-ray source, a target of carbon, and a detector at scattering angle, θ.

One can measure the *intensity* of the scattered X-rays as a function of θ. In addition, we can imagine fixing the scattering angle θ and *tuning* the detector to measure X-rays of varying wavelengths. When Compton did this, he found a result that looked qualitatively like Figure 2.5. That is, Compton saw two peaks in the intensity spectrum as a function of wavelength. The peak at λ, labeled "primary," occurred at the same wavelength as the approximately monochromatic source; the additional peak, labeled "secondary," occurred with $\lambda' > \lambda$. Compton explained his results by assuming a particle picture for the X-ray beam, as had Einstein, and by assuming energy and momentum conservation during the collision of the "photon," the particle of light, and the electron.

The kinematics of this collision can be pictured as in Figure 2.6. One assumes that in this frame of reference, the electron is initially at rest. It then acquires a momentum, P, and kinetic energy, T, from the photon. We know from Einstein's special theory of relativity that the energy of a particle of mass m and momentum P is

$$E = \sqrt{m^2c^4 + P^2c^2}. \tag{2.6}$$

We also know from this theory that the speed of light, c, is unattainable for any *material* particle; however, for a particle with zero mass, the speed of light is the *required* velocity. Under these circumstances, which

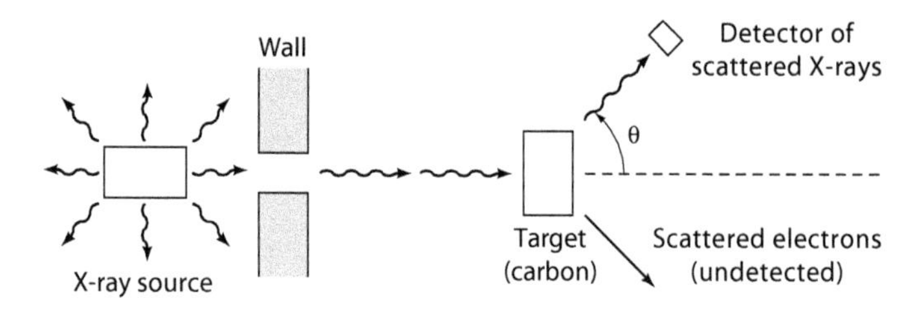

FIGURE 2.4 Schematic representation of the Compton experiment.

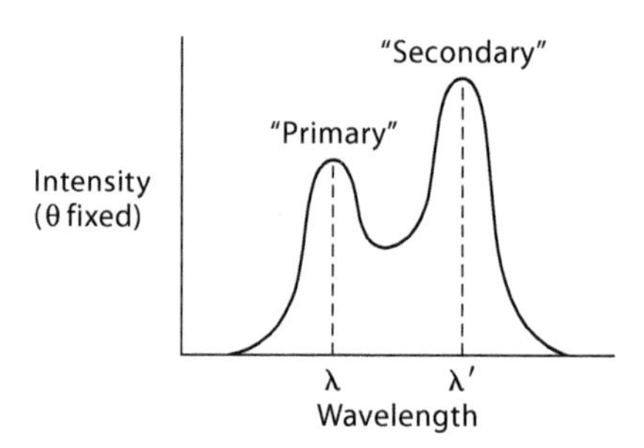

FIGURE 2.5 The two-peak result for intensity versus wavelength from the scattering of X-rays from electrons.

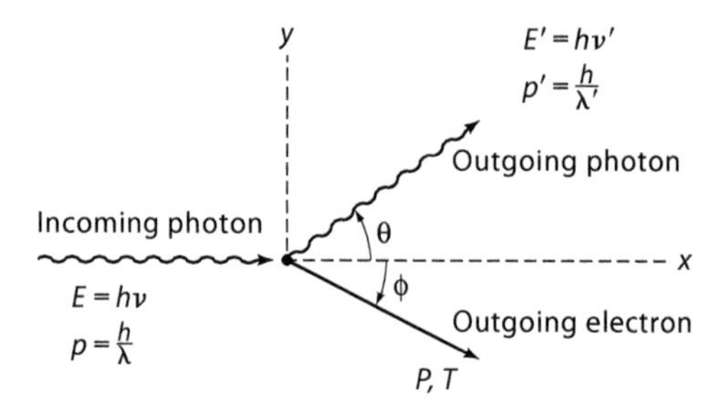

FIGURE 2.6 Kinematic diagram for the Compton effect. θ is the scattering angle for the X-ray.

apply to the photon, the relationship between E and photon momentum, p, from Equation 2.6 is

$$E = pc, \tag{2.7}$$

where $p = |\vec{p}|$ Then using Equation 2.1 we may write

$$p = \frac{h}{\lambda}. \tag{2.8}$$

Writing momentum conservation in the x and y directions for Figure 2.6 tells us

$$x:\ p = p'\cos\theta + P\cos\phi, \tag{2.9}$$

$$y:\ 0 = p'\sin\theta - P\sin\phi. \tag{2.10}$$

Therefore, we have

$$P^2\cos^2\phi = (p - p'\cos\theta)^2, \tag{2.11}$$

$$P^2\sin^2\phi = (p'\sin\theta)^2. \tag{2.12}$$

Adding Equations 2.11 and 2.12 gives

$$P^2 = p^2 + p'^2 - 2pp'\cos\theta. \tag{2.13}$$

We now write conservation of energy as

$$E - E' = T, \tag{2.14}$$

or

$$p - p' = \frac{T}{c}. \tag{2.15}$$

Squaring this gives

$$\left(\frac{T}{c}\right)^2 = p'^2 + p^2 - 2pp'. \tag{2.16}$$

Let us now subtract Equation 2.16 from Equation 2.13:

$$P^2 - \left(\frac{T}{c}\right)^2 = 2pp'(1 - \cos\theta). \tag{2.17}$$

The *relativistic* connection between P and T is *not* $T = P^2/2m$. We write the relation between the electron's energy and momentum in Equation 2.6 as

$$E_e^2 = P^2c^2 + m^2c^4. \tag{2.18}$$

The definition of kinetic energy T is as follows:

$$E_e = T + mc^2. \tag{2.19}$$

Substituting Equation 2.19 in Equation 2.18 and solving for P^2 then gives

$$P^2 = \left(\frac{T}{c}\right)^2 + 2Tm, \tag{2.20}$$

or

$$P^2 - \left(\frac{T}{c}\right)^2 = 2Tm. \tag{2.21}$$

Comparing Equations 2.21 and 2.17 we now have

$$2Tm = 2pp'(1 - \cos\theta). \tag{2.22}$$

However, from Equation 2.15, this now becomes

$$2mc(p - p') = 2pp'(1 - \cos\theta), \tag{2.23}$$

or

$$\left(\frac{1}{p'} - \frac{1}{p}\right) = \frac{1}{mc}(1 - \cos\theta). \tag{2.24}$$

Now using Equation 2.8 we find that

$$\lambda' - \lambda = \frac{h}{mc}(1 - \cos\theta). \tag{2.25}$$

Numerically,

$$\frac{h}{mc} = 2.43 \times 10^{-10}\,\text{cm}, \tag{2.26}$$

a quantity that is called the electron *Compton wavelength*. The shifted wavelength, λ', corresponds to the secondary intensity peak seen by Compton. Compton explained the primary peak as elastic scattering from the atom as a whole, leading to only a very tiny shift in wavelength due to the carbon atom's large mass relative to the electron's. Although a more detailed theory would be needed to explain the complete intensity spectrum shown earlier, the previously discussed experiment supplied rather convincing evidence of the particle nature of light.*

* The wavelength shift calculated holds for the scattering of free electrons, but of course the electrons in an atom are bound to a nucleus. However, because the X-ray photon energies are so much larger than the electronic binding energy, the aforementioned treatment is still very accurate.

2.3 UNCERTAINTY RELATION FOR PHOTONS

We now know that light particles are described by $p = h/\lambda$. Let us examine the consequence of this statement in view of the fact that the wave nature of light results in *diffraction*.

Consider the Fraunhofer (named after Joseph Fraunhofer) single-slit device in Figure 2.7. Viewing the incoming light rays as a stream of particles, we see that there is an uncertainty in the y-position of an individual photon in passing through the diffraction slit. We say that

$$\Delta y = a, \tag{2.27}$$

where Δ means "uncertainty in." Now, we expect to see the first diffraction minimum, indicated earlier, when the conditions in Figure 2.8 are obtained.

From the figure, we see that if $L \gg a$, we will have an interference minimum between rays in the "upper" half of the triangle and the "lower" half when

$$\frac{a}{2}\sin\theta = \frac{\lambda}{2}, \tag{2.28}$$

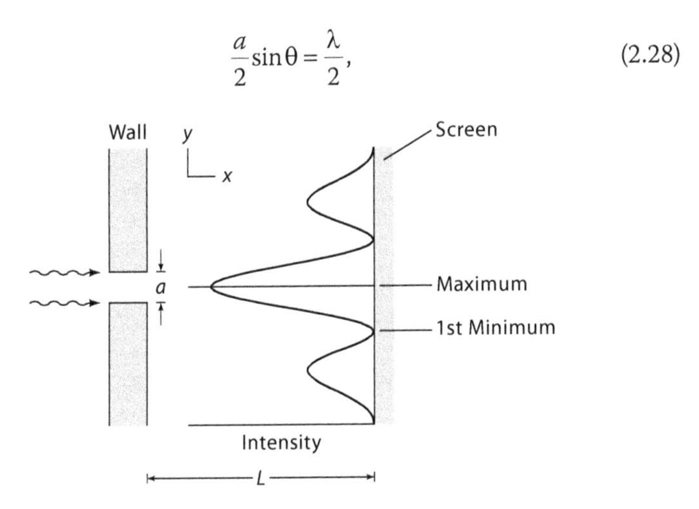

FIGURE 2.7 Schematic representation of a diffraction intensity pattern arising from a single slit.

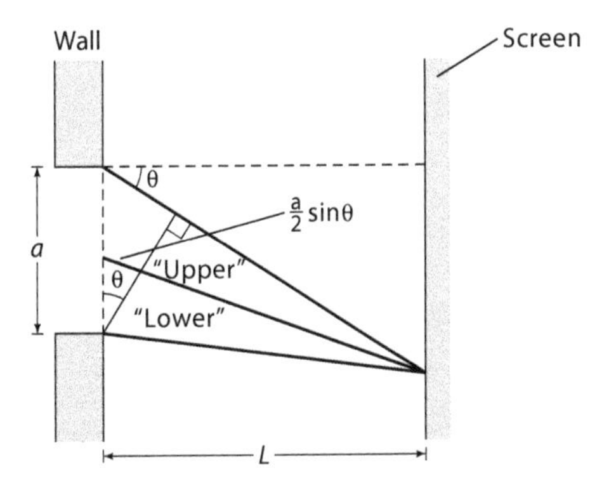

FIGURE 2.8 Geometry associated with locating the angular position of the first diffraction minimum from a single slit.

for then they will be completely out of phase. Now we know that *most* of the photons fall within the central maximum of the pattern. This means, in particle terms, there is also an uncertainty in y-momentum of individual photons passing through the slit. If we define this uncertainty so as to roughly correspond to the diffraction minimum, we have

$$\frac{\Delta p_y}{p} \approx \frac{L \sin\theta}{L} = \sin\theta. \tag{2.29}$$

Therefore,

$$\Delta y \; \Delta p_y \approx a(p \sin\theta) = a\left(p \frac{\lambda}{a}\right) = p\lambda. \tag{2.30}$$

Now using $p = h/\lambda$ this tells us

$$\Delta y \, \Delta p_y \approx h. \tag{2.31}$$

The product of the uncertainties in position and momentum of an individual photon is on the order of Planck's constant. Equation 2.31 is an example of the *Heisenberg uncertainty principle* as applied to photons. However, relations such as Equation 2.31 also hold for material particles that can be brought to rest in a laboratory, such as protons and electrons. Assuming energy and momentum conservation, the Compton effect discussed earlier demonstrates this, for if we could measure the initial and final position and momentum of the electron, we could determine the position and momentum of the photon to an arbitrary accuracy, contradicting Equation 2.31. A more general and rigorous statement of the uncertainty relation for material particles like an electron, involving x and p_x, will be seen (in Chapter 4) to be

$$\Delta x \; \Delta p_x \geq \frac{\hbar}{2}. \tag{2.32}$$

A discussion of Equation 2.32 in a mathematical context will be given in Section 2.6.

2.4 STABILITY OF GROUND STATES

If uncertainty relations apply to electrons, then we can immediately understand the reason for the stability of the ground state (i.e., lowest energy state) of the hydrogen atom, which, according to classical ideas, should quickly decay. The electron's energy is

$$E = \frac{\vec{p}^{\,2}}{2m} - \frac{e^2}{r}, \tag{2.33}$$

where $\vec{p}$ and m refer to the electron's momentum and mass. Now the hydrogen atom's ground state has zero angular momentum, so the $\vec{p}$ in

Equation 2.33 can be replaced by p_r, and we have essentially a one-dimensional problem in r, the radial coordinate. We hypothesize that

$$rp_r \sim \hbar, \tag{2.34}$$

for the hydrogen atom ground state; that is, we suppose the ground state is also close to being a minimum uncertainty state. Then using Equation 2.34 in Equation 2.33 gives

$$E = \frac{1}{2m}\left(\frac{\hbar}{r}\right)^2 - \frac{e^2}{r}. \tag{2.35}$$

A plot of Equation 2.34 looks like Figure 2.9.

We find E_0 by setting

$$\frac{\partial E}{\partial r} = 0, \tag{2.36}$$

which gives

$$a_0 = \frac{\hbar^2}{me^2}, \tag{2.37}$$

and

$$E_0 = -\frac{me^4}{2\hbar^2} (= -13.6 \text{ eV}). \tag{2.38}$$

The negative sign in Equation 2.38 means the same thing as for a classical mechanics system—that the system is *bound*. The value for E_0 turns out to be very close to the actual ground state energy of the real hydrogen atom. Our calculation is actually a bit of a swindle because Equation 2.34 is only a rough guess. The quantity a_0 in Equation 2.37 is called the "Bohr radius" and is numerically equal to 0.53×10^{-8} cm. Thus, the uncertainty principle implies the existence of a ground state and gives a rough value of the associated binding energy.

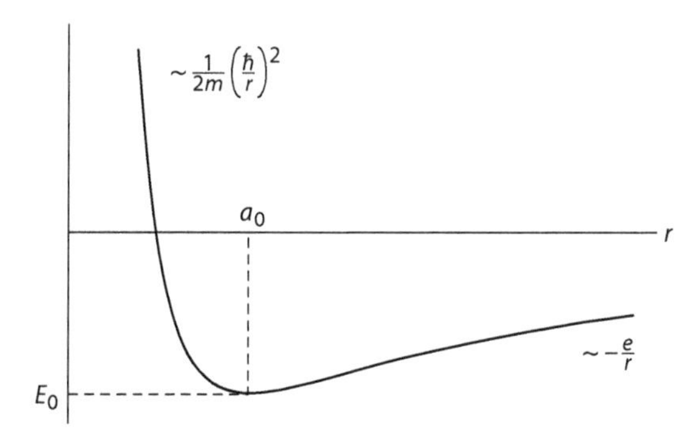

FIGURE 2.9 Effective ground state energy of hydrogen from the uncertainty principle.

2.5 BOHR MODEL

The famous Bohr model from which the term "Bohr radius" derives was a significant step forward in our understanding of atomic systems and yields an important insight into the nature of such systems. One way to recover the content of this model is to assume that the relation Equation 2.8, written for the photon, holds also for electrons and other material particles. That is, we are assuming a wavelike nature for entities we usually think of as particles, just as we previously were led to assume both wave and particle characteristics for light. If indeed objects such as electrons have wave characteristics, then they should exhibit constructive and destructive interference under appropriate circumstances. Consider therefore a simplified model of a wavelike electron trying to fit in the circular orbit as shown in Figure 2.10.

For the electron wave to avoid a situation where destructive interference would not permit it to persist in a stable configuration, we should have

$$2\pi a_n = n\lambda, \tag{2.39}$$

where $n=1, 2, 3, \ldots$. We then have

$$a_n = n\frac{\lambda}{2\pi} = n\frac{\hbar}{p}. \tag{2.40}$$

Multiplying both sides of this by p, we get another interpretation of Equation 2.39 from

$$a_n p = n\hbar. \tag{2.41}$$

The left-hand side of Equation 2.39 represents *orbital* angular momentum, which the right-hand side now says is quantized in units of $\hbar$. Minimizing the energy

$$\begin{aligned} E &= \frac{\vec{p}^2}{2m} - \frac{e^2}{a_n}, \\ &= \frac{1}{2m}\left(\frac{n\hbar}{a_n}\right)^2 - \frac{e^2}{a_n}, \end{aligned} \tag{2.42}$$

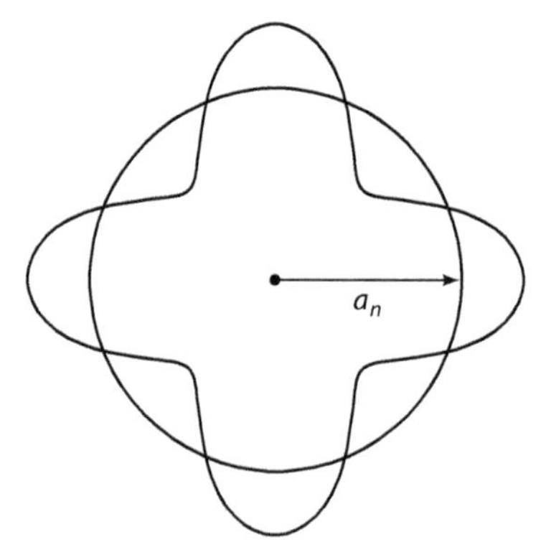

FIGURE 2.10 Schematic representation of the phase of an electron in a circular orbit around a proton.

with respect to a_n, we find that

$$\frac{\partial E}{\partial a_n} = 0 = -\frac{n^2\hbar^2}{ma_n^3} + \frac{e^2}{a_n^2}, \tag{2.43}$$

so that

$$a_n = \frac{n^2\hbar^2}{me^2}. \tag{2.44}$$

Plugging this back into Equation 2.40 now gives

$$E_n = -\frac{e^2}{2a_n} = -\frac{me^4}{2n^2\hbar^2}. \tag{2.45}$$

Setting $n=1$ in Equation 2.45 just gives Equation 2.38. The energy levels in Equation 2.45 agree closely with what is seen experimentally. The Bohr model leads us to believe that material particles can also have wavelike characteristics. Thus, the stability of the system is guaranteed by the uncertainty principle, and the discrete nature of energy levels is due to the assumed wave nature of particles. The ultimate verification of such a hypothesis would be to actually observe the effects of diffraction on a beam of supposedly particle-like electrons. That such a phenomenon can exist was hypothesized by Louis de Broglie in his doctoral dissertation (1925), and this was experimentally verified by Davisson and Germer (1927). The wavelength associated with electrons (called their *de Broglie wavelength*) according to Equation 2.8, which we originally applied only to photons, would in most instances be very tiny. For this reason, Davisson and Germer needed the close spacing of the atomic planes of a crystal to see electron diffraction take place.

The model Bohr presented of the behavior of the electron in a hydrogen atom is correct in its perception of the wave nature of particles, but is wrong in the assignment of the planetary-like orbits to electrons. In fact, as pointed out earlier, the hydrogen atom ground state actually has *zero* angular momentum, as opposed to the single unit of $\hbar$. assigned to it by Equation 2.39.

Our minds may rebel at the thought of something that shares both wave and particle characteristics. This seems like a paradox. It remains an experimental fact, however, that electrons and photons show either aspect under appropriately designed circumstances. We must, therefore, think of particle and wave characteristics as being not paradoxical, but *complementary*. This is the essence of what is called *Bohr's principle of complementary*.

We will have to build this dualism into the structure of the theory we hope to construct. The next step we take in this direction is the realization that the appropriate mathematical tool to use is the Fourier transform.

2.6 FOURIER TRANSFORM AND UNCERTAINTY RELATIONS

Leaving physics behind a minute and specializing to a single dimension, let us inquire into the Fourier transform of a function as in Figure 2.11.

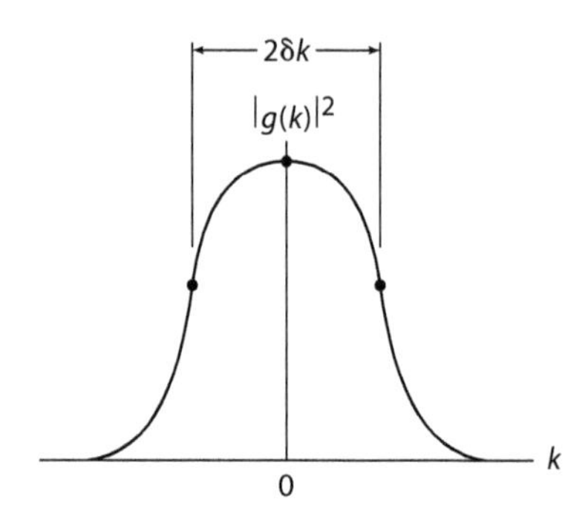

FIGURE 2.11 A function, $g(k)$, whose absolute square, $|g(k)|^2$, is peaked near the origin in k-space with an approximate full width of $2\delta k$.

That is, we are Fourier transforming a function $g(k)$ that is nonzero only when

$$-\delta k \le k \le \delta k. \tag{2.46}$$

We define the Fourier transform of $g(k)$ to be

$$f(x) = \frac{1}{\sqrt{2\pi}} \int_{-\infty}^{\infty} g(k) e^{ikx} \, dk. \tag{2.47}$$

What will the magnitude of the resulting function, $|f(x)|^2$, look like? We answer this question as follows:

Choose x such that	e^{ikx} is	$\|f(x)\|^2$ value is
$x\delta k \ll 1$	~ 1	*Maximum*
$x\delta k \sim 1$	e^{-i} to e^{i} ($\sim 1/\pi$ of a phase rotation)	*Lowered* because of partial destructive interference
$x\delta k \gg 1$	Rapidly oscillating phase	*Small* because of strong destructive interference

Qualitatively then, we would expect $|f(x)|^2$ to look like the result in Figure 2.12, where

$$\delta x \delta k \approx 1. \tag{2.48}$$

Of course, the detailed appearance of $|f(x)|^2$ depends on the exact mathematical form we assume for $g(k)$. To get some experience, let us consider a specific example. Let

$$g(k) = e^{-\alpha k^2}, \tag{2.49}$$

which is a "Gaussian" centered around $k = 0$. Then $f(x)$ is given by

$$f(x) = \frac{1}{\sqrt{2\pi}} \int_{-\infty}^{\infty} e^{-\alpha k^2} e^{ikx} \, dk. \tag{2.50}$$

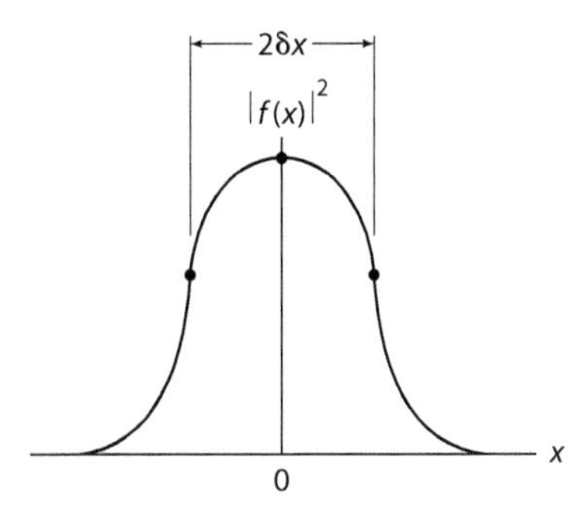

FIGURE 2.12 A qualitative plot of absolute square of the function, *f*(*x*), which results from the Fourier transform of the *g*(*k*) function shown in Figure 2.11.

We can write

$$-\alpha k^2 + ikx = -\alpha\left(k - \frac{ix}{2\alpha}\right)^2 - \frac{x^2}{4\alpha}. \tag{2.51}$$

This is called "completing the square." Then Equation 2.50 reads

$$f(x) = \frac{1}{\sqrt{2\pi}} e^{-x^2/4\alpha} \int_{-\infty}^{\infty} e^{-\alpha(k - ix/2\alpha)2} dk. \tag{2.52}$$

Now it is justified to let $k' = k - ix/2\alpha$ and still keep the integral along the real axis. The remaining integral we must do is

$$\int_{-\infty}^{\infty} e^{-\alpha k'^2} dk' = \frac{1}{\sqrt{\alpha}} \int_{-\infty}^{\infty} e^{-x^2} dx \equiv \frac{I}{\sqrt{\alpha}}. \tag{2.53}$$

I^2 can be evaluated as follows:

$$\begin{aligned} I^2 &= \int_{-\infty}^{\infty} dx e^{-x^2} \int_{-\infty}^{\infty} dy e^{-y^2} = \int_{-\infty}^{\infty} dx dy \, e^{-(x^2+y^2)} \\ &= \int_0^{\infty} r dr \int_0^{2\pi} d\theta e^{-r^2} = \pi \int_0^{\infty} dr^2 e^{-r^2} = \pi. \end{aligned} \tag{2.54}$$

Therefore, $I = \int_{-\infty}^{\infty} e^{-x^2} dx = \sqrt{\pi}$, and we have

$$f(x) = \frac{1}{\sqrt{2\pi}} e^{-x^2/4\alpha} \sqrt{\frac{\pi}{\alpha}} = \frac{1}{\sqrt{2\alpha}} e^{-x^2/4\alpha}. \tag{2.55}$$

(In this special example, *f*(*x*) turns out to be real, but in general it is complex.) Equation 2.55 has the advertised properties. That is, the half-width of the Gaussian in *k*-space is $\delta k \approx 1/(2\sqrt{\alpha})$ (this makes $|g(k = 1/(2\sqrt{\alpha})/g(k=0)|^2 = e^{-1/2}$) while the half-width of the *x*-space Gaussian, in the same sense, is $\delta x \approx \sqrt{\alpha}$. Therefore, the product of these two quantities is

$$\delta x \delta k \approx \sqrt{\alpha} \cdot \frac{1}{2\sqrt{\alpha}} = \frac{1}{2}. \tag{2.56}$$

Thus, we can make δx or δk separately as small as we wish, but then the other distribution will be spread out so as to satisfy Equation 2.56.

The argument leading to Equation 2.56 was mathematical. The argument leading to Equation 2.31 was a physical one. We now realize, however, that the uncertainty product in Equation 2.31 will be guaranteed to hold if we were to identify

$$k = \frac{p_x}{\hbar}, \tag{2.57}$$

because then Equation 2.56 would read (compare with Equation 2.32)

$$\Delta x \Delta p_x \approx \frac{\hbar}{2}, \tag{2.58}$$

where we are now interpreting the x and p_x distribution widths as uncertainties, for which we use the "Δ" symbol. Thus, the Fourier transform plus the identification Equation 2.57 accomplishes the goal of ensuring that uncertainty relations involving position and momenta are built into the theory.

2.7 SCHRÖDINGER EQUATION

Equation 2.47 says that the $f(x)$ distribution function or "wave packet" is actually a superposition of functions given by e^{ikx} with continuous k-values. Notice that (k can be a positive or negative quantity)

$$e^{ik(x+2\pi/|k|)} = e^{ikx}, \tag{2.59}$$

which says that e^{ikx} is a periodic function with a wavelength $2\pi/|k|$. Figure 2.13 displays the real and imaginary parts of e^{ikx} as a function of x.

The relation $k = p_x/\hbar$ is consistent with the statement that $2\pi/|k$ represents a particle wavelength since

$$\lambda = \frac{2\pi}{|k|} = \frac{2\pi\hbar}{|p_x|} = \frac{h}{|p_x|}, \tag{2.60}$$

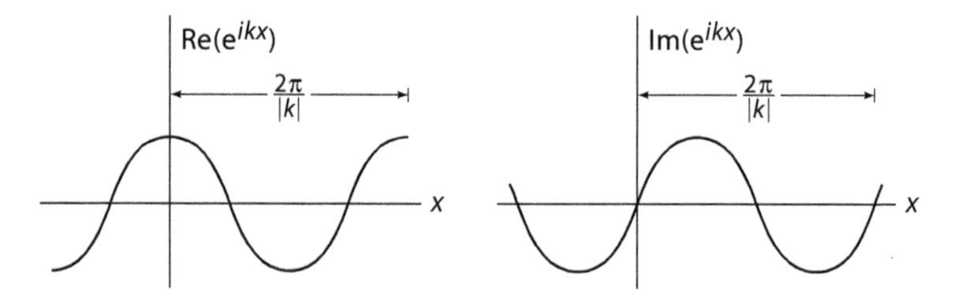

FIGURE 2.13 The real and imaginary parts of e^{ikx} as a function of x.

which gives back the previous form of the de Broglie wavelength. The function e^{ikx} (or $e^{i\vec{k}\cdot\vec{x}}$ in three dimensions) is usually called a "plane wave" because points of constant phase of this function form a plane in three dimensions.

Still dealing with only a single spatial dimension, let us now notice that the function ($t = \text{time}$)

$$e^{i(kx-\omega t)}$$

represents a *traveling* plane wave. We can understand the velocity of the motion by following a point of constant phase in the wave, for which

$$e^{i(kx-\omega t)} = \text{constant}. \tag{2.61}$$

As a function of time, these positions of constant phase then must satisfy (if $x = 0$ at $t = 0$)

$$kx = \omega t \Rightarrow x = \frac{\omega}{k}t. \tag{2.62}$$

Therefore, the *phase velocity* of the traveling plane wave is just ω/k. We then see that these phases move in the $+x$ direction if $k > 0$ and in the $-x$ direction if $k < 0$. The quantity ω represents the *angular frequency* of the moving wave. This is easy to see by fixing the position, x. Then $e^{i\omega t}$ tells us how many waves pass our observation point per unit time. We have

$$e^{-i\omega(t+2\pi/\omega)} = e^{-i\omega t}, \tag{2.63}$$

so the period is given by the usual formula

$$T = \frac{2\pi}{\omega}. \tag{2.64}$$

Free particles always have a positive definite angular frequency, ω.

For photons, the de Broglie wavelength is the same as the usual wavelength concept, and the relationship between ω and $|k|$ is given simply by

$$\omega = \frac{E}{\hbar} = \frac{|p_x|c}{\hbar} = |k|c, \tag{2.65}$$

using Equations 2.1, 2.7, and 2.57. Equation 2.65 becomes $\omega = |\vec{k}|c$ in three dimensions. (This simple linear relationship is strictly true only for light rays traveling in free space.) The phase speed is then

$$\frac{\omega}{|k|} = c, \tag{2.66}$$

as it should be. Since $\omega/|k|$ is a constant independent of $|k|$ itself, the phases of all wavelengths travel at the same velocity.

The equation that describes the time propagation of free photons is now easy to find. Based on the preceding observations, we generalize Equation 2.47 to account for time dependence:

$$f(x,t) = \frac{1}{\sqrt{2\pi}} \int_{-\infty}^{\infty} g(k) e^{i(kx-\omega t)} \, dk. \tag{2.67}$$

The differential equation that Equation 2.67 obeys is easy to construct. Observe that

$$\frac{1}{c^2}\frac{\partial^2 f(x,t)}{\partial t^2} = \frac{1}{\sqrt{2\pi}} \int_{-\infty}^{\infty} g(k) e^{ik(x-ct)} (-k^2) \, dk, \tag{2.68}$$

and that

$$\frac{\partial^2 f(x,t)}{\partial x^2} = \frac{1}{\sqrt{2\pi}} \int_{-\infty}^{\infty} g(k) e^{ik(x-ct)} (-k^2) \, dk, \tag{2.69}$$

so that

$$\frac{1}{c^2}\frac{\partial^2 f(x,t)}{\partial t^2} = \frac{\partial^2 f(x,t)}{\partial x^2}. \tag{2.70}$$

Equation 2.70 is called the "wave equation." In three dimensions the $\partial^2/\partial x^2$ operator would be changed to $\vec{\nabla}^2$. (We are ignoring the phenomenon of light polarization in this discussion.)

Let us now go through a similar discussion for a nonrelativistic particle in order to get the analog of Equation 2.70. The crucial step here is the assumption that the relation $\omega = E/\hbar$, found to hold for photons, also holds for material particles, where ω retains its meaning as the de Broglie angular frequency. This is a very reasonable supposition since we know that for any wave motion, frequency is a conserved quantity in time. (Think about Snell's law, for example.) Since energy is also conserved in the motion of a free particle, it is natural to assume that these quantities are proportional. With this hypothesis, we find for a free nonrelativistic particle

$$\omega = \frac{E}{\hbar} = \frac{p_x^2}{2m\hbar} = \frac{\hbar k^2}{2m}, \tag{2.71}$$

which would read $\omega = \hbar \vec{k}^2/2m$ in three dimensions.

The equation analogous to Equation 2.67 now becomes (one dimension again)

$$f(x,t) = \frac{1}{\sqrt{2\pi}} \int_{-\infty}^{\infty} g(k) e^{i(kx-\omega(|k|)t)} \, dk, \tag{2.72}$$

where the relation between ω and $|k|$ is given in Equation 2.71. The phase speed of these particles is

$$\frac{\omega}{|k|} = \frac{\hbar|k|}{2m} = \frac{|p_x|}{2m}. \tag{2.73}$$

This is just half the value of the mechanical speed of propagation, $|p_x|/m$. Therefore, we conclude that the phase speed of de Broglie waves is not the same thing as the actual propagation speed of the particle. Equation 2.73 also differs from Equation 2.66 in that the phases of different de Broglie wavelengths travel at different velocities. This is called *dispersion*, and its effects in an example will be studied later. For a slowly varying function $g(k)$ in Equation 2.72 (corresponding, say, to a sufficiently peaked function in position space), most of the contribution to the rapidly varying exponential integral will come from the integration domain

$$\frac{\partial(kx - \omega(|k|)t)}{\partial k} \approx 0,$$

(this is called the *stationary phase approximation*), which identifies the *average* particle propagation velocity as

$$\frac{p_x}{m} = \frac{\partial\omega}{\partial k}. \tag{2.74}$$

However, remember that a given wave packet contains a continuous range of $\vec{k}$ or $\vec{p}$ values (in three dimensions) so that a particle's velocity can only be defined in an average sense. The vector quantity $\partial\omega / \partial\vec{k}$ is called *group velocity*. The group speed of light in free space is

$$\left|\frac{\partial\omega}{\partial\vec{k}}\right| = c, \tag{2.75}$$

the same as its phase speed.

Using Equation 2.72, we find that

$$\frac{\partial f(x,t)}{\partial t} = \frac{1}{\sqrt{2\pi}}\int_{-\infty}^{\infty} g(k)\mathrm{e}^{i(kx-\omega(|k|)t)}\left(\frac{-i\hbar k^2}{2m}\right)\mathrm{d}k, \tag{2.76}$$

and

$$\frac{\partial^2 f(x,t)}{\partial x^2} = \frac{1}{\sqrt{2\pi}}\int_{-\infty}^{\infty} g(k)\mathrm{e}^{i(kx-\omega(|k|)t)}(-k^2)\mathrm{d}k, \tag{2.77}$$

so that

$$i\hbar\frac{\partial f(x,t)}{\partial t} = -\frac{\hbar^2}{2m}\frac{\partial^2 f(x,t)}{\partial x^2}. \tag{2.78}$$

Equation 2.78 is a special case of the celebrated *Schrödinger equation* written in one spatial dimension for a free particle.

2.8 SCHRÖDINGER EQUATION EXAMPLE

I will now show that

$$\psi_g(x,t) = \frac{1}{(2\pi)^{1/4}\sqrt{\delta x + \dfrac{i\hbar t}{2m\delta x}}}\cdot\exp\left(\frac{i}{\hbar}\left(\bar{p}x - \frac{\bar{p}^2}{2m}t\right) - \frac{1}{4\delta x}\cdot\frac{\left(x - \dfrac{\bar{p}}{m}t\right)^2}{\delta x + \dfrac{i\hbar t}{2m\delta x}}\right), \tag{2.79}$$

is a solution to Equation 2.78. The left-hand side of Equation 2.78 involves $\partial\psi_g(x,t)/\partial t$, which can be written as

$$\frac{\partial\psi_g(x,t)}{\partial t} = \psi_g(x,t)\cdot\left\{\begin{array}{l} \dfrac{-\dfrac{1}{2}\left(\dfrac{i\hbar}{2m\delta x}\right)}{\left(\delta x + \dfrac{i\hbar t}{2m\delta x}\right)} - \dfrac{i}{\hbar}\dfrac{\bar{p}^2}{2m} \\ -\dfrac{1}{4\delta x}\left[\dfrac{-2\left(x - \dfrac{\bar{p}}{m}t\right)\dfrac{\bar{p}}{m}}{\left(\delta x + \dfrac{i\hbar t}{2m\delta x}\right)} - \dfrac{\dfrac{i\hbar}{2m\delta x}\left(x - \dfrac{\bar{p}}{m}t\right)^2}{\left(\delta x + \dfrac{i\hbar t}{2m\delta x}\right)^2}\right] \end{array}\right\}. \tag{2.80}$$

Some necessary algebra is

$$\begin{aligned} &-\frac{i}{\hbar}\frac{\bar{p}^2}{2m} + \frac{\bar{p}}{2m\delta x}\frac{\left(x - \dfrac{\bar{p}}{m}t\right)}{\left(\delta x + \dfrac{i\hbar t}{2m\delta x}\right)} + \frac{i\hbar}{8m\delta x^2}\frac{\left(x - \dfrac{\bar{p}}{m}t\right)^2}{\left(\delta x + \dfrac{i\hbar t}{2m\delta x}\right)^2} \\ &= -\frac{i}{2m\hbar}\frac{\left(\bar{p}\delta x + \dfrac{i\hbar}{2\delta x}x\right)^2}{\left(\delta x + \dfrac{i\hbar t}{2\delta x}\right)^2}, \end{aligned} \tag{2.81}$$

so we can write

$$i\hbar\frac{\partial\psi_g(x,t)}{\partial t} = \frac{\hbar^2}{2m}\psi_g(x,t)\cdot\left(\frac{1}{2\left(\delta x^2 + \dfrac{i\hbar t}{2m}\right)} + \frac{1}{\hbar^2}\frac{\left(\bar{p}\delta x + \dfrac{i\hbar}{2\delta x}x\right)^2}{\left(\delta x + \dfrac{i\hbar t}{2m\delta x}\right)^2}\right). \tag{2.82}$$

For the other side of the equation, we find

$$\frac{\partial\psi_g(x,t)}{\partial x}=\psi_g(x,t)\cdot\left\{-\frac{1}{2\delta x}\frac{\left(x-\frac{\bar{p}}{m}t\right)}{\left(\delta x+\frac{i\hbar t}{2m\delta x}\right)}+\frac{i}{\hbar}\bar{p}\right\},\tag{2.83}$$

and so

$$\frac{\partial^2\psi_g(x,t)}{\partial x^2}=\psi_g(x,t)\cdot\left\{\begin{array}{l}-\frac{1}{2\delta t}\frac{1}{\left(\delta x+\frac{i\hbar t}{2m\delta x}\right)}+\frac{1}{4\delta x^2}\frac{\left(x-\frac{\bar{p}}{m}t\right)^2}{\left(\delta x+\frac{i\hbar t}{2m\delta x}\right)^2}\\ -\frac{\bar{p}^2}{\hbar^2}-\frac{i\bar{p}}{\hbar\delta x}\frac{\left(x-\frac{\bar{p}}{m}t\right)}{\left(\delta x+\frac{i\hbar t}{2m\delta x}\right)}\end{array}\right\}.\tag{2.84}$$

We recognize the last three terms in Equation 2.84 as the left-hand side of Equation 2.81, apart from an overall factor. This gives us

$$-\frac{\hbar^2}{2m}\frac{\partial^2\psi_g(x,t)}{\partial x^2}=-\frac{\hbar^2}{2m}\psi_g(x,y)\cdot\left\{-\frac{1}{2\left(\delta x^2+\frac{i\hbar t}{2m}\right)}-\frac{1}{\hbar^2}\frac{\left(\bar{p}\delta x+\frac{i\hbar}{2\delta x}x\right)^2}{\left(\delta x+\frac{i\hbar t}{2m\delta x}\right)^2}\right\}.\tag{2.85}$$

The right-hand sides of Equations 2.82 and 2.85 are now seen to be the same, which proves that $\psi_g(x,t)$ is a solution of Equation 2.78.

Let us now try to interpret this solution. At $t=0$ we have

$$\psi_g(x,0)=\frac{1}{\sqrt{\sqrt{2\pi}\ \delta x}}\exp\left(\frac{i}{\hbar}\bar{p}x-\frac{x^2}{4\delta x^2}\right),\tag{2.86}$$

so

$$|\psi_g(x,0)|^2=\frac{1}{\sqrt{2\pi}\,|\delta x|}\exp\left(-\frac{x^2}{2\delta x^2}\right).\tag{2.87}$$

Equation 2.87 is a Gaussian in x like we just finished studying. At time $t>0$ from Equation 2.79 we have

$$|\psi_g(x,t)|^2=\frac{1}{\sqrt{2\pi}\,|\delta x(t)|}\exp\left(-\frac{\left(x-\frac{\bar{p}}{m}t\right)^2}{2\delta x(t)^2}\right),\tag{2.88}$$

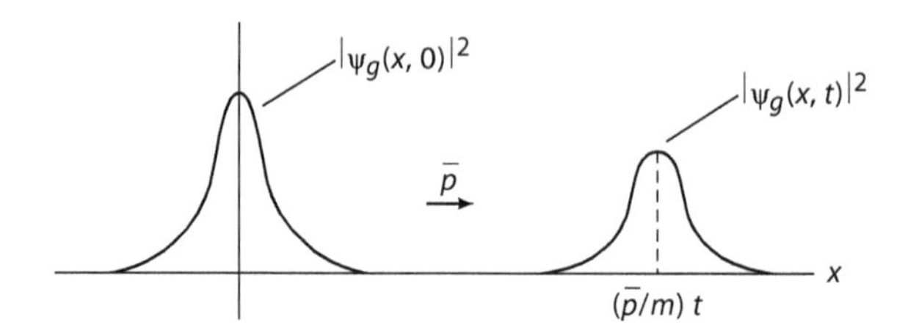

FIGURE 2.14 Representation of the evolution in time of a Gaussian wave packet.

where we have defined

$$\delta x(t)^2 \equiv \delta x^2 + \left(\frac{\hbar t}{2m\delta x}\right)^2. \tag{2.89}$$

Comparing Equation 2.88 with 2.87 gives us a picture of the time evolution of a Gaussian *wave packet*; see Figure 2.14.

The comparison referred to shows that the peak of the position-space wave packet $|\psi_g(x,t)|^2$ moves with velocity $\bar{p}/m$. You will show in an exercise that this is the expectation value of the wave packet's momentum divided by mass and thus corresponds to the usual notion of particle velocity. In addition, it does not maintain its same shape but *spreads* in time because of dispersion in values of momentum. We shall see momentarily that the magnitude squared of the momentum-space distribution does *not* change with time. Interpreted as uncertainties, this means that the product $\Delta x\ \Delta p_x$ grows in time. This is consistent with the $\geq$ sign in the uncertainty relation (Equation 2.32). The behavior of the wave packet described by Equation 2.72 is in contrast with the function $f(x,t)$ in Equation 2.67, which obeys $f(x,0) = f(x+ct,t)$ (for $g(k) = 0$, $k < 0$) and therefore propagates undistorted in time with velocity, c.

In quantum mechanics there is also an uncertainty relation for energy and time similar to that for momentum and position. Consider the Gaussian free wave function in the case $\langle p_x \rangle = \bar{p} = 0$. (This will not limit the generality of our conclusions.) We know that this wave function spreads in time according to Equation 2.89. The amount of time it takes for the wave function to evolve into a considerably wider form can be characterized approximately as

$$\frac{\hbar t_c}{2m\delta x} \approx \delta x,$$

or

$$t_c \approx \frac{2m\delta x^2}{\hbar}. \tag{2.90}$$

The Gaussian's uncertainty in energy (we will have to wait until Chapter 4 to define this; see Problem 4.7.3) is

$$\Delta E = \frac{1}{4\sqrt{2}m}\frac{\hbar^2}{\delta x^2}, \tag{2.91}$$

which, combined with Equation 2.90, gives

$$\Delta E\ t_c \approx \frac{\hbar}{2\sqrt{2}}. \tag{2.92}$$

Thus, if t_c is large, the uncertainty in energy of the Gaussian is quite small and vice versa. (If $\bar{p} \neq 0$, then the given product is larger than the right-hand side of Equation 2.92.) The generally accepted statement of the energy–time uncertainty relation is

$$\Delta E\ \Delta t \gtrsim \hbar. \tag{2.93}$$

Equation 2.93 is very similar to the Heisenberg uncertainty principle $\Delta p_x \Delta x \geq \hbar/2$. However, we will see in Chapter 4 that the Heisenberg relation can be *derived* since p_x and x are both operator quantities. We will not be able to do the same for Equation 2.93 since the time, t, is a parameter, not an operator, in nonrelativistic quantum mechanics. (In relativistic theories, both x and t are parameters.) In this form the energy–time uncertainty relation is inferred rather than derived.* This does not mean it is any less applicable to the real world, however. Its meaning is completely general if Δt is interpreted as a *correlation time* between wave functions during some transition that takes place.

2.9 DIRAC DELTA FUNCTIONS

It will behoove us in what is to come to study a little bit about *Dirac delta functions* now. As stated in Equation 2.45, the Fourier transform of $g(k)$ is defined as

$$f(x) \equiv \frac{1}{\sqrt{2\pi}} \int_{-\infty}^{\infty} g(k) e^{ikx}\, dk, \tag{2.94}$$

while the $g(k)$ distribution can be shown as

$$g(k) = \frac{1}{\sqrt{2\pi}} \int_{-\infty}^{\infty} f(x) e^{-ikx} dx. \tag{2.95}$$

Substituting Equation 2.95 in Equation 2.94, we get

$$\begin{aligned} f(x) &= \frac{1}{\sqrt{2\pi}} \int_{-\infty}^{\infty} dk\, e^{ikx} \int_{-\infty}^{\infty} dx' f(x') e^{-ik'x'} \\ &= \int_{-\infty}^{\infty} dx' f(x') \left[\int_{-\infty}^{\infty} \frac{dk}{(2\pi)} e^{ik(x-x')} \right]. \end{aligned} \tag{2.96}$$

We define

$$\delta(x - x') \equiv \int_{-\infty}^{\infty} \frac{dk}{(2\pi)} e^{ik(x-x')}, \tag{2.97}$$

* For a more rigorous approach to the energy–time uncertainty principle, see K. Gottfried and T.-M. Yan, *Quantum Mechanics: Fundamentals*, 2nd ed., (Springer, 2003), Section 2.4(d).

so

$$f(x)=\int_{-\infty}^{\infty} dx' f(x')\delta(x-x')\cdot \tag{2.98}$$

Let us set $x=0$ in Equation 2.98. Then

$$f(0)=\int_{-\infty}^{\infty} dx' f(x')\delta(x')\cdot \tag{2.99}$$

Setting $f(x)=1$ in Equation 2.99 tells us also that

$$1=\int_{-\infty}^{\infty} dx'\delta(x')\cdot \tag{2.100}$$

Equation 2.100 tells us that the area under the curve $\delta(x')$ is unity, but Equation 2.99 tells us that the only nonzero value of $\delta(x')$ is at $x'=0$. Such a function only exists in a limiting sense. In this limit, $\delta(x')$ is an even function of x' (this can be established from Equation 2.97):

$$\delta(x')=\delta(-x'). \tag{2.101}$$

We can show various properties of the Dirac delta function based on Equation 2.101. For example, we have

$$\delta(ay)=\delta(-ay), \tag{2.102}$$

where a is a constant. Therefore, we have

$$\delta(ay)=\delta(|a|y)\cdot \tag{2.103}$$

We can now write (the integral limits are understood to include the point $y=0$)

$$\begin{aligned}\int dy\, f(y)\delta(ay) &= \int dy\, f(y)\delta(|a|y) \\ &= \frac{1}{|a|}\int dz\, f\left(\frac{z}{|a|}\right)\delta(z)=\frac{1}{|a|}f(0)\cdot\end{aligned} \tag{2.104}$$

On the other hand

$$\frac{1}{|a|}f(0)=\int dy\, f(y)\left(\delta(y)\frac{1}{|a|}\right), \tag{2.105}$$

so that

$$\delta(ay)=\frac{1}{|a|}\delta(y)\cdot \tag{2.106}$$

Equation 2.106, like all identities satisfied by Dirac delta functions, is really understood to be true in the context of integration. The functions $f(x)$ must be sufficiently smooth, but are otherwise completely arbitrary.

The form (Equation 2.97) is not the only explicit representation of the Dirac delta function. An easy way of inventing representations of $\delta(x)$ is to set

$$\delta(x)=\int_{-\infty}^{\infty}\frac{dk}{(2\pi)}e^{ikx}K(k), \tag{2.107}$$

following Equation 2.97, requiring for $K(k)$ that

$$K(k) = \begin{cases} \sim 1, & \text{below a cutoff in } |k| \\ \sim 0, & \text{above a cutoff in } |k| \end{cases}. \tag{2.108}$$

The Dirac delta function is then defined in the limit of the cutoff going to infinity.

For example, consider

$$K(k) = \lim_{\varepsilon \to 0^+} e^{-(\varepsilon k^2/4)}, \tag{2.109}$$

where $\varepsilon \to 0^+$ denotes the limit as ε goes to zero through positive values. We then have

$$K(k) = \begin{cases} \sim 1, & \text{for } k \lesssim \dfrac{1}{\sqrt{\varepsilon}} \\ \sim 0, & \text{for } k \geq \dfrac{1}{\sqrt{\varepsilon}} \end{cases} \tag{2.110}$$

Putting Equation 2.109 into Equation 2.107 and interchanging the limit and the integral, we then find that $\delta(x)$ can be represented as

$$\delta(x) = \frac{1}{\sqrt{\pi}} \lim_{\varepsilon \to 0^+} \frac{1}{\sqrt{\varepsilon}} e^{-(x^2/\varepsilon)}. \tag{2.111}$$

Equation 2.111 is a Gaussian peaked at $x = 0$ that becomes increasingly narrow as $\varepsilon \to 0^+$, but which continues to have unit area. Using this technique, we can construct many forms of the Dirac delta function.

2.10 WAVE FUNCTIONS AND PROBABILITY

Let us return to interpreting the Gaussian wave packet solution of the Schrödinger equation. We already found that the magnitude of this solution spreads in time. The k-space (or "momentum space" since $k = p_x/\hbar$) transform of $\psi_g(x,0)$ is also a Gaussian, the absolute square of which has a width given by

$$\delta p_x = \frac{\hbar}{2\delta x}. \tag{2.112}$$

(The width of the Gaussian is defined the same as in Section 2.6). We can then write

$$\delta x(t)^2 = \delta x^2 + (\delta v t)^2, \tag{2.113}$$

where $\delta v \equiv \delta p_x / m$. Thus, the wave packet is spreading because of the initial velocity distribution at $t = 0$. Our Gaussian wave packet seems to be spreading out not because the particle itself is spreading, but because there was an initial uncertainty in the particle's velocity. We are being led to the point of view that the magnitude of the function $\psi_g(x,t)$ at the point x somehow represents the *probability* that the particle is in a given location. The simplest possible positive semidefinite quantity we can form out of a solution $\psi(x,t)$ of the Schrödinger equation is $|\psi(x,t)|^2$ (remember, $\psi(x,t)$ is in general complex), so we hypothesize that

$$\left.\begin{array}{l}\textbf{The probability that the single}\\ \textbf{particle descibed by the wave}\\ \textbf{packet } \psi(x,t) \textbf{ may be found}\\ \textbf{between positions } x \textbf{ and } x+\mathrm{d}x\\ \textbf{at time } t\end{array}\right\} = |\psi(x,t)|^2\,\mathrm{d}x. \quad (2.114)$$

Thus $|\psi(x,t)|^2$ is more properly called a *probability density*. Because of the interpretation in Equation 2.114, we will often require that

$$\int_{-\infty}^{\infty} \mathrm{d}x\,|\psi(x,t)|^2 = 1, \quad (2.115)$$

stating that the probability that the particle is somewhere is unity. Equation 2.115 is called a *normalization* condition. (We do not always explicitly require Equation 2.115 for solutions of Schrödinger's equation, as we will see.) For the integral in Equation 2.115 to exist, it is clear that we must have $\psi(x,t) \to 0$ as $|x| \to \infty$. In fact, we must have $|\psi(x,t)|$ decreasing faster than $|x|^{-1/2}$ as $|x| \to \infty$.

We now ask the question: What does Equation 2.115 imply for the momentum-space distribution of the wave packet? Referring back to Equation 2.47, we define the momentum-space wave packet $\psi(p_x)$ implicitly through*

$$\psi(x,0) \equiv \frac{1}{\sqrt{2\pi\hbar}} \int_{-\infty}^{\infty} \mathrm{d}p_x \psi(p_x) \mathrm{e}^{ip_x x/\hbar}. \quad (2.116)$$

In fact, we may generalize Equation 2.116 to

$$\psi(x,t) = \frac{1}{\sqrt{2\pi\hbar}} \int_{-\infty}^{\infty} \mathrm{d}p_x \psi(p_x,\, t) \mathrm{e}^{ip_x x/\hbar}, \quad (2.117)$$

using Equation 2.67 as a model, where we have defined ($\omega = E/\hbar$, where $E = p_x^2/2m$ for a free particle)

$$\psi(p_x,t) \equiv \psi(p_x)\mathrm{e}^{-i\omega t}. \quad (2.118)$$

From Equation 2.118 it is clear that the function $|\psi(p_x, t)|^2$ is time-independent for our Gaussian wave packet, which means it does not spread out in time, contrary to the position-space distribution. Using Equation 2.116 in Equation 2.115, we have the statement that at time t

* The x, p_x arguments of ψ imply different functions in this notation.

$$\int dx\,|\psi(x,t)|^2 = \int dx\,\psi(x,t)\psi^*(x,t)$$

$$= \frac{1}{2\pi\hbar}\int dx \int dp_x \psi(p_x,t)e^{ip_x x/\hbar}\int dp'_x \psi^*(p'_x,t)e^{-ip'_x x/\hbar} \quad (2.119)$$

$$= \int\int dp_x dp'_x \psi(p_x,t)\psi^*(p'_x,t)\left[\frac{1}{\sqrt{2\pi\hbar}}\int dx\, e^{-ix(p_x-p'_x)/\hbar}\right].$$

In the preceding integrals and others following where the limits of integration are not explicitly stated, we are to understand these to be from $-\infty$ to $+\infty$ on both position and momentum. From our recent discussion of Dirac delta functions, we realize that the quantity in square brackets in Equation 2.119 is just a delta function in momentum,

$$\delta(p_x - p'_x) = \frac{1}{2\pi\hbar}\int_{-\infty}^{\infty} dx\, e^{ix(p_x-p'_x)/\hbar}. \quad (2.120)$$

Therefore, we have

$$\int dx|\psi(x,t)|^2 = \int dp_x\,|\psi(p_x,t)|^2. \quad (2.121)$$

This result is known as *Parseval's theorem*. Thus, if we normalize using Equation 2.115, Equation 2.121 suggests an interpretation of $|\psi(p_x,t)|^2$ similar to Equation 2.114 for $|\psi(x,t)|^2$:

The probability that the single particle described by the wave packet $\psi(p_x,t)$ may be found with momentum values between p_x and $p_x + dp_x$ at time t $= |\psi(p_x,t)|^2\, dp_x$. (2.122)

The Schrödinger equation, Equation 2.78, is written as a differential statement involving the position-space function $f(x,t) = \psi(x,t)$. The analogous statement, which can be derived using Equations 2.78 and 2.117, for the momentum-space distribution $\psi(p_x,t)$ is

$$i\hbar\frac{\partial\psi(p_x,t)}{\partial t} = \frac{p_x^2}{2m}\psi(p_x,t). \quad (2.123)$$

This is called the *momentum-space Schrödinger equation*. In the form of Equation 2.123 it is still referring to a free particle in a single dimension. The general solution to Equation 2.123 is just Equation 2.118:

$$\psi(p_x,t) = \psi(p_x)e^{-i\frac{p_x^2}{2m\hbar}t}. \quad (2.124)$$

2.11 PROBABILITY CURRENT

It is crucial for this discussion that probability be a conserved quantity. That is, once we have imposed Equation 2.115 at $t=0$, it must continue to be true that the total probability is one. Our one-dimensional Schrödinger equation is

$$i\hbar\frac{\partial\psi(x,t)}{\partial t}=-\frac{\hbar^2}{2m}\frac{\partial^2\psi(x,t)}{\partial x^2}. \tag{2.125}$$

The complex conjugate of this is

$$-i\hbar\frac{\partial\psi^*(x,t)}{\partial t}=-\frac{\hbar^2}{2m}\frac{\partial^2\psi^*(x,t)}{\partial x^2}. \tag{2.126}$$

Using Equations 2.125 and 2.126, we find (we suppress the arguments of $\psi(x,t)$

$$\begin{aligned} i\hbar\frac{\partial}{\partial t}(\psi^*\psi) &= i\hbar\left(\frac{\partial\psi^*}{\partial t}\psi+\psi^*\frac{\partial\psi}{\partial t}\right)\\ &= \frac{\hbar^2}{2m}\left(\frac{\partial^2\psi^*}{\partial x^2}\psi-\psi^*\frac{\partial^2\psi}{\partial x^2}\right). \end{aligned} \tag{2.127}$$

This last statement, called the *continuity equation*, can be written as

$$\frac{\partial}{\partial t}(\psi^*\psi)+\frac{\partial}{\partial x}j(x,t)=0, \tag{2.128}$$

where

$$j(x,t)=\frac{i\hbar}{2m}\left[\frac{\partial\psi^*}{\partial x}\psi-\psi^*\frac{\partial\psi}{\partial x}\right], \tag{2.129}$$

$$=\frac{\hbar}{m}\mathrm{Im}\left[\psi^*\frac{\partial\psi}{\partial x}\right]. \tag{2.130}$$

$j(x,t)$ is called the "probability flux" and represents the flux or change in probability at a given position and time. Notice from Equation 2.130 that it is a *real* quantity, as it should be. Integrating Equation 2.128 over all positions, we then come to the conclusion that

$$\frac{\partial}{\partial t}\int_{-\infty}^{\infty}\mathrm{d}x|\psi(x,t)|^2=-j(x,t)|_{-\infty}^{\infty}. \tag{2.131}$$

Now we argued earlier that the requirement Equation 2.115 on $\psi(x,t)$ meant that it had to decrease faster than $|x|^{-1/2}$ as $|x|\to\infty$. This means that $j(x,t)$ must decrease faster than $|x|^{-2}$ as $|x|\to\infty$, from Equation 2.130. Therefore, $j(x,t)$ vanishes at the limits $x=\pm\infty$ and we conclude that

$$\frac{\partial}{\partial t}\int_{-\infty}^{\infty}\mathrm{d}x|\psi(x,t)|^2=0. \tag{2.132}$$

Thus, the existence of the continuity equation, Equation 2.128, plus the requirement that the norm of $\psi(x,t)$ be bounded, Equation 2.115, ensures a probabilistic interpretation for $|\psi(x,t)|^2$.

2.12 TIME SEPARABLE SOLUTIONS

We now try to find solutions of Equation 2.125 that are *separable*, that is, that can be written in the form

$$\psi(x,t) = u(x)T(t). \tag{2.133}$$

Notice that our *general* solution for $\psi(x,t)$, Equation 2.117 with Equation 2.118, is *not* of this form. We will give a physical interpretation of this in a moment. Substituting Equation 2.133 into Equation 2.125, we find that

$$i\hbar u(x)\frac{\mathrm{d}T}{\mathrm{d}t} = -\frac{\hbar^2}{2m}\frac{\mathrm{d}^2u}{\mathrm{d}x^2}T(t). \tag{2.134}$$

Dividing by $u(x)T(t)$, we get

$$i\hbar\,\frac{\mathrm{d}T/\mathrm{d}t}{T(t)} = -\frac{\hbar^2}{2m}\frac{\mathrm{d}^2u/\mathrm{d}x^2}{u(x)}. \tag{2.135}$$

Since x and t are independent variables, this equation can only be satisfied if both sides are equal to a constant, which with foresight we call E. The solution of

$$i\hbar\frac{\mathrm{d}T}{\mathrm{d}t} = ET(t), \tag{2.136}$$

is

$$T(t) = C\mathrm{e}^{-iEt/\hbar}. \tag{2.137}$$

The equation for $u(x)$ is then (we may label $u(x)$ as $u_E(x)$ if desired)

$$-\frac{\hbar^2}{2m}\frac{\mathrm{d}^2}{\mathrm{d}x^2}u(x) = E\,u(x). \tag{2.138}$$

The two linearly independent solutions of Equation 2.138 are of the form

$$C\mathrm{e}^{\pm ik'x}, \tag{2.139}$$

where $k' = \sqrt{2mE}/\hbar$ is an undetermined positive quantity. Thus, separable solutions to Equation 2.123 are of the form

$$\psi(x,t) \rightarrow C\mathrm{e}^{i(k'x - Et/\hbar)},\quad C\mathrm{e}^{i(-k'x - Et/\hbar)}, \tag{2.140}$$

where E is allowed by Equation 2.138 to take on all positive values. (The solutions to Equation 2.138 for $E<0$ are unbounded in x and do not lend themselves to a probabilistic interpretation.) We now realize that the forms of Equation 2.140 for $\psi(x,t)$ result from taking $\psi(p_x) = C'\delta(p_x - \hbar k')$

or $\psi(p_x) = C'\delta(p_x + \hbar k')$ in our general solution for $\psi(x,t)$, Equations 2.117 and 2.118. Thus, the plane waves of Equation 2.140 represent *specific* solutions to the Schrödinger equation that have a given kinetic energy, E. We have just found in fact that there are two such solutions, specified by $p_x = \pm\hbar k'$. The fact that our *general* solution Equation 2.117 is not separable in x and t means, therefore, that it is not a solution with a unique value of E. The Gaussian wave packet, for example, contained a continuous distribution of p_x (and therefore E) values. It makes physical sense that solutions to Equation 2.138 exist for all positive values of E since the kinetic energy of a free particle takes on positive values. We will see in the following chapter, however, that the generalization of Equation 2.138 that takes a quantum mechanical potential into account in general *does* restrict the values of the particle's total energy, often becoming discrete. Equation 2.138 is the *time-independent Schrödinger equation* for a free particle in one dimension.

2.13 COMPLETENESS FOR PARTICLE STATES

We have succeeded in deducing a probability-conserving differential equation for a material particle that has the position-momentum uncertainty relation built into it. We have done this without reference to our earlier discussion involving spin. However, I would now like to show that the Schrödinger equation is a natural outgrowth of the formalism we developed in the last chapter. I also wish to provide the connections between the differential equation point of view in this chapter and the operator formalism of the prior chapter.

The foundation on which we built the operator concept was the existence of what we were calling a *basis* of measurement symbols. Such a basis must allow a complete specification of all possible outcomes of a given experimental apparatus. This is the physical content of the mathematical expression of the *completeness* concept, Equation 1.91 (see Chapter 1). The basis there was discrete. In the case of spin 1/2 it consisted of two states $|+\rangle$ and $|-\rangle$. In the case of a free particle (considered spinless for now), an appropriate basis would be a specification of all possible locations or momentum values, which we assume take on continuous values. We can imagine doing all the operations discussed in the last chapter—selection, modulation, and transitions—but in this case on a "beam" of particles taking on various values of position or momenta. (A selection experiment on particle position might be realized by a diffraction experiment, for example.) Therefore, in analogy to the previous discrete specifications of completeness, we postulate the continuum statements

$$\int dx' |x'\rangle\langle x'| = 1, \tag{2.141}$$

and

$$\int dp'_x |p'_x\rangle\langle p'_x| = 1, \tag{2.142}$$

as expressing the completeness of a physical description based on continuous positions, Equation 2.141, or momentum values, Equation 2.142. The right-hand sides of Equations 2.141 and 2.142 are not the number one but the unity operator. To be consistent, we must now have (we again write $\langle x'|\cdot|x''\rangle \equiv \langle x'|x''\rangle$ and recognize this product is an ordinary number)

$$1 = 1\cdot 1 = \int dx'dx''|x'\rangle\langle x'|x''\rangle\langle x''|, \tag{2.143}$$

from which we realize that

$$\langle x' \mid x''\rangle = \delta(x' - x''), \tag{2.144}$$

for then

$$\begin{aligned}\int dx'dx''|x'\rangle\langle x'|x''\rangle\langle x''| &= \int dx'|x'\rangle\int dx''\delta(x' - x'')\langle x''| \\ &= \int dx'|x'\rangle\langle x'| = 1.\end{aligned} \tag{2.145}$$

Likewise, we have that

$$\langle p_x'|p_x''\rangle = \delta(p_x' - p_x''). \tag{2.146}$$

Equations 2.144 and 2.146 are the expressions of *orthonormality* in position and momentum space. (Compare with Chapter 1, Equation 1.163, where the Kronecker delta has just been replaced by the Dirac delta function.) In addition, we expect there to be *operators* for position and momentum just as we constructed an operator for S_z. Our model in such a construction is Equation 1.101, where, however, we would expect the discrete sum there to be replaced with an integral over continuous positions or momentums. Thus, as a natural outgrowth of our earlier experiences with discrete systems, we expect a representation for position and momentum operators by

$$x = \int dx'x'|x'\rangle\langle x'|, \tag{2.147}$$

and

$$p_x = \int dp_x'p_x'|p_x'\rangle\langle p_x'|, \tag{2.148}$$

where, in this context, the x' and p_x' are numbers and the x and p_x are our more abstract operator quantities. We now check whether Equation 1.102 is holding:

$$\begin{aligned}x|x'\rangle &= \left(\int dx''x''|x''\rangle\langle x''|\right)\cdot|x'\rangle \\ &= \int dx''x''|x''\rangle\delta(x'' - x') = x'|x'\rangle,\end{aligned} \tag{2.149}$$

and

$$p_x|p_x'\rangle = \left(\int dp_x'' p_x''|p_x''\rangle\langle p_x''|\right)\cdot|p_x'\rangle$$
$$= \int dp_x'' p_x''|p_x''\rangle\delta(p_x'' - p_x') = p_x'|p_x'\rangle. \tag{2.150}$$

Let us also assume the existence of the mathematical adjoint operation, denoted by "†," which connects our new bra and ket states. That is, we assume that

$$(|x'\rangle)^\dagger = \langle x'|, \tag{2.151}$$

$$(|p_x'\rangle)^\dagger = \langle p_x'|. \tag{2.152}$$

Now the physical outcomes, x' and p_x' in Equations 2.147 and 2.148, represent the result of position or momentum measurements in a one-dimensional space. They are necessarily real and this has the consequence that

$$x^\dagger = \left(\int dx' x'|x'\rangle\langle x'|\right)^\dagger = \int dx' x'|x'\rangle\langle x'| = x, \tag{2.153}$$

$$p_x^\dagger = \left(\int dp_x' p_x'|p_x'\rangle\langle p_x'|\right)^\dagger = \int dp_x' p_x'|p_x'\rangle\langle p_x'| = p_x. \tag{2.154}$$

Such a construction always results in a physical property that is Hermitian. As pointed out at in Chapter 1, Section 1.18, Hermitian operators have real expectation values. Using Equations 2.153 and 2.154, we then find that

$$\langle x'|x = x'\langle x'|, \tag{2.155}$$

$$\langle p_x'|p_x = p_x'\langle p_x'|, \tag{2.156}$$

where we have used the adjoint in Equations 2.149 and 2.150. Equations 2.155 and 2.156 may be independently verified using the definition of x and p_x, Equations 2.147 and 2.148. We will also refer to equations in the form of Equation 2.155 and Equation 2.156 as eigenvalue equations.

We make another important realization in the development of our x, p_x formalism by considering Equation 2.146. Using the x-representation of the unity operator, we write the left-hand side of this equation as (inserting the unit operator in the x-basis, Equation 2.141)

$$\langle p_x'|p_x''\rangle = \int dx' \langle p_x'|x'\rangle\langle x'|p_x''\rangle, \tag{2.157}$$

which, because

$$\langle p_x'|x'\rangle = (\langle x'|p_x'\rangle)^\dagger = \langle x'|p_x'\rangle^*, \tag{2.158}$$

we can write as

$$\langle p'_x|p''_x\rangle = \int dx' \langle x'|p'_x\rangle^* \langle x'|p''_x\rangle. \tag{2.159}$$

Therefore, Equation 2.146 reads

$$\int dx' \langle x'|p'_x\rangle^* \langle x'|p''_x\rangle = \delta(p'_x - p''_x). \tag{2.160}$$

By comparing the left-hand side of Equation 2.160 with an explicit representation of the delta function (see Equation 2.120; remember, the Dirac delta function is even in its argument)

$$\frac{1}{2\pi\hbar}\int dx' e^{-ix'(p'_x - p''_x)/\hbar} = \delta(p'_x - p''_x), \tag{2.161}$$

we see that it is consistent to choose

$$\langle x'|p'_x\rangle = \frac{1}{\sqrt{2\pi\hbar}} e^{ix'p'_x/\hbar}. \tag{2.162}$$

We will take Equation 2.162 as the *definition* of the bra–ket product $\langle x'|p'_x\rangle$. We will test the consistency of this definition shortly.

2.14 PARTICLE OPERATOR PROPERTIES

We are now in a position to make a crucial realization. We introduce the "Hamiltonian" operator

$$H \equiv \frac{p_x^2}{2m}, \tag{2.163}$$

which is the obvious operator quantity to represent the energy of a free particle. In analogy to the eigenvalue equations for spin, position, and momentum, we postulate a similar equation for H:

$$H\,|a'\rangle = E_{a'}\,|a'\rangle. \tag{2.164}$$

The a''s are *labels* of the allowed energy values $E_{a'}$. (In the case of the free particle, a' represents a *continuous* label.) We now multiply both sides of Equation 2.164 on the left by the bra $\langle x'|$:

$$\langle x'|\frac{p_x^2}{2m}|a'\rangle = E_{a'}\langle x'|a'\rangle. \tag{2.165}$$

To show the connection of Equation 2.165 to our previous results, consider the quantity $\langle x'|p_x|a'\rangle$. By inserting a complete set of p'_x states, we see that

$$\begin{aligned}\langle x'|p_x|a'\rangle &= \int dp'_x \langle x'|p'_x\rangle\langle p'_x|p_x|a'\rangle \\ &= \frac{1}{\sqrt{2\pi\hbar}}\int dp'_x e^{ix'p'_x/\hbar}\, p'_x \langle p'_x|a'\rangle,\end{aligned} \tag{2.166}$$

where we have used Equations 2.156 and 2.162. We may now write

$$\int dp'_x e^{ix'p'_x/\hbar}\, p'_x \langle p'_x|a'\rangle = \frac{\hbar}{i}\frac{\partial}{\partial x'}\int dp'_x e^{ix'p'_x/\hbar}\, \langle p'_x|a'\rangle. \tag{2.167}$$

Working backward, it is clear that

$$\begin{aligned}\frac{1}{\sqrt{2\pi\hbar}}\int dp'_x e^{ix'p'_x/\hbar}\langle p'_x|a'\rangle &= \int dp'_x \langle x'|p'_x\rangle\langle p'_x|a'\rangle \\ &= \langle x'|a'\rangle.\end{aligned} \tag{2.168}$$

Putting Equations 2.166 through 2.168 together yields the statement that

$$\langle x' \,|\, p_x \,|\, a'\rangle = \frac{\hbar}{i}\frac{\partial}{\partial x'}\langle x'|a'\rangle. \tag{2.169}$$

Since Equation 2.169 is supposed to be true for *any* $|a'\rangle$, this then implies ("stripping off the $|a'\rangle$")

$$\langle x' \,|\, p_x = \frac{\hbar}{i}\frac{\partial}{\partial x'}\langle x'|. \tag{2.170}$$

In other words, the momentum operator p_x acting on a position bra is the same as a differential operator $(\hbar/i)(\partial/\partial x')$ acting on the same bra. Please do *not* think this means that $p_x = (\hbar/i)(\partial/\partial x')$. It only means that the operator quantity p_x (defined in Equation 2.148) is replaced by the differential operator $(\hbar/i)(\partial/\partial x')$ when acting in position space. (When p_x acts in momentum space, it gives the number p'_x.) By taking the Hermitian adjoint of Equation 2.170 we then find that

$$p_x|x'\rangle = -\frac{\hbar}{i}\frac{\partial}{\partial x'}|x'\rangle. \tag{2.171}$$

(Remember, p_x is Hermitian.) Likewise, by considering the quantity $\langle p'_x|x|a'\rangle$, it is possible in the same manner to show that

$$\langle p'_x|x = -\frac{\hbar}{i}\frac{\partial}{\partial p'_x}\langle p'_x|, \tag{2.172}$$

and

$$x|p'_x\rangle = \frac{\hbar}{i}\frac{\partial}{\partial p'_x}|p'_x\rangle. \tag{2.173}$$

Let us test the consistency of these conclusions along with the statement Equation 2.162. Consider the quantity $\langle x'|x|p'_x\rangle$. By allowing x to act first to the left, we find that

$$\langle x'|x|p'_x\rangle = x'\langle x'|p'_x\rangle. \tag{2.174}$$

On the other hand, by allowing x to act first on $|p_x'\rangle$ we find

$$\begin{aligned}\langle x'|x|p_x'\rangle &= \frac{\hbar}{i}\langle x'|\frac{\partial}{\partial p_x'}|p_x'\rangle = \frac{\hbar}{i}\frac{\partial}{\partial p_x'}\langle x'|p_x'\rangle \\ &= \frac{\hbar}{i}\frac{1}{\sqrt{2\pi\hbar}}\frac{\partial}{\partial p_x'}\mathrm{e}^{ip_x'x'/\hbar} = \frac{x'}{\sqrt{2\pi\hbar}}\mathrm{e}^{ip_x'x'/\hbar} \\ &= x'\langle x'|p_x'\rangle.\end{aligned} \tag{2.175}$$

The right-hand sides of Equations 2.174 and 2.175 agree, as they should. This confirms the consistency of Equation 2.162, up to an overall constant.

Let us now go back and apply what we have learned to Equation 2.165. We now recognize that

$$\langle x'|\,p_x{}^2 = \frac{\hbar}{i}\frac{\partial}{\partial x'}\langle x'|\,p_x = -\hbar^2\frac{\partial}{\partial x'^2}\langle x'|, \tag{2.176}$$

so that Equation 2.165 reads

$$-\frac{\hbar^2}{2m}\frac{\partial^2}{\partial x'^2}\langle x'|a'\rangle = E_{a'}\langle x'|a'\rangle. \tag{2.177}$$

We recognize Equation 2.177 as just the free particle time-independent Schrödinger equation, Equation 2.138, that we originally motivated from a differential equation point of view. We now see that our differential and operator viewpoints will connect if we take

$$u_{a'}(x') = \langle x'|a'\rangle. \tag{2.178}$$

That is, we have come to the realization that the time-independent Schrödinger equation is just a position-space statement of the eigenvalue equation for the Hamiltonian, and the functions $u_{a'}(x')$ are *wave functions* that express the transition amplitude from the energy basis to the position basis. That is, along with a characterization of the unit operator in position and momentum space as in Equations 2.141 and 2.142, we also assume that there is also an *energy* characterization*:

$$\left\{\sum_{\substack{p_x'\geq 0,\\ p_x'<0}}\int_0^\infty \mathrm{d}E_{a'} \quad \text{or} \quad \sum_{a'}\right\}|a'\rangle\langle a'| = 1. \tag{2.179}$$

I am suggesting in Equation 2.179 that in some occasions we will find a discrete *spectrum* of energy values, $E_{a'}$, when the particle is no longer free. You should go back to the discussion in Chapter 1 to refresh yourself on

---

* Comment on Equation 2.179: What we are seeing here for the first time, in the case of the continuous energy values of the free particle, is a case of energy *degeneracy* of the energy eigenkets, $|a'\rangle$. Specifying their energy, $E_{a'}$ still leaves open the question of whether the particle has positive or negative momentum. Thus, in the case of the continuum statement of completeness in Equation 2.179, it is necessary to add the sum $\sum_{p_x'\geq 0, p_x'<0}$ in order to satisfy completeness. We will talk more about degeneracies of systems in Chapter 4.

the concept of a wave function as a transition amplitude. Also compare Equation 2.178 with Equation 1.221 from Chapter 1.

Interpreting $|u_{a'}(x')|^2$ as a probability density is also consistent with our earlier finding in the case of spin that probabilities are the absolute squares of transition amplitudes. That is, we may use completeness in position space to write energy-space orthonormality assuming the energies are discrete, $\langle a'|a''\rangle = \delta_{a'a''}$, as

$$\int dx'\langle a'|x'\rangle\langle x'|a''\rangle = \int dx' u^*_{a'}(x')u_{a''}(x') = \delta_{a'a''}. \tag{2.180}$$

When $a' = a''$ we get

$$\int_{-\infty}^{\infty} dx'\,|u_{a'}(x')|^2 = 1, \tag{2.181}$$

which is the same as Equation 2.115 when $\psi(x,t)$ refers to an energy eigenstate, $\psi(x,t) \to u_{a'}(x')e^{-iE_{a'}t/\hbar}$. Projecting Equation 2.164 into $\langle p'_x|$ and considering a general functional form $H = H(x, p_x)$, we have

$$\langle p'_x|H(x,p_x)|a'\rangle = E_{a'}\langle p'_x|a'\rangle. \tag{2.182}$$

Using Equation 2.172 then gives

$$H\left(-\frac{\hbar}{i}\frac{\partial}{\partial p'_x}, p'_x\right)\langle p'_x|a'\rangle = E_{a'}\langle p'_x|a'\rangle. \tag{2.183}$$

Equation 2.183 is the new time-independent momentum-space (differential) Schrödinger equation. The position-space energy wave functions $u_{a'}(x') = \langle x'|a'\rangle$ can be related to the $\langle p'_x|a'\rangle$ by the use of momentum-space completeness:

$$\begin{aligned} u_{a'}(x') = \langle x'|a'\rangle &= \int dp'_x\langle x'|p'_x\rangle\langle p'|a'\rangle \\ &= \frac{1}{\sqrt{2\pi\hbar}}\int dp'_x e^{ip'_x x'/\hbar}\langle p'_x|a'\rangle. \end{aligned} \tag{2.184}$$

Comparing Equation 2.184 with Equation 2.116 when $\psi(x,0) = u_{a'}(x')$, we see that

$$v_{a'}(p'_x) \equiv \langle p'_x|a'\rangle \tag{2.185}$$

should be interpreted as the momentum-space energy wave function.

2.15 OPERATOR RULES

It is important that we get off on the right foot in developing the formalism for quantum calculations. Let us carefully go over some important mathematical distinctions, then do a seminal mathematical calculation.

We have encountered two types of "operators":

- Quantum mechanical: $\mathbf{O}_Q$ (Examples: p_x, x, S_z)
- Differential: $\mathbf{O}_D$ (Examples: $(\partial/\partial p_x'), (\partial/\partial x')$)

By "quantum mechanical" operators, I mean the mathematical objects such as momentum, position, or spin that act in Dirac bra or ket space and which themselves are products of bras and kets (see Equations 1.136, 2.147, and 2.148). (The type of product leading to an operator, formed as $|a'\ a''| \equiv |a'\rangle\langle a''|$ in Chapter 1, will be called an "outer product" in Chapter 4.)

The commutative rules of these objects are quite simple:

Rule 1: Constants commute with $\mathbf{O}_Q$ and $\mathbf{O}_D$.
If C is a constant,
Rule 1a: $C\mathbf{O}_Q = \mathbf{O}_Q C$.
Rule 1b: $C\mathbf{O}_D = \mathbf{O}_D C$.
Of course, if C is actually a function of x' and/or p_x', Rule 1b can fail. (Example: $x'(\partial/\partial x') \neq (\partial/\partial x')x'$.) However, this is the usual result from calculus.

Rule 2: Operator types $\mathbf{O}_Q$ and $\mathbf{O}_D$ commute:

$$\mathbf{O}_Q\mathbf{O}_D = \mathbf{O}_D\mathbf{O}_Q$$

Again, if $\mathbf{O}_Q$ is a function of x' and p_x', Rule 2 can fail also. (Example: $x'1(\partial/\partial x') \neq (\partial/\partial x')x'1$, where 1 is the unit operator. This is just the same example as above in a different context.)

As an example of employing these rules, let us consider the evaluation of the structure

$$\underset{\text{"Term 1"}}{\underset{\uparrow}{(xp_x}} - \underset{\text{"Term 2"}}{\underset{\uparrow}{p_x x)}}|x'\rangle.$$

I will first do this using previous results from this chapter and the aforementioned rules, and then again in a limiting sense for confirmation. We have

$$\text{Term 1} \equiv xp_x|x'\rangle = \underset{\text{Equation 2.171}}{\underset{\uparrow}{-\frac{\hbar}{i}x\frac{\partial}{\partial x'}|x'\rangle}} = \underset{\text{Rule 2}}{\underset{\uparrow}{-\frac{\hbar}{i}\frac{\partial}{\partial x'}x|x'\rangle}}$$

$$\underset{\text{Equation 2.149}}{\underset{\uparrow}{=}} -\frac{\hbar}{i}\frac{\partial}{\partial x'}x'|x'\rangle = i\hbar|x'\rangle - \frac{\hbar}{i}x'\frac{\partial}{\partial x'}|x'\rangle. \quad (2.186)$$

$$\text{Term } 2 \equiv -p_x x|x'\rangle = -p_x x'|x'\rangle = -x' p_x|x'\rangle = \frac{\hbar}{i} x' \frac{\partial}{\partial x'}|x'\rangle. \tag{2.187}$$

$$\uparrow \qquad\qquad \uparrow \qquad\qquad \uparrow$$

Equation 2.149 Rule 1a Equations 2.171

Adding Terms 1 and 2 thus results in

$$(xp_x - p_x x)|x'\rangle = i\hbar|x'\rangle. \tag{2.188}$$

Since this holds for all x' and the basis $|x'\rangle$ is complete, we can conclude that

$$[x, p_x] = i\hbar, \tag{2.189}$$

where the commutator from Equation 1.137 is being employed.

This result may be confirmed as follows. Using the meaning of a derivative as a limit, one may define (see again in the adjoint of Equation 4.152 in Chapter 4)

$$p_x|x'\rangle \equiv -\frac{\hbar}{i} \lim_{\delta x' \to 0^+} \left(\frac{|x' + \delta x'\rangle - |x'\rangle}{\delta x'} \right).$$

The average of the limits for $\delta x' > 0$ ("forward difference") and $\delta x' < 0$ ("backward difference") gives the "symmetrical difference,"

$$p_x|x'\rangle \equiv -\frac{\hbar}{i} \lim_{\delta x' \to 0^+} \left(\frac{|x' + \delta x'\rangle - |x' - \delta x'\rangle}{2\delta x'} \right). \tag{2.190}$$

Armed with this, we may re-evaluate Term 1 as

$$\begin{aligned} &\text{Term } 1 \\ &= -\frac{\hbar}{i} x \lim_{\delta x' \to 0^+} \left(\frac{|x' + \delta x'\rangle - |x' - \delta x'\rangle}{2\delta x'} \right), \\ &= -\frac{\hbar}{i} \lim_{\delta x' \to 0^+} \left(\frac{(x' + \delta x')|x' + \delta x'\rangle - (x' - \delta x')|x' - \delta x'\rangle}{2\delta x'} \right), \\ &= -\frac{\hbar}{i} x' \lim_{\delta x' \to 0^+} \left(\frac{|x' + \delta x'\rangle - |x' - \delta x'\rangle}{2\delta x'} \right) + \frac{i\hbar}{2} \lim_{\delta x' \to 0^+} \left(|x' + \delta x'\rangle + |x' - \delta x'\rangle \right), \\ &= -\frac{\hbar}{i} x' \frac{\partial}{\partial x'}|x'\rangle + i\hbar|x'\rangle. \end{aligned} \tag{2.191}$$

This is in agreement with Equation 2.186. Term 2 is similar:

$$\begin{aligned} \text{Term } 2 &= -\frac{\hbar}{i} x' \lim_{\delta x' \to 0^+} \left(\frac{|x' + \delta x'\rangle - |x' - \delta x'\rangle}{2\delta x'} \right), \\ &= \frac{\hbar}{i} x' \frac{\partial}{\partial x'}|x'\rangle. \end{aligned} \tag{2.192}$$

The addition of these two terms again gives Equations 2.188 and 2.189.

Some further thoughts and comparisons of different Dirac notations, important for the beginning student, are given in Appendix A.

2.16 TIME EVOLUTION AND EXPECTATION VALUES

Notice that there is as yet no reference to time development in our operator formalism, as opposed to the wave packet discussion, where the Schrödinger equation described the evolution in time of our Gaussian wave packet, for example. However, we receive an important hint on one way to incorporate time development from Equation 2.133, which tells us how the energy eigenvalue wave functions $u_{a'}(x) = \langle x|a'\rangle$ evolve in time. In bra–ket notation, Equation 2.133 can be written as

$$\langle x'|a',t\rangle = e^{-iE_{a'}t/\hbar}\,\langle x'|a'\rangle, \tag{2.193}$$

where we have defined the time-evolved state $|a',t\rangle$. But notice that

$$\langle x'|e^{-iHt/\hbar}|a'\rangle = e^{-iE_{a'}t/\hbar}\langle x'|a'\rangle. \tag{2.194}$$

Therefore,

$$\langle x'|a',t\rangle = \langle x'|e^{-iHt/\hbar}|a'\rangle. \tag{2.195}$$

Equation 2.195, being true for all $\langle x'|$, then implies that

$$|a',t\rangle = e^{-iHt/\hbar}|a'\rangle. \tag{2.196}$$

The quantity $e^{-iHt/\hbar}$ is called the *time evolution operator*. It provides the key to understanding the time development of particle states. We notice that this operator, like the operators that describe rotations, is unitary. That is, given that H is Hermitian we have that

$$(e^{-iHt/\hbar})^{\dagger} = e^{iH^{\dagger}t/\hbar} = e^{iHt/\hbar}. \tag{2.197}$$

Now we take the time derivative of Equation 2.196. This yields

$$i\hbar\frac{\partial}{\partial t}|a',t\rangle = H\,e^{-iHt/\hbar}\,|a'\rangle = H\,|a',t\rangle \tag{2.198}$$

By projection into a position bra, $|x'\rangle$, and considering a general functional form $H = H(x, p_x)$ using Equation 2.170, we have

$$i\hbar\frac{\partial}{\partial t}\langle x'|a',t\rangle = H(x',\frac{\hbar}{i}\frac{\partial}{\partial x'})\langle x'|a',t\rangle. \tag{2.199}$$

Equation 2.199 is the new time-dependent position-space (differential) Schrödinger equation. As a continuation of our earlier notation, we will write $u_{a'}(x',t) \equiv \langle x'|a',t\rangle = e^{-iE_{a'}t/\hbar}u_{a'}(x')$. Equation 2.198 projected into momentum space also gives the time-dependent momentum-space Schrödinger equation with $v_{a'}(p'_x,t) \equiv \langle p'_x|a',t\rangle = e^{-iE_{a'}t/\hbar}v_{a'}(p'_x)$. Assuming a discrete set of energy states, the most general solution in position space is

$$\psi(x',t) = \sum_{a'} C_{a'}u_{a'}(x',t) = \sum_{a'} C_{a'}e^{-iE_{a'}t/\hbar}\,u_{a'}(x'), \tag{2.200}$$

where the $C_{a'}$ are an arbitrary set of constants. Introducing the notation

$$|\psi,t> \equiv \sum_{a'} C_{a'}|a',t\rangle, \tag{2.201}$$

for the most general linear combination of ket states, we have that

$$\psi(x',t) = \langle x'|\psi,t\rangle. \tag{2.202}$$

In the same way, defining $\psi(p'_x,t) \equiv \langle p'_x|\psi,t\rangle$, the corresponding momentum-space wave function is

$$\psi(p'_x,t) = \sum_{a'} C_{a'} v_{a'}(p'_x,t) = \sum_{a'} C_{a'} e^{-iE_{a'}t/\hbar} v_{a'}(p'_x). \tag{2.203}$$

The quantities $\psi(x',t)$ and $\psi(p'_x,t)$ are, of course, Fourier transforms of each other, the general connection being Equation 2.117. An alternative treatment of time development will be presented in Chapter 4.

In the context of our coordinate space discussion, if we take $A = A(x)$, expectation values are given as

$$\begin{aligned}\langle A(x)\rangle_{\psi,t} &= \langle \psi,t|A(x)|\psi,t\rangle \\ &= \int dx' dx'' \langle \psi,t|x'\rangle\langle x'|A(x)|x''\rangle\langle x''|\psi,t\rangle.\end{aligned} \tag{2.204}$$

Now

$$A(x)|x''\rangle = A(x'')|x''\rangle, \tag{2.205}$$

and so

$$\langle x'|A(x)|x''\rangle = A(x'')\langle x'|x''\rangle = A(x'')\delta(x'-x''), \tag{2.206}$$

which results in

$$\langle A(x)\rangle_{\psi,t} = \int dx' A(x')|\psi(x',t)|^2. \tag{2.207}$$

Since $|\psi(x',t)|^2$ is the probability density, we see that $\langle A(x)\rangle_{\psi,t}$ is obtained as a probability density-weighted integral and is explicitly real for Hermitian operators.

In the same manner, if $A = A(p_x)$, one can show that

$$\langle A(p_x)\rangle_{\psi,t} = \int dp'_x A(p'_x)|\psi(p'_x,t)|^2. \tag{2.208}$$

2.17 WRAP-UP

We have now recovered the basic dynamical equation of wave mechanics, the Schrödinger equation, from our earlier, spin-inspired, operator formalism. We have done this by applying the lessons we learned in the simpler spin case by analogy to particles in coordinate space. (For a Process Diagram–inspired recovery of this operator structure as a limiting process, please see Appendix B.)

Our understanding of the mathematics of the underlying operator formalism is still quite incomplete. However, we have reached a point where, using what we have learned, we can solve some simple one-dimensional problems in quantum mechanics. This is what we will do in the next chapter. Following that, I will try to fill in some of the gaps in our understanding of the operator formalism in Chapter 4.

PROBLEMS

2.2.1 Let us say we tried to use visible light instead of X-rays in a Compton-like scattering experiment (photons scattering from electrons). In this case, show that the fractional change in the frequency, ν, of the scattered light is given approximately by

$$\frac{|\Delta\nu|}{\nu} \approx \frac{2h\nu}{mc^2}\sin^2\frac{\theta}{2}.$$

2.3.1 Let us return to the original S–G setup. We now recognize that the condition $\delta p_z \delta z \geq \hbar$ is also responsible for the diffraction or spreading out of the atomic beam as it passes through the slit in the wall. (See Chapter 1, Figure 1.11.)

(a) Find the approximate value of the slit width, δz, that causes the magnetically split S–G beam to "wash out" due to diffraction. Evaluate δz numerically for our usual values:

$$m_{\text{Ag}} = 1.79 \times 10^{-22}\ \text{gm}, \quad \frac{\partial H}{\partial z} = 10^4\ \text{gauss}\cdot\text{cm}^{-1},$$

$$L = 10\ \text{cm}, \ |\gamma| \approx 10^7\ \text{gauss}^{-1}\text{sec}^{-1}, \ \frac{1}{2}mv^2 = \frac{3}{2}kT,$$

$$T = 10^{3\,\circ}\text{K}.$$

(b) Now replace the silver atom's mass, m_{Ag}, with the electron's mass in this calculation. Find the new slit width, δz, which causes diffraction washout.

2.4.1 Use the uncertainty relation to show that the potential

$$V(r) = \frac{-k}{r^{2+\varepsilon}}, \quad k, \varepsilon > 0,$$

is unstable for zero angular momentum states. (Remember the argument for the stability of the hydrogen atom. Also remember that setting $\partial E / \partial r = 0$ can pick out either a maximum or a minimum.)

2.4.2 Use the Heisenberg uncertainty principle to estimate the ground state energy of a one-dimensional harmonic oscillator with energy,

$$E = \frac{p_x^2}{2m} + \frac{1}{2}m\omega^2 x^2.$$

2.5.1 Consider a model of a heavy-light molecular system where the potential energy between the attractive heavy molecule (of infinite mass) and the light orbiting molecule (of mass m) is given by

$$V(r) = \frac{1}{2}m\omega^2 r^2,$$

where r is the separation distance between the two molecules and ω is a constant. The angular momentum of the electrons in the Bohr atom was quantized in units of $\hbar$. Assuming that the angular momentum of the light molecule is similarly quantized, and that only circular orbits are possible, find:

(a) The radius, r_n, of the light molecule's orbit in the nth angular momentum state.

(b) The total energy, E_n, of the nth angular momentum state.

2.6.1 Consider a wave packet defined by Equation 2.47 with $g(k)$ given by

$$g(k) = \begin{cases} 0, & k < -K \\ N, & -K < k < K \\ 0 & K < k \end{cases}$$

(a) Find the form $f(x)$ and plot it. [Answer: $f(x) = (2N/\sqrt{2\pi})(\sin(Kx)/x)$.]

(b) Show that a reasonable definition of δx for (a) yields

$$\delta k\ \delta x \sim 1.$$

2.7.1 Find the wave function equation, similar to the Schrödinger equation, for the function,

$$f(x,t) = \frac{1}{2\pi}\int_{-\infty}^{\infty} dk\ g(k)e^{i(kx-\omega t)},$$

which has an energy (E)–wavenumber (k) relationship:

$$\omega = \frac{E}{\hbar} = c\sqrt{k^2 + \left(\frac{mc}{\hbar}\right)^2}.$$

2.8.1 Evaluate the "correlation coefficient,"

$$C(t) \equiv \int_{-\infty}^{\infty} dx'\psi_g(x',0)\psi_g^*(x',t),$$

for the *stationary* Gaussian wave packet ($\overline{p}=0$). Define the correlation time, t_c, by setting $|C(t_c)|=1/2$. Compare your answer to Equation 2.90.

2.9.1 Considering the wave packet in Problem 2.6.1, find the value of N (up to a phase factor) for which

$$\int_{-\infty}^{\infty} \mathrm{d}x |f(x)|^2 = 1.$$

[Hint: Relate $f(x)$ to $g(k)$ and think delta function.]

2.9.2 Derive the delta function rule

$$\delta(f(x)) = \sum_i \frac{1}{\left|\frac{\mathrm{d}f}{\mathrm{d}x}(x_i)\right|}\delta(x-x_i),$$

for a function, $f(x)$, which has simple zeros located at $x=x_i$. $(\mathrm{d}f/\mathrm{d}x)(x_i)$ is the derivative of $f(x)$ at $x=x_i$. [Hint: Assume that near each zero one may write $f(x) \approx (\mathrm{d}f/\mathrm{d}x)(x_i)(x-x_i)$, then use Equation 2.106.]

2.10.1 (a) Given Equation 2.117, show that

$$\psi(p_x,t) = \frac{1}{\sqrt{2\pi\hbar}}\int_{-\infty}^{\infty} \mathrm{d}x\ \psi(x,t)\mathrm{e}^{-ip_x x/\hbar}.$$

(b) Find the momentum space wave function $\psi_g(p_x,0)$ corresponding to the $t=0$ coordinate space Gaussian wave function $\psi_g(x,0)$, given in Equation 2.86. What value of p_x maximizes $|\psi_g(p_x,0)|^2$? Show that the width of the probability distribution $|\psi_g(p_x,0)|^2$, in the same sense as Section 2.6, is

$$\delta p_x = \frac{\hbar}{2\delta x}.$$

$$\left[\text{Partial answer: } \psi_g(p_x,0) = \left(\frac{\sqrt{2\delta x}}{(2\pi\hbar^2)^{1/4}}\right) \exp\left(-\frac{\delta x^2}{\hbar^2}(\overline{p}-p_x)^2\right).\right]$$

2.10.2 Defining the expectation value of p_x as an integral over the probability distribution $|\psi(p_x,0)|^2$ (see Section 2.16 and the Problem 2.10.1 answer)

$$\langle p_x\rangle \equiv \int_{-\infty}^{\infty} \mathrm{d}p_x p_x |\psi(p_x,0)|^2,$$

show for the Gaussian wave packet that

$$\frac{\langle p_x\rangle}{m} = \frac{\overline{p}}{m},$$

and therefore $\langle p_x\rangle/m$ "corresponds to the usual notion of particle velocity."

2.10.3 Starting with the *coordinate space* free particle Schrödinger equation (Equation 2.78) for $f(x,t) = \psi(x,t)$, show that the *momentum space* Schrödinger equation is given by Equation 2.123.

2.15.1 Try reducing or completing the equality (p_x, x quantum mechanical operators):

$$\langle x'|(p_x x - x p_x) = ?$$

Justify all your steps. [Note: One easy way is to take the Hermitian conjugate of Equation 2.188, but please treat this problem as if that equation were not derived!]

2.16.1 Argue that, in addition to Equation 2.208, we may also write

$$\langle A(p_x)\rangle_{\psi,t} = \int_{-\infty}^{\infty} dx'\, \psi^*(x',t)\, A\left(\frac{\hbar}{i}\frac{\partial}{\partial x'}\right)\psi(x',t),$$

for the expectation value of $A(p_x)$, if $A(p_x)$ is a power series in p_x.

2.16.2 Using the result of Problem 2.16.1 (or any other means), establish that

$$\frac{\langle p_x\rangle_{\psi,t}}{m} = \int_{-\infty}^{\infty} dx'\, j(x',t),$$

where

$$j(x',t) = \frac{i\hbar}{2m}\left(\frac{\partial \psi^*}{\partial x'}\psi - \psi^*\frac{\partial \psi}{\partial x'}\right),$$

is the probability current.

2.16.3 Given the momentum space free-particle wave function ($\bar{x}$ and a are just constants) at $t = 0$,

$$\psi(p_x) = A\frac{\exp\left(\frac{i}{\hbar}\bar{x}p_x\right)}{\sqrt{p_x^2 + a^2}},$$

(a) Find the value of the constant A up to an overall phase.
(b) Find the expectation value of the momentum, $\langle p_x\rangle$.
(c) Find the momentum wave function at all later times, $\psi(p_x,t)$.

2.16.4 Consider the following experiment, represented in Figure 2.15, done with spin 1/2 particles (in this case thermal neutrons would probably work the best) on a flat tabletop. $S_y' = \hbar/2$ is first selected by magnet "1," then this beam is split into two parts (with equal amplitudes) that travel along the (equal length) paths shown. Before reaching the beam splitter, the particles have velocity v and the beam has

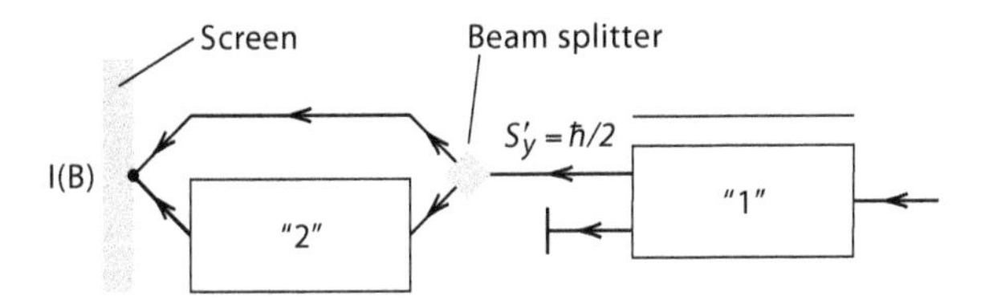

FIGURE 2.15 A selection/beam-splitting tabletop experiment done with spin 1/2 particles.

intensity I_0. A second magnet with a *uniform* magnetic field pointing along the z-axis is also positioned along one of the beam paths, as shown, with a magnetic field strictly confined to the particle path. Assume that the beams constructively interfere with one another at the screen when magnet "2" is turned off, that is, $I(B=0)=I_0$.

(a) Show that the state that emerges from magnet "1" ($S'_y = +\hbar/2$) is

$$|\bar{+}\rangle = \frac{1}{\sqrt{2}}(|+\rangle + i|-\rangle).$$

$$\left(S_z|\pm\rangle = \pm\frac{\hbar}{2}|\pm\rangle \text{ as usual}\right).$$

(b) Find an expression for the time-evolved state that emerges from magnet "2," given that $H = -\vec{\mu}\cdot\vec{B} = -\gamma B S_z$:

$$|\bar{+},t\rangle = ?$$

(c) Find an expression for the intensity of the beam spot, $I(B)$, as a function of the magnetic field of magnet "2." I will give you a choice of three methods, only one of which is correct:

Method 1:

$$|\psi\rangle = \frac{1}{2}(|\bar{+},t\rangle + |\bar{+}\rangle),$$

$$I(B) = I_0\ |\langle\psi|\psi\rangle|^2.$$

Method 2:

$$I(B) = I_0\left|\langle\bar{+},t|\bar{+}\rangle\right|^2.$$

Method 3:

$$I(B) = I_0 \text{ always}.$$

Some One-Dimensional Solutions to the Schrödinger Equation

Synopsis: At this point we are equipped to solve some simple but fundamental one-dimensional problems with the Schrödinger equation. After setting the stage with an introduction, four quantum mechanical problems are specified and solved. Along the way, we encounter important concepts such as parity in the case of the infinite well and harmonic oscillator problems, transmission and reflection coefficients in the case of the finite barrier, and energy bands and gaps in a model of conduction in solids. Raising and lowering ladder operators are introduced in the harmonic oscillator and square well cases. I conclude with some qualitative comments regarding expected types of solutions in one-dimensional problems.

3.1 INTRODUCTION

The discussion in the last chapter centered upon the case of a free particle in a single space dimension. The Schrödinger equation can be written as

$$i\hbar\frac{\partial}{\partial t}|\psi,t\rangle = H|\psi,t\rangle, \tag{3.1}$$

where

$$H = \frac{p_x^2}{2m}. \tag{3.2}$$

It is natural to assume that more general forms for H are possible. The form of Equation 3.2, which is an operator statement, is very classical looking. We hypothesize that the interaction of a quantum mechanical particle with an external potential $V(x)$ can also be represented by its classical form:

$$H = \frac{p_x^2}{2m} + V(x). \tag{3.3}$$

The crucial thing that must be checked in writing down Equation 3.3 is that the probability density interpretation given to $|\psi(x,t)|^2$ in

Equation 2.114 (Chapter 2), which was based on the existence of a conserved probability current, still holds. You will provide this check in a problem. By projecting Equations 3.1 and 3.3 into a coordinate ket, $<x|$, we obtain

$$i\hbar\frac{\partial}{\partial t}\psi(x,t) = -\frac{\hbar^2}{2m}\frac{\partial^2\psi(x,t)}{\partial x^2} + V(x)\,\psi(x,t). \tag{3.4}$$

This is still separable in space and time, which implies the time evolution of the energy eigenvalue states as in Equation 2.196 (Chapter 2) and that $e^{-iHt/\hbar}$ is still the evolution operator.

Of course, not *every* $V(x)$ in Equation 3.3 has a physical significance. The energies of our system must be real, which implies that

$$\langle a'|H|a'\rangle = E_{a'}\langle a'|a'\rangle = E_{a'}, \tag{3.5}$$

$$E_a^* = \langle a'|H|a'\rangle^* = \langle a'|H|a'\rangle^\dagger = \langle a'|H^\dagger|a'\rangle. \tag{3.6}$$

Comparing Equations 3.5 and 3.6 for any state a' implies in general that

$$H^\dagger = H. \tag{3.7}$$

For Equation 3.3 this means we must have

$$V(x)^\dagger = V(x). \tag{3.8}$$

That is, the potential operator must be Hermitian.

Since the Schrödinger Equation 3.4 is separable, we can define a time-evolved energy eigenstate

$$|a',t\rangle \equiv e^{-iHt/\hbar}|a'\rangle = |a'\rangle e^{-iE_{a'}t/\hbar}, \tag{3.9}$$

as we did in the last chapter for the free particle. Completeness of the $|a'\rangle$ (we assume a discrete form)

$$\sum_{a'}|a'\rangle\langle a'| = 1, \tag{3.10}$$

then implies for any state $|\psi\rangle$ that

$$|\psi\rangle = \sum_{a'}|a'\rangle\langle a'|\psi\rangle. \tag{3.11}$$

When projected into position space, this becomes (identical to Equation 2.200 at $t=0$)

$$\psi(x,0) = \sum_{a'}u_{a'}(x)C_{a'} \tag{3.12}$$

where we have $C_{a'} = \langle a'|\psi\rangle$. Since we know the time development of the $u_{a'}(x)$, we then have that

$$\psi(x,t) = \sum_{a'}u_{a'}(x)e^{-iE_{a'}t/\hbar}C_{a'}. \tag{3.13}$$

Equation 3.13 indicates that the knowledge of the energy eigenfunctions $u_{a'}(x)$ and eigenvalues $E_{a'}$ provides a way of constructing all possible functions $\psi(x,t)$ that solve the Schrödinger equation. For this reason, the solution to the time-independent Schrödinger equation

$$\left[-\frac{\hbar^2}{2m}\frac{d^2}{dx^2}+V(x)\right]u(x)=E\,u(x), \tag{3.14}$$

where we have included a potential $V(x)$, is of paramount importance in quantum mechanics.

We will study the solution to Equation 3.14 in this chapter for a simple set of potentials for which complete analytic solutions are possible. The four problems we will study here will be as follows:

- The infinite square well
- The finite potential barrier
- The harmonic oscillator
- The attractive Kronig–Penney model

We will continue to limit the discussion to a single dimension of space for now.

3.2 THE INFINITE SQUARE WELL: DIFFERENTIAL SOLUTION

We will take the potential to be as in Figure 3.1. That is, we are assuming $V(x)=0$ for $-a<x<a$, but $V(x)\rightarrow\infty$ for $|x|>a$. A consistent way of interpreting this potential is to say that there is zero probability of the particle to escape from the interior region of the well into the shaded region. This will be ensured if we take the boundary conditions

$$u(x)|_{x=\pm a}=0. \tag{3.15}$$

The easiest way to solve this problem is in the coordinate space representation of the wave function. Thus, we need to solve

$$-\frac{\hbar^2}{2m}\frac{d^2}{dx^2}u(x)=E\,u(x), \tag{3.16}$$

subject to the boundary conditions of Equation 3.15. Equation 3.16 can be written as

$$-\frac{d^2}{dx^2}u(x)=k^2u(x), \tag{3.17}$$

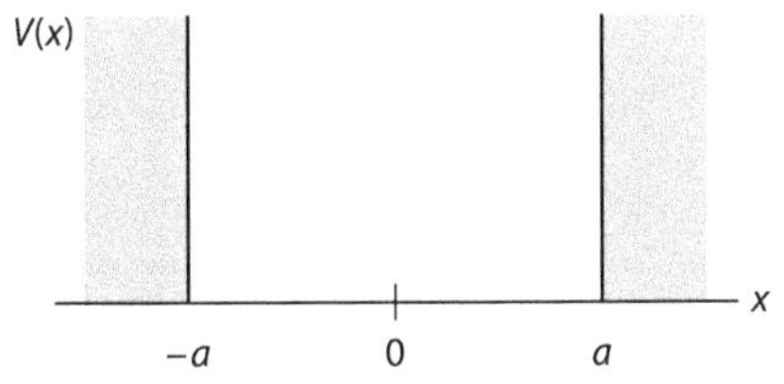

FIGURE 3.1 The infinite square well potential.

where (k is now a magnitude only)

$$k \equiv \left(\frac{2mE}{\hbar^2}\right)^{1/2}. \tag{3.18}$$

We have left the usual subscript off $u(x)$ in anticipation of a labeling scheme for the energy eigenvalues. Of course, the linearly independent solutions to Equation 3.17 are

$$u(x) = A\sin(kx) \tag{3.19}$$

or

$$u(x) = A'\cos(kx). \tag{3.20}$$

If we apply the boundary conditions of Equation 3.15 to the solutions of Equation 3.19, we find that this means

$$\sin(\pm ka) = 0, \tag{3.21}$$

which implies that

$$ka = n\pi, \tag{3.22}$$

for $n = 1, 2, 3, \ldots$. $n = 0$ is a trivial solution and $n = -1, -2, -3, \ldots$ are not linearly independent. Equation 3.22 tells us the allowed energy levels associated with the odd-space wave functions $\sin(kx)$ are discrete:

$$E_{n-} = \frac{1}{2m}\left(\frac{\hbar n\pi}{a}\right)^2. \tag{3.23}$$

(The $n-$ notation means the nth odd energy level.) Likewise, for the even solutions, Equation 3.20, we have

$$\cos(\pm ka) = 0, \tag{3.24}$$

which means

$$ka = \left(n - \frac{1}{2}\right)\pi, \tag{3.25}$$

where $n = 1, 2, 3, \ldots$ ($n = 0, -1, -2, -3, \ldots$ are not linearly independent). The energies of the even-space wave functions $\cos(kx)$ are thus

$$E_{n+} = \frac{1}{2m}\left(\frac{\hbar(n - 1/2)\pi}{a}\right)^2. \tag{3.26}$$

Qualitative plots of the lowest few odd and even wave functions are given in Figures 3.2 and 3.3.

The lowest energy solution is given by E_{1+}. If we did not know its exact value, we could guess it approximately from the uncertainty principle, assuming it is a minimum uncertainty state. We have that

$$\Delta p_x \Delta x \approx \hbar, \tag{3.27}$$

for such a state. If we say that $\Delta x \approx 2a$ and that $p_x \approx \Delta p_x$, then one obtains the estimate

$$E_{\text{lowest}} = \frac{p_x^2}{2m} \approx \frac{\hbar^2}{8ma^2}, \tag{3.28}$$

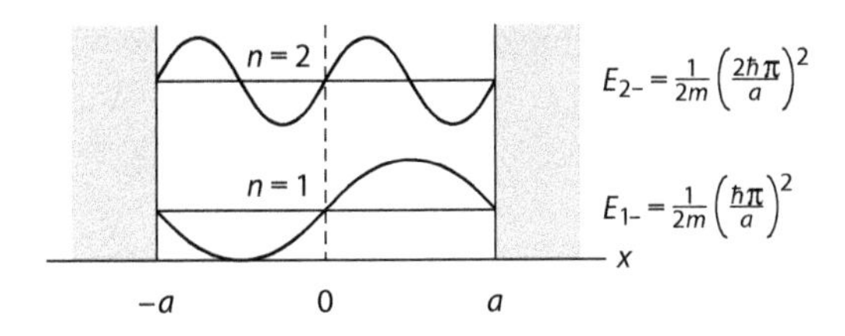

FIGURE 3.2 Qualitative plot of the wave functions for the first two odd square well solutions.

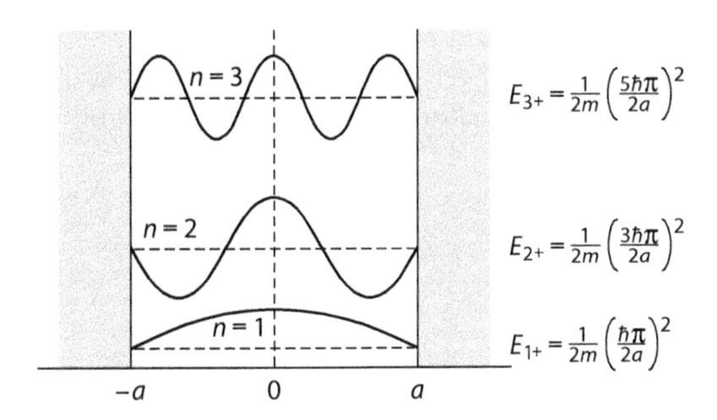

FIGURE 3.3 Qualitative plot of the wave functions for the first three even square well solutions.

which is to be compared with the actual value, $E_{1+} = \hbar^2\pi^2/8ma^2$.

We now wish to normalize the solutions Equations 3.19 and 3.20. We use the notation

$$u_{n-}(x) = \langle x|n-\rangle = A\sin(k_{n-}x), \quad (3.29)$$

$$u_{n+}(x) = \langle x|n+\rangle = A'\cos(k_{n+}x), \quad (3.30)$$

and set

$$1 = \langle n-|n-\rangle = \int_{-a}^{a} \mathrm{d}x\; u_{n-}^*(x)u_{n-}(x), \quad (3.31)$$

and

$$1 = \langle n+|n+\rangle = \int_{-a}^{a} \mathrm{d}x\; u_{n+}^*(x)u_{n+}(x). \quad (3.32)$$

Doing the integral in Equation 3.31 gives

$$\int_{-a}^{a} \mathrm{d}x\; u_{n-}^*(x)u_{n-}(x) = |A|^2\, a, \quad (3.33)$$

and similarly

$$\int_{-a}^{a} \mathrm{d}x\; u_{n+}^*(x)u_{n+}(x) = |A'|^2\, a. \quad (3.34)$$

The overall phase of wave functions is arbitrary, so we may choose A, A' as real and positive:

$$A, A' = \frac{1}{\sqrt{a}}. \tag{3.35}$$

One of the major tenants of quantum mechanics is orthogonality of states. Verification in the case of the product $\langle n-|n'+\rangle$ is easy:

$$\langle n-|n'+\rangle = \frac{1}{a}\int_{-a}^{a} dx\, \sin(k_{n-}x)\cos(k_{n'+}x) = 0. \tag{3.36}$$

(The integral of an odd × even = odd function over an even interval is zero.) The product $\langle n+|n'+\rangle$ should also be zero for $n \neq n'$. We can see this as follows:

$$\begin{aligned}\langle n+|n'+\rangle &= \frac{1}{a}\int_{-a}^{a} dx\, \cos(k_{n+}x)\cos(k_{n'+}x)\\ &= \frac{1}{a}\left\{ \frac{\sin\left[(k_{n+}-k_{n'+})x\right]}{2(k_{n+}-k_{n'+})}\Bigg|_{-a}^{a} + \frac{\sin\left[(k_{n+}+k_{n'+})x\right]}{2(k_{n+}+k_{n'+})}\Bigg|_{-a}^{a} \right\} = 0.\end{aligned} \tag{3.37}$$

This comes about since

$$\left.\begin{aligned} ak_{n+} &= \left(n-\frac{1}{2}\right)\pi,\\ &\Rightarrow a(k_{n+}-k_{n'+}) = (n-n')\pi\\ &\Rightarrow a(k_{n+}+k_{n'+}) = (n+n'-1)\pi, \end{aligned}\right\}. \tag{3.38}$$

Likewise, one can show that $\langle n-|n'-\rangle = 0$ for $n \neq n'$. By defining a label P that takes on values $\pm$ (labeling even/odd functions of x), we may summarize the statement of orthonormality here by

$$\langle nP \mid n'P'\rangle = \delta_{nn'}\delta_{PP'}. \tag{3.39}$$

That the solution of the Schrödinger equation forms a complete orthogonal set of functions is a theorem that can be proven, so we are seeing here a special case of a very general situation. (We will continue to sharpen our understanding of the mathematical meaning of completeness in the next chapter.)

The most general wave function consistent with the boundary conditions can now be written as

$$|\psi\rangle = \sum_{n=1}^{\infty}\left[C_{n+}|n+\rangle + C_{n-}|n-\rangle\right], \tag{3.40}$$

where C_{n+} and C_{n-} are sets of constants. These constants are not totally arbitrary since we must have

$$1 = \langle\psi|\psi\rangle, \tag{3.41}$$

which means that

$$\sum_{n=1}^{\infty}[|C_{n+}|^2+|C_{n-}|^2]=1. \tag{3.42}$$

Equation 3.42 suggests that $|C_{nP}|^2$ be interpreted as the probability that the general state $|\psi\rangle$ is in the energy eigenstate $|nP\rangle$. As pointed out below Equation 3.12, these constants are given by $C_{a'} = \langle a'|\psi\rangle$, which in this specific case means that (see also Chapter 2, Equation 2.178)

$$C_{nP} = \langle nP\,|\psi\rangle = \int_{-a}^{a} \mathrm{d}x\; u_{nP}^{*}(x)\psi(x). \tag{3.43}$$

In an analogy with a three-dimensional vector space, C_{nP} is like the projection of an arbitrary vector on a given axis or direction. Our interpretation of $|C_{nP}|^2$ as a probability is indeed consistent with our spin discussion of Chapter 1, the main difference being that the number of spin states was finite ("up" and "down" only for spin 1/2), whereas here the number of possible states, nP, is infinite. (It is a "countable infinity" in mathematician's jargon.)

Now while $|nP\rangle$ has sharp energy eigenvalues, we should realize that $|\psi\rangle$ does not. It is a coherent mixture of states with different energies. $|\psi\rangle$ does, however, have a well-defined *average* energy, given by its expectation value:

$$\langle H\rangle_{\psi} = \langle\psi|H|\psi\rangle = \sum_{n'=1}^{\infty}[C_{n'+}^{*}\langle n'+| + C_{n-}^{*}\langle n'-|] \cdot H \cdot \sum_{n=1}^{\infty}[C_{n+}|n+\rangle + C_{n-}|n-\rangle], \tag{3.44}$$

$$= \sum_{n=1}^{\infty}[E_{n+}|C_{n+}|^2 + E_{n-}|C_{n-}|^2]. \tag{3.45}$$

Equation 3.45 is consistent with our interpretation of $|C_{nP}|^2$ as the probability that $|\psi\rangle$ is in the state $|nP\rangle$.

Given the energies E_{nP}, it is easy to write down the time-evolved state $|\psi, t\rangle$:

$$|\psi,t\rangle = \sum_{n=1}^{\infty}[C_{n+}|n+, t\rangle + C_{n-}|n-, t\rangle], \tag{3.46}$$

or

$$|\psi, t\rangle = \sum_{n=1}^{\infty}[C_{n+}|n+\rangle\, \mathrm{e}^{-iE_{n+}t/\hbar} + C_{n-}|n-\rangle \mathrm{e}^{-iE_{n-}t/\hbar}]. \tag{3.47}$$

Of course, we still have

$$\langle\psi,t|\psi,t\rangle=\sum_{n'=1}^{\infty}[C^{*}_{n'+}e^{iE_{n'+}t/\hbar}\langle n'+|+C^{*}_{n'-}e^{iE_{n'-}t/\hbar}\langle n'-|]$$
$$\cdot\sum_{n=1}^{\infty}[C_{n+}e^{-iE_{n+}t/\hbar}|n+\rangle+C_{n-}e^{-iE_{n-}t/\hbar}|n-\rangle] \tag{3.48}$$
$$=\sum_{n=1}^{\infty}[|C_{n+}|^{2}+|C_{n-}|^{2}]=1,$$

so that probability is conserved.

The quantity P in $|nP\rangle$ is called "parity." It simply categorizes whether a wave function is even ($P=+$) or odd ($P=-$) under the substitution $x\rightarrow -x$. Let us define a parity operator by

$$\mathbb{P}|x\rangle\equiv|-x\rangle. \tag{3.49}$$

This implies that

$$\langle x|\mathbb{P}^{\dagger}=\langle -x|. \tag{3.50}$$

We have that

$$\langle x|(\overset{\leftarrow}{\mathbb{P}^{\dagger}}-\overset{\rightarrow}{\mathbb{P}})|x'\rangle=\langle -x|x'\rangle-\langle x|-x'\rangle \tag{3.51}$$
$$=\delta(x+x')-\delta(x+x')=0.$$

Since Equation 3.51 is true for all $\langle x|$, $|x'\rangle$, we have that

$$\mathbb{P}^{\dagger}=\mathbb{P}. \tag{3.52}$$

That is, the parity operator is Hermitian. We then have that

$$\langle x|\mathbb{P}\,|nP\rangle=\underbrace{\langle -x|nP\rangle}_{u_{nP}(-x)}=P\langle x|nP\rangle. \tag{3.53}$$

Since Equation 3.53 is true for all $\langle x|$, we may remove it to reveal that

$$\mathbb{P}\,|nP\rangle=P|nP\rangle, \tag{3.54}$$

$$\Rightarrow\langle nP|\mathbb{P}=P\langle nP|. \tag{3.55}$$

We thus learn that the $|nP\rangle$ are eigenstates of $\mathbb{P}$ with eigenvalues P. We also notice that

$$\mathbb{P}H|nP\rangle=E_{nP}\mathbb{P}|nP\rangle=PE_{nP}|nP\rangle, \tag{3.56}$$

and

$$H\mathbb{P}|nP\rangle=PH|nP\rangle=PE_{nP}|nP\rangle, \tag{3.57}$$

so that

$$[H,\mathbb{P}]|nP\rangle = 0. \tag{3.58}$$

Equation 3.58 being true for all states $|nP\rangle$ means that

$$[H,\mathbb{P}]=0. \tag{3.59}$$

Thus, the Hamiltonian and the parity operator commute. The reason that Equation 3.59 is significant is because we will learn in the next chapter that any operator that does not explicitly depend on time and that commutes with the Hamiltonian has expectation values that are a constant of the motion. In particular this means the parity, P, of a state $|nP\rangle$ does not change with time.

Note that the characterization of the states as $|nP\rangle$, where P is the parity eigenvalue, depends upon the choice of origin. In this case, we chose the origin to be in the center of the well, which gave an even symmetry to the potential. Besides the pedagogical value of such a choice, this characterization allows anticipation of the effects on states when small perturbations, such as those that will be considered in Chapter 5, are introduced.

3.3 THE INFINITE SQUARE WELL: OPERATOR SOLUTION

We have found the energies of the infinite square well by solving a differential equation with given boundary conditions. As an illustration of techniques that does not involve the solution of a differential equation, I will now solve for the energies of this system using operator methods.

First, let me introduce a more convenient notation for the states:

$$u_n(x) \equiv u_{n-}(x) = \frac{1}{\sqrt{a}}\sin\left(\frac{n\pi x}{a}\right), \tag{3.60}$$

$$u_{n-1/2}(x) \equiv u_{n+}(x) = \frac{1}{\sqrt{a}}\cos\left(\frac{(n-1/2)\pi x}{a}\right), \tag{3.61}$$

where $n = 1, 2, 3, \ldots$ for both. The advantage of this notation is that it makes it unnecessary to carry the parity designation to distinguish between states. When the label on the wave function is an integer, as in Equation 3.60, these are odd parity states; when it is half-integer, it has even parity. The other advantage is that the wave function label reminds us of the energy ordering. We will also define $u_n(x) \equiv \langle x|n\rangle$, $u_{n-1/2}(x) \equiv \langle x|n-1/2\rangle$ as usual.

Notice that we may write

$$\begin{aligned}\frac{1}{\sqrt{a}}\sin\left(\frac{n\pi x'}{a}\right) &= \frac{1}{\sqrt{a}}\sin\left[\frac{(n-1/2)\pi x'}{a}+\frac{\pi x'}{2a}\right]\\ &= \frac{1}{\sqrt{a}}\left[\sin\left(\frac{\pi x'}{2a}\right)\cos\left(\frac{(n-1/2)\pi x'}{a}\right)+\cos\left(\frac{\pi x'}{2a}\right)\sin\left(\frac{(n-1/2)\pi x'}{a}\right)\right].\end{aligned} \tag{3.62}$$

Also writing

$$\sin\left(\frac{(n-1/2)\pi x'}{a}\right) = -\frac{a}{(n-1/2)\pi}\frac{\mathrm{d}}{\mathrm{d}x'}\cos\left(\frac{(n-1/2)\pi x'}{a}\right) \tag{3.63}$$

says that we have the connection

$$\langle x'|n\rangle = \left[\sin\left(\frac{\pi x'}{2a}\right) - \cos\left(\frac{\pi x'}{2a}\right)\frac{a}{(n-1/2)\pi}\frac{\mathrm{d}}{\mathrm{d}x'}\right]\langle x'|n-1/2\rangle, \tag{3.64}$$

in terms of $\langle x|n\rangle$ and $\langle x|n-1/2\rangle$. Now using

$$\langle x'|x = x'\langle x'|, \tag{3.65}$$

and

$$\langle x'|p_x = \frac{\hbar}{i}\frac{\partial}{\partial x'}\langle x'|, \tag{3.66}$$

Equation 3.64 may be written as

$$\langle x'|n\rangle = \langle x'|R_{n-1/2}|n-1/2\rangle, \tag{3.67}$$

where

$$R_{n-1/2} \equiv \sin\left(\frac{\pi x}{2a}\right) - \left(\frac{ia}{(n-1/2)\pi\hbar}\right)\cos\left(\frac{\pi x}{2a}\right)p_x. \tag{3.68}$$

Let me emphasize that x and p_x in Equation 3.68 are operators, not numbers. Since Equation 3.67 is true for all $\langle x'|$, we have the operator statement

$$|n\rangle = R_{n-1/2}|n-1/2\rangle. \tag{3.69}$$

Similarly, we have that

$$\begin{aligned}\frac{1}{\sqrt{a}}\cos\left(\frac{(n-1/2)\pi x'}{a}\right) &= \frac{1}{\sqrt{a}}\cos\left(\frac{n\pi x'}{a} - \frac{\pi x'}{2a}\right)\\ &= \frac{1}{\sqrt{a}}\left[\sin\left(\frac{\pi x'}{2a}\right)\sin\left(\frac{n\pi x'}{a}\right) + \cos\left(\frac{\pi x'}{2a}\right)\cos\left(\frac{n\pi x'}{a}\right)\right].\end{aligned} \tag{3.70}$$

For the last term, I write

$$\cos\left(\frac{n\pi x'}{a}\right) = \frac{a}{n\pi}\frac{\mathrm{d}}{\mathrm{d}x'}\sin\left(\frac{n\pi x'}{a}\right), \tag{3.71}$$

which gives the connection,

$$\langle x'|n-1/2\rangle = \left[\sin\left(\frac{\pi x'}{2a}\right) + \cos\left(\frac{\pi x'}{2a}\right)\frac{a}{n\pi}\frac{\mathrm{d}}{\mathrm{d}x'}\right]\langle x'|n\rangle. \tag{3.72}$$

This can be written as

$$\langle x'|n-1/2\rangle = \langle x'|L_n|n\rangle \tag{3.73}$$

or

$$\left|n-1/2\right\rangle = L_n\left|n\right\rangle, \tag{3.74}$$

where I have introduced the operator,

$$L_n \equiv \sin\left(\frac{\pi x}{2a}\right)+\left(\frac{ia}{n\pi\hbar}\right)\cos\left(\frac{\pi x}{2a}\right)p_x. \tag{3.75}$$

In a like manner,

$$\left|n+1/2\right\rangle = -R_n\left|n\right\rangle, \tag{3.76}$$

$$\left|n-1\right\rangle = -\,L_{n-1/2}\left|n-1/2\right\rangle. \tag{3.77}$$

Thus, given the ground state $|1/2\rangle$, we can generate all the higher n states by repeated applications of $R_{n+1/2}$ and $-R_n$. For example, we have

$$|2\rangle = -R_{3/2}R_1R_{1/2}\left|1/2\right\rangle. \tag{3.78}$$

The R_n or L_n (for various allowed n values) are called raising or lowering "ladder" operators, respectively.

Let us say we know the ladder operators and their effect on states, but suppose we did not know the energy eigenvalues of the system. It is possible to find the energies as follows. Let us evaluate the quantity

$$[H, R_{n-1/2}]\left|n-1/2\right\rangle.$$

It is necessary to do some operator algebra, which should provide us good practice. We shall have to deal with the commutator

$$[H, R_{n-1/2}] = \left[\frac{p_x^2}{2m}, \sin\left(\frac{\pi x}{2a}\right)-\left(\frac{ia}{(n-1/2)\pi\hbar}\right)\cos\left(\frac{\pi x}{2a}\right)p_x\right]. \tag{3.79}$$

Before considering Equation 3.79 explicitly, let us work out some simpler things. First, consider the quantity

$$\langle x'|\left[p_x, f(x)\right] = \langle x'|\left(p_x f(x) - f(x)p_x\right). \tag{3.80}$$

We then find that

$$\begin{aligned}\langle x'|\,[p_x, f(x)] &= \frac{\hbar}{i}\frac{\partial}{\partial x'}(\langle x'|)f(x) - f(x')\langle x'|\,p_x \\ &= \frac{\hbar}{i}f(x')\,\frac{\partial}{\partial x'}\langle x'| + \frac{\hbar}{i}\frac{df(x')}{dx'}\langle x'| - \frac{\hbar}{i}f(x')\,\frac{\partial}{\partial x'}\langle x'| \\ &= \frac{\hbar}{i}\frac{df(x')}{dx'}\langle x'| = \frac{\hbar}{i}\langle x'|\,\frac{df(x)}{dx}.\end{aligned} \tag{3.81}$$

Being true for all $\langle x'|$, Equation 3.81 implies that

$$[p_x, f(x)] = \frac{\hbar}{i}\frac{df(x)}{dx}. \tag{3.82}$$

Likewise, one may show that

$$[x, f(p_x)] = -\frac{\hbar}{i}\frac{\mathrm{d}f(p_x)}{\mathrm{d}p_x}. \tag{3.83}$$

Using Equation 3.82 we have

$$\left[p_x, \cos\left(\frac{\pi x}{2a}\right)\right] = i\hbar\frac{\pi}{2a}\sin\left(\frac{\pi x}{2a}\right) \tag{3.84}$$

and

$$\left[p_x, \sin\left(\frac{\pi x}{2a}\right)\right] = -i\hbar\frac{\pi}{2a}\cos\left(\frac{\pi x}{2a}\right), \tag{3.85}$$

for example. Also useful are the following commutator identities:

$$[A,B] = -[B,A], \tag{3.86a}$$

$$[A+B,C] = [A,C] + [B,C], \tag{3.86b}$$

$$[AB,C] = A[B,C] + [A,C]B. \tag{3.86c}$$

Using Equation 3.86c we can now write

$$\begin{aligned}[H, R_{n-1/2}] &= \frac{p_x}{2m}\left[p_x, \sin\left(\frac{\pi x}{2a}\right) - \left(\frac{ia}{(n-1/2)\pi\hbar}\right)\cos\left(\frac{\pi x}{2a}\right)p_x\right] \\ &+ \left[p_x, \sin\left(\frac{\pi x}{2a}\right) - \left(\frac{ia}{(n-1/2)\pi\hbar}\right)\cos\left(\frac{\pi x}{2a}\right)p_x\right]\frac{p_x}{2m}.\end{aligned} \tag{3.87}$$

Using Equations 3.84 and 3.85 this becomes

$$\begin{aligned}[H, R_{n-1/2}] = \frac{p_x}{2m}&\left(-i\hbar\frac{\pi}{2a}\cos\left(\frac{\pi x}{2a}\right) + \frac{1}{2(n-1/2)}\sin\left(\frac{\pi x}{2a}\right)p_x\right) \\ &+ \left(-i\hbar\frac{\pi}{2a}\cos\left(\frac{\pi x}{2a}\right) + \frac{1}{2(n-1/2)}\sin\left(\frac{\pi x}{2a}\right)p_x\right)\frac{p_x}{2m}.\end{aligned} \tag{3.88}$$

We want to try to combine various types of terms together. To do this, let us try to move all the p_x operators in Equation 3.88 to the right of the factors involving x. We must be careful in doing this because x and p_x do not commute. We have that

$$\begin{aligned}p_x\sin\left(\frac{\pi x}{2a}\right) &= \sin\left(\frac{\pi x}{2a}\right)p_x + \left[p_x, \sin\left(\frac{\pi x}{2a}\right)\right] \\ &= \sin\left(\frac{\pi x}{2a}\right)p_x - i\hbar\frac{\pi}{2a}\cos\left(\frac{\pi x}{2a}\right),\end{aligned} \tag{3.89}$$

$$p_x \cos\left(\frac{\pi x}{2a}\right) = \cos\left(\frac{\pi x}{2a}\right) p_x + \left[p_x, \cos\left(\frac{\pi x}{2a}\right)\right]$$
$$= \cos\left(\frac{\pi x}{2a}\right) p_x + i\hbar \frac{\pi}{2a} \sin\left(\frac{\pi x}{2a}\right). \tag{3.90}$$

Therefore, Equation 3.88 can be written as

$$[H, R_{n-1/2}] = \frac{\hbar^2}{2m}\left(\frac{\pi}{2a}\right)^2 \sin\left(\frac{\pi x}{2a}\right) - 2\cdot\frac{i\hbar}{2m}\left(\frac{\pi}{2a}\right)\cos\left(\frac{\pi x}{2a}\right)p_x$$
$$- \frac{i\hbar}{4m(n-1/2)}\left(\frac{\pi}{2a}\right)\cos\left(\frac{\pi x}{2a}\right)p_x \tag{3.91}$$
$$+ 2\cdot\frac{1}{4m\,(n-1/2)}\sin\left(\frac{\pi x}{2a}\right)p_x^2.$$

This gives

$$[H, R_{n-1/2}]|n-1/2\rangle$$
$$= \left\{\begin{array}{l} \dfrac{\hbar^2}{2m}\left(\dfrac{\pi}{2a}\right)^2 \sin\left(\dfrac{\pi x}{2a}\right) - \dfrac{i\hbar}{2m}\left(\dfrac{\pi}{a}\right)\left(1+\dfrac{1}{4(n-1/2)}\right)\cos\left(\dfrac{\pi x}{2a}\right)\dfrac{ip_x}{\hbar} \\ + \dfrac{1}{2m(n-1/2)}\sin\left(\dfrac{\pi x}{2a}\right)p_x^2 \end{array}\right\}|n-1/2\rangle. \tag{3.92}$$

In the last term in Equation 3.92, we may make the replacement

$$p_x^2|n-1/2\rangle = 2mE_{n-1/2}\,|n-1/2\rangle, \tag{3.93}$$

for the unknown $E_{n-1/2}$. As for the other terms in Equation 3.92, the following consideration will be of use. We know from Equation 3.69 that

$$|n\rangle = \left[\sin\left(\frac{\pi x}{2a}\right) - \left(\frac{ia}{(n-1/2)\pi\hbar}\right)\cos\left(\frac{\pi x}{2a}\right)p_x\right]|n-1/2\rangle, \tag{3.94}$$

and from Equation 3.77 that

$$|n-1\rangle = \left[-\sin\left(\frac{\pi x}{2a}\right) - \left(\frac{ia}{(n-1/2)\pi\hbar}\right)\cos\left(\frac{\pi x}{2a}\right)p_x\right]|n-1/2\rangle. \tag{3.95}$$

Subtracting Equation 3.95 from Equation 3.94 gives

$$|n\rangle - |n-1\rangle = 2\sin\left(\frac{\pi x}{2a}\right)|n-1/2\rangle. \tag{3.96}$$

Adding Equations 3.94 and 3.95 gives

$$|n\rangle + |n-1\rangle = -\frac{2ia}{n\pi\hbar}\cos\left(\frac{\pi x}{2a}\right)p_x|n-1/2\rangle. \tag{3.97}$$

Substituting Equations 3.96 and 3.97 into Equation 3.92 and simplifying gives

$$[H,R_{n-1/2}]|n-1/2\rangle = \left\{\left(\frac{\hbar^2\pi^2}{16ma^2}+\frac{E_{n-1/2}}{2(n-1/2)}\right)[|n\rangle-|n-1\rangle] + \frac{\hbar^2\pi^2}{4ma^2}(n-1/2)\left(1+\frac{1}{4(n-1/2)}\right)[|n\rangle+|n-1\rangle]\right\}. \tag{3.98}$$

Now, project into $\langle n-1|$. On the left-hand side of Equation 3.98 we simply have

$$\langle n-1|[H,R_{n-1/2}]|n-1/2\rangle = 0, \tag{3.99}$$

by the nature of the raising operator. However, explicitly evaluating the right-hand side tells us that

$$0 = -\frac{\hbar^2\pi^2}{16ma^2}-\frac{E_{n-1/2}}{2(n-1/2)}+\frac{\hbar^2\pi^2}{4ma^2}(n-1/4),$$
$$\Rightarrow E_{n-1/2} = \frac{1}{2m}\left(\frac{\hbar\pi(n-1/2)}{a}\right)^2. \tag{3.100}$$

Thus, we have evaluated the energies of the positive parity states simply from a knowledge of the properties of the ladder operators. We may find the energies of the negative parity states from similar considerations.

The evaluation of the positive parity energies has been an extended exercise in the use of operator commutators. The initial solution for the energies using the coordinate space Schrödinger equation was in fact much easier. We will soon look at a problem, the harmonic oscillator, where the opposite is true. That is, the solution of the coordinate space differential equation is much more difficult than the solution of the problem using operator methods. We will only use operator methods there because of this.

3.4 THE FINITE POTENTIAL BARRIER STEP POTENTIAL

The next potential we consider is shown in Figure 3.4. We have broken the x space into three regions in which we will separately solve the Schrödinger equation.

In Regions I and III, we simply have the free space time-independent Schrödinger equation to solve. The general solution to

$$-\frac{\hbar^2}{2m}\frac{d^2u}{dx^2} = E\,u(x), \tag{3.101}$$

can be written as*

$$u_{\mathrm{I}}(x) = Ae^{ik_1x}+Be^{-ik_1x}, \tag{3.102}$$

$$u_{\mathrm{III}}(x) = Ee^{ik_1x}+Fe^{-ik_1x}, \tag{3.103}$$

* Forgive me for using E as both an energy in Equation 3.101 and an amplitude in Equation 3.103! The context will help keep things clear, however.

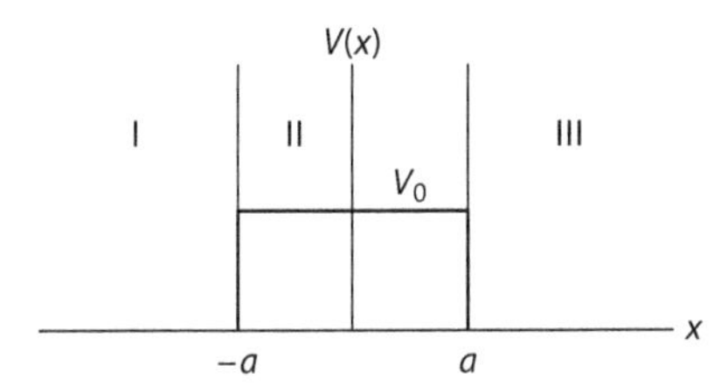

FIGURE 3.4 The finite potential barrier step potential.

where $k_1 = \sqrt{2mE}\,/\,\hbar$. We saw in Chapter 2 that $e^{i(kx-\omega t)}$ represents a wave with momentum $p_x = k\hbar$ traveling in the $+x$ direction. Therefore, we interpret the coefficient A in Equation 3.102 as the amplitude of a plane wave incident from the left. (There is no time dependence in Equations 3.102 and 3.103 because we are interested in time independent or stationary solutions in this chapter.) We will take as a boundary condition the fact that the incident waves come from the left. However, it would be wrong to conclude from this that we could choose $B = 0$ here because we expect that there will be *reflected* waves from the potential steps at $x = \pm a$. Because we are choosing only waves incident from the left, we *must* choose $F = 0$ in Region III.

In Region II, we must solve

$$-\frac{\hbar^2}{2m}\frac{d^2u}{dx^2} = (E - V_0)u(x). \tag{3.104}$$

The solution of Equation 3.104 depends on whether $E > V_0$ or $E < V_0$. When $E > V_0$, we have

$$u_{II}(x) = Ce^{ik_2x} + De^{-ik_2x}, \tag{3.105}$$

where $k_2 = \sqrt{2m(E - V_0)}\,/\,\hbar$.

Our solution so far consists of Equations 3.102 and 3.103 (with $F = 0$), and Equation 3.105. How are we to find the five coefficients A, B, C, D, and E? First, it should be clear that one condition on these coefficients is an overall (arbitrary) normalization. Applying Equation 2.129 to the "A" term in Equation 3.102, we have for the incident probability current

$$(j(x,t))^A = \frac{\hbar k_1}{m}|A|^2, \tag{3.106}$$

from which it is seen that $|A|^2$ corresponds to a constant density of incoming particles. Thus, we really have only to determine four out of these five unknown coefficients. It will be convenient, therefore, to solve only for the four dimensionless ratios

$$\frac{B}{A}, \frac{C}{A}, \frac{D}{A}, \text{ and } \frac{E}{A}.$$

To do this, we must bring out an underlying requirement of solutions of the Schrödinger equation. To see this requirement, let us integrate the time-independent Schrödinger equation over an infinitesimal region surrounding a point, x:

$$\int_{x_0-\varepsilon}^{x_0+\varepsilon} dx \left[-\frac{\hbar^2}{2m}\frac{d^2u}{dx^2} + V(x)\,u(x) = Eu(x) \right]. \tag{3.107}$$

Assuming $u(x)$ and $V(x)$ are bounded, this says that in the limit $\varepsilon \to 0^+$,

$$\left.\frac{du}{dx}\right|_+ = \left.\frac{du}{dx}\right|_-, \tag{3.108}$$

or, in other words, that the first derivative of $u(x)$ evaluated in a limiting sense from points on the left or right of x_0 is continuous. This eliminates a wave function that has a derivative discontinuity, as in Figure 3.5.

By integrating Equation 3.107 a second time, we may show that $u(x)$ is also continuous. This is fortunate, because one can show that a discontinuity in $u(x)$ or $u'(x)$ (the prime indicates the first derivative) is not consistent with our interpretation of $|u(x)|^2$ as a probability density. The only exceptions to having a continuous $u'(x)$ come when unbounded potentials are considered. We have already seen an example of this in the infinite square well where, in fact, we have a discontinuity in $u'(x)$ at the edges of the well, $x = \pm a$. Another example of an unbounded potential where a discontinuity in $u'(x)$ is allowed is a delta-function potential. We will study such a situation shortly in the attractive Kronig–Penney model.

For our problem, we must require the continuity of our wave functions at *all* positions. In particular, this means our wave functions and their first derivatives must be continuous in the neighborhood of the joining positions of Regions I, II, and III. This gives us four conditions on our coefficients, which is just enough to determine the four unknown ratios from earlier. At $x = -a$, we find that

$$Ae^{-ik_1a} + Be^{ik_1a} = Ce^{-ik_2a} + De^{ik_2a}, \tag{3.109}$$

from continuity of $u(x)$, and

$$Ak_1e^{-ik_1a} - Bk_1e^{ik_1a} = k_2Ce^{-ik_2a} - k_2De^{ik_2a}, \tag{3.110}$$

from continuity of $u'(x)$. At $x = a$, we get the equations

$$Ce^{ik_2a} + De^{-ik_2a} = Ee^{ik_1a} \tag{3.111}$$

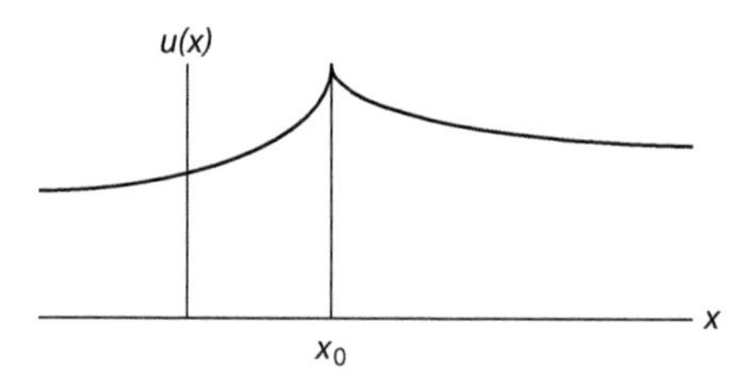

FIGURE 3.5 A wave function, $u(x)$, with a derivative discontinuity at $x = x_0$.

and

$$k_2Ce^{ik_2a} - k_2De^{-ik_2a} = k_1Ee^{ik_1a} \tag{3.112}$$

from continuity of $u(x)$ and $u'(x)$. We now define the transmission and reflection coefficients:

$$T \equiv \left|\frac{E}{A}\right|^2, \tag{3.113}$$

$$R \equiv \left|\frac{B}{A}\right|^2. \tag{3.114}$$

We use the absolute squares of the ratios of amplitudes in order that the results be interpretable as probabilities. Note that

$$T + R = 1, \tag{3.115}$$

as one must have if probabilities are conserved. Solving Equations 3.109 through 3.112 gives us

$$\frac{B}{A} = \frac{\frac{i}{2}\left(\frac{k_2}{k_1} - \frac{k_1}{k_2}\right)e^{-2ik_1a}\sin(2k_2a)}{\cos(2k_2a) - \frac{i}{2}\left(\frac{k_1}{k_2} + \frac{k_2}{k_1}\right)\sin(2k_2a)}, \tag{3.116}$$

$$\frac{C}{A} = \frac{\frac{1}{2}\left(1 + \frac{k_1}{k_2}\right)e^{-ik_2a}e^{-ik_1a}}{\cos(2k_2a) - \frac{i}{2}\left(\frac{k_1}{k_2} + \frac{k_2}{k_1}\right)\sin(2k_2a)}, \tag{3.117}$$

$$\frac{D}{A} = \frac{\frac{1}{2}\left(1 - \frac{k_1}{k_2}\right)e^{-ik_2a}e^{-ik_1a}}{\cos(2k_2a) - \frac{i}{2}\left(\frac{k_1}{k_2} + \frac{k_2}{k_1}\right)\sin(2k_2a)}, \tag{3.118}$$

$$\frac{E}{A} = \frac{e^{-2ik_1a}}{\cos(2k_2a) - \frac{i}{2}\left(\frac{k_1}{k_2} + \frac{k_2}{k_1}\right)\sin(2k_2a)}. \tag{3.119}$$

The transmission coefficients T and R from Equations 3.119 and 3.116, respectively, are then found to be

$$T = \frac{1}{\cos^2(2k_2a) + \frac{1}{4}\left(\frac{k_1}{k_2} + \frac{k_2}{k_1}\right)^2\sin^2(2k_2a)}, \tag{3.120}$$

and

$$R = \frac{\frac{1}{4}\left(\frac{k_2}{k_1} - \frac{k_1}{k_2}\right)^2\sin^2(2k_2a)}{\cos^2(2k_2a) + \frac{1}{4}\left(\frac{k_1}{k_2} + \frac{k_2}{k_1}\right)^2\sin^2(2k_2a)}. \tag{3.121}$$

Thus, even though in the case we are studying the plane wave has an energy in excess of V_0, the classical energy necessary to overcome the potential barrier, there is in general a nonzero probability that a particle will reflect from the barrier. Equation 3.115 may be verified from these equations.

Let us examine the solution in the case $E < V_0$ now. The wave function in Regions I and III is as before, but now the general solution in Region II is

$$u_{II}(x) = Ce^{-Kx} + De^{Kx}, \tag{3.122}$$

where $K = \sqrt{2m(V_0 - E)}/\hbar$. We notice that the only difference now is the fact that we are working with real exponentials. Therefore, rather than reworking out T and R from the start, it is only necessary to make the substitution $k_2 \to +iK$ everywhere. We find in the $E < V_0$ case that

$$T = \frac{1}{\cosh^2(2Ka) + \frac{1}{4}\left(\frac{K}{k_1} - \frac{k_1}{K}\right)^2 \sinh^2(2Ka)} \tag{3.123}$$

and

$$R = \frac{\frac{1}{4}\left(\frac{K}{k_1} + \frac{k_1}{K}\right)\sinh^2(2Ka)}{\cosh^2(2Ka) + \frac{1}{4}\left(\frac{K}{k_1} - \frac{k_1}{K}\right)^2 \sinh^2(2Ka)}. \tag{3.124}$$

Putting Equations 3.120 and 3.123 together, we qualitatively plot the result for $T(E)$ in Figure 3.6.

There are several aspects to remark upon on this graph. The most obvious thing to observe is that $T \neq 0$ even when $E < V_0$. Classically, such a thing could never happen. That is, if we had a classical particle with an energy $E < V_0$, there would be zero probability that the particle would be able to overcome the potential barrier. The nonzero transmission here is just a manifestation of the uncertainty principle, Equation 2.32. Note that under the replacement $k_2 \to i\,K$, we have

$$e^{\pm ik_2x} \to e^{\mp Kx}, \tag{3.125}$$

so that the forward (backward) traveling wave is changed into a decreasing (increasing) exponential for x large and positive. This suggests

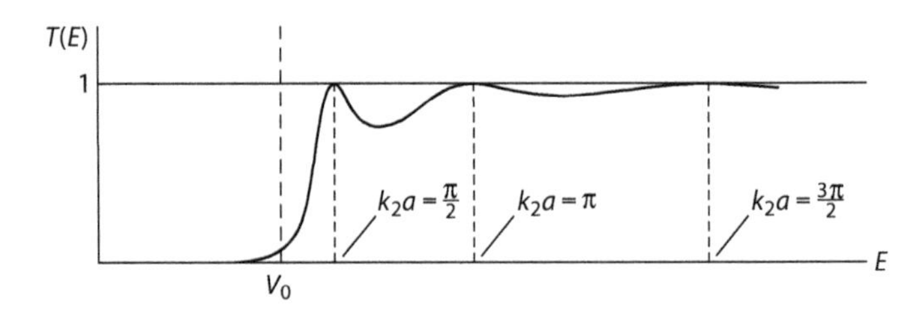

FIGURE 3.6 Qualitative plot of the transmission coefficient, $T(E)$, as a function of incoming energy for the step potential.

for a wide barrier ($2a \gg 1/(2K)$) that the probability of transmission of the $2a$-length barrier, given a small reflection at the other end, is $T \sim |e^{-2Ka}|^2 = e^{-4Ka}$, which is in agreement with Equation 3.123. To overcome the height of the barrier, the particle's momentum must have an uncertainty of $\Delta p \sim \hbar K$, leading to an approximate barrier penetration length of

$$\Delta x \sim \frac{1}{2K}, \tag{3.126}$$

from the uncertainty principle, Equation 2.32, and roughly in agreement with a simple estimate from T when $2a \to \Delta x$. This phenomenon is known as *tunneling*.

The other remarkable thing about the graph of $T(E)$ is the fact that there is complete transmission ($T = 1$) when the condition

$$2k_2 a = n\pi, \quad n = 1, 2, 3, \ldots \tag{3.127}$$

is fulfilled. This is a one-dimensional form of *resonance*. That Equation 3.127 leads to maxima in $T(E)$ versus E is easy to understand. The path difference between the incident wave and an internally reflected wave in Region II, which has transversed the barrier and back, is $4a$. These two waves will interfere constructively whenever the path difference is an integral multiple of the wavelength in Region II. Alternatively, the path difference of $4a$ leads to destructive interference between the wave reflected at $x = -a$ and a reflected wave from $x = a$ in Region I, given that the reflected wave at $x = a$ undergoes a phase shift of π. However, this simple argument does not tell us that $T = 1$ at these positions, simply that we should expect local maxima there.

When $V_0 < 0$ the potential barrier changes into a potential well. This system contains discrete states with negative energy in addition to the continuum of positive energy states. This system will be discussed and its discrete energy and eigenvalues evaluated in an unusual way in Chapter 11, Section 11.5.

3.5 THE HARMONIC OSCILLATOR

One of the most important problems in classical mechanics is the simple harmonic oscillator, described by the equation of motion,

$$m\ddot{x} + kx = 0. \tag{3.128}$$

If we integrate this equation, we get an equation for the energy of the system:

$$E = \frac{1}{2}m\dot{x}^2 + \frac{1}{2}kx^2 \ (= \text{constant}). \tag{3.129}$$

Let us study the same problem in quantum mechanics. That is, we will take

$$H = \frac{p_x^2}{2m} + \frac{1}{2}kx^2, \tag{3.130}$$

where this is now an operator equation. To simplify life, let us introduce some new variables that will eliminate the quantities m and k from appearing explicitly in our equations. First, define

$$\omega \equiv \sqrt{\frac{k}{m}} \tag{3.131}$$

and rewrite Equation 3.130 as (we have divided by $1/\hbar\omega$)

$$\frac{H}{\hbar\omega} = \frac{p_x^2}{2m\hbar\omega} + \frac{m\omega}{2\hbar}x^2. \tag{3.132}$$

Now define the dimensionless variables

$$\mathcal{H} \equiv \frac{H}{\hbar\omega}, \quad p \equiv \frac{p_x}{\sqrt{m\omega\hbar}}, \quad q \equiv \sqrt{\frac{m\omega}{\hbar}}x. \tag{3.133}$$

Then the eigenvalue problem we wish to solve may be written as

$$\mathcal{H}|n\rangle = \mathcal{E}_n|n\rangle, \tag{3.134}$$

$$\mathcal{H} = \frac{1}{2}(p^2 + q^2), \tag{3.135}$$

where $n = 0, 1, 2, 3, \ldots$ is (initially) just a labeling of the expected discrete energy states of this system and $\mathcal{E}_n = E_n/\hbar\omega$. Thus, we let $n = 0$ label the ground state, $n = 1$ pick out the first excited state, and so on.

We can easily verify that

$$[q, p] = i \tag{3.136}$$

for these new variables, and that

$$\left(\frac{q+ip}{\sqrt{2}}\right)\left(\frac{q-ip}{\sqrt{2}}\right) = \frac{q^2+p^2}{2} + \frac{i}{2}[p,q], \tag{3.137}$$

$$\Rightarrow \mathcal{H} = \left(\frac{q+ip}{\sqrt{2}}\right)\left(\frac{q-ip}{\sqrt{2}}\right) - \frac{1}{2}. \tag{3.138}$$

Let us define

$$A \equiv \frac{q+ip}{\sqrt{2}}, \tag{3.139}$$

which, since q and p are Hermitian, means that

$$A^\dagger = \frac{q-ip}{\sqrt{2}}. \tag{3.140}$$

Therefore, we may write

$$\mathcal{H} = AA^\dagger - \frac{1}{2}. \tag{3.141}$$

Now let us consider the commutator $[\mathcal{H}, A]$. We have that

$$\begin{aligned}[\mathcal{H}, A] &= \frac{1}{2\sqrt{2}}[q^2, q+ip] + \frac{1}{2\sqrt{2}}[p^2, q+ip] \\ &= \frac{i}{2\sqrt{2}}[q^2, p] + \frac{1}{2\sqrt{2}}[p^2, q] \\ &= \frac{i}{2\sqrt{2}}\{q[q,p]+[q,p]\,q\} + \frac{1}{2\sqrt{2}}\{p[p,q]+[p,q]\,p\} \\ &= -\frac{(q+ip)}{\sqrt{2}}.\end{aligned} \tag{3.142}$$

or

$$[\mathcal{H}, A] = -A. \tag{3.143}$$

Now notice

$$[\mathcal{H}, A]|n\rangle = -A|n\rangle, \tag{3.144}$$

so

$$\mathcal{H}A|n\rangle - A\underbrace{\mathcal{H}|n\rangle}_{\mathcal{E}_n|n\rangle} = -A|n\rangle \tag{3.145}$$

and

$$\mathcal{H}(A|n\rangle) = (\mathcal{E}_n - 1)(A|n\rangle). \tag{3.146}$$

Therefore, $A|n\rangle$ is also an eigenstate of $\mathcal{H}$, but with a lower value of energy than the state $|n\rangle$. Let us assume this is just the next lowest state, $|n-1\rangle$, outside of an unknown multiplicative constant

$$A|n\rangle = C_n|n-1\rangle, \tag{3.147}$$

for $n = 1, 2, 3, \ldots$. This implies that the energy levels of the system are all equally spaced. We can always choose the C_n in Equation 3.147 to be real and positive by associating any phase factor that arises with the state $|n-1\rangle$. Equation 3.147 then implies

$$\langle n|A^\dagger = C_n\langle n-1|, \tag{3.148}$$

from taking the adjoint of both sides.

The adjoint of Equation 3.143 is

$$[\mathcal{H}, A^\dagger] = A^\dagger. \tag{3.149}$$

Equation 3.149 then implies, the same way we derived Equation 3.146, that

$$\mathcal{H}(A^\dagger|n\rangle) = (\mathcal{E}_n + 1)(A^\dagger|n\rangle). \tag{3.150}$$

Therefore, $A^\dagger|n\rangle$ is an eigenstate of $\mathcal{H}$ with the next highest energy to $|n\rangle$ since the energy levels are equally spaced:

$$A^\dagger|n\rangle = C'_n|n+1\rangle. \tag{3.151}$$

There is no immediate reason why the C'_n should be real, since in picking C_n real we defined the phase of the states. The operators A and $A^\dagger$ are seen to be ladder operators for the states, but unlike the ladder operators for the infinite square well, there is no dependence of the A or $A^\dagger$ on the state label, n.

Now notice that

$$\begin{aligned}\mathcal{H}|n\rangle &= \left(AA^\dagger - \frac{1}{2}\right)|n\rangle = C'_n\, A|n+1\rangle - \frac{1}{2}|n\rangle \\ &= \left(C'_n\, C_{n+1} - \frac{1}{2}\right)|n\rangle = \mathcal{E}_n|n\rangle.\end{aligned} \tag{3.152}$$

Since $\mathcal{H}^\dagger = \mathcal{H}$, we have that the energies of the system must be real. (Remember the argument in Chapter 1, Section 1.18.) Therefore, since we chose the C_n real, the C'_n must also be real from the above.

We are assuming in this discussion that this system has a lowest energy state, which, because of the Heisenberg uncertainty principle, has a positive value. (We estimated its value in Chapter 2, Problem 2.4.2.) That is, we can only consistently maintain that $|0\rangle$ is the state of lowest energy if we take

$$A|0\rangle = 0, \tag{3.153}$$

which means, from Equation 3.147, that we must take

$$C_0 = 0. \tag{3.154}$$

Equation 3.154 will be useful in a moment.

To complete this argument let us consider the quantity

$$Q \equiv \langle 0 | A^n (A^\dagger)^n |0\rangle. \tag{3.155}$$

By allowing the A^n $(A^\dagger)^n$ operators to act one at a time to the right, we learn that

$$Q = \underbrace{\langle 0|0\rangle}_{1}\; C_1 \ldots C_n C'_{n-1} \ldots C'_0, \tag{3.156}$$

or

$$Q = (C_1 C'_0)\,(C_2 C'_1)\,\ldots\,(C_n C'_{n-1}). \tag{3.157}$$

Now, by allowing A^n to act to the left while $(A^\dagger)^n$ is still acting to the right, reveals that Q may also be written as

$$Q = C'_0 \ldots C'_{n-1}\, \underbrace{\langle n|n\rangle}_{1}\, C'_{n-1} \ldots C'_0. \tag{3.158}$$

In writing Equation 3.158, we are using the fact that the C'_n are real, plus the adjoint of Equation 3.151, which tells us that

$$\langle n | A = C'_n \langle n+1|. \tag{3.159}$$

Thus, Equation 3.158 gives us the alternative evaluation:

$$Q = (C_0')^2 (C_1')^2 \dots (C_{n-1}')^2. \tag{3.160}$$

Comparing Equations 3.157 and 3.160 in the case of $n = 1$ tells us that

$$C_1 = C_0'. \tag{3.161}$$

Therefore, in the case $n = 2$, we conclude that

$$C_2 = C_1'. \tag{3.162}$$

By mathematical induction (see the procedure in Problem 3.5.2) we may show then in general that

$$C_n = C_{n-1}'. \tag{3.163}$$

From the equal spacing of the energy levels

$$\mathcal{E}_{n-1} = \mathcal{E}_n - 1, \tag{3.164}$$

we then have, from Equations 3.152 and 3.163, that

$$C_{n+1}^2 = C_n^2 + 1. \tag{3.165}$$

Using Equation 3.154 in 3.165 now tells us that

$$C_n^2 = n. \tag{3.166}$$

Remember, we choose C_n to be positive so that

$$C_n = \sqrt{n}. \tag{3.167}$$

The dimensionless energies of the system are now given as

$$\mathcal{E}_n = C_{n+1}^2 - \frac{1}{2} = n + \frac{1}{2}. \tag{3.168}$$

Putting the dimensions back in, we have

$$E_n = \hbar\omega\mathcal{E}_n = \hbar\omega\left(n + \frac{1}{2}\right). \tag{3.169}$$

Notice that the lowest energy is nonzero, $E_0 = (1/2)\hbar\omega$. As already pointed out, this is a consequence of the Heisenberg uncertainty principle for momentum and position. This lowest energy of the simple harmonic oscillator is called its *zero point energy*. Actually, the zero point energy is not observable, since the energy of a system is arbitrary up to an additive constant. However, changes in the zero-point energy are uniquely defined and can be observed in laboratory experiments.

The next thing we do will be to get explicit expressions for the coordinate space wave functions, $\langle q'|n\rangle$. First, we know that

$$A^\dagger|n-1\rangle = \sqrt{n}|n\rangle, \tag{3.170}$$

so that we may write

$$\begin{aligned}|n\rangle &= \frac{1}{\sqrt{n}}A^\dagger\,|n-1\rangle = \frac{1}{\sqrt{n}}\frac{1}{\sqrt{n-1}}(A^\dagger)^2\,|n-2\rangle \\ &= \cdots = \frac{1}{\sqrt{n(\,n-1)\,\ldots\,1}}\,(A^\dagger)^n|0\rangle.\end{aligned} \tag{3.171}$$

Therefore, we may write the energy eigenkets as

$$|n\rangle = \frac{(A^\dagger)^n}{\sqrt{n!}}|0\rangle. \tag{3.172}$$

Since

$$(A^\dagger)^n = \left(\frac{q-ip}{\sqrt{2}}\right)^n, \tag{3.173}$$

Equation 3.172 becomes

$$|n\rangle = \frac{1}{\sqrt{2^n n!}}\,(q-ip)^n|0\rangle. \tag{3.174}$$

Our coordinate space wave functions are then

$$u_n(q') = \langle q'\,|\,n\rangle = \frac{1}{\sqrt{2^n n!}}\langle q'\,|(q-ip)^n\,|\,0\rangle, \tag{3.175}$$

where the $\langle q'|$ state is just a relabeling of the state $\langle x'|$. Using our previous results from Chapter 2, we have that

$$\langle q'|q = q'\langle q'|, \tag{3.176}$$

and

$$\langle q'|p = \frac{1}{i}\frac{\partial}{\partial q'}\langle q'|. \tag{3.177}$$

We may now use

$$\begin{aligned}\langle q'\,|(q-ip)^n\,|\,0\rangle &= \langle q'\,|(q-ip)(q-ip)^{n-1}\,|\,0\rangle \\ &= \left(q' - \frac{\mathrm{d}}{\mathrm{d}q'}\right)\langle q'|(q-ip)^{n-1}|0\rangle \\ &= \left(q' - \frac{\partial}{\partial q'}\right)^2\langle q'|(q-ip)^{n-2}|0\rangle \\ &= \ldots\left(q' - \frac{\mathrm{d}}{\mathrm{d}q'}\right)^n\langle q'|0\rangle\end{aligned} \tag{3.178}$$

in Equation 3.175. Therefore,

$$u_n(q') = \frac{1}{\sqrt{2^n n!}}\left(q' - \frac{\mathrm{d}}{\mathrm{d}q'}\right)^n \langle q' \mid 0\rangle. \tag{3.179}$$

The question now is: What is the ground state wave function $\langle q'|0\rangle$? If we can find this, then Equation 3.179 gives all the rest of the wave functions. We may find this wave function by solving a differential equation. We know that $A|0\rangle = 0$, but

$$\langle q'|A|0\rangle = \frac{1}{\sqrt{2}}\left(q' + \frac{\mathrm{d}}{\mathrm{d}q'}\right)\langle q'|0\rangle, \tag{3.180}$$

so

$$\left(q' + \frac{\mathrm{d}}{\mathrm{d}q'}\right)\langle q'|0\rangle = 0. \tag{3.181}$$

The solution to Equation 3.181 is

$$\langle q' \mid 0\rangle = C\mathrm{e}^{-q'^2/2}, \tag{3.182}$$

where C is an unknown constant. Normalizing this to unity, we find

$$|C|^2 \int_{-\infty}^{\infty} \mathrm{d}q'\, \mathrm{e}^{-q'^2} = |C|^2 \sqrt{\pi} = 1, \tag{3.183}$$

so we may choose

$$C = \frac{1}{\pi^{1/4}}. \tag{3.184}$$

Therefore,

$$u_0(q') = \langle q' \mid 0\rangle = \frac{1}{\pi^{1/4}}\, \mathrm{e}^{-q'^2/2}, \tag{3.185}$$

and we have that

$$u_n(q') = \frac{1}{\sqrt{\sqrt{\pi}\, 2^n n!}}\left(q' - \frac{\mathrm{d}}{\mathrm{d}q'}\right)^n \mathrm{e}^{-q'^2/2}. \tag{3.186}$$

Let us simplify the result Equation 3.186 a bit in order to connect up with some standard results. It is possible to show that

$$\left(\frac{\mathrm{d}}{\mathrm{d}q'}\right)^n \mathrm{e}^{-q'^2/2} f(q') = \mathrm{e}^{-q'^2/2}\left(\frac{\mathrm{d}}{\mathrm{d}q'} - q'\right)^n f(q'), \tag{3.187}$$

for an arbitrary function $f(q')$ by the use of induction. Now by choosing $f(q') = \mathrm{e}^{-q'^2/2}$, this means that

$$\left(\frac{\mathrm{d}}{\mathrm{d}q'}\right)^n \mathrm{e}^{-q'^2} = \mathrm{e}^{-q'^2/2}(-1)^n\left(q' - \frac{\mathrm{d}}{\mathrm{d}q'}\right)^n \mathrm{e}^{-q'^2/2}. \tag{3.188}$$

Using Equation 3.188 in 3.179 gives us the alternative form

$$u_n(q') = \frac{(-1)^n}{\sqrt{\sqrt{\pi}2^n n!}} e^{q'^2/2} \left(\frac{\mathrm{d}}{\mathrm{d}q'}\right)^n e^{-q'^2}. \tag{3.189}$$

Equation 3.189 is more familiar when the definition of *Hermite polynomials*,

$$H_n(q') \equiv (-1)^n e^{q'^2} \left(\frac{\mathrm{d}}{\mathrm{d}q'}\right)^n e^{-q'^2}, \tag{3.190}$$

is introduced. Equation 3.190 in 3.189 gives us

$$u_n(q') = \frac{1}{\sqrt{\sqrt{\pi}2^n n!}} e^{-q'^2/2} H_n(q'). \tag{3.191}$$

The first few Hermite polynomials are

$$H_0 = 1, \quad H_1 = 2q', \quad H_2 = 4q'^2 - 2. \tag{3.192}$$

The order of the nth Hermite polynomial is n. Also, the number of zeros in the wave function $u_n(q')$ is also n (excluding the points at infinity). The first three $u_n(q')$ are shown in Figure 3.7.

The orthonormality of the states $|n\rangle$ may be demonstrated as follows. We know that

$$\langle n|n'\rangle = \langle 0|\frac{A^n}{\sqrt{n!}} \frac{(A^\dagger)^{n'}}{\sqrt{n'!}}|0\rangle. \tag{3.193}$$

Now we may write

$$A(A^\dagger)^{n'} = (A^\dagger)^{n'} A + [A, (A^\dagger)^{n'}]. \tag{3.194}$$

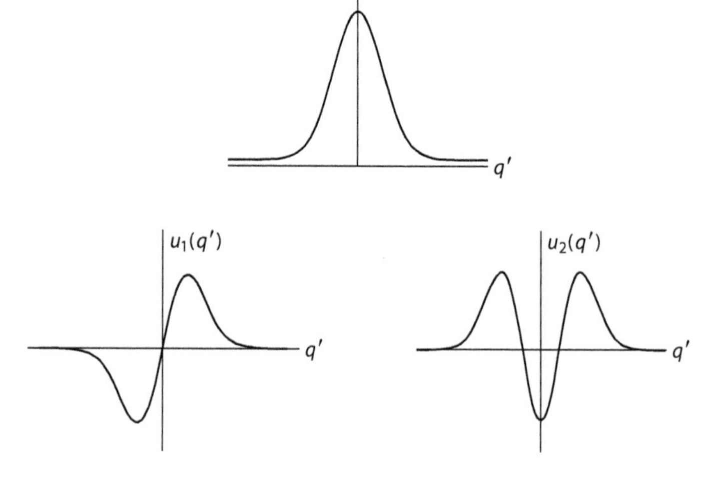

FIGURE 3.7 Qualitative plot of the first three wave functions of the harmonic oscillator potential.

In a problem, you will evaluate the commutator in Equation 3.194 as

$$[A,(A^\dagger)^{n'}]=n'(A^\dagger)^{n'-1}. \tag{3.195}$$

Using Equations 3.195 and 3.194 in 3.193, and using $A|0\rangle=0$, we now find

$$\begin{aligned}\langle n|n'\rangle &= \frac{n'}{\sqrt{nn'}}\langle 0|\frac{A^{n-1}}{\sqrt{(n-1)!}}\frac{(A^\dagger)^{n'-1}}{\sqrt{(n'-1)!}}|0\rangle \\ &= \frac{n'}{\sqrt{nn'}}\langle n-1|n'-1\rangle.\end{aligned} \tag{3.196}$$

We have three possible cases. If $n=n'$, then

$$\langle n|n\rangle=\langle n-1|n-1\rangle=\cdots=\langle 0|0\rangle=1. \tag{3.197}$$

However, if $n>n'$

$$\langle n|n'\rangle=\frac{n'}{\sqrt{nn'}}\langle n-1|n'-1\rangle=\cdots=\text{constants }\langle (n-n')|0\rangle, \tag{3.198}$$

but

$$\langle (n-n')|0\rangle=\langle 0|\frac{A^{n-n'}}{\sqrt{(n-n')!}}|0\rangle=0, \tag{3.199}$$

so $\langle n\,|\,n'\rangle=0$ for $n>n'$. Similarly, $\langle n|n'\rangle=0$ when $n<n'$. Therefore, we have shown that

$$\langle n|n'\rangle=\delta_{nn'}. \tag{3.200}$$

In this discrete context, completeness of the energy eigenfunctions reads

$$\sum_n |n\rangle\langle n|=1. \tag{3.201}$$

You will find the momentum space wave function, $\langle p'|n\rangle$, in a problem.

We notice that here, as in the square well problem, that the parity operator commutes with $\mathcal{H}$. We have that

$$\langle q'\,|\,\mathbb{P}\,|\,n\rangle=\langle -q'\,|\,n\rangle=u_n(-q'). \tag{3.202}$$

But since, from Equation 3.190

$$H_n(-q')=(-1)^n H_n(q'), \tag{3.203}$$

we have from Equation 3.191 that

$$u_n(-q')=(-1)^n u_n(q'). \tag{3.204}$$

Therefore, Equation 3.202 reads

$$\langle q'|\mathbb{P}\,|n\rangle=(-1)^n\langle q'|n\rangle. \tag{3.205}$$

Being true for all $\langle q'|$ tells us that

$$\mathbb{P}|n\rangle = (-1)^n |n\rangle. \tag{3.206}$$

It is now easy to show that

$$\mathbb{P}\mathcal{H}|n\rangle = \mathbb{P}E_n|n\rangle = E_n(-1)^n|n\rangle, \tag{3.207}$$

$$\mathcal{H}\mathbb{P}|n\rangle = (-1)^n\mathcal{H}|n\rangle = (-1)^n E_n|n\rangle, \tag{3.208}$$

and thus that

$$[\mathcal{H},\mathbb{P}]|n\rangle = 0, \tag{3.209}$$

or, again, since this is true for all $|n\rangle$, that

$$[\mathcal{H},\mathbb{P}] = 0. \tag{3.210}$$

Thus, the parity of the states $u_n(q')$ does not change when they are time-evolved.

The wave functions $u_n(q')$ are dimensionless and normalized so that

$$\int_{-\infty}^{\infty} dq'\, u_n(q')u_n(q') = 1. \tag{3.211}$$

If we wish to work with the physically dimensionful quantity x', then we should use the wave function

$$u_n(x') = \left(\frac{m\omega}{\hbar}\right)^{1/4} u_n\left(q' \to \sqrt{\frac{m\omega}{\hbar}}\,x'\right), \tag{3.212}$$

which is normalized so that

$$\int_{-\infty}^{\infty} dx' u_n(x')u_n(x') = 1. \tag{3.213}$$

3.6 THE ATTRACTIVE KRONIG–PENNEY MODEL

Next, we will study some of the physics of electrons in conductors and insulators. Of course, we will have to simplify the situation a great deal in order to be able to reduce the complexity of this problem. The first idealization is the reduction of the real situation to one spatial dimension. Then we might expect, qualitatively, the potential experienced by a single electron in the material to look somewhat like Figure 3.8.

Notice that the potentials are attractive, as should be the case for electrons in the vicinity of atoms with a positively charged core. Also notice that the potential at the edge of the material rises to a constant level as one is getting farther away from the attractive potentials of the interior atoms.

Even the potential shown in Figure 3.8 is too difficult for us to consider here. We will make two additional simplifications. First, we totally ignore all surface effects by assuming we are dealing with an infinite collection of atoms arrayed in one dimension. Second, we model the attractive Coulomb potentials by Dirac delta-function spikes. The result is Figure 3.9.

We now have a well-defined potential for which we can solve for the allowed energies. Notice we have chosen our zero of potential in Figure 3.9 to correspond to the constant potential felt by the electrons in the "interior" of the metal, away from the positions of the "atoms." In the following, we will restrict our attention to solving for the allowed positive energy levels, although states with $E<0$ also exist. We will call the $E>0$ states as *conductance* electrons and the $E<0$ states as *valence* electrons.

The potential in Equation 3.14 is now determined as

$$V(x) = -\frac{\hbar^2}{2m}\frac{\lambda}{a}\sum_{n=-\infty}^{\infty}\delta(x-na). \tag{3.214}$$

The positive constant λ is dimensionless. This follows since the dimensions of the delta function in Equations 3.214 from

$$\int dx\ \delta(x-x')=1, \tag{3.215}$$

are [1/length], and $\hbar^2/2ma$ has dimensions of [energy · length]. The equation we want to solve is

$$-\left[\frac{\partial^2}{\partial x^2}+\frac{\lambda}{a}\sum_n\delta(x-na)\right]u(x)=\frac{2mE}{\hbar^2}u(x). \tag{3.216}$$

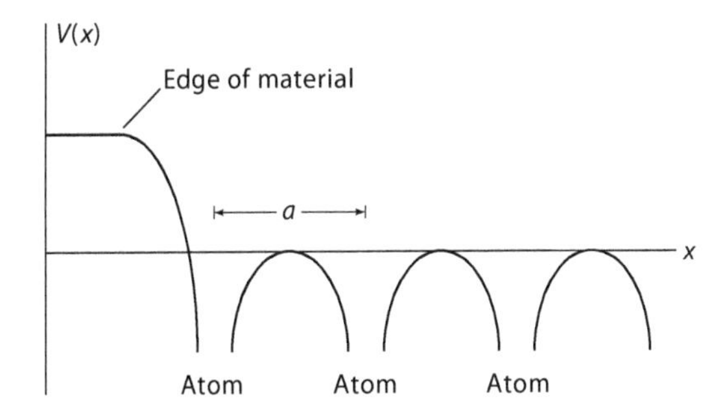

FIGURE 3.8 Qualitative representation of the potential inside a conductor.

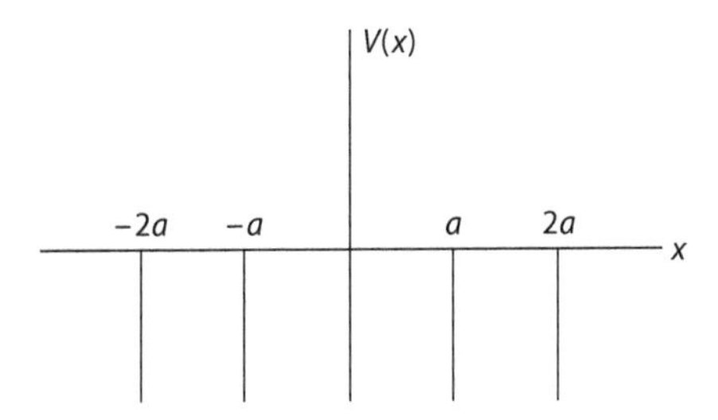

FIGURE 3.9 Attractive Dirac potential model of the interior of a conductor.

Let us simplify this equation a bit by introducing the dimensionless variable $y = x/a$ so that, by using Equation 2.105, we have

$$\delta(x - na) = \frac{1}{a}\delta(y - n). \tag{3.217}$$

Let us also set $k = \sqrt{2mE}\,/\,\hbar$ as usual. Then we have that

$$\left[\frac{\partial^2}{\partial y^2} + \lambda \sum_n \delta(y - n)\right] u(y) = -(ka)^2 u(y) \tag{3.218}$$

for the positive energy solutions. At any positions $y \neq n$, the solutions to

$$\frac{\partial^2}{\partial y^2} u(y) = -(ka)^2 u(y), \tag{3.219}$$

are simply $\sin(kay)$ and $\cos(kay)$. The general solution to $u(y)$ in the region $(n - 1) < y < n$ is then given by

$$u(y) = A_n \sin(ka(y - n)) + B_n \cos(ka(y - n)), \tag{3.220}$$

where A_n and B_n are (complex) coefficients, and the phase factors, $-(ka)n$, in the sine and cosine are chosen for convenience. We have learned that for bounded potentials, our probability interpretation for wave functions holds since $u(x)$ and du/dx are continuous functions. Of course here our potential contains delta functions and so does not satisfy the condition of being bounded. What continuity conditions must we therefore impose on the wave functions for this problem? It is clear that the troublesome positions are the locations of the delta functions at $y = n$. To find the requirements on the wave functions at these locations, let us repeat the argument based on Equation 3.107. Here we have that

$$\int_{n-\varepsilon}^{n+\varepsilon} dy \left\{\left[\frac{\partial^2}{\partial y^2} + \lambda \sum \delta(y - n)\right] u(y) = -(ka)^2 u(y)\right\}, \tag{3.221}$$

where we have integrated over a small neighborhood around $y = n$. Then, in the limit $\varepsilon \to 0^+$ we have

$$\left.\frac{\partial u}{\partial y}\right|_{y=n+} - \left.\frac{\partial u}{\partial y}\right|_{y=n-} = -\lambda u(n). \tag{3.222}$$

Equation 3.222, in contrast to Equation 3.108, says that a discontinuity in $\partial u/\partial y$ is *required*. We assume that the wave functions themselves are continuous across $y = n$; i.e.,

$$u(n)|_+ = u(n)|_- . \tag{3.223}$$

Let us now apply Equations 3.222 and 3.223 to 3.220. The condition Equation 3.223 applied at $y = n$ (between the wave functions defined in the regions $(n - 1) \leq y \leq n$ and $n \leq y \leq n + 1$) says that

$$B_n = -A_{n+1}\sin(ka) + B_{n+1}\cos(ka), \tag{3.224}$$

while Equation 3.222 implies

$$ka(A_{n+1}\cos(ka) - A_n) + kaB_{n+1}\sin(ka) = -\lambda B_n. \tag{3.225}$$

These last two equations require that

$$A_{n+1} = A_n\cos(ka) - \left(\frac{\lambda}{ka}\cos(ka) + \sin(ka)\right)B_n, \tag{3.226}$$

$$B_{n+1} = \left(\cos(ka) - \frac{\lambda}{ka}\sin(ka)\right)B_n + \sin(ka)\,A_n, \tag{3.227}$$

which are recursion relations for the A_{n+1} and B_{n+1} given A_n and B_n.

Equations 3.226 and 3.227 are not sufficient to complete the description of this system. There must be further relations between wave functions in different spatial regions. Let us notice that under the substitution $x \to x + a$ (equivalent to $y \to y + 1$) we have that

$$V(x) \to V(x+a) = V(x), \tag{3.228}$$

and

$$\frac{\mathrm{d}}{\mathrm{d}x} \to \frac{\mathrm{d}}{\mathrm{d}(x+a)} = \frac{\mathrm{d}}{\mathrm{d}x}. \tag{3.229}$$

This means that the differential equation we are solving, Equation 3.216, is *invariant* or unchanged under this substitution. We will, therefore, demand that all observables are also invariant under this change. This means that the probability densities in neighboring regions must be identical.

$$|u(y+1)|^2 = |u(y)|^2. \tag{3.230}$$

This says in general that

$$u(y+1) = \mathrm{e}^{i\phi}u(y), \tag{3.231}$$

where ϕ is some unknown phase. (From the general form Equation 3.220, this phase cannot be y-dependent.)

We have gone as far as we can without specifying boundary conditions. We will use the condition

$$u(y+N) = u(y), \tag{3.232}$$

where $N = 1, 2, 3, \ldots$. (If there are ~10^{23} atoms present, there will be $10^8 \approx (10^{23})^{1/3}$ atomic planes in any one direction.) These are called *periodic boundary conditions.* This condition couples the electrons on one side of the material to the other, resulting in what can be thought of as a ring of atoms. The phase angle ϕ in Equation 3.231 is now determined as

$$\mathrm{e}^{i\phi N} = 1 \tag{3.233}$$

$$\Rightarrow \phi = \frac{2\pi}{N}m, \tag{3.234}$$

where $m = 0, \pm 1, \pm 2, \ldots, +N/2$ (N even case) or $\pm(N-1)/2$ (N odd case), giving N values of ϕ. Equation 3.234 gives the dimensionless *quasi-momentum* values in our model. In terms of the coefficients in Equation 3.220, this says that

$$A_{n+1} = e^{i\phi} A_n, \tag{3.235}$$

$$B_{n+1} = e^{i\phi} B_n. \tag{3.236}$$

Let us now substitute Equations 3.235 and 3.236 into Equations 3.226 and 3.227. We get

$$e^{i\phi} A_n = A_n \cos(ka) - \left(\frac{\lambda}{ka} \cos(ka) + \sin(ka) \right) B_n, \tag{3.237}$$

$$e^{i\phi} B_n = \left(-\frac{\lambda}{ka} \sin(ka) + \cos(ka) \right) B_n + \sin(ka)\, A_n. \tag{3.238}$$

These may be put into the form

$$(e^{i\phi} - \cos(ka)) A_n = -\left(\frac{\lambda}{ka} \cos(ka) + \sin(ka) \right) B_n, \tag{3.239}$$

$$\sin(ka)\, A_n = -\left(-\frac{\lambda}{ka} \sin(ka) + \cos(ka) - e^{i\phi} \right) B_n. \tag{3.240}$$

For Equations 3.239 and 3.240 to be consistent, we must have that

$$\frac{e^{i\phi} - \cos(ka)}{\sin(ka)} = \frac{(\lambda / ka)\cos(ka) + \sin(ka)}{-(\lambda / ka)\sin(ka) + \cos(ka) - e^{i\phi}}. \tag{3.241}$$

This equation can be reduced to

$$\cos\phi = \cos(ka) - \frac{\lambda}{2} \frac{\sin(ka)}{ka}. \tag{3.242}$$

Imagine letting $N \to \infty$ in Equation 3.234. Then ϕ essentially becomes a continuous variable and $\cos\phi$ can take on any value between 1 and –1. This is a type of *dispersion relation*. This equation determines the energies of the system, given by $E = (\hbar k)^2/2m$, in terms of the allowed quasi-momentum values of the system, ϕ.

Plots of the right-hand side of Equation 3.242 follow for the cases $\lambda < 4$ and $\lambda > 4$ in Figures 3.10 and 3.11, respectively. Now notice that since the left-hand side of Equation 3.242 is bounded by 1 and –1, there are no solutions to Equation 3.242 in the regions marked "Forbidden." Otherwise we have a continuum of solutions (at least in the $N \to \infty$ limit). These continuum solutions are called *energy bands*. The forbidden zones, or energy *gaps*, are a result of destructive interference between waves reflected off the various delta-function potentials. We saw such a phenomenon before in our discussion of the finite potential barrier, where the transmission of the waves through the potential barrier was reduced for some values of the momentum because of destructive interference. Now, the positive constant λ in some sense represents the strength of the attractive delta function potentials.

By mapping all the quasi-momentum values in Figures 3.10 and 3.11 into the interval $-\pi < \phi < \pi$, we get the so-called *reduced-zone* description of energy eigenstates. What we see in Figures 3.12 and 3.13 is that when the attraction between the electrons and atoms is weak, the conductance energy bands come all the way down to $E=0$, that is, to the top of the valence band. We see that, however, when λ becomes large enough, an

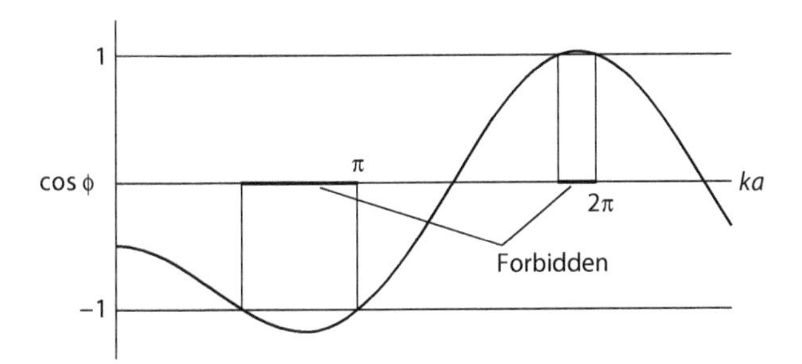

FIGURE 3.10 Qualitative plot of Equation 3.242 for the $\lambda < 4$ case.

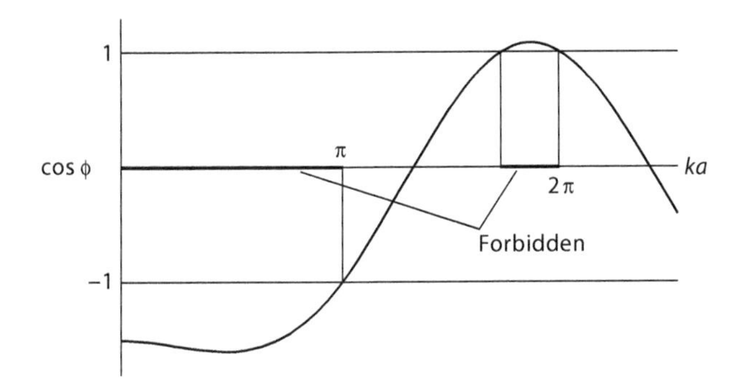

FIGURE 3.11 Qualitative plot of Equation 3.242 for the $\lambda > 4$ case.

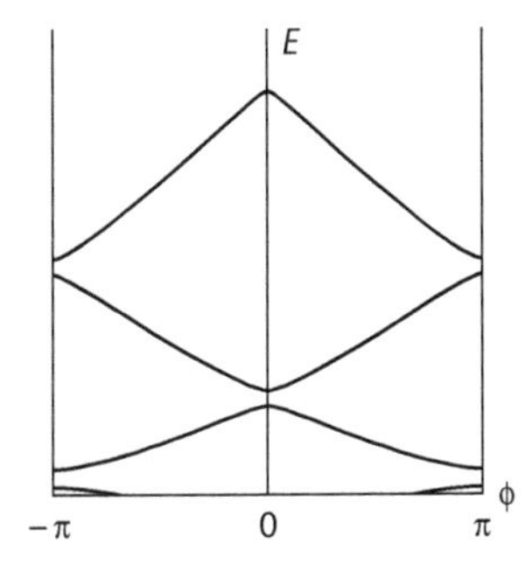

FIGURE 3.12 Reduced-zone map of the quasi-momentum values in Figure 3.10 for $\lambda < 4$.

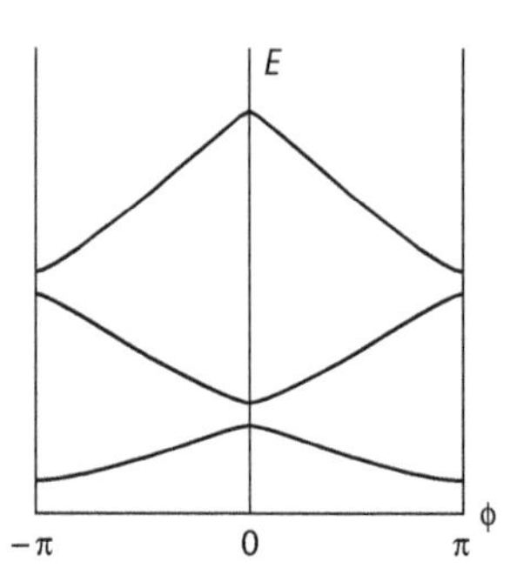

FIGURE 3.13 Reduced-zone map of the quasi-momentum values in Figure 3.11 for $\lambda > 4$.

energy gap forms between $ka = 0$ and $ka = \pi$. This is a very rough model of the band structures in conductors (λ small) and insulators (λ large). In conductors, where the interaction between the electrons and atoms is relatively weak, there is no energy gap, and valence electrons can easily occupy states in the conductance band where they are able to transport charge through the material. On the other hand, insulators hold on tightly to their electrons, and they generally have an energy gap, as in this simplified model. An external electric field is then not effective in moving electrons into the conductance band, and no flow of electricity results. Temperature can play an important role in exciting some electrons upward in energy into the conductance band in some materials. Actually, conductors usually have a partially filled conductance band at room temperatures and therefore conduct electricity readily.

Although we have succeeded in forming a simple model of conductor and insulator energy levels, I hasten to observe that this model is incomplete without an understanding that the electron states involved cannot be multiply occupied. A stable system always seeks its lowest energy, and without a rule to forbid multiple particles in a state, all electrons would tend to occupy their lowest energy in the valence band, although there would always be some occupancy in higher levels at nonzero temperature. There must be some principle limiting the occupancy of each energy state we have found to produce the physics of conductors and insulators. This issue is also related to the missing electronic component of specific heat, discussed in Chapter 1, Section 1.1. We will see the same issue emerge in the nuclear and atomic spectroscopy of Chapter 7. We will have to defer an explanation of this mysterious principle to Chapter 9.

Even with this understanding, there are many simplifications inherent in this one-dimensional model of the interior of conductors and insulators. In reality, because one is working with a finite system instead of the infinitely long system considered earlier, the continuous energy conductance bands we have found are really in a collection of extremely closely spaced discrete levels. In addition, there are interesting effects having specifically to do with the surface of such a material, which we have not considered here. The Kronig–Penney model does correspond to reality in the illustration of the formation of energy gaps, and it was for this reason that I have presented it here.

3.7 BOUND STATE AND SCATTERING SOLUTIONS

We have solved the time-independent Schrödinger equation in one spatial dimension for a variety of potentials. We have seen two basic types of solutions, *bound state* (as in the infinite square well and the simple harmonic oscillator) and *scattering* solutions (as for the finite barrier and the Kronig–Penney model). Bound state solutions have discrete energies and their wave functions can be normalized in space. In scattering solutions, energy values are contiguous and the full wave functions are not normalizable in space (they would be infinite) but instead are normalized in a much more sophisticated manner we will begin to learn about in Chapters 6 and 11.

Generalizing from the experience we have gained in this chapter, we would expect to find the following classification of energy states for the examples in Figures 3.14 and 3.15.

In Figure 3.14 or 3.15, for $E > V_2$, we would expect the particles to be "free" in the sense of having oscillatory wave functions whose wavelengths, however, would be a function of position, just like we saw for energies higher than the top of the potential in the finite barrier problem. In Figure 3.14, when $V_1 < E < V_2$, we expect states that will describe waves that completely reflect off of the potential barrier which extends to $x = \infty$. This does not mean, however, that the particle will not penetrate into the right-hand side region (the shaded region) where, classically, the particle would not be allowed to go. We saw in the finite potential barrier problem that this indeed can happen. However, because the particle's wave function will be damped in this region, one has a smaller probability

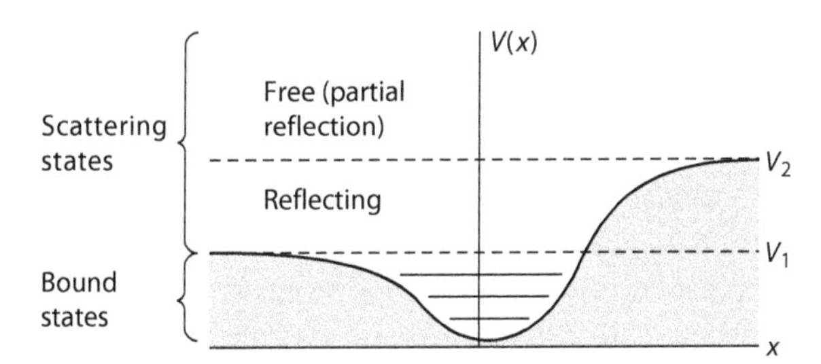

FIGURE 3.14 Energy state potential example.

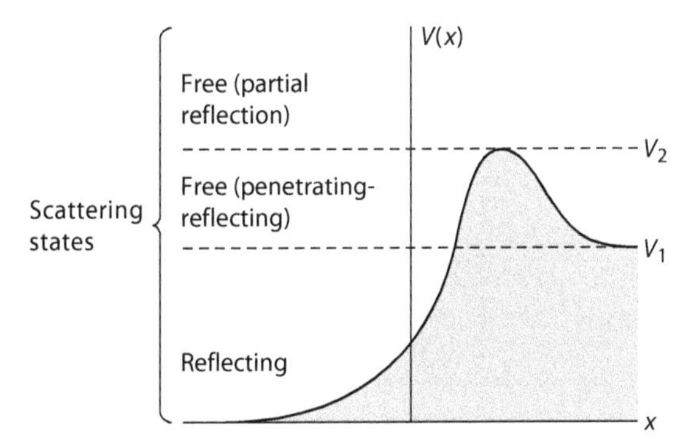

FIGURE 3.15 Another energy state potential example.

of detecting a particle with such an energy as we move farther to the right in this figure. *All* of the particles in this case must *eventually* reflect because in a steady-state situation the flux of particles at $x = \infty$ is zero. In Figure 3.15, for $V_1 < E < V_2$, the particle has only a finite length potential barrier to overcome and it can "tunnel" from one side of the barrier to the other resulting in both penetration and reflection from the barrier. In Figure 3.14, for $E < V_1$, we see that the particles are trapped in a potential well. Then, because of destructive interference between waves, we would expect only a discrete set of energy values to be allowed in a steady state. In Figure 3.15, for $E < V_1$, we again expect complete reflection from, but partial penetration of, the barrier which we again model as extending to $x = \infty$. The key to understanding these qualitative behaviors is the wave picture of particles as de Broglie waves developed in the last chapter.

PROBLEMS

3.1.1 Show for the Schrödinger equation with an arbitrary potential, $V(x)$, that the continuity equation, Equation 2.128, still holds true. Assume that $V(x)$ is Hermitian.

3.2.1 A square well wave function at $t = 0$ is given by

$$|\psi, 0\rangle = \frac{1}{\sqrt{13}}[2|1+\rangle + 3i|2-\rangle].$$

Find $\langle \mathbb{P} \rangle (\equiv \langle \psi, t | \mathbb{P} | \psi, t \rangle)$ as a function of time.

3.2.2 A particle is confined to a one-dimensional box, with walls at $x = 0$, $x = L$. (Note the different origin, box length, in this problem.) At $t = 0$, the particle's wave function is

$$\psi(x,0) = A\left(\sin\left(\frac{\pi x}{L}\right) - \frac{1}{10}\sin\left(\frac{2\pi x}{L}\right)\right).$$

(a) Normalize $\psi(x,0)$ and find the value of the constant A.
(b) What is the probability the particle is in the interval $(0, L/2)$?
(c) What is the probability the particle has energy E_2 $(=4\hbar^2 \pi^2/2mL^2)$?
(d) Find $\psi(x,t)$ for $t > 0$.

3.3.1 Show Equations 3.76 and 3.77.

3.3.2 Evaluate the double commutators:
(a) $[p_x,[p_x, f(x)]] = ?$
(b) $[x, [x, f(p_x)]] = ?$
(c) $[x, [p_x, f(x, p_x)]] = ?$
(d) $[p_x,[x, f(x, p_x)]] = ?$

3.3.3 Show Equations 3.86a, b, c are true.

3.3.4 Find the energies of the negative parity states from the ladder operators by considering the quantity, $\langle n - 1/2|[H, R_n]|n\rangle$.

3.4.1 We are given a wave function, $u(x)$, which solves the coordinate space Schrödinger equation (with a piecewise continuous potential $V(x)$). Assume that this $u(x)$ has a kink in it at $x=x_0$, as illustrated in Figure 3.5. Give a convincing mathematical argument that such a derivative discontinuity would imply that probability is not conserved.

***3.4.2** Derive the result, Equation 3.119, for the E/A coefficient for the finite potential barrier problem described there. [Hints: Organize your calculation carefully. Notice that you can get some of the coefficients to drop out by adding or subtracting multiples of these equations.]

***3.4.3** Consider the scattering problem shown in Figure 3.16. The energy, E, of the incoming particle is greater than the top of the step potential: $E > V_0$. The incoming particle originates from the left (i.e., from $x<0$). Define

$$k_1 \equiv \frac{\sqrt{2m(E-V_0)}}{\hbar}, \quad k_2 \equiv \frac{\sqrt{2mE}}{\hbar}.$$

(a) Write the general solution to the energy eigenvalue equation in Region I.
(b) Write the general solution to the energy eigenvalue equation in Region II.
(c) Show that the reflection coefficient, R, is given by

$$R=\left(\frac{k_1-k_2}{k_1+k_2}\right)^2.$$

(d) Derive the transmission coefficient, T. You may *not* just assume $R+T=1$. [Hint: Be careful! It might not be what you think.]

***3.5.1** (a) From

$$A|n\rangle=\sqrt{n}|n-1\rangle,$$

derive

$$\frac{\mathrm{d}}{\mathrm{d}q'}H_n(q')=2n\,H_{n-1}(q').$$

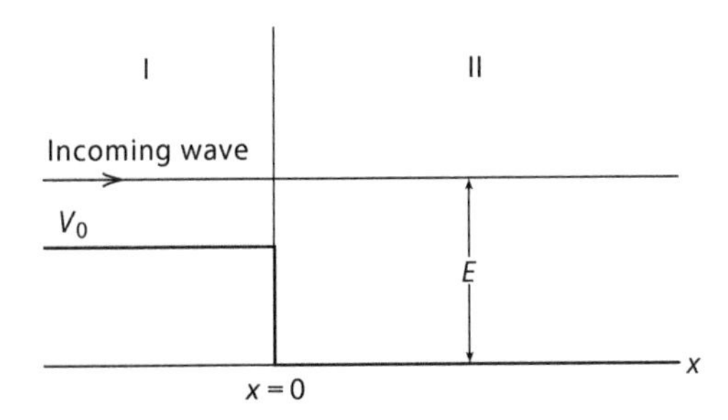

FIGURE 3.16 A single-step scattering problem.

(b) From

$$A^{\dagger}|n\rangle = \sqrt{n+1}|n+1\rangle,$$

derive

$$\left(2q' - \frac{\mathrm{d}}{\mathrm{d}q'}\right)H_n(q') = H_{n+1}(q').$$

(c) Use (a) and (b) to show that $H_n(q')$ satisfies the differential equation:

$$\left(\frac{\mathrm{d}^2}{\mathrm{d}q'^2} - 2q'\frac{\mathrm{d}}{\mathrm{d}q'} + 2n\right)H_n(q') = 0.$$

3.5.2 Find some way of showing Equation 3.195 holds true. [Hint: Use proof by induction. That is, show it is true for $n = 1$ (starting point), *assume* it is true for the case n, then show that the $n+1$ case follows.]

***3.5.3** Given the harmonic oscillator eigenvalue equation, $H|n\rangle = E_n|n\rangle$, where $E_n = \hbar\omega\,(n+1/2)$ and

$$H = \frac{p_x^2}{2m} + \frac{1}{2}m\omega^2 x^2.$$

(a) Derive the position-space energy eigenvalue equation for $u_n(x') \equiv \langle x'|n\rangle$ ("the Schrödinger equation in position space").
(b) Derive the momentum-space energy eigenvalue equation for $v_n(p_x') \equiv \langle p_x'|n\rangle$ ("the Schrödinger equation in momentum space").
(c) Find a simple substitution of variables that changes the differential equation in (a) into the equation in (b). (This therefore gives $v_n(p_x')$, outside of an overall normalization, if $u_n(x')$ is known.)

***3.5.4** Find the *momentum-space* harmonic oscillator wave functions, $v_n(p') \equiv \langle p'|n\rangle$, normalized such that

$$\int_{-\infty}^{\infty} \mathrm{d}p' |v_n(p')|^2 = 1.$$

p is the dimensionless position variable related to p_x by

$$p \equiv \frac{p_x}{\sqrt{m\omega\hbar}}.$$

[Hint: One way is to take Equation 3.174 and project it into $\langle p'|$ instead of $\langle q'|$. Another way is to follow up on the comment in Problem 3.5.3(c).]

3.5.5 Evaluate the harmonic oscillator quantities:

(a) $\langle n+2|q^2|n\rangle = ?$

(b) $\langle n|p^2|n\rangle = ?$

3.6.1 Consider the Kronig–Penney model with periodic *repulsive* delta-function potentials, that is, $\lambda < 0$.

(a) Given the dispersion relation,

$$\cos\phi = \cos(ka) - \frac{\lambda}{2}\frac{\sin(ka)}{ka},$$

draw (qualitatively) the appropriate $\cos\phi$ versus ka graph (be sure to indicate any energy gaps).

(b) Give the *reduced-zone* energy graph corresponding to your answer to part (a). Show at least the lowest two energy bands.

(c) Show that there are no energy eigenvalue solutions to the part (a) dispersion relation when both $\lambda < 0$ and $E < 0$. Does this make sense?

3.6.2 The Kronig–Penney model dispersion relation for *negative* energies is given by

$$\cos\phi = \cosh(\kappa a) - \frac{\lambda}{2}\frac{\sinh(\kappa a)}{\kappa a},$$

where $\kappa = \sqrt{-2mE}/\hbar$.

(a) Draw (qualitatively) the appropriate $\cos\phi$ vs. κa graph. Be sure to indicate any energy gaps; assume $\lambda > 4$.

(b) Give the *reduced-zone* energy graph corresponding to your answer to part (a); again assume $\lambda > 4$.

(c) What value of ϕ is associated with the ground state (i.e., most negative) energy? *Explain how to solve for this energy.* (You do not actually have to do it.)

Hilbert Space and Unitary Transformations

Synopsis: The mathematics behind the deceptively simple ideas encountered in the spin measurements from the first chapter is further developed. Technically, we limit ourselves to finite dimensional Hilbert spaces, although exceptions are pointed out along the way. The operator/matrix analogy is developed, and important physical and mathematical concepts are defined and refined. These concepts include operator compatibility, uncertainty relations, and unitary transformations. Unitary transformations are applied to coordinate transformations and used to develop the Heisenberg time-development picture in quantum mechanics.

4.1 INTRODUCTION AND NOTATION

Although we have been progressing steadily in our understanding of the laws of nature in the microscopic world, I have not been very systematic in the development of the mathematics behind the physics. I hope I have convinced you of the utility of the bra–ket notation of Dirac, but now I owe you a deeper discussion of the mathematical properties of bras, kets, and operators. In other words, I think it would be best at this stage to consolidate our knowledge of formalism a bit before moving on to more complicated problems.

First, a little mathematical orientation before we move on. The mathematical arena in which quantum mechanics operates is called "Hilbert space," named after David Hilbert, the great German mathematician. Hilbert spaces are complex vector spaces that generalize the idea of Euclidean space vectors, familiar from classical mechanics. Hilbert spaces may contain a finite or infinite number of degrees of freedom. These may be discrete and finite, like spin; discrete and infinite, like the harmonic oscillator; or continuous and infinite, like position. We will limit consideration to just the first case in order to simplify the discussions. However, we will concern ourselves with results that are general in character and straightforwardly carry over to the important cases of infinite dimensionality as well as to position and space measurements. If a continuum of states is considered, such spaces are limited to functions that are "square-integrable."*

* Basically, this means functions f(x) such that $\int_{-\infty}^{\infty} dx\,|f(x)|^2$ is finite. A so-called rigged Hilbert space can accommodate more general functions, such as oscillating exponentials, $e^{ipx/\hbar}$.

Next, some words about notation and terminology. We will continue to let

$$|a\rangle,\ |a'\rangle,\ |a_i\rangle \qquad \text{(Case 1)}$$

or

$$|i\rangle,\ |n\rangle \qquad \text{(Case 2)}$$

represent the eigenkets of some operator A. In Case 1, we are labeling the eigenket by the actual *eigenvalue*; in Case 2 we are labeling these in some unspecified order by an integer. Because we are limiting ourselves here to a finite number of states or physical outcomes, the upper limit on all such labels is finite.

Let me also remind you of the meaning of a "basis." When we were discussing spin 1/2, the basis consisted of the states $|\sigma_3'\rangle, \sigma_3' = \pm 1$. A basis is a set of linearly independent vectors in Hilbert space that completely characterize the space in the same way that the usual unit vectors, $\hat{e}_1$, $\hat{e}_2$, $\hat{e}_3$ completely characterize 3-D space. (When I refer to a "vector" in Hilbert space, I mean either a bra or a ket.) Just as any 3-D vector $\vec{v}$ may be expanded as

$$\vec{v} = \sum_{i=1}^{n} v_i \hat{e}_i, \tag{4.1}$$

any state $|\psi\rangle$ may be expanded as

$$|\psi\rangle = \sum_{i=1}^{n} C_i |X_i\rangle, \tag{4.2}$$

where I will be using the notation $|X_i\rangle$ to denote a basis ket. The C_i in Equation 4.2 are some (in general) complex constants. We say that a basis "spans" the Hilbert space. According to the technical limitation we have imposed, the number of basis states in Equation 4.2 is a finite number, n.

What does linear independence of basis kets mean mathematically? In three-dimensional Euclidean vector language, we say that two non-null vectors, $\vec{v}_1$ and $\vec{v}_2$, are linearly independent if there is no constant C such that

$$\vec{v}_1 = C\vec{v}_2. \tag{4.3}$$

In the same way, we say that the non-null kets $|X_1\rangle$ and $|X_2\rangle$ are linearly independent if there is no constant C such that

$$|X_1\rangle = C|X_2\rangle. \tag{4.4}$$

The $|X_i\rangle$, therefore, are like vectors "pointing" in various independent directions in Hilbert space.

The completeness relation, first seen in Equation 1.91 (Chapter 1), may be deduced in this context as follows. A general ket $|\psi\rangle$ may be expanded as in Equation 4.2. Multiplying on the left of Equation 4.2 by $\langle X_j|$ and assuming orthonormality (see Equation 4.14), this tells us that the expansion coefficients C_i in Equation 4.2 are just

$$C_i = \langle X_i | \psi \rangle. \tag{4.5}$$

Plugging this back into Equation 4.2 then gives

$$|\psi\rangle = \sum_i |X_i\rangle\langle X_i|\psi\rangle = \left(\sum_i |X_i\rangle\langle X_i|\right)|\psi\rangle. \tag{4.6}$$

Since $|\psi\rangle$ in Equation 4.6 is arbitrary, this tells us that

$$\sum_i |X_i\rangle\langle X_i| = 1. \tag{4.7}$$

This sheds a different, more mathematical, light on the meaning of completeness.

4.2 INNER AND OUTER OPERATOR PRODUCTS

We will be assuming the existence of the Hermitian conjugation operation that takes bras into kets and vice versa. For some arbitrary state $|\psi\rangle$ we have

$$(|\psi\rangle)^\dagger = \langle\psi|. \tag{4.8}$$

Bras and kets are independent objects, but the existence of a state $|\psi\rangle$ implies the existence of $\langle\psi|$ and the other way around. The relationship between these two objects is called "dual." We are assuming the distributive property of this operation, so that

$$(|\psi_1\rangle + |\psi_2\rangle)^\dagger = \langle\psi_1| + \langle\psi_2|. \tag{4.9}$$

When C is a complex number, we also have

$$(C|\psi\rangle)^\dagger = C^*\langle\psi| = \langle\psi|C^*. \tag{4.10}$$

Of course, given the state $|\psi\rangle$ in Equation 4.2, the state $\langle\psi|$ is implied to be

$$\langle\psi| = \sum_{i=1}^{n} C_i^*\langle X_i|. \tag{4.11}$$

Given the bra $\langle X_i|$ and the ket $|X_j\rangle$, we may form two types of quantities, the *inner product*, $\langle X_i|X_j\rangle$ (introduced previously in Chapter 1), and the *outer product*, $|X_j\rangle\langle X_i|$ (previously $|X_jX_i|$). Just as the elements of $\hat{e}_i$ are orthogonal in 3-D, as expressed by

$$\hat{e}_i \cdot \hat{e}_j = 0, \quad i \neq j, \tag{4.12}$$

we choose a basis such that the inner product must satisfy

$$\langle X_i|X_j\rangle = 0, \quad i \neq j. \tag{4.13}$$

For convenience, we often normalize the basis so that (called an orthonormal basis)

$$\langle X_i|X_j\rangle = \delta_{ij}, \quad (\text{all } i, j), \tag{4.14}$$

where δ_{ij} is the Kronecker delta. A vector $|\psi\rangle$ is said to have a squared "length," $\langle\psi|\psi\rangle$ in Hilbert space given by

$$\langle\psi|\psi\rangle = \sum_{i,j} C_i^* C_j \langle X_i|X_j\rangle = \sum_i |C_i|^2 \geq 0. \tag{4.15}$$

It is in this sense that a bra or a ket is said to be "non-null" (as expressed earlier in Equation 4.3).

The outer product $|X_j\rangle\langle X_i|$ is another name for what we have been calling an operator or a measurement symbol. The most general expression for an operator, A, in our finite Hilbert space is

$$A = \sum_{i,j} A_{ij}\,|X_i\rangle\langle X_j|. \tag{4.16}$$

The coefficients A_{ij} are just the elements of an $n \times n$ matrix representation of A. This is easy to see since

$$\begin{aligned}\langle X_k|A|X_\ell\rangle &= \sum_{i,j} A_{ij}\langle X_k|X_i\rangle\langle X_j|X_\ell\rangle \\ &= \sum_{i,j} A_{ij}\delta_{ki}\delta_{j\ell} = A_{k\ell}.\end{aligned} \tag{4.17}$$

When acting on A in Equation 4.16, Hermitian conjugation has the effect

$$A^\dagger = \sum_{i,j} A_{ij}^*\,|X_j\rangle\langle X_i| = \sum_{i,j} (A^*)_{ij}^t\,|X_i\rangle\langle X_j|, \tag{4.18}$$

where "t" is the transpose operation. Thus we see that the operator statement

$$A = A^\dagger \tag{4.19}$$

is equivalent to the matrix statement

$$A = (A^*)^t. \tag{4.20}$$

(The transpose and complex conjugation operations commute.) As remarked on before in Chapter 1, we often denote the complex transpose of a matrix as "†." The meaning of "†" is determined by its context.

Consider the operator product AB. This can be written as

$$\begin{aligned}AB &= \sum_{i,j} A_{ij}\,|X_i\rangle\langle X_j| \;\sum_{k,\ell} B_{k\ell}\,|X_k\rangle\langle X_\ell| \\ &= \sum_{i,j,\ell} A_{ij}B_{j\ell}\,|X_i\rangle\langle X_\ell|.\end{aligned} \tag{4.21}$$

The Hermitian conjugate is given by

$$\begin{aligned}(AB)^\dagger &= \sum_{i,j,\ell} A_{ij}^* B_{j\ell}^*\,|X_\ell\rangle\langle X_i| \\ &= \sum_{i,j,\ell} A_{\ell j}^* B_{ji}^*\,|X_i\rangle\langle X_\ell|,\end{aligned} \tag{4.22}$$

whereas we also have that

$$\begin{aligned}B^\dagger A^\dagger &= \sum_{i,j} B_{ji}^*\,|X_i\rangle\langle X_j| \cdot \sum_{k,\ell} A_{\ell k}^*\,|X_k\rangle\langle X_\ell| \\ &= \sum_{i,j,\ell} A_{\ell j}^* B_{ji}^*\,|X_i\rangle\langle X_\ell|,\end{aligned} \tag{4.23}$$

which proves in general that

$$(AB)^{\dagger} = B^{\dagger}A^{\dagger}. \tag{4.24}$$

This is the same rule encountered in the spin algebra of Chapter 1.

4.3 OPERATOR–MATRIX RELATIONSHIP

The operator–matrix relationship may be elaborated more fully as follows. We know that the product $A|\psi\rangle$, where A and $|\psi\rangle$ are arbitrary, is another ket:

$$|\psi'\rangle = A|\psi\rangle. \tag{4.25}$$

Using the forms Equations 4.16 and 4.2 for A and $|\psi\rangle$, we get

$$\begin{aligned} A|\psi\rangle &= \sum_{i,j} A_{ij}\,|X_i\rangle\langle X_j|\cdot\sum_k C_k|X_k\rangle \\ &= \sum_{i,j} A_{ij}C_j|X_i\rangle. \end{aligned} \tag{4.26}$$

On the other hand we know that $|\psi'\rangle$ is another ket and so may be expanded in basis kets:

$$|\psi'\rangle = \sum_i B_i|X_i\rangle. \tag{4.27}$$

Thus, the operator–ket statement (Equation 4.25) is equivalent to the matrix–vector statement

$$B_i = \sum_j A_{ij}C_j, \tag{4.28}$$

which is often written with understood indices as

$$B = AC, \tag{4.29}$$

where B and C are interpreted as *column* matrices. Likewise, given the statement

$$\langle\psi'| = \langle\psi|A, \tag{4.30}$$

and the expansions

$$\langle\psi'| = \sum_i B_i'\langle X_i|, \tag{4.31}$$

$$\langle\psi| = \sum_i C_i'\langle X_i|, \tag{4.32}$$

the analogous matrix statement to Equation 4.30 is

$$(B')^t = (C')^t A, \tag{4.33}$$

where $(B')^t$ and $(C')^t$ are now considered *row* matrices.*

Our operator–matrix analogy stands as follows:

Operator	$\leftrightarrow$	**Matrix**
Ket	$\leftrightarrow$	**Column matrix**
Bra	$\leftrightarrow$	**Row matrix**
Complex conjugation ⊕ bra, ket interchange	$\leftrightarrow$	**Complex conjugation ⊕ transpose**

The last line gives the meaning of Hermitian conjugation in the two cases.

4.4 HERMITIAN OPERATORS AND EIGENKETS

A distinguishing characteristic of a Hermitian operator is that its physical outcomes, which mathematically speaking are its eigenvalues, are real. We can show this as follows. Let $|a_i\rangle$ be any nonzero eigenket of A,

$$A|a_i\rangle = a_i|a_i\rangle. \tag{4.34}$$

Then

$$\langle a_i|A|a_i\rangle = a_i\langle a_i|a_i\rangle. \tag{4.35}$$

The adjoint of Equation 4.34 is

$$\langle a_i|A^\dagger = a_i^*\langle a_i|, \tag{4.36}$$

which then gives

$$\langle a_i \mid A^\dagger \mid a_i\rangle = a_i^*\langle a_i|a_i\rangle. \tag{4.37}$$

If $A = A^\dagger$, then Equations 4.35 and 4.37 demand that

$$a_i = a_i^*, \tag{4.38}$$

and the eigenvalues are real. (A quicker proof is to recall that in Chapter 1, Section 1.18 we proved that the expectation value of Hermitian operators is real. Since the left-hand side of Equation 4.35 is just such a quantity and because $\langle a_i|a_i\rangle \geq 0$, we must have that the a_i are real.)

It is easy to prove that eigenkets of a Hermitian operator corresponding to distinct eigenvalues are orthogonal. Equation 4.34 implies that

$$\langle a_i|A|a_j\rangle = a_j\langle a_i|a_j\rangle, \tag{4.39}$$

* Technically, it is not necessary to put the t, indicating the transpose, on these quantities, since their placement next to A, an n-by-n matrix, automatically indicates they are row vectors. However, just to be clear in all contexts, I will leave the transpose symbol.

and Equation 4.36 (with $A = A^\dagger$, $a_i = a_i^*$) implies

$$\langle a_i|A|a_j\rangle = a_i\langle a_i|a_j\rangle. \tag{4.40}$$

Therefore, from Equations 4.39 and 4.40 we have

$$(a_i - a_j)\langle a_i|a_j\rangle = 0, \tag{4.41}$$

which says that $\langle a_i|a_j\rangle = 0$ if $a_i \neq a_j$. (The operator A in Equation 4.39 can be visualized as operating to the right, whereas the A in Equation 4.40 acts to the left.)

Now there are n independent basis kets $|X_i\rangle$. The eigenkets of a Hermitian operator can, of course, be expanded in such a basis. The number of linearly independent eigenkets $|a_i\rangle$ cannot exceed n, the number of basis kets, for if they did then the $|X_i\rangle$ would not span the space contrary to our definition. Therefore, the maximum number of distinct eigenvalues is always less than or equal to n. If all n of them *are* distinct, then it is clear that we may choose the $|a_i\rangle$ as a (not necessarily normalized) basis for our description. This is a most useful result since it provides a means of finding a basis. However, when the number of distinct eigenvalues of a Hermitian operator is less than n, the number of *linearly independent* eigenkets is still n and can, therefore, be chosen as a basis. I will be content to just lay out the bare bones of this explanation since the details involve some mathematical technicalities. In doing so, the connection we have found between operators and matrices will be extremely useful.

Let us say we have a Hermitian operator A, and we want to find its eigenvalues and eigenkets. Let A be represented in terms of an arbitrary basis as in Equation 4.13. We want to find all possible $|\psi\rangle$ such that

$$A|\psi\rangle = a|\psi\rangle, \tag{4.42}$$

where a is an eigenvalue of A. Expanding $|\psi\rangle$ in our basis as

$$|\psi\rangle = \sum_i y_i|X_i\rangle, \tag{4.43}$$

we know that Equation 4.42 is equivalent to the matrix statement

$$Ay = ay. \tag{4.44}$$

Showing the explicit matrix indices, Equation 4.44 may be written as

$$\sum_k (A_{jk} - a\delta_{jk})y_k = 0. \tag{4.45}$$

It is a well-known result out of linear algebra that the necessary and sufficient condition that Equation 4.45 has a nontrivial solution is that

$$\det(A - aI) = 0, \tag{4.46}$$

where I is the unit matrix. Equation 4.46 is called the *characteristic equation* of the matrix A. It is easily shown that the left-hand side of

Equation 4.46 is just a nth order polynomial in a, and therefore has n solutions or roots. These roots constitute the totality of eigenvalues, a_i. After these have been found by solving Equation 4.46, we find the associated eigenvectors $y^{(i)}$ by solving

$$Ay^{(i)} = a_i y^{(i)}, \tag{4.47}$$

for each value a_i. (We actually went through this process for the spin matrix $\bar{\sigma}_3$ in Chapter 1, Section 1.17.) If the n solutions to Equation 4.46 are all distinct, then the corresponding eigenkets span the space. However, the solution of Equation 4.46 may involve repeated roots. A root that appears k times in the solution of the characteristic equation is called a kth order *degeneracy*. (There is no first order degeneracy, however!) In this case, we cannot use the previous argument to establish that the nonzero eigenkets or eigenvectors of these repeated roots are all orthogonal. However, it may be shown that the number of linearly independent eigenvectors corresponding to a kth order root of the characteristic equation of a Hermitian matrix is exactly k.* Given this fact, this means that there will be exactly n linearly independent eigenkets or eigenvectors, and these can now be used as a basis.

4.5 GRAM–SCHMIDT ORTHOGONALIZATION PROCESS

Although we are guaranteed that the number of linearly independent eigenkets of a Hermitian operator is n, this does not mean that any such set will satisfy orthogonality. Those eigenvectors or eigenkets that correspond to *distinct* eigenvalues are orthogonal from the previous argument. We only have to worry about the eigenvectors or eigenkets corresponding to each kth order eigenvalue degeneracy. However, there is a procedure called the *Gram–Schmidt orthogonalization process* that allows us to construct an orthogonal (but not normalized) set of vectors or kets from any linearly independent set. Let us say that we have k linearly independent vectors $|Y_i\rangle$ corresponding to the single eigenvalue a:

$$A|Y_i\rangle = a|Y_i\rangle, \quad i = 1,\ldots,k \tag{4.48}$$

(The order of labeling of these objects is arbitrary.) We can construct an orthogonal set by the following procedure.

First, choose

$$|X_1\rangle = |Y_1\rangle \tag{4.49}$$

as the first of our orthogonal set. $|Y_1\rangle$ is actually *any* of the original set of nonorthogonal kets.

Now form

$$|X_2\rangle = |Y_2\rangle - \frac{\langle X_1|Y_2\rangle}{\langle X_1|X_1\rangle}|X_1\rangle. \tag{4.50}$$

* D.C. Lay, *Linear Algebra and Its Applications*, 3rd ed. (Addison-Wesley, 2005), p. 452.

Notice that

$$\langle X_1|X_2\rangle = \langle X_1|Y_2\rangle - \frac{\langle X_1|Y_2\rangle}{\langle X_1|X_1\rangle}\langle X_1|X_1\rangle = 0. \tag{4.51}$$

Thus, the second term in Equation 4.50 has been chosen in such a way as to remove the overlap between $|X_1\rangle$ and $|X_2\rangle$. The next vector is

$$|X_3\rangle = |Y_3\rangle - \frac{\langle X_1|Y_3\rangle}{\langle X_1|X_1\rangle}|X_1\rangle - \frac{\langle X_2|Y_3\rangle}{\langle X_2|X_2\rangle}|X_2\rangle. \tag{4.52}$$

Again notice

$$\langle X_1|X_3\rangle = \langle X_1|Y_3\rangle - \frac{\langle X_1|Y_3\rangle}{\langle X_1|X_1\rangle}\langle X_1|X_1\rangle - \frac{\langle X_2|Y_3\rangle}{\langle X_2|X_2\rangle}\langle X_1|X_2\rangle = 0. \tag{4.53}$$

$$\langle X_2|X_3\rangle = \langle X_2|Y_3\rangle - \frac{\langle X_1|Y_3\rangle}{\langle X_1|X_1\rangle}\langle X_2|X_1\rangle - \frac{\langle X_2|Y_3\rangle}{\langle X_2|X_2\rangle}\langle X_2|X_2\rangle = 0. \tag{4.54}$$

In general, the procedure is to pick $|X_1\rangle = |Y_1\rangle$, after which we have

$$|X_i\rangle = |Y_i\rangle - \sum_{j=1}^{i-1}\frac{\langle X_j|Y_i\rangle}{\langle X_j|X_j\rangle}|X_j\rangle, \quad i \geq 2. \tag{4.55}$$

We can now normalize these in a Hilbert space, so that they will be orthonormal.

The end result of the preceding considerations is this: The eigenkets of $A = A^\dagger$ can all be chosen as orthonormal and therefore represents a possible basis for the space.

4.6 COMPATIBLE OPERATORS

The aforementioned information is a very useful result. However, the situation is not really satisfactory yet. We have seen that the eigenvalues of a Hermitian operator do not necessarily specify or classify all the eigenkets of the system, and therefore the states of a particle, because of eigenvalue degeneracy. In our analogy with vectors in 3-D, this is like labeling the unit vectors identically although we know they are linearly independent. The process just described, which produces an orthogonal set of kets or vectors corresponding to the same eigenvalue, has an element of arbitrariness in it because of the random choice $|X_1\rangle = |Y_1\rangle$. It represents a formal mathematical way of producing an orthogonal set of basis kets, but there must be a more physical way of doing the same thing so that known physical properties are associated with each and every ket. A way of doing this is contained in the following theorem.

THEOREM

If $A = A^\dagger$ and $B = B^\dagger$, a necessary and sufficient condition that $[A,B] = 0$ is that A and B possess a common complete set of orthonormal eigenkets.

The proof of sufficiency is as follows.

We are assuming that A and B possess a common complete set of orthonormal basis vectors, which we will label as $|a_i,b_i\rangle$. Thus

$$A|a_i,b_i\rangle = a_i|a_i,b_i\rangle, \tag{4.56}$$

$$B|a_i,b_i\rangle = b_i|a_i,b_i\rangle. \tag{4.57}$$

Then

$$AB|a_i,b_i\rangle = b_iA|a_i,b_i\rangle = b_ia_i|a_i,b_i\rangle, \tag{4.58}$$

$$BA|a_i,b_i\rangle = a_iB|a_i,b_i\rangle = a_ib_i|a_i,b_i\rangle. \tag{4.59}$$

Therefore

$$[A,B]\,|a_i,b_i\rangle = 0. \tag{4.60}$$

This statement holds for all i, which then implies that*

$$[A,B] = 0. \tag{4.61}$$

The proof of necessity in this theorem is trickier and will not be presented here. (I will ask you, however, to prove a restricted version of the necessary condition in a problem.)

Notice that we did not require all the eigenvalues of either A or B in this theorem to be unique. Repeated eigenvalues are termed degenerate; this is the same idea as introduced in Section 4.4. Hence, if we have a situation where several of the eigenkets of A have the same eigenvalue, these can also be chosen as eigenkets of B as long as $[A, B] = 0$. Then, the eigenvalues of B may possibly serve to distinguish between them. If B does not completely distinguish between them, there may be another Hermitian operator C with $[C, A] = [C, B] = 0$ that *will* do so and so on if the eigenkets are still not physically distinguished from one another. A set of Hermitian operators $A, B, C, \ldots$ whose common complete eigenkets can be characterized completely by their eigenvalues is said to constitute a *complete set* of observables (or operators). Such a set, however, may or may not be minimal and may also not be unique. An example is the states of the 1-D free particle we studied in Chapter 2. Such states are uniquely characterized by the eigenvalues of the momentum operator, p_x. Thus, this operator constitutes the complete set, which in this case is unique. We may add to this the particle energy $H = p_x^2/2m$ since $[H, p_x] = 0$, and the list is still complete but no longer minimal. The H and p_x operators therefore possess a common complete set of orthogonal† eigenkets, $|a', p'_x\rangle$ such that

$$H|a',p'_x\rangle = E_{a'}|a',p'_x\rangle, \tag{4.62}$$

$$p_x|a',p'\rangle = p'_x|a',p'_x\rangle. \tag{4.63}$$

---

* At a number of points here, and before, we have assumed that $O_1|X_i\rangle = O_2|X_i\rangle$ for all $|X_i\rangle$ implies that $O_1 = O_2$. Can you show this?

† Note that I state "orthogonal" instead of "orthonormal" here because the continuum normalization of momentum states follows a different rule; see Chapter 2, Equation 2.146.

Of course just H itself is not complete, because there is a remaining two-fold momentum sign degeneracy. A nonredundant but still nonminimal set in distinguishing the energy states of the free particle could have included $\mathbb{P}$, the parity operator from Chapter 3, Section 3.2, to resolve the two-fold degeneracy since we can show that $[H, \mathrm{P}] = 0$. Such states would be labeled as $|a', P\rangle$, say, with $P = \pm 1$:

$$H|a', P\rangle = E_{a'}|a', P\rangle, \tag{4.64}$$

$$\mathbb{P}|a', P\rangle = P|a', P\rangle. \tag{4.65}$$

The $|a', p'_x\rangle$ and $|a', P\rangle$ are just linear combinations of each other. Notice we *cannot* choose our eigenkets to be common to H, p'_x, *and* $\mathbb{P}$ since $\mathbb{P}$ and p'_x do not commute. In fact we can show $\{\mathbb{P}, p'_x\} = 0$, where $\{A, B\} \equiv AB + BA$ is defined to be the *anticommutator* of operators A and B. It is common to call Hermitian operators A and B for which $[A, B] = 0$ "compatible observables." In terms of Process Diagrams, this says that the order in which the operations A and B are carried out is immaterial.

4.7 UNCERTAINTY RELATIONS AND INCOMPATIBLE OPERATORS

On the other side of the coin are Hermitian operators for which $[A,B] \neq 0$. I will now prove an extremely important theorem that will make more concrete some of my prior statements concerning uncertainty relations.

Let us say that $|\psi_1\rangle$ and $|\psi_2\rangle$ represent arbitrary kets in some finite dimensional Hilbert space. Let us set

$$|\psi\rangle = |\psi_1\rangle + \lambda|\psi_2\rangle, \tag{4.66}$$

where λ is some arbitrary complex number. By Equation 4.15 we have that

$$\langle\psi|\psi\rangle \geq 0, \tag{4.67}$$

so that

$$(\langle\psi_1| + \lambda^*\langle\psi_2|)\cdot(|\psi_1\rangle + \lambda|\psi_2\rangle) \geq 0. \tag{4.68}$$

Writing this out in full we get

$$\langle\psi_1|\psi_1\rangle + |\lambda|^2\langle\psi_2|\psi_2\rangle + 2\mathrm{Re}[\lambda\langle\psi_1|\psi_2\rangle] \geq 0, \tag{4.69}$$

where "Re" means the real part of the argument. Since Equation 4.69 must be true for any λ, we may choose

$$\lambda = -\frac{\langle\psi_2|\psi_1\rangle}{\langle\psi_2|\psi_2\rangle}, \tag{4.70}$$

assuming that $|\psi_2\rangle$ is a nonzero ket. Substitution of Equation 4.70 into Equation 4.69 yields

$$\langle\psi_1|\psi_1\rangle + \frac{|\langle\psi_2|\psi_1\rangle|^2}{\langle\psi_2|\psi_2\rangle} + 2\mathrm{Re}\left[-\frac{|\langle\psi_2|\psi_1\rangle|^2}{\langle\psi_2|\psi_2\rangle}\right] \geq 0. \tag{4.71}$$

But the argument of Re is purely real, so Equation 4.71 implies that

$$\langle\psi_1|\psi_1\rangle \geq \frac{|\langle\psi_2|\psi_1\rangle|^2}{\langle\psi_2|\psi_2\rangle}, \tag{4.72}$$

or

$$\langle\psi_1|\psi_1\rangle\langle\psi_2|\psi_2\rangle \geq |\langle\psi_2|\psi_1\rangle|^2. \tag{4.73}$$

Equation 4.73 is called the *Schwartz inequality.* Let us let $|\psi_1\rangle$ and $|\psi_2\rangle$ in Equation 4.73 be given by

$$(A-\langle A\rangle_\psi)|\psi\rangle = |\psi_1\rangle, \tag{4.74}$$

$$(B-\langle B\rangle_\psi)|\psi\rangle = |\psi_2\rangle, \tag{4.75}$$

where A and B are Hermitian and the expectation values $\langle A\rangle_\psi$ and $\langle B\rangle_\psi$ are as usual given by

$$\langle A\rangle_\psi = \langle\psi|A|\psi\rangle, \tag{4.76}$$

$$\langle B\rangle_\psi = \langle\psi|B|\psi\rangle. \tag{4.77}$$

It is understood in Equations 4.74 and 4.75 that in an operator context $\langle A\rangle_\psi = \langle A\rangle_\psi \cdot 1$ where 1 is the unit operator. We choose $|\psi\rangle$ to be normalized, that is, $\langle\psi|\psi\rangle = 1$. Notice that

$$\begin{aligned}\langle\psi_1|\psi_1\rangle &= (\langle\psi|(A-\langle A\rangle_\psi))\;((A-\langle A\rangle_\psi)|\psi\rangle) \\ &= \langle\psi|(A-\langle A\rangle_\psi))^2|\psi\rangle.\end{aligned} \tag{4.78}$$

Similarly for $\langle\psi_2|\psi_2\rangle$. We now get that

$$\langle\psi|(A-\langle A\rangle_\psi)^2|\psi\rangle\langle\psi|(B-\langle B\rangle_\psi)^2|\psi\rangle \geq |\langle\psi|(A-\langle A\rangle_\psi)\;(B-\langle B\rangle_\psi)|\psi\rangle|^2. \tag{4.79}$$

The quantities $\langle\psi|(A-\langle A\rangle_\psi)^2|\psi\rangle = \langle\psi_1|\psi_1\rangle$ and $\langle\psi|(B-\langle B\rangle_\psi)^2|\psi\rangle = \langle\psi_2|\psi_2\rangle$ are intrinsically positive or zero. We now define the *uncertainties* in the operators A and B in the state $|\psi\rangle$ to be

$$\Delta A \equiv \sqrt{\langle\psi|(A-\langle A\rangle_\psi)^2|\psi\rangle} \geq 0. \tag{4.80}$$

$$\Delta B \equiv \sqrt{\langle\psi|(B-\langle B\rangle_\psi)^2|\psi\rangle} \geq 0. \tag{4.81}$$

Since

$$\langle\psi|\langle A\rangle_\psi|\psi\rangle = \langle A\rangle_\psi\langle\psi|\psi\rangle = \langle A\rangle_\psi, \tag{4.82}$$

we may also write

$$\left.\begin{aligned}\Delta A &= \sqrt{\langle A^2\rangle_\psi - \langle A\rangle_\psi^2} \\ \Delta B &= \sqrt{\langle B^2\rangle_\psi - \langle B\rangle_\psi^2}\end{aligned}\right\}. \tag{4.83}$$

Now let us deal with the right-hand side of Equation 4.79. We have

$$\begin{aligned}&\langle\psi|(A-\langle A\rangle_\psi)(B-\langle B\rangle_\psi)|\psi\rangle\\&=\langle\psi|(AB-\langle A\rangle_\psi B-A\langle B\rangle_\psi+\langle A\rangle_\psi\langle B\rangle_\psi)|\psi\rangle\\&=\langle\psi|(AB-\langle A\rangle_\psi\langle B\rangle_\psi)|\psi\rangle.\end{aligned}\tag{4.84}$$

By adding and subtracting the quantity 1/2 BA, we may also write

$$\begin{aligned}&\langle\psi|(AB-\langle A\rangle_\psi\langle B\rangle_\psi)|\psi\rangle\\&=\langle\psi|\left(\frac{1}{2}(AB+BA)+\frac{1}{2}(AB-BA)-\langle A\rangle_\psi\langle B\rangle_\psi\right)|\psi\rangle\\&=\langle\psi|\left(\frac{1}{2}\{A,B\}-i\left(\frac{i}{2}[A,B]\right)-\langle A\rangle_\psi\langle B\rangle_\psi\right)|\psi\rangle,\end{aligned}\tag{4.85}$$

where we have introduced both the commutator, $[A,B]$, and the anticommutator, $\{A,B\}$, of A and B. Let us define the new operators

$$x\equiv\frac{1}{2}\{A,B\}-\langle A\rangle_\psi\langle B\rangle_\psi,\tag{4.86}$$

$$y\equiv\frac{i}{2}[A,B].\tag{4.87}$$

Since A and B are Hermitian, it is easy to show that x and y are also Hermitian. Therefore, our inequality, Equation 4.79, now reads

$$(\Delta A)^2\,(\Delta B)^2\geq|\langle\psi|x|\psi\rangle-i\langle\psi|y|\psi\rangle|^2.\tag{4.88}$$

Since x and y are Hermitian, we have that the expectation values

$$\langle\psi|x|\psi\rangle=\langle x\rangle_\psi,\tag{4.89}$$

$$\langle\psi|y|\psi\rangle=\langle y\rangle_\psi,\tag{4.90}$$

are real numbers. Therefore, we must have that*

$$|\langle x\rangle_\psi-i\langle y\rangle_\psi|^2=\langle x\rangle_\psi^2+\langle y\rangle_\psi^2\geq\langle y\rangle_\psi^2.\tag{4.91}$$

Using Equation 4.91 in Equation 4.88 now gives us that

$$(\Delta A)^2(\Delta B)^2\geq\langle y\rangle_\psi^2=\left\langle\frac{i}{2}[A,B]\right\rangle_\psi^2.\tag{4.92}$$

Since both sides of Equation 4.92 are intrinsically positive or zero, this means that

$$\Delta A\Delta B\geq\left|\left\langle\frac{i}{2}[A,B]\right\rangle_\psi\right|,\tag{4.93}$$

* Note that $\langle x\rangle_\psi\neq 0$ can also be used to establish a lower bound on $\Delta A\Delta B$ even if $\langle y\rangle_\psi=0$. However, this situation corresponds to a numerical correlation in the simultaneous measurement of A and B, and is a standard statistical effect.

where the absolute value sign is used on the right-hand side because while $\langle y \rangle_\psi = \langle i/2[A, B] \rangle_\psi$ is guaranteed real, it may not be positive. Equation 4.93 says that if the Hermitian operators A and B are not compatible, there will in general be an uncertainty relation connecting them. Equations 4.73 and 4.93 were derived in a situation where the dimensionality of the Hilbert space is finite, but they also hold where the dimensionality increases without limit. In that case, we can apply Equation 4.93 to the incompatible observables x and p_x, which tells us that

$$\Delta x \Delta p_x \geq \frac{\hbar}{2}, \tag{4.94}$$

since $[x, p_x] = i\hbar$. Equation 4.94 is the promised relation Equation 2.32 (see Chapter 2). It is clear that we cannot use this type of derivation to establish the energy–time uncertainty relation since the time is a parameter, not an operator, as I also pointed out in Section 2.8.

4.8 SIMULTANEOUSLY MEASUREABLE OPERATORS

We will now show the proposition

$$\langle y \rangle_\psi = 0 \text{ for all } |\psi\rangle \Rightarrow [A,B] = 0. \tag{4.95}$$

Because of the tie-in to the uncertainty product $\Delta A \Delta B$, such operators are called *simultaneous observables*. (The ideas of compatible Hermitian operators and simultaneous observables are synonymous.) In order to prove this proposition, let us define $\mathcal{O} = AB - BA$ and let $|X_i\rangle$, $|X_j\rangle$ be any elements of a basis that spans the space. Consider

$$|\psi\rangle = |X_i\rangle + |X_j\rangle. \tag{4.96}$$

Using Equation 4.96 to evaluate $\langle y \rangle_\psi$ gives us the starting point

$$\langle X_i|\mathcal{O}|X_i\rangle + \langle X_j|\mathcal{O}|X_j\rangle + \langle X_i|\mathcal{O}|X_j\rangle + \langle X_j|\mathcal{O}|X_i\rangle = 0. \tag{4.97}$$

The first two terms of Equation 4.97 are zero because of the proposition in Equation 4.95. Thus, since we may write

$$\langle X_j|\mathcal{O}|X_i\rangle = \langle X_i|\mathcal{O}^\dagger|X_j\rangle^*, \tag{4.98}$$

and

$$\mathcal{O}^\dagger = (AB - BA)^\dagger = (BA - AB) = -\mathcal{O}, \tag{4.99}$$

we have from Equation 4.97 that

$$\langle X_i|\mathcal{O}|X_j\rangle - \langle X_i|\mathcal{O}|X_j\rangle^* = 0, \tag{4.100}$$

which is the same as saying

$$\text{Im}\left(\langle X_i|\mathcal{O}|X_j\rangle\right) = 0. \tag{4.101}$$

Likewise, consider

$$|\psi\rangle = |X_i\rangle + i|X_j\rangle. \tag{4.102}$$

Substituting Equation 4.102 into the proposition tells us

$$i\langle X_i|\mathcal{O}|X_j\rangle - i\langle X_j|\mathcal{O}|X_i\rangle = 0, \tag{4.103}$$

which, with the repeated use of Equations 4.98 and 4.99, says

$$i\left(\langle X_i|\mathcal{O}|X_j\rangle + \langle X_i|\mathcal{O}|X_j\rangle^*\right) = 0, \tag{4.104}$$

or

$$\mathrm{Re}\left(\langle X_i|\mathcal{O}|X_j\rangle\right) = 0. \tag{4.105}$$

Equations 4.101 and 4.105 together imply that

$$\langle X_i|\mathcal{O}|X_j\rangle = 0, \tag{4.106}$$

for all i, j, so that

$$\mathcal{O} = [A,B] = 0. \tag{4.107}$$

Thus, the statement that $\langle y\rangle_\psi = 0$ for all $|\psi\rangle$ is equivalent to saying $[A,B] = 0$. However, if $\langle y\rangle_\psi = 0$ holds for just some particular states, then $AB \neq BA$ in general. Note also the rather subtle point that saying $\langle y\rangle_\psi = 0$, and therefore that A and B commute, is *not* the same as saying $\Delta A \Delta B = 0$. This is because of the additional, independent $\langle x\rangle_\psi$ term present in Equation 4.88. Given that $[A,B] \neq 0$, the $|\psi\rangle$ for which the equality holds in Equation 4.93 are minimum uncertainty states called *coherent states*. We saw an example of such a state in the 1-D Gaussian wave packet of Chapter 2, evaluated at $t = 0$. (At later times $\Delta x(t) = \delta x(t)$, where $\delta x(t)$ is given by Equation 2.89, so that the wave packet does not remain coherent.)

4.9 UNITARY TRANSFORMATIONS AND CHANGE OF BASIS

I have pointed out an analogy between Hilbert space and our ordinary 3-D world. I have stated that a normalized basis, $\{|X_i\rangle\}$, is like a set of unit vectors in ordinary space. Now the description of a general vector in terms of orthogonal unit vectors is not unique. There is always the freedom of a *choice* of basis. A different choice description cannot, of course, change the length of a vector. If we let x_i represent the projections of an arbitrary vector upon three orthogonal directions and $\bar{x}_i$ represent the projections of the same vector upon another set of mutually orthogonal 3-D unit vectors, then we must have

$$\sum_i \bar{x}_i^2 = \sum_i x_i^2. \tag{4.108}$$

The transformation equations relating $\bar{x}$ to x can be written as

$$\left.\begin{aligned}\bar{x}_1 &= x_1\lambda_{11} + x_2\lambda_{21} + x_3\lambda_{31},\\ \bar{x}_2 &= x_1\lambda_{12} + x_2\lambda_{22} + x_3\lambda_{32},\\ \bar{x}_3 &= x_1\lambda_{13} + x_2\lambda_{23} + x_3\lambda_{33},\end{aligned}\right\} \tag{4.109}$$

or more compactly as

$$\bar{x}_i = \sum_j x_j\lambda_{ji}. \tag{4.110}$$

The requirement (Equation 4.108) says that

$$\sum_{i,j,k} \lambda_{ji}\lambda_{ki}x_j x_k = \sum_i x_i^2, \tag{4.111}$$

which is only satisfied if

$$\sum_i \lambda_{ji}\lambda_{ki} = \delta_{jk}, \tag{4.112}$$

where δ_{jk} is the Kronecker delta symbol. In matrix notation Equations 4.110 and 4.112 read

$$\bar{x} = x\lambda, \tag{4.113}$$

and

$$\lambda\lambda^t = 1, \tag{4.114}$$

respectively, where "t" denotes the transpose. Comparing Equation 4.114 with the definition of λ^{-1},

$$\lambda\lambda^{-1} = 1 \tag{4.115}$$

means that

$$\lambda^t = \lambda^{-1}, \tag{4.116}$$

and Equation 4.114 may also be written as

$$\lambda^t\lambda = 1. \tag{4.117}$$

Any nonsingular transformation parity that satisfies Equation 4.116 preserves the length of vectors. This includes both rotations and parity inversions of the coordinates.

Now let us attempt to do the same thing for state vectors in Hilbert space. We have seen that the quantity $\langle\psi|\psi\rangle$ is a real, positive quantity for non-null vectors $|\psi\rangle$ (Equation 4.15). Interpreting this quantity as the square of the "length," we require that

$$\langle\bar{\psi}|\bar{\psi}\rangle = \langle\psi|\psi\rangle, \tag{4.118}$$

under a change of basis. The analog of the linear transformation (Equation 4.113) is

$$\langle\bar{\psi}| = \langle\psi|U, \tag{4.119}$$

where U is an operator in the Hilbert space. Equation 4.118 now gives us that

$$\langle\psi|UU^{\dagger}|\psi\rangle = \langle\psi|\psi\rangle, \tag{4.120}$$

similar to Equation 4.111 above. The requirement that Equation 4.120 holds for all states $|\psi\rangle$ then results in

$$UU^{\dagger} = 1. \tag{4.121}$$

(The reasoning that yields Equation 4.121 from Equation 4.120 is essentially the same as the way we showed $[A, B] = 0$ follows from $\langle\psi|[A, B]|\psi\rangle = 0$.) Comparison of Equation 4.121 with

$$UU^{-1} = 1, \tag{4.122}$$

tells us that

$$U^{\dagger} = U^{-1}, \tag{4.123}$$

similar to Equation 4.116. Therefore, we may also write Equation 4.121 as

$$U^{\dagger}U = 1. \tag{4.124}$$

We recognize Equation 4.123 as the definition of a *unitary transformation*, first seen in the discussion of spin 1/2 in Chapter 1. Thus, what was seen there as a special case is revealed as being general. A unitary transformation describes a change of basis that preserves the length of state vectors in Hilbert space. Because of the strong analogy between real space vectors and Hilbert space vectors, I will sometimes refer to Equation 4.119 as a "rotation" in Hilbert space.

We have been regarding Equation 4.113 as describing the point of view where the $\bar{x}_i$ are the components of a fixed vector in a rotated coordinate system. This is called a *passive* rotation of the vector. However, an equally valid interpretation of Equation 4.113 is that the $\bar{x}_i$ represent the components of a rotated vector in a fixed coordinate system, provided this rotation is taken in the opposite direction to the passive one. This is called an *active* rotation of the vector. We have also been regarding the analogous Hilbert space statement, Equation 4.119, in a passive sense. That is, we have taken Equation 4.119 as describing a situation where the bra vector is fixed but the basis is rotated. Just as for real space vectors, however, we could just as well view Equation 4.119 as an active rotation on the bra in the opposite direction to the passive one. I am using the passive terminology here mainly to connect smoothly with the discussion of unitary transformations in Chapter 1. You should be aware that either point of view is equally valid. We will use the point of view that is most convenient at the time. Once we have chosen an interpretation, however, we must strive for consistency.

Of course, completeness and orthonormality are preserved under a unitary transformation. That is, given

$$\sum_i |X_i\rangle\langle X_i| = 1, \tag{4.125}$$

and

$$\langle X_i|X_j\rangle = \delta_{ij}, \tag{4.126}$$

we have that

$$\sum_i |\bar{X}_i\rangle\langle\bar{X}_i| = U^\dagger\left(\sum_i |X_i\rangle\langle X_i|\right)U = 1, \tag{4.127}$$

and

$$\langle\bar{X}_i|\bar{X}_j\rangle = \langle X_i|UU^\dagger|X_j\rangle = \delta_{ij}. \tag{4.128}$$

Hermiticity of operators is also preserved under unitary transformations. Given the rotation, either passive or active, on the bra vectors as in Equation 4.119, the same rotation applied to an arbitrary operator A is

$$\bar{A} = U^\dagger A U. \tag{4.129}$$

(The rotation in the opposite direction is given by $UAU^\dagger$.) If $A = A^\dagger$, then

$$(\bar{A})^\dagger = (U^\dagger A U)^\dagger = U^\dagger A U = \bar{A}. \tag{4.130}$$

Also, given $A|a'\rangle = a'|a'\rangle$, we have

$$\bar{A}|\bar{a'}\rangle = U^\dagger A U(U^\dagger|a'\rangle) = a'|\bar{a'}\rangle, \tag{4.131}$$

so that eigenvalues are not changed, and

$$\langle\bar{\psi}|\bar{A}|\bar{\psi}\rangle = \langle\psi|UU^\dagger AUU^\dagger|\psi\rangle = \langle\psi|A|\psi\rangle, \tag{4.132}$$

so that expectation values are also unchanged.

A theorem can be proven that any unitary operator U can be written as

$$U = e^{iA}, \tag{4.133}$$

where $A = A^\dagger$.* The operator e^x is defined as its power series,

$$e^x \equiv 1 + x + \frac{x^2}{2} + \cdots + = \sum_{n=0}^{\infty} \frac{x^n}{n!}. \tag{4.134}$$

From Equation 4.134, it is easy to see that

$$(e^x)^\dagger = e^{x^\dagger}, \tag{4.135}$$

from which Equation 4.133 gives us

$$UU^\dagger = e^{iA}(e^{-iA}) = 1, \tag{4.136}$$

* See E. Merzbacher, *Quantum Mechanics*, 2nd ed. (John Wiley, 1970), p. 323.

as it should. By the way, in general we have

$$e^A e^B \neq e^{A+B}, \tag{4.137}$$

when A and B do not commute. To convince yourself of this, just expand both sides in powers of the arguments of the exponents. There is an equality sign in Equation 4.137 only in general if A and B commute. (See Problem 4.9.1.)

We can use unitary transformations to represent a change in basis due to coordinate displacements, rotations (in two or more spatial dimensions) momentum boosts, and other purposes. (In Chapter 12 they will be used on wave functions to do "gauge transformations.") In addition, we will see that these transformations also provide an alternate means of viewing the time development of a quantum system.

4.10 COORDINATE DISPLACEMENTS AND UNITARY TRANSFORMATIONS

As the simplest of these possibilities, let us consider the unitary representation of coordinate displacements. (We will deal with the equally simple case of velocity boosts in a problem.) The appropriate operator is

$$U = e^{-ix' p_x/\hbar}, \tag{4.138}$$

where x' is a number (with dimensions of length), and x and p_x are the usual position and momentum operators. This unitary transformation cannot be represented by a finite matrix since the number of eigenvalues x' of $x|x'\rangle = x'|x'\rangle$ is infinite. We will not worry about this subtlety and will treat it as if the space were finite. Let us first consider the quantity

$$\bar{x} = U^\dagger x U, \tag{4.139}$$

with U given by Equation 4.138. To find the effect of U on x, let us construct a differential equation for $\bar{x}$. We have that

$$\frac{\hbar}{i}\frac{d\bar{x}}{dx'} = e^{ix' p_x/\hbar}[p_x x - x p_x] e^{-ix' p_x/\hbar}, \tag{4.140}$$

where we see that the commutator $[p_x, x] = \hbar/i$ has arisen. Then we have

$$\frac{\hbar}{i}\frac{d\bar{x}}{dx'} = \frac{\hbar}{i} e^{ix' p_x/\hbar} e^{-ix' p_x/\hbar} = \frac{\hbar}{i}, \tag{4.141}$$

$$\Rightarrow \frac{d\bar{x}}{dx'} = 1. \tag{4.142}$$

The solution to Equation 4.142 is simply

$$\bar{x} = x' + \mathcal{O}, \tag{4.143}$$

where $\mathcal{O}$ is an unknown "constant" independent of x'. Both sides of Equation 4.143 actually have an operator character. We should understand

x' in Equation 4.143 to mean $x'1$, where 1 is the unit operator. The value of the operator $\mathcal{O}$ is specified by letting $x' = 0$ in Equation 4.139, which means that

$$\bar{x}(x' = 0) = x, \tag{4.144}$$

which implies that

$$\mathcal{O} = x. \tag{4.145}$$

Therefore, we have that

$$x + x' = e^{ix'p_x/\hbar} x e^{-ix'p_x/\hbar}. \tag{4.146}$$

The effect of this unitary transformation on coordinate states can be found as follows. We know from Equation 4.146 that

$$e^{ix''p_x/\hbar} x = (x + x'')e^{ix''p_x/\hbar}. \tag{4.147}$$

Therefore,

$$\begin{aligned}(\langle x'|e^{ix''p_x/\hbar})x &= \langle x'|(x + x'')e^{ix''p_x/\hbar} \\ &= (\langle x'|e^{ix''p_x/\hbar})(x' + x'').\end{aligned} \tag{4.148}$$

This means that

$$\langle x'|e^{ix''p_x/\hbar} = C\langle x' + x''|, \tag{4.149}$$

where C is some proportionality factor. Because unitary transformations preserve length in Hilbert space, we must have $|C|^2 = 1$. We may then *choose*

$$\langle x'|e^{ix''p_x/\hbar} = \langle x' + x''|. \tag{4.150}$$

Therefore, the unitary operator $e^{ix''p_x/\hbar}$ generates a spatial translation when acting on bra coordinate states.

Using Equation 4.150 we can recover our previous result for $\langle x'|p_x$. To first order in $\delta x''$ we have the approximate statement

$$\langle x'|\left(1 + i\frac{\delta x'' p_x}{\hbar}\right) \approx \langle x' + \delta x''|. \tag{4.151}$$

This gives

$$\langle x'|p_x \approx \frac{\hbar}{i}\left(\frac{\langle x' + \delta x''| - \langle x'|}{\delta x''}\right). \tag{4.152}$$

Taking the limit of both sides of Equation 4.152 as $\delta x'' \to 0$ now yields the exact statement

$$\langle x'|p_x = \frac{\hbar}{i}\frac{\partial}{\partial x'}\langle x'|, \tag{4.153}$$

the same as Equation 2.170 (see Chapter 2).

4.11 SCHRÖDINGER AND HEISENBURG PICTURES OF TIME EVOLUTION

As our second example of a unitary transformation, consider

$$U(t) = e^{iHt/\hbar}. \tag{4.154}$$

We saw this operator (or rather, its adjoint) in Chapter 2 where it was termed the time evolution operator. In Equation 2.196 we saw that

$$|a',t\rangle = e^{-iHt/\hbar}|a'\rangle \tag{4.155}$$

where $|a'\rangle$ is an energy eigenstate (also called a stationary state) of H at $t=0$ and $|a',t\rangle$ is the time evolved state. However, Equation 4.155 also holds for any linear combination of such eigenstates or continuous superposition of energy states, $|\psi,t\rangle$, as long as the Hamiltonian is not an explicit function of time, $H \neq H(t)$.

Under these conditions, the time evolution of a general operator, A, is given by

$$\frac{d\langle A\rangle_{\psi,t}}{dt} = \frac{\partial\langle\psi,t|}{\partial t}A|\psi,t\rangle + \langle\psi,t|A\frac{\partial|\psi,t\rangle}{\partial t} + \left\langle\frac{\partial A}{\partial t}\right\rangle_{\psi,t}. \tag{4.156}$$

Consider the case of an operator with no explicit time dependence, $\partial A/\partial t = 0$. We may use the time-dependent Schrödinger equations,

$$i\hbar\frac{\partial|\psi,t\rangle}{\partial t} = H|\psi,t\rangle, \tag{4.157}$$

$$\Rightarrow -i\hbar\frac{\partial\langle\psi,t|}{\partial t} = \langle\psi,t|H, \tag{4.158}$$

to give

$$\frac{\hbar}{i}\frac{d\langle A\rangle_{\psi,t}}{dt} = \langle[H,A]\rangle_{\psi,t}. \tag{4.159}$$

Let me take this opportunity to point out another way of dealing with time evolution using the unitary operator $U(t)$. Let us again consider the expectation value of some physical property A at time t. This can be written as

$$\langle A\rangle_{\psi,t} \equiv \langle\psi,t|A|\psi,t\rangle = \langle\psi|U(t)AU^{\dagger}(t)|\psi\rangle, \tag{4.160}$$

with $U(t)$ as given earlier. This point of view assigns the time evolution to the states. However, from Equation 4.160 we see that it is equally valid to assign all time dependence to the operator A. That is, we may set

$$\langle A\rangle_{\psi,t} \equiv \langle A(t)\rangle_{\psi}, \tag{4.161}$$

where

$$A(t) \equiv U(t)AU^{\dagger}(t). \tag{4.162}$$

The problem now is to find the dynamical equation satisfied by $A(t)$ so that its time behavior can be determined. Consider, therefore, the time derivative of $A(t)$. One finds that

$$\frac{\hbar}{i}\frac{\mathrm{d}A(t)}{\mathrm{d}t} = U(t)[HA - AH]U^{\dagger}(t) + U(t)\frac{\partial A}{\partial t}U^{\dagger}(t). \qquad (4.163)$$

The three terms in Equation 4.163 come from the time dependence in $U(t)$, $U^{\dagger}(t)$, and a possible explicit time dependence in the operator A. Now we may write

$$U(t)[HA - AH]U^{\dagger}(t) = [HA(t) - A(t)H] = [H, A(t)], \qquad (4.164)$$

since in the simple case we are studying, $U(t)$ and H commute. Again, consider the case where there is no explicit time dependence in A. Then Equation 4.163 becomes

$$\frac{\hbar}{i}\frac{\mathrm{d}A(t)}{\mathrm{d}t} = [H, A(t)]. \qquad (4.165)$$

This is the operator equation of motion satisfied by $A(t)$, analogous to Equation 4.159, but without the expectation value. Equation 4.165 is called the *Heisenberg equation of motion*. Taking the expectation value of both sides of Equation 4.161, we learn that

$$\frac{\hbar}{i}\frac{\mathrm{d}\langle A(t)\rangle_{\psi}}{\mathrm{d}t} = \langle [H, A(t)]\rangle_{\psi}, \qquad (4.166)$$

in an arbitrary state $|\psi\rangle$. (The subscript, ψ, now emphasizes that the expectation value is taken with respect to a time-independent state.) Equations 4.166 and 4.159 are completely equivalent.

Notice that if $[H, A] = 0$, this implies $[H, A(t)] = 0$ because $U(t)$ and H commute. Therefore, when H and A commute, A is a *constant of the motion*. In this case, Equations 4.159 and 4.166 give rise to the equivalent statements

$$\frac{\mathrm{d}\langle A\rangle_{\psi,t}}{\mathrm{d}t} = 0, \quad \frac{\mathrm{d}\langle A(t)\rangle_{\psi}}{\mathrm{d}t} = 0, \qquad (4.167)$$

respectively. This says that expectation values of A are a constant in time. This makes sense because if A is assumed to commute with H, then it is clear that

$$\langle A(t)\rangle_{\psi} = \langle\psi | U(t)AU^{\dagger}(t) | \psi\rangle = \langle\psi | A | \psi\rangle,$$

for all t. In particular, if at $t = 0$ the wave function is an eigenvector of A with eigenvalue a', this will continue to hold true at a later time t. a' is called a *good quantum number*, and A can be chosen as one of the complete set of observables that characterize eigenvectors. In addition to energy eigenvalues, we previously saw an example of a constant of the motion in the parity operator, $\mathbb{P}$, in the case of the infinite square well and the simple harmonic oscillator. A state with a given $\langle\mathbb{P}\rangle_{\Psi}$ value will keep this quantity fixed in time if $[H, \mathbb{P}] = 0$.

The representation for $U(t)$ in Equation 4.154 is in general only true when the Hamiltonian is not an explicit function of the time. Other forms for the evolution operator hold when $H = H(t)$.* We will not deal with time-dependent Hamiltonians here (but see Chapter 10).

Note that the commutation properties of operators are preserved in the Heisenberg picture if these are interpreted as *equal time* relations. For any unitary transformation $U(t)$ we have, for example,

$$\begin{aligned} &U(t)\,[x,p_x]\,U^\dagger(t) = i\hbar, \\ &\Rightarrow U(t)xp_xU^\dagger(t) - U(t)p_xxU^\dagger(t) = i\hbar, \\ &\Rightarrow x(t)p_x(t) - p_x(t)x(t) = i\hbar, \\ &\Rightarrow [x(t),\ p_x(t)] = i\hbar. \end{aligned} \tag{4.168}$$

We thus have an alternate and equivalent way of viewing the time dynamics of quantum systems. Previously, we were taking the operators as static and viewing the states $|\psi\rangle$ as evolving either actively or passively in time. This point of view leads to the Schrödinger equation and is called the *Schrödinger picture*. We have now learned that we may instead time-evolve the operators actively or passively and let the states $|\psi\rangle$ be static. This point of view leads to the Heisenberg equations of motion and is called the *Heisenberg picture*. These approaches are entirely equivalent.

4.12 FREE GAUSSIAN WAVE PACKET IN THE HEISENBERG PICTURE

As an example of the use of the Heisenberg picture, let us reexamine the free Gaussian wave packet of Chapter 2. The Gaussian distribution $|\psi_g(x,t)|^2$ spread with time but maintained its shape. It is easy to check that the expectation values of position and momentum in the Schrödinger picture are

$$\langle x\rangle_{\psi,t} = \frac{\bar{p}}{m}t, \tag{4.169}$$

$$\langle p_x\rangle_{\psi,t} = \bar{p}, \tag{4.170}$$

from which we have that

$$\langle x\rangle_{\psi,t} = \frac{\langle p_x\rangle_{\psi,t}}{m}t. \tag{4.171}$$

Let us try to recover Equation 4.171 from the Heisenberg picture. We have that

$$\frac{\hbar}{i}\frac{\mathrm{d}x(t)}{\mathrm{d}t} = [H, x(t)] = U(t)\left[\frac{p_x^2}{2m}, x\right]U^\dagger(t). \tag{4.172}$$

* See the discussion in J.J. Sakurai and J. Napolitano, *Modern Quantum Mechanics*, 2nd ed. (Addison-Wesley, 2011), Chapters 2 and 5.

The commutator in Equation 4.172 is

$$\left[\frac{p_x^2}{2m}, x\right] = \frac{1}{2m}\{p_x[p_x, x] + [p_x, x]p_x\} = \frac{\hbar}{i}\frac{p_x}{m}. \tag{4.173}$$

We now find

$$\frac{\hbar}{i}\frac{dx(t)}{dt} = \frac{\hbar}{i}U(t)\frac{p_x}{m}U^\dagger(t) = \frac{\hbar}{i}\frac{p_x(t)}{m}, \tag{4.174}$$

so we have the *operator* statement

$$\frac{dx(t)}{dt} = \frac{p_x(t)}{m}. \tag{4.175}$$

Of course, we also have that

$$\frac{\hbar}{i}\frac{dp_x(t)}{dt} = U(t)\left[\frac{p_x^2}{2m}, p_x\right]U^\dagger(t) = 0, \tag{4.176}$$

so that

$$\frac{dp_x(t)}{dt} = 0, \tag{4.177}$$

and $p_x(t)$ is a constant operator in time ($p_x(t) = p_x$). Therefore, from Equation 4.175 we get by integration

$$x(t) = \frac{p_x}{m}t + C, \tag{4.178}$$

where C is a constant operator. We know from $x(t) = U(t)xU^\dagger(t)$ that $x(0) = x$, so $C = x$. Taking the expectation value of both sides of Equation 4.178 finally gives us

$$\langle x(t)\rangle_\psi = \frac{\langle p_x\rangle_\psi}{m}t, \tag{4.179}$$

since $\langle x\rangle_\psi = 0$. In Equation 4.171 the time dependence of the expectation values is in the state, whereas in Equation 4.179 the time dependence is in the operator. We get identical results either way. Notice that the relation Equation 4.171 was derived for a particular wave packet, whereas Equation 4.179 shows that this relation holds for any particle wave packet as long as $\langle x\rangle_\psi = 0$.

4.13 POTENTIALS AND THE EHRENFEST THEOREM

The preceding application of the Heisenberg picture was to the case of free particles. We can go a step beyond this in applying this formalism in the case where an unspecified potential is present. Let us let

$$H = \frac{p_x^2}{2m} + V(x). \tag{4.180}$$

Then we have

$$\frac{\hbar}{i}\frac{\mathrm{d}p_x(t)}{\mathrm{d}t} = U(t)[H, p_x]U^\dagger(t), \tag{4.181}$$

$$[H, p_x] = [V(x), p_x]. \tag{4.182}$$

From Equation 3.82 recall that

$$[p_x, f(x)] = \frac{\hbar}{i}\frac{\mathrm{d}f(x)}{\mathrm{d}x}, \tag{4.183}$$

which is an operator statement. For the commutator in Equation 4.182, we have, therefore,

$$[V(x), p_x] = -\frac{\hbar}{i}\frac{\mathrm{d}V(x)}{\mathrm{d}x}, \tag{4.184}$$

from which we have

$$\frac{\mathrm{d}p_x(t)}{\mathrm{d}t} = -U(t)\frac{\mathrm{d}V(x)}{\mathrm{d}x}U^\dagger(t). \tag{4.185}$$

We also have that

$$\frac{\hbar}{i}\frac{\mathrm{d}x(t)}{\mathrm{d}t} = U(t)\,[H, x]U^\dagger(t) = \frac{1}{2m}U(t)\,[p_x^2, x]U^\dagger(t), \tag{4.186}$$

where $[p_x^2, x] = (2\hbar/i)\,p_x$. Therefore

$$\frac{\mathrm{d}x(t)}{\mathrm{d}t} = \frac{p_x(t)}{m}, \tag{4.187}$$

the same as Equation 4.175. Taking another derivative in Equation 4.187 now gives us

$$\frac{\mathrm{d}^2x(t)}{\mathrm{d}t^2} = \frac{1}{m}\frac{\mathrm{d}p_x(t)}{\mathrm{d}t} = -\frac{1}{m}U(t)\,\frac{\mathrm{d}V(x)}{\mathrm{d}x}U^\dagger(t), \tag{4.188}$$

where we have used Equation 4.185. If $V(x)$ is a power series in x, then

$$U(t)\,\frac{\mathrm{d}V(x)}{\mathrm{d}x}U^\dagger(t) = \frac{\mathrm{d}V(x(t))}{\mathrm{d}x(t)} \tag{4.189}$$

and we have

$$m\frac{\mathrm{d}^2x(t)}{\mathrm{d}t^2} = -\frac{\mathrm{d}V(x(t))}{\mathrm{d}x(t)}, \tag{4.190}$$

which has the appearance of Newton's second law but written for operators. We may, if we wish, take the expectation value of both sides of Equation 4.190 in an arbitrary state. This yields

$$m\frac{\mathrm{d}^2\langle x(t)\rangle_\psi}{\mathrm{d}t^2} = -\left\langle \frac{\mathrm{d}V(x(t))}{\mathrm{d}x(t)} \right\rangle_\psi. \tag{4.191}$$

The analogous statement in the Schrödinger language writes this as

$$m\frac{d^2\langle x\rangle_{\psi,t}}{dt^2} = -\left\langle\frac{dV(x)}{dx}\right\rangle_{\psi,t}. \tag{4.192}$$

Equation 4.191 or 4.192 is called the *Ehrenfest theorem*. In other words, it says that the expectation value of the position operator moves like a classical particle subjected to a "force" given by

$$-\left\langle\frac{dV}{dx}\right\rangle.$$

PROBLEMS

4.4.1 *Prove*: In a finite Hilbert space of dimensionality N, given the N eigenvectors of A can all be chosen orthonormal and the associated eigenvalues are all real, then $A = A^\dagger$ (it is Hermitian).

4.4.2 (a) Given the 2×2 matrix,

$$A = \begin{pmatrix} 1 & i \\ -i & 2 \end{pmatrix},$$

show that A is Hermitian.

(b) Find the eigenvalues and normalized eigenvectors (in the form of column matrices) of A.

4.6.1 *Prove*: (Restricted form of the necessary condition for the theorem in Section 4.6.) Given that A and B are Hermitian and the eigenvalues of A (or B) are all distinct, and if $[A,B] = 0$, then A and B possess a common set of orthonormal eigenkets.

4.6.2 Show for the parity operator, $\mathbb{P}$, that ($\{A,B\} \equiv AB + BA$ is the anticommutator)

(a) $\{\mathbb{P}, p_x\} = 0$.

(b) $\{\mathbb{P}, x\} = 0$.

(c) $\mathbb{P}\mathbb{P}^\dagger = 1$.

4.7.1 *Prove*:

$$|\langle\psi_1|A|\psi_2\rangle|^2 \leq \langle\psi_1|A|\psi_1\rangle\langle\psi_2|A|\psi_2\rangle$$

for any $A = A^\dagger$ such that $\langle\psi|A|\psi\rangle \geq 0$ for all $|\psi\rangle$.

4.7.2 Using the definition,

$$\Delta A \equiv \sqrt{\langle A^2\rangle_\psi - \langle A\rangle_\psi^2},$$

for the Gaussian wave function, $\psi_g(x,0)$, as in Equation 2.86 (Chapter 2), find:
(a) $\Delta x = ?$
(b) $\Delta p_x = ?$
[Note: A useful integral is

$$\int_{-\infty}^{\infty} dx\, x^2\, e^{-\alpha x^2} = -\frac{d}{d\alpha}\int_{-\infty}^{\infty} dx\, e^{-\alpha x^2}$$

$$\text{where} \quad \int_{-\infty}^{\infty} dx\, e^{-\alpha x^2} = \sqrt{\frac{\pi}{\alpha}}.$$

The partial answer to 2.10.1 may also be useful.]

***4.7.3** Continue the calculation in Problem 4.7.2 by showing that the uncertainty in the Gaussian's energy, $E = p_x^2/2m$, at $t=0$ is

$$\Delta E = \frac{\hbar}{2m|\delta x|}\sqrt{\frac{\hbar^2}{8\delta x^2} + \bar{p}^2}.$$

[Relevant to the discussion in Chapter 2, Section 2.8. Note that this is actually time independent; see comments in Section 2.10.]

4.7.4 Using the raising and lowering operators of Chapter 3, show that

(a) $(\Delta p_x)^2 = \langle n | p^2 | n \rangle = \hbar m\omega\left(n + \frac{1}{2}\right)$,

(b) $(\Delta x)^2 = \langle n | x^2 | n \rangle = \frac{\hbar}{m\omega}\left(n + \frac{1}{2}\right)$,

for the squares of the uncertainties of momentum and position in the nth state for the harmonic oscillator, and therefore that

$$\Delta x \Delta p_x = \left(n + \frac{1}{2}\right)\hbar.$$

[Comment: We see that the ground state is a minimum uncertainty or coherent state, in the terminology of Section 4.8.]

4.8.1 Prove: If $\langle\psi|A|\psi\rangle$ is real for all $|\psi\rangle$, then $A = A^\dagger$. (A is Hermitian.)

***4.9.1** Show that

$$e^A e^B = e^{A+B} e^{1/2[A,B]},$$

given that operators A and B each commute with $[A,B]$. Take as proven the "Baker–Hausdorff lemma,"

$$e^A B e^{-A} = B + [A,B] + \frac{1}{2!}[A,[A,B]] + \frac{1}{3!}[A,[A,[A,B]]] + \cdots.$$

[Hint: Using the above, I built and solved a differential equation for dg/dn, where $g(n) \equiv e^{nA}e^{nB}e^{-n(A+B)}$.]

4.9.2 Use the result in Problem 4.9.1 to show that we may write

$$[e^{iax}, e^{ibp_x}] = -2ie^{i(ax+bp_x)}\sin\left(\frac{\hbar ab}{2}\right),$$

where p_x and x are operators, and a and b are numbers. Notice when $\hbar ab = 2\pi n$, $n = \pm 1, \pm 2, \ldots$, these exponentials commute, although x and p_x do not!

4.10.1 (a) Show that (set $\hbar = 1$ here for simplicity)

$$e^{-ip'_x x}\, p_x\, e^{ip'_x x} = p_x + p'_x,$$

where x and p_x are the usual position, momentum operators, and p'_x is a (real) number. [Hint: Construct a differential equation in p'_x.]

(b) Use part (a) to argue that

$$\langle p'_x | x = i\frac{\partial}{\partial p'_x}\langle p'_x|.$$

4.11.1 (a) For a free particle ($V(x) = 0$) in one dimension, show that there is an uncertainty relation connecting x (the position operator) and $x(t)$ (the Heisenberg time-evolved position operator):

$$\Delta x\, \Delta x(t) \geq \left|\frac{\hbar t}{2m}\right|.$$

(b) Express in words what this uncertainty relation means in terms of experiments that measure a particle's position.

***4.13.1** (a) By considering the time derivative of the quantity $(f(x)p_x + p_x f(x))$ in a *stationary state* of $H = p_x^2/2m + V(x)$ for a general function $f(x)$, obtain the generalized one-dimensional virial theorem in quantum mechanics. Show that

$$\left\langle f(x)\frac{dV(x)}{dx}\right\rangle = \frac{1}{4m}\left\langle p_x^2\frac{df(x)}{dx} + 2p_x\frac{df(x)}{dx}p_x + \frac{df(x)}{dx}p_x^2\right\rangle.$$

(b) Use (a) to obtain for example,

$$\left\langle\frac{dV(x)}{dx}\right\rangle = 0,$$

and

$$\left\langle x \frac{\mathrm{d}V(x)}{\mathrm{d}x} \right\rangle = 2\langle T \rangle,$$

where T is the kinetic energy operator. Does the first result contradict Equation 4.192?

4.13.2 Derive Equation 4.192 in the Schrödinger picture.

Three Static Approximation Methods

Synopsis: It is rare that exact solutions for physical quantum systems are available, and likely that one or more types of perturbation theories need to be applied. Three time-independent, or static, methods are discussed. Using exact energy shift formulas, a result correct to second order for the perturbed system energy is derived and applied to a two-state system and a harmonic oscillator example. The WKB semiclassical approximation is motivated, and applications to barrier penetration and various types of bound-state problems are made. The third class of methods gives upper limit estimates of system ground-state properties based on energy functional variations.

5.1 INTRODUCTION

It is in the nature of physical systems that the closer they are observed, the more detail there is to see; the closer we come to mathematically describing these additional details, the more intricate our mathematical considerations often then become; the more intricate the mathematical description is, the more difficult it becomes to solve our system of equations in some *exact* analytical way. Sometimes, the additional physical details we wish to incorporate into the theory are sufficiently small compared to other quantities already in the theory that their incorporation does not involve an entirely new solution or an entirely new starting point but can be treated as a perturbation of the old solution or starting point. What I want to talk about here are three methods for carrying out this program.

The first technique, called *time-independent perturbation theory* or *Rayleigh–Schrödinger perturbation theory*, is mainly useful when additional small time-independent interactions are added to a system for which an exact analytical solution is already available. This is often the case in the intricate interactions that occur in atomic and nuclear physics. The results that it gives are very general and are not necessarily limited to a nonrelativistic domain. Another technique, called the WKB (named after Gregor Wentzel, Hendrik Kramers, and Leon Brillouin) or JWKB (add Harold Jeffries) *semiclassical approximation*, is based upon the use of a "classical" version of the Schrödinger equation as a new starting point for the description of situations where particles are subjected to slowly varying potentials in space. The third method, called the *Rayleigh–Ritz variational method* or technique, gives upper limit estimates for the ground-state energies and wave functions of quantum systems. The treatment of time-independent perturbation theory is independent of any assumptions concerning our space dimensionality. However, we will maintain our limitation to a single spatial dimension for purposes of simplicity in the WKB and variational methods discussions. We will discuss *time-dependent* perturbation theory as part of a study of the time development of quantum systems in Chapter 10.

5.2 TIME-INDEPENDENT PERTURBATION THEORY*

To begin with, let us consider a Hamiltonian that is dependent on some real parameter, λ. (It could, for example, represent an interaction with an electric or magnetic field.) The basic eigenvalue–eigenvector statement is given by

$$(H(\lambda) - E(\lambda) | E\lambda\rangle = 0, \tag{5.1}$$

where in general the energies and the states also have a λ dependence. (λ is simply a parameter in $|E\lambda\rangle$, not an eigenvalue.) Taking the derivative of Equation 5.1 with respect to λ gives

$$(H - E)\frac{\partial}{\partial\lambda} | E\lambda\rangle + \left(\frac{\partial H}{\partial\lambda} - \frac{\partial E}{\partial\lambda}\right) | E\lambda\rangle = 0. \tag{5.2}$$

Now project both terms in Equation 5.2 into the state $\langle E\lambda|$. Since we know that $\langle E\lambda|(H - E) = 0$, we get (*Feynman–Hellman theorem*)

$$\langle E\lambda | \frac{\partial H}{\partial\lambda} | E\lambda\rangle = \frac{\partial E}{\partial\lambda}, \tag{5.3}$$

which is an exact statement. This equation is useful on occasions when $E(\lambda)$ is known and we wish to evaluate certain operator expectation values.

We now wish to solve Equation 5.2 for $(\partial/\partial\lambda)|E\lambda\rangle$. Assuming the inverse of the operator $(H - E)$ exists, the general solution is

$$\frac{\partial}{\partial\lambda}|E\lambda\rangle = iC(\lambda) | E\lambda\rangle + \frac{1}{E - H}\left(\frac{\partial H}{\partial\lambda} - \frac{\partial E}{\partial\lambda}\right) | E\lambda\rangle, \tag{5.4}$$

where $C(\lambda)$ is an arbitrary real constant. We can see why this term is allowed because if we try to reproduce Equation 5.2 from Equation 5.4 by operating on both sides by $(E - H)$, we see that the term proportional to $C(\lambda)$ will project to zero. It actually arises because of the freedom of choice of a λ-dependent phase in the definition of the state $|E\lambda\rangle$. We will put $C(\lambda) = 0$ in the following, but this will not limit the generality of the results.

We will now specialize to problems that have discrete, nondegenerate energy eigenvalues. Completeness can be written as

$$\sum_{E'} | E'\lambda\rangle\langle E'\lambda | = 1. \tag{5.5}$$

Using Equation 5.5 we may write

$$\left(\frac{\partial H}{\partial\lambda} - \frac{\partial E}{\partial\lambda}\right) | E\lambda\rangle = \sum_{E'} |E'\lambda\rangle\langle E'\lambda | \left(\frac{\partial H}{\partial\lambda} - \frac{\partial E}{\partial\lambda}\right) | E\lambda\rangle, \tag{5.6}$$

* Much of this section is based on my 1979 Schwinger notes.

$$= \sum_{E' \neq E} |E'\lambda\rangle\langle E'\lambda| \frac{\partial H}{\partial \lambda} | E\lambda\rangle. \tag{5.7}$$

We know by Equation 5.3 that when $E' = E$ in the sum in Equation 5.6 that the matrix element $\langle E\lambda|(\partial H/\partial\lambda) - (\partial E/\partial\lambda)|E\lambda\rangle$ vanishes. That is why the sum in Equation 5.7 leaves out this term. Once this single term is eliminated, we know that the $\partial E/\partial\lambda$ term in Equation 5.6 does not contribute because of the orthogonality of the states $\langle E'\lambda|$ and $|E\lambda\rangle$. This term is zero and the result is then Equation 5.7. Replacing Equation 5.7 in Equation 5.4 gives us

$$\frac{\partial}{\partial \lambda} | E\lambda\rangle = \frac{1}{E - H} \sum_{E' \neq E} | E'\lambda\rangle\langle E'\lambda| \frac{\partial H}{\partial \lambda} | E\lambda\rangle. \tag{5.8}$$

Now we postulate on the basis of Equation 1.105 (Chapter 1) that

$$f(H) | E'\lambda\rangle = f(E') | E'\lambda\rangle, \tag{5.9}$$

so that

$$(E - H)^{-1} | E'\lambda\rangle = (E - E')^{-1} | E'\lambda\rangle. \tag{5.10}$$

Equation 5.8 now becomes

$$\frac{\partial}{\partial \lambda} | E\lambda\rangle = \sum_{E' \neq E} | E'\lambda\rangle \frac{\langle E'\lambda | \partial H / \partial \lambda | E\lambda\rangle}{(E - E')}. \tag{5.11}$$

Let us go back to Equation 5.3 and work out the second derivative of $E(\lambda)$:

$$\frac{\partial^2 E}{\partial \lambda^2} = \langle E\lambda | \frac{\partial^2 H}{\partial \lambda^2} | E\lambda\rangle + \langle E\lambda | \frac{\partial H}{\partial \lambda} \left(\frac{\partial}{\partial \lambda} |E\lambda\rangle \right) + \left(\frac{\partial}{\partial \lambda} \langle E\lambda| \right) \frac{\partial H}{\partial \lambda} | E\lambda\rangle. \tag{5.12}$$

The last two terms in Equation 5.12 are in fact just complex conjugates of each other, so that

$$\frac{\partial^2 E}{\partial \lambda^2} = \langle E\lambda | \frac{\partial^2 H}{\partial \lambda^2} | E\lambda\rangle + 2\,\mathrm{Re} \left[\langle E\lambda| \frac{\partial H}{\partial \lambda} \underbrace{\left(\frac{\partial}{\partial \lambda} |E\lambda\rangle \right)}_{\text{Replace}} \right]. \tag{5.13}$$

Now let us use Equation 5.11 in Equation 5.13. We get

$$\frac{\partial^2 E}{\partial \lambda^2} = \langle E\lambda | \frac{\partial^2 H}{\partial \lambda^2} | E\lambda\rangle + 2 \sum_{E' \neq E} \frac{|\langle E\lambda | \partial H / \partial \lambda | E'\lambda\rangle|^2}{(E - E')}, \tag{5.14}$$

where we have dropped the real part restriction because the quantity in brackets in Equation 5.13 *is* real. Equation 5.14 is also an exact formula.

Now let us do a Taylor series for $E(\lambda)$. We have (assuming the series exists)

$$E(\lambda) = E(0) + \lambda \frac{\partial E(0)}{\partial \lambda} + \frac{\lambda^2}{2} \frac{\partial^2 E(0)}{\partial \lambda^2} + \cdots \tag{5.15}$$

where the partials with respect to λ are evaluated at $\lambda = 0$. Using Equations 5.3 and 5.14 in Equation 5.15 now reveals that

$$E(\lambda) = E(0) + \lambda\langle E|\frac{\partial H(0)}{\partial\lambda}|E\rangle + \frac{\lambda^2}{2}\langle E|\frac{\partial^2 H(0)}{\partial\lambda^2}|E\rangle$$
$$+\lambda^2\sum_{E'\neq E}\frac{|\langle E|\frac{\partial H(0)}{\partial\lambda}|E'\rangle|^2}{(E-E')}+\ldots. \tag{5.16}$$

Equation 5.16 is usually applied to the situation where the Hamiltonian is given by

$$H = H_0 + H_1. \tag{5.17}$$

H_0 represents a Hamiltonian for which an exact solution is known and H_1 represents the "perturbation." Instead of Equation 5.17 we may formally write

$$H = H_0 + \lambda H_1, \tag{5.18}$$

and then evaluate the Taylor series, Equation 5.16, when $\lambda = 1$ to get the effect of the perturbation H_1 on the energy levels. The result is

$$E = E_0 + \langle E_0|H_1|E_0\rangle + \sum_{E_0'\neq E_0}\frac{|\langle E_0|H_1|E_0'\rangle|^2}{(E_0 - E_0')} + \cdots \tag{5.19}$$

where I have labeled the unperturbed energies as E_0. Equation 5.19 says the leading correction to the E_0 energy level is just the diagonal element of the perturbation matrix. Because the perturbation H_1 appears linearly in the diagonal term, this is the first-order correction to the energy. The next term, where the H_1 matrix element appears squared, is the second-order correction, and so on. Corresponding to these corrections in the energies are corrections to the energy wave functions. For a system with a discrete, nondegenerate and complete set of states, Problem 5.2.6 gives the first-order perturbative result,

$$|\Psi_n\rangle \simeq |\psi_n\rangle + \sum_{n'\neq n}|\psi_{n'}\rangle\frac{\langle\psi_{n'}|H_1|\psi_n\rangle}{(E_n - E_{n'})}, \tag{5.20}$$

where the $|\Psi_n\rangle$ represent the new eigenvectors and $|\psi_n\rangle$ the old ones.

It sometimes happens that the leading order correction in Equation 5.19 vanishes for certain perturbations, but the second-order term does not. Notice that if E_0 represents the ground-state energy, then the effect of the second-order correction is such as to *lower* the energy of the ground state because $(E_0 - E_0') < 0$ for all $E_0' \neq E_0$ by definition. This cannot be said for higher lying states. There we see that the second-order correction tends to produce a *repulsion* between neighboring energy levels. The sign of the overall energy shift, however, is not determined.

Finally, also note that although we are discussing time-independent perturbation theory in the context of nonrelativistic quantum mechanics, there are no nonrelativistic assumptions above, just the assumption that an eigenvalue/eigenvector statement of the problem exists.

5.3 EXAMPLES OF TIME-INDEPENDENT PERTURBATION THEORY

As a simple example in a matrix context, consider a two-state problem with a Hamiltonian,

$$H = H_0 + H_1, \tag{5.21}$$

$$H_0 = \begin{pmatrix} E_1 & 0 \\ 0 & E_2 \end{pmatrix}, \tag{5.22}$$

$$H_1 = \begin{pmatrix} 0 & a \\ a^* & 0 \end{pmatrix}, \tag{5.23}$$

with $|a| \ll |E_2 - E_1|$ and $E_2 > E_1$. The H_1 Hamiltonian is the "perturbing" one. The normalized H_0 eigenvectors are as follows:

$$\Psi_1^{(0)} = \begin{pmatrix} 1 \\ 0 \end{pmatrix}, \quad \Psi_2^{(0)} = \begin{pmatrix} 0 \\ 1 \end{pmatrix}, \tag{5.24}$$

and the new energies from first-order perturbation theory,

$$E_1' = E_1 + (\Psi_1^{(0)})^\dagger H_1 \, \Psi_1^{(0)} = E_1, \tag{5.25}$$

$$E_2' = E_1 + (\Psi_2^{(0)})^\dagger H_1 \, \Psi_2^{(0)} = E_2, \tag{5.26}$$

are unchanged. We must go to second order, using Equation 5.19, to see an energy shift. Then

$$E_1' = E_1 + \frac{|(H)_{12}|^2}{(E_1 - E_2)}. \tag{5.27}$$

Identifying $(H_1)_{12} = a$, this gives

$$E_1' = E_1 - \frac{|a|^2}{(E_2 - E_1)}. \tag{5.28}$$

Similarly,

$$E_2' = E_2 + \frac{|a|^2}{(E_2 - E_1)}. \tag{5.29}$$

Notice the E_1' energy has been pushed down by the perturbation, as expected from the discussion at the end of the last section. The exact energies and eigenvectors can be found for this problem and will be compared to the perturbative solution in Problem 5.3.1.

Let us look at another perturbative example in a position-space context. We will reexamine the simple harmonic oscillator with dimensionless Hamiltonian

$$H_0 = \frac{1}{2}(p^2 + q^2), \tag{5.30}$$

and energies

$$E_0 = n + \frac{1}{2}, \quad n = 0,1,2,\ldots \tag{5.31}$$

We will take the perturbation as

$$H_1 = \gamma q^3, \quad (\gamma \text{ dimensionless}), \tag{5.32}$$

making the system *anharmonic*. The new dimensionless energies of the system are given approximately by

$$E_n \approx \left(n + \frac{1}{2}\right) + \langle \gamma q^3 \rangle_n + \sum_{n' \neq n} \frac{|\langle n|\gamma q^3|n'\rangle|^2}{(n-n')}. \tag{5.33}$$

Now, the first-order energy correction vanishes in the unperturbed states because positive and negative position values occur symmetrically in $\psi_n^*(q')\psi_n(q')$. Another way of arguing this is to say that the q^3 operator changes the parity of the state $|n\rangle$. We can work out the necessary matrix elements of q^3 for the second-order term as follows. Remember that

$$q = \frac{A + A^\dagger}{\sqrt{2}}, \tag{5.34}$$

where

$$A|n\rangle = \sqrt{n}\,|n-1\rangle, \tag{5.35}$$

$$A^\dagger|n\rangle = \sqrt{n+1}\,|n+1\rangle. \tag{5.36}$$

We now find successively:

$$q|n\rangle = \frac{1}{\sqrt{2}}\left[\sqrt{n}\,|n-1\rangle + \sqrt{n+1}\,|n+1\rangle\right], \tag{5.37}$$

$$\begin{aligned} q^2|n\rangle = q[q|n\rangle] = \frac{1}{2}\Big[&\sqrt{n(n-1)}|n-2\rangle \\ &+(2n+1)|n\rangle + \sqrt{(n+1)(n+2)}\,|n+2\rangle\Big], \end{aligned} \tag{5.38}$$

$$\begin{aligned} q^3|n\rangle = q[q^2|n\rangle] = \frac{1}{2\sqrt{2}}\Big[&\sqrt{n(n-1)(n-2)}|n-3\rangle \\ &+\sqrt{n}(3n)|n-1\rangle + \sqrt{(n+1)}\,(3n+3)|n+1\rangle \\ &+\sqrt{(n+1)(n+2)(n+3)}|n+3\rangle\Big]. \end{aligned} \tag{5.39}$$

Therefore, we have from Equation 5.33 that

$$E_n \approx \left(n + \frac{1}{2}\right) + \frac{\gamma^2}{8}\left[\frac{n(n-1)(n-2)}{3} + 9n^3 - 9(n+1)^3 + \frac{(n+1)(n+2)(n+3)}{-3}\right], \tag{5.40}$$

or

$$E_n = \left(n + \frac{1}{2}\right) - \frac{\gamma^2}{8}(30n(n+1) + 11). \tag{5.41}$$

What has happened to the energy levels? Notice that the correction term in Equation 5.41 is always negative, lowering all of the energies. This lowering in energy increases in magnitude as n increases. In fact, for neighboring energy levels we have

$$E_{n+1} - E_n = 1 - \frac{15}{2}\gamma^2(n+1). \tag{5.42}$$

Equation 5.42 implies that there is a value of n for which the difference in energies is zero. This is a backward way of finding out that our treatment of the perturbing Hamiltonian can hardly be valid under these conditions. Equation 5.42 shows that our perturbative treatment of H_1 must break down when

$$\gamma^2 n \sim 1. \tag{5.43}$$

Why has this happened? At higher energy levels, the system is "sampling" larger q' (position) values. However, for any fixed value of γ in Equation 5.32 there will be values of q' for which $\gamma q'^3 > q'^2/2$ for larger q'. Under these conditions the "perturbation" will in fact be the dominant term in the energy and a perturbative treatment is bound to be inadequate.

5.4 ASPECTS OF DEGENERATE PERTURBATION THEORY

We have left out a large class of problems in deriving the result in Equation 5.19. We have specified that the energy levels of our systems be nondegenerate. Many physical systems have such degeneracies. (The hydrogen atom is one such system we will study in Chapters 7 and 8.) Let us assume that we are trying to solve for the energy levels of a Hamiltonian of the form Equation 5.17, but that there exists a k-fold degeneracy (meaning k degenerate states, but $k=1$ is not degenerate!) of the unperturbed energy levels. In addition to the unshifted energy label, E_0, there will now be another label that will distinguish between these k states. Let us label such a state as $|E_0 a\rangle$ where $a=1, \ldots, k$. Of course the point is to find a representation that diagonalizes the full Hamiltonian, $H=H_0+H_1$, the diagonal elements being the energy eigenvalues. Now, it is reasonable (and justifiable) to assume that the first-order effect on the k members of the unperturbed (degenerate) energy spectrum will just come from those states that are elements of the degenerate subspace; that is, we neglect the effect of any "distant" energy states. If this is so, then it is only necessary to diagonalize the perturbation, H_1, in the degenerate subspace of dimension k. Then, the shifted energy levels will of course be given by the eigenvalue/eigenvector problem

$$\left.\begin{aligned} \det(H_1 - \lambda) &= 0, \\ H_1\psi^{(a)} &= \lambda^{(a)}\psi^{(a)}, \\ E_a &= E_0 + \lambda^{(a)}. \end{aligned}\right\} \tag{5.44}$$

where $a=1, \ldots, k$. Thus, this is just a standard eigenvalue/eigenvector problem, but carried out entirely within the originally k-fold degenerate subspace. Then, if the diagonal elements of this matrix are all distinct, the degeneracy will have been lifted and we will have k distinct energy levels where before there was only one. However, it may happen that not all the diagonal elements are distinct after H_1 is diagonalized, that is, some energy degeneracies remain. To proceed beyond this point in perturbation theory, it is necessary to use second-order degenerate perturbation theory (which takes into account the effects of "distant" states). We will not pursue this subject further here as it occurs rather infrequently.*

5.5 WKB SEMICLASSICAL APPROXIMATION†

Let us now move on to talk about another useful approximation method: the WKB semiclassical approximation. The Schrödinger equation for an arbitrary potential in one spatial dimension is

$$\left[-\frac{\hbar^2}{2m}\frac{d^2}{dx^2}+V(x)\right]u(x)=Eu(x) \tag{5.45}$$

or

$$\left[\frac{d^2}{dx^2}+\frac{2m}{\hbar^2}(E-V(x))\right]u(x)=0. \tag{5.46}$$

When we were solving Equation 5.45 or 5.46 for flat potentials, as in the finite potential barrier problem, we defined the constant (for $E > V_0$, say)

$$k^2=\frac{2m}{\hbar^2}(E-V_0), \tag{5.47}$$

for which the solutions to the Schrödinger equation were

$$u(x)\sim e^{\pm ikx}. \tag{5.48}$$

The wave number, k, is related to the de Broglie wavelength by

$$k=\frac{2\pi}{\lambda}\equiv\frac{1}{\bar{\lambda}}. \tag{5.49}$$

Following this lead, let us define a position-dependent wave number by

$$k^2(x)\equiv\frac{2m}{\hbar^2}(E-V(x)), \tag{5.50}$$

when $(E - V(x)) > 0$. If we think of the potential $V(x)$ in Equation 5.50 as changing sufficiently slowly with x, then we might expect to get solutions of the form

$$u(x)\sim e^{\pm i\int^x dx' k(x')}, \tag{5.51}$$

where the lower limit on the integral is not yet specified. With this $u(x)$ we have‡

$$\frac{1}{i}\frac{d}{dx}u(x)=\pm\, k(x)u(x), \tag{5.52}$$

* See K. Gottfried, and T.-M. Yan. *Quantum Mechanics: Fundamentals*, 2nd ed. (Springer, 2003), Sect. 3.7(b) for a good example along these lines.

† This section is again based on my Schwinger lecture notes.

‡ Here we are using the rule (Leibnitz): $(d/dt)\int_{a(t)}^{b(t)} f(x)dx = f(b(t))(db/dt) - f(a(t))(da/dt)$.

and therefore

$$\frac{d^2}{dx^2}u(x) = -k^2(x)u(x) \pm i\frac{dk(x)}{dk}u(x), \tag{5.53}$$

which is just Equation 5.46 if

$$\frac{1}{k^2(x)}\left|\frac{dk(x)}{dx}\right| \ll 1, \tag{5.54}$$

or, using an obvious definition of $\lambdabar(x)$:

$$\left|\frac{d\lambdabar(x)}{dx}\right| \ll 1. \tag{5.55}$$

In words, Equation 5.55 says that the change in the reduced wavelength because of the varying potential must be small compared to a change in x. This should be the case in many semiclassical applications where the particle energies are large compared to the potential.

To improve upon Equation 5.51, let us define a new slowly varying wave function $\phi(x)$ such that

$$u(x) \equiv e^{\pm i\int^x dx' k(x')}\phi(x). \tag{5.56}$$

We then have

$$\frac{d}{dx}u(x) = e^{\pm i\int^x dx' k(x')}\left[\frac{d}{dx} \pm ik(x)\right]\phi(x), \tag{5.57}$$

and similarly for the second derivative. The Schrödinger equation now takes the form

$$\left[\left(\frac{d}{dx} \pm 2ik(x)\right)\left(\frac{d}{dx}\right) \pm i\frac{dk(x)}{dx}\right]\phi(x) = 0, \tag{5.58}$$

which is still exact. Since $\phi(x)$ is supposedly a slowly varying function of x, let us neglect the second derivative term in Equation 5.58. Then we have

$$\frac{(d\phi/dx)}{\phi} + \frac{(dk/dx)}{2k} = 0. \tag{5.59}$$

Integrating indefinitely gives

$$\phi(x) = \frac{C}{\sqrt{k(x)}}, \tag{5.60}$$

where C is an unspecified constant. Thus, we have approximately that

$$u(x) \simeq \frac{C}{\sqrt{k(x)}}e^{\pm i\int^x dx' k(x')}, \tag{5.61}$$

for $(E - V(x)) > 0$. The plus sign in Equation 5.61 represents a wave traveling to the right and the minus sign a wave traveling to the left.

Now let us think about the case $(E - V(x)) < 0$. We know that classical particles cannot penetrate into such regions because they have insufficient energy. Quantum mechanically, however, we know that non-zero wave functions are allowed and that they are given by real exponentials. We saw that these solutions are given by making the substitution $k \to \pm iK$, where $K = \sqrt{2m(V_0 - E)} / \hbar$. Letting $K(x) = \sqrt{2m(V(x) - E)} / \hbar$ in Equation 5.61, we find the WKB solutions

$$u(x) = \frac{C'}{\sqrt{K(x)}} e^{\pm \int^x dx' K(x')}, \tag{5.62}$$

valid for $(E - V(x)) < 0$.

We now have formulas for slowly varying potentials when $(E-V(x)) \gtrless 0$. However, the basis for the WKB description of wave functions breaks down near classical turning points, that is, positions near where $(E - V(x)) = 0$. Consider a de Broglie wave of energy E interacting with a potential barrier in the situation shown in Figure 5.1. In this situation, we see that the local reduced wavelength, $\lambda\!\!\!^{-}(x)$, becomes infinite at $x = x_2$, and the condition Equation 5.55 is violated. However, we do have approximate solutions far to the left and right of the turning point in Equations 5.61 and 5.62, respectively. It is a question of trying to approximately solve the Schrödinger equation in the vicinity of the turning point and then matching this middle wave function to the WKB solutions on either side. The approximate Schrödinger equation near the turning point is given by assuming an expansion of the potential

$$V(x) \approx E + (x - x_2) \left. \frac{\partial V}{\partial x} \right|_{x=x_2}, \tag{5.63}$$

near $x = x_2$. Then the Schrödinger equation becomes

$$\left[\frac{d^2}{dx^2} + \frac{2m}{\hbar^2} (x - x_2) \left. \frac{\partial V}{\partial x} \right|_{x=x_2} \right] u(x) = 0. \tag{5.64}$$

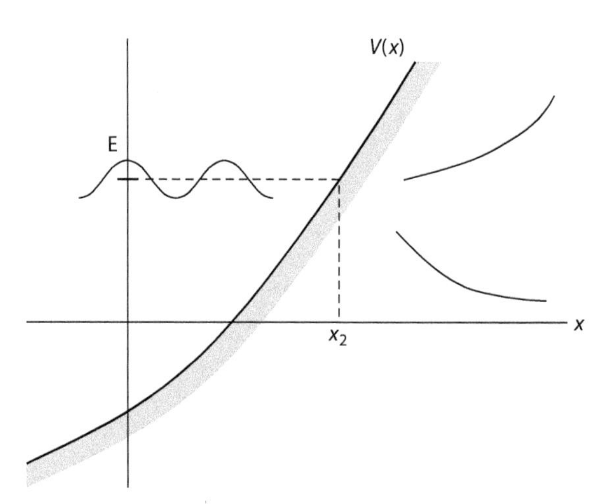

FIGURE 5.1 Illustrating a wave function of energy E interacting with a potential barrier. The classical turning point, x_2, is shown. The part of the potential where $E > V(x)$ is to the left of the turning point.

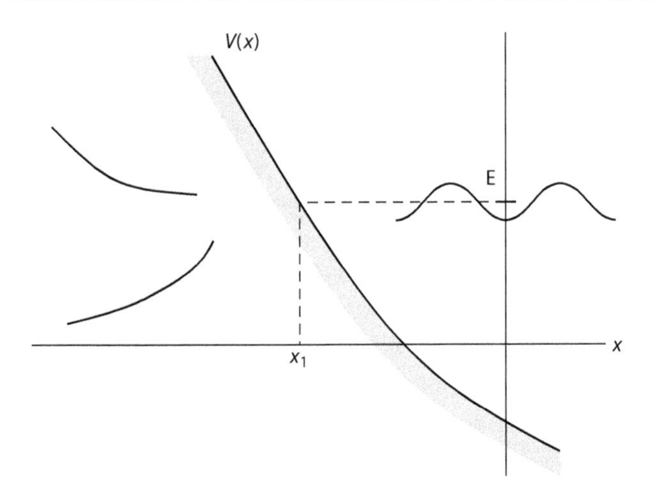

FIGURE 5.2 Same as Figure 5.1, except the part of the potential where $E > V(x)$ is to the right of the turning point, labeled as x_1.

The details of the mathematics that describe the solution of this equation and the matching of wave functions and their first derivatives at the turning points will not be recorded here.* The results of such an analysis tell us that the connections between wave functions in the two regions are given by

$$\underline{x < x_2} \qquad\qquad\qquad \underline{x > x_2}$$

$$\frac{1}{\sqrt{k(x)}} 2\cos\left[\int_{x_2}^{x} dx' k(x') + \frac{\pi}{4}\right] \leftrightarrow \frac{1}{\sqrt{K(x)}} e^{-\int_{x_2}^{x} dx' K(x')}, \tag{5.65}$$

$$\frac{1}{\sqrt{k(x)}} \sin\left[\int_{x_2}^{x} dx' k(x') + \frac{\pi}{4}\right] \leftrightarrow -\frac{1}{\sqrt{K(x)}} e^{\int_{x_2}^{x} dx' K(x')}. \tag{5.66}$$

Likewise, if we have a situation as in Figure 5.2, then we have the connections

$$\underline{x > x_1} \qquad\qquad\qquad \underline{x < x_1}$$

$$\frac{1}{\sqrt{k(x)}} 2\cos\left[\int_{x_1}^{x} dx' k(x') - \frac{\pi}{4}\right] \leftrightarrow \frac{1}{\sqrt{K(x)}} e^{-\int_{x}^{x_1} dx' K(x')}, \tag{5.67}$$

$$-\frac{1}{\sqrt{k(x)}} \sin\left[\int_{x_1}^{x} dx' k(x') - \frac{\pi}{4}\right] \leftrightarrow \frac{1}{\sqrt{K(x)}} e^{\int_{x}^{x_1} dx' K(x')}. \tag{5.68}$$

These sets of equations are called the *WKB connection formulas*. The double arrows mean that a solution of one form in the given spatial region corresponds to the other form in the neighboring region.

---

* See E. Merzbacher, *Quantum Mechanics*, 3rd ed. (John Wiley & Sons, 1998), Chapter 7, for example.

5.6 USE OF THE WKB APPROXIMATION IN BARRIER PENETRATION

There are several interesting applications of these formulas. One of these comes from considering the situation in Figure 5.3. Here we are imagining an incident wave from the left impinging on the potential but having insufficient energy classically to overcome the barrier. However, in quantum mechanics we know that there will be a finite probability that the particle reaches the region to the right of $x = x_2$ because of tunneling. Because of Equations 5.61 and 5.62 the approximate solutions in Regions I and III are as follows:

$$u_{\mathrm{I}}(x) = \frac{A}{\sqrt{k(x)}}\cos\left[\int_{x_1}^{x} dx'k(x') + \frac{\pi}{4}\right] + \frac{B}{\sqrt{k(x)}}\sin\left[\int_{x_1}^{x} dx'k(x') + \frac{\pi}{4}\right], \tag{5.69}$$

$$u_{\mathrm{III}}(x) = \frac{E}{\sqrt{k(x)}}\cos\left[\int_{x_2}^{x} dx'k(x') - \frac{\pi}{4}\right] + \frac{F}{\sqrt{k(x)}}\sin\left[\int_{x_2}^{x} dx'k(x') - \frac{\pi}{4}\right]. \tag{5.70}$$

Our requirement of waves impinging from the left requires $F = iE$, since $u_{\mathrm{III}}(x)$ must be of the form of Equation 5.61 with the upper, positive sign. Now, instead of matching wave functions and their first derivatives at $x = x_1$ and x_2, it is only necessary to use the connection formulas. Again, an overall normalization determines one of the constants, so we can, for example, divide everything through by A to isolate two ratios, E/A and B/A. We must use the connection formulas from Regions I and III to get two expressions for $u_{\mathrm{II}}(x)$. Requiring consistency of these two expressions gives us the ratios

$$\frac{E}{A} = \frac{i}{2\theta} \tag{5.71}$$

and

$$\frac{B}{A} = \frac{-i}{4\theta^2}, \tag{5.72}$$

where

$$\theta \equiv e^{\int_{x_1}^{x_2} dx'K(x')}. \tag{5.73}$$

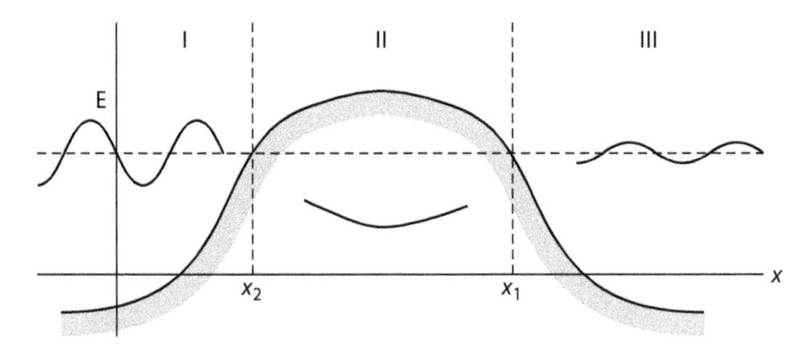

FIGURE 5.3 A wave function of energy *E*, with wave incident from the left, interacting with a finite potential hill.

Now the transmission coefficient is defined as the ratio of the absolute square of the coefficients of the right-traveling waves in Regions III and I. One can show that this means

$$T = \left| \frac{-2i(E/A)}{(1 - i(B/A))} \right|^2. \tag{5.74}$$

Applying Equation 5.74 to the barrier problem of Chapter 3, Section 3.4, we find

$$T \approx e^{-4Ka}, \tag{5.75}$$

where we have neglected the $-1/(4\theta^2)$ term in the denominator of Equation 5.74 for $Ka \gg 1$. Equation 3.123 gives

$$T \approx \left[\frac{4}{1 + \frac{1}{4}\left(\frac{K}{k_1} - \frac{k_1}{K} \right)^2} \right] e^{-4Ka}, \tag{5.76}$$

in the same limit. We see that the WKB method got the exponential part of T correct but has missed the overall constant in front. However, a rectangular potential does not satisfy well the WKB requirement for a slowly varying potential.

5.7 USE OF THE WKB APPROXIMATION IN BOUND STATES

Another application of the WKB method is to get approximate bound-state energies. Let us say that our potential looks like Figure 5.4.

The energy specified corresponds to a bound state. In Region I, we must have an exponentially decreasing wave function:

$$u_{\mathrm{I}}(x) = \frac{C}{\sqrt{k(x)}} e^{-\int_x^{x_1} dx' K(x')}. \tag{5.77}$$

By the connection formula Equation 5.67, the solution in Region II is

$$u_{\mathrm{II}}(x) = \frac{C}{\sqrt{k(x)}} 2\cos\left[\int_{x_1}^{x} dx' k(x') - \frac{\pi}{4} \right]. \tag{5.78}$$

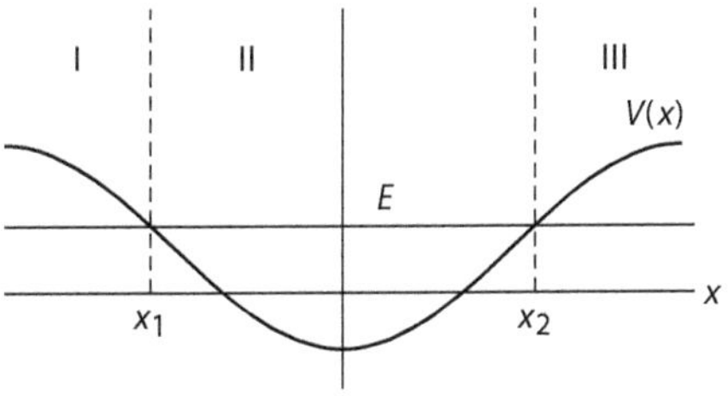

FIGURE 5.4 The two turning points, x_1 and x_2, for a particle of energy E in a finite potential well with possible bound states.

On the other hand, we must also have a decreasing exponential solution in Region III:

$$u_{\text{III}}(x) = \frac{C'}{\sqrt{k(x)}} e^{-\int_{x2}^{x} dx' K(x')}. \tag{5.79}$$

Using the connection formula Equation 5.65, this then gives

$$u_{\text{II}}(x) = \frac{C'}{\sqrt{k(x)}} 2\cos\left[\int_{x2}^{x} dx' k(x') + \frac{\pi}{4}\right]. \tag{5.80}$$

We see that Equations 5.78 and 5.80 are only compatible when

$$\int_{x1}^{x2} dx' k(x') = \left(n + \frac{1}{2}\right)\pi \tag{5.81}$$

and

$$\frac{C}{C'} = (-1)^n, \tag{5.82}$$

where $n = 0, 1, 2, \ldots$. Equation 5.81 determines approximately the bound-state energies of the system.

When one turning point intercepts an infinite step-function wall, as in Figure 5.5, one may show by joining approximate solutions in Regions I and II, that

$$\int_{x1}^{x2} dx' k(x') = \left(n + \frac{3}{4}\right)\pi. \quad (n = 0, 1, 2, \ldots). \tag{5.83}$$

In addition, for the case where both turning points intercept infinite step-function walls, as in Figure 5.6, we have

$$\int_{x1}^{x2} dx' k(x') = (n+1)\pi. \quad (n = 0, 1, 2, \ldots). \tag{5.84}$$

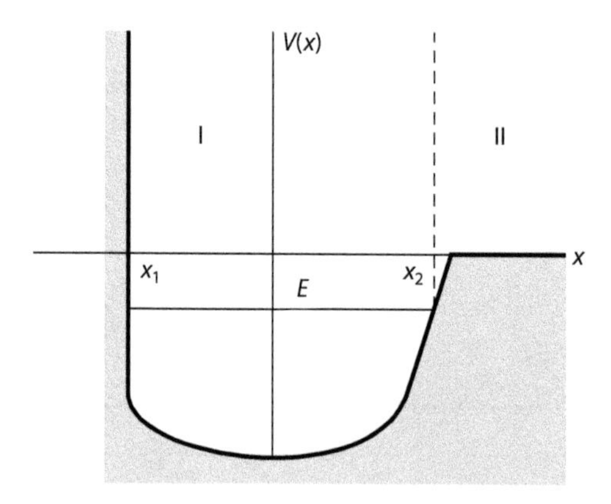

FIGURE 5.5 The two turning points, x_1 and x_2, for a bound particle of energy E in a potential with one infinite step wall. One turning point is at the position of the wall.

Let us do an example using Equation 5.83. Consider the one-dimensional potential, $V(x)$, of Figure 5.7, where

$$V(x) = \begin{cases} +\infty, & x < 0 \\ -\dfrac{\alpha}{x}, & x > 0. \end{cases} \tag{5.85}$$

We will examine the $E < 0$ case for bound states. We have

$$k^2(x) = \frac{2m}{\hbar^2}\left(E + \frac{\alpha}{x}\right), \tag{5.86}$$

and the turning points for the integral are as follows:

$$x_1 = 0, \quad x_2 = -\frac{\alpha}{E}. \tag{5.87}$$

Equation 5.83 then gives

$$\sqrt{\frac{2m|E|}{\hbar^2}} \int_0^{-\alpha/E} \mathrm{d}x \sqrt{\frac{\alpha}{|E|x} - 1} = \left(n + \frac{3}{4}\right)\pi. \tag{5.88}$$

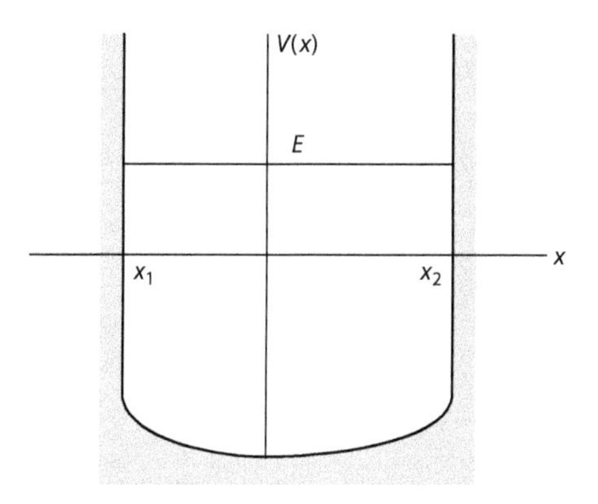

FIGURE 5.6 The two turning points, x_1 and x_2, for a bound particle of energy E in a potential with two infinite step walls. Both turning points are at the position of the wall.

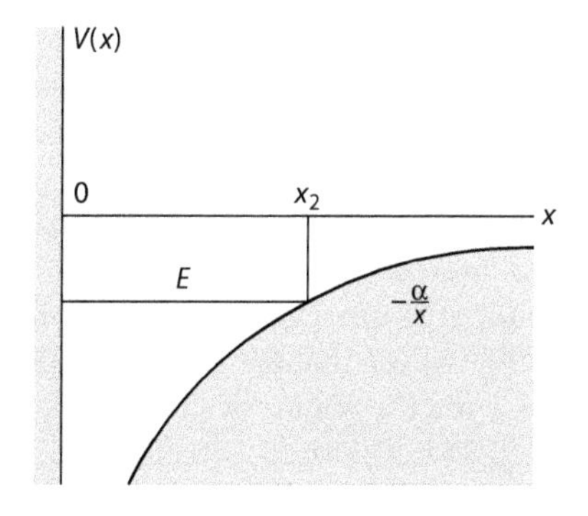

FIGURE 5.7 The potential of Equation 5.85. One turning point is on the vertical wall at $x = 0$, the other is at x_2.

Define $y \equiv (|E|x)/\alpha$. This now gives

$$\int_0^1 dy \sqrt{\frac{1-y}{y}} = \frac{\pi\hbar\left(n+\frac{3}{4}\right)}{\alpha}\sqrt{\frac{|E|}{2m}}. \tag{5.89}$$

The dimensionless integral on the left is

$$\int_0^1 dy \sqrt{\frac{1-y}{y}} = \tan^{-1}\left(\sqrt{\frac{y}{1-y}}\right)\Bigg|_0^1 = \frac{\pi}{2}. \tag{5.90}$$

Therefore, from Equation 5.89,

$$E^{\text{WKB}} = -\frac{m\alpha^2}{2\hbar^2\left(n+\frac{3}{4}\right)^2}, \tag{5.91}$$

where $n = 0, 1, 2, \ldots$. The exact solution for the potential of Equation 5.85 is available from the three-dimensional Coulomb solution we will examine in Chapter 7 (Equation 7.245), which has the same radial differential equation and boundary conditions as the one-dimensional problem for $\ell = 0$. (ℓ is an angular momentum quantum number.) In this case, the solution reads

$$E^{\text{exact}} = -\frac{m\alpha^2}{2\hbar^2(n+1)^2}, \tag{5.92}$$

again for $n = 0, 1, 2, \ldots$. The ratio is

$$\frac{E^{\text{WKB}}}{E^{\text{exact}}} = \left(\frac{n+1}{n+\frac{3}{4}}\right)^2, \tag{5.93}$$

which slightly overestimates the bounding energy, but improves quickly as $n \gg 1$.

A change of variables can sometimes yield an improved WKB approximation for problems such as the above. If one can find a new coordinate variable that maps the $x > 0$ region into a new, unrestricted range, the condition Equation 5.83 can be replaced by Equation 5.84. One would then expect the WKB method applied to the new, smoother "potential" to be more accurate. This technique, applied to the radial Schrödinger equation, does indeed yield better results.*

What about applications to potentials where a sharp, but not infinite, step potential is encountered on one side or other of a potential well? Technically, one should still use Equation 5.81 to find the energy levels. However, we should expect that for deeply bound states (this would require $-mV_0a^2/\hbar^2 \gg 1$ for the square well) that the lower levels would obey Equation 5.83 or 5.84 better than Equation 5.81.

---

* See J. Schwinger, *Quantum Mechanics: Symbolism of Atomic Measurements*, edited by B.-G. Englert (Springer, 2001), Problems 8–9a.

5.8 VARIATIONAL METHODS

Consider a system with a complete, nondegenerate energy spectrum, $|n\rangle$, labeled as the subscript of the energy E_n, with a spectrum $E_0 < E_1 < E_2 \ldots$, where E_0 is the ground state. Form the energy functional (H is the Hamiltonian)

$$\mathcal{E}_0 = \frac{\langle\psi|H|\psi\rangle}{\langle\psi|\psi\rangle}, \tag{5.94}$$

from a state, $|\psi\rangle$, which satisfies the same boundary conditions as the ground-state wave function. Since the $|n\rangle$ are complete, we may write

$$|\psi\rangle = \sum_{n=0}^{\infty} C_n |n\rangle, \tag{5.95}$$

where the C_n are constants. Then using this in Equation 5.94 gives

$$\mathcal{E}_0 = \frac{\sum_{n=0}^{\infty} |C_n|^2 (E_n - E_0)}{\sum_{n=0}^{\infty} |C_n|^2} + E_0 \geq E_0. \tag{5.96}$$

Technically, $\mathcal{E}_0$ is just an upper limit to the true energy but can be regarded as an estimate of the true E_0. The variational technique now consists of using a trial wave function with adjustable parameters, $|\psi(a_i)\rangle$, for $i = 1, 2, \ldots, n$, and solving the n equations,

$$\frac{\partial \mathcal{E}_0(a_i)}{\partial a_i} = 0, \quad i = 1, \ldots, n, \tag{5.97}$$

to produce a minimum in the estimate. In carrying this out, one may use a normalized wave function, $\langle\psi(a_i)|\psi(a_i)\rangle = 1$, or introduce a functional with a "Lagrange multiplier," λ (also used in Chapter 9, Section 9.11):

$$\mathcal{E}_0^*(a_i) \equiv \langle\psi(a_i)|H|\psi(a_i)\rangle - \lambda(\langle\psi(a_i)|\psi(a_i)\rangle - 1). \tag{5.98}$$

The solution now involves $n + 1$ equations,

$$\frac{\partial \mathcal{E}_0^*(a_i)}{\partial a_i} = 0, \quad \langle\psi(a_i)|\psi(a_i)\rangle = 1. \tag{5.99}$$

Lagrange multiplier methods are described in many physics and mathematics books that discuss variational methods.* The general rule is: multiply the constraint equation by a Lagrange multiplier, subtract from the original functional, and vary the new quantity as if there were no constraint, treating the constraint as a side condition. The Lagrange

---

* See, for example, Chapter 6 of my Open Text Project book, *Modern Introductory Mechanics* (Bookboon.com).

multiplier in this case has the value $\lambda = (\mathcal{E}_0(a_i))_{\min}$, so its value is just the energy we are looking for! (See Problem 5.8.1.)

There is definitely an art to choosing good wave functions! A guide to choosing good forms is to try to imitate large and small x behaviors from the appropriate Schrödinger equation solutions. For example, the harmonic oscillator equation reads

$$\left(-\frac{\hbar^2 d^2}{2m\,dx^2} - \frac{1}{2}m\omega^2 x^2\right)\psi(x) = E\psi(x). \tag{5.100}$$

One would expect, for even solutions that $\psi_e(x) \sim$ constant near the origin, and approximate normalizable exponential behavior, $\psi_e(x) \sim \exp(-m^2\omega^2x^2/2)$ at large distances. The natural ansatz is thus $\psi_e(x) = A\exp(-m^2\omega^2x^2/2)$ (A a constant), which is exact for the even parity ground state. Likewise, one might guess $\psi_o(x) = A'\,x\exp(-m^2\omega^2x^2/2)$ (A' another constant) for the lowest energy odd parity state, which would also result in an exact estimate. Note that one can use the variational method on even and odd parity ground states separately since they have different boundary conditions. (The odd parity states have $\psi_o(0) = 0$.)

I will give two examples of the use of the variational method. Exact solutions are available for both, so that I can illustrate the accuracy of the method. I will use examples that will shed light on the relative accuracy of perturbative corrections (Sections 5.2 and 5.3) and the WKB estimates (Section 5.7) for ground states.

The first example is the *asymmetric harmonic oscillator*,

$$H = \frac{1}{2}(p^2 + q^2) + \mathbb{E}q \tag{5.101}$$

I am using the dimensionless operators, p and q, of Equation 3.133 (Chapter 3), and have introduced an extra term, $\mathbb{E}q$, where the constant $\mathbb{E}$ can be thought of as an external electric field. The energy spectrum (see Problem 5.2.5) is given exactly by

$$\mathcal{E}_n = \frac{E_n}{\hbar\omega} = \left(n + \frac{1}{2}\right) - \frac{\mathbb{E}^2}{2}. \tag{5.102}$$

Let $|n\rangle$ represent the usual harmonic oscillator wave functions (Section 3.5). The result of first-order perturbation theory for the new, modified ground-state wave function, $|\tilde{0}\rangle$, is (see Equation 5.20)

$$|\tilde{0}\rangle \approx |0\rangle + \sum_{n' \neq 0}^{\infty} |n'\rangle \frac{\langle n'|\mathbb{E}q|0\rangle}{(E_0 - E_{n'})} \tag{5.103}$$

Using Equation 5.34, only the $n' = 1$ term will contribute and we have

$$|\tilde{0}\rangle \approx |0\rangle - \frac{\mathbb{E}}{\sqrt{2}}|1\rangle. \tag{5.104}$$

These perturbative results are supposed to hold for $|\mathbb{E}| \ll 1$.

Based upon this result, let us simply assume that

$$|\psi(a)\rangle = |0\rangle + a|1\rangle, \tag{5.105}$$

for the variational wave function, where a is a parameter we can vary. I choose to vary the normalized energy functional Equation 5.94, for which I find

$$\mathcal{E}_0(a) = \frac{\frac{1}{2} + \frac{3}{2}|a|^2 + \sqrt{2}\mathbb{E}\,\mathrm{Re}\{a\}}{1+|a|^2}. \tag{5.106}$$

Varying both Re{a} and Im{a} independently, continuity when $\mathbb{E} = 0$ now requires Im{a} = 0. Designating $a_r \equiv \mathrm{Re}\{a\}$, then

$$\frac{\partial \mathcal{E}_0(a_r)}{\partial a_r} = 0 \Rightarrow 2a_r - \sqrt{2}\mathbb{E}a_r^2 + \sqrt{2}\mathbb{E} = 0, \tag{5.107}$$

$$\Rightarrow a_r = \frac{1}{\sqrt{2}\mathbb{E}}\left(1 \pm \sqrt{1+2\mathbb{E}^2}\right). \tag{5.108}$$

In order to connect to the correct solution for $\mathbb{E} = 0$, we must choose the negative sign in Equation 5.108. This gives

$$\mathcal{E}_0^V = \frac{3 - \sqrt{1+2\mathbb{E}^2} + \frac{3}{2\mathbb{E}^2}(1-\sqrt{1+2\mathbb{E}^2})}{2 + \frac{1}{\mathbb{E}^2}(1-\sqrt{1+2\mathbb{E}^2})}. \tag{5.109}$$

I examine the numerical accuracy of Equation 5.109 by forming the ground-state energy *shift* ratio $R(\mathcal{E}) \equiv (\mathbb{E}_0^V - 1/2)/(\mathbb{E}_0 - 1/2)$, where $\mathcal{E}_0 = 1/2 - \mathbb{E}^2/2$. The result is presented graphically in Figure 5.8.

Note that this ratio is less than one, which, because the energy shift is negative, is indeed an upper limit on the total energy. In addition, for $\mathbb{E} \ll 1$, the variational and exact results agree quite well, as guaranteed by our perturbative-type approach. However, even for $\mathbb{E} = 1/2$, which gives an energy shift of one-fourth of the original ground-state energy, although the estimate in the energy shift is 10.1% too small, the actual energy value is only 3.4% high. To get a better model, we would have to mix other states

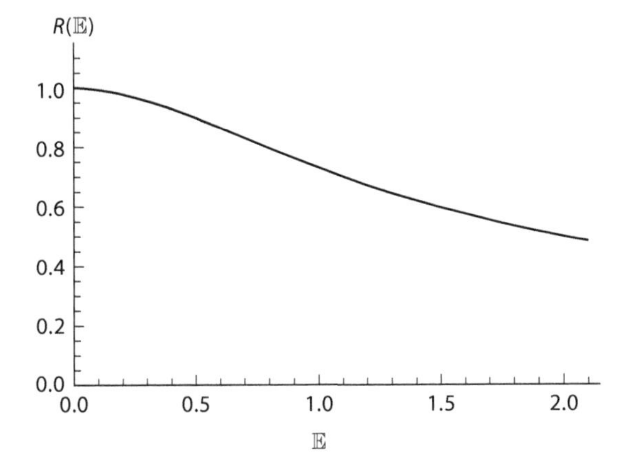

FIGURE 5.8 The energy shift ratio, R defined in the text, for the asymmetric harmonic oscillator.

besides the first excited state in Equation 5.105. (Can you figure out the exact form of the new wave functions? See Problem 5.2.5 again.)

Another interesting application of the variational method sheds light on the WKB energy estimation results for the ground state of the "bouncing ball" potential,

$$V(x) = \begin{cases} +\infty, & x < 0 \\ mgx, & x > 0' \end{cases} \tag{5.110}$$

where g is the acceleration due to gravity. Here, it is a little harder to model the correct large-x behavior,* and to be consistent with the boundary conditions at the origin and spatial infinity, I will simply assume

$$\psi(x) = Axe^{-\beta x}. \tag{5.111}$$

All the necessary integrals may be evaluated from

$$\int_0^\infty dx\, e^{-2\beta x} = \frac{1}{2\beta}, \tag{5.112}$$

and derivatives with respect to β. Normalization yields

$$|A| = \sqrt{4\beta^3}. \tag{5.113}$$

I find

$$\mathcal{E}_0(\beta) = \frac{\hbar^2}{2m}\beta^2 + \frac{3mg}{2\beta}, \tag{5.114}$$

$$\frac{\partial \mathcal{E}_0(\beta)}{\partial \beta} = 0 \Rightarrow \beta = \left(\frac{3m^2 g}{2\hbar^2}\right)^{1/3}. \tag{5.115}$$

The final estimate for the variational energy is then given as

$$\mathcal{E}_0(\beta) = \underbrace{\frac{3^{5/3}}{2^{4/3}}}_{\approx 2.476} \left(\frac{mg^2\hbar^2}{2}\right)^{1/3}. \tag{5.116}$$

The estimate of 2.476 (error of 5.9%) should be compared to the one from the WKB method in the book by Sakurai and Napolitano† of 2.320 (error of −0.8%). The energies are proportional to the zeros of the nonsingular Airy function, Ai(x), and the true proportionality constant above is approximately 2.338 to four decimal places. Although the variational estimate here is clearly not as accurate as the WKB one for a comparable amount of work, the advantage is that it can be improved with the addition of a function with additional parameters in it or a better initial guess.

---

* Ignoring all nonexponential behavior, one actually expects $\psi(x) \sim \exp(-(2m^2 g/\hbar^2)^{1/2} x^{3/2})$; see Problem 5.8.2.

† J.J. Sakurai and J. Napolitano, *Modern Quantum Mechanics*, 2nd ed. (Addison-Wesley, 2011), p. 115.

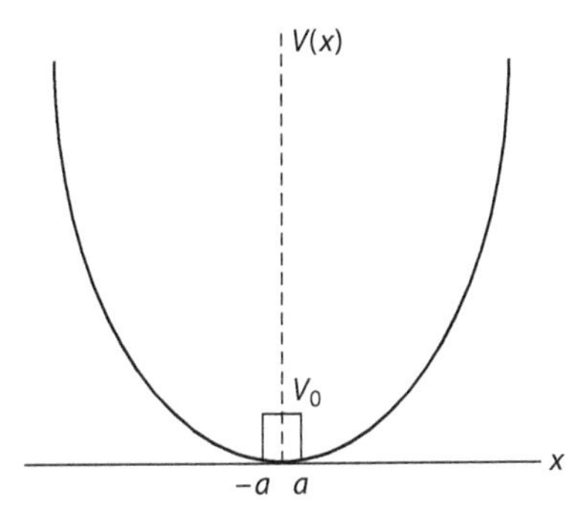

FIGURE 5.9 Harmonic oscillator potential with a small, flat bump centered at $x = 0$.

The variational method is often used in circumstances where previous methods have already given good estimates, but additional accuracy is still desired. The classic example of this is the ground state of helium, in which the central nucleon electric charge is changed into a parameter, which can be thought of as representing the shielding of the proton's charge by one of the electrons.*

PROBLEMS

5.2.1 Given the harmonic oscillator Hamiltonian,

$$H = \frac{p_x^2}{2m} + \frac{1}{2}m\omega^2 x^2,$$

and energies,

$$E_n = \hbar\omega\left(n + \frac{1}{2}\right), \quad n = 0, 1, 2, \ldots$$

(a) Evaluate $\langle x^2 \rangle_n$ using Equation 5.3. [Hint: Carefully choose the parameter λ to be one of the constants in H.]

(b) Now evaluate $\langle p_x^2 \rangle_n$.

(c) Show the sum rule

$$\langle x^2 \rangle_n = 2\left(\frac{m\omega}{\hbar}\right)\sum_{n' \neq n} \frac{|\langle n|\, x^2 \,|n'\rangle|^2}{n' - n}.$$

5.2.2 Prove that the constant $C(\lambda)$ in Equation 5.4 is *real*.

5.2.3 You are given a harmonic oscillator potential with a small bump at its bottom, as in Figure 5.9.

$$H = \frac{p_x^2}{2m} + \frac{1}{2}m\omega^2 x^2 + V(x),$$

$$V(x) = \begin{cases} V_0, & |x| \leq a \\ 0, & \text{elsewhere.} \end{cases}$$

* Sakurai and Napolitano, *Modern Quantum Mechanics*. p. 455.

Make the simplifying assumption that $a^2 \ll \hbar/m\omega$ in this problem. The *unperturbed* energy eigenfunctions for this problem are (see Chapter 3, Equation 3.212)

$$u_n(x) = \left(\frac{m\omega}{\hbar}\right)^{1/4} \frac{1}{\sqrt{\sqrt{\pi}2^n n!}} e^{-m\omega x^2/2\hbar} H_n\left(\sqrt{\frac{m\omega}{\hbar}}x\right),$$

and the *unperturbed* energies are as follows:

$$E_n = \hbar\omega\left(n + \frac{1}{2}\right), \quad n = 0,1,2,....$$

(a) Show that the first-order effect of this perturbation on the lowest energy state is

$$E'_{n=0} \simeq \frac{\hbar\omega}{2} + 2aV_0\sqrt{\frac{m\omega}{\pi\hbar}}.$$

(b) Show that the first-order effect of this perturbation on the expectation values of x^2 and p_x^2 in the ground state is given by

$$\langle x^2 \rangle_{n=0} \simeq \frac{\hbar}{2m\omega} + \frac{aV_0}{\sqrt{\pi\hbar m\omega^3}},$$

$$\langle p_x^2 \rangle_{n=0} \simeq \frac{\hbar m\omega}{2} - aV_0\sqrt{\frac{m^3\omega}{\pi\hbar}}.$$

[Notice, we do *not* know what the new *eigenstates* are. Consider derivatives of $E'_{n=0}$ using Equation 5.3.]

5.2.4 Consider adding to the one-dimensional Hamiltonian,

$$H_0 = \frac{p_x^2}{2m} + V(x),$$

where the potential $V(x)$ is an even function of x ($V(x) = V(-x)$), the perturbation (γ = constant)

$$H_1 = \gamma x.$$

(Physically, this could represent adding an electric field.)

(a) On the basis of some physical property, argue that the new energies, to lowest order in γ, must be given by

$$E_n = E_n^{(0)} + \alpha_n \gamma^2,$$

where the α_n are constants and $E_n^{(0)}$ are the unperturbed energies. (Notice there is no *linear* term $\sim\gamma$ in E_n.)

(b) Show that the constants α_n are given by the sum

$$\alpha_n = \sum_{n' \neq n} \frac{|\langle n|x|n'\rangle|^2}{E_n^{(0)} - E_{n'}^{(0)}}.$$

5.2.5 (a) Evaluate α_n from Problem 5.2.4 for the *ground state* $(n=0)$ of a harmonic oscillator. The energies are $E_n^{(0)} = \hbar\omega\left(n+(1/2)\right)$. [Hint: Remember that (see Chapter 3, Section 3.5)

$$x = \left(\frac{\hbar}{m\omega}\right)^{1/2} \frac{A + A^\dagger}{\sqrt{2}},$$

where

$$A\,|\,n\rangle = \sqrt{n}|n-1\rangle,$$

$$A^\dagger\,|\,n\rangle = \sqrt{n+1}\,|\,n+1\rangle.$$

(b) Show that the (a) result is actually *exact* for all n! [Hint: Consider a shift in coordinates, $x \to x$ + constant in the Hamiltonian.]

5.2.6 The eigenvalue equation for a system with a discrete, nondegenerate and complete set of states may be written formally as (Dirac notation; assume $\langle\psi_n|\psi_n\rangle = 1$)

$$H_0|\psi_n\rangle = E_n|\psi_n\rangle,$$

where H_0 is the unperturbed Hamiltonian, and the old eigenenergies are given by E_n. A small perturbation, H_1, is added to H_0, $H = H_0 + H_1$. Show that the new eigenvectors, $|\Psi_n\rangle$, are given approximately by Equation 5.20:

$$|\,\Psi_n\rangle \simeq |\,\psi_n\rangle + \sum_{n'\neq n} |\,\psi_{n'}\rangle \frac{\langle\psi_{n'}\,|\,H_1\,|\,\psi_n\rangle}{(E_n - E_{n'})}.$$

[Hint: One way is to use Equation 5.11 and a Taylor series expansion. There are many others.]

***5.2.7** Apply Equation 5.20 to reconfirm the results in Problem 5.2.3(b).

***5.3.1** (a) Find the exact eigenvalues for the matrix problem of Section 5.3. Show that for $|a| \ll |E_2 - E_1|$ the results agree with Equations 5.28 and 5.29, derived from perturbation theory.

(b) Find the exact eigenvectors of this same problem. Show that for $|a| \ll |E_2 - E_1|$ the results agree with those derived from the perturbative expression, Equation 5.20.

5.4.1 Consider again the two-state matrix problem but with degenerate diagonal elements, $E_1 = E_2 = E$:

$$H = \begin{pmatrix} E & a \\ a^* & E \end{pmatrix}.$$

Find the eigenvalues and eigenvectors.

5.5.1 In deriving Equation 5.60, I dropped a second derivative term from Equation 5.58. Use the method of successive approximations to find a more accurate expression for $\phi(x)$ that takes this term into account. [The "method of successive approximations" means to estimate the unknown term from the previous solution. Answer:

$$\phi(x) \simeq \frac{C}{\sqrt{k(x)}} \exp\left(\mp \frac{i}{2} \int^x dx' \left[\frac{1}{(k(x'))^{1/2}} \frac{d^2}{dx'^2} \frac{1}{(k(x'))^{1/2}} \right]\right).\Bigg]$$

5.6.1 Show Equations 5.71 and 5.72.

5.7.1 Apply the WKB semiclassical method to approximate the energy levels of the harmonic oscillator. That is, apply the estimate (x_1 and x_2 are the turning points; $k(x) = \sqrt{2m(E - V(x))} / \hbar$)

$$\int_{x_1}^{x_2} dx' k(x') = \left(n + \frac{1}{2}\right)\pi, \quad n = 0,1,2,3,...$$

to the Hamiltonian

$$H = \frac{p_x^2}{2m} + \frac{1}{2} m\omega^2 x^2.$$

Is your result surprising?

5.7.2 Show Equations 5.81 and 5.82 follow from Equations 5.78 and 5.80. [Hint: Try to write the arguments of the cosines the same except for a "leftover" part, then use double-angle formulas for the cosine.]

5.7.3 Verify Equation 5.83.

5.7.4 Verify that Equation 5.84 gives the correct energy levels of an infinite square well (Chapter 3, Sections 3.2 and 3.3).

5.8.1 Show that the Lagrange multiplier, λ, in Equation 5.98 represents $(\mathcal{E}_0(a_i))_{\text{min}}$, the minimum value of the energy functional, Equation 5.94.

***5.8.2** Use the trial wave function

$$\psi(x) = Cx \exp\left(-\alpha \left(\frac{2m^2 g}{\hbar^2}\right)^{1/2} x^{3/2}\right),$$

where C and α are parameters, to estimate the ground-state energy of the bouncing ball potential of Equation 5.110. The following formula* for the necessary integrals,

* I.S. Gradshteyn and I.M. Ryzhik, *Table of Integrals, Series, and Products* (Academic Press, 1965), #3.478(1).

$$\int_0^\infty dx\, x^{\nu-1}\exp(-\mu x^p) = \frac{1}{|p|}\mu^{-(\nu/p)}\Gamma\left(\frac{\nu}{p}\right), \quad [\mathrm{Re}(\mu)>0, \mathrm{Re}(\nu)>0],$$

follows from the defining equation integral for $\Gamma(z)$, the famous gamma function.* After a great churning of integrals and a single variational derivative with respect to α (C cancels out), I find

$$(\mathcal{E}_0)_{\min} = \left[\frac{3}{2^{2/3}}(\gamma\delta^2)^{1/3}\right]\left(\frac{m\hbar^2 g^2}{2}\right)^{1/3},$$

where

$$\gamma = \frac{15}{8}\frac{\Gamma\left(\frac{5}{3}\right)}{\Gamma(2)} - \frac{9}{16}\frac{\Gamma\left(\frac{8}{3}\right)}{\Gamma(2)}, \quad \delta = \frac{\Gamma\left(\frac{8}{3}\right)}{\Gamma(2)}.$$

Numerically,

$$(\mathcal{E}_0)_{\min} = [2.34723]\left(\frac{m\hbar^2 g^2}{2}\right)^{1/3},$$

which is about 0.38% higher compared to the exact answer.

5.8.3 Consider a Hamiltonian,

$$H = H_0 + \lambda H_1.$$

We are given

$$H_0|\psi_0\rangle = E_0|\psi_0\rangle,$$

where E_0 and $|\psi_0\rangle$ are the nondegenerate ground-state energy and eigenket of H_0. Assuming $\langle\psi_0|H_1|\psi_0\rangle = 0$, argue that

$$E \le E_0,$$

from a variational point of view, where E is the true ground-state energy of H. This means that the energy is always lowered by the addition of H_1. This is the same conclusion as in Section 5.2, but this current argument does not use perturbation theory. (Note that after establishing an initial downward trajectory of the ground-state energy for λ small, Equation 5.14 guarantees this downward behavior will continue. Often, H_1 can be taken to represent an electric field. I would like to thank Byron Jennings and Richard Woloshyn for discussions leading to these conclusions.)

* The gamma function is a famous classical mathematical function explained, for example, in M. Abramowitz and I. Stegun, eds., *Handbook of Mathematical Functions*, (U.S. Department of Commerce, 1972), Section 6.

Generalization to Three Dimensions

Synopsis: Direct product spaces are encountered in the generalization to higher dimensions, and the connection to wave function separability is discussed. The spherical coordinate basis is introduced in preparation for discussing spherically symmetric potentials. Angular momentum operators and properties are examined, and a complete set of compatible observables is determined. The angular momentum eigenvalues and eigenstates are determined using ladder operators, and the connection to spherical harmonics is made. The stage is set for understanding the Schrödinger radial equation.

6.1 CARTESIAN BASIS STATES AND WAVE FUNCTIONS IN THREE DIMENSIONS

Our first step is to generalize some obvious results to 3-D. A fundamental starting point is to postulate that

$$\vec{x}\,|\vec{x}'\rangle = \vec{x}'\,|\vec{x}'\rangle, \quad \vec{p}\,|\vec{p}'\rangle = \vec{p}'\,|\vec{p}'\rangle. \tag{6.1}$$

The *Cartesian* bases (we will also encounter a *spherical* basis) $|\vec{x}'\rangle$ and $|\vec{p}'\rangle$ are direct products of the basis states in the three orthogonal directions: (One sometimes writes $|\vec{x}'\rangle = |x_1'\rangle \otimes |x_2'\rangle \otimes |x_3'\rangle$.)

$$\left.\begin{aligned} |\vec{x}'\rangle &= |x_1'\rangle\,|x_2'\rangle\,|x_3'\rangle, & |\vec{p}'\rangle &= |p_1'\rangle\,|p_2'\rangle\,|p_3'\rangle \\ &\equiv |\,x_1', x_2', x_3'\,\rangle & &\equiv |\,p_1', p_2', p_3'\,\rangle \end{aligned}\right\}. \tag{6.2}$$

(I will try to consistently label the three orthogonal space directions as 1, 2, and 3 rather than as x, y, and z from now on.) We also have (see Chapter 2, Equation 2.162)

$$\langle \vec{x}'\,|\,\vec{p}'\rangle = \frac{1}{(2\pi\hbar)^{3/2}}\,e^{i\vec{x}'\cdot\vec{p}'/\hbar}, \tag{6.3}$$

$$\langle \vec{x}'\,|\,\vec{x}''\rangle = \delta^3(\vec{x}' - \vec{x}''), \quad \langle \vec{p}'\,|\,\vec{p}''\rangle = \delta^3(\vec{p}' - \vec{p}''), \tag{6.4}$$

where

$$\begin{aligned} \delta^3(\vec{x}' - \vec{x}'') &\equiv \delta(x_1' - x_1'')\,\delta(x_2' - x_2'')\,\delta(x_3' - x_3''), \\ \delta^3(\vec{p}' - \vec{p}'') &\equiv \delta(p_1' - p_1'')\,\delta(p_2' - p_2'')\,\delta(p_3' - p_3''). \end{aligned} \tag{6.5}$$

Also

$$1=\int d^3x'\,|\vec{x}'\rangle\langle\vec{x}'|,\quad 1=\int d^3p'\,|\vec{p}'\rangle\langle\vec{p}'|, \tag{6.6}$$

with

$$d^3x' = dx_1'\,dx_2'\,dx_3',\quad d^3p' = dp_1'\,dp_2'\,dp_3'\,. \tag{6.7}$$

The formal energy eigenvalue problem is still stated as

$$H\,|a'\rangle = E_{a'}\,|a'\rangle, \tag{6.8}$$

where the $|a'\rangle$ are a complete, orthogonal set of states:

$$\sum_{a'}|a'\rangle\langle a'|=1,\quad \langle a'|a''\rangle=\delta_{a'a''}. \tag{6.9}$$

Equation 6.9 assumes that the energy eigenvalues are discrete and non-degenerate. Wave functions are given by the projections (see Chapter 2, Equation 2.178)

$$u_{a'}(\vec{x}') = \langle\vec{x}'|a'\rangle, \tag{6.10}$$

which satisfy (using Equation 6.6)

$$\int d^3x\, u_{a'}^*(\vec{x})u_{a'}(\vec{x}) = 1. \tag{6.11}$$

Equation 6.11 tells us the engineering dimensions of the $u_{a'}(\vec{x})$ are

$$[u_{a'}(\vec{x})] \sim [\text{length}]^{-3/2}. \tag{6.12}$$

For continuous spectra, we usually use a momentum rather than an energy basis to completely specify the state of the particle. Then defining*

$$u_{\vec{p}}(\vec{x}') \equiv \langle\vec{x}'|\vec{p}\rangle, \tag{6.13}$$

we have (again from Equation 6.6)

$$\int d^3x\, u_{\vec{p}}^*(\vec{x})u_{\vec{p}'}(\vec{x}) = \delta^3(\vec{p}-\vec{p}'). \tag{6.14}$$

We will limit ourselves to consideration of Hamiltonians of the form

$$H = \frac{\vec{p}^{\,2}}{2m} + V(\vec{x}), \tag{6.15}$$

for which the three-dimensional time-independent Schrödinger equation reads

$$\left[-\frac{\hbar^2}{2m}\vec{\nabla}^2 + V(\vec{x})\right]u(\vec{x}) = Eu(\vec{x}). \tag{6.16}$$

The form of the potential, $V(\vec{x})$, will determine the nature of the spatial basis to be used. For example, for $V(\vec{x})=F(x_1)+F(x_2)+F(x_3)$ (as for the 3-D harmonic oscillator) we would use a Cartesian basis; for $V(\vec{x})=V(r)$, where $r=|\vec{x}|$, one would use a spherical basis. We must

* Many books use $u_{\vec{k}}(\vec{x})=\hbar^{3/2}u_{\vec{p}}(\vec{x})$ as the momentum eigenfunction, in which case it is dimensionless. I will switch to the dimensionless form in Chapter 11.

use a basis in which the time-independent Schrödinger equation separates; for example, $u_{a'}(\vec{x}) = u_1(x_1)u_2(x_2)u_3(x_3)$ in Cartesian coordinates or $u_{a'}(\vec{x}) = u_1(r)u_2(\theta)u_3(\phi)$ in spherical coordinates. The time-independent Schrödinger equation is, of course, just Equation 6.8 projected into an explicit basis.

We will continue to assume that

$$[x_i, p_i] = i\hbar, \tag{6.17}$$

for *each* $i = 1, 2, 3$, where x_i and p_i are *operators*. However, what about $[x_i, x_j]$ for $i \neq j$? In the Cartesian basis

$$[x_1, x_2]|\vec{x}'\rangle = (x_1x_2 - x_2x_1)|x_1', x_2', x_3'\rangle = 0. \tag{6.18}$$

Since Equation 6.18 is true for any $|\vec{x}'\rangle$,

$$\Rightarrow [x_1, x_2] = 0. \tag{6.19}$$

Similarly, $[x_2, x_3] = [x_1, x_3] = 0$. Therefore,

$$[x_i, x_j] = 0, \tag{6.20}$$

for all i, j. We learned earlier in Chapter 4 that $[A, B]$ is a measure of the "compatibility" of the operators A and B. Equation 6.19 tells us that measurements of x_i do not limit the precision of measurements of x_j $(i \neq j)$ for a particle. Thus the x_i are simultaneously measurable for all states.

6.2 POSITION/MOMENTUM EIGENKET GENERALIZATION

Let us generalize the unitary displacement operator we had before, also in Chapter 4:

$$\langle x_1'|e^{ix_1'' p_1/\hbar} = \langle x_1' + x_1''|. \tag{6.21}$$

Now we have

$$\langle x_1', x_2', x_3'|e^{ix_1'' p_1/\hbar} = \langle x_1' + x_1'', x_2', x_3'|, \tag{6.22}$$

and similarly for displacements in the 2 and 3 directions. Now consider the displacements shown in Figure 6.1.

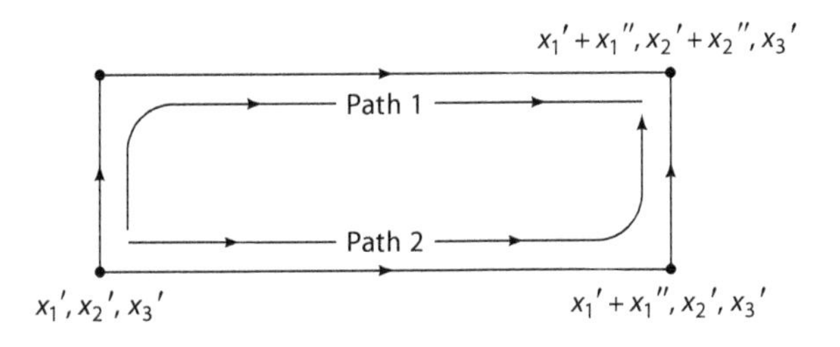

FIGURE 6.1 Two planar displacement paths involving the same initial and final points.

We may get from (x_1', x_2', x_3') to $(x_1' + x_1'', x_2' + x_2'', x_3')$ along paths 1 or 2. Along path 1:

$$\langle x_1', x_2', x_3' | e^{ix_2'' p_2/\hbar} e^{ix_1'' p_1/\hbar} = \langle x_1' + x_1'', x_2' + x_2'', x_3' |. \tag{6.23}$$

Along path 2:

$$\langle x_1', x_2', x_3' | e^{ix_1'' p_1/\hbar} e^{ix_2'' p_2/\hbar} = \langle x_1' + x_1'', x_2' + x_2'', x_3' |. \tag{6.24}$$

The equivalence of these two operators tells us that

$$[p_1, p_2] = 0. \tag{6.25}$$

This can obviously be done for the sets $[p_2, p_3]$ and $[p_1, p_3]$ also. The conclusion is

$$[p_i, p_j] = 0, \tag{6.26}$$

for all i, j. Thus the p_i are simultaneously measurable in all states. Notice that since Equation 6.26 is true, we have

$$e^{ix_1' p_1/\hbar}\, e^{ix_2' p_2/\hbar}\, e^{ix_3' p_3/\hbar} = e^{i\vec{x}'\cdot\vec{p}/\hbar}, \tag{6.27}$$

so that a general displacement can be written as

$$\langle \vec{x}' | e^{i\vec{x}''\cdot\vec{p}/\hbar} = \langle \vec{x}' + \vec{x}'' |. \tag{6.28}$$

What about mixed objects like $[x_1, p_2]$? Consider the infinitesimal change

$$\langle \vec{x}' | \left(1 + i d\vec{x}''\cdot\vec{p}/\hbar\right) = \langle \vec{x}' + d\vec{x}'' |. \tag{6.29}$$

Multiplying both sides by the operator $\vec{x}$, we have

$$\langle \vec{x}' | \left(1 + i\, d\vec{x}''\cdot\vec{p}/\hbar\right)\vec{x} = \langle \vec{x}' + d\vec{x}'' | \left(\vec{x}' + d\vec{x}''\right). \tag{6.30}$$

Now do these two operations in the opposite order:

$$\begin{aligned}\langle \vec{x}' | \vec{x}\left(1 + i\, d\vec{x}''\cdot\vec{p}/\hbar\right) &= \langle \vec{x}' | \left(1 + i d\vec{x}''\cdot\vec{p}/\hbar\right)\vec{x}' \\ &= \langle \vec{x}' + d\vec{x}'' | \vec{x}'.\end{aligned} \tag{6.31}$$

Therefore, we have that

$$\langle \vec{x}' | [(1 + i d\vec{x}''\cdot\vec{p}/\hbar), \vec{x}] = \langle \vec{x}' + d\vec{x}'' | d\vec{x}'', \tag{6.32}$$

$$\Rightarrow \frac{i}{\hbar}[d\vec{x}''\cdot\vec{p}, \vec{x}] = d\vec{x}''. \tag{6.33}$$

This statement becomes more transparent in component language:

$$\frac{i}{\hbar}\sum_k dx_k''[p_k, x_j] = dx_j'', \tag{6.34}$$

$$\Rightarrow [p_k, x_j] = \frac{\hbar}{i}\delta_{kj}. \tag{6.35}$$

From the 1-D statement,

$$\langle x'|e^{i\vec{x}''p_x/\hbar} = \langle x' + x''|, \tag{6.36}$$

I derived the result (Equation 4.153)

$$\langle x'|\, p_x = \frac{\hbar}{i}\frac{\partial}{\partial x'}\langle x'|. \tag{6.37}$$

Likewise in our 3-D Cartesian basis, given in Equation 6.27, it is easy to show that

$$\langle \vec{x}'|\, \vec{p} = \frac{\hbar}{i}\vec{\nabla}'\langle \vec{x}'|, \tag{6.38}$$

where $\vec{\nabla}'$ is the usual gradient (differential) operator:

$$\vec{\nabla}' = \sum_i \hat{e}_i \frac{\partial}{\partial x_i'}. \tag{6.39}$$

Likewise, one can show that

$$\langle \vec{p}'|\, \vec{x} = -\frac{\hbar}{i}\vec{\nabla}_{p'}\langle \vec{p}'|, \tag{6.40}$$

where

$$\vec{\nabla}_{p'} = \sum_i \hat{e}_i \frac{\partial}{\partial p_i'}. \tag{6.41}$$

6.3 EXAMPLE: THREE-DIMENSIONAL INFINITE SQUARE WELL

Let us apply our Cartesian basis to a simple problem. Consider the 3-D square well, with positively infinite values of the potential at the walls and $V = 0$ inside, shown in Figure 6.2.

The boundary conditions are

$$\left.\begin{aligned} u(\pm a, x_2, x_3) &= 0,\\ u(x_1, \pm a, x_3) &= 0,\\ u(x_1, x_2, \pm a) &= 0. \end{aligned}\right\} \tag{6.42}$$

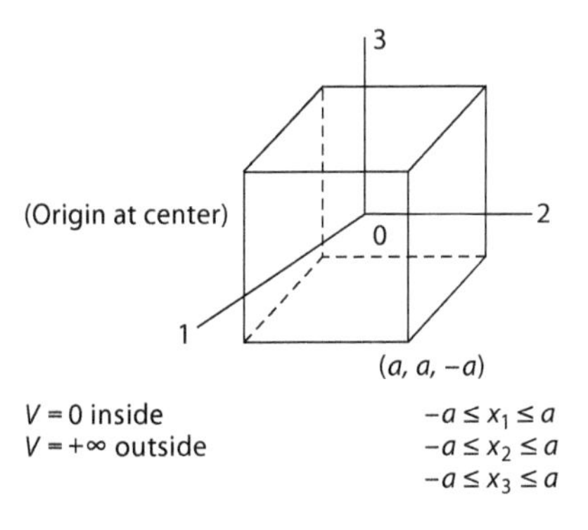

FIGURE 6.2 The infinite potential three-dimensional square well box geometry.

(In other words $u|_{\text{surface}} = 0$). Projected into the Cartesian basis, the energy eigenvalue condition $H|a'\rangle = E_{a'}|a'\rangle$ becomes

$$-\frac{\hbar^2}{2m}\left(\frac{\partial^2}{\partial x_1^2}+\frac{\partial^2}{\partial x_2^2}+\frac{\partial^2}{\partial x_3^2}\right)u_{a'}(\vec{x}) = E_{a'}u_{a'}(\vec{x}). \tag{6.43}$$

This is obviously separable in x_1, x_2, and x_3. Let

$$u_{a'}(\vec{x}) = u_1(x_1)u_2(x_2)u_3(x_3). \tag{6.44}$$

Then Equation 6.43 may be put into the form $(u_1'' \equiv \partial^2 u/\partial x_1^2, \text{etc.})$

$$\frac{u_1''}{u_1}+\frac{u_2''}{u_2}+\frac{u_3''}{u_3} = -\frac{2mE_{a'}}{\hbar^2}, \tag{6.45}$$

which means we may set

$$\left.\begin{aligned} -\frac{\hbar^2}{2m}u_1'' &= E_1u_1(x_1),\\ -\frac{\hbar^2}{2m}u_2'' &= E_2u_2(x_2),\\ -\frac{\hbar^2}{2m}u_3'' &= E_3u_3(x_3), \end{aligned}\right\} \tag{6.46}$$

where

$$E_1 + E_2 + E_3 = E_{a'}. \tag{6.47}$$

Equations 6.46 and 6.42 ensure that the solution in each direction is identical to the one-dimensional case solved in Chapter 3, Section 3.2. Let me remind you of these solutions:

$$u_{n-}(x) = \langle x\,|\,n-\rangle = \frac{1}{\sqrt{a}}\sin(k_{n-}x), \tag{6.48}$$

$$u_{n+}(x) = \langle x\,|\,n+\rangle = \frac{1}{\sqrt{a}}\cos(k_{n+}x), \tag{6.49}$$

where $(n = 1, 2, 3, \ldots)$

$$k_{n-} = \frac{n\pi}{a} \tag{6.50}$$

$$k_{n+} = \frac{(n-1/2)\pi}{a}, \tag{6.51}$$

and $E = \hbar^2k^2/2m$. Thus in 3-D the solutions are

$$\begin{aligned} u_{a'}(\vec{x}) &= \langle \vec{x}'\,|a'\rangle = \langle x_1\,|E_1\rangle\langle x_2\,|E_2\rangle\langle x_3\,|E_3\rangle \\ &= u_{n_1P_1}(x_1)u_{n_2P_2}(x_2)u_{n_3P_3}(x_3), \end{aligned} \tag{6.52}$$

where each of the $u_{n_i P_i}$ is given in Equations 6.48 and 6.49 with $P_{1,2,3} = \pm$ giving the parities of the state. Equation 6.47 says that the *total* energy is given by the sum of the E_1, E_2, and E_3 eigenenergies. This means that there are energy degeneracies. For example, consider energy levels for which $P_1 = P_2 = P_3 = -$,

$$E_{n_1-,n_2-,n_3-} = \frac{\pi^2\hbar^2}{2ma^2}\left(n_1^2 + n_2^2 + n_3^2\right). \tag{6.53}$$

Although the lowest such energy is specified by $n_1 = n_2 = n_3 = 1$, the first excited state is threefold degenerate in energy: ($n_1 = 2$, $n_2 = 1$, $n_3 = 1$), ($n_1 = 1$, $n_2 = 2$, $n_3 = 1$), and ($n_1 = 1$, $n_2 = 1$, $n_3 = 2$). Of course, the overall ground state is the one with $P_1 = P_2 = P_3 = +$ and $n_1 = n_2 = n_3 = 1$.

6.4 SPHERICAL BASIS STATES

A much more useful coordinate basis in physics is a *spherical* basis in which the position of a particle is specified by the three numbers r, θ, and ϕ:

$$\langle x_1', x_2', x_3'| \rightarrow \langle r,\theta,\phi|. \tag{6.54}$$

Based on what we have seen before, we expect that

$$1 = \int \mathrm{d}^3r\, |r,\theta,\phi\rangle\langle r,\theta,\phi|, \tag{6.55}$$

where

$$\mathrm{d}^3r = r^2 \sin\theta\, \mathrm{d}r\mathrm{d}\theta\mathrm{d}\phi. \tag{6.56}$$

The range of these variables is as usual

$$\left.\begin{aligned} r&:\ 0 \rightarrow \infty, \\ \theta&:\ 0 \rightarrow \pi, \\ \phi&:\ 0 \rightarrow 2\pi, \end{aligned}\right\} \tag{6.57}$$

which picks out all points in coordinate space.

Just as the 3-D Cartesian basis,

$$\langle x_1', x_2', x_3'| = \langle x_1'|\langle x_2'|\langle x_3'| \tag{6.58}$$

is a direct product of three Hilbert spaces, we expect that the spherical basis,

$$\langle r,\theta,\phi| = \langle r|\langle\theta|\langle\phi|, \tag{6.59}$$

is also a direct product of separate Hilbert spaces. And just as we have completeness in each Cartesian subspace,

$$\left.\begin{aligned} 1_{x_1} &= \textstyle\int_{-\infty}^{\infty} \mathrm{d}x_1'\, |x_1'\rangle\langle x_1'|, \\ 1_{x_2} &= \textstyle\int_{-\infty}^{\infty} \mathrm{d}x_2'\, |x_2'\rangle\langle x_2'|, \\ 1_{x_3} &= \textstyle\int_{-\infty}^{\infty} \mathrm{d}x_3'\, |x_3'\rangle\langle x_3'|, \end{aligned}\right\} \tag{6.60}$$

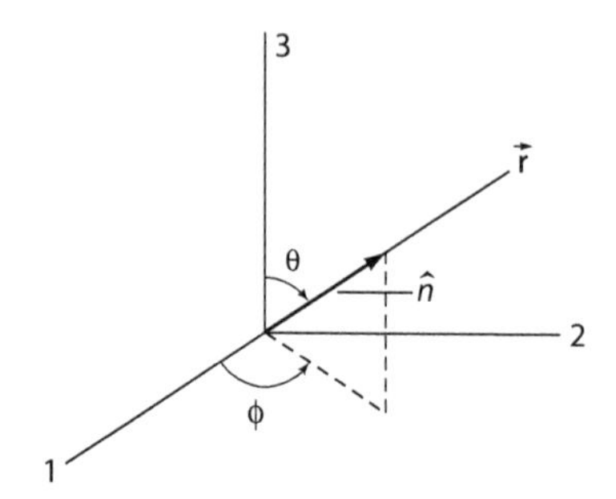

FIGURE 6.3 Spherical polar, θ, and azimuthal, ϕ, angles, leading to the unit radial direction vector, $\hat{n}$.

we demand that

$$\left.\begin{aligned} 1_r &= \int_0^\infty \mathrm{d}r\, r^2 |r\rangle\langle r|, \\ 1_\theta &= \int_0^\pi \mathrm{d}\theta \sin\theta\, |\theta\rangle\langle\theta|, \\ 1_\phi &= \int_0^{2\pi} \mathrm{d}\phi\, |\phi\rangle\langle\phi|, \end{aligned}\right\} \tag{6.61}$$

so that

$$1 = 1_1 \cdot 1_\theta \cdot 1_\phi = \int \mathrm{d}^3 r\, |r,\theta,\phi\rangle\langle r,\theta,\phi|. \tag{6.62}$$

We will label the angular position as

$$|\hat{n}\rangle \equiv |\theta,\phi\rangle, \tag{6.63}$$

where $\hat{n}$ is a unit vector pointing in the $\vec{r}$ direction, as shown in Figure 6.3. Of course, there are many other bases possible, corresponding to cylindrical coordinates, elliptical coordinates, etc. It is clear that we have to be consistent in a given problem to stick with an initial choice, but other than this one is free to switch between various bases to simplify derivations and expressions. I will generally use $|\vec{x}\rangle$ to denote a Cartesian basis and $|\vec{r}\rangle$ to denote a spherical one.

6.5 ORBITAL ANGULAR MOMENTUM OPERATOR

Let us now introduce the quantum mechanical operator representing *orbital* angular momentum:*

$$\vec{L} \equiv \vec{x} \times \vec{p}. \tag{6.64}$$

* This is surely the classical form of the angular momentum, but an equivalent classical form is $\vec{L} = -\vec{p} \times \vec{x}$. Luckily, both of these produce the same quantum mechanical operator since only orthogonal components of $\vec{x}, \vec{p}$ are multiplied together. This is not always the case, however, and sometimes gives rise to "operator-ordering ambiguities." The ambiguity is usually resolved by averaging the possibilities.

Component-wise, we have

$$\left.\begin{aligned} L_1 &= x_2 p_3 - x_3 p_2, \\ L_2 &= x_3 p_1 - x_1 p_3, \\ L_3 &= x_1 p_2 - x_2 p_1. \end{aligned}\right\}. \tag{6.65}$$

Notice that since $[x_i, p_j] = 0$ $(i \neq j)$, the order of the operators in Equation 6.65 does not matter. Also notice that

$$\vec{L} = \vec{L}, \tag{6.66}$$

that is, it is Hermitian and, therefore, has real eigenvalues.

The various L_i do not commute. To see this, consider

$$\begin{aligned} \left[L_1, L_2\right] &= \left[x_2 p_3 - x_3 p_2, x_3 p_1 - x_1 p_3\right] \\ &= \left[x_2 p_3, x_3 p_1\right] + \left[x_3 p_2, x_1 p_3\right] \\ &= x_2 p_1 \left[p_3, x_3\right] + p_2 x_1 \left[x_3, p_3\right] \\ &= i\hbar\left(x_1 p_2 - x_2 p_1\right) = i\hbar L_3. \end{aligned} \tag{6.67}$$

Likewise

$$\left[L_1, L_3\right] = -i\hbar L_2, \tag{6.68}$$

$$\left[L_2, L_3\right] = i\hbar L_1. \tag{6.69}$$

The general statement is that

$$\left[L_i, L_j\right] = i\hbar \sum_k \varepsilon_{ijk} L_k, \tag{6.70}$$

is the general statement. (ε_{ijk} is the usual permutation symbol, which is completely antisymmetric). This shows that the three components, L_i, are mutually incompatible observables. A quantum mechanical state cannot, in general, be in an eigenstate of *both* L_1 and L_2. This is distinctly different from *linear* momentum for which we have seen

$$[p_i, p_j] = 0, \tag{6.71}$$

for all i, j.

6.6 EFFECT OF ANGULAR MOMENTUM ON BASIS STATES

We will now try to find the effect of the L_i on a state $|\vec{x}'\rangle$. Consider

$$\left(1 - i\left(\frac{\delta\phi}{\hbar}\right)L_3\right)|\vec{x}'\rangle = \left(1 - i\left(\frac{\delta\phi}{\hbar}\right)(p_2 x_1' - p_1 x_2')\right)|\vec{x}'\rangle, \tag{6.72}$$

where $\delta\phi$ is a positive, infinitesimal quantity. Remember that

$$e^{-i\vec{x}''\cdot\vec{p}/\hbar}|\vec{x}'\rangle = |\vec{x}'+\vec{x}''\rangle. \tag{6.73}$$

Let us choose $\vec{x}'' = (-\delta x'',0,0)$ where $\delta x''$ is also a positive, infinitesimal quantity. Then Equation 6.73 implies that

$$\left(1+i\frac{\delta x''}{\hbar}p_1\right)|\vec{x}'\rangle = |x_1'-\delta x'', x_2', x_3'\rangle. \tag{6.74}$$

Likewise for $\vec{x}'' = (0,\delta x'',0)$, we get

$$\left(1-i\frac{\delta x''}{\hbar}p_2\right)|\vec{x}'\rangle = |x_1', x_2'+\delta x'', x_3'\rangle. \tag{6.75}$$

Since we may write

$$\left(1-i\left(\frac{\delta\phi}{\hbar}\right)L_3\right) = \left(1-i\left(\frac{\delta\phi}{\hbar}\right)x_1'p_2\right)\left(1+i\left(\frac{\delta\phi}{\hbar}\right)x_2'p_1\right), \tag{6.76}$$

(because $\delta\phi$ is infinitesimal) we get that

$$\left(1-i\left(\frac{\delta\phi}{\hbar}\right)L_3\right)|\vec{x}'\rangle = |x_1'-\delta\phi x_2', x_2'+\delta\phi x_1', x_3'\rangle. \tag{6.77}$$

The right-hand side of Equation 6.77 reveals that a rotation about the 3-axis has been performed. (See Figure 6.4.) We are adopting the convention that this represents an *active* rotation of the physical system itself (rather than a passive rotation of the coordinate system in the opposite direction). The rotation shown is defined to have $\delta\phi > 0$.

I have used the Cartesian basis to make these conclusions. In terms of a spherical basis, the effect of this operator is clearly

$$\left(1-i\left(\frac{\delta\phi}{\hbar}\right)L_3\right)|r,\theta,\phi\rangle = |r,\theta,\phi+\delta\phi\rangle. \tag{6.78}$$

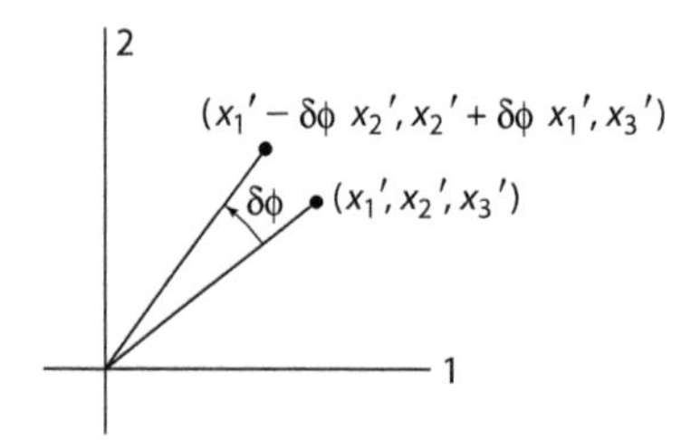

FIGURE 6.4 The displacement described in the text by the positive angle $\delta\phi$ occurring about the third axis.

Since $\delta\phi$ is infinitesimal, we have

$$|r,\theta,\phi+\delta\phi\rangle = |r,\theta,\phi\rangle + \delta\phi\frac{\partial}{\partial\phi}|r,\theta,\phi\rangle. \tag{6.79}$$

Matching the coefficient of $\delta\phi$ on both sides of Equation 6.78, we conclude that

$$L_3\,|r,\theta,\phi\rangle = i\hbar\frac{\partial}{\partial\phi}|r,\theta,\phi\rangle, \tag{6.80}$$

or since r and θ play no role here, that

$$L_3\,|\phi\rangle = i\hbar\frac{\partial}{\partial\phi}|\phi\rangle. \tag{6.81}$$

Equivalently,

$$\langle\phi|\,L_3 = -i\hbar\frac{\partial}{\partial\phi}\langle\phi|\,. \tag{6.82}$$

Finite relations can also be produced using L_3. Any finite rotation, ϕ, can always be imagined to consist of N identical partial rotations by an amount ϕ/N. But in the limit $N\to\infty$ each of these partial rotations becomes infinitesimal. Thus, a finite rotation is accomplished by

$$\lim_{N\to\infty}\left(1-i\left(\frac{\phi/N}{\hbar}\right)L_3\right)^N.$$

Applying the formula

$$\lim_{N\to\infty}\left(1+\frac{x}{N}\right)^N = \mathrm{e}^x, \tag{6.83}$$

to the above gives

$$\lim_{N\to\infty}\left(1-i\left(\frac{\phi/N}{\hbar}\right)L_3\right)^N = \mathrm{e}^{-iL_3\phi/\hbar}, \tag{6.84}$$

as the operator which performs finite rotations about the third axis. That is

$$\mathrm{e}^{-iL_3\phi'/\hbar}\,|r,\theta,\phi\rangle = |r,\theta,\phi+\phi'\rangle. \tag{6.85}$$

Since L_3 is Hermitian, we recognize $\mathrm{e}^{\pm iL_3\phi/\hbar}$ as a unitary operator. L_3 is called the *generator* of rotations about the third axis.

We will now find the effect of L_1 and L_2 on the $\langle\vec{r}|$ basis by a more cookbook-type approach. We have that

$$\begin{aligned}\langle\vec{r}\,|\vec{L} &= \langle\vec{r}\,|\vec{x}\times\vec{p} = \langle\vec{r}\,|\vec{r}\times\vec{p}\\ &= \vec{r}\times(\langle\vec{r}\,|\vec{p}) = \vec{r}\times\left(\frac{\hbar}{i}\vec{\nabla}_r\langle\vec{r}\,|\right).\end{aligned} \tag{6.86}$$

Now since our basis is spherical, the gradient operator must be stated in spherical variables (which is symbolized by $\vec{\nabla}_r$):

$$\vec{\nabla}_r = \hat{e}_r \frac{\partial}{\partial r} + \hat{e}_\phi \frac{1}{r \sin\theta} \frac{\partial}{\partial \phi} + \hat{e}_\theta \frac{1}{r} \frac{\partial}{\partial \theta}, \tag{6.87}$$

where $\hat{e}_r$, $\hat{e}_\phi$, and $\hat{e}_\theta$ are unit vectors pointing in the instantaneous r, ϕ, and θ directions.

Figure 6.5 informs us that

$$\left.\begin{aligned} \hat{e}_r \times \hat{e}_\theta &= \hat{e}_\phi, \\ \hat{e}_\theta \times \hat{e}_\phi &= \hat{e}_r, \\ \hat{e}_\phi \times \hat{e}_r &= \hat{e}_\theta, \end{aligned}\right\} \tag{6.88}$$

so that

$$\langle \vec{r} | \vec{L} = \frac{\hbar}{i} \left[\hat{e}_\phi \frac{\partial}{\partial \theta} - \hat{e}_\theta \frac{1}{\sin\theta} \frac{\partial}{\partial \phi} \right] \langle \vec{r} |. \tag{6.89}$$

The $\hat{e}_r$, $\hat{e}_\theta$, and $\hat{e}_\phi$ can be related to unit vectors along $\hat{e}_1$, $\hat{e}_2$, and $\hat{e}_3$ in Figure 6.5 by

$$\hat{e}_r = \sin\theta \cos\phi \, \hat{e}_1 + \sin\theta \sin\phi \, \hat{e}_2 + \cos\theta \, \hat{e}_3, \tag{6.90}$$

$$\hat{e}_\phi = -\sin\phi \, \hat{e}_1 + \cos\phi \, \hat{e}_2, \tag{6.91}$$

$$\hat{e}_\theta = \cos\theta \cos\phi \, \hat{e}_1 + \cos\theta \sin\phi \, \hat{e}_2 - \sin\theta \, \hat{e}_3. \tag{6.92}$$

So, when the basis in Equation 6.89 is expressed in terms of the $\hat{e}_i$, we find

$$\langle \vec{r} | L_1 = \frac{\hbar}{i} \left[-\sin\phi \frac{\partial}{\partial \theta} - \cos\phi \cot\theta \frac{\partial}{\partial \phi} \right] \langle \vec{r} |, \tag{6.93}$$

$$\langle \vec{r} | L_2 = \frac{\hbar}{i} \left[\cos\phi \frac{\partial}{\partial \theta} - \sin\phi \cot\theta \frac{\partial}{\partial \phi} \right] \langle \vec{r} |, \tag{6.94}$$

$$\langle \vec{r} | L_3 = \frac{\hbar}{i} \frac{\partial}{\partial \phi} \langle \vec{r} |. \tag{6.95}$$

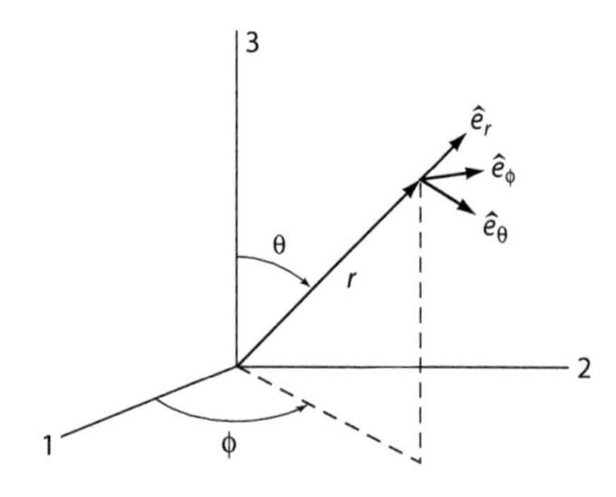

FIGURE 6.5 The unit vectors $\hat{e}_r$, $\hat{e}_\phi$, and $\hat{e}_\theta$ in a spherical coordinate basis.

Equation 6.95 is, of course, consistent with Equation 6.82. We can also show from the preceding that

$$\langle \vec{r}\,|\,\vec{L}^2 = \langle \vec{r}\,|\left(L_1^2 + L_2^2 + L_3^2\right)$$

$$= -\hbar^2\left[\frac{1}{\sin^2\theta}\frac{\partial^2}{\partial\phi^2} + \frac{1}{\sin\theta}\frac{\partial}{\partial\theta}\left(\sin\theta\frac{\partial}{\partial\theta}\right)\right]\langle \vec{r}\,|, \tag{6.96}$$

$$\equiv L_{\text{op}}^2\langle \vec{r}\,|, \tag{6.97}$$

where we have defined the *differential* operator (as opposed to the Hilbert space operator, $\vec{L}^2$)

$$L_{\text{op}}^2 \equiv -\hbar^2\left[\frac{1}{\sin^2\theta}\frac{\partial^2}{\partial\phi^2} + \frac{1}{\sin\theta}\frac{\partial}{\partial\theta}\left(\sin\theta\frac{\partial}{\partial\theta}\right)\right]. \tag{6.98}$$

Apart from an overall factor of $-1/(r\hbar)^2$, this will be seen to be the angular part of the spherical $\vec{\nabla}_r^2$ operator (see Equation 6.117). Equations 6.93 through 6.95 could also have been stated in terms of the angular basis $\langle \hat{n}|$ since the purely radial part of the basis plays no role in these considerations.

6.7 ENERGY EIGENVALUE EQUATION AND ANGULAR MOMENTUM

I will now prove a useful identity for $\vec{L}^2$ in preparation for applying it to the energy eigenvalue problem. We know that

$$L_i = \sum_{j,k}\varepsilon_{ijk}x_j p_k \tag{6.99}$$

where the order of the operators x_j, p_k does not matter since $[x_j, p_k] = 0$ for $j \neq k$. Therefore,

$$\vec{L}^2 = \sum_i L_i^2 = \sum_i\left(\sum_{j,k}\varepsilon_{ijk}x_j p_k \sum_{\ell,m}\varepsilon_{i\ell m}x_\ell p_m\right)$$

$$= \sum_{j,k,\ell,m}\left(\sum_i \varepsilon_{ijk}\varepsilon_{i\ell m}\right)x_j p_k x_\ell p_m. \tag{6.100}$$

One has

$$\sum_i \varepsilon_{ijk}\varepsilon_{i\ell m} = \left(\delta_{km}\delta_{j\ell} - \delta_{k\ell}\delta_{jm}\right), \tag{6.101}$$

so that we may write

$$\vec{L}^2 = \sum_{j,k,\ell,m}\left(\delta_{j\ell}\delta_{km}x_j\overbrace{\left(x_\ell p_k - i\hbar\delta_{\ell k}\right)}^{p_k x_\ell}p_m - \delta_{k\ell}\delta_{jm}x_j p_k\underbrace{\left(p_m x_\ell + i\hbar\delta_{m\ell}\right)}_{x_\ell p_m}\right) \tag{6.102}$$

or

$$\vec{L}^2 = \vec{x}^2\vec{p}^2 - 2i\hbar\vec{x}\cdot\vec{p} - (\vec{x}\cdot\vec{p})(\vec{p}\cdot\vec{x}). \tag{6.103}$$

But (can you show it?)

$$\vec{p}\cdot\vec{x} = \vec{x}\cdot\vec{p} - 3i\hbar, \tag{6.104}$$

so

$$\vec{L}^2 = \vec{x}^2\vec{p}^2 - (\vec{x}\cdot\vec{p})^2 + i\hbar\vec{x}\cdot\vec{p}. \tag{6.105}$$

Notice that if $\vec{x}$ and $\vec{p}$ were regarded as numbers, the last term in Equation 6.105 would not be present.

Now we can try to construct the differential equation implied by

$$H\,|a'\rangle = E_{a'}\,|a'\rangle, \tag{6.106}$$

in a problem with spherical symmetry, $H = \vec{p}^2/2m + V(r)$. Project both sides of Equation 6.106 into the spherical basis $\langle\vec{r}|$:

$$\langle\vec{r}\,|\,H\,|\,a'\rangle = E_{a'}\langle\vec{r}\,|\,a'\rangle, \tag{6.107}$$

$$\langle\vec{r}\,|\frac{\vec{p}^2}{2m}|\,a'\rangle + V(r)\langle\vec{r}\,|\,a'\rangle = E_{a'}\langle\vec{r}\,|\,a'\rangle, \tag{6.108}$$

where $u_{a'}(\vec{r}) \equiv \langle\vec{r}\,|\,a'\rangle$. We now have from Equation 6.105 that

$$\langle\vec{r}\,|\,\vec{x}^2\vec{p}^2\,|\,a'\rangle = \langle\vec{r}\,|\,\vec{L}^2\,|\,a'\rangle + \langle\vec{r}\,|\,(\vec{x}\cdot\vec{p})^2\,|\,a'\rangle - i\hbar\langle\vec{r}\,|\,\vec{x}\cdot\vec{p}\,|\,a'\rangle. \tag{6.109}$$

We have that

$$\langle\vec{r}\,|\,\vec{x}\cdot\vec{p}\,|\,a'\rangle = \vec{r}\cdot(\langle\vec{r}\,|\,\vec{p}\,|\,a'\rangle) = \vec{r}\cdot\left(\frac{\hbar}{i}\vec{\nabla}_r\langle\vec{r}\,|\,a'\rangle\right), \tag{6.110}$$

with $\vec{\nabla}_r$ given by Equation 6.87. Therefore,

$$\langle\vec{r}\,|\,\vec{x}\cdot\vec{p}\,|\,a'\rangle = \frac{\hbar}{i}r\frac{\partial}{\partial r}\langle\vec{r}\,|\,a'\rangle = \frac{\hbar}{i}r\frac{\partial}{\partial r}u_{a'}(\vec{r}). \tag{6.111}$$

Likewise

$$\begin{aligned}\langle\vec{r}\,|\,(\vec{x}\cdot\vec{p})^2\,|\,a'\rangle &= \langle\vec{r}\,|\,(\vec{x}\cdot\vec{p})(\vec{x}\cdot\vec{p})\,|\,a'\rangle \\ &= \left(\frac{\hbar}{i}r\frac{\partial}{\partial r}\right)\left(\frac{\hbar}{i}r\frac{\partial}{\partial r}\right)\langle\vec{r}\,|\,a'\rangle \\ &= -\hbar^2\left(r^2\frac{\partial^2}{\partial r^2} + r\frac{\partial}{\partial r}\right)u_{a'}(\vec{r}),\end{aligned} \tag{6.112}$$

and

$$\langle\vec{r}\,|\,\vec{x}^2\vec{p}^2\,|\,a'\rangle = r^2\langle\vec{r}\,|\,\vec{p}^2|a'\rangle. \tag{6.113}$$

Using Equations 6.111 through 6.113 in Equation 6.109, we find that (dividing by r^2)

$$\langle \vec{r} \mid \vec{p}^2 | a' \rangle = -\hbar^2 \left(\frac{\partial^2}{\partial r^2} + \frac{2}{r}\frac{\partial}{\partial r} \right) u_{a'}(\vec{r}) + \frac{1}{r^2} \langle \vec{r} \mid \vec{L}^2 | a' \rangle. \tag{6.114}$$

Now using Equation 6.97, we get

$$\langle \vec{r} \mid \vec{p}^2 | a' \rangle = \left[-\frac{\hbar^2}{r^2}\frac{\partial}{\partial r}\left(r^2 \frac{\partial}{\partial r} \right) + \frac{L_{op}^2}{r^2} \right] u_{a'}(\vec{r}). \tag{6.115}$$

Since

$$\langle \vec{r} \mid \vec{p}^2 | a' \rangle = -\hbar^2 \, \vec{\nabla}_r^2 \, u_{a'}(\vec{r}), \tag{6.116}$$

all we have really accomplished in Equation 6.115 is to find an explicit expression for the $\vec{\nabla}_r^2$ operator in spherical coordinates (in an especially interesting way, however). So, using Equation 6.98,

$$\vec{\nabla}_r^2 = \frac{1}{r^2}\frac{\partial}{\partial r}\left(r^2 \frac{\partial}{\partial r} \right) + \frac{1}{r^2 \sin^2\theta}\frac{\partial^2}{\partial \phi^2} + \frac{1}{r^2 \sin\theta}\frac{\partial}{\partial \theta}\left(\sin\theta \frac{\partial}{\partial \theta} \right). \tag{6.117}$$

Therefore, returning to Equation 6.108, we have the explicit radial Schrödinger equation:

$$\left[-\frac{\hbar^2}{2mr^2}\left(\frac{\partial}{\partial r}\left(r^2 \frac{\partial}{\partial r} \right) - \left(\frac{\vec{L}_{op}}{\hbar} \right)^2 \right) + V(r) \right] u_{a'}(\vec{r}) = E_{a'} u_{a'}(\vec{r}). \tag{6.118}$$

6.8 COMPLETE SET OF OBSERVABLES FOR THE RADIAL SCHRÖDINGER EQUATION

We could now proceed in a standard way to separate variables in Equation 6.118 and find the eigenvalues and eigenvectors of the angular part of the problem using purely differential operator techniques.* Instead, let us proceed by considering what the aforementioned set of quantum numbers $\{a'\}$ consists of. Notice

$$\left. \begin{aligned} \langle \vec{r}' | V(r) \vec{L} &= V(r') \langle \vec{r}' | \vec{L} = V(r') \vec{L}_{\rm op} \langle \vec{r}' |, \\ \langle \vec{r}' | \vec{L} V(r) &= \vec{L}_{\rm op} V(r') \langle \vec{r}' | = V(r') \vec{L}_{\rm op} \langle \vec{r}' |, \end{aligned} \right\} \tag{6.119}$$

where $\vec{L}_{\rm op}$ is a differential operation (see Equation 6.86). Therefore,

$$[\vec{L}, V(r)] = 0. \tag{6.120}$$

* For example, see E. Merzbacher, *Quantum Mechanics*, 3rd ed. (John Wiley & Sons, 1998), Chapter 11.

Also we have that

$$\left[\vec{p}^2, L_j\right] = \left[\sum_i p_i^2, \sum_{k,\ell} \varepsilon_{jk\ell} x_k p_\ell\right] = \sum_{i,k,\ell} \left[p_i^2, x_k p_\ell\right] \varepsilon_{jk\ell}. \tag{6.121}$$

Now

$$\begin{aligned} [p_i^2, x_k p_\ell] &= p_i[p_i, x_k p_\ell] + [p_i, x_k p_\ell] p_i \\ &= p_i(x_k \underbrace{[p_i, p_\ell]}_{0} + \underbrace{[p_i, x_k]}_{\frac{\hbar}{i}\delta_{ik}} p_\ell) \\ &\quad + (x_k \underbrace{[p_i, p_\ell]}_{0} + \underbrace{[p_i, x_k]}_{\frac{\hbar}{i}\delta_{ik}} p_\ell) p_i, \end{aligned} \tag{6.122}$$

so that

$$\left[p_i^2, x_k p_\ell\right] = 2\frac{\hbar}{i} p_i p_\ell \delta_{ik}. \tag{6.123}$$

Using this in Equation 6.121 gives

$$\left[\vec{p}^2, L_j\right] = \sum_{i,k,\ell} \varepsilon_{jk\ell} 2\frac{\hbar}{i} p_i p_\ell \delta_{ik} = 2\frac{\hbar}{i} \sum_{k,\ell} \varepsilon_{jk\ell} p_k p_\ell. \tag{6.124}$$

Notice in Equation 6.124 that we have an object that is symmetric in two summed indices (k,ℓ) multiplied into an object that is antisymmetric in the same two indices. The result is zero. We can see this in general as follows. Let us say we have the two index objects A_{ij} and B_{ij} and that

$$\left.\begin{aligned} A_{ij} &= -A_{ji}, \\ &\text{and} \\ B_{ij} &= B_{ji}. \end{aligned}\right\} \tag{6.125}$$

Then we have

$$\sum_{i,j} A_{ij} B_{ij} = -\sum_{i,j} A_{ji} B_{ji}. \tag{6.126}$$

But since we are free to rename our indices, this means

$$\sum_{i,j} A_{ij} B_{ij} = -\sum_{i,j} A_{ij} B_{ij}. \tag{6.127}$$

Anything that is equal to minus itself is zero, and so it is for the right-hand side of Equation 6.124:

$$\left[\vec{p}^2, \vec{L}\right] = 0. \tag{6.128}$$

Since $H = \vec{p}^2 / 2m + V(r)$, Equations 6.120 and 6.128 imply

$$\left[H, \vec{L}\right] = 0. \tag{6.129}$$

Thus the $\vec{L}$ gives rise to good quantum numbers. (See discussion in Chapter 4, Section 4.11). Equation 6.129 implies of course that

$$[H,\vec{L}^2]=0. \tag{6.130}$$

What's more

$$[\vec{L}^2,L_i]=\left[\sum_j L_j^2,L_i\right]=\sum_j (L_j[L_j,L_i]+[L_j,L_i]L_j), \tag{6.131}$$

and using Equation 6.70 we get

$$[\vec{L}^2,L_i]=i\hbar\sum_{j,k}\varepsilon_{jik}(L_jL_k+L_kL_j). \tag{6.132}$$

This is again a situation in which a sum over two indices (j,k) is being performed on a symmetric object $(L_jL_k+L_kL_j)$ and an antisymmetric one (ε_{jik}). Therefore,

$$[\vec{L}^2,\vec{L}]=0. \tag{6.133}$$

Now one of the theorems I talked about, and partially proved in Chapter 4, Section 4.6, said essentially that: $(A=A^\dagger, B=B^\dagger)$

A,B possess a common complete set of
orthonormal eigenkets $\Leftrightarrow [A,B]=0$.

Therefore, from Equations 6.129, 6.130, and 6.133, I may choose to characterize the set of quantum numbers $\{a'\}$ in Equation 6.118 (which is just an explicit version of Equation 6.106 in a spherical basis) as eigenvalues of the set,

$$\{H,\vec{L}^2,L_3\}.$$

I could not, for example, add L_1 or L_2 to this list since $[L_{1,2}, L_3]\neq 0$. I did not have to choose L_3 in the above set; L_1 or L_2 would have done just as well. However, the choice of L_3 is simpler and conventional. The above represents a complete set of observables for a spinless particle subjected to a spherically symmetric potential. (This is very nearly the case for an electron in the hydrogen atom.)

6.9 SPECIFICATION OF ANGULAR MOMENTUM EIGENSTATES

Given the aforementioned choice of commuting observables, we take

$$|a'\rangle=|n,a,b\rangle, \tag{6.134}$$

where n is a radial quantum number that depends on the nature of the potential (more about this later), and a and b are eigenvalues of $\vec{L}^2$ and L_3:

$$\vec{L}^2|a,b\rangle = \hbar^2 a|a,b\rangle, \tag{6.135}$$

$$L_3|a,b\rangle = \hbar b|a,b\rangle, \tag{6.136}$$

where factors of $\hbar$ have been inserted for convenience. (a and b are real since $\vec{L}^2$ and L_3 are Hermitian.)

Let us find the eigenvalues and eigenvectors of $\vec{L}^2$ and L_3. Let us introduce (the same form as the "very special operator" of Chapter 1, Section 1.15)

$$L_\pm = L_1 \pm iL_2, \tag{6.137}$$

$$L_+^\dagger = L_- \text{ (not Hermitian)}. \tag{6.138}$$

We can now show that (same proof as in Problem 1.15.1a of Chapter 1)

$$[L_+, L_-] = 2\hbar L_3, \tag{6.139}$$

$$[L_3, L_\pm] = \pm\hbar L_\pm. \tag{6.140}$$

Also

$$[\vec{L}^2, L_\pm] = 0. \tag{6.141}$$

Now consider

$$\begin{aligned} L_3(L_\pm|a,b\rangle) &= L_\pm(L_3 \pm \hbar)|a,b\rangle \\ &= \hbar(b \pm 1)(L_\pm|a,b\rangle), \end{aligned} \tag{6.142}$$

and

$$\vec{L}^2(L_\pm\,|\,a,b\rangle) = L_\pm\hbar^2 a\,|\,a,b\rangle = \hbar^2 a(L_\pm\,|\,a,b\rangle). \tag{6.143}$$

Therefore, the $L_\pm$ are ladder operators in the b space; they raise or lower the value of this quantum number by one unit (similar to the operators A and $A^\dagger$ in the harmonic oscillator problem). The conclusion is that

$$L_\pm|a,b\rangle = C_\pm|a,b \pm 1\rangle, \tag{6.144}$$

or

$$\langle a,b|L_\mp = C_\pm^*\langle a,b \pm 1|, \tag{6.145}$$

where the $C_\pm$ are unknown constants.

Now we have that

$$\begin{aligned} L_- L_+ &= (L_1 - iL_2)(L_1 + iL_2) \\ &= L_1^2 + L_2^2 + i[L_1, L_2] \\ &= \vec{L}^2 - L_3^2 - \hbar L_3, \end{aligned} \tag{6.146}$$

so that

$$\begin{aligned} L_- L_+\,|a,b\rangle &= (\hbar^2 a - \hbar^2 b^2 - \hbar(\hbar b))|a,b\rangle \\ &= \hbar^2[a - b(b+1)]|a,b\rangle. \end{aligned} \tag{6.147}$$

But since

$$\langle a,b|L_-L_+|a,b\rangle = |C_+|^2, \tag{6.148}$$

we have

$$|C_+|^2 = \hbar^2[a - b(b+1)]. \tag{6.149}$$

We *choose* the arbitrary phase to be such that

$$C_+ = \hbar\sqrt{a - b(b+1)}. \tag{6.150}$$

Likewise

$$L_+L_- = \vec{L}^2 - L_3^2 + \hbar L_3, \tag{6.151}$$

so that

$$L_+L_-|a,b\rangle = \hbar^2[a - b(b-1)]|a,b\rangle, \tag{6.152}$$

and since

$$|C_-|^2 = \langle a,b|L_+L_-|a,b\rangle, \tag{6.153}$$

we have

$$|C_-|^2 = \hbar^2\left[a - b(b-1)\right], \tag{6.154}$$

so

$$C_- = e^{i\phi}\hbar\sqrt{a - b(b-1)}, \tag{6.155}$$

where ϕ is an unknown phase. Actually ϕ is fixed from our previous choice for C_+ in Equation 6.150. To see this, consider

$$L_+\left[L_-|a,b\rangle = e^{i\phi}\hbar\sqrt{a - b(b-1)}\,|a,b-1\rangle\right], \tag{6.156}$$

$$\begin{aligned}\Rightarrow L_+L_-|a,b\rangle &= e^{i\phi}\hbar^2\sqrt{a - b(b-1)}\sqrt{a-(b-1)b}\,|a,b\rangle \\ &= e^{i\phi}\hbar^2\left[a - b(b-1)\right]|a,b\rangle.\end{aligned} \tag{6.157}$$

Comparing Equation 6.157 with Equation 6.152 implies that $e^{i\phi} = 1$. Therefore, we have found that

$$L_\pm|a,b\rangle = \hbar\sqrt{a - b(b\pm 1)}\,|a,b\pm 1\rangle. \tag{6.158}$$

Now, what are the allowed values of a and b? We can easily show that the expectation value of the square of a Hermitian operator is always nonnegative. Therefore, since

$$L_1^2 + L_2^2 = \vec{L}^2 - L_3^2, \tag{6.159}$$

we have

$$\left(L_1^2 + L_2^2\right)|a,b\rangle = \hbar^2\left(a - b^2\right)|a,b\rangle, \tag{6.160}$$

and since, when multiplying on the left by $\langle a,b|$, the left-hand side of Equation 6.160 is guaranteed to be nonnegative, we have that

$$a - b^2 \geq 0. \tag{6.161}$$

Now we know that the $L_\pm$ raise or lower the value of b in $|a,b\rangle$ by ± 1 unit while keeping the value of a unchanged. Hence, for a fixed a value, we have

$$-\sqrt{a} \le b \le \sqrt{a}. \tag{6.162}$$

Therefore, for a given a value there is a largest value of b; let us call this b_{max}. (b_{max} does not necessarily equal $\sqrt{a}$ since there may exist more restrictive conditions. Equation 6.162 only shows b is bounded.) Thus by definition we must have

$$L_+ \,|\, a, b_{max}\rangle = 0. \tag{6.163}$$

Likewise, call the minimum value of b, for a given a, b_{min}. Therefore,

$$L_- \,|\, a, b_{min}\rangle = 0. \tag{6.164}$$

From Equations 6.146 and 6.151 we have

$$L_\mp L_\pm = \vec{L}^2 - L_3\left(L_3 \pm \hbar\right). \tag{6.165}$$

Thus, applying Equation 6.165 to Equations 6.163 and 6.164, we find

$$\left.\begin{aligned} L_-L_+\,|a,b_{max}\rangle &= \hbar^2\left[a - b_{max}\left(b_{max}+1\right)\right]|\, a,b_{max}\rangle = 0,\\ L_+L_-\,|a,b_{min}\rangle &= \hbar^2\left[a - b_{min}\left(b_{min}-1\right)\right]|\, a,b_{min}\rangle = 0.\end{aligned}\right\} \tag{6.166}$$

The Equations 6.166 together imply that

$$b_{max}(b_{max}+1) = b_{min}(b_{min}-1), \tag{6.167}$$

or

$$(b_{max}+b_{min})(b_{max}-b_{min}+1) = 0. \tag{6.168}$$

But since $b_{max} \ge b_{min}$, we get that

$$b_{max} = -b_{min}. \tag{6.169}$$

Let us say it takes 2ℓ steps (2ℓ is a positive or zero integer) to go from $b = b_{min}$ to $b = b_{max}$ in steps of one:

$$b_{max} = b_{min} + 2\ell, \tag{6.170}$$

where the possible ℓ values are

$$\ell = 0\ \overset{\textbf{X}}{\frac{1}{2}},\ 1,\ \overset{\textbf{X}}{\frac{3}{2}},\ldots \tag{6.171}$$

(The reason for the "**X**" above $\ell = 1/2,\ 3/2,\ldots$ will be discussed shortly.) Then because $b_{min} = -b_{max}$, we have

$$b_{max} = \ell, \tag{6.172}$$

and from either of the equations in 6.166, we get

$$a = \ell(\ell+1). \tag{6.173}$$

So, for a given ℓ value, we have the possible b values:

$$\left.\begin{array}{l} b = b_{\min}, b_{\min}+1, \ldots, b_{\max}-1, b_{\max} \\ \text{or} \\ b = -\ell, -\ell+1, \ldots, \ell-1, \ell \end{array}\right\}. \tag{6.174}$$

It is more conventional to relabel b as m, called the "magnetic quantum number." As an example, let us choose $\ell = 3$. Then m can take on $2\ell+1=7$ values:

$$\ell = 3 \left\{ \begin{array}{rl} m=3 & \downarrow L_- \\ 2 & \\ 1 & \\ 0 & \\ -1 & \\ -2 & \\ -3 & \uparrow L_+. \end{array} \right.$$

There is a subtlety involved in the labeling of the eigenstates of $\vec{L}^2$ and L_3. Since $\vec{L}^2$ and L_3 commute (and, therefore, are simultaneously measurable), we may regard the state $|\ell,m\rangle$ as a type of direct product, which would seem to imply that

$$|\ell,m\rangle \sim |\ell\rangle \otimes |m\rangle.$$

However, because $\vec{L}^2$ does not project entirely into θ space (see Equation 6.96), the value of m enters the eigenvalue equation for $\vec{L}^2$ (see upcoming Equation 6.188). Therefore, we will define*

$$|\ell,m\rangle = \underset{\substack{\text{Projects into}\\ \theta\text{ space}}}{|\ell(m)\rangle} \otimes \underset{\substack{\text{Projects into}\\ \phi\text{ space}}}{|m\rangle}.$$

$|\ell,m\rangle$ is just a way of *labeling* the more proper object on the right-hand side. The $\ell(m)$ notation is supposed to indicate that ℓ is the quantum number associated with the θ eigenvalue equation, but that m enters this equation as a parameter. The same sort of subtlety affects the labeling of the Hilbert space description of the radial eigenstates; that is, the value of ℓ enters the radial eigenvalue equation (see Equation 6.253), and we shall define

$$|n,\ell,m\rangle \equiv |n(\ell)\rangle \otimes |\ell,m\rangle.$$

* Technically speaking, these states are separable but not true direct products.

These Hilbert spaces are such that

$$\langle r,\theta,\phi|n,\ell,m\rangle = \langle r|n(\ell)\rangle\langle\theta|\ell(m)\rangle\langle\phi|m\rangle$$
$$= u_{n\ell}(r)u_{\ell m}(\theta)u_m(\phi).$$

The eigenvalue equations for $u_m(\phi)$, $u_{\ell m}(\theta)$, and $u_{n\ell}(r)$ are given by Equations 6.180, 6.188, and 6.253, respectively.

Let us relabel our states as

$$|a,b\rangle \to |\ell,m\rangle. \tag{6.175}$$

We have, therefore, found the eigenvalues of $\vec{L}^2$ and L_3 as

$$\vec{L}^2|\ell,m\rangle = \hbar^2\ell(\ell+1)|\ell,m\rangle, \tag{6.176}$$

$$L_3|\ell,m\rangle = \hbar m|\ell,m\rangle, \tag{6.177}$$

and

$$L_\pm|\ell,m\rangle = \hbar\sqrt{(\ell \mp m)(\ell \pm m+1)}\,|\ell,m\pm 1\rangle. \tag{6.178}$$

Actually, "half-integer" values of ℓ ($\ell = 1/2, 3/2, 5/2, \ldots$) are not allowed. One way to see this is as follows. The eigenvalue equation for L_3 is

$$L_3|m\rangle = m\hbar|m\rangle. \tag{6.179}$$

Projecting both sides of Equation 6.179 into $\langle\phi|$ and calling $u_m(\phi) = \langle\phi|m\rangle$, we get

$$\frac{\hbar}{i}\frac{\partial}{\partial\phi}u_m(\phi) = m\hbar u_m(\phi). \tag{6.180}$$

The solution to Equation 6.180 is

$$u_m(\phi) = \frac{1}{\sqrt{2\pi}}e^{im\phi}, \tag{6.181}$$

which is normalized so that

$$\int_0^{2\pi} d\phi\,|u_m|^2 = 1. \tag{6.182}$$

Now consider half-integer values of ℓ. From Equation 6.181 it would seem that

$$\langle\phi|m\rangle = \frac{1}{\sqrt{2\pi}}e^{im\phi}, \tag{6.183}$$

where m is also required, by Equation 6.174, to take on half-integer values. Therefore, given Equation 6.183, we have

$$\langle\phi+2\pi|m\rangle = -\frac{1}{\sqrt{2\pi}}e^{im\phi}. \tag{6.184}$$

But $\langle\phi|$ and $\langle\phi+2\pi|$ pick out the *same* point in coordinate space. Therefore, the spatial wave functions of half-integer ℓ are not single valued, a condition we must require for the transformation between the $|\phi\rangle$ and $|m\rangle$ bases.

There are other arguments as to why half-integer ℓ values are not allowed. The end result is to limit ℓ to the values

$$\ell = 0,1,2,3,\ldots$$

and m to

$$m = -\ell, -\ell+1, \ldots, \ell-1, \ell$$

giving $(2\ell+1)$ m values for each ℓ.

6.10 ANGULAR MOMENTUM EIGENVECTORS AND SPHERICAL HARMONICS

Now we know the eigenvalues of $\vec{L}^2$ and L_3. We would like to find the explicit eigenvectors also. We write (the $Y_{\ell m}$ are called "spherical harmonics")

$$\begin{aligned} Y_{\ell m}(\theta,\phi) &\equiv \langle\theta,\phi|\ell,m\rangle \\ &= \langle\theta|\ell(m)\rangle\langle\phi|m\rangle \\ &\equiv u_{\ell m}(\theta)u_m(\phi). \end{aligned} \tag{6.185}$$

The eigenvalue equation for $u_m(\theta)$ is written in Equation 6.180, and its normalized solution is Equation 6.181. The eigenvalue equation for $u_{\ell m}(\theta)$ comes from

$$\vec{L}^2|\ell,m\rangle = \hbar^2\ell(\ell+1)|\ell,m\rangle. \tag{6.186}$$

We have to start out with $|\ell,m\rangle$ in Equation 6.186 (and not $|\ell(m)\rangle$) since we do not know the effect of $\vec{L}^2$ on $\langle\theta|$ but only on $\langle\theta,\phi|$ from Equations 6.97 and 6.98.

Projected into $\langle\theta,\phi|$ space using Equations 6.97 and 6.98 gives

$$\begin{aligned} &-\left[\frac{1}{\sin^2\theta}\frac{\partial^2}{\partial\phi^2} + \frac{1}{\sin\theta}\frac{\partial}{\partial\theta}\left(\sin\theta\frac{\partial}{\partial\theta}\right)\right]u_{\ell m}(\theta)u_m(\phi) \\ &= \ell(\ell+1)u_{\ell m}(\theta)u_m(\phi), \end{aligned} \tag{6.187}$$

or making the replacement $(\partial^2/\partial\phi^2) \to -m^2$ from Equation 6.180 and then dividing both sides by $u_m(\phi)$ we get

$$-\left[\frac{-m^2}{\sin^2\theta} + \frac{1}{\sin\theta}\frac{\partial}{\partial\theta}\left(\sin\theta\frac{\partial}{\partial\theta}\right)\right]u_{\ell m}(\theta) = \ell(\ell+1)u_{\ell m}(\theta). \tag{6.188}$$

The solution to this equation determines the eigenvectors $u_{\ell m}(\theta)$.

Rather than trying to solve Equation 6.188 directly, we will generate the solutions by operator techniques, using the ladder operators $L_\pm$.

We can construct all of the $|\ell,m\rangle$ by considering $|\ell,\ell\rangle$ and then applying L_- $(\ell-m)$ times:

$$\begin{aligned}(L_-)^{\ell-m}|\ell,\ell\rangle &= \hbar\sqrt{2\ell}(L_-)^{(\ell-m)-1}|\ell,\ell-1\rangle\\ &= \hbar^2\sqrt{2\ell}\sqrt{(2\ell-1)2}(L_-)^{(\ell-m)-2}|\ell,\ell-2\rangle\\ &= \hbar^3\sqrt{2\ell}\sqrt{(2\ell-1)2}\sqrt{(2\ell-2)3}(L_-)^{(\ell-m)-3}|\ell,\ell-3\rangle,\end{aligned} \tag{6.189}$$

or, in general, after $(\ell-m)$ applications of L_-:

$$(L_-)^{(\ell-m)}|\ell,\ell\rangle = (\hbar)^{\ell-m}\underbrace{\sqrt{2\ell}\sqrt{(2\ell-1)2}\ldots\sqrt{(2\ell-(\ell-m-1))(\ell-m)}}_{(\ell-m\text{ factors})}|\ell,m\rangle. \tag{6.190}$$

Let us look at some of the individual pieces that make up the overall factor in Equation 6.190. We recognize the combination,

$$(2\ell)(2\ell-1)\cdots\underbrace{(2\ell-(\ell-m-1))}_{(\ell+m+1)}$$

in Equation 6.190, which we can write as

$$\frac{(2\ell)(2\ell-1)\cdots(2)(1)}{(\ell+m)(\ell+m-1)\cdots(2)(1)} = \frac{(2\ell)!}{(\ell+m)!}. \tag{6.191}$$

We also have the combination

$$(1)(2)\cdots(\ell-m)=(\ell-m)! \tag{6.192}$$

Therefore, we may write Equation 6.190 as

$$(L_-)^{\ell-m}|\ell,\ell\rangle = (\hbar)^{\ell-m}\sqrt{\frac{(2\ell)!(\ell-m)!}{(\ell+m)!}}|\ell,m\rangle, \tag{6.193}$$

or, solving for $|\ell,m\rangle$,

$$|\ell,m\rangle = \frac{1}{(\hbar)^{\ell-m}}\sqrt{\frac{(\ell+m)!}{(2\ell)!(\ell-m)!}}(L_-)^{\ell-m}|\ell,\ell\rangle. \tag{6.194}$$

I remind you of the effect of L_1 and L_2 (compare with Equations 6.93 and 6.94; here I am using the $\langle\hat{n}| = \langle\theta,\phi|$ notation):

$$\langle\hat{n}|L_1 = \frac{\hbar}{i}\left(-\sin\phi\frac{\partial}{\partial\theta} - \cos\phi\cot\theta\frac{\partial}{\partial\phi}\right)\langle\hat{n}|, \tag{6.195}$$

$$\langle\hat{n}|L_2 = \frac{\hbar}{i}\left(\cos\phi\frac{\partial}{\partial\theta} - \sin\phi\cot\theta\frac{\partial}{\partial\phi}\right)\langle\hat{n}|. \tag{6.196}$$

So therefore

$$\langle\hat{n}|L_+ = \frac{\hbar}{i}\left(\overbrace{(-\sin\phi+i\cos\phi)}^{ie^{i\phi}}\frac{\partial}{\partial\theta} \quad -\overbrace{(\cos\phi+i\sin\phi)}^{e^{i\phi}}\cot\theta\frac{\partial}{\partial\phi}\right)\langle\hat{n}| \tag{6.197}$$

or

$$\langle\hat{n}|L_+ = \hbar e^{i\phi}\left(\frac{\partial}{\partial\theta}+i\cot\theta\frac{\partial}{\partial\phi}\right)\langle\hat{n}|. \tag{6.198}$$

Also

$$\langle\hat{n}|L_- = \frac{\hbar}{i}\left((-\sin\phi-i\cos\phi)\frac{\partial}{\partial\theta} \quad -(\cos\phi-i\sin\phi)\cot\theta\frac{\partial}{\partial\phi}\right)\langle\hat{n}| \tag{6.199}$$

or

$$\langle\hat{n}|L_- = -\hbar e^{-i\phi}\left(\frac{\partial}{\partial\theta}-i\cot\theta\frac{\partial}{\partial\phi}\right)\langle\hat{n}|. \tag{6.200}$$

Now consider

$$\langle\hat{n}|\,L_-\,|\ell,m\rangle = -\hbar e^{-i\phi}\left(\frac{\partial}{\partial\theta}-i\cot\theta\frac{\partial}{\partial\phi}\right)\langle\hat{n}|\ell,m\rangle, \tag{6.201}$$

where $\langle\hat{n}|\ell,m\rangle \equiv Y_{\ell m}(\hat{n})$. We know that

$$Y_{\ell m}(\theta,\phi) = u_{\ell m}(\theta)\frac{1}{\sqrt{2\pi}}e^{im\phi}, \tag{6.202}$$

where $m = 0, \pm1, \pm2, \ldots, \pm\ell$ ($2\ell+1$ values). Therefore

$$\langle\hat{n}|\,L_-\,|\ell,m\rangle = -\frac{\hbar}{\sqrt{2\pi}}e^{i(m-1)\phi}\left(\frac{d}{d\theta}+m\cot\theta\right)u_{\ell m}(\theta). \tag{6.203}$$

Consider the identity:

$$\begin{aligned}
&(\sin\theta)^{1-m}\frac{d}{d\cos\theta}[(\sin\theta)^m F(\theta)] \\
&= (\sin\theta)^{1-m}\underbrace{\frac{d\theta}{d\cos\theta}}_{-\frac{1}{\sin\theta}}\underbrace{\frac{d}{d\theta}[(\sin\theta)^m F(\theta)]}_{\left[(\sin\theta)^m\frac{dF(\theta)}{d\theta}+m(\sin\theta)^{m-1}\cos\theta F(\theta)\right]} \\
&= -\left(\frac{d}{d\theta}+m\cot\theta\right)F(\theta).
\end{aligned} \tag{6.204}$$

But the right-hand side of Equation 6.204 is the same as the structure in Equation 6.203, so we may make the replacement

$$\langle\hat{n}|\,L_-|\ell,m\rangle = \frac{\hbar}{\sqrt{2\pi}}e^{i(m-1)\phi}\sin(\theta)^{1-m}\frac{d}{d\cos\theta}[(\sin\theta)^m u_{\ell m}(\theta)]. \tag{6.205}$$

Now consider

$$\langle\hat{n}|\,L_-L_-\,|\ell,m\rangle = -\hbar e^{-i\phi}\left(\frac{\partial}{\partial\theta} - i\cot\theta\frac{\partial}{\partial\phi}\right)\langle\hat{n}|\,L_-\,|\ell,m\rangle$$

$$= \frac{-(\hbar)^2}{\sqrt{2\pi}}e^{i(m-2)\phi}\left(\frac{d}{d\theta} + (m-1)\cot\theta\right) \tag{6.206}$$

$$\cdot\left[(\sin\theta)^{1-m}\frac{d}{d\cos\theta}((\sin\theta)^m u_{\ell m}(\theta))\right].$$

Employing Equation 6.204 again in Equation 6.206 (with $m \to m-1$) gives

$$\langle\hat{n}|\,(L_-)^2\,|\ell,m\rangle = \frac{(\hbar)^2}{\sqrt{2\pi}}e^{i(m-2)\phi}(\sin\theta)^{2-m}\frac{d^2}{d\cos\theta^2}[(\sin\theta)^m u_{\ell m}(\theta)]. \tag{6.207}$$

Seeing the pattern that seems to have developed, we may now prove by induction that

$$\langle\hat{n}|(L_-)^k\,|\ell,m\rangle = \frac{(\hbar)^k}{\sqrt{2\pi}}e^{i(m-k)\phi}(\sin\theta)^{k-m}\left(\frac{d}{d\cos\theta}\right)^k[(\sin\theta)^m u_{\ell m}(\theta)]. \tag{6.208}$$

Let us set $m = \ell$ and $k = \ell - m$ in Equation 6.208:

$$\langle\hat{n}\,|(L_-)^{\ell-m}|\ell,\ell\rangle = \frac{(\hbar)^{\ell-m}}{\sqrt{2\pi}}e^{im\phi}(\sin\theta)^{-m}\left(\frac{d}{d\cos\theta}\right)^{\ell-m}[(\sin\theta)^{\ell} u_{\ell\ell}(\theta)]. \tag{6.209}$$

But from Equation 6.194

$$\langle\hat{n}|\ell,m\rangle = (\hbar)^{m-\ell}\sqrt{\frac{(\ell-m)!}{(2\ell)!(\ell-m)!}}\,\langle\hat{n}|(L_-)^{\ell-m}|\,\ell,\ell\rangle, \tag{6.210}$$

and so we find

$$Y_{\ell m}(\theta,\phi) = \frac{e^{im\phi}}{\sqrt{2\pi}}\sqrt{\frac{(\ell+m)!}{(2\ell)!(\ell-m)!}}(\sin\theta)^{-m}\left(\frac{d}{d\cos\theta}\right)^{\ell-m}[(\sin\theta)^{\ell} u_{\ell\ell}(\theta)]. \tag{6.211}$$

Therefore, we will have a general expression for *all* the $Y_{\ell m}(\hat{n})$ if we can find the explicit expression for $u_{\ell\ell}(\theta)$. Now remember that

$$L_+|\ell,\ell\rangle = 0, \tag{6.212}$$

so that

$$\langle\hat{n}|L_+\,|\,\ell,\ell\rangle = \hbar e^{i\phi}\left(\frac{\partial}{\partial\theta} + i\cot\theta\frac{\partial}{\partial\phi}\right)\underbrace{\langle\hat{n}\,|\ell,\ell\rangle}_{\frac{e^{i\ell\phi}}{\sqrt{2\pi}}u_{\ell\ell}(\theta)}, \tag{6.213}$$

which gives us the first-order differential equation:

$$\left[\frac{d}{d\theta} - \ell\cot\theta\right]u_{\ell\ell}(\theta) = 0. \tag{6.214}$$

It is easy to check that the solution to Equation 6.214 is

$$u_{\ell\ell}(\theta) = C_\ell(\sin\theta)^\ell. \tag{6.215}$$

(Do it.) C_ℓ is an unknown constant that is determined by the normalization condition

$$\int d\Omega_{\hat{n}} \left| u_{\ell\ell}(\theta)\frac{e^{i\ell\phi}}{\sqrt{2\pi}} \right|^2 = 1. \tag{6.216}$$

(Equation 6.216 can be viewed as saying the probability of seeing the particle somewhere in angular space is unity.) Explicitly, this gives

$$|C_\ell|^2 \int_0^\pi d\theta \sin\theta(\sin\theta)^{2\ell} = 1, \tag{6.217}$$

or

$$|C_\ell|^2 \int_{-1}^1 d(\cos\theta)(\sin\theta)^{2\ell} = 1. \tag{6.218}$$

Setting

$$x = \cos\theta, \tag{6.219}$$

we then get

$$|C_\ell|^2 \int_{-1}^1 (dx)(1-x^2)^\ell = 1. \tag{6.220}$$

The integral in Equation 6.220 can be done by parts (this will be a homework problem) to yield

$$|C_\ell|^2 \left(2\frac{(2^\ell \ell!)^2}{(2\ell+1)!} \right) = 1. \tag{6.221}$$

It is conventional to choose the phase such that

$$C_\ell = \frac{(-1)^\ell}{2^\ell \ell!}\sqrt{\frac{(2\ell+1)!}{2}}, \tag{6.222}$$

and so

$$u_{\ell\ell}(\theta) = \frac{(-1)^\ell}{2^\ell \ell!}\sqrt{\frac{(2\ell+1)!}{2}}(\sin\theta)^\ell. \tag{6.223}$$

Using Equation 6.223 in Equation 6.211 now gives the general result

$$Y_{\ell m}(\theta,\phi) = \frac{(-1)^\ell e^{im\phi}}{2^\ell \ell!}\sqrt{\frac{(2\ell+1)(\ell+m)!}{4\pi(\ell-m)!}}(\sin\theta)^{-m}\left(\frac{d}{d\cos\theta}\right)^{\ell-m}[(\sin\theta)^{2\ell}]. \tag{6.224}$$

6.11 COMPLETENESS AND OTHER PROPERTIES OF SPHERICAL HARMONICS

Because of the normalization of the spherical basis, Equation 6.62, we have

$$\langle \ell',m' | \left[\int d\Omega_{\hat{n}} |\hat{n}\rangle\langle\hat{n}| = 1 \right] |\ell,m\rangle, \tag{6.225}$$

which means that

$$\int d\Omega_{\hat{n}} Y^{*}_{\ell' m'}(\hat{n}) Y_{\ell m}(\hat{n}) = \delta_{\ell\ell'}\delta_{mm'} \tag{6.226}$$

Equation 6.226 is completeness for spherical harmonics in angular space. We also have

$$\langle \hat{n} | \left[\sum_{\ell,m} |\ell,m\rangle\langle \ell,m| = 1 \right] |\hat{n}'\rangle, \tag{6.227}$$

or

$$\sum_{\ell,m} Y_{\ell m}(\hat{n}) Y^{*}_{\ell m}(\hat{n}') = \langle \hat{n} | \hat{n}'\rangle, \tag{6.228}$$

which expresses completeness of the $|\ell,m\rangle$ basis states. The $\langle \hat{n}|\hat{n}'\rangle$ is a *spherical* Dirac delta function. We can get an explicit form for it by requiring that

$$\begin{aligned} 1_{\hat{n}} &= \int d\Omega_{\hat{n}} \, |\hat{n}\rangle\langle \hat{n}| \int d\Omega_{\hat{n}'} |\hat{n}'\rangle\langle \hat{n}'|, \\ &= \int d\Omega_{\hat{n}} d\Omega_{\hat{n}'} |\hat{n}\rangle\langle \hat{n}|\hat{n}'\rangle\langle \hat{n}'|, \\ &= \int d(\cos\theta) d(\cos\theta') d\phi d\phi' |\theta,\phi\rangle\langle\theta,\phi|\theta',\phi'\rangle\langle\theta',\phi'|. \end{aligned} \tag{6.229}$$

Comparing this with the original expression

$$1_{\hat{n}} = \int d(\cos\theta) d\phi |\theta,\phi\rangle\langle\theta,\phi|, \tag{6.230}$$

means that we may take

$$\langle \hat{n} | \hat{n}'\rangle = \delta(\cos\theta - \cos\theta')\delta(\phi - \phi'). \tag{6.231}$$

[If we wish, we may use the delta function rule (see Chapter 2, Problem 2.9.2)

$$\delta(f(x)) = \sum_i \frac{1}{\left| \frac{df}{dx}(x_i) \right|} \delta(x - x_i),$$

where the sum is over the simple zeros of $f(x)$, located at $x = x_i$, to write

$$\delta(\cos\theta - \cos\theta') = \frac{1}{\sin\theta'} \delta(\theta - \theta'),$$

where both θ and θ' are assumed to be in the range from 0 to π.] Therefore

$$\sum_{\ell=0}^{\infty} \sum_{m=-\ell}^{\ell} Y_{\ell m}(\theta,\phi) Y^{*}_{\ell m}(\theta',\phi') = \delta(\cos\theta - \cos\theta')\delta(\phi - \phi'). \tag{6.232}$$

Notice we have not really *proven* either Equation 6.226 or 6.232; the proofs require more sophisticated analysis.

A useful connection is

$$Y_{\ell 0}(\theta,\phi)=\frac{(-1)^{\ell}}{2^{\ell}\ell!}\sqrt{\frac{(2\ell+1)}{4\pi}}\left(\frac{\mathrm{d}}{\mathrm{d}\cos\theta}\right)^{\ell}(\sin\theta)^{2\ell}$$
$$\equiv\sqrt{\frac{(2\ell+1)}{4\pi}}P_{\ell}(\cos\theta), \tag{6.233}$$

where $P_{\ell}(x)$ is called a "Legendre polynomial." (Notice that when $m=0$, there is no ϕ dependence in $Y_{\ell 0}(\theta,\phi)$.) From Equation 6.233, we deduce that

$$P_{\ell}(x)=\frac{1}{2^{\ell}\ell!}\left(\frac{\mathrm{d}}{\mathrm{d}x}\right)^{\ell}(x^2-1)^{\ell}. \tag{6.234}$$

I am not going to go through all of the explicit steps, but in the same way that I showed Equation 6.194 to be true by operating $(\ell-m)$ times with L_- on $|\ell,\ell\rangle$ to give $|\ell,m\rangle$, we can also start at the other end and operate $(\ell+m)$ times with L_+ on $|\ell,-\ell\rangle$ to give $|\ell,m\rangle$. The result is

$$|\ell,m\rangle=(\hbar)^{-(\ell+m)}\sqrt{\frac{(\ell-m)!}{(2\ell)!(\ell+m)!}}(L_+)^{\ell+m}|\ell,-\ell\rangle. \tag{6.235}$$

We can also show that

$$\langle\hat{n}|(L_+)^{\ell+m}|\ell,-\ell\rangle=\frac{(\hbar)^{(\ell+m)}}{\sqrt{2\pi}}(-1)^{\ell+m}\mathrm{e}^{im\phi}(\sin\theta)^{m}\left(\frac{\mathrm{d}}{\mathrm{d}\cos\theta}\right)^{\ell+m}[(\sin\theta)^{\ell}u_{\ell-\ell}(\theta)]. \tag{6.236}$$

From our earlier $Y_{\ell m}(\theta,\phi)$ expression, Equation 6.224, we can show that

$$Y_{\ell-\ell}(\theta,\phi)=\frac{\mathrm{e}^{-i\ell\phi}}{2^{\ell}\ell!}\sqrt{\frac{(2\ell+1)!}{4\pi}}(\sin\theta)^{\ell}, \tag{6.237}$$

from which we can identify

$$u_{\ell-\ell}(\theta)=\frac{1}{2^{\ell}\ell!}\sqrt{\frac{(2\ell+1)!}{2}}(\sin\theta)^{\ell}. \tag{6.238}$$

(Notice that unlike Equation 6.223 there is no factor of $(-1)^{\ell}$ here.) The preceding steps then lead in the same manner as before to the alternate expression

$$Y_{\ell m}(\theta,\phi)=\frac{(-1)^{m+\ell}\mathrm{e}^{im\phi}}{2^{\ell}\ell!}\sqrt{\frac{(2\ell+1)(\ell-m)!}{4\pi(\ell+m)!}}(\sin\theta)^{m}\left(\frac{\mathrm{d}}{\mathrm{d}\cos\theta}\right)^{\ell+m}[(\sin\theta)^{2\ell}]. \tag{6.239}$$

You should now go back to Equation 6.224 and carefully compare it to Equation 6.239. By using the expression in Equation 6.224 when $m\geq 0$ and the expression Equation 6.239 when $m\leq 0$, one may also write, for example

$$Y_{\ell m}(\theta,\phi)$$
$$=\frac{(-1)^{(m-|m|)/2}(-1)^{\ell}}{2^{\ell}\ell!}\mathrm{e}^{im\phi}\sqrt{\frac{(2\ell+1)(\ell+|m|)!}{4\pi(\ell-|m|)!}}(\sin\theta)^{-|m|}\left(\frac{\mathrm{d}}{\mathrm{d}\cos\theta}\right)^{\ell-|m|}[(\sin\theta)^{2\ell}]. \tag{6.240}$$

We can read off from Equation 6.240 the symmetry property

$$(-1)^m Y_{\ell-m}(\theta,-\phi) = Y_{\ell m}(\theta,\phi), \tag{6.241}$$

but since

$$Y^*_{\ell m}(\theta,\phi) = Y_{\ell m}(\theta,-\phi), \tag{6.242}$$

we may write Equation 6.241 as

$$(-1)^m Y^*_{\ell-m}(\theta,\phi) = Y_{\ell m}(\theta,\phi). \tag{6.243}$$

We can tie this discussion into the parity operator (introduced in Section 3.2) for which, by definition

$$\langle \hat{n} | \mathbb{P} = \langle -\hat{n} |, \tag{6.244}$$

where we may take

$$\langle -\hat{n} | = \begin{cases} \langle \pi-\theta, \phi+\pi |, & \text{if } \phi < \pi \\ \langle \pi-\theta, \phi-\pi |, & \text{if } \phi \geq \pi \end{cases}. \tag{6.245}$$

Now we have that

$$\begin{aligned} \cos(\pi-\theta) &= -\cos\theta, \\ \sin(\pi-\theta) &= \sin\theta, \end{aligned} \tag{6.246}$$

which helps us to see from Equation 6.240 that

$$Y_{\ell m}(\pi-\theta, \phi \pm \pi) = Y_{\ell m}(\theta,\phi) \underbrace{e^{\pm im\pi}(-1)^{\ell-|m|}}_{(-1)^\ell}. \tag{6.247}$$

Therefore

$$\langle \hat{n} | \mathbb{P} | \ell, m\rangle = \langle -\hat{n} | \ell, m\rangle = (-1)^\ell \langle \hat{n} | \ell, m\rangle. \tag{6.248}$$

Since Equation 6.248 is true for all $\langle \hat{n} |$, we have that

$$\mathbb{P}|\ell,m\rangle = (-1)^\ell |\ell,m\rangle. \tag{6.249}$$

The words that go with Equation 6.249 say "the parity of the state $|\ell, m\rangle$ is $(-1)^\ell$."

6.12 RADIAL EIGENFUNCTIONS

We return to the Schrödinger equation in spherical coordinates, Equation 6.118. We now know that

$$u_{a'}(\vec{r}) = u_{n\ell}(r) Y_{\ell m}(\theta,\phi), \tag{6.250}$$

and also that

$$L^2_{\text{op}} \langle \hat{n} | \ell, m\rangle = \langle \hat{n} | \vec{L}^2 | \ell, m\rangle = \hbar^2 \ell(\ell+1) \langle \hat{n} | \ell, m\rangle, \tag{6.251}$$

or

$$L_{\text{op}}^2 Y_{\ell m}(\theta,\phi) = \hbar^2 \ell(\ell+1) Y_{\ell m}(\theta,\phi), \tag{6.252}$$

so that Equation 6.118 is equivalent to

$$\left[-\frac{\hbar^2}{2mr^2}\left(\frac{\partial}{\partial r}\left(r^2\frac{\partial}{\partial r}\right) - \ell(\ell+1)\right) + V(r)\right] u_{n\ell}(r) = E_{n\ell}\, u_{n\ell}(r). \tag{6.253}$$

Notice that the magnetic quantum number, m, does not enter in Equation 6.253. This equation determines the energy levels of the system; therefore, the energies are independent of m for a problem with spherical symmetry and we have a $2\ell + 1$ fold degeneracy (at least) of each energy level labeled by $(n\,\ell)$. Also note that, as stated earlier, ℓ enters this equation simply as a *parameter*; the *quantum number* determined by this equation is n. In the next chapter, we will examine solutions to Equation 6.253 for various forms for the potential $V(r)$.

PROBLEMS

6.1.1 The Hamiltonian of a three-dimensional harmonic oscillator is

$$H = \frac{p_1^2 + p_2^2 + p_3^2}{2m} + \frac{1}{2} m\omega^2 (x_1^2 + x_2^2 + x_3^2).$$

(a) Given the wave function and energy levels of the one-dimensional harmonic oscillator as given in Chapter 3, find the wave functions and allowed energies for the three-dimensional case. [Hint: The Schrödinger equation is separable in the three dimensions. Just *use*, do not *redo*, the Chapter 3 calculation.]

(b) What is the degree of degeneracy of the first three energy levels?

(c) Find a formula that expresses the degree of degeneracy of the nth energy level. [Hint: The summation formula,

$$\sum_{i=1}^{n} i = \frac{n}{2}(n+1),$$

is useful here.]

6.2.1 The three-dimensional time-dependent Schrödinger equation for the Hamiltonian Equation 6.15 is

$$i\hbar \frac{\partial \Psi(\vec{x},t)}{\partial t} = -\frac{\hbar^2}{2m}\vec{\nabla}^2 \psi(\vec{x},t) + V(\vec{x})\psi(x,t).$$

Use this to show that the three-dimensional particle probability current,

$$\vec{J}(\vec{x},t) \equiv \frac{\hbar}{m}\text{Im}(\psi^*(\vec{x},t)\vec{\nabla}\psi(\vec{x},t)),$$

satisfies

$$\frac{\partial}{\partial t}(\psi^*\psi) + \vec{\nabla} \cdot \vec{J}(\vec{x},t) = 0.$$

(This generalizes Equations 2.128 through 2.130.)

6.3.1 Find a complete set of commuting operators that uniquely specifies the independent states of the three-dimensional infinite square well.

6.5.1 Can one measure a particle's momentum, $\vec{p}$, and angular momentum, $\vec{L}$, along the same axis? What quantity must you compute to answer this question? Compute it!

6.5.2 Show, from the fundamental commutation relations,
(a) $[L_1, p_1] = 0$,
(b) $[L_1, p_2] = i\hbar p_3$.
(c) $[L_1, x_1] = 0$,
(d) $[L_1, x_2] = i\hbar x_3$.
[The cyclic summary of these is $[L_i, p_j] = i\hbar\varepsilon_{ijk}\, p_k$ and $[L_i, x_j] = i\hbar\varepsilon_{ijk}\, x_k$, where there is an understood sum on the k index.]

***6.6.1** Using Equations 6.93 through 6.95, show that Equation 6.96 is true.

6.6.2 (a) Show that:

$$[L_3, \phi] = -i\hbar.$$

(ϕ is an operator whose eigenvalue is the spherical azimuthal angle:

$$\phi|\phi'\rangle = \phi'|\phi'\rangle.)$$

(b) Apply (a) to evaluate the quantity:

$$e^{-iL_3\phi'/\hbar}\,\phi\, e^{iL_3\phi'/\hbar} = ?$$

(ϕ' is a number.)

6.8.1 Show the angular analog of the Ehrenfest theorem (from Chapter 4. Section 4.13):

$$\frac{d\langle\vec{L}\rangle}{dt} = -\langle\vec{x}\times\vec{\nabla}V(\vec{x})\rangle.$$

[Hints: Notice we are assuming $V(\vec{x})$ and not $V(r)$. First argue that $i\hbar(d\langle\vec{L}\rangle/dt) = -\langle[H,\vec{L}]\rangle$, then evaluate the commutator.]

6.9.1 (a) Given an operator U with the properties

$$[\vec{L}^2, U] = 0$$
$$[L_3, U] = -\ell\hbar U,$$

show that

$$U\,|\,\ell,\ell\rangle = \text{const.}\,|\,\ell,0\rangle.$$

(b) Given an operator V with the properties

$$[L_+, V] = 0,$$
$$[L_3, V] = \hbar V,$$

show that

$$V\,|\,\ell,\ell\rangle = \text{const.}\,|\,\ell+1,\ell+1\rangle.$$

(c) Given an operator W with the properties,

$$[L_-, W] = 0,$$
$$[L_3, W] = -\hbar W,$$

find:

$$W\,|\,\ell,-\ell\rangle = ?$$

6.9.2 *Prove*: The expectation value of the square of a Hermitian operator is nonnegative. (This is essentially a one-line proof.)

6.9.3 Given $[L_1, A] = [L_2, A] = 0$, show that $[L_3, A] = 0$.

6.9.4 Assume a particle has an orbital angular momentum with $L_3' = \hbar m$ and $\vec{L}^{2\prime} = \hbar^2 \ell(\ell+1)$. Show that in this state we have the expectation values:
(a) $\langle L_1 \rangle = \langle L_2 \rangle = 0$,
(b) $\langle L_1^2 \rangle = \langle L_2^2 \rangle = \frac{1}{2}\hbar^2(\ell(\ell+1) - m^2)$.

6.10.1 Prove Equation 6.208 by induction. (Remember the meaning of this term from Problem 3.5.2 in Chapter 3.)

6.10.2 Do the integral in Equation 6.220.

6.10.3 (a) Using Equation 6.224, construct the explicit forms for the spherical harmonics:

$$Y_{00},\ Y_{11},\ Y_{10},\ Y_{1-1}.$$

Obviously, you should show your work!

$$Y_{00} = \left(\frac{1}{4\pi}\right)^{1/2},$$

$$Y_{11} = -\left(\frac{3}{8\pi}\right)^{1/2} \sin\theta\, e^{i\phi},$$

$$Y_{10} = \left(\frac{3}{4\pi}\right)^{1/2} \cos\theta,$$

$$Y_{1-1} = \left(\frac{3}{8\pi}\right)^{1/2} \sin\theta\, e^{-i\phi}.$$

(b) Do some others for fun:

$$Y_{22} = \left(\frac{15}{32\pi}\right)^{1/2} \sin^2\theta\, e^{2i\phi},$$

$$Y_{21} = -\left(\frac{15}{8\pi}\right)^{1/2} \sin\theta \cos\theta e^{i\phi},$$

$$Y_{20} = \left(\frac{5}{16\pi}\right)^{1/2} (3\cos^2\theta - 1),$$

$$Y_{2-1} = \left(\frac{15}{8\pi}\right)^{1/2} \sin\theta \cos\theta\, e^{-i\phi},$$

$$Y_{2-2} = \left(\frac{15}{32\pi}\right)^{1/2} \sin^2\theta\, e^{-2i\phi}.$$

6.10.4 The wave function of a bound particle is given by

$$\Psi(\vec{r},0) = x_1 x_3\, f(r),$$

where $f(r)$ is a given function of $r = |\vec{r}\,|$.

(a) If $\vec{L}^2$ is measured at $t = 0$, what value is found?

(b) What possible values of L_3 will measurement find at $t = 0$, and with what probability will they occur? [Hint: Consult the $Y_{\ell m}$'s from Problem 6.10.3.]

6.11.1 Show Equation 6.235 for $|\ell, m\rangle$ by operating $(\ell + m)$ times with L_+ on $|\ell, -\ell\rangle$.

II Quantum Particles

The Three-Dimensional Radial Equation

Synopsis: Attention is now turned to a number of spherically symmetric problems. After a recap, readers are guided through the mathematics associated with free-particle wave functions. A simple nuclear shell model is constructed using the infinite spherical well. Two-body formalism is introduced in the context of the finite spherical well, viewed as a crude neutron–proton (deuteron) potential. Additional applications are made to the Coulomb potential for atomic systems and the confined Coulombic potential as a model of the heavy–light quark system.

7.1 RECAP OF THE SITUATION

Let us recap the situation. We started out as usual by projecting

$$H\,|a'\rangle = E_{a'}\,|a'\rangle, \tag{7.1}$$

where

$$H = \frac{\vec{p}^{\,2}}{2m} + V(r), \tag{7.2}$$

into a spherical basis:

$$\langle \vec{r}\,|\left(\frac{\vec{p}^{\,2}}{2m} + V(r)\right)|\,a'\rangle = E_{a'}\langle \vec{r}\,|\,a'\rangle. \tag{7.3}$$

Using the definition

$$u_{a'}(\vec{r}) \equiv \langle \vec{r}\,|\,a'\rangle \tag{7.4}$$

and the result

$$\langle \vec{r}\,|\,\vec{p}^{\,2}\,|\,a'\rangle = \left[-\frac{\hbar^2}{r^2}\frac{\partial}{\partial r}\left(r^2\frac{\partial}{\partial r}\right) + \frac{L_{op}^2}{r^2}\right]u_{a'}(\vec{r}), \tag{7.5}$$

we got

$$\left[-\frac{\hbar^2}{2mr^2}\left(\frac{\partial}{\partial r}\left(r^2\frac{\partial}{\partial r}\right) - \left(\frac{L_{op}^2}{\hbar^2}\right)\right) + V(r)\right]u_{a'}(\vec{r}) = E_{a'}u_{a'}(\vec{r}). \tag{7.6}$$

We know the solution to Equation 7.6 can be written as

$$u_{a'}(\vec{r}) = u_{n\ell}(r)Y_{\ell m}(\theta,\phi) \tag{7.7}$$

and the radial eigenvalue equation we are to solve is

$$\left[-\frac{\hbar^2}{2mr^2}\left(\frac{\partial}{\partial r}\left(r^2\frac{\partial}{\partial r}\right) - \ell(\ell+1)\right) + V(r)\right]u_{n\ell}(r) = E_{n\ell}u_{n\ell}(r). \tag{7.8}$$

We can cast this equation into a more convenient form by introducing

$$u_{n\ell}(r) = \frac{R_{n\ell}(r)}{r}, \tag{7.9}$$

for which we get

$$\left[-\frac{\hbar^2}{2m}\left(\frac{\mathrm{d}^2}{\mathrm{d}r^2} - \frac{\ell(\ell+1)}{r^2}\right) + V(r)\right]R_{n\ell}(r) = E_{n\ell}R_{n\ell}(r). \tag{7.10}$$

In this form, the radial 3-D eigenequation looks very much like a 1-D problem (see Chapter 3, Equation 3.14) with an *effective potential*

$$V_{\text{eff}}(r) = \begin{cases} V(r) + \dfrac{\hbar^2}{2m}\dfrac{\ell(\ell+1)}{r^2}, & r > 0 \\ \infty, & r \leq 0. \end{cases} \tag{7.11}$$

The "centrifugal energy" term, $\hbar^2\ell(\ell+1)/2mr^2$, should be familiar to you from classical mechanics, where it has the form $\vec{\ell}^2/2mr^2$, where $\vec{\ell}$ is the classical relative angular momentum. If this *were* a one-dimensional problem, the fact that the potential becomes infinite at $r = 0$ would imply that the particle would be unable to penetrate to the $r < 0$ region and we would expect that

$$R(0) = 0 \tag{7.12}$$

would be the correct boundary condition. Actually, the correct boundary condition (based on the requirement that H be Hermitian) is somewhat more complicated. Equation 7.12 will be sufficient for our purposes here.*

We are going to discuss the solution to Equation 7.8 for the following five problems:

- The free particle,

$$V(r) = 0.$$

- The infinite spherical well,

$$V(r) = \begin{cases} \infty, & r \geq a \\ 0, & r < a. \end{cases}$$

- The "deuteron" (really a finite potential well),

$$V(r) = \begin{cases} 0, & r \geq a \\ -V_0, & r < a. \end{cases}$$

- The Coulomb potential (via the 2-D harmonic oscillator),

$$V(r) = -\frac{Ze^2}{r}.$$

- The "confined Coulombic model" where,

$$V(r) = \begin{cases} \infty, & r \geq a \\ -\dfrac{\xi}{r}, & r < a. \end{cases}$$

* For more discussion, see E. Merzbacher, *Quantum Mechanics*, 3rd ed. (John Wiley & Sons, 1998), Section 12.4.

7.2 THE FREE PARTICLE

The description of a free particle in spherical coordinates is not trivial. We talked extensively about the description of the 1-D free particle in Chapter 2. The solution to the 1-D time-independent Schrödinger equation

$$-\frac{\hbar^2}{2m}\frac{\mathrm{d}^2 u_{p'_x}(x)}{\mathrm{d}x^2} = E u_{p'_x}(x), \tag{7.13}$$

with $u_{p'_x}(x) = \langle x|p'_x\rangle$ subject to the normalization

$$\left.\begin{aligned} &\int_{-\infty}^{\infty} \mathrm{d}p'_x\, |p'_x\rangle\langle p'_x| = 1, \\ \Rightarrow &\int_{-\infty}^{\infty} \mathrm{d}p'_x\, u_{p'_x}(x) u^*_{p'_x}(x') = \delta(x - x'), \end{aligned}\right\} \tag{7.14}$$

is just

$$u_{p'_x}(x) = \frac{1}{\sqrt{2\pi\hbar}} \mathrm{e}^{ixp'_x/\hbar}, \tag{7.15}$$

where $p'_x > 0$ solutions represent particles moving in the $+x$ direction, and similarly for $p'_x < 0$. [We could also, if we wish, adopt the energy normalization condition (recall Equation 2.179 and let $s = \mathrm{sign}(p'_x)$)

$$\sum_{s=\pm}\int_0^{\infty} \mathrm{d}E\,|E,s\rangle\langle E,s| = 1 \Rightarrow \sum_{s=\pm}\int_0^{\infty} \mathrm{d}E\; u_{E,s}(x) u^*_{E,s}(x') = \delta(x - x'),$$

which gives (k defined in Equation 7.17)

$$u_{E,s}(x) = \left(\frac{m}{16\pi^2\hbar^2 E}\right)^{1/4} \mathrm{e}^{iskx}.]$$

In spherical coordinates, we want to find solutions to (we relabel $n \to k$ in this problem)

$$-\frac{\hbar^2}{2m}\left[\frac{\mathrm{d}^2}{\mathrm{d}r^2} - \frac{\ell(\ell+1)}{r^2}\right] R_{k\ell}(r) = \pm\frac{\hbar^2 k^2}{2m} R_{k\ell}(r), \tag{7.16}$$

subject to the boundary condition Equation 7.12 and where $\ell = 0, 1, 2, \ldots$. We have defined

$$k \equiv \frac{\sqrt{2mE}}{\hbar}. \tag{7.17}$$

The $\pm$ sign on the right-hand side of Equation 7.16 corresponds to $E = \pm\hbar^2 k^2/2m$; that is, to whether E is positive or negative. Defining a new dimensionless variable

$$p \equiv kr, \tag{7.18}$$

we may write Equation 7.16 as

$$\left[\frac{\mathrm{d}^2}{\mathrm{d}p^2} - \frac{\ell(\ell+1)}{p^2} \pm 1\right] R_{k\ell}(p) = 0. \tag{7.19}$$

We can eliminate the possibility of negative energies here. Consider the case $\ell = 0$, $E < 0$:

$$\left[\frac{\mathrm{d}^2}{\mathrm{d}p^2} - 1\right] R_{k\ell}(p) = 0. \tag{7.20}$$

The solutions to Equation 7.20 are real exponentials:

$$R_{k\ell}(p) \sim \mathrm{e}^{\pm p}. \tag{7.21}$$

But these solutions cannot simultaneously satisfy the boundary condition Equation 7.12 and $R(\infty) = 0$, which would be appropriate for square-integrable bound states; similarly for the other $\ell \neq 0$ values. Thus, the physically relevant solutions in this problem have $E \geq 0$ and the equation we must solve is (except for $E = 0$ exactly)

$$\left[\frac{\mathrm{d}^2}{\mathrm{d}p^2} - \frac{\ell(\ell+1)}{p^2} + 1\right] R_{k\ell}(p) = 0. \tag{7.22}$$

Let us look at the $\ell = 0$ case of Equation 7.22:

$$\left[\frac{\mathrm{d}^2}{\mathrm{d}p^2} + 1\right] R_{k0}(p) = 0. \tag{7.23}$$

The solutions to Equation 7.23 are of course

$$R_{k0}(p) \sim \sin p, \quad \cos p. \tag{7.24}$$

However, only the first possibility on the right satisfies the boundary condition Equation 7.12.

To find the solutions for $\ell \neq 0$, let us define

$$R_{k\ell}(p) \equiv \left(-1\right)^{\ell} p^{\ell+1} X_{k\ell}(p). \tag{7.25}$$

Let us find the equation that $X_{k\ell}(p)$ satisfies. We have that

$$\frac{\mathrm{d}R_{k\ell}}{\mathrm{d}p} = (-1)^{\ell} \left\{ (\ell+1) p^{\ell} X_{k\ell} + p^{\ell+1} \frac{\mathrm{d}X_{k\ell}}{\mathrm{d}p} \right\}, \tag{7.26}$$

and

$$\frac{\mathrm{d}^2 R_{k\ell}}{\mathrm{d}p^2} = (-1)^{\ell} \left\{ (\ell+1) p^{\ell-1} X_{k\ell} + 2(\ell+1) p^{\ell} \frac{\mathrm{d}X_{k\ell}}{\mathrm{d}p} + p^{\ell+1} \frac{\mathrm{d}^2 X_{k\ell}}{\mathrm{d}p^2} \right\}. \tag{7.27}$$

Therefore, we have

$$\begin{aligned}
&\left[\frac{\mathrm{d}^2}{\mathrm{d}p^2} - \frac{\ell(\ell+1)}{p^2} + 1\right] R_{k\ell}(p) \\
&= (-1)^{\ell} \left\{ \ell(\ell+1) p^{\ell-1} X_{k\ell} + 2(\ell+1) p^{\ell} \frac{\mathrm{d}X_{k\ell}}{\mathrm{d}p} \right. \\
&\quad \left. + p^{\ell+1} \frac{\mathrm{d}^2 X_{k\ell}}{\mathrm{d}p^2} - \frac{\ell(\ell+1)}{p^2} p^{\ell+1} X_{k\ell} + p^{\ell+1} X_{k\ell} \right\} \qquad (7.28) \\
&= (-1)^{\ell} p^{\ell+1} \left[\frac{\mathrm{d}^2 X_{k\ell}}{\mathrm{d}p^2} + \frac{2(\ell+1)}{p} \frac{\mathrm{d}X_{k\ell}}{\mathrm{d}p} + X_{k\ell} \right].
\end{aligned}$$

Thus, the equation satisfied by $X_{k\ell}$ is

$$\left[\frac{\mathrm{d}^2}{\mathrm{d}p^2}+\frac{2(\ell+1)}{p}\frac{\mathrm{d}}{\mathrm{d}p}+1\right]X_{k\ell}(p)=0. \tag{7.29}$$

Differentiating Equation 7.29 yields

$$X'''_{k\ell}+\frac{2(\ell+1)}{p}X''_{k\ell}-\frac{2(\ell+1)}{p^2}X'_{k\ell}+X'_{k\ell}=0, \tag{7.30}$$

where the primes denote differentiation with respect to p. Now let us substitute

$$\left.\begin{aligned} X'_{k\ell}&=p\psi_{k\ell},\\ X''_{k\ell}&=\psi_{k\ell}+p\psi'_{k\ell},\\ X'''_{k\ell}&=2\psi'_{k\ell}+p\psi''_{k\ell},\end{aligned}\right\} \tag{7.31}$$

into Equation 7.30. We get

$$(2\psi'_{k\ell}+p\psi''_{k\ell})+\frac{2(\ell+1)}{p}(\psi_{k\ell}+p\psi'_{k\ell})-\frac{2(\ell+1)}{p^2}(p\psi_{k\ell})+p\psi_{k\ell}=0, \tag{7.32}$$

or

$$\psi''_{k\ell}+\frac{2(\ell+2)}{p}\psi'_{k\ell}+\psi_{\ell}=0. \tag{7.33}$$

Comparing with Equation 7.29, we thus conclude that we may choose

$$\psi_{k\ell}(p)=X_{k\ell+1}(p). \tag{7.34}$$

From the first of the equations in 7.31, we then have

$$X_{k\ell+1}(p)=\frac{1}{p}\frac{\mathrm{d}}{\mathrm{d}p}X_{k\ell}(p). \tag{7.35}$$

Therefore, by induction

$$X_{k\ell}(p)=\left(\frac{1}{p}\frac{\mathrm{d}}{\mathrm{d}p}\right)^{\ell}X_{k0}(p). \tag{7.36}$$

This implies that (from Equation 7.25)

$$R_{k\ell}(p)=(-1)^{\ell}p^{\ell+1}\left(\frac{1}{p}\frac{\mathrm{d}}{\mathrm{d}p}\right)^{\ell}X_{k0}(p). \tag{7.37}$$

But from Equations 7.24 and 7.25 (for $\ell=0$), we have

$$X_{k0}(p)\sim\frac{\sin p}{p}. \tag{7.38}$$

Equations 7.38 and 7.37 determine, outside of normalization, the full set of solutions to Equation 7.22 with the boundary condition Equation 7.12.

To connect to standard definitions, let us introduce the "spherical Bessel functions," $j_\ell(p)$ (which are solutions of Equation 7.8):

$$j_\ell(p) \equiv (-p)^\ell \left(\frac{1}{p}\frac{\mathrm{d}}{\mathrm{d}p}\right)^\ell \left[\frac{\sin p}{p}\right]. \tag{7.39}$$

Note that $j_0(p)$ goes to unity as p goes to zero, whereas the other $j_\ell(p)$ are zero at $p=0$.

What we have accomplished is to solve Equation 7.8

$$-\frac{\hbar^2}{2m}\left[\frac{1}{r^2}\frac{\partial}{\partial r}\left(r^2\frac{\partial}{\partial r}\right)-\frac{\ell(\ell+1)}{r^2}\right]u_{k\ell}(r)=\frac{\hbar^2k^2}{2m}u_{k\ell}(r), \tag{7.40}$$

subject to the boundary condition Equation 7.12. More abstractly (strictly speaking, a better notation for the right-hand side is $\langle r|k(\ell)\rangle$)

$$u_{k\ell}(r)=\langle r\,|\,k\ell\rangle. \tag{7.41}$$

Equation 7.40 is equivalent to

$$\langle r\,|\left(\frac{p_r^2}{2m}+\frac{\hbar^2\ell(\ell+1)}{2mr^2}\right)|k\ell\rangle\;=\frac{\hbar^2k^2}{2m}\langle r\,|\,k\ell\rangle, \tag{7.42}$$

where we have defined the Hilbert space operator p_r

$$\langle r\,|\,p_r \equiv \frac{\hbar}{i}\frac{1}{r}\frac{\partial}{\partial r}r\,\langle r\,|. \tag{7.43}$$

Since Equation 7.42 holds for all $\langle r\,|$, we then have

$$\left(\frac{p_r^2}{2m}+\frac{\hbar^2\ell(\ell+1)}{2mr^2}\right)|\,k\ell\rangle\;=\frac{\hbar^2k^2}{2m}|\,k\ell\rangle, \tag{7.44}$$

as the eigenvalue equation for $|\,k\ell\rangle$. The first operator on the left corresponds to the radial kinetic energy; the second is the angular kinetic energy. Notice that for a given value of k (and therefore the energy, E), there is an *infinite* degeneracy in ℓ and m. Physically, this corresponds to the fact that a particle with a given energy in 3-D has an infinite number of directions in which to travel, as opposed to the twofold degeneracy in 1-D space.

The $|\,k\ell\rangle$ are assumed to give a complete description and so we assume

$$\int_0^\infty \mathrm{d}k\,k^2\,|\,k\ell\rangle\langle k\ell\,|=1_r, \tag{7.45}$$

in spherical radial space for each ℓ value. Other normalizations of the states $|\,k\ell\rangle$ are possible. Equation 7.45 is written in analogy to the first of the equations in 6.61 (see Chapter 6), just like $\int \mathrm{d}p_1'\;|\,p_1'\;\rangle\langle p_1'\;|=1_1$ is the momentum space version of $\int \mathrm{d}x_1'\;|\,x_1'\;\rangle\langle x_1'\;|=1_1$. The consequence of Equation 7.45 is that

$$\begin{aligned}1_r\cdot 1_r &= \int \mathrm{d}k'\,k'^2\,|\,k'\ell\rangle\langle k'\ell\,|\int \mathrm{d}k\,k^2\,|\,k\ell\rangle\langle k\ell\,| \\ &= \int \mathrm{d}k'\,\mathrm{d}k\,k'^2k^2\,|\,k'\ell\rangle\langle k'\ell\,|\,k\ell\rangle\langle k\ell\,|.\end{aligned} \tag{7.46}$$

Comparison of Equations 7.46 and 7.45 reveals that

$$\langle k'\ell \,|\, k\ell\rangle = \frac{1}{k^2}\delta(k-k'). \tag{7.47}$$

In the same way from

$$1_r = \int_0^\infty \mathrm{d}r\, r^2 \,|\, r\rangle\langle r\,|, \tag{7.48}$$

we have

$$\langle r' \,|\, r\rangle = \frac{1}{r^2}\delta(r-r'). \tag{7.49}$$

Thus, the complete set of radial and angular eigenkets for the free particle is given by (being more precise in the r and θ space notation):

$$|\,k,\ell,m\rangle \equiv \underbrace{|\,k(\ell)\,\rangle}_{r\text{-space}} \otimes \underbrace{|\,\ell(m)\rangle}_{\theta\text{-space}} \otimes \underbrace{|\,m\rangle}_{\phi\text{-space}} \begin{pmatrix} 0<k<\infty \\ \ell = 0,1,2,\ldots \\ m=-\ell,\ldots,\ell \end{pmatrix}.$$

Of course, as pointed out at the beginning of this chapter, there is the alternate complete set,

$$|\,p_1',p_2',p_3'\,\rangle \equiv |\,p_1'\,\rangle \otimes |\,p_2'\,\rangle \otimes |\,p_3'\,\rangle, \quad (-\infty < p_1', p_2', p_3' < \infty)$$

based on a Cartesian description. Because both sets are complete, we should be able to expand one in terms of another after projection into the appropriate coordinate space. We will explore this connection further in Chapter 11, Section 11.8.

If we adopt Equation 7.45 as our normalization (similar to Equation 7.14 in the 1-D case), then we require

$$\left.\begin{aligned} &\langle r|\left[\int_0^\infty \mathrm{d}k\, k^2 \,|\, k\ell\rangle\langle k\ell| = 1\right]|\, r'\rangle \\ &\Rightarrow \int_0^\infty \mathrm{d}k k^2 u_{k\ell}(r) u_{k\ell}^*(r') = \frac{1}{r^2}\delta(r-r') \end{aligned}\right\}. \tag{7.50}$$

We have found that

$$u_{k\ell}(r) = C_\ell j_\ell(kr), \tag{7.51}$$

where C_ℓ is an unknown normalization factor. Putting Equation 7.51 into 7.50 gives

$$\int_0^\infty \mathrm{d}k\, k^2 \,|\, C_\ell\,|^2\, j_\ell(kr) j_\ell(kr') = \frac{1}{r^2}\delta(r-r'). \tag{7.52}$$

Equation 7.52 determines the normalization factor C_ℓ. The C_ℓ cannot be a function of k, since by the normalization condition (Equation 7.50), the $u_{k\ell}(r)$ are dimensionless. To get an explicit expression, it is necessary to perform the integration of the left of Equation 7.52. The mathematics involved in doing this is beyond the level of this discussion. I will simply quote the necessary result:*

$$\int_0^\infty dk\, k^2 j_\ell(kr) j_\ell(kr') = \frac{\pi}{2r^2}\delta(r-r'). \tag{7.53}$$

(You will, however, be asked to confirm this result for the $\ell = 0$ case. Equation 7.53 is, mathematically, only true for the case $r, r' > 0$.)

Comparing Equations 7.53 and 7.52, we find that

$$|C_\ell|^2 = \frac{2}{\pi}, \tag{7.54}$$

for which we choose

$$C_\ell = \sqrt{\frac{2}{\pi}}. \tag{7.55}$$

This implies finally that

$$u_{k\ell}(r) = \langle r \,|\, k\ell\rangle = \sqrt{\frac{2}{\pi}}\, j_\ell(kr). \tag{7.56}$$

The complete free-particle wave function is then

$$u_{k\ell m}(\vec{r}) \equiv \langle \vec{r} \,|\, k\ell, \ell, m\rangle = \langle r \,|\, k\ell\rangle\langle \hat{r} \,|\, \ell, m\rangle = \sqrt{\frac{2}{\pi}}\, j_\ell(kr) Y_{\ell m}(\theta,\phi). \tag{7.57}$$

The entire content of their completeness can be stated as

$$\begin{aligned} &\langle \vec{r}\,|\left[\int dk\, k^2 \sum_{\ell,m} |\, k\ell,\ell,m\rangle\langle k\ell,\ell,m\,| = 1\right]|\,\vec{r}'\rangle, \\ &\Rightarrow \int_0^\infty dk\, k^2 \sum_{\ell,m} u_{k\ell m}(\vec{r})\, u^*_{k\ell m}(\vec{r}\,') = \delta(\vec{r} - \vec{r}\,'), \end{aligned} \tag{7.58}$$

and as

$$\begin{aligned} &\langle k\ell', \ell', m'\,|\left[\int d^3r |\vec{r}\rangle\langle \vec{r}| = 1\right]|\, k\ell, \ell, m\rangle, \\ &\Rightarrow \int d^3r\, u^*_{k'\ell' m'}(\vec{r}) u_{k\ell m}(\vec{r}) = \delta_{\ell\ell'}\, \delta_{mm'}\, \frac{1}{k^2}\delta(k-k'). \end{aligned} \tag{7.59}$$

7.3 THE INFINITE SPHERICAL WELL POTENTIAL

The solutions for the free particle resulted in a continuous spectrum, as it should. We now imagine putting a particle into a spherical well that has a potential that rises to ∞ at the surface. This means the particle is trapped in the region $r < a$ and has zero probability of escaping from this spherical region. In the 1-D case we subjected the free-particle solutions,

$$u(x) \sim \sin(kx), \cos(kx),$$

* See K. Gottfried, *Quantum Mechanics* (W.A. Benjamin, 1966), Section 11.

to the boundary conditions

$$u(x)|_{x=\pm a}=0 \tag{7.60}$$

to find the allowed energies. We do a similar thing here in our radial 3-D description. The free-particle solutions are given by Equation 7.57, and our boundary condition

$$u_{klm}(\vec{r})|_{r=a}=0, \tag{7.61}$$

determines the energy levels as being solutions of

$$j_\ell(ka)=0. \tag{7.62}$$

Equation 7.62 allows only certain discrete values of ka as solutions. These are called *zeros* of the Bessel junction $j_\ell(ka)$. Let us look at the $\ell=0$ case:

$$j_0(ka)=0, \Rightarrow \frac{\sin(ka)}{ka}=0, \tag{7.63}$$

$$\Rightarrow ka=n\pi, \quad n=1,2,3,\ldots \tag{7.64}$$

In the $\ell=1$ case, however, we have

$$j_1(ka)=0 \Rightarrow \sin(ka)=(ka)\cos(ka). \tag{7.65}$$

The transcendental Equation 7.65 can be solved numerically to yield approximately

$$ka=4.493,7.725,10.90,\ \ldots \tag{7.66}$$

for the first three zeros. Tables of the zeros of Bessel functions are, for example, published in the National Bureau of Standards' *Handbook of Mathematical Functions*. The energies of these states are then given by

$$E=\frac{\hbar^2(ka)^2}{2ma^2}. \tag{7.67}$$

A convenient label that categorizes the solutions is the number of nodes (or zeros) in the radial wave function (for $r\neq 0$). The lowest energy solution for each ℓ value has a single node at $r=a$ (the surface). The next highest energy solution has two nodes, the next has three nodes, and so on. We will use $n=1, 2, 3, \ldots$ to label these solutions. (Notice that only in the case $\ell=0$ is ka proportional to n.) Defining

$$u_{n\ell}(r)\equiv N_\ell j_\ell(k_n r), \tag{7.68}$$

where k_n represents a solution with n nodes and N is a normalization factor determined by

$$\int_0^a dr\, r^2\,|u_{n\ell}(r)|^2=1, \tag{7.69}$$

we may give a schematic representation of these wave functions as in Figure 7.1. (Notice that only for $\ell=0$ states is $u_{n\ell}(0)\neq 0$.) The normalization factor in Equation 7.68 can be shown to be given by

$$N_\ell^2=\frac{2}{a^3 j_{\ell+1}^2(k_n a)}. \tag{7.70}$$

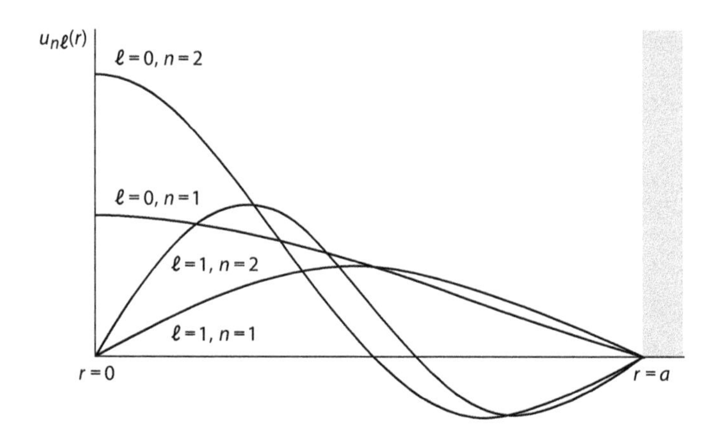

FIGURE 7.1 Schematic representation of the radial behavior of the four lowest energy wave functions for the infinite spherical well potential.

Some conventional terminology to denote these states is given next.

$$\left.\begin{array}{l}\ell=0\ s \text{ state "sharp"} \\ \ell=1\ p \text{ state "principle"} \\ \ell=2\ d \text{ state "diffuse"} \\ \ell=3\ f \text{ state}\end{array}\right\} \text{from the days of experimental spectral analysis.}$$

For $\ell=4, 5, \ldots$, the letter designation becomes alphabetical. In addition, one also sees the notation:

$\ell=0, n=1$: 1s *shell*
$\ell=0, n=2$: 2s *shell*
$\ell=1, n=1$: 1p *shell*
$\ell=1, n=2$: 2p *shell.*

This is called spectroscopic notation. The number in front of the letter in this case is just n, the number of radial ($r \neq 0$) nodes. It also indicates the energy level *order.* That is, the 1s state is the lowest energy $\ell=0$ state, the 2s state has the next highest energy for $\ell=0$, and so on.

Since the energies of the states are proportional to the dimensionless quantity $(ka)^2$, a listing of the first few levels gives an idea of the separations involved.

State	**1s**	**1p**	**1d**	**2s**	**1f**
$(ka)^2$	9.87	20.14	33.21	39.48	48.83
	2p	**1g**	**2d**	**1h**	**3s**
	59.68	66.96	82.72	87.53	88.83

Each energy level, of course, is in general degenerate: remember the $(2\ell+1)$ degeneracy in m of the ℓth level.

One of the reasons the spherical well is so interesting is because it turns out to represent a crude, and yet fairly accurate, model of the nucleus. Such models assume that the interaction between a single nucleon (either a proton or neutron) and all the other nucleons can be represented by an external potential, $V(r)$. This is called the *single-particle* or *shell* model of the nucleus and is described in all introductory nuclear physics books. This picture is, of course, a gross phenomenological oversimplification, but such simplifications are in fact necessary because of the extreme complexity of the problem.

Using the mysterious assumption that we can only place at most two neutrons or two protons in each energy level, known by early workers as *zweideutigkeit*, German for "two-valuedness," we can calculate the number of nucleons held by each shell. This assumption will have to wait until Chapter 9 to be explained. We came upon a similar assumption for electrons previously in the model of conductors and insulators in Chapter 3, Section 3.6.

Shell	# Protons $(Z) = 2(2\ell + 1)$	# Neutrons $(N) = 2(2\ell + 1)$	Additive Total
1s $(\ell = 0)$	2	2	2
1p $(\ell = 1)$	6	6	$2+6=8$
1d $(\ell = 2)$	10	10	$8+10=18$
2s $(\ell = 0)$	2	2	$18+2=20$
1f $(\ell = 3)$	14	14	$20+14=34$
2p $(\ell = 1)$	6	6	$34+6=40$
1g $(\ell = 4)$	18	18	$40+18=58$

Because of the fact that jumps in energy occur when a given shell is filled, we might expect that nuclei with the number of protons, Z, or number of neutrons, N, equal to the numbers in the right-hand column in the preceding table should be particularly stable. This is similar to the closed-shell effect for the Nobel gases in atomic physics. On the basis of the aforementioned simple model then, we might expect to have particularly stable nuclei whenever Z or N equals

$$2, 8, 18, 20, 34, 40, 58, \ldots.$$

Instead, experiment shows that the favored stable nuclei have Z or N equal to

$$2, 8, 20, 28, 50, 82, \text{ and } 126.$$

These Z values correspond to helium, oxygen, calcium, nickel, tin, and lead nuclei. (A $Z = 126$ nucleus is experimentally unknown.) These are called the nuclear "magic numbers." Nuclei that have *both* Z and N equal to a magic number are especially stable. An example is a type of calcium nucleus, ${}^{40}_{20}Ca$, where the top number represents the nucleon number, $A = Z + N$, and the bottom is Z. The Z value determines which element, but each element can have a variety of N values, called *isotopes*. (Lead with $Z = 82$ and $N = 126$ is also particularly stable.)

Of course, the assumption of an infinite spherical potential is partly to blame for the fact that the aforementioned set of predicted and observed magic numbers do not agree. A more realistic model is to assume some sort of *finite* potential well. We will study this in Section 7.4. It is also important to allow for the known exponential tail in the nucleon distribution rather than postulate a sharp cutoff at some spherical radius, a. Another important ingredient in a successful theory of the atomic nucleus is an assumption known as *LS coupling* or *spin-orbit coupling*. This phenomenon will be explained when we come to the hydrogen atom in the next chapter. The end result of such calculations is an energy-level diagram, which can be interpreted as the filling order of the last added nucleon. A schematic representation of such an energy level diagram is shown in Figure 7.2. The spectroscopic notation on the right-hand side of this figure (in the case of spin-orbit coupling) will also be explained in Chapter 8. It represents the effect of adding angular momentum ℓ to a spin 1/2 from a neutron or proton. The magic numbers can be explained from this type of model.

A real, fundamental theory of strong interactions has existed since about 1972 and is called *quantum chromodynamics* (QCD). The observed particles in a nucleus, the proton and neutron, are actually composite states made up of three *quarks*. QCD describes the interactions of quarks and the force-mediating particles called *gluons*. Given this theory, we should be able to calculate all internuclear forces from QCD. In practice, it is very difficult because of the complicated composite nature of the systems involved. However, a computer-based numerical technique for solving QCD, known as *lattice QCD*, is making inroads on these challenging problems. All this and more will be explored further in Chapter 12.

7.4 THE "DEUTERON"

Up to now, we have only been investigating the behavior of a single, isolated particle subject to some simple external potential. More realistically, the simplest systems in nature are two-particle systems. Continuing the earlier discussion of the atomic nucleus, we will concentrate here on a phenomenological description of the simplest nuclear subsystem, the deuteron, which consists of a single proton and neutron. Of course, from the more fundamental point of view of quark dynamics, this is already an extremely complicated system consisting of six quarks. To get a practical description, we will ignore this reality and treat the neutron and proton as fundamental objects. Physics is a multilayered discipline. It turns out for many applications that we can ignore the deeper layers of reality (usually at higher energies) if we wish to describe a system at some reasonable level of accuracy.

Since we are considering a system consisting of two independent particles, it is reasonable to suppose we can introduce independent position and momentum operators for each particle. That is, we introduce momentum operators $\vec{p}_1, \vec{p}_2$ and position operators $\vec{x}_1, \vec{x}_2$ such that

$$\left[\vec{p}_\alpha, \vec{p}_\beta\right] = 0, \quad \left[\vec{x}_\alpha, \vec{x}_\beta\right] = 0 \tag{7.71}$$

where α and β are now *particle* labels. Our direct product states can be taken as

$$|\vec{p}_1', \vec{p}_2'\rangle = |\vec{p}_1'\rangle \otimes |\vec{p}_2'\rangle, \tag{7.72}$$

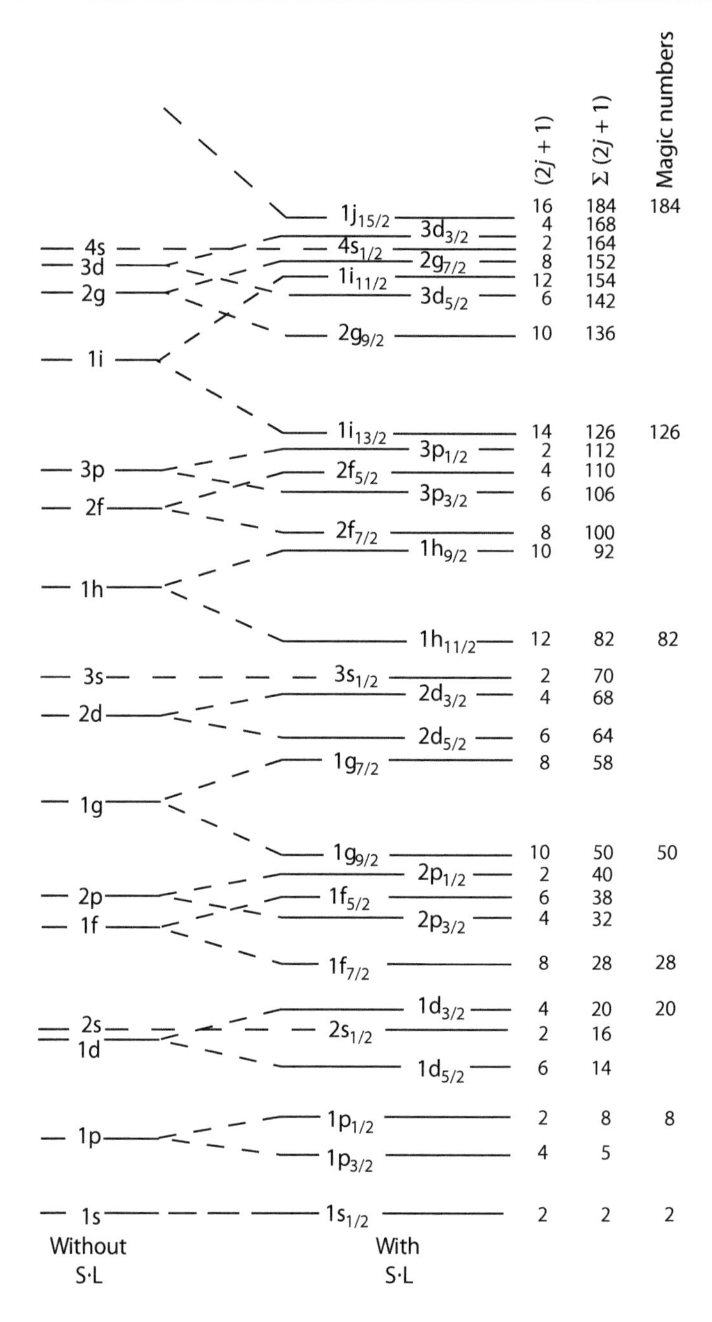

FIGURE 7.2 The energy level filling order of rounded edge square well potentials ("Without **S.L**") is shown on the left. The right part shows the filling order, which arises when a realistic spin-orbit interaction ("With **S.L**") is added. The occupancy of each level is given by the first numerical column, $2J+1$, where the J value is read from the subscript on the spin-orbit level. The cumulative total up to that level is given by the second column. The "magic numbers" appear as significant energy gaps associated with the cumulative numbers on the right. This figure is analogous to the atomic one in Figure 7.9; there are also relatively small ordering level differences between proton and neutron. The spin-orbit spectroscopic notation will be explained in Chapter 8. (From R. Eisberg and R. Resnick, *Quantum Physics*, Wiley, Figures 15–18, 538 (1985). With permission.)

$$|\vec{x}_1',\vec{x}_2'\rangle = |\vec{x}_1'\rangle \otimes |\vec{x}_2'\rangle. \tag{7.73}$$

We will also assume that position and momentum operators from different particles commute:

$$\left[p_{\alpha i}, x_{\beta j}\right] = \frac{\hbar}{i}\delta_{ij}\delta_{\alpha\beta}. \tag{7.74}$$

Just as in classical dynamics, we can imagine introducing a different set of coordinates, based on the center of mass of the system that locates the positions of our two particles. Given particles at positions $\vec{x}_1'$ and $\vec{x}_2'$ relative to a fixed coordinate system, we can also locate their positions given the center-of-mass vector, $\vec{X}'$, and relative position, $\vec{x}'$. The relationship between these coordinates is illustrated in Figure 7.3.

Based upon these relationships, we define operators for these quantities as follows

$$\vec{x} \equiv \vec{x}_1 - \vec{x}_2, \tag{7.75}$$

$$\vec{X} \equiv \frac{m_1\vec{x}_1 + m_2\vec{x}_2}{(m_1 + m_2)}. \tag{7.76}$$

Just as we introduced a new basis to represent a particle's position in spherical coordinates, we imagine an alternate description based upon knowledge of $\vec{X}'$ and $\vec{x}'$:

$$\langle\vec{x}_1', \vec{x}_2'| \rightarrow \langle\vec{x}', \vec{X}'|. \tag{7.77}$$

Such a description is only allowed if $\vec{x}'$ and $\vec{X}'$ may be simultaneously measured, that is, if $\vec{x}$ and $\vec{X}$ commute. This is easy to show using Equation 7.71:

$$\left[X_i, x_j\right] = \left[\frac{(m_1x_{1i} + m_2x_{2i})}{(m_1 + m_2)}, x_{1j} - x_{2j}\right] = 0. \tag{7.78}$$

Thus, we may take

$$\langle\vec{x}',\vec{X}'| \equiv \langle\vec{x}'|\ \langle\vec{X}'|, \tag{7.79}$$

(the right-hand side is a direct product) where we define (we imagine Cartesian bases for now)

$$\langle\vec{x}'|\vec{x} \equiv \langle\vec{x}'|\vec{x}', \tag{7.80}$$

$$\langle\vec{X}'|\vec{X} \equiv \langle\vec{X}'|\vec{X}'. \tag{7.81}$$

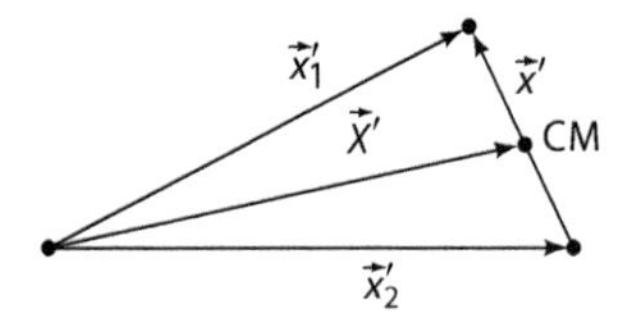

FIGURE 7.3 The relationship of the various coordinates for the two-body problem.

Along with this coordinate description, we introduce center-of-mass momentum operators

$$\vec{P} \equiv \vec{p}_1 + \vec{p}_2, \tag{7.82}$$

$$\vec{p} \equiv \frac{m_2 \vec{p}_1 - m_1 \vec{p}_2}{(m_1 + m_2)}. \tag{7.83}$$

We have

$$[p_i, P_j] = \left[\frac{m_2 p_{1i} - m_1 p_{2i}}{(m_1 + m_2)}, p_{1j} - p_{2j} \right] = 0, \tag{7.84}$$

from Equation 7.71. Therefore, a new momentum description

$$\langle \vec{p}_1', \vec{p}_2' | \rightarrow \langle \vec{p}', \vec{P}' |, \tag{7.85}$$

where

$$\langle \vec{p}', \vec{P}' | \equiv \langle \vec{p}' | \langle \vec{P}' |, \tag{7.86}$$

(the right-hand side is again a direct product) and

$$\langle \vec{p}' | \vec{p} = \langle \vec{p}' | \vec{p}', \tag{7.87}$$

$$\langle \vec{P}' | \vec{P} = \langle \vec{P}' | \vec{P}', \tag{7.88}$$

is allowed. We assume both the center-of-mass coordinate description (Equation 7.79) and the center-of-mass momentum description (Equation 7.86) are complete.

Based upon Equation 7.74, we also have that

$$\left[P_i, x_j \right] = \left[X_i, p_j \right] = 0. \tag{7.89}$$

However, we will also get

$$\begin{aligned} \left[P_i, X_j \right] &= \left[p_{1i} + p_{2i}, \frac{m_1 x_{1j} + m_2 x_{2j}}{(m_1 + m_2)} \right] \\ &= \frac{m_1}{m_1 + m_2} \left[p_{1i}, x_{1j} \right] + \frac{m_1}{m_1 + m_2} \left[p_{2i}, x_{2j} \right] = \frac{\hbar}{i} \delta_{ij}, \end{aligned} \tag{7.90}$$

$$\begin{aligned} \left[p_i, x_j \right] &= \left[\frac{m_2 p_{1i} - m_1 p_{2i}}{(m_1 + m_2)}, x_{1j} - x_{2j} \right] \\ &= \frac{m_1}{m_1 + m_2} \left[p_{1i}, x_{1j} \right] + \frac{m_1}{m_1 + m_2} \left[p_{2i}, x_{2j} \right] = \frac{\hbar}{i} \delta_{ij}. \end{aligned} \tag{7.91}$$

Based on Equations 7.90 and 7.91, we may show that

$$\langle \vec{X}' | \vec{P} = \frac{\hbar}{i} \vec{\nabla}_{X'} \langle \vec{X}' |, \tag{7.92}$$

$$\langle \vec{x}' | \vec{p} = \frac{\hbar}{i} \vec{\nabla}_{x'} \langle \vec{x}' |, \tag{7.93}$$

by reasoning similar to that in Chapter 4 leading up to Equation 4.153.

Solving for $\vec{p}_1$ and $\vec{p}_2$ from Equations 7.82 and 7.83, we find

$$\vec{p}_1 = \frac{m_1}{m_1 + m_2}\vec{P} + \vec{p}, \tag{7.94}$$

$$\vec{p}_2 = \frac{m_2}{m_1 + m_2}\vec{P} - \vec{p}, \tag{7.95}$$

so that for the kinetic energy we get

$$\frac{\vec{p}_1^{\,2}}{2m_1} + \frac{\vec{p}_2^{\,2}}{2m_2} = \frac{\vec{P}^2}{2M} + \frac{\vec{p}^{\,2}}{2\mu}, \tag{7.96}$$

where $M = m_1 + m_2$ and $\mu = m_1 m_2 / M$ (the usual formula for "reduced mass"). Assuming the interaction of the two particles depends only upon the distance between the two particles, the Hamiltonian becomes

$$H = \frac{\vec{P}^2}{2M} + \frac{\vec{p}^{\,2}}{2\mu} + V(r), \tag{7.97}$$

where, in a spherical coordinate description, r represents an operator giving the magnitude of the distance between the particles. Let us call (external Hamiltonian)

$$H_{\text{ext}} = \frac{\vec{P}^2}{2M}, \tag{7.98}$$

and (internal Hamiltonian)

$$H_{\text{int}} = \frac{\vec{p}^{\,2}}{2\mu} + V(r). \tag{7.99}$$

Notice that

$$[H_{\text{ext}}, H_{\text{int}}] = \left[\frac{\vec{P}^2}{2M}, \frac{\vec{p}^{\,2}}{2\mu} + V(r)\right] = 0, \tag{7.100}$$

so that our system can be taken to have simultaneous eigenvalues of H_{ext} and H_{int}. The eigenvalue equation for H_{ext} is just

$$H_{\text{ext}} |\vec{P}'\rangle = \frac{\vec{P}'^2}{2M} |\vec{P}'\rangle, \tag{7.101}$$

so that

$$\langle \vec{X}' | H_{\text{ext}} | \vec{P}' \rangle = \frac{\vec{P}'^2}{2M} \langle \vec{X}' | \vec{P}' \rangle, \tag{7.102}$$

or

$$-\frac{\hbar^2}{2M} \vec{\nabla}_{X'}^2 \, u_{\vec{P}'}(\vec{X}') = \frac{\vec{P}'^2}{2M} u_{\vec{P}'}(\vec{X}'), \tag{7.103}$$

where $u_{\vec{P}'}(\vec{X}') \equiv \langle \vec{X}' | \vec{P}' \rangle$ and we have used Equation 7.92. The solution to Equation 7.103 is just

$$u_{\vec{P}'}(\vec{X}') = \frac{1}{(2\pi\hbar)^{3/2}} e^{i\vec{P}' \cdot \vec{X}'/\hbar}, \tag{7.104}$$

that is, just free-particle plane waves normalized such that

$$\int d^3X \, u^*_{\vec{P}'}(\vec{X}) u_{\vec{P}''}(\vec{X}) = \delta\left(\vec{P}' - \vec{P}''\right), \tag{7.105}$$

as usual. This shows that the center of mass of the system behaves like a free particle, taking on continuous energy values.

The interesting part of the two-body problem is the eigenvalue equation for H_{int}. We switch to a spherical polar description for this sector

$$\langle \vec{x}' | \to \langle \vec{r} |, \tag{7.106}$$

and write the eigenvalue equation as

$$\langle \vec{r} | H_{\text{int}} | a' \rangle = E_{a'} \langle \vec{r} | a' \rangle, \tag{7.107}$$

or using the form for H_{int} (Equation 7.99) we have

$$\left[-\frac{\hbar^2}{2\mu} \vec{\nabla}_r^2 + V(r) \right] u_{a'}(\vec{r}) = E_{a'} u_{a'}(\vec{r}), \tag{7.108}$$

where $u_{a'}(\vec{r}) = \langle \vec{r} \, | a' \rangle$. This is of the same form as the one-body Schrödinger equation we have already considered, except that m is replaced by μ.

The preceding has been a short introduction to two-body formalism and has not been directed toward the deuteron problem as such. Let me now recite some of the known experimental facts concerning the deuteron. It has been determined that the binding energy of the deuteron is

$$E = -2.225 \, \text{MeV}. \tag{7.109}$$

We have that

$$1\,\text{eV} = \text{Energy given to an electronic charge accelerated through } 1\,\text{V}$$

or

$$1\,\text{eV} = 1.602 \times 10^{-12} \text{erg}.$$

1 MeV = 10^6 eV, of course. Since energies can be measured in MeV, mass can be measured in units of MeV/c^2. Equation 7.109 implies that it takes 2.226 MeV of energy to separate the proton and neutron in deuterium to infinity. In addition, it is known that there are no excited energy states of the deuteron, that is, there is only a single bound state of this system. One (nonfundamental) theory says the force between nucleons (or at least the long range part of the potential) is due to *pion exchange* between them (see Chapter 10, Figure 10.18). The energy–time uncertainty principle (Chapter 2, Equation 2.93) reads

$$\Delta E \, \Delta t \gtrsim \hbar. \tag{7.110}$$

The energy of a system of two nucleons fluctuates when we consider events that take place over short periods of time, but observable events never violate energy conservation. Because of this, one nucleon can spontaneously give rise to a pion, which has a rest mass energy

$$m_\pi c^2 \simeq 140 \text{ MeV}. \tag{7.111}$$

According to Equation 7.110, an energy fluctuation will result in an unobservable violation of energy conservation if it occurs in a time scale, T, such that

$$T \lesssim T_{\max} = \frac{\hbar}{\Delta E}. \tag{7.112}$$

Setting $\Delta E = m_\pi c^2$, and assuming the maximum velocity of the pion is the speed of light, then we learn that this pion has a maximum range, $R_{\max} = c\, T_{\max}$, of

$$R_{\max} = \frac{\hbar}{m_\pi c}. \tag{7.113}$$

The right-hand side of Equation 7.113 is just the pion's reduced (i.e., divided by 2π) Compton wavelength (Chapter 2, Section 2.2) that has a numerical value of

$$\frac{\hbar}{m_\pi c} \approx 1.41 \times 10^{-15}\, m = 1.41 \text{ fm}. \tag{7.114}$$

(fm = 1 fermi = 10^{-15} m = 10^{-13} cm.) A useful numerical relationship is

$$1 \text{ fm}^{-1} = (197.3 \text{ MeV}) / \hbar c.$$

Therefore, if the second nucleon is within this approximate range, the pion can be reabsorbed by it and a force between the two nucleons can be transmitted. This picture of the origin of the strong nuclear force is due to Hidiki Yukawa. It seems to explain the short-range nature of the strong nuclear force. In its details it cannot be completely correct since we now know that all of these particles, protons, neutrons, and pions are made up of more fundamental objects. (In the simplest possible interpretation, a pion is just a bound state of a quark and an antiquark.) The emitted and absorbed pion in the discussion is called a "virtual" particle. This particle exchange picture of nucleon interactions explains the long-range portion of the strong force quite well. Historically, however, this argument was made in the opposite order. Instead of knowing the particle mass and deducing the implied range, the range of the nuclear force was known approximately, and the exchange picture was used to deduce the particle mass. At the time, this range was estimated be about 2 fm from scattering experiments, and Yukawa predicted the existence of a particle with a mass of about 100 MeV/c^2. This prediction was confirmed by the discovery of the pion in 1947.

Assuming a spherically symmetric force, the radial equation we need to solve is (just Equation 7.10 with $m \to \mu$)

$$\left[-\frac{\hbar^2}{2\mu}\left(\frac{d^2}{dr^2} - \frac{\ell(\ell+1)}{r^2}\right) + V(r)\right] R(r) = ER(r). \tag{7.115}$$

(I will sometimes leave the quantum number labels off of the radial eigenfunctions until it is clear what they are.) What about the potential, $V(r)$? The remarks about the range of the nuclear force suggest the simplest possible picture, shown in Figure 7.4. That is

$$V(r) = \begin{cases} 0, & r \geq a \\ -V_0, & r < a. \end{cases} \tag{7.116}$$

(Note that we have $V_0 > 0$ here for a potential well, as opposed to the negative sign in Chapter 3, Section 3.4.) We expect that the well radius, a, is on the order of 1.41 fm. To produce a binding energy of about 2.2 MeV, and to have only a single bound state, what should the well depth, V_0, be?

We have labeled the region of space with $r \leq a$ as "I" and $r \geq a$ as "II" in Figure 7.4. In I, we need to solve

$$-\frac{\hbar^2}{2\mu}\left(\frac{d^2}{dr^2} - \frac{\ell(\ell+1)}{r^2}\right)R(r) = (E + V_0)R(r). \tag{7.117}$$

This is the same form as the free-particle problem (Equation 7.16) except that $E \rightarrow E + V_0$ here. Following the same reasoning as before, we learn that $E + V_0$ must be a positive quantity and that the solutions are of the form $j_\ell(Ka)$, where

$$K \equiv \frac{\sqrt{2\mu(E + V_0)}}{\hbar}. \tag{7.118}$$

In the region labeled as II, we must instead solve

$$-\frac{\hbar^2}{2\mu}\left(\frac{d^2}{dr^2} - \frac{\ell(\ell+1)}{r^2}\right)R(r) = ER(r), \tag{7.119}$$

but now because we are specifically interested in a bound state we require $E < 0$. (There are also $E > 0$ solutions. These describe scattering.) To find the appropriate solutions to Equation 7.119 in this case, let us for the moment specialize to $\ell = 0$. In dimensionless form with

$$\left.\begin{aligned} \rho &\equiv K'r, \\ K' &\equiv \frac{\sqrt{-2\mu E}}{\hbar} = \frac{\sqrt{2\mu|E|}}{\hbar}, \end{aligned}\right\} \tag{7.120}$$

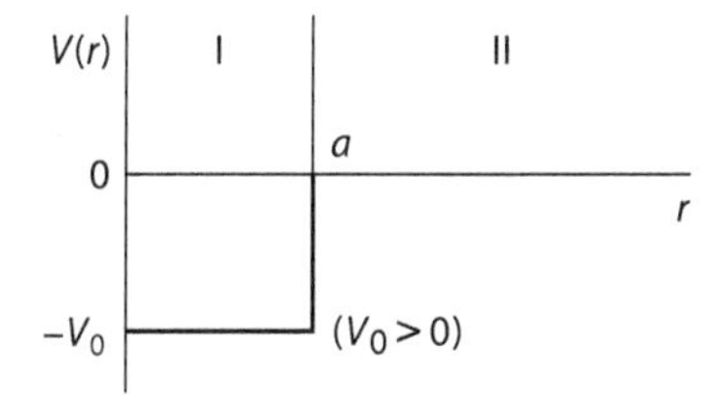

FIGURE 7.4 The finite spherical well potential in radial coordinates.

the $\ell = 0$ version of Equation 7.119 reads

$$\left[\frac{d^2}{d\rho^2} - 1\right] R(\rho) = 0. \tag{7.121}$$

We encountered this equation before (Equation 7.20) and found that its solutions were of the form

$$R(\rho) \sim e^{\pm\rho}. \tag{7.122}$$

These were discarded because they could not have $R(0)$ and $R(\infty)$ simultaneously zero. However, we are now solving Equation 7.121 in Region II, which does not include the origin. Region II does, however, include spatial infinity where the wave function must vanish,

$$R(\infty) = 0. \tag{7.123}$$

This is actually a necessary but not sufficient condition that the wave function must satisfy. The real requirement on the wave function is that it be square integrable, which here means that $R(\rho)$ must fall off faster than $\rho^{-1/2}$ as $\rho \to \infty$. (Review the similar 1-D discussion in Section 2.10.) The condition Equation 7.123 applied to Equation 7.122 is enough to pick out the unique solution

$$R(\rho) \sim e^{-\rho}. \tag{7.124}$$

This brings the present discussion up to the same point as the free-particle discussion leading to the unique solution $R_{k0}(\rho) \sim \sin\rho$. We can repeat the rest of that discussion, which leads to the solutions of the form (see Equation 7.37)

$$R(\rho) \sim \rho^{\ell+1} \left(\frac{1}{\rho}\frac{d}{d\rho}\right)^{\ell} X_{\ell=0}(\rho). \tag{7.125}$$

Earlier for the free particle we had $X_{\ell=0}(\rho) \sim \sin\rho/\rho$, which led to the use of the spherical Bessel functions

$$j_\ell(\rho) = (-\rho)^\ell \left(\frac{1}{\rho}\frac{d}{d\rho}\right)^{\ell} \left(\frac{\sin\rho}{\rho}\right), \tag{7.126}$$

as solutions of the radial equation. However, in this case, we have $X_{\ell=0}(\rho) \sim e^{-\rho}/\rho$, and we can take our solutions to be proportional to the *spherical Hankel functions of the first kind* (with imaginary arguments), $h_\ell^{(1)}(i\rho)$, defined as

$$h_\ell^{(1)}(i\rho) \equiv -(i\rho)^\ell \left(\frac{1}{\rho}\frac{d}{d\rho}\right)^{\ell} \left(\frac{e^{-\rho}}{\rho}\right). \tag{7.127}$$

Of course, the solutions in Regions I and II have to be joined smoothly, which means in this case that $R(\rho)$ and its first derivative must connect continuously at $r = a$. These conditions arise here in three dimensions completely analogously to the continuity conditions discussed in Chapter 3,

Section 3.4 in one dimension. Therefore, our explicit solutions are (A and B are normalization constants)

$$u(r)=\begin{cases} Aj_\ell(Kr), & r\le a, \\ Bh_\ell^{(1)}(iK'r), & r\ge a, \end{cases} \tag{7.128}$$

and our requirements of continuity in $u(r)$ and $u'(r)$ at $r=a$ read

$$Aj_\ell(Ka)=Bh_\ell^{(1)}(iK'a), \tag{7.129}$$

$$A\frac{\mathrm{d}j_\ell(Kr)}{\mathrm{d}r}\bigg|_{r=a}=B\frac{\mathrm{d}h_\ell^{(1)}(iK'r)}{\mathrm{d}r}\bigg|_{r=a}. \tag{7.130}$$

Dividing Equation 7.130 by 7.129 gives us the statement

$$\frac{\left.\dfrac{\mathrm{d}j_\ell(Kr)}{\mathrm{d}r}\right|_{r=a}}{j_\ell(Ka)}=\frac{\left.\dfrac{\mathrm{d}h_\ell^{(1)}(iK'r)}{\mathrm{d}r}\right|_{r=a}}{h_\ell^{(1)}(K'a)}. \tag{7.131}$$

Equation 7.131 is the eigenvalue equation for the energies, and for finite well depth, V_0, has only a finite number of solutions. Let us explicitly investigate Equation 7.131 in the case $\ell=0$. Since

$$j_0(Ka)=\frac{\sin Ka}{Ka}, \tag{7.132}$$

$$h_0^{(1)}(iK'a)=-\frac{\mathrm{e}^{-K'a}}{K'a}, \tag{7.133}$$

Equation 7.131 reads

$$\frac{\left.\dfrac{\mathrm{d}}{\mathrm{d}r}\left(\sin\dfrac{Kr}{r}\right)\right|_{r=a}}{\left(\sin\dfrac{Ka}{a}\right)}=\frac{\left.\dfrac{\mathrm{d}}{\mathrm{d}r}\left(\dfrac{\mathrm{e}^{-K'r}}{r}\right)\right|_{r=a}}{\left(\dfrac{\mathrm{e}^{-K'a}}{a}\right)}, \tag{7.134}$$

or

$$(Ka)\cot Ka=-(K'a). \tag{7.135}$$

Notice also that from Equations 7.118 and 7.120

$$(Ka)^2+(K'a)^2=\frac{2\mu V_0 a^2}{\hbar^2}. \tag{7.136}$$

Therefore, we may solve the eigenvalue problem, given a value of V_0, by plotting Equations 7.135 and 7.136 simultaneously on axes labeled (Ka) and $(K'a)$. This is shown in Figure 7.5.

Notice that, based on the figure, we can determine the number of bound s-states for a given well depth quite easily. When we have, for example,

$$\frac{2\mu V_0 a^2}{\hbar^2}<\left(\frac{\pi}{2}\right)^2, \tag{7.137}$$

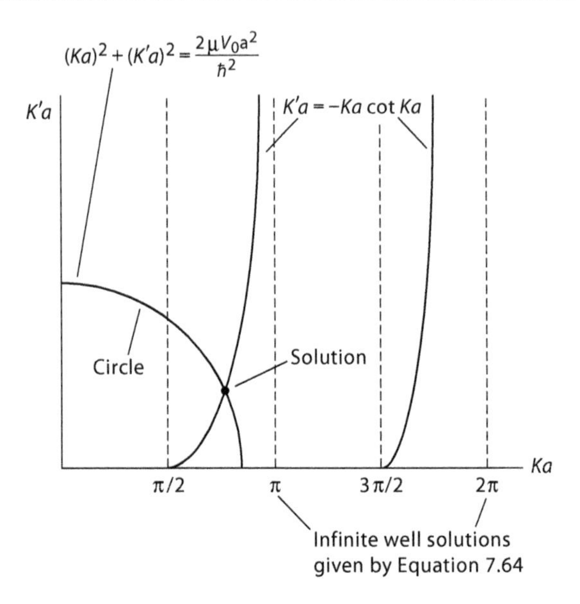

FIGURE 7.5 A qualitative simultaneous plot of *K′a* as a function of *Ka* from Equations 7.135 and 7.136. The points of intersection are the solutions. The infinite well solutions ($V_0 \to \infty$) are also indicated.

then there will be *no* bound s-states because the two sets of curves in the figure never intersect. If

$$\left(\frac{\pi}{2}\right)^2 \leq \frac{2\mu V_0 a^2}{\hbar^2} < \left(\frac{3\pi}{2}\right)^2, \tag{7.138}$$

there will be a single bound s-state. In general, it is easy to see that

$$\left(n-\frac{1}{2}\right)^2 \pi^2 \leq \frac{2\mu V_0 a^2}{\hbar^2} < \left(n+\frac{1}{2}\right)^2 \pi^2 \tag{7.139}$$

is the condition on the well depth such that n bound s-states exist. Of course, for the deuteron, we would like to choose V_0 such that a single bound state exists. This would seem to mean that we should require Equation 7.138 to hold for the well depth, V_0. However, this is only a condition to have a single bound s-state; there can also be bound states with $\ell = 1, 2, 3, \ldots$. One can show that the first $\ell = 1$ state becomes bound when

$$\frac{2\mu V_0 a^2}{\hbar^2} \geq \pi^2. \tag{7.140}$$

One can also show that if $(V_0)_\ell$ represents the minimum well depth to bind a state of angular momentum ℓ, then*

$$(V_0)_0 < (V_0)_1 < (V_0)_2 \ldots \tag{7.141}$$

* The $(V_0)_\ell$ can be related to zeros of spherical Bessel functions. One may show that the $E \to 0^-$ roots of Equation 7.131 (for $\ell \geq 1$) are such that $j_{\ell-1}\left((x)_{\ell-1}\right) = 0$, where $(x)_{\ell-1} = \sqrt{\left(2\mu(V_0)_\ell / \hbar^2\right)}\, a$. Then proving the above binding order comes down to showing that $(x_0)_{\ell-1} < (x_0)_\ell$, where the $(x_0)_\ell$ are the smallest positive roots of this equation for each ℓ.

Combining Equations 7.138 and 7.140, the condition to produce a single bound state is

$$\left(\frac{\pi}{2}\right)^2 \le \frac{2\mu V_0 a^2}{\hbar^2} < \pi^2. \tag{7.142}$$

In terms of our estimate that $a \simeq 1.41$ fm, Equation 7.81 says that ($\mu \simeq 1/2\, m_{\text{proton}}$, $m_{\text{proton}} = 940\ \text{MeV}/c^2$)

$$52\ \text{MeV} < V_0 < 206\ \text{MeV}. \tag{7.143}$$

According to our simple theory, this single bound state must have $\ell = 0$. In reality, because the neutron–proton potential is not exactly spherically symmetric (because of the so-called *tensor force*), the deuteron is actually a mixture of $\ell = 0$ and a small amount of $\ell = 2$. (The angular momentum $\vec{L}^2$ does not give rise to a good quantum number.)

The wave function of the lowest bound s-state looks qualitatively like the wave function in Figure 7.6. Notice the exponential tail in the classically excluded Region II.

We have not found explicit formulas for the normalization factors A and B in Equation 7.128, but they are, as usual, determined from

$$\int_0^\infty \mathrm{d}r\, r^2 \left|u(r)\right|^2 = 1. \tag{7.144}$$

Counting the point at infinity as a node (since $u(r) = 0$ there), the lowest bound s-state has one radial node, the next highest (if bound) has two nodes, and so on. Similarly, for the higher ℓ values. Therefore, the bound-state quantum numbers can be taken to be $n\ell$ (and the $2\ell + 1$ m values), where n represents the number of radial nodes, as in the infinite spherical well also. This is the first explicit case we have studied where the eigenenergies have *both* a discrete ($E < 0$) and a continuous ($E > 0$) spectrum. We will study the scattering solutions later in Chapter 11, Section 11.11. If we desired to write completeness for these sets of radial wave functions, we would have to have both a discrete sum over the bound states as well as an integral over the scattering states as in

$$\sum_{\text{bound}} |n\ell\rangle\langle n\ell| + \int_0^\infty \mathrm{d}k\, k^2\, |k\ell\rangle\langle k\ell| = 1, \tag{7.145}$$

(k has the same meaning as in the earlier free-particle problem).

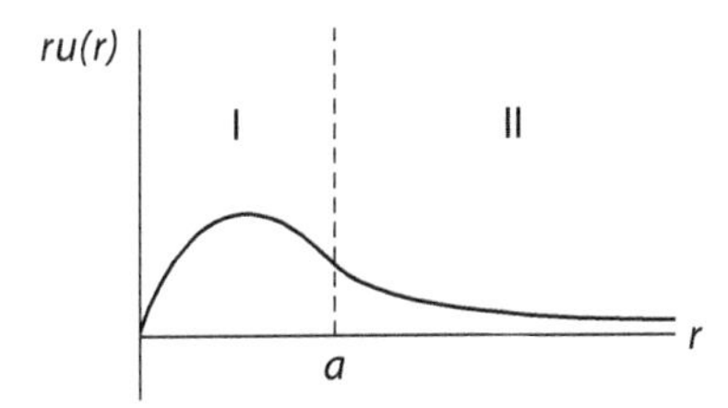

FIGURE 7.6 Qualitative representation of the $r\,u(r)$ wave function for the lowest energy solution of the finite square well potential of radius a.

Of course the model considered here for the deuteron is very crude and cannot fit all the experimental facts. In particular, the characterization of the potential as a finite spherical well with the estimates Equations 7.113 and 7.143 is unrealistic. Better models are based upon the use of a potential that turns repulsive at small enough distances (the "repulsive core"), has an exponential tail at large distances (to simulate the finite range pion exchange potential), and includes the previously mentioned tensor force.

7.5 THE COULOMB POTENTIAL: INITIAL CONSIDERATIONS

The static potential between two electric charges e_1 and e_2 is

$$V(r) = \frac{e_1 e_2}{r}, \tag{7.146}$$

where r is the distance between them. If we take $e_1 = -e$ and $e_2 = Ze$ ($e = |e|$ is the magnitude of charge on an electron), we have

$$V(r) = -\frac{Ze^2}{r}. \tag{7.147}$$

Z is the number of protons in the atomic nucleus. Such a potential describes "hydrogen-like" systems. The radial eigenvalue equation is given by Equation 7.115:

$$\left[\frac{\hbar^2}{2\mu}\left(-\frac{d^2}{dr^2} + \frac{\ell(\ell+1)}{r^2}\right) - \frac{Ze^2}{r}\right] R(r) = ER(r). \tag{7.148}$$

Let us consider solutions to Equation 7.148 in the neighborhood of $r = 0$. For r sufficiently small, the equation $R(r)$ satisfies is

$$\left[-\frac{d^2}{dr^2} + \frac{\ell(\ell+1)}{r^2}\right] R(r) \approx 0. \tag{7.149}$$

Actually, any potential such that $\lim_{r\to 0} r^2 V(r) = 0$ will satisfy Equation 7.149 near $r = 0$. (This includes all the cases that will be studied in this chapter.) Let us try a power law solution to Equation 7.149:

$$u(r) = Cr^a. \tag{7.150}$$

This implies that

$$a(a-1) = \ell(\ell+1). \tag{7.151}$$

We therefore have two possible solutions:

$$a = \ell+1, -\ell. \tag{7.152}$$

The solutions $\sim 1/r^\ell$ do not satisfy the boundary condition Equation 7.12. Therefore near the origin, we expect the $R(r)$ function to behave like $r^{\ell+1}$.

Actually, for the Coulomb potential, the $\ell = 0$ case has to be handled separately to confirm that $R(r) \sim r$ near the origin. For $\ell = 0$, we have

$$\left[\frac{\hbar^2}{2\mu}\frac{d^2}{dr^2} + \frac{Ze^2}{r}\right]R(r) \approx 0, \tag{7.153}$$

near the origin. Assuming $R(r) = Cr$ in this vicinity, the first-order correction is then determined by

$$\frac{d^2 R_{\text{corr.}}(r)}{dr^2} \simeq -\frac{2Z}{a_0 r}(Cr), \tag{7.154}$$

where $a_0 = \hbar^2/\mu e^2$. (The "Bohr radius" of Equation 2.37, but with $m \to \mu$.) The correction induced by Equation 7.154 is clearly of order r^2, giving us

$$R(r) \simeq Cr\left(1 - \frac{Zr}{a_0}\right). \tag{7.155}$$

Clearly, this process of correcting the lowest order solution can be continued to higher orders, showing that the assumption $R(r) \approx Cr$ is self-consistent.

Now let us consider the opposite extreme, $r \to \infty$, in Equation 7.148. In this case, we expect that

$$-\frac{\hbar^2}{2\mu}\frac{d^2 R(r)}{dr^2} \approx ER(r). \tag{7.156}$$

We will choose to look at the bound-state solutions for which $E = -|E|$. (Scattering solutions also exist for $E > 0$ as for the deuteron.) The solutions to Equation 7.156 are of the form

$$R(r) \sim e^{-kr}, e^{kr}. \tag{7.157}$$

Only the first possibility satisfies the boundary condition, Equation 7.123. ($k = \sqrt{2\mu|E|}/\hbar$, as usual.) Summing up what we have learned about solutions to Equation 7.148, we can state that

$$\left.\begin{aligned} r \to 0:&\ R(r) \sim r^{\ell+1} \\ r \to \infty:&\ R(r) \sim e^{-kr} \end{aligned}\right\}. \tag{7.158}$$

Based on Equation 7.158, we guess for the $\ell = 0$ wave function that

$$R(r) = Cre^{-kr}. \tag{7.159}$$

Substituting Equation 7.159 into 7.148, we learn that we indeed have a solution if

$$k = \frac{Z}{a_0}. \tag{7.160}$$

Expressing k in terms of E, we can show that this means

$$E = -\frac{\hbar^2 Z^2}{2\mu a_0^2} = -\frac{\mu e^4 Z^2}{2\hbar^2}. \tag{7.161}$$

We estimated this result before in Section 2.4 (Chapter 2) when $Z = 1$. The only difference between the above and our previous estimate is the use of the reduced mass, μ, in place of the electron mass, m.

A conventional way of stating the energy of atomic states is to give the E/hc value, which has units of inverse distance. Then

$$\frac{E}{hc} = -\frac{\alpha^2 \mu c Z^2}{2h}. \tag{7.162}$$

The right-hand side of Equation 7.162 is closely related to the *Rydberg constant*, which is the most accurately measured of all fundamental constants. It is denoted as R_∞ and is defined by

$$R_\infty \equiv \frac{\alpha^2 m_e c}{2h}, \tag{7.163}$$

where m_e is the electron mass and $\alpha = e^2/\hbar c$ is the fine structure constant. Numerically, the value is*

$$R_\infty = 1.0973731568508(65) \times 10^7 \mathrm{m}^{-1},$$

where the numbers in parentheses give the uncertainty in the last two significant digits. R_∞ represents the idealized ground-state energy of the hydrogen ($Z = 1$) atom as if the proton's mass was infinite (which explains the "∞" subscript on R). Actually, the real hydrogen ground-state energy receives corrections from higher-order electrodynamic and relativistic effects. The value of R_∞ is determined from extremely accurate measurements of transition frequencies in hydrogen and "deuterium," (the atomic levels associated with the deuteron nucleus, studied in Section 7.4) and comparing experiment to theoretical expressions. (The Lamb shift from Chapter 8, Section 8.6 is included in these.) It is reassuring to know that atomic physics is on such a solid basis!

Returning to Equation 7.159, wave function normalization requires

$$1 = \int_0^\infty \mathrm{d}r\, R^2(r) = C^2 \int_0^\infty \mathrm{d}r\, r^2 \mathrm{e}^{-2Zr/a_0}, \tag{7.164}$$

but

$$\int_0^\infty \mathrm{d}x\, x^2 \mathrm{e}^{-ax} = 2a^{-3}, \tag{7.165}$$

so that we may take

$$C = 2\left(\frac{Z}{a_0}\right)^{3/2}. \tag{7.166}$$

Therefore

$$u(r) = R(r) r^{-1} = 2\left(\frac{Z}{a_0}\right)^{3/2} \mathrm{e}^{-Zr/a_0}. \tag{7.167}$$

* 2014 CODATA recommended values, http://www.physics.nist.gov.

The full $\ell = 0$ wave function ($Y_{00} = (1/\sqrt{4\pi})$) is

$$u(r)Y_{00} = \frac{1}{\sqrt{\pi}}\left(\frac{Z}{a_0}\right)^{3/2} \mathrm{e}^{-Zr/a_0}. \tag{7.168}$$

7.6 THE COULOMB POTENTIAL: 2-D HARMONIC OSCILLATOR COMPARISON

We will now solve for the wave functions in general. We will do this using a method I learned about from Julian Schwinger.* We begin in a seemingly strange place. Let us remind ourselves of the basic facts concerning the one-dimensional harmonic oscillator (covered in Chapter 3, Section 3.5). The Hamiltonian is

$$H = \frac{p_x^2}{2m} + \frac{1}{2}m\omega^2 x^2. \tag{7.169}$$

By introducing dimensionless variables

$$\left.\begin{aligned} \mathcal{H} &= \frac{H}{\hbar\omega}, \\ p &= \frac{p_x}{\sqrt{m\omega\hbar}}, \\ q &= \sqrt{\frac{m\omega}{\hbar}}x, \end{aligned}\right\} \tag{7.170}$$

the Hamiltonian becomes

$$\mathcal{H} = \frac{1}{2}(p^2 + q^2). \tag{7.171}$$

The eigenvalue equation is ($\langle q|p = (1/\mathrm{i})(\mathrm{d}/\mathrm{d}q)\langle q|$)

$$\left[-\frac{1}{2}\frac{\mathrm{d}^2}{\mathrm{d}q^2} + \frac{1}{2}q^2\right]u_n(q) = \frac{E_n}{\hbar\omega}u_n(q), \tag{7.172}$$

where

$$E_n = \hbar\omega\left(n + \frac{1}{2}\right), \quad n = 0,1,2,\ldots. \tag{7.173}$$

The normalization we chose was (see the comments at the end of Section 3.5 regarding switching variables to x instead of q)

$$\int_{-\infty}^{\infty} \mathrm{d}q' u_n(q') u_n(q') = 1, \tag{7.174}$$

* See J.S. Schwinger, *Quantum Mechanics: Symbolism of Atomic Measurements*, edited by B.-G. Englert (Springer, 2001), Section 8.1.

and we found explicitly that

$$u_n(q) = \frac{1}{\sqrt{\sqrt{\pi}2^n n!}} e^{q^2/2} \left(-\frac{d}{dq} \right)^n e^{-q^2}. \tag{7.175}$$

A Taylor series for a general function $f(q)$ may be written as

$$f(q+q') = \sum_{n=0}^{\infty} \frac{1}{n!} \left(q' \frac{\partial}{\partial q} \right)^n f(q). \tag{7.176}$$

Now multiply both sides of Equation 7.175 by $\lambda^n / \sqrt{n!}$ (where λ is a real number) and sum over n:

$$\sum_{n=0}^{\infty} \frac{\lambda^n}{\sqrt{n!}} u_n(q) = \frac{1}{\pi^{1/4}} e^{q^2/2} \sum_{n=0}^{\infty} \frac{1}{n!} \left(-\frac{\lambda}{\sqrt{2}} \frac{d}{dq} \right)^n e^{-q^2}. \tag{7.177}$$

Comparing the right-hand side of Equation 7.177 to that of Equation 7.176, we can identify $q' = -\lambda / \sqrt{2}$, and we find that

$$\sum_{n=0}^{\infty} \frac{\lambda^n}{\sqrt{n!}} u_n(q) = \frac{1}{\pi^{1/4}} e^{q^2/2} e^{-(q-\lambda/\sqrt{2})^2}, \tag{7.178}$$

$$= \frac{1}{\pi^{1/4}} e^{-q^2/2+\sqrt{2}\lambda q - \lambda^2/2}. \tag{7.179}$$

The right-hand side of Equations 7.178 or 7.179 is called a *generating function* for the $u_n(q)$ because by doing a specified mathematical operation, we may *generate* all the $u_n(q)$'s. Reversing the reasoning that leads to these results, we find that

$$u_n(q) = \frac{1}{\pi^{1/4}} \frac{1}{\sqrt{n!}} \left(\frac{d}{d\lambda} \right)^n \left[e^{-q^2/2+\sqrt{2}\lambda q - \lambda^2/2} \right]_{\lambda=0}. \tag{7.180}$$

Now that we have that under our belts, let us consider the *two*-dimensional harmonic oscillator. Using the same dimensionless form as before, our Hamiltonian is

$$\mathcal{H} = \frac{1}{2}(p_1^2 + p_2^2) + \frac{1}{2}(q_1^2 + q_2^2). \tag{7.181}$$

Defining

$$\mathcal{H}_1 = \frac{1}{2}(p_1^2 + q_1^2), \quad \mathcal{H}_2 = \frac{1}{2}(p_1^2 + q_2^2), \tag{7.182}$$

we easily see that

$$[\mathcal{H}_1, \mathcal{H}_2] = 0. \tag{7.183}$$

The simultaneous eigenkets of $\mathcal{H}_1$ and $\mathcal{H}_2$ are

$$|n_1, n_2\rangle \equiv |n_1\rangle \otimes |n_2\rangle, \tag{7.184}$$

where

$$\left.\begin{aligned}\mathcal{H}_1 \,|\, n_1\rangle &= \left(n_1 + \frac{1}{2}\right)|\, n_1\rangle \\ \mathcal{H}_2 \,|\, n_2\rangle &= \left(n_2 + \frac{1}{2}\right)|\, n_2\rangle\end{aligned}\right\}, \tag{7.185}$$

so that (n_1, n_2 = 0, 1, 2, … independently)

$$\mathcal{H}\,|n_1, n_2\rangle = (n_1 + n_2 + 1)|n_1, n_2\rangle. \tag{7.186}$$

Calling $n = n_1 + n_2$, the energy degeneracy of the energy level labeled by n is clearly $n + 1$. The eigenvalue Equation 7.186 projected into a coordinate basis is

$$\left[-\frac{1}{2}\left(\frac{\partial^2}{\partial q_1^2} + \frac{\partial^2}{\partial q_2^2}\right) + \frac{1}{2}(q_1^2 + q_2^2)\right] u_{n_1 n_2}(q_1, q_2) = (n+1) u_{n_1 n_2}(q_1, q_2), \tag{7.187}$$

where of course

$$\langle q_1, q_2 \,|\, n_1, n_2\rangle = \langle q_1 \,|\, n_1\rangle\langle q_2 \,|\, n_2\rangle, \tag{7.188}$$

or

$$u_{n_1 n_2}(q_1, q_2) = u_{n_1}(q_1) u_{n_2}(q_2). \tag{7.189}$$

Let us now do the same thing for the two-dimensional harmonic oscillator that we just got through doing for the one-dimensional case, that is, find a generating function for the wave functions. It is now easy to see that the statement analogous to Equation 7.179 is

$$\begin{aligned}&\sum_{n_1, n_2 = 0}^{\infty} \frac{\lambda_1^{n_1} \lambda_2^{n_2}}{\sqrt{n_1! n_2!}} u_{n_1 n_2}(q_1, q_2) \\ &= \frac{1}{\pi} \exp\left[-\frac{1}{2}(q_1^2 + q_2^2) + \sqrt{2}(q_1\lambda_1 + q_2\lambda_2) - \frac{1}{2}(\lambda_1^2 + \lambda_2^2)\right].\end{aligned} \tag{7.190}$$

Although these wave functions are a complete solution to the problem, we wish to find the eigenfunctions in terms of the plane polar coordinates ρ and ϕ, defined in Figure 7.7, in preparation for solving the Coulomb problem.

The transformation equations are

$$\left.\begin{aligned}q_1 &= \rho \cos\phi \\ q_2 &= \rho \sin\phi\end{aligned}\right\}. \tag{7.191}$$

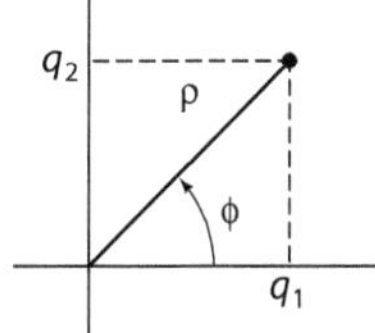

FIGURE 7.7 Showing polar coordinates ρ and ϕ in the q_1, q_2 plane.

(In terms of the physical coordinates x_1 and x_2, $\rho^2 = (m\omega/\hbar)(x_1^2 + x_2^2)$.) Let us introduce the complex numbers,

$$\left.\begin{aligned} \lambda_+ &\equiv \frac{1}{\sqrt{2}}(\lambda_1 - i\lambda_2), \\ \lambda_- &\equiv \frac{1}{\sqrt{2}}(\lambda_1 + i\lambda_2). \end{aligned}\right\}, \tag{7.192}$$

where λ_1 and λ_2 are real. Notice that $\lambda_+^* = \lambda_-$ and that

$$\lambda_+\lambda_- = \frac{1}{2}\left(\lambda_1^2 + \lambda_2^2\right). \tag{7.193}$$

The generating function (Equation 7.190) can be written in terms of ρ and ϕ as

$$\begin{aligned} &\frac{1}{\sqrt{\pi}}\exp\left[-\frac{1}{2}(q_1^2 + q_2^2) + \sqrt{2}(\lambda_1 q_1 + \lambda_2 q_2) - \frac{1}{2}\left(\lambda_1^2 + \lambda_2^2\right)\right] \\ &= \frac{1}{\sqrt{\pi}}\exp\left[-\frac{1}{2}\rho^2 + \rho + (\lambda_+ e^{i\phi} + \lambda_- e^{-i\phi}) - \lambda_+\lambda_-\right]. \end{aligned} \tag{7.194}$$

By expanding the left-hand side of Equation 7.194 in a Taylor series in λ_1 and λ_2, we have learned that the coefficients, outside of a numerical factor, *are* the wave functions in the q_1, q_2 basis. We now do the same thing in λ_+ and λ_-. That is, we expand the right-hand side of Equation 7.194 in powers of λ_+ and λ_-, thereby defining a complete set of solutions in the ρ, ϕ basis:

$$\frac{1}{\sqrt{\pi}}\exp\left[-\frac{1}{2}\rho^2 + \rho(\lambda_+ e^{i\phi} + \lambda_- e^{-i\phi}) - \lambda_+\lambda_-\right] \equiv \sum_{n_+, n_- = 0}^{\infty} \frac{\lambda_+^{n_+}\lambda_-^{n_-}}{\sqrt{n_+! n_-!}} u_{n_+ n_-}(\rho, \phi). \tag{7.195}$$

Since the left-hand side of Equation 7.195 is equal to the right-hand side of Equation 7.190, we have that

$$\sum_{n_1, n_2 = 0}^{\infty} \frac{\lambda_1^{n_1}\lambda_2^{n_2}}{\sqrt{n_1! n_2!}} u_{n_1 n_2}(q_1, q_2) = \sum_{n_+, n_- = 0}^{\infty} \frac{\lambda_+^{n_+}\lambda_-^{n_-}}{\sqrt{n_+! n_-!}} u_{n_+ n_-}(\rho, \phi). \tag{7.196}$$

But, actually, we can say more than this. Let us imagine picking out all the terms of the left of Equation 7.196 such that $n_1 + n_2 = n$, where n is fixed. (There will be $n + 1$ such terms.) These terms must be equal to the terms on the right of Equation 7.196 such that $n_+ + n_- = n$ also, since λ_+ and λ_- are just linear combinations of λ_1 and λ_2. Therefore, we have the more specific statement that

$$\sum_{n_1 + n_2 = n} \frac{\lambda_1^{n_1}\lambda_2^{n_2}}{n_1! n_2!} u_{n_1 n_2}(q_1, q_2) = \sum_{n_+ + n_- = n} \frac{\lambda_+^{n_+}\lambda_-^{n_-}}{n_+! n_-!} u_{n_+ n_-}(\rho, \phi). \tag{7.197}$$

(The notation, $\sum_{n_1 + n_2 = n}$, means to sum over all values of n_1, $n_2 = 0, 1, 2, \ldots$ such that $n_1 + n_2 = n$, where n is fixed.) Equation 7.197 says that the $u_{n_1 n_2}$ (q_1, q_2) and the $u_{n_+ n_-}(\rho, \phi)$ are just linear combinations of each other. Since we know that the $u_{n_1 n_2}(q_1, q_2)$ represent a complete set of basis functions, it follows that the $u_{n_+n_-}(\rho, \phi)$ are also a complete set but in a different coordinate basis. I prove in Appendix C that they are also orthonormal.

We now will find explicit expressions for the $u_{n_+n_-}$. By operating on both sides of Equation 7.195 with $(1/\sqrt{n_+!})(d/d\lambda_+)^{n_+}$, evaluated at $\lambda_+ = 0$, we get

$$\sum_{n_-=0}^{\infty} \frac{\lambda_-^{n_-}}{\sqrt{n_-!}} u_{n_+n_-} = \frac{1}{\sqrt{\pi n_+!}} \exp\left(-\frac{1}{2}\rho^2 + \rho e^{-i\phi}\lambda_-\right)(\rho e^{i\phi} - \lambda_-)^{n_+}. \quad (7.198)$$

Now we operate with $(1/\sqrt{n_-!})(d/d\lambda_-)^{n_-}$, evaluated at $\lambda_- = 0$, on both sides of Equation 7.198 to get

$$u_{n_+n_-} = \frac{1}{\sqrt{\pi n_+! n_-!}} e^{-\rho^2/2} \left(\frac{d}{d\lambda_-}\right)^{n_-} \left[e^{\rho e^{-i\phi}\lambda_-}(\rho e^{i\phi} - \lambda_-)^{n_+}\right]\Big|_{\lambda_-=0}. \quad (7.199)$$

Let us start to simplify this expression by defining a new parameter z such that

$$\lambda_- \equiv \rho e^{i\phi}(1 - z). \quad (7.200)$$

Replacing λ_- by z in Equation 7.199, we get

$$\begin{aligned} u_{n_+n_-} &= \frac{1}{\sqrt{\pi n_+! n_-!}} e^{-1/2\rho^2} \left(-\frac{1}{(\rho e^{i\phi})} \frac{d}{dz}\right)^{n_-} \left[e^{\rho^2(1-z)}(\rho e^{i\phi} z)^{n_+}\right]\Big|_{z=1} \\ &= \frac{1}{\sqrt{\pi n_+! n_-!}} e^{1/2\rho^2} (\rho e^{i\phi})^{n_+ - n_-} \left(-\frac{d}{dz}\right)^{n_-} \left[e^{-z\rho^2} z^{n_+}\right]\Big|_{z=1}. \end{aligned} \quad (7.201)$$

Notice that we may write

$$\begin{aligned} \left(-\frac{d}{dz}\right)^{n_-} \left[z^{n_+} e^{-\rho^2 z}\right]\Big|_{z=1} &= \rho^{-2n_+} \left(-\frac{d}{dz}\right)^{n_-} \left[(\rho^2 z)^{n_+} e^{-\rho^2 z}\right]\Big|_{z=1} \\ &= \rho^{2(n_- - n_+)} \left(-\frac{d}{d(\rho^2 z)}\right)^{n_-} \left[(\rho^2 z)^{n_+} e^{-\rho^2 z}\right]\Big|_{z=1}. \end{aligned} \quad (7.202)$$

We can now set $z = 1$ in the last expression to get

$$\left(-\frac{d}{dz}\right)^{n_-} \left[z^{n_+} e^{-\rho^2 z}\right]\Big|_{z=1} = \rho^{2(n_- - n_+)} \left(-\frac{d}{d\rho^2}\right)^{n_-} \left[\rho^{2n_+} e^{-\rho^2}\right]. \quad (7.203)$$

Our simplified expression for $u_{n_+n_-}$ is

$$u_{n_+n_-}(\rho, \phi) = \frac{1}{\sqrt{\pi n_+! n_-!}} e^{i(n_+ - n_-)\phi} \rho^{n_- - n_+} e^{1/2\rho^2} \left(-\frac{d}{d\rho^2}\right)^{n_-} \left[\rho^{2n_+} e^{-\rho^2}\right]. \quad (7.204)$$

Let us introduce the quantity

$$L_n^{\alpha}(x) \equiv x^{-\alpha} e^{x} \frac{1}{n!} \left(\frac{d}{dx}\right)^{n} \left[x^{n+\alpha} e^{-x}\right], \quad (7.205)$$

where n is an integer (0, 1, 2, …) and α is an arbitrary real quantity. These are called *Laguerre polynomials* when $\alpha = 0$ (we define $L_n(x) \equiv L_n^0(x)$) or *associated Laguerre polynomials* for $\alpha \neq 0$. They are polynomials of order n.

We can write our $u_{n_+n_-}$ functions in terms of the associated Laguerre polynomials as follows:

$$u_{n_+n_-}(\rho,\phi) = \frac{1}{\sqrt{2\pi}} e^{i(n_+-n_-)\phi} P_{n_+n_-}(\rho), \tag{7.206}$$

where

$$P_{n_+n_-}(\rho) \equiv \sqrt{2\frac{n_-!}{n_+!}} e^{-\rho^2/2} (-1)^{n_-} \rho^{n_+-n_-} L_{n_-}^{n_+-n_-}(\rho^2). \tag{7.207}$$

The quantity $(1/\sqrt{2\pi})e^{i(n_+-n_-)\phi}$ in Equation 7.206 is the two-dimensional analog of the spherical harmonics. Notice also that the argument of the Laguerre function is ρ^2, not ρ. The associated Laguerre polynomials have the symmetry property that*

$$L_n^{\alpha}(x) = \frac{(\alpha+n)!}{n!}(-1)^{\alpha} x^{-\alpha} L_{\alpha+n}^{-\alpha}(x), \tag{7.208}$$

as long as $n = 0, 1, 2, \ldots$ and $\alpha = -n, -n+1, -n+2, \ldots$. Using Equation 7.208 in 7.207, we get an alternative form for $P_{n_+n_-}(\rho)$:

$$P_{n_+n_-}(\rho) = \sqrt{2\frac{n_+!}{n_-!}} e^{-\rho^2/2} (-1)^{n_+} \rho^{n_--n_+} L_{n_+}^{n_--n_+}(\rho^2). \tag{7.209}$$

Just like when we had two alternative expressions for the $Y_{\ell m}(\theta,\phi)$ and combined them into a third, more convenient, expression, we now do the same thing here with the alternative expressions, Equations 7.207 and 7.209. We choose to use Equation 7.207 when $n_+ - n_- \geq 0$ and Equation 7.209 when $n_+ - n_- \leq 0$. We can now write

$$P_{n_+n_-}(\rho) = \sqrt{2\frac{\left(\frac{n_++n_-}{2} - \frac{|n_+-n_-|}{2}\right)!}{\left(\frac{n_++n_-}{2} + \frac{|n_+-n_-|}{2}\right)!}} e^{-\rho^2/2} (-1)^{\frac{n_++n_-}{2} - \frac{|n_+-n_-|}{2}} \times \rho^{|n_+-n_-|} L_{\frac{n_++n_-}{2} - \frac{|n_+-n_-|}{2}}^{|n_+-n_-|}(\rho^2) \tag{7.210}$$

The expression, Equation 7.210, is somewhat awkward. Let us define

$$\left.\begin{aligned} m &\equiv n_+ - n_-, \\ n_r &\equiv \frac{n_++n_-}{2} - \frac{|m|}{2}, \end{aligned}\right\} \tag{7.211}$$

and relabel $P_{n_+n_-}$ as $P_{n_r m}$. We find

$$P_{n_r m}(\rho) = P_{n_r,-m}(\rho) = \sqrt{2\frac{n_r!}{(n_r+|m|)!}} e^{-\rho^2/2} (-1)^{n_r} \rho^{|m|} L_{n_r}^{|m|}(\rho^2). \tag{7.212}$$

---

* Schwinger, *Quantum Mechanics*, Section 7.3 and Problem 7-13a.

The quantum number n_r gives the number of nodes or zeros in $L_{n_r}^{|m|}(\rho)$ (or $P_{n_r m}(\rho)$ excluding the point at infinity). Our relabeled, complete radial wave functions are thus

$$u_{n_r m}(\rho,\phi) = \frac{1}{\sqrt{2\pi}} e^{im\phi} P_{n_r m}(\rho), \tag{7.213}$$

with $P_{n_r m}(\rho)$ given earlier.

We now have two different basis sets in which to uniquely characterize the states of our two-dimensional system. The explicit forms for the Cartesian wave functions are given in Equations 7.175 and 7.189, and their quantum numbers are n_1 and n_2, where each can take on the values 0, 1, 2, … independently. Likewise, the explicit radial wave functions are given in Equations 7.212 and 7.213, and their quantum numbers are specified by n_r and m, where $n_r = 0, 1, 2, \ldots$ and m is any positive or negative integer (or zero). Thus, our Hilbert space is spanned either by the basis

$$|n_1, n_2\rangle,$$

which has energies

$$H|n_1, n_2\rangle = \hbar\omega(n_1 + n_2 + 1)|n_1, n_2\rangle, \tag{7.214}$$

or by the basis

$$|n_r, m\rangle,$$

which has energies

$$H|n_r, m\rangle = \hbar\omega(2n_r + |m| + 1)|n_r, m\rangle. \tag{7.215}$$

(We have from Equation 7.197 that $n_+ + n_- = 2n_r + |m|$ characterizes the energy of the states.) Although both characterizations are complete, in general there is not a one-to-one correspondence between these different basis states because of the degeneracies in energy. In either characterization, the possible energies of the system are $E = \hbar\omega(n+1)$, where $n = 0$, 1, 2, …, and the degeneracy of the nth level is given by $n + 1$. An explicit energy diagram of the different basis states is given next.

Cartesian basis		**Radial basis**
$n_1=0, n_2=2$		$n_r=0, m=2$
$n_1=1, n_2=1$		$n_r=0, m=-2$
$n_1=2, n_2=0$	$E=3\hbar\omega$	$n_r=1, m=0$
$n_1=0, n_2=1$		$n_r=0, m=1$
$n_1=1, n_2=0$	$E=2\hbar\omega$	$n_r=0, m=-1$
$n_1=n_2=0$	$E=\hbar\omega$	$n_r=m=0$

In general, the $(n+1)$ Cartesian or radial states for which $E = \hbar\omega(n+1)$ are linear combinations of each other. These are given by Equation 7.197. (There is, of course, a one-to-one correspondence between the two nondegenerate ground states.)

Just to get a feeling for the correspondence of the two sets of states, let us work some of the relationships out explicitly. The Cartesian basis ground state is given by

$$u_{00}(q_1,q_2)=u_0(q_1)u_0(q_2), \tag{7.216}$$

where (from Equation 7.175)

$$u_0(q)=\frac{1}{\pi^{1/4}}\mathrm{e}^{-q^2/2}. \tag{7.217}$$

Therefore,

$$u_{00}(q_1,q_2)=\frac{1}{\sqrt{\pi}}\mathrm{e}^{-(q_1^2+q_2^2)/2}\left(=\frac{1}{\sqrt{\pi}}\mathrm{e}^{-\rho^2/2}\right). \tag{7.218}$$

On the other hand, we have

$$u_{00}(\rho,\phi)=\frac{1}{\sqrt{\pi}}\mathrm{e}^{-\rho^2/2}\,L_0^0(\rho^2), \tag{7.219}$$

but it is easy to show that

$$L_0^{\alpha}(\rho^2)=1, \tag{7.220}$$

for any α, so that

$$u_{00}(\rho,\phi)=\frac{1}{\sqrt{\pi}}\mathrm{e}^{-\rho^2/2}. \tag{7.221}$$

Equations 7.218 and 7.221 are identical.

For the $E=2\hbar\omega$ energy levels, we have the linearly independent Cartesian set

$$\left.\begin{aligned}u_{10}(q_1,q_2)&=u_1(q_1)u_0(q_2),\\ u_{01}(q_1,q_2)&=u_0(q_1)u_1(q_2).\end{aligned}\right\} \tag{7.222}$$

From Equation 7.175 we have that

$$u_1(q)=\frac{\sqrt{2}}{\pi^{1/4}}q\mathrm{e}^{-q^2/2}, \tag{7.223}$$

so that

$$\left.\begin{aligned}u_{10}(q_1,q_2)&=\sqrt{\frac{2}{\pi}}q_1\mathrm{e}^{-(q_1^2+q_2^2)/2}\left(=\sqrt{\frac{2}{\pi}}\rho\cos\phi\,\mathrm{e}^{-\rho^2/2}\right),\\ u_{01}(q_1,q_2)&=\sqrt{\frac{2}{\pi}}q_1\mathrm{e}^{-(q_1^2+q_2^2)/2}\left(=\sqrt{\frac{2}{\pi}}\rho\sin\phi\,\mathrm{e}^{-\rho^2/2}\right).\end{aligned}\right\} \tag{7.224}$$

The radial basis set is

$$\left.\begin{aligned}u_{01}(\rho,\phi)&=\frac{1}{\sqrt{2\pi}}\mathrm{e}^{i\phi}\sqrt{2}\mathrm{e}^{-\rho^2/2}\rho L_0^1(\rho^2),\\ &=\frac{1}{\sqrt{\pi}}\rho\mathrm{e}^{i\phi}\mathrm{e}^{-\rho^2/2},\\ u_{0-1}(\rho,\phi)&=\frac{1}{\sqrt{2\pi}}\mathrm{e}^{-i\phi}\sqrt{2}\mathrm{e}^{-\rho^2/2}\rho L_0^1(\rho^2),\\ &=\frac{1}{\sqrt{\pi}}\rho\mathrm{e}^{-i\phi}\mathrm{e}^{-\rho^2/2}.\end{aligned}\right\} \tag{7.225}$$

The relationship between these two sets is given by Equation 7.197 with $n=1$:

$$\sum_{n_1+n_2=1}\frac{\lambda_1^{n_1}\lambda_2^{n_2}}{\sqrt{n_1!n_2!}}u_{n_1n_2}=\sum_{n_++n_-=1}\frac{\lambda_+^{n_+}\lambda_-^{n_-}}{\sqrt{n_+!n_-!}}u_{n_rm},\tag{7.226}$$

which gives

$$\lambda_1u_{10}(q_1,q_2)+\lambda_2u_{01}(q_1,q_2)=\lambda_+u_{01}(\rho,\phi)+\lambda_-u_{0-1}(\rho,\phi).\tag{7.227}$$

However, expressing λ_+ and λ_- in terms of λ_1 and λ_2 using Equation 7.192 and matching coefficients of λ_1 and λ_2 on both sides of Equation 7.227, we find that

$$\left.\begin{aligned}u_{10}(q_1,q_2)&=\frac{1}{\sqrt{2}}(u_{01}(\rho,\phi)+u_{0-1}(\rho,\phi)),\\ u_{01}(q_1,q_2)&=\frac{i}{\sqrt{2}}(u_{0-1}(\rho,\phi)-u_{01}(\rho,\phi)).\end{aligned}\right\}\tag{7.228}$$

Plugging in our explicit expressions above, we see that the equations in 7.228 are identically satisfied.

What explicit equations do the $u_{n_rm}(\rho,\phi)$ solve? Our two-dimensional Schrödinger equation reads

$$\left[-\frac{1}{2}\left(\frac{\partial^2}{\partial q_1^2}+\frac{\partial^2}{\partial q_2^2}\right)+\frac{1}{2}\left(q_1^2+q_2^2\right)\right]u_{n_1n_2}=(n_1+n_2+1)u_{n_1n_2},\tag{7.229}$$

in Cartesian coordinates. We change variables to ρ and ϕ:

$$\left.\begin{aligned}&\frac{\partial^2}{\partial q_1^2}+\frac{\partial^2}{\partial q_2^2}=\frac{\partial^2}{\partial\rho^2}+\frac{1}{\rho}\frac{\partial}{\partial\rho}+\frac{1}{\rho^2}\frac{\partial^2}{\partial\phi^2},\\ &q_1^2+q_2^2=\rho^2.\end{aligned}\right\}\tag{7.230}$$

The $u_{n_rm}(\rho,\phi)$ satisfy

$$\left[-\frac{1}{2}\left(\frac{\partial^2}{\partial\rho^2}+\frac{1}{\rho}\frac{\partial}{\partial\rho}+\frac{1}{\rho^2}\frac{\partial^2}{\partial\phi^2}\right)+\frac{1}{2}\rho^2\right]u_{n_rm}=(2n_r+|m|+1)u_{n_rm}.\tag{7.231}$$

The variables ϕ and ρ separate in Equation 7.231, and according to Equation 7.213, we may replace

$$\frac{1}{\rho^2}\frac{\partial^2}{\partial\phi^2}\to-\frac{m^2}{\rho^2}.\tag{7.232}$$

In Equation 7.231, we also replace $u_{n_rm}(\rho,\phi)$ by $P_{n_rm}(\rho)$. Thus $P_{n_rm}(\rho)$ satisfies

$$\left[-\frac{1}{2}\left(\frac{d^2}{d\rho^2}+\frac{1}{\rho}\frac{d}{d\rho}-\frac{m^2}{\rho^2}\right)+\frac{1}{2}\rho^2\right]P_{n_rm}(\rho)=(2n_r+|m|+1)P_{n_rm}(\rho),\tag{7.233}$$

or

$$\left[\frac{d^2}{d\rho^2}+\frac{1}{\rho}\frac{d}{d\rho}-\frac{m^2}{\rho^2}+2(2n_r+|m|+1)-\rho^2\right]P_{n_rm}(\rho)=0.\tag{7.234}$$

Let us now come to the point. The Coulomb problem, Equation 7.148, can be cast into the form:

$$\left[\frac{\mathrm{d}^2}{\mathrm{d}r^2} - \frac{\ell(\ell+1)}{r^2} + \frac{2Z}{a_0 r} - \frac{2\mu}{\hbar^2}|E|\right] R(r) = 0, \tag{7.235}$$

where we have set $E = -|E|$ for bound states. To establish a connection between Equations 7.234 and 7.235, let us let

$$\rho^2 = \lambda r, \tag{7.236}$$

in the two-dimensional oscillator equation. It is easy to establish that

$$\left.\begin{aligned} \frac{\mathrm{d}^2}{\mathrm{d}\rho^2} &= \frac{4}{\lambda}\left[\frac{1}{2}\frac{\mathrm{d}}{\mathrm{d}r} + r\frac{\mathrm{d}^2}{\mathrm{d}r^2}\right], \\ \frac{1}{\rho}\frac{\mathrm{d}}{\mathrm{d}\rho} &= \frac{2}{\lambda}\frac{\mathrm{d}}{\mathrm{d}r}. \end{aligned}\right\} \tag{7.237}$$

Then, multiplying by $\lambda/4r$, we find that Equation 7.234 may be cast into the form

$$\left[\frac{\mathrm{d}^2}{\mathrm{d}r^2} + \frac{1}{r}\frac{\mathrm{d}}{\mathrm{d}r} - \frac{m^2}{4r^2} + \frac{\lambda(2n_r + |m| + 1)}{2r} - \frac{\lambda^2}{4}\right] P_{n_r m}\left(\sqrt{\lambda r}\right) = 0. \tag{7.238}$$

Equation 7.238 is of the same form of Equation 7.235 except for the $(1/r)(\mathrm{d}/\mathrm{d}r)$ term. We can get rid of it by writing the equation satisfied by $\sqrt{r}P_{n_r m}$:

$$\left[\frac{\mathrm{d}^2}{\mathrm{d}r^2} - \frac{m^2 - 1}{4r^2} + \frac{\lambda(2n_r + |m| + 1)}{2r} - \frac{\lambda^2}{4}\right]\left[\sqrt{r}P_{n_r m}\left(\sqrt{\lambda r}\right)\right] = 0. \tag{7.239}$$

Casting our eyes back on Equation 7.235, we see that since we know the explicit solutions to Equation 7.239, we can also work out the solutions to Equation 7.235 given the following correspondence:

2-D oscillator	**Coulomb**
$(m^2 - 1)/4$	$\ell(\ell+1)$
$\lambda(2n_r + \|m\| + 1)/2$	$2Z/a_0$
$\lambda^2/4$	$2\mu\|E\|/\hbar^2$

$(a_0 = \hbar^2/\mu e^2)$ The first line tells us what $|m|$ corresponds to; the last two lines tell us what λ and n_r correspond to. Thus, from the first line we get

$$m^2 \to 4\ell(\ell+1) + 1 = (2\ell+1)^2,$$

so that

$$|m| \to |2\ell + 1| = 2\ell + 1. \tag{7.240}$$

The arrow symbol "→" means "corresponds to." We should view Equation 7.240 simply as a correspondence between parameters; that is, as far as Equations 7.235 and 7.239 are concerned, $|m|$ and ℓ are just parameters that can take on any values. (Separate eigenvalue equations determine the values of m and ℓ.) From the third line of the correspondence, we get

$$\lambda \to \frac{\sqrt{8\mu|E|}}{\hbar}. \tag{7.241}$$

(λ must, from Equation 7.236, be a positive quantity.) Then, from the second line above, we get

$$\lambda n_r \to \frac{2Z}{a_0} - (\ell+1)\lambda,$$

from Equation 7.240. Then, using Equation 7.241, we find the correspondence

$$n_r \to \frac{2Z\hbar}{a_0\sqrt{8\mu|E|}} - (\ell+1). \tag{7.242}$$

Now, according to the $\vec{L}^2$ eigenvalue equation, Equation 6.188, ℓ takes on values 0, 1, 2, …. Also, n_r is *required* by its eigenvalue equation (either Equation 7.234, 7.238, or 7.239) to take on values 0, 1, 2, … also. Therefore, the quantity $2Z\hbar/(a_0\sqrt{8\mu|E|})$ is also restricted to integer values such that Equation 7.242 is satisfied. In other words we have $(n = n_r + \ell + 1)$

$$\frac{2Z\hbar}{a_0\sqrt{8\mu|E|}} = n, \tag{7.243}$$

where, for a fixed ℓ, we must have n $n \geq \ell+1$ for the integer n (i.e., $n_r \geq 0$). Equation 7.243 determines the energy levels of the Coulomb problem! Solving for E, we get

$$E = -|E| = -\frac{Z^2\hbar^2}{2\mu a_0^2 n^2}, \tag{7.244}$$

$$= -\frac{Z^2 e^2}{2a_0 n^2}. \tag{7.245}$$

The integer n is called the *principal quantum number* since it completely determines the energy value. Reinserting n into Equations 7.241 and 7.242, we find that the complete correspondence between the two-dimensional harmonic oscillator and the (three-dimensional) Coulomb problem is specified by

$$\left.\begin{aligned} |m| &\to 2\ell+1, \\ \lambda &\to \frac{2Z}{a_0 n}, \\ n_r &\to n-(\ell+1). \end{aligned}\right\} \tag{7.246}$$

Equation 7.246 is very useful because it is only necessary to make the above substitutions in the quantity $\sqrt{r}P_{n_rm}(\sqrt{\lambda}r)$(see Equation 7.239) to read off the Coulomb radial eigenfunctions. (The angular eigenfunctions are, of course, just the $Y_{\ell m}(\theta,\phi)$ as in any central force problem.) Making the substitutions (Equation 7.246) into the explicit form for P_{n_rm}, Equation 7.212, we find for these eigenfunctions

$$R_{nl}(r) = N\sqrt{\frac{a_0 n}{Z}\frac{(n-\ell-1)!}{(n+\ell)!}}(-1)^{n-\ell-1}e^{-Zr/a_0n}\left(\frac{2Zr}{a_0n}\right)^{\ell+1} L_{n-\ell-1}^{2\ell+1}\left(\frac{2Zr}{a_0n}\right). \tag{7.247}$$

N is an unknown normalization constant, which will be evaluated shortly. The phase factor $(-1)^{n-\ell-1}$ is unimportant and may be discarded if desired. (Because of the symmetry property, Equation 7.208, there is an alternate form of these wave functions that can be written down.)

We can evaluate the constant N as follows. We require that (remember that $u(r) = R(r)/r$ from Equation 7.9)

$$\int_0^\infty dr|R_{n\ell}(r)|^2 = 1. \tag{7.248}$$

We may also effectively evaluate the integral on the left-hand side of Equation 7.248 by appealing to the two-dimensional harmonic oscillator. Consider the integral

$$I \equiv \int_0^\infty d\rho\rho^3(P_{n_rm}(\rho))^2. \tag{7.249}$$

This integral may be evaluated using the result of Equation 5.3 (Chapter 5):

$$\left\langle\frac{\partial H}{\partial\lambda}\right\rangle_E = \frac{\partial E}{\partial\lambda}. \tag{7.250}$$

The two-dimensional harmonic oscillator Hamiltonian is

$$H = \frac{p_1^2 + p_2^2}{2m} + \frac{m\omega^2}{2}\left(x_1^2 + x_2^2\right), \tag{7.251}$$

and our orthonormality condition was (see Equation C.11 of Appendix C and Equations 7.206 and 7.212)

$$\int_0^\infty d\rho\rho P_{n_rm}(\rho)P_{n_r'm}(\rho) = \delta_{n_rn_r'}, \tag{7.252}$$

when expressed in polar coordinates. (Remember, $\rho^2 = (m\omega/\hbar)(x_1^2 + x_2^2)$.) Picking $\lambda = \omega$ for use in Equation 7.250, we find that

$$\frac{1}{\hbar}\left\langle\frac{\partial H}{\partial\omega}\right\rangle_{n_rm} = \left(\rho^2\right)_{n_rm}, \tag{7.253}$$

but

$$\left\langle\rho^2\right\rangle_{n_rm} = I, \tag{7.254}$$

and

$$\frac{1}{\hbar}\frac{\partial E}{\partial\omega} = (2n_r + |m| + 1). \tag{7.255}$$

Putting Equations 7.253, 7.254, and 7.255 together, we find explicitly that

$$\int_0^\infty d\rho\rho^3 (P_{n_r m}(\rho))^2 = 2n_r + |m| + 1. \tag{7.256}$$

Now making the substitutions $\rho = (\lambda r)^{1/2}$ and Equation 7.246 into 7.256, we get the statement that

$$\frac{1}{2}\left(\frac{2Z}{a_0 n}\right)^2 \int_0^\infty dr\, r\left(P_{\substack{n_r = n-(\ell+1)\\ m=2\ell+1}}\left(\sqrt{\frac{2Zr}{a_0 n}}\right)\right)^2 = 2n. \tag{7.257}$$

However, in Equation 7.247, we have defined

$$R_{n\ell}(r) \equiv N r^{1/2} P_{\substack{n_r = n-(\ell+1)\\ m=2\ell+1}}\left(\sqrt{\frac{2Zr}{a_0 n}}\right), \tag{7.258}$$

so that Equation 7.257 implies that

$$\int_0^\infty dr R_{n\ell}^2(r) = \frac{nN^2}{(Z/a_0 n)^2}. \tag{7.259}$$

To reconcile Equation 7.259 with 7.248, we can choose

$$N = \frac{Z}{a_0 n^{3/2}}. \tag{7.260}$$

Thus, the fully normalized radial Coulomb wave functions are determined as

$$R_{n\ell}(r) = \sqrt{\frac{Z}{a_0 n^2}\frac{(n-\ell-1)!}{(n+\ell)!}}(-1)^{n-\ell-1} e^{-Zr/a_0 n}\left(\frac{2Zr}{a_0 n}\right)^{\ell+1} L_{n-\ell-1}^{2\ell+1}\left(\frac{2Zr}{a_0 n}\right), \tag{7.261}$$

and the three-dimensional wave functions are as follows:

$$u_{n\ell m}(\vec{r}) = \frac{R_{n\ell}(r)}{r} Y_{\ell m}(\theta,\phi). \tag{7.262}$$

They satisfy orthonormality:

$$\int d^3 r\, u_{n\ell m}^*(\vec{r}) u_{n'\ell' m'}(\vec{r}) = \delta_{nn'}\delta_{\ell\ell'}\delta_{mm'}. \tag{7.263}$$

However, because there are scattering solutions ($E > 0$) that we have not included, the set in Equation 7.262 is *not* complete.

We can connect with our earlier special solution for the ground state (Equation 7.168) now by putting $n = 1$ and $\ell = 0$ in Equations 7.262 and 7.261. We get

$$\begin{aligned} u_{100}(\vec{r}) &= 2\left(\frac{Z}{a_0}\right)^{3/2} e^{-Zr/a_0} L_0^1 \frac{1}{\sqrt{4\pi}} \\ &= \frac{1}{\sqrt{\pi}}\left(\frac{Z}{a_0}\right)^{3/2} e^{-Zr/a_0}, \end{aligned} \tag{7.264}$$

exactly as we had before.

Notice that since the energy levels, Equation 7.245, are given by *n alone*, we now have a degeneracy in both m and ℓ for the Coulomb potential. This is called an *accidental degeneracy*. (This term seems to be reserved for situations where the *dynamics* and not the *symmetry* determines the degeneracy. Remember, it was the fact that we are working with a spherically symmetric central force that caused the m degeneracy.) Because $\ell+1 \leq n$, we have the following classifications of the first few energy levels:

n	ℓ	Notation	#States	Total
1	0	1s	1	1
2	0	2s	1	4
	1	2p	3	
3	0	3s	1	
	1	3p	3	9
	2	3d	5	

One can easily show that the degeneracy of the nth energy level is n^2. Schematically, the Coulomb energy levels appear like Figure 7.8 $(\omega_0 = \mu Z^2 e^4 / 2\hbar^2)$.

Using the same mysterious zweideutigkeit rule as before, applied to electrons now ("at most two electrons to each energy level"), we can, on the basis of the Coulomb solution (neglecting the repulsive interactions of the electrons among themselves), begin to get a crude idea of the structure of atomic energy levels. We use the spectroscopic notation introduced before with the addition of a superscript to tell us how many electrons there are in a given atomic shell. (The principle quantum number, n, does *not* represent the number of nodes in the radial wave function here but simply indicates the ordering of the energy levels for an ideal hydrogen atom. The actual number of radial nodes is given by $n - \ell - 1$ since the wave function, Equation 7.262, is proportional to $e^{-Zr/a_0 n}\, L_{n-\ell-1}^{2\ell+1}(2Zr/a_0 n)$. I am not counting the point at $r = \infty$ as a node.) The *atomic configurations* of the first 19 elements are given in Table 7.1.

There are obviously regularities in the order in which these energy levels are filled. In order to try to explain these, I will refer to the schematic energy level diagram in Figure 7.9.

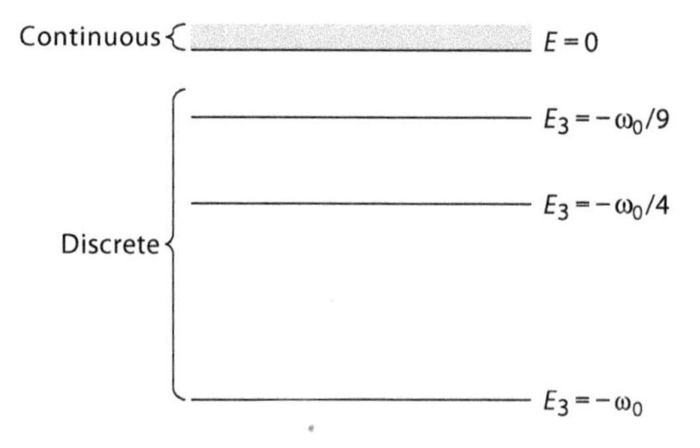

FIGURE 7.8 Representation of the Coulomb energy levels in a hydrogen-like system. The energy levels above $E = 0$ are continuous.

TABLE 7.1 Atomic Configurations of the First 19 Elements, Showing the Total Configuration and Last Electron States

Element	Z	Total Config	Last Electron
H	1	1s	1s
He (inert)	2	$(1s)^2$	1s
Li	3	He(2s)	2s
Be	4	$He(2s)^2$	2s
B	5	Be(2p)	2p
C	6	$Be(2p)^2$	2p
N	7	$Be(2p)^3$	2p
0	8	$Be(2p)^4$	2p
F	9	$Be(2p)^5$	2p
Ne (inert)	10	$Be(2p)^6$	2p
Na	11	Ne(3s)	3s
Mg	12	$Ne(3s)^2$	3s
Al	13	Mg(3p)	3p
Si	14	$Mg(3p)^2$	3p
P	15	$Mg(3p)^3$	3p
S	16	$Mg(3p)^4$	3p
Cl	17	$Mg(3p)^5$	3p
Ar (inert)	18	$Mg(3p)^6$	3p
K	19	Ar(4s)	4s

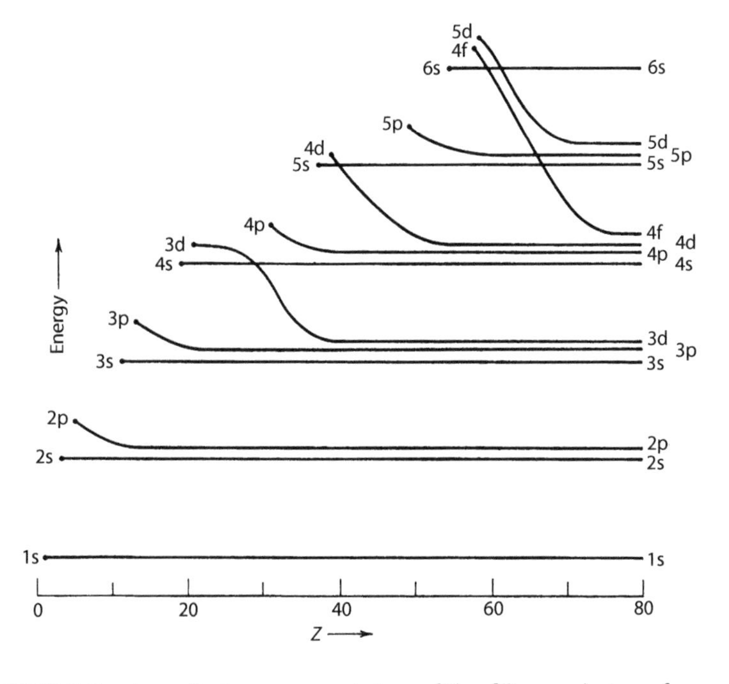

FIGURE 7.9 A qualitative representation of the filling ordering of energy levels in a neutral atom with atomic number Z, up to $Z=80$ (mercury). Each curve begins at the actual Z for which the subshell begins to be occupied. (From R. Eisberg and R. Resnick, *Quantum Physics*, 2nd ed., Wiley, Figures 9–14, 333 (1985). Reproduced with permission.)

Notice from Table 7.1 that the 2s shell is filled before the 2p shell and the 3s shell is filled before the 3p shell. These facts are also brought out in Figure 7.9. In our ideal Coulomb solution, the 2s, 2p and the 3s, 3p energy levels are degenerate. In real atoms, with their complicated interactions between the many particles involved, these energy degeneracies are split. We can get a qualitative idea of the ordering of the split levels from the following considerations. If we imagine adding electrons to an atomic nucleus with a fixed charge, the first electrons added will be the most strongly bound and therefore will tend to have wave functions that are largest near the nucleus. As more loosely bound electrons are added, the inner electrons tend to *shield* some of the nucleus's electric charge from the outer electrons. Loosely speaking, this means that for electrons in outer shells, a wave function that has a larger magnitude near the nucleus will, in effect, sample a larger attractive central charge. Thus, the 2s energy levels will be filled before the 2p shell since on the average the 2s electron's wave function has a smaller separation from the nucleus than the 2p electron does. Likewise for the 3s and 3p energy levels. These splittings are between levels that have the same value of n. This shielding effect eventually means that some energy levels with higher values of n will fill up before levels with a lower n value. The first example of this is potassium (K) in which the outer electron is in a 4s state rather than 3d.

It is the order in filling up atomic shells that puts the "period" in the periodic table. Obviously, there are many other things to observe about the atomic structure of the various elements, but we will not dwell on them here. We will, however, have more to say about the energy levels of the real hydrogen atom in the next chapter.

7.7 THE CONFINED COULOMBIC MODEL*

Let us contrast the schematic energy level diagrams found in the Coulomb and infinite spherical well cases. Plotting the energies and the assumed potentials simultaneously, we get the diagrams shown in Figure 7.10. (We generalize the Coulomb solution to the form $V(r) = -\xi / r$; let us keep an open mind as to the meaning of the constant ξ.)

In the Coulomb case, we supply the boundary condition that $R(r) \underset{r\to\infty}{\longrightarrow} 0$, and in the infinite spherical well case, we require that $R(a) = 0$. For both, we have that $R(0) = 0$. In the Coulomb problem, the discrete energies, which are negative, are given by (Equation 7.245 with $Ze^2 \to \xi$)

$$E_n = -\frac{\xi^2 \mu}{2\hbar^2 n^2}. \tag{7.265}$$

In the infinite well case the energies are positive and are given by (Equation 7.67).

$$E = \frac{\hbar^2 (ka)^2}{2ma^2}, \tag{7.266}$$

* W. Wilcox, O.V. Maxwell, and K.A. Milton, *Phys. Rev.* D **31**, 1081 (1985).

where the dimensionless quantity ka is given by the condition (Equation 7.62). We used the infinite spherical well as a starting point for describing nuclear dynamics and we developed the Coulomb solution as a way to begin to understand some physics of atomic systems. I would like to point out now that there is a way of connecting these two seemingly different situations as special cases of another model, which I am calling the confined Coulombic model. The assumed potential of this composite model looks qualitatively like Figure 7.11.

Just like the spherical well problem, we are imagining an infinite confining wall to exist at $r = a$, where we will assume that the radial wave function vanishes, $R(a) = 0$. Inside the wall, instead of there being a flat potential, we are postulating an attractive Coulombic potential, $V(r) = -\xi / r$. Because of the confining boundary condition, we expect that the eigenenergies will always be discrete like the spherical well. Then, in the limit $a \to \infty$, we would also expect that the allowed energies become more Coulomb-like. That is for a large but finite, although the $E > 0$ energies remain discrete, they should become denser. Likewise, the $E < 0$ discrete states will be better and better reproduced as a becomes larger. On the other hand, as a becomes smaller we would expect that the energies of the states to be essentially due to the kinetic energy of confinement. That is, as we squeeze a quantum mechanical particle into a smaller volume, we would, as a consequence of having decreased the

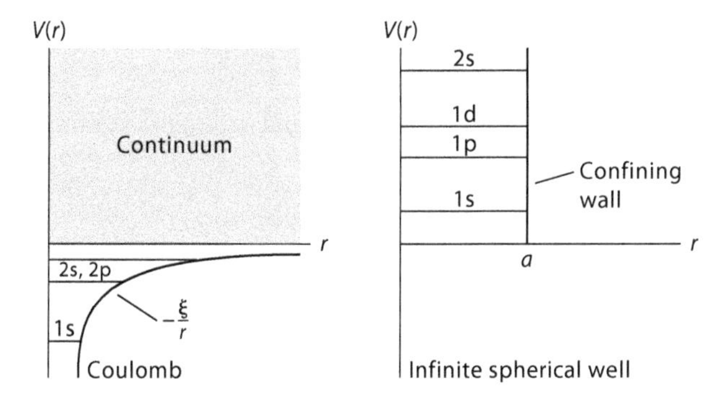

FIGURE 7.10 Simultaneous plot of the energies, superimposed on the potentials, of the Coulomb (left) and infinite spherical well (right) potentials.

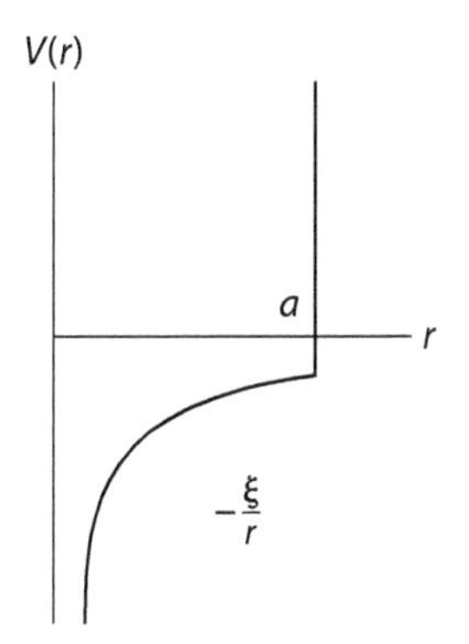

FIGURE 7.11 A representation of the confined Coulombic potential.

particle's position uncertainty, also expect the particle's momentum (and energy) to increase. This means, for small enough a, that the value of the energies will be essentially independent of the Coulomb potential and will approach the infinite spherical well energies arbitrarily closely. We may well ask what use this model is in the real physical world; I will touch upon this later.

The equation we need to solve is Equation 7.10,

$$\left[-\frac{\hbar^2}{2m}\left(\frac{\mathrm{d}^2}{\mathrm{d}r^2}-\frac{\ell(\ell+1)}{r^2}\right)+V(r)\right]R(r)=ER(r), \tag{7.267}$$

where the potential is

$$V(r)=\begin{cases}\infty, & r\geq a\\ -\dfrac{\xi}{r}, & r<a,\end{cases} \tag{7.268}$$

and the boundary conditions are, essentially, that $R(0)=0$ and $R(a)=0$. In general, there are both positive and negative energy solutions to Equation 7.267. (However, if the confinement radius becomes small enough, there will only be positive energy solutions.) We will only examine the $E>0$ solutions here. Defining $\rho=kr$ with $k=\sqrt{2mE}/\hbar$ as usual, we may write Equation 7.267 in the form

$$\left[\frac{\mathrm{d}^2}{\mathrm{d}\rho^2}-\frac{\ell(\ell+1)}{\rho^2}+\frac{\xi k}{E\rho}+1\right]R(\rho)=0. \tag{7.269}$$

We can cast Equation 7.269 into a more standard form by defining a function $\omega(\rho)$ such that*

$$R(\rho)=\rho^{\ell+1}\mathrm{e}^{-i\rho}\omega(\rho). \tag{7.270}$$

We can now work out the first and second derivative as follows:

$$\frac{\mathrm{d}R}{\mathrm{d}\rho}=\rho^{\ell+1}\mathrm{e}^{-i\rho}\left(\frac{(\ell+1)}{\rho}\omega-i\omega+\frac{\mathrm{d}\omega}{\mathrm{d}\rho}\right), \tag{7.271}$$

$$\frac{\mathrm{d}^2R}{\mathrm{d}\rho^2}=\rho^{\ell+1}\mathrm{e}^{-i\rho}\left(\frac{\ell(\ell+1)}{\rho^2}\omega-\frac{2i(\ell+1)}{\rho}\omega-\omega+\frac{2(\ell+1)}{\rho}\frac{\mathrm{d}\omega}{\mathrm{d}\rho}-2i\frac{\mathrm{d}\omega}{\mathrm{d}\rho}+\frac{\mathrm{d}^2\omega}{\mathrm{d}\rho^2}\right). \tag{7.272}$$

The differential equation satisfied by $\omega(\rho)$ can therefore be written as

$$\frac{\mathrm{d}^2\omega}{\mathrm{d}\rho^2}+2\left(\frac{(\ell+1)}{\rho}-i\right)\frac{\mathrm{d}\omega}{\mathrm{d}\rho}+\left(\frac{\xi k}{E\rho}-\frac{2i(\ell+1)}{\rho}\right)\omega=0. \tag{7.273}$$

We now define a new x such that $x=i\rho$. Equation 7.273 can now be written as

$$-\frac{\mathrm{d}^2\omega}{\mathrm{d}x^2}+2\left(\frac{-(\ell+1)}{x}+1\right)\frac{\mathrm{d}\omega}{\mathrm{d}x}+\left(\frac{i\xi k}{Ex}+\frac{2(\ell+1)}{x}\right)\omega=0 \tag{7.274}$$

* Motivation: small and large ρ behavior given by the prefactors to $\omega(\rho)$.

or as

$$\frac{x}{2}\frac{\mathrm{d}^2\omega}{\mathrm{d}x^2}+(\ell+1-x)\frac{\mathrm{d}\omega}{\mathrm{d}x}+\left(-(\ell+1)-\frac{i\xi k}{2E}\right)\omega=0. \tag{7.275}$$

We compare Equation 7.275 with the differential equation satisfied by the *confluent hypergeometric functions*:

$$z\frac{\mathrm{d}^2F}{\mathrm{d}z^2}+(b-z)\frac{\mathrm{d}F}{\mathrm{d}z}-aF=0. \tag{7.276}$$

One often writes the solution to Equation 7.276 as $F(a,b,z)$. The solution to Equation 7.276, which is finite at $z=0$, is given by the series solution

$$F(a,b,z)=1+\frac{az}{b}+\frac{a(a+1)z^2}{b(b+1)2!}+\frac{a(a+1)(a+2)}{b(b+1)(b+2)}\frac{z^3}{3!}+\cdots \tag{7.277}$$

It is understood in Equation 7.277 that b is not zero or a negative integer; $F(a,b,z)$ is undefined for these values. Because Equation 7.276 is a *second-order* differential equation, there is another, linearly independent solution that, however, we will not be interested in because it will not be able to satisfy the boundary condition $R(0)=0$. There is a more compact notation for writing Equation 7.277. Let us define

$$\left.\begin{aligned}(a)_n&\equiv a(a+1)(a+2)\ldots(a+n-1),\\(a)_0&\equiv 1.\end{aligned}\right\} \tag{7.278}$$

Then, Equation 7.277 can be written as

$$F(a,b,z)=\sum_{n=0}^{\infty}\frac{(a)_n z^n}{(b)_n n!}. \tag{7.279}$$

The Equations 7.275 and 7.276 become identical upon making the identifications

$$\left.\begin{aligned}z&=2x,\\a&=\ell+1+\frac{i\xi k}{2E},\\b&=2\ell+2.\end{aligned}\right\} \tag{7.280}$$

The solution to Equation 7.273 is, therefore, of the form

$$\omega(\rho)=CF\left(\ell+1+\frac{i\xi k}{2E},2\ell+2,2i\rho\right), \tag{7.281}$$

and the radial wave function, $R(r)$, for this problem looks like

$$R(r)=N(kr)^{\ell+1}\mathrm{e}^{-ikr}F\left(\ell+1+\frac{i\xi k}{2E},2\ell+2,2ikr\right). \tag{7.282}$$

(Amazingly, the function $\mathrm{e}^{-ika}F(\ell+1+i\xi k/2E,2\ell+2,2ikr)$ is completely real!) The quantity N is the normalization factor, which we will not determine here. The condition that determines the positive energy levels, similar to, but more complicated than Equation 7.62, is thus

$$e^{-ika}F\left(\ell+1+\frac{i\xi k}{2E},2\ell+2,2ika\right)=0. \tag{7.283}$$

Unfortunately, the energy eigenvalue condition cannot be simplified more than this. Now there are two dimensionless quantities in the arguments of F in Equation 7.283. Taking ka as one of these quantities, since we may write

$$\frac{\xi k}{2E}=\frac{(\xi ma/\hbar^2)}{ka}, \tag{7.284}$$

we learn that $\xi ma/\hbar^2$ is the other dimensionless quantity. Choosing a value of $\xi ma/\hbar^2$, Equation 7.283 then determines the possible values of ka, and therefore the energies. Qualitatively, a plot of ka as a function of $\xi ma/\hbar^2$, looks like Figure 7.12 for the 1s state.

The place where $ka=0$ in this graph is where the condition (Equation 7.283) specializes to $k\to 0$. In fact, one can show that (just use the form of Equation 7.279)

$$\lim_{k\to 0}F\left(\ell+1+i\frac{\xi m/\hbar^2}{k},2\ell+2,2ika\right)\to$$
$$(2\ell+1)!\,J_{2\ell+1}\left(2\sqrt{\frac{2ma\xi}{\hbar^2}}\right)\left(\frac{2ma\xi}{\hbar^2}\right)^{-\ell-1/2}, \tag{7.285}$$

where $J_n(x)$ ($n=2\ell+1$ and $x=2\sqrt{2ma\xi/\hbar^2}$ above) is a *Bessel function* of order n. Do not get them confused with the earlier spherical Bessel functions we discussed in the free-particle and spherical-well cases. The relationship between these two types of Bessel functions is given by

$$j_n(x)=\sqrt{\frac{\pi}{2x}}J_{n+1/2}(x). \tag{7.286}$$

A formula for the Bessel function $J_n(x)$ is

$$J_n(x)=\left(\frac{1}{2}x\right)^n\sum_{k=0}^{\infty}\frac{\left(-\frac{x^2}{4}\right)^k}{k!(n+k)!}. \tag{7.287}$$

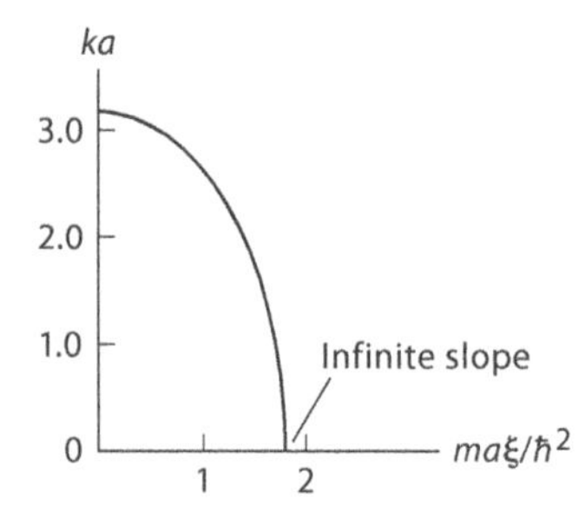

FIGURE 7.12 Qualitative plot of ka as a function of $\xi ma/\hbar^2$ as determined by Equation 7.283.

A more general formula for the order n not restricted to zero or positive integers is also available (Equation 9.1.10 of the *National Bureau of Standards*), but will not be dealt with here. Thus, the condition Equation 7.283, restricted to $k=0$ (i.e., solutions with exactly zero energy) becomes

$$J_{2\ell+1}\left(2\sqrt{\frac{2ma\xi}{\hbar^2}}\right)=0. \tag{7.288}$$

The first zero of $J_1(x)$ for $x \neq 0$ occurs when

$$x=3.83171\ldots \tag{7.289}$$

which means the curve in Figure 7.12 crosses the $\xi ma/\hbar^2$ axis when

$$\frac{\xi ma}{\hbar^2}=1.83525\ldots \tag{7.290}$$

One can also show that the slope on this graph at this point becomes infinite. (This fact is true in general for the other energy states.)

As stated at the beginning of this section, the confined Coulombic model provides a connection between the Coulomb and spherical well solutions in the limits $a \to \infty$ and $a \to 0$, respectively. Since we are working with $E>0$ here, let us consider the $a \to 0$ limit of the eigenvalue condition (Equation 7.283). Let us imagine fixing the values of m and ξ in the $a \to 0$ limit. Then we have that

$$\lim_{a\to 0}\frac{ma\xi}{\hbar^2}=0, \tag{7.291}$$

but from the qualitative behavior seen in Figure 7.12, we would expect that

$$\lim_{a\to 0} ka = \text{fixed number}. \tag{7.292}$$

(Can you understand why Equation 7.292 is really just a statement of the uncertainty relation in the small a limit for this system?) This equation may seem somewhat mysterious to you until you realize that k is actually an implicit function of a; the eigenvalue condition (Equation 7.283) determines this dependence. Thus, when a is made small enough to make

$$\frac{\xi ma}{\hbar^2} \ll ka \tag{7.293}$$

(the system does not have to become relativistic to do this), then we get

$$\lim_{a\to 0} F\left(\ell+1+\frac{i\xi ma/\hbar^2}{ka}, 2\ell+2, 2ika\right) \to F(\ell+1, 2\ell+2, 2ika). \tag{7.294}$$

However, from an identity, Equation 13.6.4 of the *NBS Handbook*, we have

$$F(\ell+1,2\ell+2,2ika)=\Gamma\left(\frac{3}{2}+\ell\right)e^{ika}\left(\frac{1}{2}ka\right)^{-\ell-1/2}\left(\frac{2ka}{\pi}\right)^{1/2} j_\ell(ka). \tag{7.295}$$

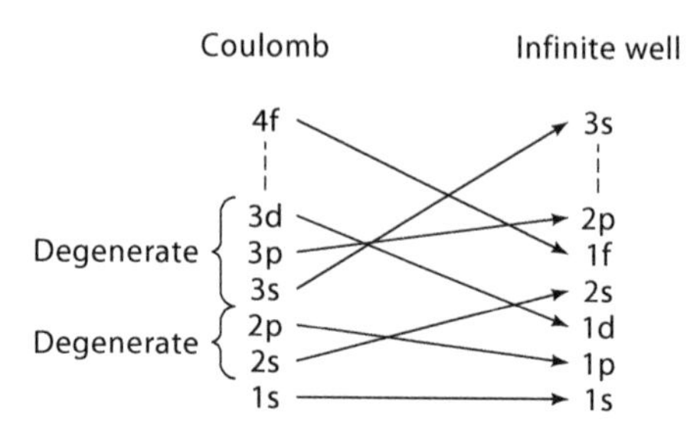

FIGURE 7.13 Correspondence between the lower energy levels in the Coulombic problem on the left ($a \to \infty$) and those in the infinite spherical well problem ($a \to 0$), as connected by the confined Coulombic model.

$\Gamma(x)$ (where $x = 3/2 + \ell$ above) is a famous classical function called the *gamma function*. Its form and value are immaterial for the argument here except to say that $\Gamma(x) \neq 0$ for $x > 0$. Thus, we learn that in the small a limit that the eigenvalue condition just becomes

$$j_\ell(ka) = 0. \tag{7.296}$$

This constitutes the recovery of the result Equation 7.62 for the infinite spherical well. One can also show, although it will not be done here, that the $a \to \infty$ limit of the eigenvalue condition for $E < 0$ (which is slightly different from Equation 7.283) gives back the Coulomb energies.*

Since one is able to recover the bound-state Coulomb energies or the infinite spherical well energies as special cases of this model, this means there must be a one-to-one correspondence between these levels. It is easy to establish what this correspondence is, based upon the values of ℓ, the angular momentum quantum number, and the number of nodes in the radial wave functions, neither of which can change as the confinement radius is adjusted. Remembering that the number of nodes in the radial Coulomb wave functions is given by $n - \ell$ (where n is the principle quantum number), we get the correspondence, shown in Figure 7.13, between the $E < 0$ Coulomb bound states and the $E > 0$ infinite spherical well solutions. (Remember, the number in front of the spectroscopic letter in the Coulomb case gives the value of the principle quantum number, while the number preceding the letter in the infinite well case just tells us the number of radial nodes.) The order of energy levels for the confined Coulombic model holds on the left for $a \to \infty$ and on the right for $a \to 0$.

There are many other aspects to this problem that I have not addressed. I will stop with my short survey of this model here, except to point out the type of system this model is supposed to portray. Earlier, I mentioned quarks as being the building blocks of protons and neutrons. Experimentally, six types or "flavors" of quarks are known (really 18 types of quarks if we count the fact that each quark has a threefold symmetry called "color"). They are called up (u), down (d), strange (s), charmed (c), bottom (b), and top (t). These

* Small deviations from the Coulomb energies can be found with techniques discussed in W. Wilcox, *Am. J. Phys.* **57**, 526 (1989), and *Annals Phys.* **279**, 65 (2000).

are listed in the order of their masses; the u quark has the lightest mass and the t quark is the most massive. The concept of the "mass" of a quark is a somewhat fuzzy concept because isolated quarks are never seen in nature. This fact of nature is called *confinement.* The model presented here is a crude mock-up of heavy–light quark–antiquark systems that consist of a c or b quark (or antiquark) paired with a $\bar{u}$ or $\bar{d}$ antiquark (or quark). The infinite potential barrier in the model presented is a way of representing the confinement of quarks, while the Coulomb potential in the interior is a way of modeling the short-range interactions between the quarks. In this model, the heavy quark, relatively unaffected by the dynamics of its light partner, resides in the center of the spherical well, giving rise to the Coulomb potential. The Coulombic coupling constant ξ introduced here is actually related to the so-called strong coupling constant in quantum chromodynamics (QCD). The model given here is, of course, nonrelativistic. A relativistic version, based upon solution of the *Dirac equation,* provides a more realistic description of such systems.

PROBLEMS

7.2.1 Get explicit forms for $j_0(p)$, $j_1(p)$, and $j_2(p)$ from Equation 7.39. Show your work! [Answer:

$$j_0(p) = \frac{\sin p}{p},$$
$$j_1(p) = \frac{\sin(p)}{p^2} - \frac{\cos(p)}{p},$$
$$j_2(p) = \left(\frac{3}{p^3} - \frac{1}{p}\right)\sin(p) - 3\frac{\cos(p)}{p^2}.]$$

7.2.2 Get explicit forms for $h_0^{(1)}(ip)$, $h_1^{(1)}(ip)$, and from Equation 7.127. Show your work! [Answer:

$$h_0^{(1)}(ip) = -\frac{e^{-p}}{p},$$
$$h_1^{(1)}(ip) = i\frac{e^{-p}}{p}\left(1 + \frac{1}{p}\right),$$
$$h_2^{(1)}(ip) = \frac{e^{-p}}{p}\left(1 + \frac{3}{p} + \frac{3}{p^2}\right).]$$

7.2.3 Using the form of Equation 7.39, derive the recurrence relation

$$\frac{dj_\ell(p)}{dp} = -j_{\ell+1}(p) + \frac{\ell}{p}j_\ell(p).$$

7.2.4 Show that we may write (see Equation 7.43)

$$\langle r'|p_r \equiv \frac{\hbar}{i}\frac{1}{r'}\frac{\partial}{\partial r'}r'\langle r'| = \langle r'|\frac{1}{2}\left[\frac{\vec{r}}{r}\cdot\vec{p} + \vec{p}\cdot\frac{\vec{r}}{r}\right].$$

[This is an example of the operator ordering ambiguity resolution process, described in a footnote in Chapter 6.]

7.2.5 Verify Equation 7.53 in the case $\ell = 0$. [Hint: Consider the quantity $[\delta(r-r') - \delta(r+r')]$ where

$$\delta(r-r') \equiv \frac{1}{2\pi}\int_{-\infty}^{\infty} dk\ e^{ik(r-r')}.]$$

7.2.6 A free-particle wave function is given at $t=0$ by the ket (I am using $|k,\ell,m\rangle$ notation as in the text),

$$|\psi(k),0\rangle \equiv C_1|k,0,0\rangle + \frac{C_2}{\sqrt{5}}(|2k,1,0\rangle + 2i|2k,1,1\rangle).$$

(a) What is $\langle\vec{r}|\psi(k),0\rangle$? (Write it out as explicitly as possible as a function of r, ϕ, θ.)

(b) What is $|\psi(k),t\rangle$?

(c) What is the expectation value of the energy for this particle?

(d) What possible values of L_3 will measurement find at $t=0$, and with what probabilities?

[Please show some intermediate steps or explain your reasoning in reaching your answers in parts (c) and (d).]

***7.3.1** This will be a rather long problem that will, hopefully, lead you to the result of Equation 7.70. The radial eigenvalue equation in spherical coordinates is ($R = r\ u(r)$, $R' = dR/dr$)

$$-\frac{\hbar^2}{2m}\left[R'' - \frac{\ell(\ell+1)}{r^2}R\right] + V(r)R = E\ R. \tag{P.1}$$

(a) First, take the derivative of Equation P.1 with respect to E and multiply by R^*; call this Equation P.2. Then complex conjugate Equation P.1 and multiply by $\partial R/\partial E$; call this Equation P.3. Subtract Equation P.2 from Equation P.3 to get

$$-\frac{\hbar^2}{2m}\left[R^*\frac{\partial R''}{\partial E} - R''^*\frac{\partial R}{\partial E}\right] = |R|^2. \tag{P.4}$$

(b) Considering that the radial normalization condition for the finite spherical well is

$$\int_0^a dr|R|^2 = 1, \tag{P.5}$$

show that Equation P.4 implies

$$\frac{\hbar^2}{2m}\left[R'^*\frac{\partial R}{\partial E} - R^*\frac{\partial R'}{\partial E}\right]_{r=a} = 1, \tag{P.6}$$

given that $R(0) = 0$ and that $R'(0)$ is finite.

(c) Now assuming

$$R(r) = N(E)\,j_\ell(k_n r)r, \tag{P.7}$$

where $k_n(=\sqrt{2mE}/\hbar)$ is determined by

$$j_\ell(k_n a)=0, \tag{P.8}$$

show that Equation P.6 specializes to

$$\frac{|N|^2a^3}{2}\left(\frac{\mathrm{d}j_\ell(x)}{\mathrm{d}x}\right)^2\bigg|_{x=k_na}=1. \tag{P.9}$$

(d) Now, using the recursion relation from Problem 7.2.3, show that Equation P.9 gives

$$|N|^2=\frac{2}{a^3 j_{\ell+1}^2(k_n a)}, \tag{P.10}$$

when Equation P.8 is also used.

7.3.2 Check the result of the last problem, Equation P.10, by finding $|N|^2$ for $\ell=0$ by doing the normalization integral explicitly. Compare with Equation P.10 specialized to $\ell=0$.

7.4.1 Show for the two-body problem that

$$\vec{L}=\vec{L}_{cm}+\vec{\ell},$$

where

$$\vec{L}=\vec{x}_1\times\vec{p}_1+\vec{x}_2\times\vec{p}_2,$$

(1,2 are particle labels)

$$\vec{L}_{cm}=\vec{X}\times\vec{P},$$

$$\vec{\ell}=\vec{x}\times\vec{p}.$$

All these quantities are operators.

7.4.2 Using the definitions in Problem 7.4.1, show that

$$[(L_{cm})_i,\ell_j]=0,$$

so that these two quantities may be specified simultaneously.

7.4.3 Find the normalization factors A and B (in Equation 7.128) for the $\ell=0$ deuteron bound state. Show that

$$|A|^2=\frac{2\kappa'\kappa^2}{1+\kappa' a},$$

$$|B|^2=\frac{2\kappa'^3(\sin\kappa a)^2 e^{2\kappa' a}}{1+\kappa' a}.$$

Be sure to require continuity in $u(r)$ at $r=a$.

***7.4.4** (a) Show that the $\ell = 1$ eigenvalue equation of the deuteron, Equation 7.131, can be written as

$$\frac{1}{(\kappa' a)^2} + \frac{1}{(\kappa' a)} = -\frac{1}{(\kappa a)^2} + \frac{\cot(\kappa a)}{(\kappa a)}.$$

(b) Argue on the basis of this equation that the well depth that binds the first $\ell = 1$ state is (see Equation 7.140)

$$\frac{2\mu V_0 a^2}{\hbar^2} = \pi^2.$$

7.4.5 (a) Write the $\ell = 1$ *infinite* square well ($V(r) = 0, r < a$; $V(r) = \infty$, $r \geq a$) energy eigenvalue condition ($E > 0$) as a function of $k = \sqrt{2mE}/\hbar$ and a.

(b) By taking some appropriate limit, show how the expression in Problem 7.4.4a yields the part (a) eigenvalue condition. (Make the finite square well become infinite.)

7.4.6 Solve for the square well depth parameter, $V_0 > 0$, for the deuteron. You are given following parameters:

$$K' = \frac{\sqrt{2\mu|E|}}{\hbar}, \quad K = \frac{\sqrt{2\mu(E + V_0)}}{\hbar},$$
$$E = -2.23 \text{ MeV}, \quad \mu = \frac{940}{2} \text{MeV}/c^2, \quad a = 1.41 \text{ fm},$$

and the $\ell = 0$ bound-state condition,

$$K'a = -(Ka)\cot(Ka).$$

[Numerical methods are called for here.]

7.4.7 An alternate argument for the same binding ordering as in Equation 7.141 can be constructed from a variational point of view for a general central potential, $V(r)$. Assume all states under discussion are bound and define

$$E_\ell \equiv -\langle \ell | H_\ell | \ell \rangle > 0,$$
$$H_\ell \equiv \frac{p_r^2}{2\mu} + \frac{\hbar^2 \ell(\ell+1)}{2\mu r^2} + \lambda V(r),$$

where $\langle \ell |$ represents the radial ground-state angular momentum wave function. Construct a *variational* argument that

$$E_{\ell+1} < E_\ell.$$

Since this ordering is invariant as the dimensionless well depth parameter, λ, is continuously changed, this shows that a result equivalent to Equation 7.141 must hold for the spherical well and other finite-range potentials.

***7.5.1** (a) Write the *radial* equation for the three-dimensional harmonic oscillator. By introducing the dimensionless variable $\rho = (m\omega/\hbar)^{1/2} r$, show that this equation may be put into the form $(V(r) = m\omega^2 r^2/2 = \hbar\omega\rho^2/2)$

$$\left[\frac{d^2}{d\rho^2} - \frac{\ell(\ell+1)}{\rho^2} - \rho^2 + 2\left(n + \frac{3}{2}\right)\right] R_{n\ell}(\rho) = 0.$$

(b) Find the approximate power ρ-dependence of $R_{n\ell}(\rho)$ near the origin, $\rho \ll 1$, and *exponential* dependence far from the origin, $\rho \gg 1$.

(c) Use (b) to predict the form of the ground-state wave function for a given ℓ-value (which turns out to be exact for all $n = \ell$ states!).

7.6.1 Show that

$$\sum_{n=0}^{\infty} \frac{\lambda^n n}{\sqrt{n!}} u_n(q) = \lambda(\sqrt{2}q - \lambda)\frac{1}{\pi^{1/4}} e^{-q^2/2 + \sqrt{2}\lambda q - \lambda^2/2}.$$

***7.6.2** (a) Show by induction that

$$\left(\frac{d}{dx}\right)^n e^{-x} f(x) = e^{-x}\left(\frac{d}{dx} - 1\right)^n f(x)$$

for an arbitrary function $f(x)$.

(b) Using (a) and given $(L_n(x) \equiv L_n^0(x))$

$$L_n(x) = \frac{e^x}{n!}\left(\frac{d}{dx}\right)^n e^{-x} x^n,$$

show that

$$L_n(x) = \frac{1}{n!}\left(\frac{d}{dx} - 1\right)^n x^n.$$

(c) Using the binomial theorem (look it up!), show that the result in (b) then implies

$$L_n(x) = \sum_{k=0}^{n} \frac{(-1)^k n!\, x^k}{(k!)^2 (n-k)!}.$$

7.6.3 (a) Show by induction that

$$\left(\frac{d}{dx} - 1\right)^{n+1} x f(x) = x\left(\frac{d}{dx} - 1\right)^{n+1} f(x) + (n+1)\left(\frac{d}{dx} - 1\right)^n f(x).$$

(b) Using (a), and given that

$$L_{n+1}(x) = \frac{1}{(n+1)!}\left(\frac{d}{dx}-1\right)^{n+1} x\, x^n,$$

show that

$$(n+1)(L_{n+1} - L_n) = x\left(\frac{d}{dx}-1\right)L_n.$$

7.6.4 (a) Show that

$$\frac{d}{dx}(L_{n+1} - L_n) + L_n = 0.$$

[Hint: We can write

$$L_{n+1} = \frac{1}{(n+1)!}\left(\frac{d}{dx}-1\right)^n \left(\frac{d}{dx}-1\right)x^{n+1}.]$$

(b) Using part (a) and Problem 7.6.3b, now show that $L_n(x)$ satisfies the differential equation

$$\left(x\frac{d^2}{dx^2} + (1-x)\frac{d}{dx} + n\right)L_n(x) = 0.$$

7.6.5 Use the differential relation*

$$\frac{d}{dx}L_{n+1}^{q-1}(x) = -L_n^q(x),$$

to show that the associated Laguerre polynomials,

$$L_n^q(x) \equiv \frac{x^{-q}}{n!}e^x\left(\frac{d}{dx}\right)^n x^{n+q}e^{-x},$$

are related to the alternate functions (q a nonnegative integer)

$$\mathcal{L}_n^q(x) \equiv (-1)^q\left(\frac{d}{dx}\right)^q e^x\left(\frac{d}{dx}\right)^{q+n} x^{q+n}e^{-x},$$

by the equation:

$$L_n^q(x) = \frac{\mathcal{L}_n^q(x)}{(n+q)!}.$$

***7.6.6** Starting with Equation 7.234, show that the two-dimensional harmonic oscillator radial equation may be written as ($\rho^2 = (m\omega/\hbar)(x^2+y^2)$)

$$\left[\frac{d^2}{d\rho^2} - \frac{(m^2-(1/4))}{\rho^2} + 2(|m|+2n_r+1) - \rho^2\right][\sqrt{\rho}\,P_{n_r m}(\rho)] = 0.$$

* See Schwinger, *Quantum Mechanics*, Section 7.3.

Compare with the three-dimensional equation in Problem 7.5.1 and establish the correspondences between the various quantum numbers in the two cases. Then, based upon the explicit form $P_{n_r m}(\rho)$, given in Equation 7.212, write the implied form for the 3-D wave function $R_{n\ell}(\rho)$. [Hint: $R_{n\ell}(\rho) \sim e^{-\rho^2/2}\rho^{\ell+1}L_{(n-\ell)/2}^{\ell+1/2}(\rho^2)$.]

7.6.7 In the text, I used the relation

$$\left\langle \frac{\partial H}{\partial \lambda} \right\rangle_E = \frac{\partial E}{\partial \lambda}$$

to evaluate an integral necessary to normalize the Coulomb solutions. Apply the same technique to the Coulomb Hamiltonian,

$$H = \frac{\vec{p}^2}{2\mu} - \frac{Ze^2}{r},$$

to show that

$$\left\langle \frac{1}{r} \right\rangle_E = \frac{Z}{a_0 n^2}.$$

7.6.8 Find the value of the expectation value of r in the state where $\ell = n-1$ (its maximum value). That is, show that

$$\langle n, \ell = n-1 | r | n, \ell = n-1 \rangle = \frac{a_0 n}{Z}\left(n + \frac{1}{2}\right).$$

[Hints: The explicit normalization condition for the radial part of the $\ell = n - 1$ hydrogenic type wave functions is given by

$$\int_0^\infty dr |R_{n,\ell=n-1}(r)|^2$$
$$= \int_0^\infty dr\, r^{2n} \left(\frac{2Z}{a_0 n}\right)^{2n+1} \exp\left(-\frac{2Zr}{a_0 n}\right) \frac{1}{2n[(2n-1)!]} = 1.$$

The integral we want is

$$\langle n, \ell = n-1 | r | n, \ell = n-1 \rangle = \int_0^\infty dr\, r |R_{n,\ell=n-1}(r)|^2.$$

The hint is to generate the integral we want by taking a derivative with respect to a parameter.]

***7.6.9** Find an alternate form of the hydrogenic wave functions given in Equation 7.261. Do this by going back to Equation 7.210 and using the following procedure:

1. Get an expression equivalent to Equation 7.210 by using Equation 7.209 when $n_+ - n_- \geq 0$, and Equation 7.207 when $n_+ - n_- \leq 0$.

2. In your expression, change to the new quantum numbers n_r and m given in Equation 7.211.
3. Use the Equations 7.258 and 7.260 to find the new form of $R_{n\ell}(r)$.

[Hint: $R_{n\ell}(r) \sim L_{n+\ell}^{-(2\ell+1)}(2Zr/a_0 n)$.]

7.6.10* A phenomenological potential for a many-electron atom is given by

$$V(r) = -\frac{Z'e^2}{r}\left(1+\frac{b}{r}\right).$$

(a) Write the radial equation to be solved for this new potential. By comparing to Equation 7.115, complete the correspondence:

Coulomb Problem		Phenomenological Potential
Z	→	
ℓ	→	
N	→	

(b) Use the correspondence in (a) and the known form of the energies in Equation 7.245 to solve for the energy levels of the new potential.

7.7.1 Find the $E<0$ eigenvalue condition for the confined Coulombic model. (This replaces Equation 7.283 for $E>0$.)

7.7.2 The first zero of the Bessel function $J_1(x)$ occurs at $x=3.83171$ .... Look up the next three zeros of this function. Give a *physical interpretation* of these zeros in the context of the confined Coulombic model.

7.7.3 Within the confined Coulombic model, use the Heisenberg uncertainty principle to *estimate* the confinement radius, r_c, which raises the energy of the 1s state of hydrogen, originally for $a \to \infty$, to $E_{1s}=0$. (Consider ξ and m fixed parameters.) Compare your estimate with the *exact* answer, given somewhere in the text. (See Chapter 2, Section 2.4 for a review of how such estimates are made for ground states.)

7.7.4 The eigenvalue condition for negative energies, $E<0$, for the confined Coulombic potential is given by (just put $k \to -ik$ in Equation 7.283):

$$F\left(\ell+1+\frac{\xi k}{2E}, 2\ell+2, 2ka\right)=0.$$

* From R.L. Liboff, *Introductory Quantum Mechanics*, 2nd ed. (Addison-Wesley, 1992), Probem 12.13.

(a) Using Equation 7.277, show that the series for the hypergeometric function $F(a,b,z)$ $(b>0)$ terminates after X terms if $a=-n_r$, where $n_r=0, 1, 2, 3 \ldots$. Find X.

(b) The series solution, Equation 7.277 for $F(\ell+1+\xi k/2E, 2\ell+2, 2ka)$ $(k\equiv\sqrt{-2mE}/\hbar)$ diverges for $a\to\infty$ and $E<0$ unless the series terminates after a finite number of terms. Given that the series terminates as in part (a), solve for the energies, E, in the $a\to\infty$ limit. Show that

$$E=-\frac{\xi^2\mu}{2\hbar^2 n^2},$$

where $n\equiv n_r+\ell+1$.

Addition of Angular Momenta

Synopsis: General operator properties of angular momenta are recalled, and a physical motivation for adding them is given. The product and total angular momentum bases are defined. It is shown how to express one basis in terms of the other, and an explicit example of adding two spin 1/2 angular momenta is given. A further example consists of adding a general orbital angular momentum value to spin 1/2. These considerations are applied to perturbations in the hydrogen atom energy spectrum, where the correct choice for calculations is the total angular momentum basis.

8.1 GENERAL ANGULAR-MOMENTUM EIGENSTATE PROPERTIES

Let me remind you of the commutation properties associated with spin 1/2. There were four independent operators in this case, which we chose to be the unit symbol 1, and the three σ_i given by

$$\left.\begin{aligned} \sigma_1 &= |-\rangle\langle+|+|+\rangle\langle-|, \\ \sigma_2 &= i(|-\rangle\langle+|-|+\rangle\langle-|), \\ \sigma_3 &= |+\rangle\langle+|-|-\rangle\langle-|. \end{aligned}\right\} \tag{8.1}$$

They have the properties (Chapter 1, Equations 1.133 through 1.135)

$$\left.\begin{aligned} &\sigma_k^2 = 1 && (k = 1,2,3), \\ &\sigma_k\sigma_\ell = -\sigma_\ell\sigma_k && (k \neq \ell), \\ &\sigma_k\sigma_\ell = i\sigma_m && (k,\ell,m \text{ cyclic}). \end{aligned}\right\} \tag{8.2}$$

I later asserted that the entire content of their algebra, given in Equation 8.2, is combined in the statement that (Chapter 1, Equation 1.139)

$$\sigma_i\sigma_j = 1\delta_{ij} + i\sum_k \varepsilon_{ijk}\sigma_k. \tag{8.3}$$

An immediate consequence of Equation 8.3 is

$$[\sigma_i,\sigma_j] = 2i\sum_k \varepsilon_{ijk}\sigma_k. \tag{8.4}$$

The "crucial connection" that allowed us to tie our Process Diagram formalism to the real-world property of electron spin was (Chapter 1, Equation 1.136)

$$S_i = \frac{\hbar}{2}\sigma_i. \tag{8.5}$$

In terms of the S_i, the commutation relation Equation 8.4 reads

$$[S_i, S_j] = i\hbar\sum_k \varepsilon_{ijk} S_k. \tag{8.6}$$

These relations are exactly the same as the commutation relations, given by Equation 6.70 (see Chapter 6), for orbital angular momentum:

$$[L_i, L_j] = i\hbar\sum_k \varepsilon_{ijk} L_k. \tag{8.7}$$

Remember, these commutation relations lead directly to the statements

$$\left.\begin{aligned} \vec{L}^2|\ell,m\rangle &= \hbar^2\ell(\ell+1)|\ell,m\rangle, \\ L_3|\ell,m\rangle &= \hbar m|\ell,m\rangle. \end{aligned}\right\} \tag{8.8}$$

The physical picture associated with the statements in Equation 8.8 is that of an orbiting particle (or an orbiting system of two particles). The allowed values of ℓ are $\ell = 0, 1, 2, \ldots$. The cases $\ell = 1/2, 3/2, 5/2, \ldots$ were not consistent with the requirement

$$\langle\phi| = \langle\phi + 2\pi|. \tag{8.9}$$

Because the spin operators, S_i, satisfy the same commutation properties as the L_i, we must have analogous statements to Equation 8.8 for $\vec{S}^2$ and S_3. Of course we know that

$$\vec{S}^2 = S_1^2 + S_2^2 + S_3^2 = \frac{\hbar^2}{4}(\sigma_1^2 + \sigma_2^2 + \sigma_3^2) = \frac{3\hbar^2}{4}1, \tag{8.10}$$

from Equation 8.3, and that

$$S_3|\sigma_3'\rangle = \frac{\hbar}{2}\sigma_3|\sigma_3'\rangle = \frac{\hbar}{2}\sigma_3'|\sigma_3'\rangle, \tag{8.11}$$

where $\sigma'_3 = \pm 1$. Relabeling our states by $m_S = (1/2)\sigma_3' = \pm 1/2$, Equations 8.10 and 8.11 lead to

$$\left.\begin{aligned} \vec{S}^2|m_S\rangle &= \hbar^2\left(\frac{3}{4}\right)|m_S\rangle, \\ S_3|m_S\rangle &= \hbar m_S|m_S\rangle. \end{aligned}\right\} \tag{8.12}$$

The structure of these equations is the same as Equation 8.8 for $\ell = 1/2$. Now, however, since the origin of spin is an intrinsic internal property of the particle, the requirement Equation 8.9 is not relevant. Thus, although there are no systems in nature for which $\ell = 1/2, 3/2, 5/2, \ldots$, there are no such restrictions on the value of the spin angular momentum. Generalizing Equation 8.12, we expect realizations in nature of particle spins such that

$$\left.\begin{aligned}\vec{S}^2|s,m_S\rangle &= \hbar^2 s(s+1)|s,m_S\rangle\\ S_3|s,m_S\rangle &= \hbar m_S|s,m_S\rangle\end{aligned}\right\}, \tag{8.13}$$

where $s = 0, 1/2, 1, 3/2, 2, \ldots$. In terms of Process Diagrams, $s = 1/2$ represents the two-physical-outcomes case, $s = 1$ represents the three-outcomes case, and so on. Only the lower intrinsic spin values, up to $s = 2$, seem to be realized by fundamental particles in nature.

The representation of the spin raising and lowering operators, S_+ and S_-, may be constructed from the statements

$$S_\pm|S,\, m_S\rangle = \hbar\sqrt{(s \mp m_S)(s \pm m_S + 1)}\,|s,\, m_S \pm 1\rangle, \tag{8.14}$$

which can be derived exactly as was Equation 6.178 in Chapter 6. (The derivation would be based on the commutation relations, Equations 1.214 through 1.216 in Chapter 1.) The explicit operator form is $(S_- = S_+^\dagger; |m_S'\rangle \equiv |s,m_S'\rangle)$

$$S_+ = \hbar \sum_{m_S' = -s+1}^{s} \sqrt{m_S'(1 - m_S') + s(1+s)}\,|m_S'\rangle\langle m_S' - 1|. \tag{8.15}$$

(Verified in Problem 8.1.1.) Then we have

$$S_+ = \hbar\sqrt{s(1-s) + s(1+s))}\,|\frac{1}{2}\rangle\langle -\frac{1}{2}| = \hbar|\frac{1}{2}\rangle\langle -\frac{1}{2}|, \tag{8.16}$$

for the spin 1/2 ($s = 1/2$) case with two physical outcomes, and

$$S_+ = \hbar\sum_{m_S'=0}^{1}\sqrt{m_S'(1-m_S')+2)}\,|m_S'\rangle\langle m_S' - 1| = \sqrt{2}\hbar(|+\rangle\langle 0| + |0\rangle\langle -|), \tag{8.17}$$

for the spin 1 ($s = 1$) case with three physical outcomes, in agreement with Equation 1.218 (Chapter 1).

Note that $[S_i, x_j] = [S_i, p_j] = 0, \Rightarrow [S_i, L_j] = 0$, which emphasizes that spin is an "internal" property of particles.

8.2 COMBINING ANGULAR MOMENTA FOR TWO SYSTEMS

Electrons and quarks have spin 1/2 ($s = 1/2$ in Equation 8.13). In the Standard Model of particle physics (studied in Chapter 12) these are considered elementary particles without composite structure. However, many particles that were originally thought to be fundamental are now known to be composite. For example, the proton is a composite structure containing three quarks. The *total* angular momentum of this particle, which involves both the intrinsic spin of the quarks as well as the relative angular momentum between them, is measured to be 1/2. To begin to understand such composite objects, it is necessary to learn how to combine or add the angular momenta of the subsystems.

Consider two systems that possess angular-momentum states given by

$$\begin{aligned}&\text{System 1: } |j_1, m_1\rangle,\\ &\text{System 2: } |j_2, m_2\rangle,\end{aligned}$$

where we consider all possible values $j_1, j_2 = 0, 1/2, 1, 3/2, \ldots$. It is immaterial for this discussion whether this angular momentum is due to spin or orbital motion. An obvious basis for the description of this composite system is given by a direct product,

$$|j_1, j_2; m_1, m_2\rangle \equiv |j_1, m_1\rangle \otimes |j_2, m_2\rangle. \tag{8.18}$$

A complete set of operators whose eigenvalues completely characterize this state are just given by $\vec{J}_1^2, J_{1z}, \vec{J}_2^2, J_{2z}$, for which, of course,

$$\left.\begin{aligned} \vec{J}_1^2|j_1, j_2; m_1, m_2\rangle &= \hbar^2 j_1(j_1+1)|j_1, j_2; m_1, m_2\rangle, \\ J_{1z}|j_1, j_2; m_1, m_2\rangle &= \hbar m_1|j_1, j_2; m_1, m_2\rangle, \\ \vec{J}_2^2|j_1, j_2; m_1, m_2\rangle &= \hbar^2 j_2(j_2+1)|j_1, j_2; m_1, m_2\rangle, \\ J_{2z}|j_1, j_2; m_1, m_2\rangle &= \hbar m_2|j_1, j_2; m_1, m_2\rangle. \end{aligned}\right\} \tag{8.19}$$

(I will often label Cartesian components x, y, z in this chapter for clarity.) Since our composite state is just a direct product of the individual states, it follows that ("1" and "2" are the particle labels)

$$[J_{1i}, J_{2j}] = 0. \tag{8.20}$$

We can imagine a situation where a measurement on systems 1 and 2 is less convenient than a measurement on the *total* system. There should be a more appropriate description for such a characterization. Let us define the operator

$$\vec{J} = \vec{J}_1 + \vec{J}_2, \tag{8.21}$$

(which really means $\vec{J} = \vec{J}_1 \otimes 1 + 1 \otimes \vec{J}_2$) and call it the "total angular momentum." Then, if the components J_{1i} and J_{2i} separately satisfy the commutation relations (like $\vec{L}$ or $\vec{S}$)

$$\left.\begin{aligned} [J_{1i}, J_{1j}] &= i\hbar \sum_k \varepsilon_{ijk} J_{1k}, \\ [J_{2i}, J_{2j}] &= i\hbar \sum_k \varepsilon_{ijk} J_{2k}, \end{aligned}\right\} \tag{8.22}$$

we also have that the total angular-momentum components satisfy the same relations (if Equation 8.20 holds):

$$[J_i, J_j] = i\hbar \sum_k \varepsilon_{ijk} J_k. \tag{8.23}$$

A consequence of Equation 8.23 is that we may characterize the composite states of the total system by the eigenvalues of $\vec{J}^2$ and J_z, just as the commutation relations in Equation 8.7 lead to Equation 8.8. Therefore, we postulate the existence of an alternate set of states $|j_1, j_2; j, m\rangle$ such that

$$\left.\begin{aligned} \vec{J}^2|j_1, j_2; j, m\rangle &= \hbar^2 j(j+1)|j_1, j_2; j, m\rangle, \\ J_z|j_1, j_2; j, m\rangle &= \hbar m|j_1, j_2; j, m\rangle. \end{aligned}\right\} \tag{8.24}$$

The notation used anticipates the fact that the operators $\vec{J}_1^2$ and $\vec{J}_2^2$ commute with $\vec{J}^2$ and J_z; that is,

$$\left.\begin{aligned} [\vec{J}_{1,2}, \vec{J}^2] &= 0, \\ [\vec{J}_{1,2}^2, J_z] &= 0. \end{aligned}\right\} \tag{8.25}$$

Thus, the eigenvalues of $\vec{J}$, J_z, $\vec{J}_1^2$, and $\vec{J}_2^2$ can all be specified simultaneously. These operators represent an alternate complete set of operators that can be used to characterize the system.

Given the values of j_1 and j_2, the statement of completeness for the $|j_1,j_2;m_1,m_2\rangle$ basis is

$$\sum_{m_1,m_2} |j_1,j_2;m_1,m_2\rangle\langle j_1,j_2;m_1,m_2| = 1. \tag{8.26}$$

In more detail, this means that

$$\sum_{m_1} |j_1,m_1\rangle\langle j_1,m_1| \sum_{m_2} |j_2,m_2\rangle\langle j_2,m_2| = 1_1 \cdot 1_2. \tag{8.27}$$

Using Equation 8.26 then, we can formally express a basis state $|j_1,j_2;j,m\rangle$ in terms of the $|j_1,j_2;m_1,m_2\rangle$ states:

$$|j_1,j_2;j,m\rangle = \sum_{m_1,m_2} |j_1,j_2;m_1,m_2\rangle\langle j_1,j_2;m_1,m_2|j_1,j_2;j,m\rangle. \tag{8.28}$$

The quantities $\langle j_1,j_2;m_1,m_2|j_1,j_2;j,m\rangle$ are called the "Clebsch–Gordan" coefficients. We will work out the values of some of these quantities in some special cases a little later. By projecting both sides of Equation 8.28 into $\langle j_1,j_2;j,m|$, we obtain

$$1 = \sum_{m_1,m_2} |\langle j_1,j_2;m_1,m_2|j_1,j_2;j,m\rangle|^2, \tag{8.29}$$

which is just a statement of Clebsch–Gordan coefficient probability conservation ("unitarity").

Now we ask: What are the relationships between the quantum numbers j_1, j_2, m_1, and m_2 (which specify the eigenvalues of the $\vec{J}_1^2, \vec{J}_2^2, J_{1z}, J_{2z}$ set of operators) and the quantum numbers j and m? We know that

$$J_z = J_{1z} + J_{2z}, \tag{8.30}$$

from taking the z-component of Equation 8.21. We therefore can write

$$\langle j_1,j_2;j,m|J_z - J_{1z} - J_{2z}|j_1,j_2;m_1,m_2\rangle = 0. \tag{8.31}$$

Allowing J_{1z} and J_{2z} to act to the right and J_z to act to the left, we then get

$$(m - m_1 - m_2)\langle j_1,j_2;j,m|j_1,j_2;m_1,m_2\rangle = 0. \tag{8.32}$$

That is, for all states that have a nonzero overlap, $\langle j_1,j_2;j,m|j_1,j_2;m_1,m_2\rangle \neq 0$ we must have that

$$m = m_1 + m_2. \tag{8.33}$$

Thus, the quantized z-components of the individual angular momenta add to give the total value, m. This would not be surprising if we imagine Equation 8.21 to be a vector statement for numbers rather than a relation between operators. Now remember that m_1 and m_2 can only take on values separated by integer intervals between j_1 and $-j_1$ or j_2 and $-j_2$, respectively. That is

$$m_1 = j_1, j_1 - 1, \ldots, -j_1 + 1, -j_1 \quad (2j_1 + 1 \text{ values}),$$
$$m_2 = j_2, j_2 - 1, \ldots, -j_2 + 1, -j_2 \quad (2j_2 + 1 \text{ values}).$$

Thus, the maximum positive values, for fixed j_1 and j_2, are

$$(m_1)_{\max} = j_1, \quad (m_2)_{\max} = j_2. \tag{8.34}$$

Equation 8.34 then implies

$$(m)_{\max} = j_1 + j_2. \tag{8.35}$$

One can show that the state associated with $(m)_{\max} = j_1 + j_2$ has $j = j_1 + j_2$ also. This corresponds to the maximum value of j. (A similar argument shows that the coupled state $|\, j_1, j_2; j, m\rangle$ with $(m)_{\min} = -(j_1 + j_2)$ also has $j = j_1 + j_2$.) The next lowest value of j that is conceivable would be $j = j_1 + j_2 - 1/2$. However, such a state would have $m = j_1 + j_2 - 1/2, j_1 + j_2 - 3/2, \ldots, -j_1 - j_2 + 1/2$, none of which are consistent with $m = m_1 + m_2$. The only allowed values of j appear to be

$$j = j_1 + j_2, j_1 + j_2 - 1, j_1 + j_2 - 2, \ldots. \tag{8.36}$$

What is the lower limit of this process? We require that the number of states in each basis (for given j_1, j_2 values) be the same. In the $|\, j_1, j_2; m_1, m_2\rangle$ basis, the number of distinct quantum states is given by $(2j_1 + 1)(2j_2 + 1)$ by counting the allowed values of m_1 and m_2. This must also be the number of states in the $|\, j_1, j_2; j, m\rangle$ basis. Fixing the value of j in $|\, j_1, j_2; j, m\rangle$, we have $(2j + 1)$ m values. Therefore, we require

$$\sum_{j=j_{\min}}^{j_{\max}} (2j+1) = (2j_1 + 1)(2j_2 + 1), \tag{8.37}$$

where $j_{\max} = j_1 + j_2$, but $j_{\min}$ is unknown. We will try to solve Equation 8.37 for $j_{\min}$. We can always write

$$\sum_{j=j_{\min}}^{j_{\max}} j = \sum_{j=1}^{j_{\max}} j - \sum_{j=1}^{j_{\min}-1} j. \tag{8.38}$$

Using the fact that

$$\sum_{i=1}^{n} i = \frac{1}{2} n(n+1), \tag{8.39}$$

we then find that (we are covering the $j_{\max,\min}$ = integer case here; try to construct the analogous statements when $j_{\max,\min}$ are half-integers)

$$\begin{aligned}\sum_{j=j_{\min}}^{j_{\max}} j &= \frac{1}{2} j_{\max}(j_{\max}+1) - \frac{1}{2}(j_{\min}-1) j_{\min} \\ &= \frac{1}{2}[j_{\max}^2 + j_{\max} - j_{\min}^2 + j_{\min}].\end{aligned} \tag{8.40}$$

We therefore get

$$\sum_{j=j_{\min}}^{j_{\max}} (2j+1) = j_{\max}^2 + 2j_{\max} - j_{\min}^2 + 1. \tag{8.41}$$

Setting Equation 8.41 equal to $(2j_1+1)(2j_2+1)$ and solving for j^2_{min}, we now find that

$$\begin{aligned} j^2_{\text{min}} &= 1-(2j_1+1)(2j_2+\ 1)+2(j_1+j_2)+(j_1+j_2)^2 \\ &= (j_1-j_2)^2. \end{aligned} \tag{8.42}$$

Since j_{min} must be a positive number, we can then write

$$j_{\text{min}} = |\, j_1 - j_2 \,|. \tag{8.43}$$

Thus, the sum in Equation 8.37 is such that

$$|\, j_1 - j_2 \,| \le j \le j_1 + j_2, \tag{8.44}$$

which is called the "triangle inequality." This result is also not surprising if we think of j_1 and j_2 as being the "magnitudes" of the "vectors" $\vec{J}_1$ and $\vec{J}_2$. Then the case $j = j_1 + j_2$ would correspond to $\vec{J}_1$ and $\vec{J}_2$ being "parallel," whereas $j = |j_1 - j_2|$ would be associated with $\vec{J}_1$ and $\vec{J}_2$ being "antiparallel."

Now that we understand how the quantum numbers j and m arise (given fixed j_1 and j_2 values), we can write the statement of completeness in the $|\, j_1, j_2; j, m\rangle$ basis, which is analogous to Equation 8.26. It is

$$\sum_{j=|j_1-j_2|}^{j_1+j_2} \sum_{m=-j}^{j} |\, j_1, j_2; j, m\rangle\langle j_1, j_2; j, m| = 1. \tag{8.45}$$

8.3 EXPLICIT EXAMPLE OF ADDING TWO SPIN-1/2 SYSTEMS

As a concrete example of the relationships between these two bases, let us consider the simple case of adding the angular momenta of two spin-1/2 objects. That is, we set $j_1 = 1/2$ and $j_2 = 1/2$. The four states that arise in the $|\, j_1, j_2; m_1, m_2\rangle$ basis are (I denote the state $|1/2, 1/2; m_1, m_2\rangle$ as $|\, m_1, m_2\rangle$ for convenience in this context):

$$\begin{aligned} &|\frac{1}{2},\frac{1}{2}\rangle, \quad |-\frac{1}{2},-\frac{1}{2}\rangle \\ &|-\frac{1}{2},\frac{1}{2}\rangle, \quad |\frac{1}{2},-\frac{1}{2}\rangle. \end{aligned}$$

The four states in the $|\, j_1, j_2; j, m\rangle$ basis are (the state $|1/2, 1/2; j, m\rangle$ is denoted here by $|\, j, m\rangle$):

$$j=1: \begin{cases} |1,1\rangle \\ |1,0\rangle \\ |1,-1\rangle \end{cases} \qquad j=0: \quad |0,0\rangle.$$

These states must be linear combinations of each other. To find the explicit connections between them, let us recall some of the results of Chapter 6. We introduced there the operators $L_\pm = L_x \pm iL_y$ and found that these were raising (L_+) or lowering (L_-) operators on the quantum number m in the state $|\ell,m\rangle$. These results depended *only* on the commutation properties of the L_i. Now since the total angular-momentum operators, J_i, satisfy the same algebra, we have identical results for the $|j,m\rangle$. That is, introducing

$$J_\pm = J_x \pm iJ_y \quad (J_x = J_{1x} + J_{2x}, J_y = J_{1y} + J_{2y}), \tag{8.46}$$

we then have

$$J_\pm|j,m\rangle = \hbar\sqrt{(j\mp m)(j\pm m+1)}\,|j,m\pm 1\rangle, \tag{8.47}$$

and similarly for $J_{1\pm} = J_{1x} \pm iJ_{1y}$ on $|j_1,m_1\rangle$ and $J_{2\pm} = J_{2x} \pm iJ_{2y}$ on $|j_2,m_2\rangle$.

Going back to Equation 8.28 and choosing $m = j_1 + j_2$ (its maximum value), we see that the sum on the right is only a single term, and we get

$$|1,1\rangle = |\frac{1}{2},\frac{1}{2}\rangle\langle\frac{1}{2},\frac{1}{2}|1,1\rangle. \tag{8.48}$$

The Clebsch–Gordan coefficient $\langle 1/2,1/2|1,1\rangle$ can be set equal to one in order to maintain the usual normalization conditions on the states. Now let us apply the operator $J_- = J_{1-} + J_{2-}$ to both sides of Equation 8.48:

$$\Rightarrow \sqrt{(1+1)(1-1+1)}\,|1,0\rangle$$

$$= \sqrt{\left(\frac{1}{2}+\frac{1}{2}\right)\left(\frac{1}{2}-\frac{1}{2}+1\right)}\,|-\frac{1}{2},\frac{1}{2}\rangle + \sqrt{\left(\frac{1}{2}+\frac{1}{2}\right)\left(\frac{1}{2}-\frac{1}{2}+1\right)}\,|\frac{1}{2},-\frac{1}{2}\rangle, \tag{8.49}$$

$$\Rightarrow |1,0\rangle = \frac{1}{\sqrt{2}}\left[|-\frac{1}{2},\frac{1}{2}\rangle + |\frac{1}{2},-\frac{1}{2}\rangle\right].$$

Now apply it again to both sides of Equation 8.49:

$$\Rightarrow \sqrt{(1+0)(1-0+1)}\,|1,-1\rangle$$

$$= \frac{1}{\sqrt{2}}\left[\sqrt{\left(\frac{1}{2}+\frac{1}{2}\right)\left(\frac{1}{2}-\frac{1}{2}+1\right)}\,|-\frac{1}{2},\frac{1}{2}\rangle + \sqrt{\left(\frac{1}{2}+\frac{1}{2}\right)\left(\frac{1}{2}-\frac{1}{2}+1\right)}\,|-\frac{1}{2},\frac{1}{2}\rangle\right], \tag{8.50}$$

$$\Rightarrow |1,-1\rangle = |-\frac{1}{2},-\frac{1}{2}\rangle.$$

We could also have *started* with Equation 8.50 and applied $J_+ = J_{1+} + J_{2+}$ to deduce Equations 8.49 and 8.48.

The state $|0,0\rangle$ is not yet determined. Given that $m = m_1 + m_2$, its most general form in this case is

$$|0,0\rangle = C_1\,|-\frac{1}{2},\frac{1}{2}\rangle + C_2\,|\frac{1}{2},-\frac{1}{2}\rangle. \tag{8.51}$$

Now the states $|0,0\rangle$ and $|1,0\rangle$ must be orthogonal. This means that

$$\langle 1,0|0,0\rangle = \frac{1}{\sqrt{2}}\left[C_1\langle -\frac{1}{2},\frac{1}{2}|-\frac{1}{2},\frac{1}{2}\rangle + C_2\langle\frac{1}{2},-\frac{1}{2}|\frac{1}{2},-\frac{1}{2}\rangle\right]$$

$$\Rightarrow 0 = \frac{1}{\sqrt{2}}[C_1 + C_2] \tag{8.52}$$

$$\Rightarrow C_1 = -C_2.$$

If we normalize such that $\langle 0,0|0,0\rangle = 1$, we can then choose $C_1 = 1/\sqrt{2}$, and finally we get

$$|0,0\rangle = \frac{1}{\sqrt{2}}\left[|\frac{1}{2},-\frac{1}{2}\rangle - |-\frac{1}{2},\frac{1}{2}\rangle\right]. \tag{8.53}$$

Collecting our results together, we have

$$\left.\begin{aligned} |1,1\rangle &= |\frac{1}{2},\frac{1}{2}\rangle, \\ |1,0\rangle &= \frac{1}{\sqrt{2}}\left[|-\frac{1}{2},\frac{1}{2}\rangle + |\frac{1}{2},-\frac{1}{2}\rangle\right] \\ |1,-1\rangle &= |-\frac{1}{2},-\frac{1}{2}\rangle, \end{aligned}\right\}, \tag{8.54}$$

for the $j = 1$ states (which are called the "triplet" states because m takes on three values) and

$$|0,0\rangle = \frac{1}{\sqrt{2}}\left[|\frac{1}{2},-\frac{1}{2}\rangle - |-\frac{1}{2},\frac{1}{2}\rangle\right], \tag{8.55}$$

for the $j = 0$ state (called a "singlet" for obvious reasons).

It is now easy to show, for this same system of two spin-1/2 objects, that the statements of completeness in Equations 8.26 and 8.45 are equivalent.

Since either basis set is complete, there is no reason at this point to prefer one description of the composite system to the other. However, when a Hamiltonian is specified for the system, there is in general no reason why the *individual* third components should be conserved. This would be expressed by $[H, J_{1z}] \neq 0$ and $[H, J_{2z}] \neq 0$. We would only expect in general that the *total* third component would be a constant of the motion, that is, that $[H, J_z] = 0$. Thus, the $|j_1, j_2; j, m\rangle$ states are usually the more relevant ones for interacting composite systems.

8.4 EXPLICIT EXAMPLE OF ADDING ORBITAL ANGULAR MOMENTUM AND SPIN 1/2

To investigate this point some more, let us now examine the addition of an *orbital* and a *spin* 1/2 angular momentum. This is the situation that occurs in the hydrogen atom. We set $\vec{J}_1 = \vec{L}$ and $\vec{J}_2 = \vec{S}$, so that

$$\vec{J} = \vec{L} + \vec{S}. \tag{8.56}$$

To set the notation, I list the mathematical properties of these operators:

$$[L_i, S_j] = 0, \tag{8.57}$$

$$\vec{L}^2 |\ell, m_\ell\rangle = \hbar^2 \ell(\ell+1) |\ell, m_\ell\rangle, \tag{8.58}$$

$$L_z |\ell, \mathrm{m}_\ell\rangle = \hbar m_\ell |\ell, m_\ell\rangle, \tag{8.59}$$

$$\vec{S}^2|m_S\rangle = \frac{3\hbar^2}{4}|m_S\rangle, \tag{8.60}$$

$$S_z\,|m_S\rangle = \hbar m_S\,|m_S\rangle. \tag{8.61}$$

The composite state—an eigenfunction of $\vec{L}^2, \vec{S}^2$, L_z, and S_z—will be denoted as follows:

$$|m_\ell, m_S\rangle \equiv |\ell, m_\ell\rangle|m_S\rangle. \tag{8.62}$$

(A more proper notation would be: $|\ell, s; m_\ell, m_s\rangle$.) The number of such states, for fixed ℓ value, is $2(2\ell+1)$. The properties of the states of total angular momentum are, of course

$$\vec{J}^2\,|j,m\rangle = \hbar^2 j(j+1)|j,m\rangle, \tag{8.63}$$

$$J_z\,|j,m\rangle = \hbar m\,|j,m\rangle, \tag{8.64}$$

where, by the triangle inequality, we have for the allowed values of j,

$$\left.\begin{array}{ll} j = \ell - \dfrac{1}{2} \text{ or } \ell + \dfrac{1}{2}, & \ell > 0 \\ j = \dfrac{1}{2}, & \ell = 0 \end{array}\right\}. \tag{8.65}$$

The number of states is also $2(2\ell+1)$ for these two cases:

$$\left.\begin{array}{ll} \#\text{States} = \left(2\left(\ell - \dfrac{1}{2}\right)+1\right) + \left(2\left(\ell + \dfrac{1}{2}\right)+1\right), & \ell > 0 \\ \#\text{States} = 2, & \ell = 0 \end{array}\right\}. \tag{8.66}$$

The $|j,m\rangle$ are eigenfunctions of $\vec{L}^2$ and $\vec{S}^2$ also in addition to $\vec{J}^2$ and J_z. (The more proper notation would be: $|\ell, s; j,m\rangle$.)

We start our investigation of the relationship between two sets of states by writing the most general possible connection given that $m = m_\ell + m_S$:

$$|\ell + \frac{1}{2}, m\rangle = C_1\,|m - \frac{1}{2}, \frac{1}{2}\rangle + C_2\,|m + \frac{1}{2}, -\frac{1}{2}\rangle, \tag{8.67}$$

$$|\ell - \frac{1}{2}, m\rangle = C_3\,|m - \frac{1}{2}, \frac{1}{2}\rangle + C_4\,|m + \frac{1}{2}, -\frac{1}{2}\rangle. \tag{8.68}$$

The normalization conditions

$$\langle \ell + \frac{1}{2}, m\,|\,\ell + \frac{1}{2}, m\rangle = \langle \ell - \frac{1}{2}, m\,|\,\ell - \frac{1}{2}, m\rangle = 1,$$

give (we can choose all these constants real)

$$C_1^2 + C_2^2 = 1, \tag{8.69}$$

$$C_3^2 + C_4^2 = 1. \tag{8.70}$$

These mean that we may set

$$C_1 = \cos\alpha,\quad C_2 = \sin\alpha, \tag{8.71}$$

$$C_3 = \cos\beta,\quad C_4 = \sin\beta. \tag{8.72}$$

We must also have $\langle \ell+1/2, m|\ell-1/2, m\rangle = 0$ (the states are orthogonal), which means that

$$\cos\alpha\cos\beta + \sin\alpha\sin\beta = 0, \tag{8.73}$$

or

$$\cos(\alpha-\beta) = 0. \tag{8.74}$$

Choosing

$$\beta = \alpha + \frac{\pi}{2}, \tag{8.75}$$

(this makes $C_3 = -\sin\alpha$ and $C_4 = \cos\alpha$) will satisfy Equation 8.74. Thus, we now only have one undetermined constant, say $\cos\alpha$, to determine.

To determine this remaining constant, let us use the raising and lowering operators already introduced when we added two spin-1/2 angular momenta. We have, of course, that

$$\vec{J}^2 = \vec{L}^2 + \vec{S}^2 + 2\vec{L}^2\cdot\vec{S}^2. \tag{8.76}$$

(We can write $\vec{L}\cdot\vec{S} = \vec{S}\cdot\vec{L}$ here since all components of $\vec{L}$ and $\vec{S}$ commute.) We can then derive the result that

$$2\vec{L}\cdot\vec{S} = 2L_zS_z + L_+S_- + L_-S_+, \tag{8.77}$$

where, as usual

$$\left.\begin{aligned} L_\pm &= L_x \pm iL_y,\\ S_\pm &= S_x \pm iS_y.\end{aligned}\right\} \tag{8.78}$$

Now operating on both sides of Equation 8.67 with $\vec{J}^2$, we find that

$$\begin{aligned}
&\left(\ell+\frac{3}{2}\right)\left(\ell+\frac{1}{2}\right)|\,\ell+\frac{1}{2}, m\rangle\\
&= \cos\alpha\left[\ell(\ell+1)+\frac{3}{4}+2\left(m-\frac{1}{2}\right)\frac{1}{2}\right]|\,m-\frac{1}{2},\frac{1}{2}\rangle\\
&\quad+\cos\alpha\left[\sqrt{\left(\ell-m+\frac{1}{2}\right)\left(\ell+m+\frac{1}{2}\right)}\sqrt{\left(\frac{1}{2}+\frac{1}{2}\right)\left(\frac{1}{2}-\frac{1}{2}+1\right)}\right]|\,m+\frac{1}{2},-\frac{1}{2}\rangle\\
&\quad+\sin\alpha\left[\ell(\ell+1)+\frac{3}{4}+2\left(m+\frac{1}{2}\right)\left(-\frac{1}{2}\right)\right]|\,m+\frac{1}{2},-\frac{1}{2}\rangle\\
&\quad+\sin\alpha\left[\sqrt{\left(\ell+m+\frac{1}{2}\right)\left(\ell-m+\frac{1}{2}\right)}\sqrt{\left(\frac{1}{2}+\frac{1}{2}\right)\left(\frac{1}{2}-\frac{1}{2}+1\right)}\right]|\,m-\frac{1}{2},-\frac{1}{2}\rangle.
\end{aligned} \tag{8.79}$$

More simply, this is the same as (dividing both sides by $(\ell+3/2)(\ell+1/2)$)

$$
\begin{aligned}
&|\ell+\frac{1}{2}, m\rangle \\
&=\frac{|m-\frac{1}{2}, \frac{1}{2}\rangle}{\left(\ell+\frac{3}{2}\right)\left(\ell+\frac{1}{2}\right)}\left\{\cos\alpha\left[\ell(\ell+1)+\frac{1}{4}+m\right]+\sin\alpha\left[\sqrt{\left(\ell+\frac{1}{2}\right)^2-m^2}\right]\right\} \\
&+\frac{|m+\frac{1}{2}, -\frac{1}{2}\rangle}{\left(\ell+\frac{3}{2}\right)\left(\ell+\frac{1}{2}\right)}\left\{\sin\alpha\left[\ell(\ell+1)+\frac{1}{4}-m\right]+\cos\alpha\left[\sqrt{\left(\ell+\frac{1}{2}\right)^2-m^2}\right]\right\}.
\end{aligned}
\tag{8.80}
$$

Comparing Equation 8.80 with the starting point, Equation 8.67, since the left-hand sides of these equations are identical, we have that (remember $C_1=\cos\alpha$, $C_2=\sin\alpha$)

$$\cos\alpha=\frac{\cos\alpha\left[\ell(\ell+1)+\frac{1}{4}+m\right]+\sin\alpha\sqrt{\left(\ell+\frac{1}{2}\right)^2-m^2}}{\left(\ell+\frac{3}{2}\right)\left(\ell+\frac{1}{2}\right)}, \tag{8.81}$$

$$\sin\alpha=\frac{\sin\alpha\left[\ell(\ell+1)+\frac{1}{4}-m\right]+\cos\alpha\sqrt{\left(\ell+\frac{1}{2}\right)^2-m^2}}{\left(\ell+\frac{3}{2}\right)\left(\ell+\frac{1}{2}\right)}. \tag{8.82}$$

We will only need one of these equations to solve for $\cos\alpha$ or $\sin\alpha$. Dividing both sides of Equation 8.81 by $\cos\alpha$ and then solving for $\tan\alpha$ gives

$$\tan\alpha=\frac{\left(\ell+\frac{3}{2}\right)\left(\ell+\frac{1}{2}\right)-\left[\ell(\ell+1)+\frac{1}{4}+m\right]}{\sqrt{\left(\ell+\frac{1}{2}\right)^2-m^2}}. \tag{8.83}$$

We can simplify this to

$$\tan\alpha=\frac{\ell-m+\frac{1}{2}}{\sqrt{\left(\ell+\frac{1}{2}\right)^2-m^2}}=\sqrt{\frac{\ell+\frac{1}{2}-m}{\ell+\frac{1}{2}+m}}. \tag{8.84}$$

Then, solving for $\cos\alpha$ we get

$$\cos^2\alpha=\frac{1}{1+\tan^2\alpha}=\frac{\ell+\frac{1}{2}+m}{2\ell+1}, \tag{8.85}$$

$$\Rightarrow\cos\alpha=\sqrt{\frac{\ell+\frac{1}{2}+m}{2\ell+1}}, \tag{8.86}$$

where, by convention, we choose the positive root. Given Equation 8.86, we then get

$$\sin\alpha = \sqrt{\frac{\ell+\frac{1}{2}-m}{2\ell+1}}, \tag{8.87}$$

where the positive root is now *determined* because $\tan\alpha \geq 0$ from Equation 8.84, and we have chosen $\cos\alpha \geq 0$ in Equation 8.86. Our explicit connections between the states of the total angular momentum and composite orbital/spin states are therefore

$$|\ell+\frac{1}{2},m\rangle = \sqrt{\frac{\ell+\frac{1}{2}+m}{2\ell+1}}\,|m-\frac{1}{2},\frac{1}{2}\rangle + \sqrt{\frac{\ell+\frac{1}{2}-m}{2\ell+1}}\,|m+\frac{1}{2},-\frac{1}{2}\rangle, \tag{8.88}$$

$$|\ell-\frac{1}{2},m\rangle = -\sqrt{\frac{\ell+\frac{1}{2}-m}{2\ell+1}}\,|m-\frac{1}{2},\frac{1}{2}\rangle + \sqrt{\frac{\ell+\frac{1}{2}+m}{2\ell+1}}\,|m+\frac{1}{2},-\frac{1}{2}\rangle. \tag{8.89}$$

By projecting these kets into the spin/angular space $\langle\theta,\phi,m_S|$ and using the explicit matrix representation of spin 1/2 discussed in Chapter 1, Section 1.16, Equations 8.88 and 8.89 are seen to be equivalent to

$$\mathcal{Y}_\ell^{j=\ell+1/2,m}(\theta,\phi) = \sqrt{\frac{\ell+\frac{1}{2}+m}{2\ell+1}}\,Y_{\ell,m-\frac{1}{2}}(\theta,\phi)\psi_+ + \sqrt{\frac{\ell+\frac{1}{2}-m}{2\ell+1}}\,Y_{\ell,m+\frac{1}{2}}(\theta,\phi)\psi_-, \tag{8.90}$$

$$\mathcal{Y}_\ell^{j=\ell-1/2,m}(\theta,\phi) = -\sqrt{\frac{\ell+\frac{1}{2}-m}{2\ell+1}}\,Y_{\ell,m-\frac{1}{2}}(\theta,\phi)\psi_+ + \sqrt{\frac{\ell+\frac{1}{2}+m}{2\ell+1}}\,Y_{\ell,m+\frac{1}{2}}(\theta,\phi)\psi_-, \tag{8.91}$$

where we have defined

$$\left[\mathcal{Y}_\ell^{j=\ell\pm1/2,m}(\theta,\phi)\right]_{m_S} \equiv \left\langle \theta,\phi,m_S \middle| \ell\pm\frac{1}{2},m\right\rangle, \tag{8.92}$$

and where the two values of $m_S = \pm 1/2$ are being used as matrix row labels. (We associate $m_S = 1/2$ with the top row and $m_S = -1/2$ with the bottom row.) The $\psi_\pm$ are the column matrices (as in Chapter 1, Equation 1.225)

$$\psi_+ = \begin{pmatrix}1\\0\end{pmatrix},\quad \psi_- = \begin{pmatrix}0\\1\end{pmatrix}. \tag{8.93}$$

More compactly, both Equations 8.90 and 8.91 may be written as

$$\mathcal{Y}_\ell^{j=\ell\pm1/2,m}(\theta,\phi) = \begin{matrix} m_S = \frac{1}{2} \\ m_S = -\frac{1}{2} \end{matrix}\begin{pmatrix} \pm\sqrt{\frac{\ell+\frac{1}{2}\pm m}{2\ell+1}}\,Y_{\ell,m-\frac{1}{2}}(\theta,\phi) \\ \sqrt{\frac{\ell+\frac{1}{2}\mp m}{2\ell+1}}\,Y_{\ell,m+\frac{1}{2}}(\theta,\phi) \end{pmatrix}. \tag{8.94}$$

The top signs go with the case $j=\ell+1/2$ and the bottom signs with $j=\ell-1/2\,(\ell\neq 0)$. The $\mathcal{Y}_\ell^{j,m}(\theta,\phi)$ are called *spin-angle functions*. We will see them again in a calculation in Chapter 10, Section 10.8.

8.5 HYDROGEN ATOM AND THE CHOICE OF BASIS STATES

We now have two complete sets of eigenfunctions in which to describe a situation where spin 1/2 and orbital angular momentum are being added. The set $|\,\ell,s;m_\ell,m_s\rangle$ ($s=1/2$ here) is an eigenvector of $\vec{L}^2$, $\vec{S}^2$, L_z, and S_z. The other set, denoted as $|\,\ell,s;j,m\rangle$ and giving rise to the aforementioned spin-angle functions, is eigenvectors of $\vec{J}^2,\vec{L}^2,\vec{S}^2$, and J_z. Which set should we use in a given problem? Mathematically, it does not matter, but computationally it makes a lot of difference. Let us go back to the problem of the hydrogen atom in order to get some experience in these matters. The following considerations will be illustrative of both the use of spin-angle basis states as well as the perturbation theory development in Chapter 5.

First, let us get a sense of the speed involved in the motion of an electron in the hydrogen atom ground state. The Bohr radius is

$$a_0=\frac{\hbar^2}{me^2}. \tag{8.95}$$

I argued back in Chapter 2, Equation 2.34, that from the uncertainty principle

$$p_r a_0 \approx \hbar, \tag{8.96}$$

for the ground state. Setting $p_r = mv$ and solving for v from Equations 8.95 and 8.96 then gives

$$\frac{v}{c}\approx\frac{e^2}{\hbar c}. \tag{8.97}$$

We came upon the combination $\alpha = e^2/\hbar c$ back in Chapter 1, Section 1.4 where it was called the "fine structure constant." It is a pure number with the approximate value

$$\alpha=\frac{1}{137.036}. \tag{8.98}$$

The result of Equation 8.97 implies a small relativistic correction to the hydrogen energy levels calculated in Chapter 7, the so-called "fine structure."

The relativistic kinetic energy is

$$E^2=\vec{p}^{\,2}c^2+m^2c^4, \tag{8.99}$$

from which we find that

$$E-mc^2\approx\frac{\vec{p}^{\,2}}{2m}-\frac{1}{2mc^2}\left(\frac{\vec{p}^{\,2}}{2m}\right)^2. \tag{8.100}$$

The first term is the usual kinetic energy term, and the second represents the lowest order relativistic correction. From Equation 8.97, we would expect this correction to be of the order

$$\frac{\frac{1}{2mc^2}\left(\frac{\vec{p}^2}{2m}\right)^2}{\frac{\vec{p}^2}{2m}} \sim \left(\frac{v}{c}\right)^2 \sim \alpha^2, \tag{8.101}$$

relative to the unperturbed ground state energy.

There is another correction to the hydrogen atom energy levels of the same order of magnitude. It comes about because the electron, moving in the electric field of the nucleus, experiences an *effective* magnetic field given by (this comes from Maxwell's equations)

$$\vec{H}_{\text{eff}} = -\frac{\vec{v}}{c} \times \vec{E}, \tag{8.102}$$

where the electric field, $\vec{E}$, is

$$\vec{E} = -\vec{\nabla}\phi(r). \tag{8.103}$$

$\phi(r)$ is the (central) electrostatic potential, which in the case of the Coulomb law, is given by

$$\phi(r) = \frac{Ze}{r}. \tag{8.104}$$

(We imagine Z protons in the nucleus; e is the *magnitude* of the electron's charge.) The electron's magnetic moment is given by Equations 1.42 and 1.44 (see Chapter 1):

$$\vec{\mu} = -\frac{e}{mc}\vec{S}. \tag{8.105}$$

(The relation Equation 8.105 is in fact not exactly true and has some small corrections, the most important of which was calculated by Julian Schwinger in 1948.) Given Equations 8.102 and 8.105, it is reasonable to expect that there will be an interaction of the form $-\vec{\mu}\cdot\vec{H}_{\text{eff}}$ (see Equation 1.29). Now, using Equations 8.102 through 8.105, one can show that this interaction can be written as

$$-\vec{\mu}\cdot\vec{H}_{\text{eff}} = -\frac{e}{m^2c^2}\frac{1}{r}\frac{d\phi}{dr}(\vec{L}\cdot\vec{S}), \tag{8.106}$$

where we have used the classical form $\vec{L} = \vec{r} \times (m\vec{v})$. The result (Equation 8.106) is actually too large by a factor of 2. The reason is that we have not been careful in using the correct relativistic kinematics in evaluating the interaction. The extra necessary factor of 1/2 is called the "Thomas precession factor" (after the English physicist L.H. Thomas). It is difficult to justify kinematically, but extremely easy to recover from the Dirac equation, which is the *relativistic* equation satisfied by electrons.

An interaction of the form of Equation 8.106 says that the electron's spin will interact with its own angular momentum. Such an effect is called *L–S coupling.* (This is the same L–S coupling mentioned in the deuteron discussion in Chapter 7.)

In the preceding discussion, the quantities $\vec{p}, \vec{L}, \vec{S}$, and so forth are classical quantities. As usual we replace these quantities by their quantum mechanical operators. (Notice that $\vec{L}\cdot\vec{S} = \vec{S}\cdot\vec{L}$ for $\vec{S}$, $\vec{L}$ as operators so there is no ambiguity in the replacement.) Therefore, we write our slightly corrected hydrogen atom Hamiltonian as

$$H = H_0 + H_{\text{rel}} + H_{LS}, \tag{8.107}$$

where $H_0 = \vec{p}^2/2\mu - Ze^2/r$ was the original Hamiltonian, and H_{rel} and H_{LS} are given by (letting $m \to \mu$ where appropriate makes little difference here)

$$H_{\text{rel}} = -\frac{1}{2mc^2}\left(\frac{\vec{p}^2}{2m}\right)^2, \tag{8.108}$$

$$H_{LS} = -\frac{e}{2m^2c^2}\frac{1}{r}\frac{d\phi}{dr}(\vec{L}\cdot\vec{S}). \tag{8.109}$$

The only thing to do with H_{rel} and H_{LS} is to treat them as perturbations as in the development in Chapter 5. We must use the degenerate perturbation theory outlined there since, in fact, each energy level specified by n has a $2n^2$-fold degeneracy. (The factor of two comes from considering the two values of electron spin, $m_s = \pm 1/2$.) From Equation 5.44, the perturbed energy levels are given by

$$E_a = E_0 + \langle E_0 a | H_1 | E_0 a\rangle, \tag{8.110}$$

to first order. Remember, the label a in Equation 8.110 was used to distinguish between states with the same energy. Remember also that the basis to be used in Equation 8.110 is one in which the perturbing Hamiltonian, H_1, is diagonal in the degenerate subspace specified by a.

In our case, this degenerate subspace is the one specified by the quantum numbers ℓ, m_ℓ and m_s from the complete set $\{\vec{L}^2, \vec{S}^2, L_z, S_z\}$ or the quantum numbers ℓ, j, and m from the alternate set $\{\vec{L}^2, \vec{S}^2, \vec{J}^2, J_z\}$. (The radial states differing by the principle quantum number n are never degenerate.) Now H_{rel} commutes with all of the operators in both sets. The other perturbation, H_{LS}, fails to commute with all of the operators in $\{\vec{L}^2, \vec{S}^2, L_z, S_z\}$, and its perturbation matrix is not diagonal in this set. However, the perturbation $H_{\text{rel}} + H_{LS}$ is diagonal in the set $\{\vec{L}^2, \vec{S}^2, \vec{J}^2, J_z\}$, and we may use Equation 8.110 directly to find the new energies. If, for example, one were unaware of the $|\ell, s; j, m\rangle$ set of states, we could still find the first-order perturbed energies, but we could not start with Equation 8.110. We would be forced to diagonalize the perturbation in the $|\ell, s; m_\ell, m_s\rangle$ basis. This would essentially repeat the work we already did in finding the explicit relationship between the two basis sets. Eventually, this would give us the same answers but with a lot of redundant work.

As J.J. Sakurai pointed out in colorful language: "You have to be either a fool or a masochist to use the L_z, S_z eigenkets as the base kets for this problem."*

8.6 HYDROGEN ATOM AND PERTURBATIVE ENERGY SHIFTS

We thus have for the perturbative energy shifts

$$\Delta E_{n\ell sjm} = \langle n\ell sjm|(H_{\text{rel}} + H_{LS})|n\ell sjm\rangle, \tag{8.111}$$

where we are defining the separable state

$$|n\ell sjm\rangle \equiv |n\ell\rangle \otimes |\ell, s; j, m\rangle. \tag{8.112}$$

The $|n\ell\rangle$ provide the hydrogen atom radial basis found in Chapter 7:

$$\frac{R_{n\ell}(r)}{r} = u_{n\ell}(r) = \langle r|n\ell\rangle. \tag{8.113}$$

Now we know that (I continue to ignore in this chapter the distinction between m and μ)

$$\left(\frac{\vec{p}^2}{2m} - \frac{Ze^2}{r}\right)|n\ell sjm\rangle = E_n|n\ell sjm\rangle, \tag{8.114}$$

gives the unperturbed energy eigenvalues. Therefore

$$\frac{\vec{p}^2}{2m}|n\ell sjm\rangle = \left(E_n + \frac{Ze^2}{r}\right)|n\ell sjm\rangle, \tag{8.115}$$

which implies that

$$\begin{aligned} \langle n\ell sjm|H_{\text{rel}}|n\ell sjm\rangle &= \langle n\ell sjm| - \frac{1}{2mc^2}\left(E_n + \frac{Ze^2}{r}\right)^2 |n\ell sjm\rangle \\ &= -\frac{1}{2mc^2}\langle n\ell|\left(E_n + \frac{Ze^2}{r}\right)^2 |n\ell\rangle. \end{aligned} \tag{8.116}$$

To evaluate (Equation 8.116), we will need to know the expectation values $\langle 1/r\rangle_{n\ell}$ and $\langle 1/r^2\rangle_{n\ell}$. Actually, from a homework problem in Chapter 7 (Problem 7.6.7), we know that

$$\left\langle \frac{1}{r}\right\rangle_{n\ell} = \frac{Z}{a_0 n^2}. \tag{8.117}$$

We can use the same technique as displayed in this problem (and the discussion of normalization of the hydrogen atom eigenfunctions in

* J.J. Sakurai and J. Napolitano, *Modern Quantum Mechanics*, 2nd ed., (Addison-Wesley, 2011), Chapter 5.

Chapter 7) to also find $\langle 1/r^2\rangle_{n\ell}$. Our unperturbed Hamiltonian in the radial eigenspace can be written as

$$H = \frac{p_r^2}{2m} + \frac{\hbar^2\ell(\ell+1)}{2mr^2} - \frac{Ze^2}{r}, \tag{8.118}$$

with the p_r operator defined as in Equation 7.43. Then, choosing $\lambda = \ell$ in Equation 5.3 (Chapter 5), we obtain

$$\left\langle \frac{\partial H}{\partial \ell} \right\rangle_{n\ell} = \frac{\partial E_n}{\partial \ell} = \frac{\partial}{\partial \ell}\left(-\frac{Z^2\hbar^2}{2ma_0^2(n_r+\ell+1)^2} \right), \tag{8.119}$$

$$\Rightarrow \left\langle \frac{\hbar^2(2\ell+1)}{2mr^2} \right\rangle_{n\ell} = \frac{Z^2\hbar^2}{a_0^2 n^3}, \tag{8.120}$$

$$\Rightarrow \left\langle \frac{1}{r^2} \right\rangle_{n\ell} = \frac{Z^2}{a_0^2} \frac{1}{\left(\ell+\frac{1}{2}\right)n^3}. \tag{8.121}$$

Putting the pieces together, we finally evaluate

$$\begin{aligned} \langle n\ell|H_{\text{rel}}|n\ell\rangle &= -\frac{1}{2mc^2}\left(\frac{Z^2e^2}{a_0 n^2}\right)^2\left(\frac{n}{\ell+\frac{1}{2}} - \frac{3}{4}\right) \\ &= \frac{Z^2E_n}{n^2}\alpha^2\left(\frac{n}{\ell+\frac{1}{2}} - \frac{3}{4}\right). \end{aligned} \tag{8.122}$$

Thus, we have from Equation 8.122 (for $n=1$, $Z=1$) that

$$\frac{\langle H_{\text{rel}}\rangle_{10}}{E_0} \sim \alpha^2, \tag{8.123}$$

confirming Equation 8.101.

We also have that

$$\vec{J}^2 = (\vec{L}+\vec{S})^2 = \vec{L}^2 + \vec{S}^2 + 2\vec{L}\cdot\vec{S}, \tag{8.124}$$

and therefore

$$\vec{L}\cdot\vec{S} = \frac{1}{2}(\vec{J}^2 - \vec{L}^2 - \vec{S}^2), \tag{8.125}$$

$$\Rightarrow \vec{L}\cdot\vec{S}|\ell,s;j,m\rangle = \frac{\hbar^2}{2}\left(j(j+1) - \ell(\ell+1) - \frac{3}{4} \right)|\ell,s;j,m\rangle. \tag{8.126}$$

This allows us to write

$$\langle n\ell sjm|H_{LS}|n\ell sjm\rangle = -\frac{e\hbar^2}{4m^2c^2}\left[j(j+1) - \ell(\ell+1) - \frac{3}{4} \right]\langle n\ell|\frac{1}{r}\frac{d\phi}{dr}|n\ell\rangle. \tag{8.127}$$

Since $\phi(r) = Ze/r$ (Equation 8.104 for the Coulomb potential), we need to find $\langle 1/r^3 \rangle_{n\ell}$ in Equation 8.127. There is an easy way of doing this. Again, from Equation 7.43,

$$\langle r|p_r = \frac{\hbar}{i}\frac{1}{r}\frac{\partial}{\partial r} r\langle r|. \tag{8.128}$$

Therefore, for any function $f(r)$ we have (I am not being careful to distinguish the use of r as an operator or eigenvalue here)

$$\begin{aligned}\langle r|p_r f(r) &= \frac{\hbar}{i}\frac{1}{r}\frac{\partial}{\partial r}(r\langle r|f(r)) \\ &= \frac{\hbar}{i}\left(\frac{f(r)}{r}\langle r| + f(r)\frac{\partial}{\partial r}\langle r| + f'(r)\langle r|\right)\end{aligned} \tag{8.129}$$

and

$$\langle r|f(r)p_r = \frac{\hbar}{i} f(r)\left(\frac{1}{r}\langle r| + \frac{\partial}{\partial r}\langle r|\right), \tag{8.130}$$

which implies that

$$\langle r|[p_r, f(r)] = \frac{\hbar}{i}\frac{\partial f(r)}{\partial r}\langle r|, \tag{8.131}$$

$$\Rightarrow [p_r, f(r)] = \frac{\hbar}{i}\frac{\partial f(r)}{\partial r}. \tag{8.132}$$

If we now take $f(r) = H(= p_r^2/2m + \hbar^2\ell(\ell+1)/2mr^2 - Ze^2/r)$ and evaluate the expectation value of the left-hand side of Equation 8.132, we get

$$\begin{aligned}\langle n\ell|[p_r, H]|n\ell\rangle &= \langle n\ell|p_r H - Hp_r|n\ell\rangle \\ &= (E_n - E_n)\langle n\ell|p_r|n\ell\rangle = 0.\end{aligned} \tag{8.133}$$

Since

$$\left\langle \frac{\partial H}{\partial r}\right\rangle_{n\ell} = -\frac{\hbar^2\ell(\ell+1)}{m}\left\langle\frac{1}{r^3}\right\rangle_{n\ell} + Ze^2\left\langle\frac{1}{r^2}\right\rangle_{n\ell}, \tag{8.134}$$

one has that

$$\left\langle\frac{1}{r^3}\right\rangle_{n\ell} = \frac{Ze^2 n}{\hbar^2\ell(\ell+1)}\left\langle\frac{1}{r^2}\right\rangle_{n\ell}, \tag{8.135}$$

or using Equation 8.121, that

$$\left\langle\frac{1}{r^3}\right\rangle_{n\ell} = \left(\frac{Z}{na_0}\right)^3 \frac{1}{\ell\left(\ell+\frac{1}{2}\right)(\ell+1)}. \tag{8.136}$$

Notice that the expectation value in Equation 8.136 diverges if we set $\ell = 0$. It is easy to understand how this comes about. We found in Chapter 7 that near the origin, the radial function $R(r)$ behaved as $R(r) \sim r^{\ell+1}$. Therefore, for the integral in Equation 8.136, we have when $\ell = 0$,

$$\lim_{r\to 0}\frac{R^2(r)}{r^3} \sim \frac{1}{r}. \tag{8.137}$$

The integral of r^{-1} diverges logarithmically when the lower limit is $r = 0$. Therefore, the perturbative treatment given here breaks down for s-states.

We now have that

$$\langle n\ell sjm|H_{LS}|n\ell sjm\rangle = -\frac{Z^2E_n}{2n}\alpha^2\frac{\left[j(j+1)-\ell(\ell+1)-\ell+\ell+\frac{3}{4}\right]}{\ell\left(\ell+\frac{1}{2}\right)(\ell+1)}, \tag{8.138}$$

confirming that the order of magnitude of the energy shift from H_{LS} is the same as from H_{rel}. Since

$$\frac{\left[j(j+1)-\ell(\ell+1)-\frac{3}{4}\right]}{\ell\left(\ell+\frac{1}{2}\right)(\ell+1)} = \begin{cases} \dfrac{1}{\left(\ell+\frac{1}{2}\right)\left(j+\frac{1}{2}\right)}, & j=\ell+\frac{1}{2} \\ -\dfrac{1}{\left(\ell+\frac{1}{2}\right)\left(j+\frac{1}{2}\right)}, & j=\ell-\frac{1}{2} \end{cases}, \tag{8.139}$$

we can combine Equations 8.122 and 8.138 as (top sign is for $j = \ell + 1/2$ and bottom sign is for $j = \ell - 1/2$)

$$\begin{aligned} \Delta E_{nj} &= \frac{Z^2E_n}{n^2}\alpha^2\left[\frac{n}{\ell+\frac{1}{2}} \mp \frac{n}{2\left(\ell+\frac{1}{2}\right)\left(j+\frac{1}{2}\right)} - \frac{3}{4}\right] \\ &= \frac{Z^2E_n}{n^2}\alpha^2\left[\frac{n}{j+\frac{1}{2}} - \frac{3}{4}\right]. \end{aligned} \tag{8.140}$$

Equation 8.140 is our final result for the fine structure splitting in a Coulomb field and holds for both cases, $j = \ell \pm 1/2$. These levels, which before were degenerate in energy, are now split by a small amount. A new notation will be introduced here where the j value is added as a subscript on the usual principle quantum number and angular-momentum quantum number combination. Thus a 2s state with $j = 1/2$ is designated as $2s_{1/2}$. This is same as the notation used in Figure 7.2 for nuclei, which also experience L–S coupling but with a sign opposite to that in Equation 8.109. For the $n = 2$ and $\ell = 0$, 1 levels (the 2s, 2p states), we have the qualitative energy diagram shown in Figure 8.1.

Actually, one can show that Equation 8.140, even in the case of $\ell = 0$, is also a result of the relativistic Dirac equation, at least to order α^2.

It was shown experimentally in 1947 by W. Lamb and R. Retherford that the hydrogen atom $2s_{1/2}$ and $2p_{1/2}$ levels are actually split by a small

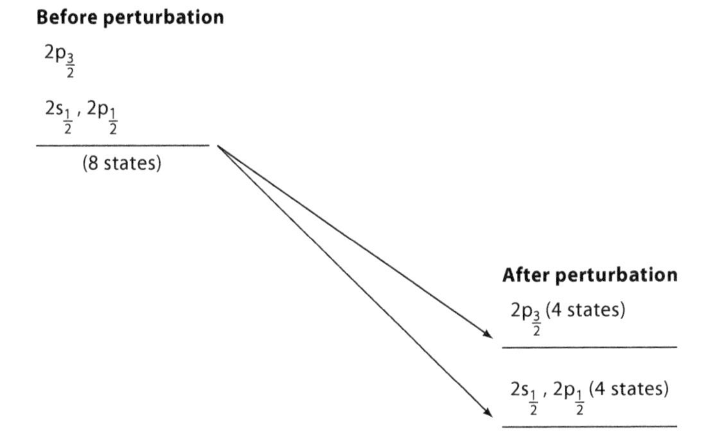

FIGURE 8.1 Qualitative energy level diagram for the $n=2$ energy levels of the hydrogen atom before the perturbations given in Equation 8.140 are added (left) and after (right). All initial levels are degenerate. After the perturbation, all levels are depressed, but the $j=1/2$ levels are lowered more than the $j=3/2$ levels.

amount. Using nonrelativistic quantum theory, Hans Bethe quickly showed that the splitting arose from an atomic electron self-energy effect. Lamb, along with N. Kroll, also published the correct relativistic derivation of the effect, which is more than justifiably called the *Lamb shift.** He won the 1955 Nobel Prize in Physics for this work.

PROBLEMS

8.1.1 (a) Using Equation 8.15,

$$S_+ = \hbar \sum_{m_s'=-s+1}^{s} \sqrt{m_s'\,(1-m_s')+s(1+s)}\;|m_s'\rangle\langle m_s'-1|,$$

verify both statements in Equation 8.14:

$$S_\pm|s,m_s\rangle = \hbar\sqrt{(s\mp m_s)(s\pm m_s+1)}|s,m_s\pm 1\rangle.$$

8.2.1 Show for $\vec{J} = \vec{J}_1 + \vec{J}_2\,([J_{1i}, J_{2j}] = 0)$ that (see Equation 8.25)

(a) $[\vec{J}_1^{\,2}, \vec{J}^{\,2}] = 0,$

(b) $[\vec{J}_1^{\,2}, J_z] = 0.$

8.2.2 Show that

$$\vec{J}_1 \cdot \vec{J}_2 = J_{1z}J_{2z} + \frac{1}{2}(J_{1+}J_{2-} + J_{1-}J_{2+}),$$

* Both Feynman and Schwinger also had the opportunity to be the first to publish the correct result for the Lamb shift. In this context, see the interesting "appropriately numbered" footnote 13 in Feynman's paper, *Phys. Rev.* **76**, p. 769 (1949). Also read the "ancient history" account, elicited from a question from Harold (Hypothetical alert reader of limitless dedication) in Section 4–11 of J. Schwinger, *Particles, Sources, and Fields*, Vol. II (Addison-Wesley, 1973).

where

$$J_{1\pm} \equiv J_{1x} \pm i\,J_{1y},$$
$$J_{2\pm} \equiv J_{2x} \pm i\,J_{2y}.$$

8.2.3 Prove that the coupled state corresponding to $m = j_1 + j_2$ (its maximum value) has $j = j_1 + j_2$. Do this as follows.

(a) First, argue that

$$\langle j_1, j_2; j, m|\vec{J}^2 - (\vec{J}_1 + \vec{J}_2)^2|j_1, j_2; m_1, m_2\rangle = 0.$$

(b) By setting $m_1 = j_1$, $m_2 = j_2$, and $m = j_1 + j_2$, now show that this expression is equivalent to

$$[j(j+1) - (j_1 + j_2)(j_1 + j_2 + 1)]\langle j_1, j_2; j, j_1 + j_2|j_1, j_2; j_1, j_2\rangle = 0,$$

and therefore conclude in general that $j = j_1 + j_2$. [I had to use the Problem 8.2.2 result on this part.]

8.2.4 Add $j_1 = 1/2$ to $j_2 = 3/2$.

(a) List the quantum numbers of the states in the uncoupled representation. How many states are there?

(b) List the quantum numbers of the states in the coupled representation. How many states are there?

(c) Write the coupled state $|3/2, 1/2; 2, 2\rangle$ in terms of uncoupled states.

8.3.1 Suppose two $\ell = 1$ states are coupled together to make a state $(\ell_1 = \ell_2 = 1; \ell = 2, m = 1)$.

(a) Assuming

$$|1,1;2,2\rangle = |1,1\rangle_1|1,1\rangle_2,$$

show that

$$|1,1;2,1\rangle = \sqrt{\frac{1}{2}}|1,1\rangle_1|1,0\rangle_2 + \sqrt{\frac{1}{2}}|1,0\rangle_1|1,1\rangle_2.$$

(b) What is the probability that $L'_{1z} = \hbar$, $L'_{2z} = 0$?

(c) Evaluate explicitly the angular-dependent amplitude (in terms of spherical harmonics):

$$\langle \theta_1, \phi_1; \theta_2, \phi_2|1,1;2,1\rangle = ?$$

8.4.1 Using Problem 8.2.2 with $\vec{J}_1 = \vec{L}_1$ and $\vec{J}_2 = \vec{L}_2$, verify the ℓ, m values of the following coupled angular-momentum eigenstate for two $\ell = 1$ atoms:

$$|1,1;2,0\rangle = \sqrt{\frac{1}{6}}|1,1\rangle_1|1,-1\rangle_2 + \sqrt{\frac{2}{3}}|1,0\rangle_1|1,0\rangle_2 + \sqrt{\frac{1}{6}}|1,-1\rangle_1|1,1\rangle_2.$$

[Note I am not asking you to *construct* this state, but simply verify the total $\ell = 2$, $m = 0$ value by operating on both sides with $\vec{L}^2 = (\vec{L}_1 + \vec{L}_2)^2$ and $L_z = L_{1z} + L_{2z}$.]

***8.4.2** (a) Given particles 1 and 2, both with spin 1/2, one may add to get coupled states with $j=1$ and $j=0$ given by Equations 8.54 and 8.55. Now add a *third* spin 1/2 (particle 3) to these results. Show, using Equations 8.88 and 8.89 with ℓ replaced by 1, that the coupled states with $j=3/2$, $m=1/2$ and $j=1/2$, $m=1/2$, are given by

$$\overset{j\quad m}{|3/2,\,1/2\rangle_{123}} = \frac{1}{\sqrt{3}}(|-1/2\rangle_1|1/2\rangle_2+|1/2\rangle_1|-1/2\rangle_2)|1/2\rangle_3 + \sqrt{\frac{1}{3}}|1/2\rangle_1|1/2\rangle_2|-1/2\rangle_3,$$

$$\overset{j\quad m}{|1/2,\,1/2\rangle_{123}} = -\frac{1}{\sqrt{6}}(|-1/2\rangle_1|1/2\rangle_2+|1/2\rangle_1|-1/2\rangle_2)|1/2\rangle_3 + \sqrt{\frac{2}{3}}|1/2\rangle_1|1/2\rangle_2|-1/2\rangle_3.$$

(b) Is there another way to make a linearly independent $|1/2,1/2\rangle_{123}$ state, other than that given above? [Hint: Two spin 1/2 states can also add up to give spin 0. Then add the third spin-1/2 particle.]

8.5.1 Using Equations 8.102 through 8.105, show that Equation 8.106 holds.

8.5.2 Show:

(a) $[\vec{L}\cdot\vec{S},L_i]=i\hbar(\vec{L}\times\vec{S})_i$,

(b) $[\vec{L}\cdot\vec{S},S_i]=-i\hbar(\vec{L}\times\vec{S})_i$.

[Note this proves that $[\vec{L}\cdot\vec{S},J_i]=0$.]

8.5.3 Given the Hamiltonian (see Equation 8.109)

$$H=V_1(r)+V_2(r)\vec{L}\cdot\vec{S},$$

which of the operators $\vec{L}^2$, $\vec{S}^2$, L_3, S_3 fail to commute with H? What is the physical meaning of this? [Remember that $[\vec{L},V(r)]=0$, as derived in Chapter 6, Equation 6.120. You may use the Problem 8.5.2 results.]

8.5.4 Imagine the electron has a small *electric dipole moment* ($\vec{p}$) in addition to its usual magnetic dipole moment ($\vec{\mu}$). If this electric dipole is proportional to spin, $\vec{p}=\gamma_p\vec{S}$, then we would expect the electron in a hydrogen atom to feel a small, additional perturbation given by

$$H'=-\vec{p}\cdot\vec{E}=-\frac{Ze}{r^3}\vec{p}\cdot\vec{r}=-\frac{Ze\gamma_p}{r^3}(\vec{S}\cdot\vec{r}).$$

$$(\vec{r}=x\hat{i}+y\hat{j}+z\hat{k}).$$

Calculate:

$$[S_z,H']=?,\quad [L_z,H']=?,\quad [J_z,H']=?$$

On the basis of your calculation, which of these operators gives rise to a conserved quantum number? [Comment: An electron electric dipole moment would actually violate parity, which was introduced in Chapter 3, and time-reversal. We will study these discrete invariances more in Chapter 12. Such an interaction has not been observed.]

8.5.5 In terms of the principal quantum number n, there is a $2n^2$ degeneracy of the nth (unperturbed) energy level in the hydrogen atom. For $n=2$, this means there are eight linearly independent states with the same energy. Specify the quantum numbers of these eight states using:
(a) The "uncoupled basis": $|\ell,s;m_\ell,m_s\rangle$.
(b) The "coupled basis": $|\ell,s;j,m\rangle$.

8.6.1 Specify the quantum numbers of all the unperturbed states of the hydrogen atom that have $n=3$ and $=1$, using:
(a) The "uncoupled basis": $|\ell,s;m_\ell,m_s\rangle$.
(b) The "coupled basis": $|\ell,s;j,m\rangle$.
(c) An additional piece of the Hamiltonian is added to $H_0\ (=\vec{p}^2/2m-Ze^2/r)$ that has the form

$$H'=\frac{eB_z}{2mc}(L_z+2S_z),$$

which represents the interaction of the atomic electron with a constant external magnetic field, B_z, pointing along z. Find the effect of H' on the degenerate $n=3$, $\ell=1$ energy levels. [You will have to make an appropriate choice of basis to do this calculation. This is called the "Zeeman effect."]

8.6.2 In addition to the *fine structure* of hydrogen discussed in the present chapter, there is an additional so-called *hyperfine interaction* caused by the interaction of the proton's spin ($\vec{S}_p$) with the electron's spin ($\vec{S}_e$). For s-states, the value of this is given by (the expectation value over the radial part of the wave function has been taken)

$$H_{hf}\equiv\frac{4\pi g_p e^2}{3m_e m_p c^2}\vec{S}_p\cdot\vec{S}_e\,|u_s(0)|^2,$$

where $|u_s(0)|^2$ is the absolute square of the s-state wave function evaluated at the origin. (m_e and m_p are the electron and proton masses, and g_p is the proton's "gyromagnetic ratio," approximately equal to 5.59.) Remember, that both electron and proton have spin 1/2. (Assume $[(S_p)_i,(S_e)_j]=0$.)

(a) Which of the states, the coupled ($|s_e, s_p; s, m\rangle$) or uncoupled ($|s_e, s_p; m_e, m_p\rangle$), are the appropriate ones to use in first-order perturbation theory? Explain your choice.
(b) Find the energy shifts caused by H_{hf} on the 1s state hydrogen energy levels. [Hint: Note that $|u_s(0)|^2 = 1/(\pi a_0^3)$, where a_0 is the Bohr radius, for the 1s state of hydrogen, according to Equation 7.264 in Chapter 7. The famous "21 cm line" (its wavelength) in astronomy originates from transitions between the two energy levels you should find.]

Spin and Statistics

Synopsis: The connection between a particle's spin and the statistics it obeys is given. Bose–Einstein and Fermi–Dirac statistics are explained and illustrated with products of wave functions. A new type of particle basis state is described, and two possibilities for the commutation properties of the associated raising ("creation") and lowering ("annihilation") ladder operators are examined, giving rise to the Bose–Einstein and Fermi–Dirac cases. The relation between particle occupation number and energy is inferred from detailed balance, and cubical enclosures of particles are studied. Two physics mysteries encountered in the earlier chapters (*zweideutigkeit* and the vanishing electronic component of specific heat) are explained. The Hartree–Fock theoretical framework for the approximate treatment of many-particle atomic systems is constructed.

9.1 THE CONNECTION BETWEEN SPIN AND STATISTICS

There are two subjects previously mentioned involving apparently unrelated phenomena that I would like to bring back to our attention. One is the *zweideutigkeit* of Chapter 7. We encountered the rule "at most two neutrons and two protons in each energy level" in connection with the simple model of the nucleus presented there. We also encountered the rule "at most two electrons to each energy level" in the atomic model presented. Related to this was the observation in the Kronig–Penney model of Chapter 3 that there was apparently a principle limiting the occupancy of conductor and insulator energy levels. The other subject was brought up at the very beginning of this course in regard to experimental indications of a need for a new type of mechanics to replace Newtonian mechanics for microscopic systems. We had defined the molar specific heat at constant volume by

$$c_V = \left(\frac{\partial \bar{E}}{\partial T}\right)_V, \tag{9.1}$$

where $\bar{E}$ was the average internal energy per mole and T the temperature.

We saw that the universal prediction of Dulong–Petit,

$$c_V = 3R, \tag{9.2}$$

($R = kN_a$) did not hold for all materials, especially diamond. The law did seem to hold true for copper and silver, at least near room temperature. However, I pointed out that there is still a paradox associated with these materials. If each atom of copper or silver gave up one or more valence electrons, we would expect there to be an electronic component to the specific heat. This is not observed at room temperatures.

The origin of these two mysteries can be explained, as we will see, by a deep connection found in nature between a particle's spin and the type of statistics obeyed by a collection of identical particles. Specifically, it has been established that:

Systems of identical particles with zero or positive integer spin have symmetrical wave functions and are said to obey Bose–Einstein statistics. Such particles are called bosons.

Systems of identical particles with half-integer spin have antisymmetrical wave functions and are said to obey Fermi–Dirac statistics. Such particles are called fermions.

This connection between the spin of a particle (which might be composite) and the wave function of the system is a cornerstone of relativistic quantum mechanics. A consistent relativistic description in fact *requires* such a connection. The connection between statistics and spin was first formulated by Wolfgang Pauli in 1940.

9.2 BUILDING WAVE FUNCTIONS WITH IDENTICAL PARTICLES

What is meant by the preceding statements regarding symmetrical and antisymmetrical wave functions? Let us consider several simple examples. First, consider a system of two identical particles that are, however, in distinct quantum states. These are individually described by the *single-particle states*:

$$\text{Particle 1}: |a'\rangle_1$$

$$\text{Particle 2}: |a''\rangle_2.$$

The labels a' and a'' ($a' \neq a''$) stand for some set of quantum numbers or properties. These could include, for example, $\vec{J}^2$, J_z, energy, etc. We have temporarily labeled the kets associated with the first or second particle with a subscript. This composite state will be denoted by

$$|a'\rangle_1 |a''\rangle_2 \equiv |a', a''\rangle. \tag{9.3}$$

However, if the particles are indistinguishable, it is not possible to know which particle is in a given state. Therefore, another possible state of the system is specified by

$$\text{Particle 1}: |a''\rangle_1$$

$$\text{Particle 2}: |a'\rangle_2.$$

The composite state is

$$|a''\rangle_1 |a'\rangle_2 \equiv |a'',a'\rangle. \tag{9.4}$$

The true physical composite state of the system, which can be neither Equation 9.3 nor 9.4 since they distinguish between the particles, must somehow be a mixture of these two possibilities. Quantum mechanics says that the physically realizable states of such a system depend on the spin of the particles involved. If we are dealing with two identical bosons, the true composite state would be symmetric under the interchange of particle labels:

$$\left|\overset{a'}{1},\overset{a''}{1}\right\rangle_b = \frac{1}{\sqrt{2}}\left(|a',a''\rangle + |a'',a'\rangle\right).$$

If we are dealing with two identical fermions, the state would be antisymmetric under label interchange:

$$\left|\overset{a'}{1},\overset{a''}{1}\right\rangle_f = \frac{1}{\sqrt{2}}\left(|a',a''\rangle - |a'',a'\rangle\right).$$

The factors of $1/\sqrt{2}$ are included for normalization. There is also, of course, an arbitrary overall phase involved. The states on the right are single-particle property states, and the states on the left are *particle occupation states.*

Now let us consider the case of three identical particles: two of which are in a quantum state a', and the other being in a state described by a''. The possible states of the system are as follows:

$$|a',a',a''\rangle, \quad |a',a'',a'\rangle, \quad |a'',a',a'\rangle.$$

The bosonic composite state associated with this example is just

$$\left|\overset{a'}{2},\overset{a''}{1}\right\rangle_b = \frac{1}{\sqrt{3}}\left(|a',a',a''\rangle + |a',a'',a'\rangle + |a'',a',a'\rangle\right).$$

This state is symmetric under the interchange of any two particle labels. When we try to construct an antisymmetrical combination for this example, we discover that it cannot be done. This is in fact what will happen any time if more than one particle is in a given state. One cannot build a completely antisymmetric state (under the interchange of any two-particle labels) from composite states, which themselves are already partly symmetric. Physically, for a system of identical fermions, this means that at most one fermion can occupy a given state of the system. This simple fact has enormous consequences in nature and is intimately related to the *zweideutigkeit* and suppressed electronic component of specific heat phenomena discussed earlier.

As a last example, consider a state of three particles in three different quantum states given by a', a'', and a''. The possible distinct combinations are six in number:

$$|a',a'',a'''\rangle, \quad |a'',a''',a'\rangle, \quad |a''',a',a''\rangle,$$
$$|a'',a',a'''\rangle, \quad |a',a''',a''\rangle, \quad |a''',a'',a'\rangle.$$

The completely symmetric and antisymmetric combinations appropriate to a system of bosons or fermions, respectively, are given by (overall phases are not important)

$$\overset{a'a''a'''}{|1,1,1\rangle_b} \equiv \frac{1}{\sqrt{6}}(|a', a'', a'''\rangle + |a'', a''', a'\rangle + |a''', a', a''\rangle$$
$$+|a'', a', a'''\rangle + |a', a''', a''\rangle + |a''', a'', a'\rangle).$$

and

$$\overset{a'a''a'''}{|1,1,1\rangle_f} \equiv \frac{1}{\sqrt{6}}(|a', a'', a'''\rangle + |a'', a''', a'\rangle + |a''', a', a''\rangle$$
$$-|a'', a', a'''\rangle - |a', a''', a''\rangle - |a''', a'', a'\rangle).$$

One can recover the results of the previous example of a symmetric combination of three identical particles in two quantum states a' and a'' by setting $a''' = a'$ in the first of these (outside of normalization). However, when we try to set two quantum states equal in the antisymmetric combination, we get a complete cancellation of terms, telling us that in fact no such state exists.

The number of possible particle occupation states of n particles in g separate single-particle quantum states is given by the expressions ("!" is the factorial symbol and $0! = 1$)

$$\binom{n+g-1}{n} = \frac{(n+g-1)!}{n!\,(g-1)!} \text{ for bosons}$$

and

$$\binom{g}{n} = \frac{(g)!}{n!(g-n)!} \text{ for fermions.}$$

The objects on the left are *binomial coefficients*. The binomial coefficient $\binom{n_1}{n_2}$ represents the number of ways n_2 objects can be chosen among n_1 objects ($n_1 \geq n_2$). This coefficient can be justified by counting the number of ways n particles can be arranged among g states, and thus $g-1$ partitionings (giving $n+g-1$ total objects), in the boson case. In the fermion case, each of the states can have only 0 or 1 particles, so the number of arrangements is just the number of ways n things can be chosen from g states. For example, the first case considered earlier has $n=2$, $g=2$, for which the number of boson states is 3 ($|2, 0\rangle_b$, $|1, 1\rangle_b$, $|0, 2\rangle_b$) and the number of fermion states is just one ($|1, 1\rangle_f$). Notice that the fermion expression makes no sense for $n > g$.

In the earlier examples we are normalizing these states consistently. The normalization factors are just given by the inverse square root of the number of distinct terms in the sums. If we let n_i = number of particles in quantum state i, and we let $n = \sum_i n_i$ be the total number of particles in the system, then clearly we have that

$$\#\text{ terms in F} - \text{D state} = n!, \tag{9.5}$$

and the normalization factor for a system of n fermions will be $1/\sqrt{n!}$. For bosons, we can show

$$\#\ \text{terms in B}-\text{E state} = \frac{n!}{\prod_i n_i!}. \tag{9.6}$$

Notice that

$$\frac{n!}{\prod_i n_i!} \leq n!. \tag{9.7}$$

The only time the equality holds is when $n_i = 1$ for all i. Equation 9.6 is a consequence of the grouping that occurs when n_i originally distinct particles are forced to be in a symmetric state. For example, the symmetric state of three particles in distinct states involves $3! = 6$ terms, as earlier. When two of the particles occupy the same state, however, we get an expression with $3!/2! = 3$ terms, which is specified by Equation 9.6 with $n = 3$, $n_1 = 2$, $n_2 = 1$. ($n = n_1 + n_2$.) Generalizing these results, one can show that the symmetric state of n identical bosons, with quantum state occupation numbers n_i, is given by

$$|n_1, n_2, \ldots\rangle_b \equiv \left(\frac{n!}{\prod_i n_i!} \right)^{-1/2} \sum_{\substack{\text{Distinct} \\ \text{permutations}}} |\overbrace{a_1, a_1, \ldots}^{n_1}, \overbrace{a_2, a_2, \ldots}^{n_2}, \ldots\rangle, \tag{9.8}$$

whereas for fermions, we can take it to be given by the completely antisymmetric expression,

$$|\overbrace{1,1,1,\ldots,1}^{n},0,0,\ldots\rangle_f \equiv \frac{1}{\sqrt{n!}} \det \begin{pmatrix} |a_1\rangle_1 & |a_1\rangle_2 & . & . & . & |a_1\rangle_n \\ |a_2\rangle_1 & |a_2\rangle_2 & . & . & . & |a_2\rangle_n \\ . & . & & & & . \\ . & . & & & & . \end{pmatrix}, \tag{9.9}$$

called the *Slater determinant.* (Technically speaking, these expressions assume that the number of allowed states of the composite system is finite.)

9.3 PARTICLE OCCUPATION BASIS

We may view these general expressions for boson or fermion wave functions as defining a new set of basis states. Instead of talking about the physical attributes of particles labeled as 1, 2, 3, … , n, one now talks about having a system of n_1 particles of type 1, n_2 particles of type 2, etc., with $\sum_i n_i = n$. That is, instead of specifying the properties of numbered particles, we count the number of particles with a specified property. We will call this new orthonormal basis a *particle occupation basis.* It is being denoted as

$$|n_1, n_2, n_3, \ldots\rangle.$$

It will be assumed to be complete in the usual sense,

$$\sum_{n_1, n_2, \ldots} |n_1, n_2, \ldots\rangle\langle n_1, n_2, \ldots| = 1,$$

and is built up out of the single-particle basis states as described before. As this form suggests, the eigenvalues of this basis are not particle *properties* but particle *numbers*.

We define an operator, $a_i^\dagger$, such that for any state $|...n_i...\rangle$ (the symbol "~" means "proportional to"),

$$a_i^\dagger \,|...n_i...\rangle \sim |...n_i+1...\rangle. \tag{9.10}$$

Taking the Hermitian conjugate of Equation 9.10 and multiplying on the right by a general state $|...\, n_i' \,...\rangle$, we see that the only nonzero result happens when $n_i' = n_i + 1$. This means that

$$a_i \,|...n_i+1...\rangle \sim |...n_i...\rangle. \tag{9.11}$$

We supplement this with the additional definition

$$a_i \,|...0_i...\rangle = 0, \tag{9.12}$$

where 0_i labels the *zero-particle* or *vacuum state* for the particle of type i.

Since the effect of the operator $a_i^\dagger$ is to increase the particle occupation number of state i by one, it is called the *creation operator*. Similarly, since from Equation 9.11 we see that a_i reduces the occupation number of state i by one, it is called the *annihilation operator*. These statements are made in the sense of $a_i^\dagger$ and a_i acting on kets. When acting on bras, we have that

$$\langle...n_i...|\, a_i \sim \langle...n_i+1...|, \tag{9.13}$$

$$\langle...n_i+1...|a_i^\dagger \sim \langle...n_i...|, \tag{9.14}$$

and

$$\langle...0_i...|\, a_i^\dagger = 0, \tag{9.15}$$

which follow by taking the Hermitian conjugate of Equations 9.10 through 9.12.

Now from Equations 9.10 and 9.11, we know that

$$a_i^\dagger a_i \,|...n_i...\rangle \sim |...n_i...\rangle. \tag{9.16}$$

For the state $n_i = 0$, we have

$$a_i^\dagger a_i \,|...0_i...\rangle = 0, \tag{9.17}$$

which follows from Equation 9.12. Based on Equations 9.16 and 9.17, we define the effect of the *number operator*,

$$N_i \equiv a_i^\dagger a_i, \tag{9.18}$$

on the occupation basis to be

$$N_i|...n_i...\rangle = n_i|...n_i...\rangle. \tag{9.19}$$

There are as many such operators as there are states in the system. The total number of operators is given by

$$N = \sum_i N_i. \tag{9.20}$$

Its eigenvalue is n, the total number of particles in the system.

Let us derive some commutation properties of these quantities. Since a_i is an annihilation operator, we have that

$$\begin{aligned} N\, a_i\,|...n_i...\rangle &= (n-1)a_i\,|...n_i...\rangle \\ &= a_i(N-1)\,|...n_i...\rangle. \end{aligned} \tag{9.21}$$

The statement Equation 9.21 being true for any state $|...n_i...\rangle$ then implies

$$[a_i, N] = a_i. \tag{9.22}$$

Since we have that

$$N^\dagger = \left(\sum_i a_i^\dagger a_i\right)^\dagger = N, \tag{9.23}$$

the adjoint of Equation 9.22 is

$$[a_i^\dagger, N] = -a_i^\dagger. \tag{9.24}$$

Now consider

$$N_j\, a_i\,|...n_i...n_j...\rangle = \begin{bmatrix} n_j \\ (n_j - 1) \end{bmatrix} a_i\,|...n_i...n_j...\rangle, \tag{9.25}$$

the top result holding for $i \neq j$ and the bottom result for $i = j$. One can write both results at once by

$$\begin{aligned} N_j\, a_i\,|...n_i...n_j...\rangle &= (n_j - \delta_{ij})a_i\,|...n_i...n_j...\rangle \\ &= a_i(N_j - \delta_{ij})\,|...n_i...n_j...\rangle. \end{aligned} \tag{9.26}$$

Equation 9.26 holding true for all occupation kets implies

$$[a_i, N_j] = a_i \delta_{ij}. \tag{9.27}$$

This is consistent with Equation 9.22 because

$$[a_i, N] = \sum_j [a_i, N_j] = \sum_j \delta_{ij} a_i = a_i. \tag{9.28}$$

Taking the adjoint of Equation 9.27 gives

$$[a_i^\dagger, N_j] = -a_i^\dagger \delta_{ij}. \tag{9.29}$$

Another test of consistency of these relations is to check to see if N_i and N_j, which are assumed to have simultaneous eigenvalues, commute:

$$\begin{aligned}[N_i, N_j] &= [a_i^\dagger a_i, N_j] = a_i^\dagger [a_i, N_j] + [a_i^\dagger, N_j] a_i \\ &= a_i^\dagger a_i \delta_{ij} - a_i^\dagger a_i \delta_{ij} = 0.\end{aligned} \tag{9.30}$$

We now require that

$$a_i^\dagger a_j^\dagger |...n_i...n_j...\rangle = \lambda_{ij} a_j^\dagger a_i^\dagger |...n_i...n_j...\rangle, \tag{9.31}$$

that is, that the states produced by $a_i^\dagger a_j^\dagger$, or alternatively by $a_j^\dagger a_i^\dagger$, be the same except for an overall numerical factor, λ_{ij}. We assume this constant is independent of the occupation numbers n_i, n_j and is symmetric in i and j.*
Then Equation 9.31 implies

$$a_i^\dagger a_j^\dagger = \lambda_{ij} a_j^\dagger a_i^\dagger. \tag{9.32}$$

Since Equation 9.32 is true for all i, j, it leads to

$$(\lambda_{ij})^2 = 1. \tag{9.33}$$

This means we have

$$[a_i, a_j] = [a_i^\dagger, a_j^\dagger] = 0, \tag{9.34}$$

when $\lambda_{ij} = 1$ and

$$\{a_i, a_j\} = \{a_i^\dagger, a_j^\dagger\} = 0, \tag{9.35}$$

when $\lambda_{ij} = -1$. In Equation 9.35, we are encountering the anticommutator:

$$\{A, B\} \equiv AB + BA. \tag{9.36}$$

Now from Equation 9.27, we may write

$$a_i(a_j^\dagger a_j) - (a_j^\dagger a_j) a_i = \delta_{ij} a_i, \tag{9.37}$$

which becomes, with the use of Equations 9.34 and 9.35,

$$[a_i, a_j^\dagger] a_j = \delta_{ij} a_i, \tag{9.38}$$

when $\lambda_{ij} = 1$ and

$$\{a_i, a_j^\dagger\} a_j = \delta_{ij} a_i, \tag{9.39}$$

when $\lambda_{ij} = -1$. Assuming that the action of the operators $[a_i, a_j^\dagger]$ and $\{a_i, a_j^\dagger\}$ does not depend on the occupation numbers of the states in the cases $\lambda_{ij} = 1$ and $\lambda_{ij} = -1$, respectively,† we conclude that

$$[a_i, a_j^\dagger] = \delta_{ij}, \tag{9.40}$$

when $\lambda_{ij} = 1$ and

$$\{a_i, a_j^\dagger\} = \delta_{ij}, \tag{9.41}$$

* Such an assumption need not be made. See the treatment in E. Merzbacher, *Quantum Mechanics*, 3rd ed. (John Wiley, 1998), Chapter 21.

† This assumption may also be avoided; see Merzbacher, *Quantum Mechanics*.

when $\lambda_{ij} = -1$.

Notice that the commutation properties of a_i and $a_i^\dagger$ in Equation 9.40 and the eigenvalue–eigenvector statement for $N_i = a_i^\dagger a_i$ in Equation 9.19 are mathematically equivalent to the ladder operators A and $A^\dagger$ that we defined for the harmonic oscillator in Chapter 3. For those operators we had

$$[A, A^\dagger] = 1, \tag{9.42}$$

$$A^\dagger A|n\rangle = n|n\rangle. \tag{9.43}$$

The difference is in the physical interpretation of these statements. Equations 9.42 and 9.43 are results for the *energy levels* of a single particle. Equations 9.19 and 9.40 refer to the *occupation numbers* of a single physical state. However, because these mathematical systems are identical, we may take over the results previously derived. In particular, we showed in Chapter 3 that

$$A|n\rangle = \sqrt{n}\ |n-1\rangle, \tag{9.44}$$

$$A^\dagger|n\rangle = \sqrt{n+1}\ |n+1\rangle, \tag{9.45}$$

and

$$|n\rangle = \frac{(A^\dagger)^n}{\sqrt{n!}}\ |0\rangle, \tag{9.46}$$

where the state $|0\rangle$ represents the ground state. Similarly, we have that

$$a_i|\ldots n_i \ldots\rangle = \sqrt{n_i}\ |\ldots n_i - 1 \ldots\rangle, \tag{9.47}$$

$$a_i^\dagger|\ldots n_i \ldots\rangle = \sqrt{n_i + 1}\ |\ldots n_i + 1 \ldots\rangle, \tag{9.48}$$

and

$$|\ldots n_i \ldots\rangle = \frac{(a_i^\dagger)^{n_i}}{\sqrt{n_i!}}\ |\ldots 0_i \ldots\rangle, \tag{9.49}$$

for each physical property labeled by i. We identify this situation ($\lambda_{ij} = 1$) as the case of Bose–Einstein statistics. The most familiar particle to which these considerations apply is the photon, but the creation and annihilation operators of any zero or integer spin particle must obey the same relations.

9.4 MORE ON FERMI–DIRAC STATISTICS

We have not encountered a system before for which anticommutation relations like Equations 9.35 and 9.41 hold. Notice that if $i = j$, Equation 9.35 implies that

$$(a_i^\dagger)^2 = 0. \tag{9.50}$$

Thus, any attempt to put two or more particles in the same quantum state fails in the case $\lambda_{ij}=-1$. We immediately recognize this situation as describing Fermi–Dirac statistics. We expect for this case that all the occupation numbers, n_i, of the state $|...n_i...\rangle$ can only take on values 0 or 1. This is confirmed from the algebra since from Equation 9.41 for $i=j$

$$a_i a_i^\dagger + a_i^\dagger a_i = 1, \tag{9.51}$$

$$\Rightarrow \quad a_i a_i^\dagger = 1 - N_i. \tag{9.52}$$

Now, by multiplying on the left by $N_i = a_i^\dagger a_i$ we learn that

$$N_i(1-N_i)=0. \tag{9.53}$$

Applying this null operator on any state $|...n_i...\rangle$ then gives us that

$$n_i = \{0,1\} \tag{9.54}$$

for all i.

To keep things simple, we will restrict our attention to the systems of one and two physical properties in the Fermi–Dirac case. We hypothesize that for the simplest case of a single physical property ($\langle 0|0\rangle = \langle 1|1\rangle = 1$)

$$a^\dagger|0\rangle = C_1|1\rangle, \tag{9.55}$$

$$\Rightarrow \quad \langle 0|a = C_1^*\langle 1|, \tag{9.56}$$

and

$$a|1\rangle = C_2|0\rangle, \tag{9.57}$$

$$\Rightarrow \quad \langle 1|a^\dagger = C_2^*\langle 0|. \tag{9.58}$$

From the number operator relation $a^\dagger a|1\rangle = |1\rangle$, we then have that

$$C_2 C_1 = 1. \tag{9.59}$$

Multiplying Equation 9.58 on the right by $a|1\rangle$ also gives us

$$|C_2|^2 = 1. \tag{9.60}$$

Ignoring a possible phase, both of these relations are satisfied if we choose

$$C_1 = C_2 = 1, \tag{9.61}$$

resulting in

$$a^\dagger|0\rangle = |1\rangle, \tag{9.62}$$

$$a|1\rangle = |0\rangle. \tag{9.63}$$

These statements imply that

$$a = |0\rangle\langle 1|, \tag{9.64}$$

$$\Rightarrow \quad a^\dagger = |1\rangle\langle 0|. \tag{9.65}$$

In our original language of measurement symbols from Chapter 1, we would say that $a = |01|$ and $a^\dagger = |10|$. However, unlike our previous applications, we are not specifying physical properties but occupation numbers in the states. Thus, this simplest fermion system is just another manifestation of the two-physical-outcomes formalism of Chapter 1. Using Equations 9.64 and 9.65 in

$$aa^\dagger + a^\dagger a = 1 \tag{9.66}$$

reveals this as just the statement of completeness for this two-level system:

$$aa^\dagger + a^\dagger a = \sum_{n=\{0,1\}} |n\rangle\langle n|. \tag{9.67}$$

Raising the complexity a notch, we now consider an identical fermion system of two physical properties. Following the above, we may choose

$$a_1^\dagger|0,0\rangle = |1,0\rangle, \tag{9.68}$$

$$a_2^\dagger|0,0\rangle = |0,1\rangle, \tag{9.69}$$

and

$$a_1|1,0\rangle = |0,0\rangle, \tag{9.70}$$

$$a_2|0,1\rangle = |0,0\rangle. \tag{9.71}$$

Notice that

$$a_1^\dagger a_2^\dagger\ |0,0\rangle = -a_2^\dagger a_1^\dagger\ |0,0\rangle, \tag{9.72}$$

because of Equation 9.35. There are now two possible definitions of the state $|1,1\rangle$:

$$|1,1\rangle = \pm a_2^\dagger a_1^\dagger\ |0,0\rangle. \tag{9.73}$$

We see that the *order* in which the states are raised or lowered is of crucial importance in this case. In general, one may choose for identical fermions that

$$a_i|\ldots n_i\ldots\rangle = \begin{cases} 0, & n_i = 0 \\ \pm\ |\ldots 0_i\ldots\rangle, & n_i = 1 \end{cases} \tag{9.74}$$

and therefore

$$a_i^\dagger|\ldots n_i\ldots\rangle = \begin{cases} \pm\ |\ldots 1_i\ldots\rangle, & n_i = 0 \\ 0, & n_i = 1 \end{cases}. \tag{9.75}$$

(Equations 9.74 and 9.75 preserve the statement $N_i\,|...n_i...\rangle = n_i\,|...n_i...\rangle$ for $n_i = 0$ or 1.) Rather than specifying an arbitrary (but eventually necessary) convention to fix the signs in Equations 9.74 and 9.75, we will find it possible to live peaceably with this ambiguity in the present study.

Fermi–Dirac statistics explains the *zweideutigkeit* of atomic and nuclear physics. In the early days of atomic spectroscopy, physicists were unaware of electron spin, and it appeared as if an unexplained rule "two electrons per energy level" was operating. Actually, since electrons as spin-1/2 particles obey Fermi–Dirac statistics, atomic systems are really built up with one electron per *state*, but the two orientations of electron spin are very nearly degenerate in energy. The fact that electrons, neutrons, and protons are all spin-1/2 particles determines to a large extent the nature of our universe. If these particles had integer spin, they would obey Bose–Einstein statistics, and there would be no restriction on the number of particles in a given state. The ground states of all such systems would closely resemble one another, removing the incredible diversity seen in atomic and nuclear systems.

9.5 INTERACTION OPERATOR AND FEYNMAN DIAGRAMS

Other useful quantities can be built out of the occupation basis creation and annihilation operators. The total particle number operator

$$N = \sum_{i=\text{all states}} a_i^\dagger a_i, \tag{9.76}$$

can be generalized to the form

$$F^{(1)} = \sum_{i,j} a_i^\dagger a_j f_{ij}, \tag{9.77}$$

to represent various additive single-particle properties. We interpret the f_{ij} as being matrix elements of some single-particle property operator, f:

$$f_{ij} = \langle i \,|f|\, j \rangle. \tag{9.78}$$

For example, if we set $f = H$ where H is the single-particle Hamiltonian, then

$$f_{ij} = \langle i \,|H|\, j \rangle = \varepsilon_i \delta_{ij}, \tag{9.79}$$

and $F^{(1)}$ will represent the total energy operator of the system:

$$F^{(1)}|\ldots n_k \ldots\rangle = \left(\sum_i \varepsilon_i n_i \right) |\ldots n_k \ldots\rangle. \tag{9.80}$$

Another useful operator in this basis that can be used to represent an additive two-particle property (like potential energy) is of the form

$$F^{(2)} = \frac{1}{2} \sum_{i,j,k,\ell} a_i^\dagger a_j^\dagger a_k a_\ell \; V_{ijk\ell}, \tag{9.81}$$

where, again, the $V_{ijk\ell}$ is a set of matrix elements. Notice the conventional factor of 1/2, which is included to avoid double counting because of a symmetry of the $V_{ijk\ell}$ (upcoming in Equation 9.86). We choose the $V_{ijk\ell}$ as matrix elements of some two-particle operator, V,

$$V_{ijk\ell} \equiv \langle i, j \,|V|\, k, \ell \rangle, \tag{9.82}$$

where as usual

$$\left.\begin{aligned} |i,j\rangle &= |i\rangle_1 \otimes \, |j\rangle_2, \\ \langle i,j| &= \langle i|_1 \otimes \, \langle j|_2. \end{aligned}\right\}. \tag{9.83}$$

We have previously used a two-particle basis in describing the deuteron in Chapter 7. A simple example of a two-body operator would be $V = V(r)$, where r is the relative distance operator between the two particles. (A more realistic but also more complicated potential in that case would also include the previously mentioned tensor force, which depends on the relative spin orientations of the two particles.) A diagrammatic way of visualizing the two-particle matrix element $V_{ijk\ell}$ is presented in Figure 9.1.

The particle to the left has been labeled as particle 1, and the dotted line in the diagram drawn perpendicularly to the time axis, represents the interaction of the two particles through the two-body potential, V. The time order of events associated with the matrix element $\langle i,j\,|\,V\,|\,k,\ell\rangle$ itself is read from right to left in this interpretation. (The same implicit time ordering of events occurs in our earlier right to left interpretation of measurement symbols in Chapter 1. The time direction in the associated Process Diagrams is also right to left, as opposed to the bottom to top time direction choice in Figure 9.1.) This picture of the interaction of two particles is called a *Feynman diagram* and can be used to represent a weak scattering event between two particles. Figure 9.1 is actually a nonrelativistic interpretation of such a diagram. In the usual relativistic diagram, the interaction, V, would not necessarily take place at a single instant in time, but would be summed over all time intervals.

As previously mentioned, these matrix elements satisfy certain identities. First, from the meaning of Hermitian conjugation, we have

$$\langle i,j|\,V|k,\ell\rangle^* = \langle k,\ell|\,V^\dagger|i,j\rangle. \tag{9.84}$$

If $V = V^\dagger$, then

$$\langle i,j|\,V|k,\ell\rangle^* = \langle k,\ell|\,V|i,j\rangle. \tag{9.85}$$

Another identity follows from the fact that a relabeling of identical particles (which we assume these are) does not affect the value of the matrix element $V_{ijk\ell}$. Thus, under the substitution $1 \leftrightarrow 2$ we have

$$\langle i,j|\,V|\,k,\ell\rangle = \langle j,i|\,V|\,\ell,k\rangle. \tag{9.86}$$

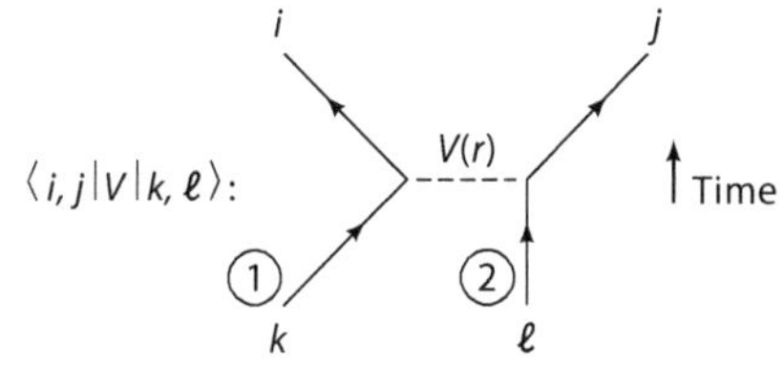

FIGURE 9.1 A scattering event between particles 1 and 2, mediated by the potential, V. The positive time axis is upward, and the interaction is considered to be instantaneous. The particle numbers are circled in this and following figures. Only the *directions* of the lines have any significance here.

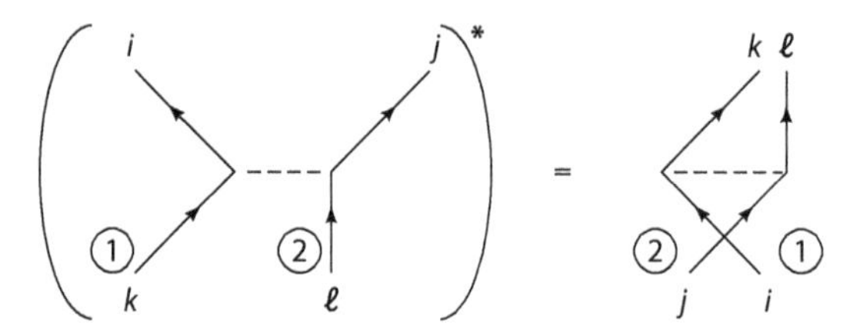

FIGURE 9.2 Diagrammatic interpretation of Equation 9.85 in terms of scattering states. The State 1 interaction side is placed to the left of 2 to mimic the matrix element involved.

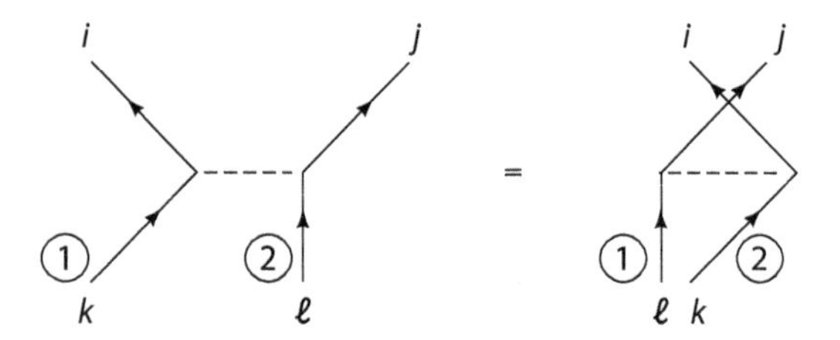

FIGURE 9.3 Similar to Figure 9.2, but illustrating Equation 9.86.

A diagrammatic interpretation of Equation 9.85 is given in Figure 9.2. The statement Equation 9.86 can be visualized as in Figure 9.3.

We are imagining that the i, j, k, ℓ include the momentum state of the particle that, if we think of these particles as being contained in a box, will be discrete.

9.6 IMPLICATIONS OF DETAILED BALANCE

We are getting closer to an application of these ideas.* Consider a gas consisting of many particles. The characteristics of such systems depend on the statistics of the particles involved. We will study such systems under the assumption of *detailed balance*. To explain what this means, consider an interaction between two distinguishable particles as illustrated in Figure 9.4 (1, 1′, 2, and 2′ are particular distinct values of the state variables i, j, k, and ℓ). In the present (classical) context, the rate for this reaction to be taking place somewhere in our gas will be proportional to the product of the number of particles of types 1 and 1′ present:

$$\text{rate}(1+1' \rightarrow 2+2') \sim n_1 n_{1'}. \tag{9.87}$$

The rate for the reversed collision, shown in Figure 9.5, is also given by

$$\text{rate}(2+2' \rightarrow 1+1') \sim n_2 n_{2'}. \tag{9.88}$$

Detailed balance implies that

$$\text{rate}(1+1' \rightarrow 2+2') = \text{rate}(2+2' \rightarrow 1+1'). \tag{9.89}$$

* The argument in this section is an expanded and illustrated version of a section taken from my 1979 Schwinger quantum mechanics notes.

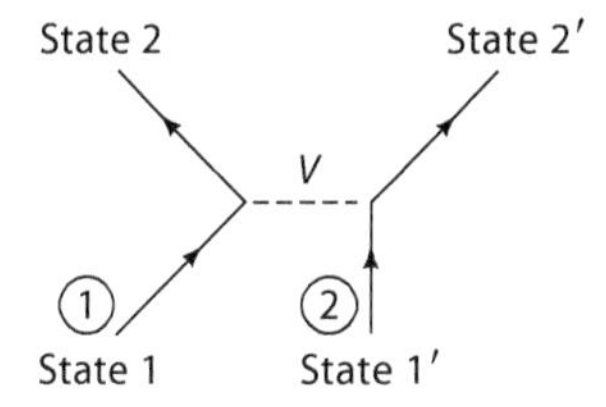

FIGURE 9.4 A scattering interaction where particle 1 goes from State 1 to State 2, while particle 2 goes from State 1′ to State 2′.

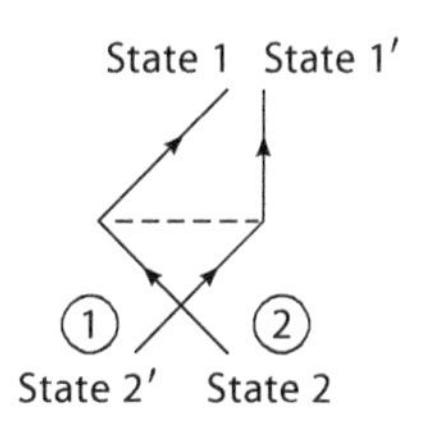

FIGURE 9.5 The reversed interaction to Figure 9.4.

That is, in equilibrium the rates of a reaction and its inverse must be equal. Assuming the same proportionality constants in Equations 9.87 and 9.88, this means that

$$n_1 n_{1'} = n_2 n_{2'}, \tag{9.90}$$

which also implies

$$\ln\frac{1}{n_1} + \ln\frac{1}{n_{1'}} = \ln\frac{1}{n_2} + \ln\frac{1}{n_{2'}}. \tag{9.91}$$

Equation 9.91 has the appearance of a conservation law for some scalar quantity. It is plausible that Equation 9.91 is an expression of energy conservation for this situation. Then, comparing Equation 9.91 with such a statement,

$$\varepsilon_1 + \varepsilon_{1'} = \varepsilon_2 + \varepsilon_{2'}, \tag{9.92}$$

the connection suggests that for any species i

$$\ln\frac{1}{n_i} = \alpha + \beta\varepsilon_i, \tag{9.93}$$

where α and β are unknown constants, or that

$$n_i = e^{-\alpha-\beta\varepsilon_i}. \tag{9.94}$$

Equation 9.94 becomes an expression of the classical statistics of distinguishable particles, Maxwell–Boltzmann statistics (see Chapter 1), if we take

$$\beta = \frac{1}{kT}, \tag{9.95}$$

where k is the Boltzmann constant and T the absolute temperature. The quantity α in Equation 9.94, called the *chemical potential*, can be thought of as a scale factor fixed by the number of particles, n, in the system:

$$\sum_i n_i = \sum_i e^{-\alpha-\beta\varepsilon_i} = n. \tag{9.96}$$

Although this argument seems sketchy, it has led to correct conclusions about the type of statistics obeyed by distinguishable particles. Actually, all we are requiring is that detailed balance and energy conservation be consistent statements. Let us see if we can repeat it for the two quantum mechanical cases of indistinguishable particles.

Using V in Equation 9.81 as an expression of the two-body potential between identical particles, we relabel

$$F^{(2)} \to H_{\text{int}} = \frac{1}{2}\sum_{i,j,k,\ell} a_i^\dagger a_j^\dagger a_k a_\ell \, \langle i,j|V|k,\ell\rangle. \tag{9.97}$$

The key to this discussion is the relation

$$\text{rate}(1 \to 2) \sim |\langle 2|H_{\text{int}}|1\rangle|^2, \tag{9.98}$$

which is a partial statement of what is called "Fermi's golden rule." This rule comes about from time-dependent perturbation theory, which we will study in detail in Chapter 10. The states $|1\rangle$ and $|2\rangle$ are the appropriate initial and final particle occupation basis states, respectively, for the reaction in question. To find the effect of H_{int} on the particle occupation states, let us go back to our previous expressions for the effects of creation and annihilation operators. From Equations 9.47 and 9.48,

$$a_2^\dagger a_1|n_1,n_2\rangle = \sqrt{n_2+1}\sqrt{n_1}\;|n_1-1,n_2+1\rangle, \tag{9.99}$$

for the Bose–Einstein case. We may make a similar statement for Fermi–Dirac statistics. We may write both statements in Equation 9.74 as

$$a_i|\ldots n_i\ldots\rangle = \pm\sqrt{n_i}|\ldots n_i - 1\ldots\rangle, \tag{9.100}$$

where $n_i = 0, 1$ only, and we see the usual Fermi–Dirac sign ambiguity. In addition, Equation 9.75 may be written as

$$a_i^\dagger|\ldots n_i\ldots\rangle = \pm\sqrt{1-n_i}|\ldots n_i + 1\ldots\rangle. \tag{9.101}$$

Putting Equations 9.100 and 9.101 together we then get

$$a_2^\dagger a_1|n_1,n_2\rangle = \pm\sqrt{1-n_2}\sqrt{n_1}|n_1-1,n_2+1\rangle, \tag{9.102}$$

similar to Equation 9.99 for Bose–Einstein particles. We write Equations 9.99 and 9.102 together as

$$a_2^\dagger a_1|n_1,n_2\rangle = (\text{S.F.})\;\sqrt{1\pm n_2}\sqrt{n_1}|n_1-1,n_2+1\rangle, \tag{9.103}$$

where the top sign is taken for bosons and the bottom sign for fermions. The quantity "(S.F.)" (meaning "sign factor") is only to be considered for fermions. Equation 9.103 implies that

$$a_2^\dagger a_{2'}^\dagger a_1 a_{1'} |n_1, n_{1'}, n_2, n_{2'}\rangle = (\text{S.F.}) \sqrt{1 \pm n_2}\sqrt{1 \pm n_{2'}} \times \sqrt{n_1}\sqrt{n_{1'}}\, |n_1 - 1, n_{1'} - 1, n_2 + 1, n_{2'} + 1\rangle, \tag{9.104}$$

where the $\pm$ signs are to be interpreted as earlier.

We are now in a position to partially determine the rate for the reaction $1 + 1' \to 2 + 2'$:

$$\text{rate}(1 + 1' \to 2 + 2') \sim |\langle n_1 - 1, n_{1'} - 1, n_2 + 1, n_{2'} + 1 | H_{\text{int}} | n_1, n_{1'}, n_2, n_{2'}\rangle|^2. \tag{9.105}$$

We are assuming in general that there are n_1, $n_{1'}$, n_2, $n_{2'}$ particles in the distinct modes 1, 1′, 2, 2′, respectively, in the initial state. Letting the labels i, j, k, and ℓ in Equation 9.97 take on all possible values, we have that

$$\begin{aligned}
&\langle n_1 - 1, n_{1'} - 1, n_2 + 1, n_{2'} + 1 | H_{\text{int}} | n_1, n_{1'}, n_2, n_{2'}\rangle \\
&= \frac{1}{2}\langle n_1 - 1, n_{1'} - 1, n_2 + 1, n_{2'} + 1 | \Big\{ a_2^\dagger a_{2'}^\dagger a_1 a_{1'}\ \langle 2,2'|V|1,1'\rangle \\
&+ a_2^\dagger a_{2'}^\dagger a_{1'} a_1\ \langle 2,2'|V|1',1\rangle + a_{2'}^\dagger a_2^\dagger a_1 a_{1'}\ \langle 2',2|V|1,1'\rangle \\
&+ a_{2'}^\dagger a_2^\dagger a_{1'} a_1\ \langle 2',2|V|1',1\rangle \Big\} |n_1, n_{1'}, n_2, n_{2'}\rangle.
\end{aligned} \tag{9.106}$$

However, from Equation 9.86, we have that

$$\langle 2,2'|V|1,1'\rangle = \langle 2',2|V|1',1\rangle \tag{9.107}$$

and

$$\langle 2,2'|V|1',1\rangle = \langle 2',2|V|1,1'\rangle. \tag{9.108}$$

Also, because we are assuming the modes 1, 1′, 2, 2′ are all distinct, we easily see that

$$a_2^\dagger a_{2'}^\dagger a_{1'} a_1 = \pm a_2^\dagger a_{2'}^\dagger a_1 a_{1'}, \tag{9.109}$$

$$a_{2'}^\dagger a_2^\dagger a_1 a_{1'} = \pm\, a_2^\dagger a_{2'}^\dagger a_1 a_{1'}, \tag{9.110}$$

where the top signs are for bosons and the bottom signs for fermions. For either case,

$$a_{2'}^\dagger a_2^\dagger a_{1'} a_1 = a_2^\dagger a_{2'}^\dagger a_1 a_{1'}. \tag{9.111}$$

The expression in Equation 9.106 now simplifies to

$$\begin{aligned}
&\langle n_1 - 1, n_{1'} - 1, n_2 + 1, n_{2'} + 1 | H_{\text{int}} | n, n_{1'}, n_2, n_{2'}\rangle \\
&= \langle n_1 - 1, n_{1'} - 1, n_2 + 1, n_{2'} + 1 | a_2^\dagger a_{2'}^\dagger a_1 a_{1'} | n_1, n_{1'}, n_2, n_{2'}\rangle \\
&\cdot [\langle 2,2'|V|1,1'\rangle \pm \langle 2,2'|V|1',1\rangle].
\end{aligned} \tag{9.112}$$

Then, using the result of Equation 9.104, we get that

$$\begin{aligned}&\langle n_1-1,n_{1'}-1,n_2+1,n_{2'}+1|H_{\text{int}}|n_1,n_{1'},n_2,n_{2'}\rangle\\&=(\text{S.F.})\ [\langle 2,2'|V|1,1'\rangle\pm\langle 2,2'|V|1',1\rangle]\\&\sqrt{(1\pm n_2)(1\pm n_{2'})n_1n_{1'}}.\end{aligned} \tag{9.113}$$

Using our previous diagrammatic conventions, the quantity in square brackets on the right of Equation 9.113 may be represented as in Figure 9.6. The diagram to the right in Figure 9.6 is often called an *exchange diagram* (or crossed diagram).

Now Fermi's golden rule, Equation 9.105, reads:

$$\begin{aligned}\text{rate}\,(1+1'\rightarrow 2+2')\sim&|\langle 2,2'|V|1,1'\rangle\pm\langle 2,2'|V|1',1\rangle\\&\times(1\pm n_2)(1\pm n_{2'})n_1n_{1'}.\end{aligned} \tag{9.114}$$

The rate for the reverse action is simply given by a relabeling of the above:

$$\begin{aligned}\text{rate}\,(2+2'\rightarrow 1\ +\ 1')\sim&|\langle 1,1'|V|2,2'\rangle\pm\langle 1,1'|V|2',2\rangle|^2\\&\times\ (1\pm n_1)(1\pm n_{1'})n_2n_{2'}.\end{aligned} \tag{9.115}$$

From the identities Equations 9.85 and 9.86 we now have that

$$\langle 2,2'|V|1,1'\rangle^*=\langle 1,1'|V|2,2'\rangle, \tag{9.116}$$

and

$$\langle 2,2'|V|1',1\rangle^*=\langle 1,1'|V|2',2\rangle \tag{9.117}$$

which reveals that

$$|\langle 2,2'|V|1,1'\rangle\pm\langle 2,2'|V|1',1\rangle|^2=|\langle 1,1'|V|2,2'\rangle\pm\langle 1,1'|V|2',2\rangle|^2. \tag{9.118}$$

The statement of detailed balance for this reaction, Equation 9.89, now gives us

$$(1\pm n_2)(1\pm n_{2'})n_1n_{1'}=(1\pm n_1)(1\pm n_{1'})n_2n_2, \tag{9.119}$$

in contrast to the classical result, Equation 9.90.

Assuming the same overall proportionality constant in the rate as in the classical result, we see that Bose–Einstein (B–E) statistics "encourages"

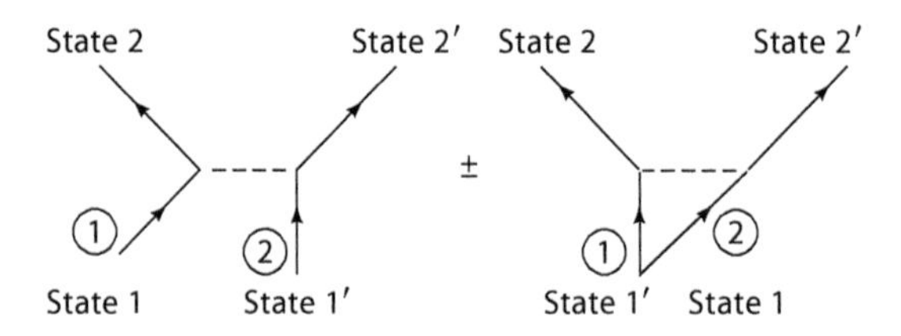

FIGURE 9.6 A diagrammatic illustration of the two terms in square brackets on the right-hand side of Equation 9.113.

the reaction relative to the classical rate, while the minus signs associated with fermions shows that Fermi–Dirac (F–D) statistics "discourages" the reaction by not permitting it if certain states are already occupied.

Performing the same mathematical steps as in the classical case, we first write Equation 9.119 as

$$\ln\left(\frac{1}{n_1} \pm 1\right) + \ln\left(\frac{1}{n_{1'}} \pm 1\right) = \ln\left(\frac{1}{n_2} \pm 1\right) + \ln\left(\frac{1}{n_{2'}} \pm 1\right), \quad (9.120)$$

which we assume is consistent with energy conservation when the system is in equilibrium. Thus, for each species, i, we now have

$$\ln\left(\frac{1}{n_i} \pm 1\right) = \alpha + \beta\varepsilon_i, \quad (9.121)$$

which gives

$$n_i = \frac{1}{e^{\alpha+\beta\varepsilon_i} + 1} \quad (9.122)$$

in the F–D case and

$$n_i = \frac{1}{e^{\alpha+\beta\varepsilon_i} - 1} \quad (9.123)$$

in the B–E case. As before, α is fixed by the total number of particles present and $\beta = (kT)^{-1}$. Notice that both Equations 9.122 and 9.123 go over to the classical result, Equation 9.94, when $e^{\alpha} \gg 1$. We assume that the energies of the nonrelativistic gas particles are given by

$$\varepsilon_i = \frac{\vec{p}_i^{\,2}}{2m}. \quad (9.124)$$

Given the results of Equations 9.94, 9.122, and 9.123 in the classical, Fermi–Dirac, and Bose–Einstein cases, respectively, we then have for the total number and energy of these systems that

$$n = \sum_i n_i, \quad (9.125)$$

$$\varepsilon = \sum_i n_i\varepsilon_i. \quad (9.126)$$

9.7 DENSITY OF STATES EXPRESSIONS

To evaluate these expressions, it will be necessary to replace the sums in Equations 9.125 and 9.126 by approximate integrals. First, let us enumerate exactly the possible momentum states of an ideal gas of noninteracting particles contained in some finite volume, V. We want to take the boundary conditions on the wave functions of the particles in the gas to be as general, and yet as simple, as possible. Let us choose a volume such that it is in the shape of a rectangular enclosure of volume $V = L_1L_2L_3$. A free-particle solution to the three-dimensional Schrödinger equation

$$-\frac{\hbar^2}{2m}\vec{\nabla}^2 u(\vec{x}) = \varepsilon u(\vec{x}), \tag{9.127}$$

normalized such that

$$\int_V d^3x |u(\vec{x})|^2 = 1, \tag{9.128}$$

is given by

$$u(\vec{x}) = \frac{1}{\sqrt{V}} e^{i\vec{p}\cdot\vec{x}/\hbar}. \tag{9.129}$$

We imagine that the volume V under consideration is actually in contact with identical volumes along its boundaries as shown in Figure 9.7.

Given this situation, the appropriate boundary conditions on the wave functions $u_i(\vec{x})$ are

$$\left.\begin{aligned} u(x_1 + L_1, x_2, x_3) &= u(x_1, x_2, x_3) \\ u(x_1, x_2 + L_2, x_3) &= u(x_1, x_2, x_3) \\ u(x_1, x_2, x_3 + L_3) &= u(x_1, x_2, x_3) \end{aligned}\right\}. \tag{9.130}$$

These are just periodic boundary conditions. (We saw these also in the Kronig–Penney model of Chapter 3.) They determine the allowed momenta values in the spatial directions. For example, in the one direction we must have

$$\frac{p_1}{\hbar}(x_1 + L_1) = \frac{p_1}{\hbar} x_1 + 2\pi n_1, \tag{9.131}$$

where $n_1 = 0, \pm 1, \pm 2, \ldots$. Therefore,

$$p_1 = \left(\frac{h}{L_1}\right) n_1. \tag{9.132}$$

Similarly in the two and three directions. Thus, in counting such states, the total number of integers in the range $\Delta n_1 \Delta n_2 \Delta n_3$ is given by

$$\Delta n_1 \Delta n_2 \Delta n_3 = \frac{V}{h^3} \Delta p_1 \Delta p_2 \Delta p_3. \tag{9.133}$$

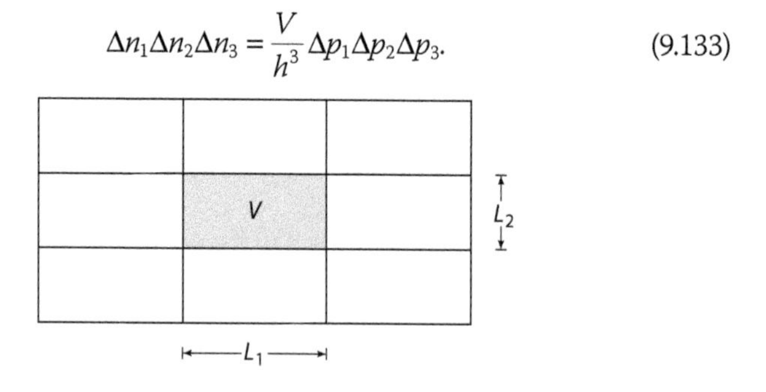

FIGURE 9.7 The rectangular volume, V, filled with a gas, in contact with identical volumes.

If in the expressions Equations 9.125 and 9.126 the sums are such that the summands vary very little over increments of n_1, n_2, and n_3 by unity, then we may make the replacement

$$\sum_{i=\text{all states}} \rightarrow \frac{V}{h^3}\int d^3p = \frac{4\pi V}{h^3}\int dp\, p^2, \tag{9.134}$$

in these expressions. Actually, in the case of particles with spin, the sum over all states must also include a sum over all values of the third component of spin, $S_3' = \hbar m_S$. For noninteracting particles, this just adds a numerical factor:

$$\sum_{i=\text{ all states}} \rightarrow D\frac{V}{h^3}\int d^3p = \frac{4\pi DV}{h^3}\int dp\, p^2 \equiv \int d\Omega_{\vec{p}}\, dE\, \rho_{[\vec{p}]}, \tag{9.135}$$

where $D \equiv 2s+1$ is a degeneracy factor that counts the number of components of S_3'. Equation 9.135 also introduces the idea of a density of states; $\rho_{[\vec{p}]}$ is the instantaneous number of states per energy in the spherical angular interval $d\Omega_{\vec{p}}$. This form is independent of the angular direction. This expression will be useful in scattering situations we will encounter in the next chapter.

Although the results in Equations 9.134 and 9.135 have been justified for very particular boundary conditions that might seem unrealistic, it can be shown that these results are really insensitive to the shape of the container the noninteracting particles are placed in and the exact surface boundary conditions as long as the average de Broglie wavelength of the particles is small compared to the physical dimensions of the container.*

9.8 CREATING AND DESTROYING PHOTONS

The most important single application of the formalism of creation and destruction operators is to the electromagnetic field, where the photons of light are to be treated with Bose-Einstein statistics. We will need to contain them in a large box, as in the last section, in order to enumerate the possible momentum states of a gas of such particles contained in some finite volume, V. We saw photons as particles in Chapter 2 in the Compton effect. Photons have spin 1, and so their creation and annihilation operators are described by Equations 9.47 and 9.48, but they also have two polarization states.†

Let us see what it takes to characterize an electromagnetic field in this context. We will need a small tutorial at this point. We will consider the portion of the box with no electromagnetic sources. (We will encounter

* I recommend the discussion in F. Reif, *Fundamentals of Statistical and Thermal Physics* (Waveland Press Inc., 2008), Section 9.10, on this point.

† I also recommend D. Griffiths, *Introduction to Electrodynamics*, 4th ed. (Cambridge University Press, 2017), Chapter 9, for background on polarization vectors and Chapter 10 for electromagnetic gauge transformations. Note the use of the *Gaussian* system here as opposed to the *Système International* (SI) of this reference for electromagnetic quantities.

such a situation in Chapter 10, Section 10.8.) The vector potential of the photon field in this case is given by a sum over a set of discrete plane waves multiplied by operators, which will describe the creation and destruction of photons in the box. The general form is

$$\vec{A}(\vec{r},t) = \sum_{\vec{k},\alpha}\left(\vec{A}_\alpha(\vec{k})e^{i(\vec{k}\cdot\vec{r}-\omega t)} + \vec{A}_\alpha^\dagger(\vec{k})e^{-i(\vec{k}\cdot\vec{r}-\omega t)}\right), \tag{9.136}$$

where the sum is over the discrete allowed set of box wave numbers, $\vec{k} = \vec{p}/\hbar$, as well as the two polarizations ($\alpha = 1,2$) of the photon with energy $E = \hbar\omega$. Each term in the sum in Equation 9.136 corresponds to a separate photon mode. Note that $\vec{A}^\dagger(\vec{r},t) = \vec{A}(\vec{r},t)$, that is, the vector field is Hermitian or self-conjugate. The quantities $\vec{A}_{\alpha'}(\vec{k}')$ and $\vec{A}_\alpha^\dagger(\vec{k})$ in box normalization are given by

$$\vec{A}_\alpha(\vec{k}) = c\sqrt{\frac{2\pi\hbar}{\omega V}}a_\alpha(\vec{k})\hat{\varepsilon}_{\alpha,\vec{k}}, \quad \vec{A}_\alpha^\dagger(\vec{k}) = c\sqrt{\frac{2\pi\hbar}{\omega V}}a_\alpha^\dagger(\vec{k})\hat{\varepsilon}_{\alpha,\vec{k}}, \tag{9.137}$$

where $a_\alpha^\dagger(\vec{k})$ and $a_\alpha(\vec{k})$ are creation and annihilation operators for photons of wave number $\vec{k}$ and polarization α. Note the volume factors V in the denominator of the square roots in Equation 9.137. The unit vectors $\hat{\varepsilon}_{\alpha,\vec{k}}$ are called *polarization vectors* and $|\vec{k}| = \omega/c$. In our case we will choose real vectors $\hat{\varepsilon}_{1,\vec{k}}$ and $\hat{\varepsilon}_{2,\vec{k}}$ such that the combination $\left(\hat{\varepsilon}_{1,\vec{k}}, \hat{\varepsilon}_{2,\vec{k}}, \hat{k}\right)$ forms a right-handed set of mutually orthogonal coordinate vectors. This type of light polarization is called *linear polarization,* and corresponds to a state where oscillations of the electric field are perpendicular to the photon momentum. Specifically, we have

$$\hat{\varepsilon}_{\alpha,\vec{k}} \cdot \hat{\varepsilon}_{\alpha'\vec{k}} = \delta_{\alpha\alpha'}, \quad \hat{\varepsilon}_{\alpha,\vec{k}} \cdot \hat{k} = 0, \quad \hat{\varepsilon}_{1,\vec{k}} \times \hat{\varepsilon}_{2,\vec{k}} = \hat{k}. \tag{9.138}$$

Note that the middle identity in Equation 9.138 results in

$$\hat{\varepsilon}_{\alpha,\vec{k}} \cdot \hat{k} = 0 \Rightarrow \vec{\nabla} \cdot \vec{A} = 0. \tag{9.139}$$

Equation 9.139 is known as the *Coulomb gauge condition* on $\vec{A}$. It turns out the vector potential in electrodynamics is not unique and Equation 9.139 represents an extra requirement that we can always impose. It also turns out that in the Coulomb gauge when there are no electric charges are present in our box, the scalar potential Φ can always be taken to vanish.

We will need to verify that the expressions for $\vec{A}_{\alpha'}(\vec{k}')$ and $\vec{A}_\alpha^\dagger(\vec{k})$ in Equation 9.137 are normalized correctly in order to give the correct photon energies. For this purpose we will evaluate the electromagnetic expression for the energy of the field, but it must be regarded as an operator in photon number space in this context. This expression is

$$H = \frac{1}{8\pi}\int d^3r\,(\vec{E}\cdot\vec{E} + \vec{B}\cdot\vec{B}), \tag{9.140}$$

where the integration is over the box. The electric and magnetic fields are (note no scalar potential term for the electric field in this context)

$$\vec{E}(\vec{r},t) = -\frac{1}{c}\frac{\partial \vec{A}}{\partial t} = \frac{i}{c}\sum_{\vec{k},\alpha}\omega\left(\vec{A}_\alpha(\vec{k})e^{i(\vec{k}\cdot\vec{r}-\omega t)} - \vec{A}_\alpha^\dagger(\vec{k})e^{-i(\vec{k}\cdot\vec{r}-\omega t)}\right)$$

$$= i\sum_{\vec{k},\alpha}\sqrt{\frac{2\pi\hbar}{\omega V}}\omega\hat{\varepsilon}_{\alpha,\vec{k}}\left(a_\alpha(\vec{k})e^{i(\vec{k}\cdot\vec{r}-\omega t)} - a_\alpha^\dagger(\vec{k})e^{-i(\vec{k}\cdot\vec{r}-\omega t)}\right), \tag{9.141}$$

$$\vec{B}(\vec{r},t) = \vec{\nabla}\times\vec{A} = \frac{i}{c}\sum_{\vec{k},\alpha}\omega\left(\vec{k}\times\vec{A}_\alpha(\vec{k})e^{i(\vec{k}\cdot\vec{r}-\omega t)} - \hat{k}\times\vec{A}_\alpha^\dagger(\vec{k})e^{-i(\vec{k}\cdot\vec{r}-\omega t)}\right)$$

$$= ic\sum_{\vec{k},\alpha}\sqrt{\frac{2\pi\hbar}{\omega V}}\vec{k}\times\hat{\varepsilon}_{\alpha,\vec{k}}\left(a_\alpha(\vec{k})e^{i(\vec{k}\cdot\vec{r}-\omega t)} - a_\alpha^\dagger(\vec{k})e^{-i(\vec{k}\cdot\vec{r}-\omega t)}\right). \tag{9.142}$$

When the square of Equations 9.141 and 9.142 are used in Equation 9.140, this introduces an independent sum over another set of momentums we shall call $\vec{k}'$. The integration rule

$$\int d^3 r\, e^{i(\vec{k}\pm\vec{k}')\cdot\vec{r}} = V\delta_{\vec{k},\mp\vec{k}'}, \tag{9.143}$$

where V is the volume factor, $V = L_1L_2L_3$, now reduces the expressions back to a single momentum sum. The student should verify the statements:

$$\frac{1}{8\pi}\int d^3 r\,\vec{E}\cdot\vec{E} = \frac{\hbar}{4}\sum_{\vec{k},\alpha,\alpha'}\omega\Big[\hat{\varepsilon}_{\alpha,\vec{k}}\cdot\hat{\varepsilon}_{\alpha'\vec{k}}\left(a_\alpha(\vec{k})a_{\alpha'}^\dagger(\vec{k}) + a_{\alpha'}^\dagger(\vec{k})a_\alpha(\vec{k})\right)$$

$$-\hat{\varepsilon}_{\alpha,\vec{k}}\cdot\hat{\varepsilon}_{\alpha',-\vec{k}}\left(a_\alpha(\vec{k})a_{\alpha'}(-\vec{k})e^{-2\omega t} + a_\alpha^\dagger(\vec{k})a_{\alpha'}^\dagger(-\vec{k})e^{2\omega t}\right)\Big], \tag{9.144}$$

$$\frac{1}{8\pi}\int d^3 r\,\vec{B}\cdot\vec{B} = \frac{\hbar}{4}\sum_{\vec{k},\alpha,\alpha'}\frac{c^2}{\omega}\Big[\vec{k}\times\hat{\varepsilon}_{\alpha,\vec{k}}\cdot\vec{k}\times\hat{\varepsilon}_{\alpha'\vec{k}}\left(a_\alpha(\vec{k})a_{\alpha'}^\dagger(\vec{k}) + a_{\alpha'}^\dagger(\vec{k})a_\alpha(\vec{k})\right)$$

$$+\vec{k}\times\hat{\varepsilon}_{\alpha,\vec{k}}\cdot\vec{k}\times\hat{\varepsilon}_{\alpha'-\vec{k}}\left(a_\alpha(\vec{k})a_{\alpha'}(-\vec{k})e^{-2\omega t} + a_\alpha^\dagger(\vec{k})a_{\alpha'}^\dagger(-\vec{k})e^{2\omega t}\right)\Big]. \tag{9.145}$$

The scalar polarization combinations that appear in Equation 9.145 can be rewritten and reduced as

$$\vec{k}\times\hat{\varepsilon}_{\alpha,\vec{k}}\cdot\vec{k}\times\hat{\varepsilon}_{\alpha',\pm\vec{k}} = \vec{k}\cdot\left(\hat{\varepsilon}_{\alpha',\pm\vec{k}}\times\left(\vec{k}\times\hat{\varepsilon}_{\alpha,\vec{k}}\right)\right)$$

$$= \vec{k}\cdot\left(\vec{k}\hat{\varepsilon}_{\alpha,\vec{k}}\cdot\hat{\varepsilon}_{\alpha',\pm\vec{k}} - \hat{\varepsilon}_{\alpha,\vec{k}}\left(\vec{k}\cdot\hat{\varepsilon}_{\alpha',\pm\vec{k}}\right)\right) \tag{9.146}$$

$$= k^2\hat{\varepsilon}_{\alpha,\vec{k}}\cdot\hat{\varepsilon}_{\alpha',\pm\vec{k}}.$$

At this point the second lines of the expressions in Equations 9.144 and 9.145 are seen to cancel, but the first lines add up. The simple result is

$$H = \sum_{\vec{k},\alpha,\alpha'}\frac{\hbar\omega}{2}\hat{\varepsilon}_{\alpha,\vec{k}}\cdot\hat{\varepsilon}_{\alpha',\vec{k}}\left(a_\alpha(\vec{k})a_{\alpha'}^\dagger(\vec{k}) + a_\alpha^\dagger(\vec{k})a_{\alpha'}(\vec{k})\right). \tag{9.147}$$

One of the polarization identities in Equation 9.138 now requires $\alpha = \alpha'$, giving

$$H = \sum_{\vec{k},\alpha} \frac{\hbar\omega}{2}\left(a_\alpha(\vec{k})a_\alpha^\dagger(\vec{k}) + a_\alpha^\dagger(\vec{k})a_\alpha(\vec{k})\right). \tag{9.148}$$

The Bose–Einstein commutator from Equation 9.40 results in

$$H = \sum_{\vec{k},\alpha} \hbar\omega\left(N_\alpha(\vec{k}) + \frac{1}{2}\right), \tag{9.149}$$

where

$$N_\alpha(\vec{k}) = a_\alpha^\dagger(\vec{k})a_\alpha(\vec{k}) \tag{9.150}$$

is the number operator for photons of wave number $\vec{k}$ and polarization α.

We have verified that the forms in Equation 9.137 result in an operator construction that correctly assigns an energy $\hbar\omega$ to each added photon in the box. However, there is a surprise here. When there are no photons, there is still an energy! This is the *zero point energy* of the electromagnetic field. We saw this circumstance first in the harmonic oscillator energy result in Equation 3.169 (Chapter 3). Mathematically, the number operator formalism here is precisely the same as the harmonic oscillator energy formalism of Section 3.5, as we pointed out previously in Section 9.3. That is, each photon degree of freedom is just behaving as an independent harmonic oscillator. One might object that the sum in Equation 9.149 is actually infinite. However, changes in this energy due to boundary conditions on the electromagnetic fields at material surfaces can effect finite changes in this energy, resulting in actual forces. This is termed the *Casimir effect*. Some of the theoretical predictions of zero-point forces in fact have been experimentally verified.* This is fascinating stuff!

We are at the starting point of a long road leading to the relativistic field theory known as quantum electrodynamics (QED), which we will examine again in Chapter 12.

9.9 MAXWELL–BOLTZMANN STATISTICS

What else can we do with this formalism? Let us go back and examine the evaluation of Equations 9.125 and 9.126, using the replacement Equation 9.135, for the cases of Maxwell–Boltzmann, Bose–Einstein, and Fermi–Dirac statistics.

Using Equation 9.94, the expressions for n and ε are as follows (we set $D = 1$, appropriate for a spinless particle):

$$n = \frac{4\pi V}{h^3}\int_0^\infty \mathrm{d}p\, p^2 \mathrm{e}^{-\alpha}\mathrm{e}^{-\beta p^2/2m}, \tag{9.151}$$

$$\varepsilon = \frac{4\pi V}{h^3}\int_0^\infty \mathrm{d}p\, p^2 \mathrm{e}^{-\alpha}\mathrm{e}^{-\beta p^2/2m}\left(\frac{p^2}{2m}\right). \tag{9.152}$$

* S. Lamoreaux, *Phys. Rev. Lett.* **78**, 1 (1997).

Introducing the dimensionless variable

$$x^2 \equiv \frac{\beta p^2}{2m}, \tag{9.153}$$

we can write Equations 9.151 and 9.152 as

$$\frac{n}{V} = \frac{4\pi e^{-\alpha}}{h^3}\left(\frac{2m}{\beta}\right)^{3/2}\int_0^\infty dx\, x^2 e^{-x^2}, \tag{9.154}$$

$$\frac{\varepsilon}{V} = \frac{4\pi e^{-\alpha}}{h^3}\left(\frac{2m}{\beta}\right)^{3/2}\frac{1}{\beta}\int_0^\infty dx\, x^4 e^{-x^2}. \tag{9.155}$$

We recall from Equations 2.53 and 2.54 (see Chapter 2) that

$$\int_0^\infty dx\, e^{-x^2} = \frac{\sqrt{\pi}}{2}. \tag{9.156}$$

Letting $x^2 \to \lambda x^2$, taking derivatives with respect to λ and then setting $\lambda = 1$, we find that

$$\int_0^\infty dx\, x^2 e^{-x^2} = \frac{\sqrt{\pi}}{4}, \tag{9.157}$$

$$\int_0^\infty dx\, x^4 e^{-x^2} = \frac{3\sqrt{\pi}}{8}. \tag{9.158}$$

We therefore have that

$$\frac{n}{V} = e^{-\alpha}\left(\frac{2\pi m}{\beta h^2}\right)^{3/2}, \tag{9.159}$$

and

$$\frac{\varepsilon}{V} = \frac{3}{2\beta}e^{-\alpha}\left(\frac{2\pi m}{\beta h^2}\right)^{3/2}. \tag{9.160}$$

Equations 9.159 and 9.160 imply

$$\varepsilon = \frac{3}{2}nkT, \tag{9.161}$$

as we would expect for an ideal gas of n particles from the equipartition theorem. Actually, Equation 9.159 is just a definition of e^α. Notice that the de Broglie wavelength, $\lambda = h/mv$, of a particle with energy kT is given by

$$\lambda = \left(\frac{h^2}{2mkT}\right)^{1/2}. \tag{9.162}$$

Solving for e^α from Equation 9.159 allows us to write

$$e^\alpha = \frac{1}{\lambda^3\left(\frac{n}{V}\right)}. \tag{9.163}$$

Since n/V is just the overall particle density and λ^3 is a volume, this says in words that

$$e^{\alpha} = \frac{1}{\begin{pmatrix}\text{Number of particles} \\ \text{in a } \lambda^3 \text{volume}\end{pmatrix}}. \tag{9.164}$$

This is the classical meaning of e^{α}. We have already observed that the quantum results, Equations 9.122 and 9.123, go over to the classical Maxwell–Boltzmann case when $e^{\alpha} \gg 1$. In terms of particle attributes, Equation 9.164 says, as we might expect, that $e^{\alpha} \gg 1$ describes a situation where the de Broglie wavelengths of particles overlap very little. As a numerical example, consider a gas of helium atoms ($m \simeq 10^{-24}$gm) at room temperature ($T \simeq 300$ K) and a particle density approximately equal to that of air at one atmosphere ($n/V \simeq 10^{19}\text{cm}^{-3}$):

$$e^{\alpha} \approx 10^5. \tag{9.165}$$

9.10 BOSE–EINSTEIN STATISTICS

Apparently, the expressions that we must evaluate in the Bose–Einstein case are (we continue to set $D = 1$)

$$n = \frac{4\pi V}{h^3}\int_0^{\infty} dp\, p^2 \frac{1}{e^{\alpha+\beta p^2/2m} - 1}, \tag{9.166}$$

$$\varepsilon = \frac{4\pi V}{h^3}\int_0^{\infty} dp\, p^2 \frac{p^2/2m}{e^{\alpha+\beta p^2/2m} - 1}. \tag{9.167}$$

We expect from the previous classical expression for e^{α} that we will begin to encounter quantum effects as we increase the de Broglie wavelength overlap by decreasing the temperature. This means lowering the value of α. On the basis of Equation 9.123, however, it is clear that α can never turn negative. Consider an $\alpha = -|\alpha|$; then, in a certain mode, i, we would have for some energies ε_i:

$$\varepsilon_i < \frac{|\alpha|}{\beta} \Rightarrow n_i < 0. \tag{9.168}$$

A negative occupation number is certainly not physical. Let us therefore consider the extreme situation that $\alpha = 0$. Then Equation 9.166 becomes

$$\frac{n}{V} = \frac{4\pi}{h^3}\int_0^{\infty} dp\, p^2 \frac{1}{e^{\beta p^2/2m} - 1}. \tag{9.169}$$

Some mathematical steps to reduce the integral in Equation 9.169 are as follows:

$$\frac{n}{V} = \frac{4\pi}{h^3}\left(\frac{2m}{\beta}\right)^{3/2}\int_0^{\infty} dx\, x^2 \frac{1}{e^{x^2} - 1}, \tag{9.170}$$

$$\int_0^\infty dx\, x^2 \frac{1}{e^{x^2}-1} = \int_0^\infty dx\, x^2 \frac{e^{-x^2}}{1-e^{-x^2}} = \int_0^\infty dx\, x^2 \sum_{n=1}^\infty e^{-nx^2}. \tag{9.171}$$

Interchanging the sum and integral in the last quantity and doing the integral (using a scaled version of Equation 9.157) then informs us that

$$\sum_{n=1}^\infty \int_0^\infty dx\, x^2 e^{-nx^2} = \frac{\sqrt{\pi}}{4}\sum_{n=1}^\infty \frac{1}{n^{3/2}} = \frac{\sqrt{\pi}}{4}\zeta\left(\frac{3}{2}\right), \tag{9.172}$$

where the "Riemann zeta function" is defined as ($x > 1$)

$$\zeta(x) \equiv \sum_{n=1}^\infty \frac{1}{n^x}. \tag{9.173}$$

In our case $\zeta(3/2) = 2.612\ldots$. One finds that

$$\frac{n}{V} \stackrel{?}{=} \left(\frac{2\pi mkT}{h^2}\right)^{3/2} \zeta\left(\frac{3}{2}\right). \tag{9.174}$$

Equation 9.174 dictates a particular relationship between the particle density, n/V, and the absolute temperature when $\alpha = 0$. We would expect to be able to approach the absolute zero of temperature to an arbitrary extent. However, once we have lowered the temperature (and α) to the point where Equation 9.174 is satisfied, our present equations give us no further guidance as to the behavior of the system. Since n and V are fixed, we may view Equation 9.174 as predicting a critical temperature, T_c, for the vanishing (or as we will see, the approximate vanishing) of α:

$$T_c \equiv \frac{h^2}{2\pi mk}\left(\frac{n/V}{\zeta\left(\frac{3}{2}\right)}\right)^{2/3}. \tag{9.175}$$

Since we know that $\alpha \geq 0$, what happens when we try to lower the temperature further? To understand this, let us reconsider the more correct discrete formula based upon Equation 9.123:

$$n = \sum_i \frac{1}{e^{\alpha+\beta\varepsilon_i}-1}. \tag{9.176}$$

To justify the expression Equation 9.166, it was necessary to assume that the change in the summand in Equation 9.176 for neighboring discrete momentum states was small. This assumption breaks down for temperature less than T_c. As $\alpha \to 0$, the single term in Equation 9.176 for $\vec{p} = 0$ diverges, and we must split this term off before replacing the sum by an integral. Thus, we replace Equation 9.166 with the more correct expression

$$n = \frac{4\pi V}{h^3}\int_0^\infty dp\, p^2 \frac{1}{e^{\alpha+\beta p^2/2m}-1} + n_0, \tag{9.177}$$

where

$$n_0 = \frac{1}{e^{\alpha}-1} \tag{9.178}$$

represents the occupation associated with the $\vec{p}=0$ state. What happens for $T<T_c$ is that the Bose particles of the gas begin to collect into this single state. We may solve approximately for n_0 as follows. From Equation 9.177, we have

$$n_0 = n - \frac{4\pi V}{h^3}\int \mathrm{d}p\, p^2 \frac{1}{\mathrm{e}^{\alpha+\beta p^2/2m}-1}. \tag{9.179}$$

To get a first-order expression for n_0, we make the zeroth order approximation that $\alpha=0$ for $T\le T_c$ in the second term of Equation 9.179. Since we already evaluated the resulting integral, we now easily see that

$$n_0 \simeq n - V\left(\frac{2\pi mkT}{h^2}\right)^{3/2}\zeta\left(\frac{3}{2}\right) = n\left(1-\left(\frac{T}{T_c}\right)^{3/2}\right). \tag{9.180}$$

Thus, a macroscopic fraction of the gas occupies the single $\vec{p}=0$ state for $T\le T_c$. This phenomenon is known as "Bose–Einstein condensation" (BEC). A plot of $n_0(T)$ looks like Figure 9.8.

Of course, the nonzero momentum-state occupation numbers obey

$$n_{\varepsilon\neq 0} = n\left(\frac{T}{T_c}\right)^{3/2}, \tag{9.181}$$

below T_c. We do not need to modify the formula Equation 9.167 for the total energy ε since $\varepsilon=0$ for the mode n_0. The assumption that $\alpha \ll 1$ when $T\le T_c$ is consistent since from Equation 9.178 we have that

$$\alpha \simeq \frac{1}{n\left(1-(T/T_c)^{3/2}\right)}. \tag{9.182}$$

BEC is a direct consequence of the statistics obeyed by integer spin particles whose available momentum states are, by assumption, changed very little by interactions between the particles. This is an idealistic situation. The question is: Is anything like this seen in nature?

In 1938, it was discovered by P. Kapitsa, J. Allen, and D. Misener that liquid ^{4}He, which is a Bose system ($A+Z=$even), undergoes a transition, called the "λ-transition" (see Figure 9.9), to a phase called "helium II" at a temperature of 2.17 K. This new phase is made up of two components called

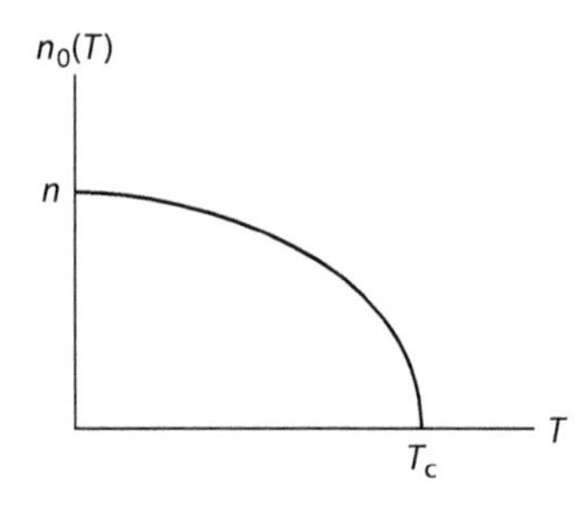

FIGURE 9.8 A qualitative plot of the density of particles in the ground state, $n_0(T)$, as a function of T for a Bose–Einstein condensate with $T<T_c$.

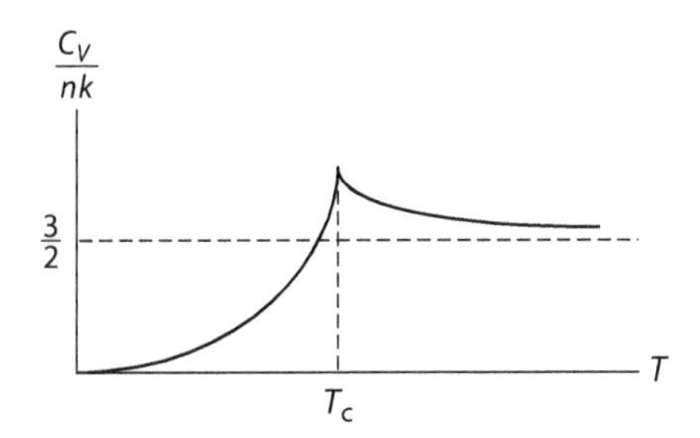

FIGURE 9.9 A qualitative plot of the expected temperature dependence of the heat capacity divided by the total number of particles times Boltzmann's constant, C_V/nk for a system that undergoes a Bose–Einstein transition in the vicinity of T_c. The λ shape of this graph is the origin for the name of this transition in ^{4}He.

the *normal fluid* and the *superfluid*, similar to the n_0 and $n_{\varepsilon\neq 0}$ components described earlier. If one simply substitutes the mass and density of liquid ^{4}He into Equation 9.175, one gets $T_c = 3.14$ K, not too far from the λ-transition temperature, but not perfect either. However, the fact that the transition is occurring in a liquid rather than a gas means the pure theory is not directly applicable. The conclusion is that Bose–Einstein condensation contributes significantly (but not exclusively) to the λ-transition in liquid ^{4}He.

True BEC has now been completely verified in gases.* A group at the University of Colorado/NIST-JILA Lab, including E. Cornell and C. Wieman, produced a gaseous state of rubidium atoms in 1995, and 4 months later W. Ketterle and coworkers at the Massachusetts Institute of Technology (MIT) produced a condensate of sodium atoms. Cornell, Wieman, and Ketterle received the Nobel Prize for this fundamental work in 2001. It has also been seen in many other types of atomic gases.

Let us also calculate Equation 9.167 when $\alpha \simeq 0$ (which we now know occurs when $T \leq T_c$). We have

$$\varepsilon = \frac{4\pi V}{h^3}\int dp\, p^2 \frac{p^2/2m}{e^{\beta p^2/2m} - 1}. \tag{9.183}$$

Running through the steps similar to Equations 9.170 through 9.174, we now find that

$$\varepsilon = \frac{3}{2}\zeta\left(\frac{5}{2}\right)\left(\frac{2\pi mkT}{h^2}\right)^{3/2} VkT \simeq .770\, n\left(\frac{T}{T_c}\right)^{3/2} kT, \tag{9.184}$$

where $\zeta(5/2) \simeq 1.341$. Equation 9.184 has a very different appearance from the Maxwell–Boltzmann result, Equation 9.161. The *total* specific heat at constant volume is given by

$$C_V \equiv \left.\frac{\partial\varepsilon}{\partial T}\right|_V \simeq 1.925\left(\frac{T}{T_c}\right)^{3/2} nk, \tag{9.185}$$

* See W. Ketterle, "Experimental Studies of Bose–Einstein Condensation," *Physics Today*, December 1999.

which is seen to vanish at $T=0$. Thus, for an ideal gas of weakly interacting B–E particles, we expect that the specific heat will initially rise as T increases until reaching a peak at $T=T_c$. After this, the value will level out to the constant value, 3/2 nk, expected at high temperatures. There is a discontinuity in C_V at $T=T_c$. This is shown qualitatively in Figure 9.9.

The very important case of Bose–Einstein statistics applied to a gas of photons in an equilibrium temperature black-body cavity is described in this formalism by putting $\alpha=0$. This makes sense since we have seen that α is determined by the (up to now) fixed number of particles present in the system. Now the number of particles present is not fixed, but is determined by the temperature and size of the box, and nothing else. Photons have spin 1, and so are described by Equations 9.166 and 9.167, but with $D=2$ (two photon polarizations), $\alpha=0$, and the photon mode energy $\varepsilon_p=pc$, where p is the magnitude of the photon's momentum (remember Equation 2.7 in Chapter 2). I will leave the treatment of this case to you in Problem 9.10.1. We will talk more about the photon, considered as an elementary particle, in Chapter 12.

9.11 FERMI–DIRAC STATISTICS

In general, we have (we set $D=2$, appropriate for a spin-1/2 particle)

$$n=\frac{8\pi V}{h^3}\int_0^\infty dp\,p^2\frac{1}{e^{\alpha+\beta p^2/2m}+1}, \tag{9.186}$$

$$\varepsilon=\frac{8\pi V}{h^3}\int_0^\infty dp\,p^2\frac{p^2/2m}{e^{\alpha+\beta p^2/2m}+1}. \tag{9.187}$$

Just as we concentrated on low temperatures in the B–E case, we will also do so here, for this is where we expect to see quantum effects. Again, α is determined from the expression for n. Before, we argued that α had to be positive, and then evaluated the expressions for n and ε in the extreme case that $\alpha=0$. Here, there is no such restriction since the expression for n_i in the F–D case, Equation 9.122, is never in any danger of turning negative. In either the B–E or F–D case, determining α by solving Equations 9.177 or 9.186, respectively, is a difficult mathematical problem. However, the situations simplify at low temperatures. In the B–E case, we found self-consistently that $\alpha \ll 1$ for $T \leq T_c$ and saw that the particles all piled up in the ground state, $\vec{p}=0$, when $T=0$. We expect a completely different situation in the F–D case at low temperatures. Because of the exclusion principle, no more than a single particle may occupy a given state of the system. Therefore, at $T=0$, we expect the identical fermions *not* to accumulate in the single $\vec{p}=0$ state, but to fill up the lowest n states of the system with single particles. The occupation number, $n(p)$, considered as a continuous function of $p^2/2m$, should appear at $T=0$ as in Figure 9.10. That is, we expect

$$\lim_{T\to 0}\frac{1}{e^{\alpha+\beta p^2/2m}+1}=\begin{cases}1, & \dfrac{p^2}{2m}<\mu\\[2mm] 0, & \dfrac{p^2}{2m}>\mu.\end{cases} \tag{9.188}$$

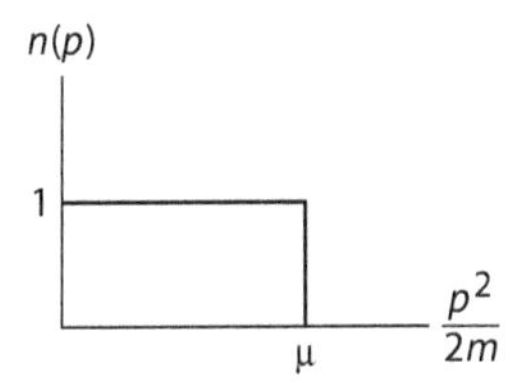

FIGURE 9.10 Occupation number as a function of kinetic energy for a Fermi–Dirac gas at zero temperature. The maximum occupied energy value, μ, is the "Fermi energy."

The quantity μ, called the "Fermi energy," is the energy of the highest occupied state and will be determined by the total number of particles present. Equation 9.188 specifies that

$$\alpha(T) = -\frac{\mu}{kT} \tag{9.189}$$

at low temperatures. Given the expected profile (Equation 9.188), Equations 9.186 and 9.187 now read (at *exactly* $T=0$)

$$n = \frac{8\pi V}{h^3}\int_0^{p_F} \mathrm{d}p\, p^2, \tag{9.190}$$

$$\varepsilon_{T=0} = \frac{8\pi V}{h^3}\int_0^{p_F} \mathrm{d}p\, p^2\left(\frac{p^2}{2m}\right), \tag{9.191}$$

where we have defined $p_F^2/2m \equiv \mu$. Doing the trivial integrals results in

$$n = \frac{8\pi V}{3}\frac{(2m\mu)^{3/2}}{h^3}, \tag{9.192}$$

$$\varepsilon_{T=0} = \frac{8\pi V}{10}\frac{(2m\mu)^{5/2}}{mh^3} = \frac{3}{5}\mu n. \tag{9.193}$$

Equation 9.192 determines μ (and also the low temperature form for $\alpha(T)$ through Equation 9.189). We get

$$\mu = \frac{h^2}{2m}\left(\frac{3}{8\pi}\frac{n}{V}\right)^{2/3}. \tag{9.194}$$

We define the "Fermi temperature,"

$$kT_F \equiv \mu. \tag{9.195}$$

Its significance will be discussed shortly.

These results are for $T=0$. Let us now look at the first-order deviations for low temperatures.* In the following, we will be concerned with the approximate evaluation of integrals of the form

$$I = \int_0^\infty \mathrm{d}\varepsilon_p \frac{f(\varepsilon_p)}{\mathrm{e}^{\alpha+\varepsilon_p/kT}+1}, \tag{9.196}$$

* This derivation is taken from my 1979 Schwinger notes.

given that $\alpha(T) \to -\mu/kT$ at low temperatures. First, rewrite Equation 9.196 exactly as

$$I = \int_0^{-\alpha kT} \mathrm{d}\varepsilon_p f(\varepsilon_p) - \int_0^{-\alpha kT} \mathrm{d}\varepsilon_p \frac{f(\varepsilon_p)}{1+\mathrm{e}^{-\alpha-\varepsilon_p/kT}} + \int_{-\alpha kT}^{\infty} \mathrm{d}\varepsilon_p \frac{f(\varepsilon_p)}{\mathrm{e}^{\alpha+\varepsilon_p/kT}+1}. \tag{9.197}$$

In the second integral, introduce the variable x through $\varepsilon_p = -\alpha kT - (kT)x$; in the third integral introduce x through $\varepsilon_p = -\alpha kT + (kT)x$. Doing the change of variables, we get (still an exact result)

$$\begin{aligned} I = \int_0^{-\alpha kT} \mathrm{d}\varepsilon_p f(\varepsilon_p) - kT \int_0^{-\alpha} \mathrm{d}x \frac{f(-\alpha kT - kTx)}{\mathrm{e}^x+1} \\ + kT \int_0^{\infty} \mathrm{d}x \frac{f(-\alpha kT + kTx)}{\mathrm{e}^x+1}. \end{aligned} \tag{9.198}$$

Again using the fact that $\alpha(T) \to -\mu/kT$ at low temperatures, we replace the upper limit in the second integral by $+\infty$ and write

$$I \simeq \int_0^{-\alpha kT} \mathrm{d}\varepsilon_p f(\varepsilon_p) + kT \int_0^{\infty} \mathrm{d}x \frac{f(-\alpha kT + kTx) - f(-\alpha kT - kTx)}{\mathrm{e}^x+1}. \tag{9.199}$$

This approximation holds when $\alpha \ll -1$, or equivalently, $T \ll T_F$. We now expand the functions $f(-\alpha kT + kTx)$ and $f(-\alpha kT - kTx)$ about their values at small x. This is allowed in Equation 9.199 because of the suppression of higher-order terms by the factor $(\mathrm{e}^x + 1)^{-1}$ at large x.

$$[f(-\alpha kT + kTx) - f(-\alpha kT - kTx)] \simeq 2kTxf'(-\alpha kT), \tag{9.200}$$

where

$$f'(-\alpha kT) \equiv \left.\frac{\mathrm{d}f(\varepsilon_p)}{\mathrm{d}\varepsilon_p}\right|_{\varepsilon_p=-\alpha kT}. \tag{9.201}$$

Our integral now reads

$$I \simeq \int_0^{-\alpha kT} \mathrm{d}\varepsilon_p f(\varepsilon_p) + 2(kT)^2 f'(-\alpha kT) \int_0^{\infty} \mathrm{d}x \frac{x}{\mathrm{e}^x+1}. \tag{9.202}$$

But

$$\int_0^{\infty} \mathrm{d}x \frac{x}{\mathrm{e}^x+1} = -\sum_{n=1}^{\infty} \frac{(-1)^n}{n^2} = \frac{\pi^2}{12}. \tag{9.203}$$

The last result can be justified from the theory of Fourier series.* Therefore, we have arrived at

$$I \underset{T \ll T_F}{\simeq} \int_0^{-\alpha kT} \mathrm{d}\varepsilon_p f(\varepsilon_p) + \frac{\pi^2}{6}(kT)^2 f'(-\alpha kT). \tag{9.204}$$

Let us apply Equation 9.204 to 9.186 and 9.187. First, we rewrite Equation 9.186 as

$$n = \frac{8\pi\sqrt{2}m^{3/2}V}{h^3} \int_0^{\infty} \mathrm{d}\varepsilon_p \frac{\varepsilon_p^{1/2}}{\mathrm{e}^{\alpha+\varepsilon_p/kT}+1}, \tag{9.205}$$

* W. Kaplan, *Advanced Calculus*, 2nd ed. (Addison-Wesley, 1973), p. 472.

where $\varepsilon_p = p^2/2m$. To use the result in Equation 9.204, we identify $f(\varepsilon_p) = \varepsilon_p^{1/2}$ for which $f'(-\alpha kT) = 1/2(-\alpha kT)^{-1/2}$. Plugging in, we find that

$$n \simeq \frac{8\pi V(2m\mu)^{3/2}}{3h^3}\left\{\left(\frac{-\alpha kT}{\mu}\right)^{3/2} + \frac{\pi^2}{8}\left(\frac{kT}{\mu}\right)^2\left(-\frac{\mu}{\alpha kT}\right)^{1/2}\right\}. \quad (9.206)$$

The first term on the right in Equation 9.206 just reproduces the result in Equation 9.192 when $T=0$, as it should. Since n and V are fixed, Equation 9.206 implies a change in the meaning of $\alpha(T)$. Setting $\alpha(T) = -(\mu/kT) + z$ where $z \ll \mu/kT$, we find self-consistently from Equation 9.206 that

$$\alpha(T) \simeq -\frac{\mu}{kT} + \frac{\pi^2}{12}\left(\frac{kT}{\mu}\right). \quad (9.207)$$

Raising the value of T is thus seen to increase $\alpha(T)$, as it should, since $\alpha(T) \geq 1$ is the classical limit.

We may employ the same technique on Equation 9.187 to show that

$$\varepsilon \simeq \frac{8\pi V}{10}\frac{(2m\mu)^{5/2}}{mh^3}\left\{\left(-\frac{\alpha kT}{\mu}\right)^{5/2} + \frac{5\pi^2}{8}\left(\frac{kT}{\mu}\right)^2\left(-\frac{\alpha kT}{\mu}\right)^{1/2}\right\}. \quad (9.208)$$

Since we now know the approximate form for $\alpha(T)$, we may rewrite this to first order as

$$\varepsilon \simeq \frac{8\pi V}{10}\frac{(2m\mu)^{5/2}}{mh^3}\left\{1 + \frac{5\pi^2}{12}\left(\frac{kT}{\mu}\right)^2\right\}. \quad (9.209)$$

Recognizing from Equation 9.193 that the overall factor in Equation 9.209 is just the energy of the gas at zero temperature, we have that

$$\varepsilon \simeq \frac{3}{5}n\mu + \frac{\pi^2}{4}n\mu\left(\frac{kT}{\mu}\right)^2. \quad (9.210)$$

We again investigate the specific heat. We have that

$$C_V = \frac{\pi^2}{2}nk^2\left(\frac{T}{\mu}\right) = \frac{\pi^2}{2}nk\left(\frac{T}{T_F}\right). \quad (9.211)$$

Thus, we see that T_F sets the scale of temperatures for the Fermi–Dirac gas.

The point of this calculation is the following. Remember that the electrons in the Kronig–Penney model are treated as a gas acted upon by the stationary atoms. If we use appropriate electronic concentrations, n/V, in Equations 9.194 and 9.195, we find that*

$$\left.\begin{aligned}(T_F)_{\mathrm{Cu}} &= 8.12\times 10^{4\circ}\mathrm{K}\\ (T_F)_{\mathrm{Ag}} &= 6.36\times 10^{4\circ}\mathrm{K}\end{aligned}\right\}, \quad (9.212)$$

* See C. Kittel, *Introduction to Solid State Physics*, 8th ed. (John Wiley, 2005), p. 139 for the values used.

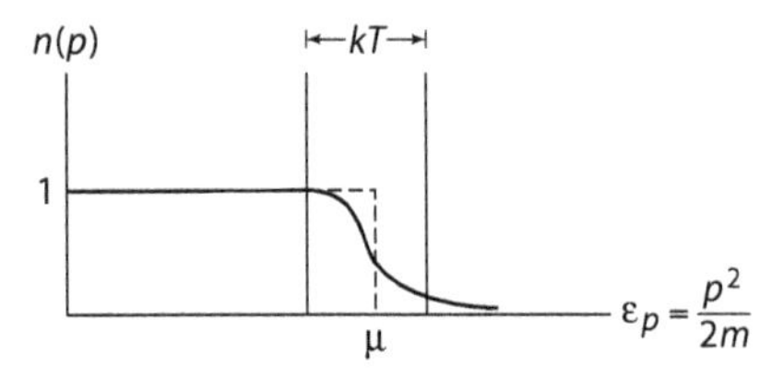

FIGURE 9.11 Occupation number as a function of kinetic energy for a Fermi–Dirac gas at a low but nonzero temperature ($T \ll T_F$). The $T = 0$ distribution is shown as a dotted line for comparison.

for copper and silver. Thus, the *electronic* component of C_V at room temperature is very small. Qualitatively, this effect can be understood from Figure 9.11, which holds at $T \ll T_F$.

From the approximate form for $n(p)$ when $T \ll T_F$,

$$n(p) \approx \frac{1}{e^{(\varepsilon_p - \mu)/kT} + 1}, \tag{9.213}$$

it is easy to see that the width of the region of particles with excited energies is of order kT. The number of particles excited is, therefore, of order $n(kT/\mu) = n(T/T_F)$. This means the excitation energy of the gas will be $[n(T/T_F)]kT = nk(T^2/T_F)$, and the specific heat will be $\simeq nk(T/T_F)$, as found earlier. This solves the mystery of the missing electronic component of specific heat since the Fermi temperature is so high compared to room temperature for these materials!

9.12 THE HARTREE–FOCK EQUATIONS

An important application of identical particle symmetry occurs in multiparticle atomic physics. The Schrödinger equation for a neutral atom with Z electrons and protons is just (the nucleus is considered infinitely heavy and $e > 0$ is the proton charge)

$$\sum_{i=1}^{Z} \left(-\frac{\hbar^2}{2m} \vec{\nabla}_i^2 u\,(\cdots) - \frac{Ze^2}{r_i} u\,(\cdots) + e^2 \sum_{i<j} \frac{u\,(\cdots)}{r_{ij}} \right) = E\, u(\cdots), \tag{9.214}$$

where $r_i \equiv |\vec{r}_i|$, $r_{ij} \equiv |\vec{r}_i - \vec{r}_j|$ and

$$u(\cdots) \equiv u\,(\vec{r}_1, \vec{r}_2, \ldots, \vec{r}_Z; \chi_1, \chi_2, \ldots, \chi_Z) \tag{9.215}$$

are the full wave functions, which contain both position ($\vec{r}_i$) and spin (χ_i) variables for all electrons. (We sum on both i and j in Equation 9.214 with $i < j$.) The first term on the left of Equation 9.214 represents the system kinetic energy, the second is the attractive potential between the electrons and the nucleus, and the third represents the repulsive electron–electron interaction. Since the variables represent identical fermions, we must have that

$$u\,(\ldots, \vec{r}_i, \vec{r}_j, \ldots; \ldots \chi_i, \chi_j, \ldots) = -u\,(\ldots, \vec{r}_j, \vec{r}_i, \ldots; \ldots \chi_j, \chi_i, \ldots), \tag{9.216}$$

for any pair ($i \neq j$). The full solution to this second-order partial differential equation in Z three-dimensional position variables and Z spins is too hard to address directly. Of course, without the electron–electron coupling term, this equation separates into Z independent equations. This suggests that an approximate treatment of this interaction term could help.

Let us go back to Equation 9.97 and find the general form of a two-body interaction for *bound*, instead of *scattered*, particles. We are of course using a particle number basis. I will for the moment limit the discussion to two different states: 1 and 1′, both of which are occupied. We have

$$\langle 1_1, 1_{1'} \,|\, H_{\text{int}} \,|\, 1_1, 1_{1'} \rangle = \frac{1}{2} \sum_{i,j,k,\ell} \langle i, j | V | \ell, k \rangle \langle 1_1, 1_{1'} \,|\, a_i^\dagger a_j^\dagger a_k a_\ell \,|\, 1_1, 1_{1'} \rangle. \tag{9.217}$$

There are four nonzero contributions to Equation 9.217. Using the identity Equation 9.86, this results in

$$\langle 1_1, 1_{1'} \,|\, H_{\text{int}} \,|\, 1_1, 1_{1'} \rangle = \langle 1, 1' \,|\, V | 1, 1' \rangle - \langle 1, 1' \,|\, V | 1', 1 \rangle. \tag{9.218}$$

To apply this result to an atomic system, let us shift back to generic state labels $1 \rightarrow i$ and $1' \rightarrow j$ in preparation for summing:

$$\langle 1_i, 1_j \,|\, H_{\text{int}} \,|\, 1_i, 1_j \rangle = \langle i, j \,|\, V | i, j \rangle - \langle i, j | V | \, j, i \rangle. \tag{9.219}$$

The right-hand side of Equation 9.219 can be graphically portrayed as in Figure 9.12, where the circled numbers are particle labels, whereas i and j are state labels.

Now define V as a two-body Coulomb potential between particles located at positions 1 and 2,

$$V(r_{12}) \equiv \frac{e^2}{r_{12}}, \tag{9.220}$$

and project the wave functions into a coordinate-spin basis with

$$\langle \vec{r}_1, \vec{r}_2; m_1, m_2 \,|\, i, j \rangle \equiv u_i(\vec{r}_1)\, u_j(\vec{r}_2)\, \chi_{m_i}^{(m_1)}\, \chi_{m_j}^{(m_2)}, \tag{9.221}$$

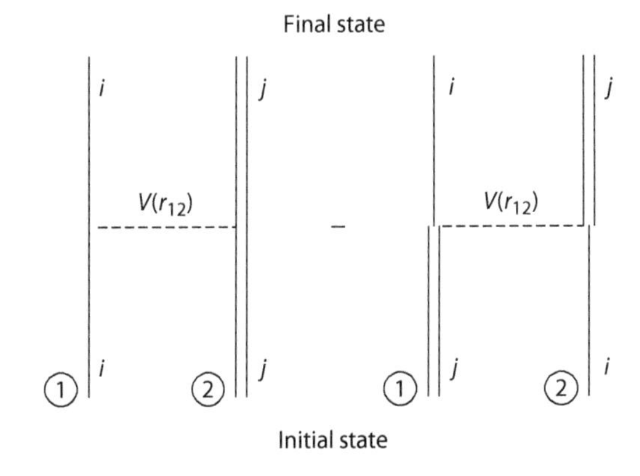

FIGURE 9.12 Diagrammatic interpretation of Equation 9.219 for a bound atomic system. The line groupings represent not scattering states but different bound states.

where the two normalized spin states are

$$\chi_{1/2} = \begin{pmatrix} 1 \\ 0 \end{pmatrix}, \quad \chi_{-1/2} = \begin{pmatrix} 0 \\ 1 \end{pmatrix}, \tag{9.222}$$

and we assume orthonormality of the wave functions,

$$\int d^3r \; u_i^*(\vec{r}) \; u_j(\vec{r}) = \delta_{ij}. \tag{9.223}$$

Inserting coordinate-spin completeness twice in each term of Equation 9.219, one finds:

$$\langle 1_i, 1_j \,|\, H_{\text{int}} \,|\, 1_i, 1_j \rangle$$

$$= e^2 \int d^3r_1 \int d^3r_2 \Bigg[\underbrace{\frac{|u_i(\vec{r}_1)|^2 |u_j(\vec{r}_2)|^2}{r_{12}}}_{\text{"direct"}} - \underbrace{(\chi_{m_i}^\dagger \chi_{m_j})^2 \frac{u_i^*(\vec{r}_1) u_j^*(\vec{r}_2) u_j(\vec{r}_1) u_i(\vec{r}_2)}{r_{12}}}_{\text{"exchange"}} \Bigg]. \tag{9.224}$$

The "direct" term is just what we would expect if we were dealing with the electrostatic interaction of two charge densities, $-e\,|u_i(\vec{r}_1)|^2$ and $-e\,|u_j(\vec{r}_2)|^2$. The "exchange" term is, however, completely nonclassical. Notice its role is to remove any overlap between the two wave functions in the integral as required by the Fermi principle. This results in complete cancellation for $i=j$, for example. In this regard, notice

$$\left(\chi_{m_i}^\dagger \, \chi_{m_j}\right)^2 = \delta_{m_i, m_j}, \tag{9.225}$$

so that spin overlap is also removed. The exchange term is very important numerically and can affect atomic system energies by 20%–30%.

We now sum over i and j, but avoid double counting an interaction by requiring $i<j$. (We could also sum on all $j \neq i$ and compensate with a factor of 1/2.) This defines a two-body potential interaction energy expectation value, $\langle U_{\text{int}} \rangle$:

$$\langle U_{\text{int}} \rangle \equiv \sum_{i<j}^{Z} \langle 1_i, 1_j | H_{\text{int}} | 1_i, 1_j \rangle. \tag{9.226}$$

In Chapter 5, we learned that we may find the approximate ground-state energy and wave function of a quantum system by varying parameters in the system energy, which is just the expectation value of the Hamiltonian operator. We may view the earlier unknown wave functions, $u_i(\vec{r})$, themselves as the parameters. Actually, we may make independent variations of the $u_i(\vec{r})$ and $u_i^*(\vec{r})$ since a general complex quantity has independent real and imaginary parts. In preparation for such a variation for an atomic system, I form

$$\langle H \rangle \equiv \langle T + U_C + U_{\text{int}} \rangle, \tag{9.227}$$

where I have added standard kinetic and nuclear Coulomb interaction terms, which results in

$$\langle H\rangle = \sum_{i=1}^{Z}\int d^3r_1\Bigg[\overbrace{-\frac{\hbar^2}{2m}u_i^*(\vec{r}_1)\,\vec{\nabla}_1^2u_i(\vec{r}_1)}^{\langle T\rangle}\overbrace{-\frac{Ze^2}{r_1}|u_i(\vec{r}_1)|^2}^{\langle U_C\rangle}$$

$$\overbrace{+e^2\sum_{j<i}\left\{|u_i(\vec{r}_1)|^2\int d^3r_2\frac{|u_j(\vec{r}_2)|^2}{r_{12}}-\delta_{m_i,m_j}u_i^*(\vec{r}_1)u_j(\vec{r}_1)\int d^3r_2\frac{u_j^*(\vec{r}_2)u_i(\vec{r}_2)}{r_{12}}\right\}}^{\langle U_{\text{int}}\rangle}\Bigg]. \tag{9.228}$$

The first term on the right represents the kinetic energy, $\langle T\rangle$. The second term is just the Coulomb interaction between the nucleus and the electron charge densities, $\langle U_C\rangle$, and the third and fourth terms are the electron–electron direct and exchange terms, respectively. The exchange term now introduces a spin interaction in the system, although the original interaction was spin independent!

Actually, one cannot vary Equation 9.228 directly with respect to the $u_i(\vec{r}_1)$ or $u_i^*(\vec{r}_1)$ to find the ground-state properties since there are Z constraints on the normalization of these functions through Equation 9.223 with $i=j$. The formalism for including such constraints involves Lagrange multipliers. (You can refresh yourself on this subject if necessary by glancing over Chapter 5, Section 5.7.) With a bit of notational sleight of hand, I define the new functional

$$H^* \equiv \langle H\rangle - \sum_{i=1}^{Z}E_i\left(\int d^3r_1|u_i(\vec{r}_1)|^2-1\right), \tag{9.229}$$

with N Lagrange multipliers, E_i. The variation of the u_i^* now produces

$$\delta_{u_i^*}H^* = \sum_{i=1}^{Z}\int d^3r_1\,\delta u_i^*(\vec{r}_1)\Bigg[-\frac{\hbar^2}{2m}\vec{\nabla}_1^2u_i(\vec{r}_1)-\frac{Ze^2}{r_1}u_i(\vec{r}_1)$$

$$+e^2\sum_{j\neq i}\left\{u_i(\vec{r}_1)\int d^3r_2\frac{|u_j(\vec{r}_2)|^2}{r_{12}}-\delta_{m_i,m_j}u_j(\vec{r}_1)\int d^3r_2\frac{u_j^*(\vec{r}_2)u_i(\vec{r}_2)}{r_{12}}\right\}-E_iu_i(\vec{r}_1)\Bigg]. \tag{9.230}$$

(Notice the $j<i$ sum has turned into a $j\neq i$ one.) The extremum condition,

$$\delta_{u_i^*}H^* = 0, \tag{9.231}$$

for arbitrary δu_i^* then results in the equations ($\vec{r}_1\to\vec{r}, \vec{r}_2\to\vec{r}\,', r_{12}\to R$ for convenience),

$$-\frac{\hbar^2}{2m}\vec{\nabla}^2u_i(\vec{r})-\frac{Ze^2}{r}u_i(\vec{r})$$

$$+e^2\sum_{j\neq i}\left\{u_i(\vec{r})\int d^3r'\frac{|u_j(\vec{r}\,')|^2}{R}-\delta_{m_i,m_j}u_j(\vec{r})\int d^3r'\frac{u_j^*(\vec{r}\,')u_i(\vec{r}\,')}{R}\right\}=E_iu_i(\vec{r}). \tag{9.232}$$

The E_i turn out to be *removal energies*, associated with electron i. The full system energy, E, is then

$$E \equiv \langle H \rangle = \frac{1}{2}\left(\sum_{i=1}^{Z} E_i + \langle T \rangle + \langle U_C \rangle \right) \tag{9.233}$$

an expression derivable from Equations 9.228 and 9.126 that correctly avoids double counting of the interaction terms.

We would just get the complex conjugate of Equations 9.232 if we varied the $u_i(\vec{r}_1)$ instead of the $u_i^*(\vec{r}_1)$. The exchange energy term has now produced a spin-dependent potential energy term affecting the force between the electrons. This gives rise to what is called an "exchange force."

Equations in 9.232 are called the *Hartree–Fock equations*, named after Douglas Hartree and Vladimir Fock. (Without the exchange term, these are called the Hartree equations.) They are a coupled set of Z second-order differential equations in Z unknowns. A set of trial wave functions, $u_1^{(0)}(\vec{r}_i)$ are assumed to begin the solution process, which then results in a new set, $u_1^{(1)}(\vec{r}_i)$, which can then be used to generate a further set. Hopefully, after N iterations, the process converges self-consistently:

$$u_i(\vec{r}) = \lim_{N \gg 1} u_i^{(N)}(\vec{r}). \tag{9.234}$$

The Hartree–Fock equations are effective "mean-field" equations, where only the lowest-order electron–electron interaction is taken into account. (They actually correspond to the lowest-order "ladder" interaction between electrons in Figure 11.12 in Chapter 11.) Other, better approximations have been formulated since 1930, when these equations were first developed. In particular, relativistic effects can be introduced. Examples of the accuracy of these equations applied to atomic systems can be found in the book by Bethe and Jackiw,* for example. These types of equations can be viewed as a foundational starting point for much of many-body atomic, molecular, and nuclear physics.

PROBLEMS

9.2.1 (a) Write all the allowed wave functions for a three-state system containing two identical bosons. How many are there?

(b) Write all the allowed wave functions for a three-state system containing two identical fermions. How many are there?

[Note: A basis element is $|i,j\rangle \equiv |i\rangle_1 |j\rangle_2$ in particle property space, where $i,j = \{a', a'', a'''\}$ represents the three possible states, and 1 and 2 are particle labels.]

9.2.2 (a) Count the possible particle occupation states of the form $|n_1, n_2\rangle$ for identical bosons when $g = 2$ and $n = 4$. Does this agree with the expression in Section 9.2?

* H. Bethe and R. Jackiw, *Intermediate Quantum Mechanics*, 2nd ed. (W.A. Benjamin, 1968).

(b) Count the possible particle occupation states of the form $|n_1,n_2,n_3,n_4\rangle$ for identical bosons when $n=2$ and $g=4$. Does this agree with the expression in Section 9.2?

9.3.1 (a) Given

$$a_i\,|\cdots n_i \cdots\rangle = C_i\,|\cdots n_i - 1\cdots\rangle,$$

where C_i is an unknown constant, and

$$a_i^\dagger a_i|\cdots n_i \cdots\rangle = n_i|\cdots n_i \cdots\rangle,$$

show that the most general form for C_i is

$$C_i = \sqrt{n_i}\,e^{i\alpha},$$

where α is an arbitrary (real) phase.

(b) Deduce from (a) that

$$a_1^\dagger|\cdots n_i - 1\cdots\rangle = e^{-i\alpha}\sqrt{n_i}\,|\cdots n_i \cdots\rangle.$$

9.3.2 (a) Evaluate the commutator $[N_i, a_i^\dagger a_i^\dagger]$, where $N_i = a_i^\dagger a_i$, for both F–D and B–E statistics.

(b) Use (a) to show that

$$N_i\, a_i^\dagger a_i^\dagger|\cdots 0_i \cdots\rangle = 2a_i^\dagger a_i^\dagger|\cdots 0_i \cdots\rangle,$$

for both F–D (trivial) and B–E statistics.

9.3.3 (a) Show for both statistics:

$$[a_i, a_j^\dagger a_k] = \delta_{ij} a_k.$$

Then derive

$$[a_i^\dagger, a_j^\dagger a_k] = -\delta_{ik} a_j^\dagger$$

by Hermitian conjugation.

(b) Apply part (a) to prove

$$[a_i^\dagger a_j, a_k^\dagger a_\ell] = \delta_{jk} a_i^\dagger a_\ell - \delta_{i\ell} a_k^\dagger a_j.$$

9.3.4 Evaluate:

(a) $e^{-i\lambda N} a_i e^{i\lambda N} = ?$

(b) $e^{-i\lambda N} a_i^\dagger e^{i\lambda N} = ?$

where $N = \sum_i N_i$ is the total number operator.

[Hint: Differentiate these quantities in λ and solve the resulting differential equations.]

9.4.1 For a two-physical-property F–D system, adopt the definition:

$$a_1^\dagger a_2^\dagger\,|0,0\rangle = a_1^\dagger\,|0,1\rangle = |1,1\rangle.$$

(a) Show that

$$a_1 = |0,0\rangle\langle 1,0| + |0,1\rangle\langle 1,1|,$$

$$a_2 = |0,0\rangle\langle 0,1| - |1,0\rangle\langle 1,1|,$$

satisfy Equations 9.68 and 9.69 and the above result for $a_1^\dagger a_2^\dagger |0,0\rangle$.

(b) Show that

$$a_i a_i^\dagger + a_i^\dagger a_i = \sum_{n_1,n_2=\{0,1\}} |n_1,n_2\rangle\langle n_1,n_2|,$$

for either $i = 1$ or 2.

9.4.2 From Problem 9.3.1 we know for F–D statistics that

$$a_i^\dagger|\cdots 0_i \cdots\rangle = e^{-i\alpha}\ |\cdots 1_i \cdots\rangle,$$

$$a_i|\cdots 1_i \cdots\rangle = e^{i\alpha}\ |\cdots 0_i \cdots\rangle.$$

Using the convention that $e^{i\alpha} = 1$ when the number of occupied one-particle states with index less than i is even, and $e^{i\alpha} = -1$ when this number is odd, evaluate:

(a) $a_3^\dagger\ |1,0,1,0,0\rangle = ?$
(b) $a_5^\dagger a_1\ |1,0,1,0,0\rangle = ?$
(c) $a_1 a_3 a_5 a_4 a_2\,|1,1,1,1,1\rangle = ?$
(d) $a_3^\dagger a_1^\dagger a_4^\dagger a_5^\dagger a_2^\dagger\ |0,0,0,0,0\rangle = ?$

The occupation numbers are given in the order:

$$|\,n_1, n_2, n_3, n_4, n_5\rangle.$$

9.4.3 Consider *fermion* creation and annihilation operators $a_i^\dagger$ and a_i for some particle property i.

(a) Find *matrix representations* for each of the following operators (use the number basis and the result of Problem 9.3.1):
   (i) a_i
   (ii) $a_i^\dagger$
   (iii) $a_i^\dagger a_i$

(b) Evaluate:
   (iv) $e^{a_i^\dagger a_i}\,|\cdots n_i \cdots\rangle = ?$
   (v) $e^{a_i a_i^\dagger}\,|\cdots n_i \cdots\rangle = ?$

9.5.1 Define the single-particle property operators:

$$F^{(1)} \equiv \sum_{i,j} a_i^\dagger a_j \langle i|f|j\rangle,$$

$$G^{(1)} \equiv \sum_{k,l} a_k^\dagger a_\ell\ \langle k|g|l\rangle.$$

Show (you may use the result of Problem 9.3.3b)

$$[F^{(1)}, G^{(1)}] = \sum_{i,j} a_i^{\dagger} a_j \; \langle i | [f, g] \, | j \rangle.$$

***9.7.1** Use (upper sign is B–E, lower sign F–D)

$$\frac{1}{e^{\alpha+\beta\varepsilon} \mp 1} \simeq e^{-\alpha-\beta\varepsilon} \pm e^{-2\alpha-2\beta\varepsilon}$$

when e^{α} is large to evaluate

$$N = \frac{DV}{(2\pi\hbar)^3} \int d^3p \frac{1}{e^{\alpha+\beta(\vec{p}^{\,2}/2m)} \mp 1}$$

and

$$E = \frac{DV}{(2\pi\hbar)^3} \int d^3p \frac{\vec{p}^{\,2}/2m}{e^{\alpha+\beta(\vec{p}^{\,2}/2m)} \mp 1}$$

approximately. Show that

$$\frac{N}{V} \simeq D\left(\frac{mkT}{2\pi\hbar^2}\right)^{3/2} e^{-\alpha}[1 \pm 2^{-3/2} e^{-\alpha}],$$

$$\frac{E}{V} \simeq \frac{3}{2} kTD\left(\frac{mkT}{2\pi\hbar^2}\right)^{3/2} e^{-\alpha}[1 \pm 2^{-5/2} e^{-\alpha}].$$

Then show that E/V and N/V are related by

$$\frac{E}{V} \simeq \frac{3}{2} kT \left(\frac{N}{V}\right)(1 \mp 2^{-5/2} e^{-\alpha}).$$

9.8.1 (a) Show the vector potential field $\vec{A}(\vec{r}, t)$ introduced in Equation 9.136 satisfies the wave equation:

$$\left[\vec{\nabla}^2 - \frac{1}{c^2}\frac{\partial^2}{\partial t^2}\right]\vec{A}(\vec{r}, t) = 0.$$

[This is the appropriate Maxwell equation for this quantity in the Coulomb gauge when no sources are present.]

(b) Show the following matrix element results:

$$\left\langle n_{\alpha,\vec{k}} + 1 \,|\, \vec{A}(\vec{r}, t) \,|\, n_{\alpha,\vec{k}} \right\rangle = c\sqrt{\frac{2\pi}{\omega V}} \sqrt{n_{\alpha,\vec{k}} + 1}\; \hat{\varepsilon}_{\alpha,\vec{k}} \, e^{-i(\vec{k}\cdot\vec{r} - \omega t)},$$

$$\left\langle n_{\alpha,\vec{k}} \,|\, \vec{A}(\vec{r}, t) \,|\, n_{\alpha,\vec{k}} + 1 \right\rangle = c\sqrt{\frac{2\pi}{\omega V}} \sqrt{n_{\alpha,\vec{k}}}\; \hat{\varepsilon}_{\alpha,\vec{k}} \, e^{i(\vec{k}\cdot\vec{r} - \omega t)}.$$

***9.9.1** Consider an ensemble of N one-dimensional harmonic oscillators whose energy levels are quantized but that obey Maxwell–Boltzmann statistics. The energy levels, as usual, are given

$$E_n = \hbar\omega\left(n + \frac{1}{2}\right).$$

(a) Given that the average occupation number of the nth state at temperature T is given by ($\beta = 1/(kT)$)

$$\langle n \rangle = \exp(-\alpha - \beta E_n),$$

find an expression for the total energy of the gas, E. [Hints: You can eliminate α by using

$$N = \sum_{n=0}^{\infty} \langle n \rangle.$$

Also, the arithmetic series

$$\sum_{n=0}^{\infty} x^n = \frac{1}{1-x},$$

($|x| < 1$) may be helpful. You should be able to do (b) and (c) even if you cannot do the sums in (a). [Answer:]

$$E = \frac{N\hbar\omega}{2} \frac{\cosh(\beta\hbar\omega/2)}{\sinh(\beta\hbar\omega/2)}.$$

(b) Find the specific heat of this collection of oscillators from

$$C_V = \left.\frac{\partial E}{\partial T}\right|_V.$$

(c) Take the limit

$$\lim_{T\to\infty} C_V = \ ?$$

[Hint: This limit should give the classical result expected from the equipartition theorem, seen in Chapter 1.]

***9.10.1** (a) Evaluate the integrals (for photons)

$$n = \frac{8\pi V}{h^3}\int_0^{\infty} dp\, p^2 \frac{1}{e^{\beta pc} - 1},$$

$$\varepsilon = \frac{8\pi V}{h^3}\int_0^{\infty} dp\, p^2 \frac{pc}{e^{\beta pc} - 1}.$$

Note that $\zeta(3) = 1.20906$, $\zeta(4) = \pi^4/90$. The second equation can be used to derive the "Stefan–Boltzmann" law for energy radiated from a black body, discussed in Chapter 2, Section 2.1.

(b) Show that we may also write

$$\frac{\varepsilon}{V} = \int_0^\infty d\nu \, u(\nu), \quad \text{where}$$

$$u(\nu) = \frac{8\pi h\nu^3}{c^3} \frac{1}{e^{h\nu/kT} - 1}.$$

Demonstrate that in the $\nu \to 0$ and $\nu \to \infty$ limits we get:

$$\lim_{\nu \to 0} u(\nu) = \frac{8\pi\nu^2}{c^3} kT, \quad \text{(A)}$$

$$\lim_{\nu \to \infty} u(\nu) = \frac{8\pi h\nu^3}{c^3} e^{-h\nu/kT}. \quad \text{(B)}$$

[Comments: Notice (A) has no factor of $\hbar$. This is the classical Rayleigh–Jeans result. (B) is the result that led Einstein to postulate that $E = h\nu = \hbar\omega$ for particles of light, confirmed in the photoelectric effect.]

9.11.1 Consider a relativistic gas ($E = pc$) of N identical Fermi–Dirac particles occupying a volume V at zero temperature ($T = 0$). Show that the total energy, E_{tot}, of the system is given by

$$E_{\text{tot}} = \frac{(9\pi)^{2/3}}{4} \hbar c N^{4/3} V^{-1/3}.$$

[Such considerations are important in stellar dynamics of white dwarfs and neutron stars, for example.]

9.11.2 Apply Equation 9.204 to Equation 9.187, and using the approximate form for $\alpha(T)$, Equation 9.207, show that (Equation 9.209)

$$\varepsilon \simeq \frac{8\pi V}{10} \frac{(2m\mu)^{5/2}}{mh^3} \left\{ 1 + \frac{5\pi^2}{12} \left(\frac{kT}{\mu} \right)^2 \right\}.$$

9.11.3 Consider a gas of n identical particles, where n (even) is not necessarily large compared to 1. The energy levels are those of the harmonic oscillator, $E_i = \hbar\omega(i + 1/2)$. Considering these as identical Fermi–Dirac particles, find μ (Fermi energy), $\varepsilon_{T=0}$ (total energy at $T = 0$), and T_F (Fermi temperature).

9.11.4 Consider a gas of n identical nonrelativistic fermions. This gas, in an enclosure of volume V, is placed in a uniform magnetic field, B_z, pointing along the positive z-axis. An atom with momentum $\vec{p}$ can have only two possible energies:

$$\varepsilon(\pm) = \frac{\vec{p}^{\,2}}{2m} \pm \lambda \, B_z,$$

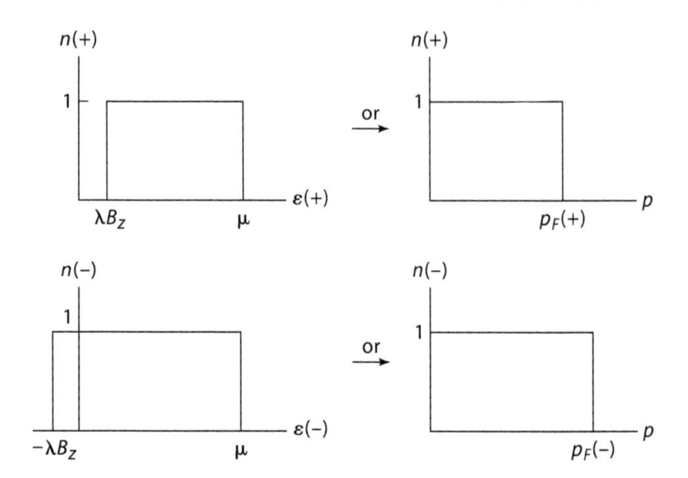

FIGURE 9.13 The occupation numbers for the gas as a function of energy on the left; occupation number as a function of momentum magnitude on the right. Top figures for the ε(+) case, bottom for the ε(−) case.

where λ is the atom's magnetic moment. At exactly $T=0$ (zero temperature), the occupation numbers $n(\pm)$ (associated with the energies $\varepsilon(\pm)$) look like Figure 9.13 (assuming $\lambda>0$; μ is the common value for the "Fermi energy"). Assuming $n(\pm)$ are given by

$$n(\pm)=\frac{4\pi V}{h^3}\int_0^{p_F(\pm)} dp\, p^2,$$

at $T=0$, show that ($n=n(-)+n(+)$)

$$n(-)-n(+)\simeq\frac{3n\lambda B_z}{2VkT_F},\quad n\simeq\frac{8\pi}{3h^3}(2m\mu)^{3/2},$$

where $kT_F=\mu$. (Make the approximation $|\lambda B_z|\ll\mu$.)

9.12.1 (a) Show Equation 9.233 holds for the system energy, $<H>$. The removal energy E_i is the sum of the kinetic, potential and interaction energies associated with electron i.

(b) Show that Equation 9.230 follows from the variation of Equation 9.229.

9.12.2 Look up the results of calculations, as well as comparisons to experiment, of some atomic energy levels using the Hartree–Fock equations in the literature.

Time-Dependent Systems

Synopsis: The time development of quantum systems with time-dependent Hamiltonians or potentials will be discussed in this chapter. First, we set up the formalism of time development in a very general way and discover that there are two basic Hilbert space descriptions of these processes. As discussed earlier, exact solutions for physical systems are rare, and we therefore find approximate ways of describing fast and slow time developments. Concentrating on two-state systems, we investigate the important role complex phases play in the slow evolution of quantum systems. In this context, we learn how to apply perturbation theory to our formalism, resulting in wonderfully practical formulas that can readily be applied in the lab, including elastic scattering and spontaneous emission rates. Finally, the important topic of time decay of systems and associated uncertainties in the energies of such states will be discussed.

10.1 TIME-DEPENDENT POTENTIALS

We have not yet addressed realistic quantum systems with nontrivial time dependence in this presentation. Quantum systems in general are not static, but change or dynamically evolve when external forces or fields perturb them. Our previous discussion in Chapter 2, Section 2.16 covered only the basics and involved system Hamiltonians that did not have explicit time dependence. It is helpful to introduce time dependence in an additive sense to build on our previous quantum formalism. We will assume a time-dependent Hamiltonian $H(t)$ of the form

$$H(t) = H_0 + u(t)V(t), \tag{10.1}$$

where $u(t)$ is a so-called step function

$$u(t) \equiv \begin{cases} 0, t < 0, \\ 1, t > 0. \end{cases} \tag{10.2}$$

This limits the explicit time dependence to $t > 0$. H_0 is a time-independent Hamiltonian that we assume has a discrete set of eigenkets $|n\rangle$ with eigenvalues E_n. This allows us to construct a Hilbert state-space for the whole system at all times as

$$H_0 |n\rangle = E_n |n\rangle, \tag{10.3}$$

which we assume is complete:

$$\sum_{n'} |n'\rangle\langle n'| = 1. \tag{10.4}$$

Although there is no explicit reference to time in these states, we have learned that energy eigenkets naturally evolve in time in order to satisfy the time-dependent Schrödinger equation,

$$i\hbar\frac{\partial}{\partial t}|\psi,t\rangle = H(t)|\psi,t\rangle. \tag{10.5}$$

The general form of a time-dependent state in this context is

$$|\psi,t\rangle = \sum_{n'} C_{n'}(t)e^{-i\omega_{n'}t}|n'\rangle, \tag{10.6}$$

where $\omega_n \equiv E_n/\hbar$. By projecting Equation 10.6 into the bra state $\langle n|$, we learn the coefficients $C_n(t)$ are formally given by

$$C_n(t) = \langle n|\psi,t\rangle e^{i\omega_n t}. \tag{10.7}$$

Let us now substitute Equation 10.6 into 10.5 in order to develop the differential behavior of these coefficients for $t > 0$. This gives (the overdot denotes a time derivative)

$$i\hbar\dot{C}_n(t) = \sum_{n'} C_{n'}(t)V_{nn'}(t)e^{i\omega_{nn'}t}, \tag{10.8}$$

where $V_{nn'}(t) \equiv \langle n|V(t)|n'\rangle$ and $\omega_{nn'} \equiv \omega_n - \omega_{n'}$. Note that the potential operator $V(t)$ may be a function of position or other variables, so a lot is hidden in this notation! Equations 10.8 are a set of coupled, first-order differential equations. Also note that the matrix elements themselves obey the identity

$$\langle n|V(t)|n'\rangle^* \left(= V^*_{nn'}(t)\right) = \langle n'|V(t)|n\rangle, \tag{10.9}$$

for $V^\dagger(t) = V(t)$. These equations may be solved in a *perturbative* manner, as we will see in Section 10.7.

The most important thing to verify before we go any further is that this formalism conserves probability. The coefficients we introduced represent a system with probabilities associated with the various eigenstates $|n\rangle$ such that

$$\sum_{n'} |C_{n'}(t)|^2 = 1. \tag{10.10}$$

The rate of change of these coefficients represents the probability flow into or out of these states, so I define the current j_n as in

$$j_n \equiv \frac{\partial}{\partial t}|C_n(t)|^2 = 2\,\mathrm{Re}\left(\dot{C}_n(t)C^*_n(t)\right). \tag{10.11}$$

Probability conservation is just the statement that the *total* flow must be zero, and can be simply stated as

$$\sum_{n'} j_{n'} = 0, \tag{10.12}$$

or equivalently

$$\mathrm{Re}\left(\sum_{n'}\dot{C}_{n'}(t)C_{n'}^{*}(t)\right)=0. \tag{10.13}$$

Using Equation 10.8 as well as completeness, Equation 10.4, one can show (see Problem 10.1.2)

$$\sum_{n'}\dot{C}_{n'}(t)C_{n'}^{*}(t)=-\frac{i}{\hbar}\langle\psi,t\,|\,V(t)\,|\,\psi,t\rangle, \tag{10.14}$$

which is purely imaginary for Hermitian $V(t)$, so Equation 10.13 is satisfied.

We can eliminate reference to diagonal element contribution in Equation 10.8 by a redefinition of the probability coefficients. Consider the new coefficients $\tilde{C}_n(t)$ such that

$$\tilde{C}_n(t)\equiv C_n(t)e^{\frac{i}{\hbar}\int_0^t dt' V_{nn}(t')}\Rightarrow|\tilde{C}_n(t)|^2=|C_n(t)|^2. \tag{10.15}$$

One may now show (informal problem for the student) that they obey the differential equations

$$i\hbar\dot{\tilde{C}}_n(t)=\sum_{n'}\tilde{C}_{n'}(t)\tilde{V}_{nn'}(t)e^{i\omega_{nn'}t}, \tag{10.16}$$

where we define the new matrix elements:

$$\tilde{V}_{nn}(t)\equiv 0;\quad \tilde{V}_{nn'}(t)\equiv e^{\frac{i}{\hbar}\int_0^t dt'(V_{nn}(t')-V_{n'n'}(t'))}V_{nn'}(t),\ n'\neq n. \tag{10.17}$$

That is, this new $\tilde{V}_{nm}(t)$ set, which zeros the on-diagonal potential terms, is just as good as the original set. This recharacterization shows there is no loss in generality in just considering off-diagonal interaction matrix elements, which is what we will do in the following.

There is another characterization of this system that comes from considering the full time-dependent Hamiltonian, $H(t)$. Again, assuming a discrete set of states, we have

$$H(t)\,|\,n,t\rangle=\bar{E}_n(t)\,|\,n,t\rangle. \tag{10.18}$$

The bar above the energy eigenvalues distinguishes these values from the H_0 set. Note that these are also time-dependent, in general. I could also have introduced a bar over the time-dependent eigenkets in Equation 10.18, but the reference to time here will be sufficient to distinguish this new set from that in Equation 10.3. Completeness has the simple form,

$$\sum_{n'}|\,n',t\rangle\langle n',t\,|=1, \tag{10.19}$$

and we characterize a general state in this basis as

$$|\,\psi,t\rangle=\sum_{n'}\bar{C}_{n'}(t)e^{i\theta_{n'}(t)}\,|\,n',t\rangle. \tag{10.20}$$

One should note that the same time-dependent Schrödinger equation, Equation 10.5, also holds here. The quantity $\theta_n(t)$ is called the *dynamical phase* associated with the nth state. It is given by

$$\theta_n(t) \equiv -\int_0^t dt'\bar{\omega}_n(t'); \quad \bar{\omega}_n(t) \equiv \frac{\bar{E}_n(t)}{\hbar}. \tag{10.21}$$

This quantity is just the analog of the time-development phase factors for the H_0 basis seen in Equation 10.6. Formally,

$$\bar{C}_n(t) = \langle n,t \mid \psi,t\rangle e^{-i\theta_n(t)}. \tag{10.22}$$

Now, putting Equation 10.20 into 10.5 we find

$$i\hbar\frac{\partial \mid \psi,t\rangle}{\partial t} = i\hbar\sum_{n'} e^{i\theta_{n'}(t)}\left[\dot{\bar{C}}_{n'}(t)\mid n',t\rangle - i\bar{\omega}_{n'}(t)\bar{C}_{n'}(t)\mid n',t\rangle + \bar{C}_{n'}(t)\frac{\partial}{\partial t}\mid n',t\rangle\right] \tag{10.23}$$

and

$$H(t)\left|\psi,t\right\rangle = \hbar\sum_{n'}\bar{\omega}_{n'}(t)\bar{C}_{n'}(t)e^{i\theta_{n'}(t)}\left|n',t\right\rangle. \tag{10.24}$$

Equating Equations 10.23 and 10.24 now gives the dynamical equation satisfied by the $\bar{C}_n(t)$ coefficients:

$$\sum_{n'}\left[\dot{\bar{C}}_{n'}(t)e^{i\theta_{n'}(t)}\mid n',t\rangle + \bar{C}_{n'}(t)e^{i\theta_{n'}(t)}\frac{\partial}{\partial t}\mid n',t\rangle\right] = 0. \tag{10.25}$$

As a final step, project Equation 10.25 into the $\langle n,t|$ space:

$$\dot{\bar{C}}_n(t) = -\sum_{n'} e^{i(\theta_{n'}(t)-\theta_n(t))}\bar{C}_{n'}(t)\langle n,t\mid\frac{\partial}{\partial t}\mid n',t\rangle. \tag{10.26}$$

This is the final differential result for these amplitudes. We will come back to this equation in the next section, where a type of slow time-development solution will be derived. However, before we do this, it is crucially important to check probability conservation for these new equations. Let us again form the current combination introduced in Equation 10.11. We start with

$$\sum_n \dot{\bar{C}}_n(t)\bar{C}_n^*(t) = -\sum_{n',n} e^{i(\theta_{n'}(t)-\theta_n(t))}\bar{C}_n^*(t)\bar{C}_{n'}(t)\langle n,t\mid\frac{\partial}{\partial t}\mid n',t\rangle. \tag{10.27}$$

Relating the $\bar{C}_n(t)$ coefficients to the wave function using Equation 10.22 and removing a redundant complete set sum gives

$$\begin{aligned}\sum_n \dot{\bar{C}}_n(t)\bar{C}_n^*(t) &= -\sum_{n',n}\langle\psi,t\mid n,t\rangle\langle n,t\mid\left(\frac{\partial}{\partial t}\mid n',t\rangle\right)\langle n',t\mid\psi,t\rangle \\ &= -\sum_{n'}\langle\psi,t\mid\left(\frac{\partial}{\partial t}\mid n',t\rangle\right)\langle n',t\mid\psi,t\rangle.\end{aligned} \tag{10.28}$$

However, completeness also implies

$$\frac{\partial}{\partial t}\left(\sum_{n'}|n',t\rangle\langle n',t|\right)=0 \;\Rightarrow\; \sum_{n'}\left(\frac{\partial}{\partial t}|n',t\rangle\right)\langle n',t| = -\sum_{n'}|n',t\rangle\frac{\partial}{\partial t}\langle n',t|. \tag{10.29}$$

As applied to the right-hand side of Equation 10.28, this gives

$$\begin{aligned}\sum_{n'}\langle\psi,t|\left(\frac{\partial}{\partial t}|n',t\rangle\right)\langle n',t|\psi,t\rangle &= -\sum_{n'}\langle\psi,t|n',t\rangle\left(\frac{\partial}{\partial t}\langle n',t|\right)|\psi,t\rangle \\ &= -\sum_{n'}\left(\langle\psi,t|\left(\frac{\partial}{\partial t}|n',t\rangle\right)\langle n',t|\psi,t\rangle\right)^*,\end{aligned} \tag{10.30}$$

which, since the complex conjugate produces an overall minus sign, reveals this quantity is purely imaginary. This implies

$$\mathrm{Re}\left(\sum_n \dot{\bar{C}}_n(t)\bar{C}_n^*(t)\right)=0, \tag{10.31}$$

and therefore probability is conserved as in Equation 10.13.

What is the relationship between the $\bar{C}_n(t)$ and the $C_n(t)$? Inserting completeness in the form of Equation 10.4 in 10.22 and using the definition of Equation 10.7 gives us

$$e^{i\theta_n(t)}\bar{C}_n(t)=\sum_{n'} e^{-i\omega_{n'}t}C_{n'}(t)\langle n,t|n'\rangle. \tag{10.32}$$

This equation will allow us to solve for the $\bar{C}_n(t)$ coefficients if the $C_n(t)$ coefficients have already been determined. We will use this approach in Section 10.3.

In preparation for the next section, let us separate out a special term in Equation 10.26. We will define the quantity

$$\omega_n^A(t)\equiv i\langle n,t|\frac{\partial}{\partial t}|n,t\rangle, \tag{10.33}$$

and will call this the *adiabatic angular velocity*. It is always a real quantity. Then our $\bar{C}_n(t)$ system of equations reads

$$\dot{\bar{C}}_n(t)=i\omega_n^A(t)\bar{C}_n(t)-\sum_{n'\neq n} e^{i(\theta_{n'}(t)-\theta_n(t))}\bar{C}_{n'}(t)\langle n,t|\frac{\partial}{\partial t}|n',t\rangle. \tag{10.34}$$

10.2 SUDDEN AND SLOW QUANTUM TRANSITIONS

Physicists love to look at extremes when it comes to calculating quantities, often because these extreme conditions have simpler and easier solutions.

First, let us consider sudden transitions. We take our Hamiltonian to be

$$H(t)=H_1+u(t)(H_2-H_1) \tag{10.35}$$

where $u(t)$ is again the step function that turns on at $t=0$. We assume discrete states given by

$$H_1|n_1\rangle=E_{n_1}|n_1\rangle;\quad H_2|n_2\rangle=E_{n_2}|n_2\rangle, \tag{10.36}$$

$$|\psi,0^-\rangle = |\psi,0^+\rangle \equiv |\psi\rangle, \tag{10.37}$$

where the original, given $|\psi\rangle$ state satisfies state normalization, $\langle\psi|\psi\rangle = 1$. The notation in Equation 10.37 indicates continuity of the initial wave function a time just before, $0^- \equiv 0 - \varepsilon$, or just after, $0^+ \equiv 0 + \varepsilon$, the transition, where ε is an infinitely small, positive quantity. The overlap amplitude for $|\psi\rangle$ being found in state $\langle n_2|$ is $\langle n_2|\psi\rangle$, leading to the probability for a transition, $|\langle n_2|\psi\rangle|^2$. The probability conservation statement

$$\sum_{n_2} |\langle n_2|\psi\rangle|^2 = 1, \tag{10.38}$$

derived by insertion of completeness in state normalization, verifies this interpretation.

It is natural to ask about energy conservation in this case. Because of the transition, one has that the expectation values of the system energies can differ by an amount

$$\Delta W \equiv \langle\psi|H_2 - H_1|\psi\rangle. \tag{10.39}$$

ΔW can be considered the external work done on the system during the transition. However, note that this is an *expectation value*, and this quantity can be either positive or negative for individual events. In the case of particle decays, this additional energy is manifested in the decay products.

One place this formalism is useful is in weak (beta) decays of nuclei. When, for example, one of the neutrons (protons) in a nucleus changes into a proton (neutron) with the emission of an electron (positron) and an antineutrino (neutrino), the bound atomic electrons are subject to a new Coulomb field with $Z \to Z+1$ ($Z \to Z-1$), where Z, the atomic number, counts the number of protons. For such a change to be considered sudden, the transition time, $h/|\Delta E_N|$, where ΔE_N is the difference in energy of the two nuclear states, must be small compared to atomic timescales, $h/|\Delta E_A|$, where ΔE_A is similarly the difference in energy of the atomic states. For these types of decays, it turns out transition times are fast for small Z but switch over to intermediate to slow for large Z where atomic transition energies approach or exceed nuclear energy differences.*

Slow changes are also termed *adiabatic*. We assume the initial state is in one of the eigenstates from the $H(t)$ basis. The appropriate formalism to use for this discussion involves the $\bar{C}_n(t)$ coefficients. Let us assume initial values $\bar{C}_a(0^+) = 1$ and $\bar{C}_n(0^+) = 0$ for $n \neq a$. Then, the approximate solution from Equation 10.34 for $\bar{C}_a(t)$ is

$$\bar{C}_a(t) \approx e^{i\gamma_a(t)}; \quad \gamma_a(t) \equiv \int_0^t dt'\, \omega_a^A(t'). \tag{10.40}$$

The state $\langle a,t|\psi,t\rangle$ therefore has two complex phases, the dynamical phase $\theta_a(t)$ and an *adiabatic phase* $\gamma_a(t)$:

$$\langle a,t|\psi,t\rangle = \bar{C}_a(t)e^{i\theta_a(t)} \approx e^{i\gamma_a(t)}e^{i\theta_a(t)}. \tag{10.41}$$

* See J. Chizma, G. Karl, and V.A. Novikov, *Phys. Rev. C* **58**, 3674 (1998). This reference gives a more detailed condition for the turnover from sudden to adiabatic based upon comparison of beta and atomic kinetic energies.

The differential equation for the other coefficients now reads

$$\dot{\bar{C}}_n(t) \approx i\omega_n^A(t)\bar{C}_n(t) - e^{i(\theta_a(t)-\theta_n(t)+\gamma_a(t))}\langle n,t|\frac{\partial}{\partial t}|a,t\rangle. \tag{10.42}$$

Note that

$$e^{i\gamma_n(t)}\frac{d}{dt}\left(\bar{C}_n(t)e^{-i\gamma_n(t)}\right) = \dot{\bar{C}}_n(t) - i\omega_n^A(t)\bar{C}_n(t). \tag{10.43}$$

This means that

$$\bar{C}_n(t)e^{-i\gamma_n(t)} \approx -\int_0^t dt'\, e^{i(\theta_a(t')-\theta_n(t')+\gamma_a(t')-\gamma_n(t'))}\langle n,t'|\frac{\partial}{\partial t'}|a,t'\rangle. \tag{10.44}$$

It is awkward to carry time arguments on the dynamical and adiabatic phases and angular velocities at this point, so in the following we simplify to $\theta_n \equiv \theta_n(t)$, $\gamma_n \equiv \gamma_n(t)$, $\bar{\omega}_n \equiv \bar{\omega}_n(t)$, and $\omega_n^A \equiv \omega_n^A(t)$ for all n. In preparation for the integration, let us write the integrand in Equation 10.44 as

$$e^{i(\theta_a-\theta_n+\gamma_a-\gamma_n)}\langle n,t|\frac{\partial}{\partial t}|a,t\rangle =$$

$$\frac{d}{dt}\left(\frac{\langle n,t|\frac{\partial}{\partial t}|a,t\rangle e^{i(\theta_a-\theta_n+\gamma_a-\gamma_n)}}{i\left(-\bar{\omega}_a+\bar{\omega}_n+\omega_a^A-\omega_n^A\right)}\right) - e^{i(\theta_a-\theta_n+\gamma_a-\gamma_n)}\frac{d}{dt}\left(\frac{\langle n,t|\frac{\partial}{\partial t}|a,t\rangle}{i\left(-\bar{\omega}_a+\bar{\omega}_n+\omega_a^A-\omega_n^A\right)}\right). \tag{10.45}$$

We have that $\langle n,t|\psi,t\rangle = e^{i\theta_n(t)}\bar{C}_n(t)$ from Equation 10.22. The first term in Equation 10.45 may now be integrated exactly, and we have

$$\langle n,t|\psi,t\rangle \approx -2\frac{\langle n,t|\frac{\partial}{\partial t}|a,t\rangle}{\left(-\bar{\omega}_a+\bar{\omega}_n+\omega_a^A-\omega_n^A\right)}e^{\frac{i}{2}(\theta_a+\theta_n+\gamma_a+\gamma_n)}\sin\left(\frac{1}{2}(\theta_a-\theta_n+\gamma_a-\gamma_n)t\right)$$

$$-ie^{i(\theta_n+\gamma_n)}\int_0^t dt'\, e^{i(\theta_a-\theta_n+\gamma_a-\gamma_n)}\frac{d}{dt'}\left(\frac{\langle n,t'|\frac{\partial}{\partial t}|a,t'\rangle}{\left(-\bar{\omega}_a+\bar{\omega}_n+\omega_a^A-\omega_n^A\right)}\right). \tag{10.46}$$

Defining

$$Q_{an} \equiv \frac{\langle n,t|\frac{\partial}{\partial t}|a,t\rangle}{\left(-\bar{\omega}_a+\bar{\omega}_n\right)}, \tag{10.47}$$

then, for adiabatic changes characterized by the three statements,

$$\left|\omega_{a,n}^A\right| \ll \left|\bar{\omega}_{a,n}\right|,\quad \left|Q_{an}\right| \ll 1,\quad \left|\frac{\dot{Q}_{an}}{\left(-\bar{\omega}_a+\bar{\omega}_n\right)}\right| \ll 1, \tag{10.48}$$

we have the remarkably simple result that

$$\langle n,t|\psi,t\rangle \approx -2\frac{\langle n,t|\frac{\partial}{\partial t}|a,t\rangle}{(-\bar{\omega}_a+\bar{\omega}_n)}e^{\frac{i}{2}(\theta_a+\theta_n+\gamma_a+\gamma_n)}\sin\left(\frac{1}{2}(\theta_a-\theta_n+\gamma_a-\gamma_n)t\right). \quad (10.49)$$

The first inequality in Equation 10.48 is just a statement that any adiabatic frequency can be made vanishingly small compared to any intrinsic system frequency. The implication is that the system stays in the original eigenstate, acquiring only an adiabatic phase change, as seen in Equation 10.40, and transitions to other states $n \neq a$ are suppressed because of the smallness (second inequality in Equation 10.48) and the relative slow time development (third inequality in Equation 10.48) of the Q_{an} coefficient. Now that we have approximate forms for *all* the coefficients $\bar{C}_n(t)$, we could go back and substitute these in Equation 10.34 to get an even better approximation! This defines a type of adiabatic series expansion. We will not pursue this here.

10.3 TWO-STATE PROBLEMS

We started this book with an examination of the simplest possible quantum system, the two-state problem. The simplest manifestation is a spin-1/2 system. Such systems can interact electromagnetically with external fields, which can drive transitions. Before examining more realistic applications, we will first look at general behaviors of such time-dependent systems as we get used to the formalism.

The following Hamiltonian describes an harmonic single-frequency system that "turns on" at $t=0$:

$$H_0=\begin{pmatrix} E_1 & 0 \\ 0 & E_2 \end{pmatrix};\quad V(t)=u(t)\begin{pmatrix} 0 & \gamma e^{-i\omega t} \\ \gamma e^{i\omega t} & 0 \end{pmatrix}, \quad (10.50)$$

$$\Rightarrow H(t)=\begin{pmatrix} E_1 & u(t)\gamma e^{-i\omega t} \\ u(t)\gamma e^{i\omega t} & E_2 \end{pmatrix}. \quad (10.51)$$

Starting with Equation 10.8, for $t>0$ we have the matrix elements and the differential equations,

$$V_{12}(t)=\gamma e^{-i\omega t};\quad V_{21}(t)=\gamma e^{i\omega t}, \quad (10.52)$$

$$i\hbar\dot{C}_1(t)=\gamma C_2(t)e^{i(\omega_{12}-\omega)t}, \quad (10.53)$$

$$i\hbar\dot{C}_2(t)=\gamma C_1(t)e^{i(\omega-\omega_{12})t}, \quad (10.54)$$

where $\omega_{12}\equiv\omega_1-\omega_2$ and γ is a real constant. We will view ω as a driving angular frequency. The characteristic roots of the second-order differential equations for $C_1(t)$ and $C_2(t)$ are (students verify!)

$$C_1(t):\quad \omega_1^{\pm}\equiv\frac{1}{2}\left[\omega_{12}-\omega\pm\kappa(\omega)\right], \quad (10.55)$$

$$C_2(t): \quad \omega_2^{\pm} \equiv \frac{1}{2}\left[\omega - \omega_{12} \pm \kappa(\omega)\right], \tag{10.56}$$

where

$$\kappa(\omega) \equiv \sqrt{(\omega_{12} - \omega)^2 + 4(\gamma/\hbar)^2}. \tag{10.57}$$

We will see that $\kappa(\omega)$ corresponds to the angular frequency at which the associated two-state probabilities oscillate. To illustrate this, let us assume that $C_1(0^+) = 1, C_2(0^+) = 0$, where we are leaving out a possible irrelevant overall phase factor on the state. The solution is

$$C_1(t) = e^{\frac{i}{2}(\omega_{12} - \omega)t}\left(\cos\left(\frac{\kappa(\omega)t}{2}\right) + i\frac{\omega - \omega_{12}}{\kappa(\omega)}\sin\left(\frac{\kappa(\omega)t}{2}\right)\right), \tag{10.58}$$

$$C_2(t) = -\frac{2i(\gamma/\hbar)}{\kappa(\omega)}e^{\frac{i}{2}(\omega - \omega_{12})t}\sin\left(\frac{\kappa(\omega)t}{2}\right). \tag{10.59}$$

The probabilities are given by the absolute squares of these, and the combination is called *Rabi's formula*:

$$|C_2(t)|^2 = \frac{4(\gamma/\hbar)^2}{\kappa(\omega)^2}\sin^2\left(\frac{\kappa(\omega)t}{2}\right); \quad |C_1(t)|^2 = 1 - |C_2(t)|^2. \tag{10.60}$$

Note the maximal initial population of the "1" state and the transition to minimal population at $t = \pi/\kappa(\omega)$. Note also that $\kappa(\omega)$ is minimized and the extremal values of $|C_1(t)|^2, |C_2(t)|^2$ become 0, 1 when $\omega = \omega_{12}$. In the following we will introduce physical pictures in which ω is an external driving frequency; this condition produces a maximal response, or resonance. We correctly find that the sum of probabilities is unity at all times; see Figure 10.1.

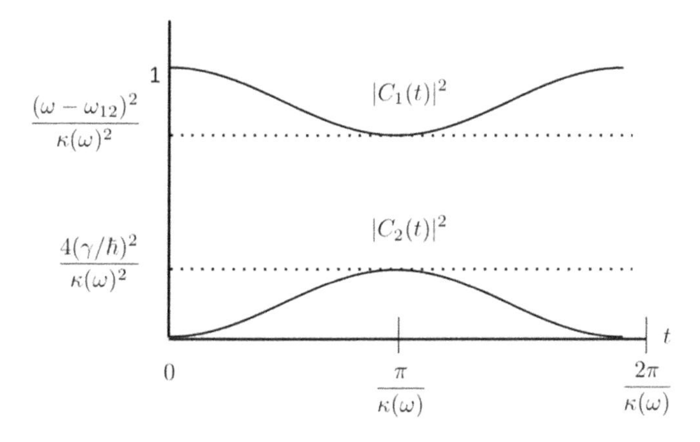

FIGURE 10.1 A qualitative illustration of the probability amplitudes $|C_1(t)|^2$ and $|C_2(t)|^2$ as functions of time for the two-state system.

One may check that the eigenvalues, $\bar{E}_{1,2} \equiv \hbar\bar{\omega}_{1,2}$, of $H(t)$ for $t>0$ are constants given by

$$\bar{\omega}_1 = \frac{1}{2}\left[\omega_1 + \omega_2 + \kappa(0)\right], \tag{10.61}$$

$$\bar{\omega}_2 = \frac{1}{2}\left[\omega_1 + \omega_2 - \kappa(0)\right]. \tag{10.62}$$

Notice these are chosen in such a way that for $\omega_1 > \omega_2$ we have $\bar{\omega}_1 \to \omega_1$ and $\bar{\omega}_2 \to \omega_2$ for $\gamma \to 0$. Since these are constants, the full $\bar{\omega}_{1,2}(t)$ notation is unnecessary. The time-dependent eigenvectors of $H(t)$ are designated $\langle n|1,t\rangle$ and $\langle n|2,t\rangle$, and are given in matrix notation as

$$\langle n|1,t\rangle \equiv \begin{pmatrix} x_1 \\ y_1 e^{i\omega t} \end{pmatrix} \begin{matrix} n=1, \\ n=2. \end{matrix} \tag{10.63}$$

$$\langle n|2,t\rangle \equiv \begin{pmatrix} x_2 \\ y_2 e^{i\omega t} \end{pmatrix} \begin{matrix} n=1, \\ n=2. \end{matrix} \tag{10.64}$$

In this context $n=1$, 2 designates the first or second row of the column matrix, and $\langle n|$ is considered the bra-space associated with H_0. We will have a comment about the distribution of time dependence between the top and bottom components, which actually is not unique, at the end of Section 10.5. Students may verify that the normalized ($x_1^2 + y_1^2 = 1$, $x_2^2 + y_2^2 = 1$) eigenvector components are

$$x_1 = -y_2 = \sqrt{\frac{\omega_1 - \omega_2 + \kappa(0)}{2\kappa(0)}};\ \ x_2 = y_1 = S_\gamma \sqrt{\frac{\omega_2 - \omega_1 + \kappa(0)}{2\kappa(0)}}\ \ (S_\gamma \equiv \text{sign}(\gamma)). \tag{10.65}$$

Note that for $\gamma \to 0$ and $\omega_1 > \omega_2$ one has $x_1 \to 1$ and $y_1 \to 0$, whereas $y_2 \to -1$ and $x_2 \to 0$. Also note $\langle 2,t|1,t\rangle = x_2 x_1 + y_2 y_1 = 0$, expressing orthogonality.

Let us solve a problem in this new basis in order to investigate adiabatic changes. Equation 10.32 gives the relationship between the H_0 and $H(t)$ amplitude coefficients,

$$e^{i\theta_n(t)}\bar{C}_n(t) = \sum_{n'} e^{-i\omega_{n'}t}\tilde{C}_{n'}(t)\,\langle n,t|n'\rangle, \tag{10.66}$$

where we have made the notational replacement $C_{n'}(t) \to \tilde{C}_{n'}(t)$ on the right-hand side to distinguish this new solution from Equations 10.58 and 10.59. Explicitly

$$e^{i\theta_1(t)}\bar{C}_1(t) = \tilde{C}_1(t)e^{-i\omega_1 t}x_1 + \tilde{C}_2(t)e^{-i(\omega_2+\omega)t}y_1, \tag{10.67}$$

$$e^{i\theta_2(t)}\bar{C}_2(t) = \tilde{C}_1(t)e^{-i\omega_1 t}x_2 + \tilde{C}_2(t)e^{-i(\omega_2+\omega)t}y_2, \tag{10.68}$$

where the complex conjugate of the components of Equations 10.63 and 10.64 were used. Let us assume the system starts in the $|2,t\rangle$ state at $t=0^+$ and follow the time evolution. This means $\bar{C}_1(0^+)=0$, $\bar{C}_2(0^+)=1$ (leaving out an irrelevant phase on $\bar{C}_2(0^+)$), which in turn gives $\tilde{C}_1(0^+) = x_2$ and $\tilde{C}_2(0^+) = y_2$. The reader may verify the new solution

$$\tilde{C}_1(t) = e^{\frac{i}{2}(\omega_{12}-\omega)t}\left(x_2 \cos\left(\frac{\kappa(\omega)t}{2}\right) + i\frac{(\omega-\omega_{12})x_2 - 2(\gamma/\hbar)y_2}{\kappa(\omega)}\sin\left(\frac{\kappa(\omega)t}{2}\right)\right), \tag{10.69}$$

$$\tilde{C}_2(t) = e^{\frac{i}{2}(\omega-\omega_{12})t}\left(y_2 \cos\left(\frac{\kappa(\omega)t}{2}\right) - i\frac{(\omega-\omega_{12})y_2 + 2(\gamma/\hbar)x_2}{\kappa(\omega)}\sin\left(\frac{\kappa(\omega)t}{2}\right)\right). \tag{10.70}$$

Using Equations 10.67 and 10.68 now gives

$$e^{i\theta_1(t)}\bar{C}_1(t) = \frac{2i\omega(\gamma/\hbar)e^{-\frac{i}{2}(\omega+\omega_1+\omega_2)t}}{\kappa(0)\kappa(\omega)}\sin\left(\frac{\kappa(\omega)t}{2}\right), \tag{10.71}$$

$$e^{i\theta_2(t)}\bar{C}_2(t) = e^{-\frac{i}{2}(\omega+\omega_1+\omega_2)t}\left(\cos\left(\frac{\kappa(\omega)t}{2}\right) + i\frac{\kappa(0)}{\kappa(\omega)}\left(1 - \frac{\omega_{12}\omega}{\kappa(0)^2}\right)\sin\left(\frac{\kappa(\omega)t}{2}\right)\right). \tag{10.72}$$

In Problem 10.3.6 I will also ask you to verify that $\bar{C}_1(t)$ has the form expected from Equation 10.49. Note that in this respect the state labeled "n" in Equation 10.49 is the state called "1" here, and the state "a" is "2" here. In the adiabatic limit, $|\omega| \ll |\omega_{1,2}|$ and $|\omega| \ll |\bar{\omega}_{1,2}|$, by recognizing that

$$-\frac{1}{2}(\omega + \omega_1 + \omega_2 - \kappa(\omega)) \approx -\bar{\omega}_2 - \frac{\omega}{2}\left(1 - \left.\frac{d\kappa(\omega)}{d\omega}\right|_{\omega=0}\right), \tag{10.73}$$

with $\left.\frac{d\kappa(\omega)}{d\omega}\right|_{\omega=0} = -\frac{\omega_{12}}{\kappa(0)}$, one has $\bar{C}_2(t) \approx e^{i\omega_2^A t}$, where

$$\omega_2^A = -\frac{\omega}{2}\left(1 + \frac{\omega_{12}}{\kappa(0)}\right). \tag{10.74}$$

This result may be verified using Equation 10.33 directly. In addition, one may show

$$\omega_1^A = -\frac{\omega}{2}\left(1 - \frac{\omega_{12}}{\kappa(0)}\right). \tag{10.75}$$

There is an interesting connection between the ω_n^A and the characteristic frequencies $\omega_n^{\pm}$ introduced earlier. We know that in general

$$\tilde{C}_2(t) = Ae^{i\omega_2^+ t} + Be^{i\omega_2^- t}, \tag{10.76}$$

where A and B are coefficients, in terms of the characteristic frequencies defined in Equation 10.56. However, from Equations 10.7, 10.19, 10.22, 10.63, and 10.64

$$\tilde{C}_2(t) = y_1\bar{C}_1(t)e^{i(\omega_2+\omega-\bar{\omega}_1)t} + y_2\bar{C}_2(t)e^{i(\omega_2+\omega-\bar{\omega}_2)t}. \tag{10.77}$$

We may now deduce A and B by setting $\omega = 0$ in Equation 10.77. In this limit one has $\bar{C}_1(t) = 0$, $\bar{C}_2(t) = 1$ by the adiabatic theorem. By comparing Equations 10.76 and 10.77 one deduces $A = y_2, B = 0$ since $\left(\omega_2 - \bar{\omega}_2\right) = \omega_2^+\big|_{\omega=0}$. One may then construct $\tilde{C}_1(t)$ from $\tilde{C}_2(t)$ using Equation 10.54. This gives the approximate values of these coefficients in the adiabatic limit:

$$\tilde{C}_2(t) \approx y_2 e^{i\omega_2^+ t}; \quad \tilde{C}_1(t) \approx x_2 e^{i\left(\omega_2^+ - \omega + \omega_{12}\right)t}. \tag{10.78}$$

This shows that these coefficients also change only by phases in this limit. Using Equation 10.76 in 10.68 now gives

$$\bar{C}_2(t) \approx e^{i\left(\bar{\omega}_2 + \omega_2^+ - \omega_2 - \omega\right)t}. \tag{10.79}$$

Expanding the exponent in Equation 10.79 in a Taylor series in ω and comparing with the expected adiabatic result, $\bar{C}_2(t) \approx e^{i\omega_2^A t}$, now identifies

$$\omega_2^A = \omega \cdot \frac{d}{d\omega}\left(\omega_2^+ - \omega\right)\bigg|_{\omega=0}. \tag{10.80}$$

In a similar fashion (Problem 10.3.8), one may show that

$$\omega_1^A = \omega \cdot \frac{d}{d\omega}\left(\omega_1^-\right)\bigg|_{\omega=0}. \tag{10.81}$$

The *adiabatic* angular velocities are thus closely related to the *characteristic* angular velocities of the system. In this regard, note the $\sim e^{i\omega t}$ dependence in the $n = 2$ (lower) components of Equations 10.63 and 10.64, as well as the lack of ω-dependence in the $n = 1$ (upper) components. Thus, in both cases ω_n^A is just the linear ω-term in the Taylor series expansion of the associated characteristic angular velocity relative to the naturally occurring ω-dependence in that amplitude.

10.4 THE BERRY PHASE

In a landmark paper,* M. V. Berry formulated the analog in quantum mechanics of the geometrical changes that occur for closed-path classical mechanical systems in their parameter spaces. A nonslipping vertical rolling hoop on a plane is one simple classical example; the point of contact changes in general after a closed path on the plane is completed. Another

* M.V. Berry, *Proc. R. Soc. Lond. A* **392**, 45 (1984). Note that this phenomenon was first discovered in the context of optical interference effects by S. Pancharatnam, *Proc. Indian Acad. Sci. A* **44**, 247 (1956).

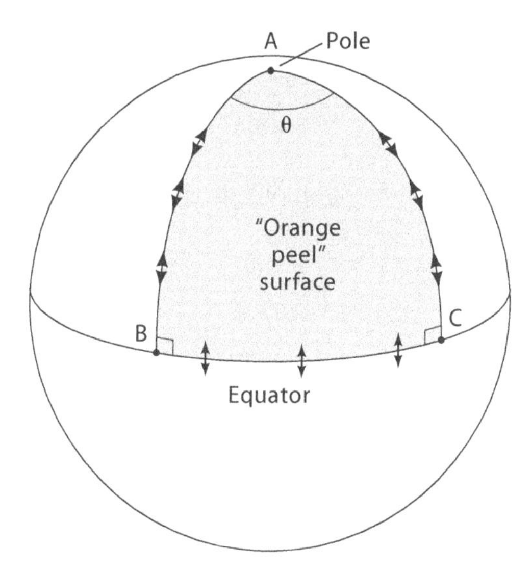

FIGURE 10.2 Transporting a plane pendulum on the given path, ABCA, moves the plane of the pendulum through the angle θ. The plane of oscillation is shown as double-headed arrows.

more advanced example is termed the Foucault plane pendulum;* in the Foucault case the plane of oscillation changes as the system is transported on a closed circular path in external space by the Earth's motion. This change comes about by the transport itself and, actually, is not a result of any noninertial forces. If, for example, the pendulum was transported along the path shown in Figure 10.2, it is clear that the angle change, θ, in the pendulum after transport would be equal to the solid angle subtended by the "orange peel" section: $\theta = \Omega_{\text{orange peel}}$. This happens because these paths are portions of great circles that maintain the relative angular relationship between pendulum and local coordinate system. Straightforward arguments (see Problem 10.4.1) then show that the angle of deviation for *any* closed path is also just the subtended solid angle: $\theta = \Omega_{\text{closed path}}$. Such systems are termed *nonholonomic*. In quantum mechanics, the *Berry phase* is a nonholonomic factor that originates in the adiabatic angular frequencies, ω_n^A (Equation 10.33). These are then integrated over the path, giving a phase (Equation 10.40). However, one can only claim the accumulated contour integral of this quantity is truly nonholonomic if, after bringing the quantum system back to the initial point or condition along a closed path in its parameter space, a nonzero phase is attached to the wave function. The Berry phase can be manifested in any type of quantum systems in which there is more than a single timescale; this happens when one of the scales is slow compared to the others and a closed path for this system also occurs. We will discuss two examples of such quantum phases in upcoming sections; one is called the *geometrical phase* and is associated with a closed spin direction path for a charged, spinning particle, and the other, called the *Aharonov–Bohm effect*, comes about when charged particles are transported around an axial magnetic flux.

* W. Wilcox, *Modern Introductory Mechanics, Part 2* (Bookboon.com), Chapter 10.

To begin with, let us back up to the harmonic two-state system of Section 10.3 and introduce the angle

$$\phi \equiv \omega t. \tag{10.82}$$

Now let the eigenvectors in Equations 10.63 and 10.64 be reexpressed as $|n,t\rangle \to |n,\phi(t)\rangle$

$$\gamma_n(\phi) = i\int_0^{\phi} d\phi' \langle n,\phi'|\frac{\partial}{\partial\phi'}|n,\phi'\rangle, \tag{10.83}$$

where we drop the reference to time in the ket $|n,\phi'\rangle$ since it is no longer necessary. We have now gone from dynamics to geometry. This is a case with a single changed variable.

In general, consider a quantum system whose time dynamics can be characterized by a single point in an associated parameter space. This is symbolized by

$$|n,t\rangle \to |n,\vec{R}(t)\rangle, \tag{10.84}$$

and we have

$$\frac{\partial}{\partial t}|n,t\rangle \to \frac{d\vec{R}(t)}{dt}\cdot\vec{\nabla}_R|n,\vec{R}\rangle, \tag{10.85}$$

where we again drop the reference to time in the ket $|n,\vec{R}\rangle$. The $\vec{R}$ are the external parameters of the system that change with time. We then have the more general statement

$$\gamma_n(\vec{R}) = i\int_0^t dt'\frac{d\vec{R}'}{dt'}\cdot\langle n,\vec{R}'|\vec{\nabla}_{R'}|n,\vec{R}'\rangle = i\int_0^{\vec{R}} d\vec{R}'\cdot\langle n,\vec{R}'|\vec{\nabla}_{R'}|n,\vec{R}'\rangle \tag{10.86}$$

for the accumulated adiabatic phase. When taken on a closed path this becomes

$$\gamma_n^C = i\oint_C d\vec{R}\cdot\langle n,\vec{R}|\vec{\nabla}_R|n,\vec{R}\rangle. \tag{10.87}$$

This is termed a Berry phase if it is nonzero in general. In the $|n'\rangle$ two-state spin basis, this becomes

$$\gamma_n^C = i\sum_{n'}\oint_C d\vec{R}\cdot\langle n,\vec{R}|n'\rangle\vec{\nabla}_R\langle n'|n,\vec{R}\rangle, \tag{10.88}$$

through the insertion of completeness.

The closed path aspect of the Berry phase equation allows one to reformulate it using Stokes' theorem.* For a smooth vector field $\vec{A}$, this is given by

$$\oint_C d\vec{R}\cdot\vec{A} = \int_S da\,\hat{n}\cdot(\vec{\nabla}\times\vec{A}). \tag{10.89}$$

* W. Wilcox and C. Thron, *Macroscopic Electrodynamics: An Introductory Graduate Treatment* (World Scientific, 2016), Section 1.4.

In applications it is important to realize that there is a global right-hand rule connection between the path direction $d\vec{R}$ and the surface normal $\check{n}$. In our case $\vec{A}_n(\vec{R}) \equiv i\langle n,\vec{R}|\vec{\nabla}_R|n,\vec{R}\rangle$, and we have

$$\gamma_n^C = i\int_S da\,\hat{n}\cdot\left[\vec{\nabla}_R\times\langle n,\vec{R}|\vec{\nabla}_R|n,\vec{R}\rangle\right]. \tag{10.90}$$

Note that since this gives a *phase*, $e^{i\gamma_n^C}$, any change in γ_n^C by $2\pi n$, $n=\pm1$, ±2, ... cannot be detected.

It was noted in Berry's original paper that the Aharonov–Bohm effect, which we will cover in Section 10.6, is an example of a nonholonomic phase. One of the early verifications of its effects involved neutron spin rotation in a twisted magnetic field.* There are several review articles available on manifestations of Berry phase in condensed matter environments.† See also the listed resource article‡ giving other reviews and seminal papers.

10.5 MAGNETIC SPIN RESONANCE AND THE GEOMETRICAL PHASE

The time-dependent Hamiltonian we will consider in this section is given by (see Equation 1.29; here we use $\vec{B}$ to represent the magnetic field)

$$H(t) = -\vec{\mu}\cdot\vec{B}, \tag{10.91}$$

where

$$\vec{\mu} = -\frac{e}{mc}\vec{S};\quad \vec{S} = \frac{\hbar}{2}\vec{\sigma}. \tag{10.92}$$

The $\vec{\sigma}$ are the Pauli matrices (see Chapter 1, Section 1.16), and e is the magnitude of the charge on an electron. The magnetic field is given as

$$\vec{B} = B_0\left[\sin\theta\cos(\omega t)\hat{e}_1 + \sin\theta\sin(\omega t)\hat{e}_2 + \cos\theta\,\hat{e}_3\right]. \tag{10.93}$$

We choose $B_0 > 0$ for definiteness. This is a two-state system with

$$H(t) = \frac{e\hbar B_0}{2mc}\begin{pmatrix} \cos\theta & e^{-i\omega t}\sin\theta \\ e^{i\omega t}\sin\theta & -\cos\theta \end{pmatrix}. \tag{10.94}$$

Let us define $\omega_0 \equiv eB_0/2mc$. We recognize that this is just a special case of the two-state system from Section 10.3 with

$$\omega_1 = \hbar\omega_0\cos\theta,\quad \omega_2 = -\hbar\omega_0\cos\theta,\quad \gamma = \hbar\omega_0\sin\theta, \tag{10.95}$$

$$\bar{\omega}_1 = \omega_0,\quad \bar{\omega}_2 = -\omega_0. \tag{10.96}$$

---

* T. Bitters and D. Dubbers, *Phys. Rev. Lett.* **59**, 251 (1987); D.J. Richardson, A.I. Kilvington, K. Green, and S.K. Lamoreaux, *Phys. Rev. Lett.* **61**, 2030 (1988).

† R. Resta, *J. Phys.: Condens. Matter* **12**, 107 (2000); D. Xiao, M.-C. Chang and Q. Niu, *Rev. Mod. Phys.* **82**, 1959 (2000).

‡ J. Anandan, J. Christian, and K. Wanelik, *Am. J. Phys.* **65**, 180 (1996).

Figure 10.3 shows the physical setup. The applied magnetic field is seen to rotate at an angular velocity ω. As we learned in Section 10.3, such a system is exactly solvable. However, this type of setup in actual laboratory applications is inconvenient. In *magnetic spin resonance* experiments, an oscillating, rather than rotating magnetic field, is actually used to produce resonance in atomic and nuclear systems in order to measure atomic electron spin energy levels or those associated with nuclear magnetic moments. If we consider an oscillating magnetic field in the 1-direction, we would have instead the Hamiltonian

$$H^{\text{osc.}}(t) = \frac{e\hbar B_0}{2mc}\begin{pmatrix} \cos\theta & \cos(\omega t)\sin\theta \\ \cos(\omega t)\sin\theta & -\cos\theta \end{pmatrix}. \tag{10.97}$$

Then we have the coupled system

$$i\hbar\dot{C}_1(t) = \frac{\gamma}{2}C_2(t)e^{i(\omega_{12}-\omega)t}\left(1+e^{2i\omega t}\right), \tag{10.98}$$

$$i\hbar\dot{C}_2(t) = \frac{\gamma}{2}C_1(t)e^{i(\omega-\omega_{12})t}\left(1+e^{-2i\omega t}\right), \tag{10.99}$$

with γ and ω_{12} values from Equation 10.95. Compare these with those in Equations 10.53 and 10.54; we see that the terms $\sim e^{\pm 2i\omega t}$ are new. These new equations can no longer be solved exactly as in the harmonic case, but, amazingly, the numerical solution is approximately the same as in Rabi's formula, Equation 10.60 (with $\gamma \to \gamma/2$), as long as the coupling is small, $|\gamma/\hbar| \ll \omega_{12}$, and the resonance condition $\omega \approx \omega_{12}$ holds. (Here we are assuming $\omega_1 > \omega_2$.) This approximation can be understood using the perturbative solution approach in Section 10.7 and will be investigated numerically in a problem at the end of the chapter (Problem 10.5.1). Basically, the extra terms in Equations 10.98 and 10.99 add a high-frequency component to the original solution, which, however, has a suppressed amplitude as long as the resonance condition and small coupling conditions hold. The approximate equivalence of an oscillating field in the laboratory with a simpler harmonic rotating field solution is called the *rotating wave approximation.*

Let us now go back to the situation in Figure 10.3 in order to investigate the implications of adiabatic changes for this system. We learned in Section 10.4 about the Berry phase. It can arise when slow changes in parameters are made. One such parameter here is the direction of the magnetic field. The eigenvectors associated with $\bar{\omega}_1$ and $\bar{\omega}_2$ from Equations 10.63 and 10.64 are

$$\langle n|1,t\rangle \equiv \begin{pmatrix} \cos\frac{\theta}{2} \\ \sin\frac{\theta}{2}e^{i\omega t} \end{pmatrix} \begin{matrix} n=1, \\ n=2. \end{matrix} \tag{10.100}$$

$$\langle n|2,t\rangle \equiv \begin{pmatrix} \sin\frac{\theta}{2} \\ -\cos\frac{\theta}{2}e^{i\omega t} \end{pmatrix} \begin{matrix} n=1, \\ n=2. \end{matrix} \tag{10.101}$$

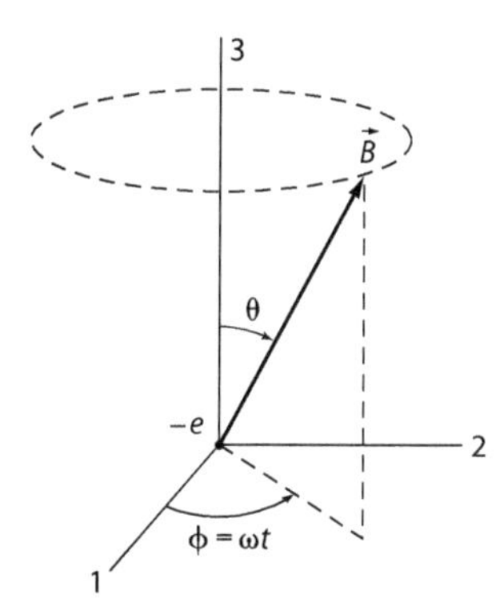

FIGURE 10.3 A stationary electron subjected to a magnetic field inclined at an angle θ whose 1 and 2 components are rotating at constant angular velocity ω. The electron spin either points along $\vec{B}$ (state "1") or opposite to it (state "2").

Also from Equations 10.75 and 10.74 we have

$$\omega_1^A = -\frac{\omega}{2}(1-\cos\theta), \quad \omega_2^A = -\frac{\omega}{2}(1+\cos\theta). \tag{10.102}$$

For a closed loop corresponding to a time period T with $\omega T = 2\pi$, we have from Equation 10.40 that

$$\gamma_1^{\text{circle}} = -\pi(1-\cos\theta) = -\frac{\Omega_{\text{top}}}{2}, \quad \gamma_2^{\text{circle}} = -\pi(1+\cos\theta) = -\frac{\Omega_{\text{bottom}}}{2}. \tag{10.103}$$

Here, we have identified Ω_{top} and Ω_{bottom} as the solid angles associated with the given circular path of the magnetic field in angular space. Of course, $\Omega_{\text{top}} + \Omega_{\text{bottom}} = 4\pi$. Since $e^{2\pi i} = 1$ we can effectively replace $-\Omega_{\text{bottom}}/2$ with $\Omega_{\text{top}}/2$. Also note that the higher energy state $\bar{E}_1 = \hbar\omega_0$ is in the magnetic eigenstate $m = 1/2$ with electron spin pointed *along* $\vec{B}$, and the lower energy state $\bar{E}_2 = -\hbar\omega_0$ has $m = -1/2$ with spin *opposite* $\vec{B}$. So, modulo 2π and replacing the labels "1" and "2" with $m = 1/2$ and $-1/2$, respectively, one has simply

$$\gamma_m^{\text{circle}}(\text{mod } 2\pi) = -m\,\Omega_{\text{top}}. \tag{10.104}$$

A remarkably simple result!

Could this solid angle result be a special case for this simple path? Now we go to the powerful geometrical point of view and use Equation 10.88. In this case

$$\vec{\nabla}_R = \frac{\hat{e}_\theta}{R}\frac{\partial}{\partial\theta} + \frac{\hat{e}_\phi}{R\sin\theta}\frac{\partial}{\partial\phi}. \tag{10.105}$$

Let us evaluate γ_1^{loop} for an arbitrary closed loop. We have

$$\gamma_1^{\text{loop}} = i\oint_{\text{loop}} d\vec{R}\cdot\begin{pmatrix}\cos\frac{\theta}{2} & \sin\frac{\theta}{2}e^{-i\phi}\end{pmatrix}\vec{\nabla}_R\begin{pmatrix}\cos\frac{\theta}{2}\\ \sin\frac{\theta}{2}e^{i\phi}\end{pmatrix}. \tag{10.106}$$

After the action of the gradient we have

$$\gamma_1^{\text{loop}} = i\oint_{\text{loop}} d\vec{R}\cdot\begin{pmatrix}\cos\frac{\theta}{2} & e^{-i\phi}\sin\frac{\theta}{2}\end{pmatrix}\begin{pmatrix}-\frac{\hat{e}_\theta}{2R}\sin\frac{\theta}{2}\\ \frac{e^{i\phi}}{R}\left(\frac{\hat{e}_\theta}{2}\cos\frac{\theta}{2}+i\hat{e}_\phi\frac{\sin\frac{\theta}{2}}{\sin\theta}\right)\end{pmatrix}. \tag{10.107}$$

Finally, we simplify this to

$$\gamma_1^{\text{loop}} = i\oint_{\text{loop}} d\vec{R}\cdot\hat{e}_\phi\,\frac{i}{R}\frac{\sin^2\frac{\theta}{2}}{\sin\theta}. \tag{10.108}$$

Using $d\vec{R}\cdot\hat{e}_\phi = R\sin\theta\, d\phi$, where θ and ϕ are no longer considered independent angles for the given loop, one then has

$$\gamma_1^{\text{loop}} = -\frac{1}{2}\oint_{\text{loop}} d\phi(1-\cos\theta). \tag{10.109}$$

Similarly,

$$\gamma_2^{\text{loop}} = -\frac{1}{2}\oint_{\text{loop}} d\phi(1+\cos\theta). \tag{10.110}$$

What is the meaning of these integrals? Let us consider Figure 10.4, which shows an oval integration path, although the shape can be generalized. The meaning of the "top" solid angle is explored topologically in Figure 10.5. It demonstrates that the right-hand rule connecting the direction of the integral loop in Equation 10.109 and the associated solid angle, Ω_{top}, is consistent under a continuous deformation of the path. We conclude that

$$\gamma_1^{\text{loop}} = -\frac{\Omega_{\text{top}}}{2},\quad \gamma_2^{\text{loop}} = -\frac{\Omega_{\text{bottom}}}{2}. \tag{10.111}$$

Therefore Equation 10.104 is unchanged but now can be applied to a *general* loop. Physically, these phases apply for the general adiabatic motion of the magnetic field in our two-state system.

We may also do an alternate calculation using Equation 10.90. Note that we may write

$$\gamma_n^C = i\oint_S da\,\hat{n}\cdot\left[\vec{\nabla}_R\times\vec{A}_n(\vec{R})\right], \tag{10.112}$$

where

$$\vec{A}_n(\vec{R}) \equiv \sum_{n'}\langle n,\vec{R}|n'\rangle\vec{\nabla}_R\langle n'|n,\vec{R}\rangle, \tag{10.113}$$

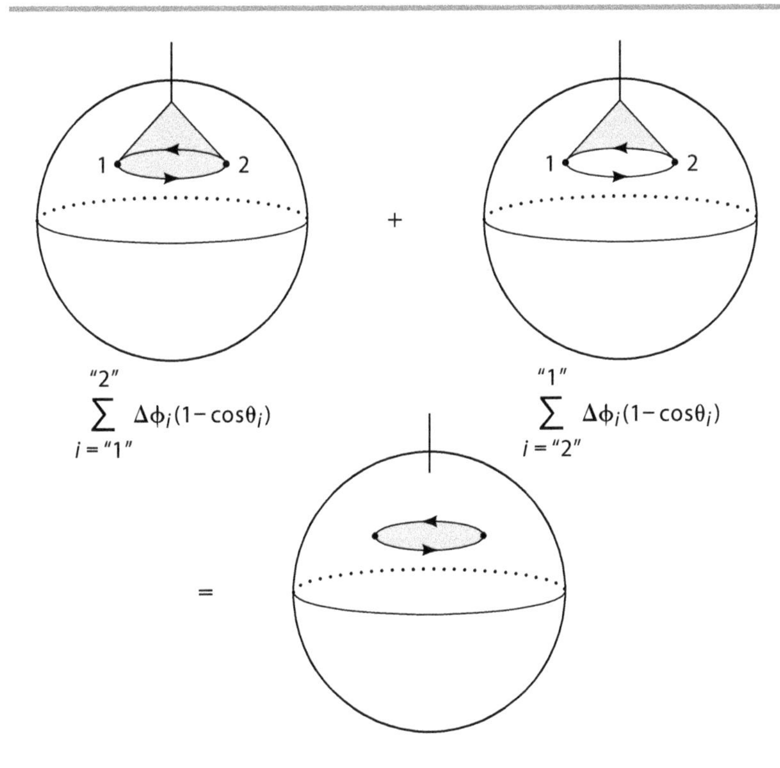

FIGURE 10.4 Evaluating Equation 10.109 for a simple loop. The integral around the loop results in the selection of the solid angle inside the loop as one integrates from 1 to 2 and back again.

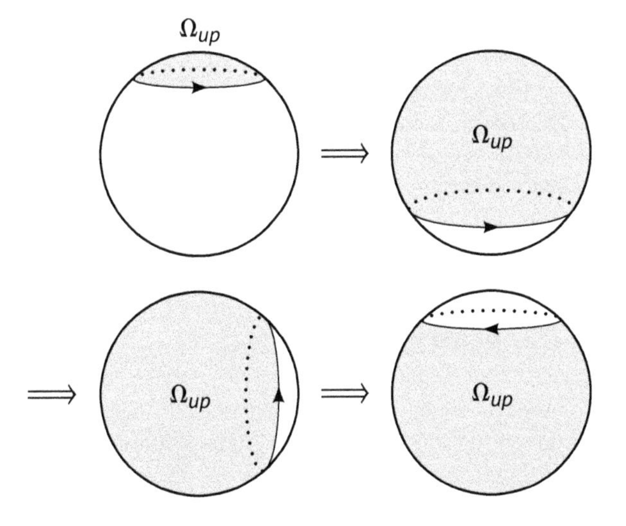

FIGURE 10.5 Showing the meaning of the "top" solid angle associated with a continuously modified circular loop.

by insertion of completeness in the original spin basis. For $n = 1$ we have already evaluated $\vec{A}_1(\vec{R})$ in the reduction from Equation 10.106 to 10.108. We have

$$\vec{A}_1(\vec{R}) = \begin{pmatrix} \cos\frac{\theta}{2} & e^{-i\phi}\sin\frac{\theta}{2} \end{pmatrix} \vec{\nabla}_R \begin{pmatrix} \cos\frac{\theta}{2} \\ e^{i\phi}\sin\frac{\theta}{2} \end{pmatrix} = \frac{i}{R}\frac{\sin^2\frac{\theta}{2}}{\sin\theta}\hat{e}_\phi. \quad (10.114)$$

The right-hand rule requires $\hat{n} = \hat{r}$ for $S =$ top in Equation 10.112. Then using the spherical curl operator gives

$$\hat{n}\cdot\left[\vec{\nabla}_R \times \vec{A}_1(\vec{R})\right] = \frac{1}{R^2}\frac{1}{\sin\theta}\left(\frac{\partial\left(A_{1\phi}(\vec{R})\sin\theta\right)}{\partial\theta}\right) = \frac{i}{2R^2}, \quad (10.115)$$

$$\Rightarrow \gamma_1^{\text{loop}} = i\oint_{\text{top}} da\, \hat{n}\cdot\left[\vec{\nabla}_R \times \vec{A}_1(\vec{R})\right] = -\frac{\Omega_{\text{top}}}{2}. \quad (10.116)$$

where $da = R^2 d\Omega$. The student is asked to confirm the result $\gamma_2^{\text{loop}} = -\frac{\Omega_{\text{bottom}}}{2}$ in Problem 10.5.2. Note that the right-hand rule requires $\hat{n} = -\hat{r}$ in this case. The factors of $\pm 1/2$ are due to this being the spin-1/2 result, for which the magnetic quantum number takes on the two values $m = \pm 1/2$. It turns out this factor would be replaced by the $(2s + 1)$ m values for arbitrary spin s, so that Equation 10.104 is even more general than we thought!

Note that when we rotate a fermion through 2π radians, such as a spin-1/2 electron, its wave function must have a sign change, as we found back in Chapter 1 (see Equations 1.222 and 1.223). If we consider a situation in which a fermion completes a rotation, we must remember to append this phase. This is the point I was alluding to when the wave functions in Equations 10.63 and 10.64 were introduced. These wave functions do not contain such a fermion phase. This removal was done for pedagogical reasons to isolate the nonholonomic part of the phase change. For a real fermion, we must remember to put this phase back in!

10.6 THE AHARONOV–BOHM EFFECT

Adiabatic phase factors can influence particles in subtle ways. The following ideas involving the effect of a magnetic solenoid on electrons was first put forward by Y. Aharonov and D. Bohm.* Let us start with the form of the Hamiltonian for a charged particle in electric and magnetic fields taken from classical mechanics

$$H = \frac{1}{2m}\left(\vec{p} - \frac{q}{c}\vec{A}(\vec{x})\right)^2 + q\Phi(\vec{x}), \quad (10.117)$$

* Y. Aharonov and D. Bohm, *Phys. Rev.* **115**, 485 (1959).

where q is the charge, while Φ and $\vec{A}$ are the scalar and vector potentials, respectively. Then with

$$\langle \vec{x} | H | \psi \rangle = i\hbar \frac{\partial}{\partial t} \langle \vec{x} | \psi \rangle, \tag{10.118}$$

$$\langle \vec{x} | \vec{p} | \psi \rangle = -i\hbar \vec{\nabla} \langle \vec{x} | \psi \rangle, \tag{10.119}$$

we have

$$i\hbar \frac{\partial}{\partial t} \psi(\vec{x},t) = \left(-\frac{\hbar^2}{2m} \left(\vec{\nabla} - i\frac{q}{\hbar c} \vec{A}(\vec{x},t) \right)^2 + q\Phi(\vec{x},t) \right) \psi(\vec{x},t), \tag{10.120}$$

which will also be constructed in Chapter 12, Section 12.6 from a gauge invariance point of view. We will specialize to static vector and scalar potentials. Let us define $\psi_0(\vec{x},t)$ such that

$$\psi(\vec{x},t) \equiv \psi_0(\vec{x},t) e^{i\frac{q}{\hbar c} \int_{\vec{x}_0}^{\vec{x}} d\vec{x}' \cdot \vec{A}(\vec{x}')}, \tag{10.121}$$

where $\vec{x}_0$ is an arbitrary reference point. Note that

$$\left(\vec{\nabla} - i\frac{q}{\hbar c} \vec{A}(\vec{x}) \right) \psi(\vec{x},t) = e^{i\frac{q}{\hbar c} \int_{\vec{x}_0}^{\vec{x}} d\vec{x}' \cdot \vec{A}(\vec{x}')} \vec{\nabla} \psi_0(\vec{x},t). \tag{10.122}$$

Then $\psi(\vec{x},t)$ is a formal solution to Equation 10.120 as long as

$$i\hbar \frac{\partial}{\partial t} \psi_0(\vec{x},t) = \left(-\frac{\hbar^2}{2m} \vec{\nabla}^2 + q\Phi(\vec{x}) \right) \psi_0(\vec{x},t). \tag{10.123}$$

Boundary conditions need to be supplied to complete the solution. These may be applied to either $\psi(\vec{x})$ or $\psi_0(\vec{x})$, as the situation demands. The phase factor in Equation 10.121 can manifest itself in unexpected ways, as we will see in the following examples.

For our first application, consider the setup in Figure 10.6. A particle with charge q is constrained to move on a circle of radius b in a field free region around a central region that contains a long solenoid that contains and confines a nonzero magnetic field along the 3-direction. The vector potential in cylindrical coordinates is given by

$$\vec{A}(\vec{x}) = \begin{cases} \dfrac{F\rho}{2\pi a^2} \hat{e}_\phi, \ \rho < a, \\ \dfrac{F}{2\pi\rho} \hat{e}_\phi, \ \rho > a, \end{cases} \tag{10.124}$$

where $F \equiv \pi a^2 B$ is the magnetic flux in the 3-direction. The curl of this gives

$$\vec{B}(\vec{x}) = \begin{cases} B\hat{e}_3, \ \rho < a, \\ 0, \quad \rho > a. \end{cases} \tag{10.125}$$

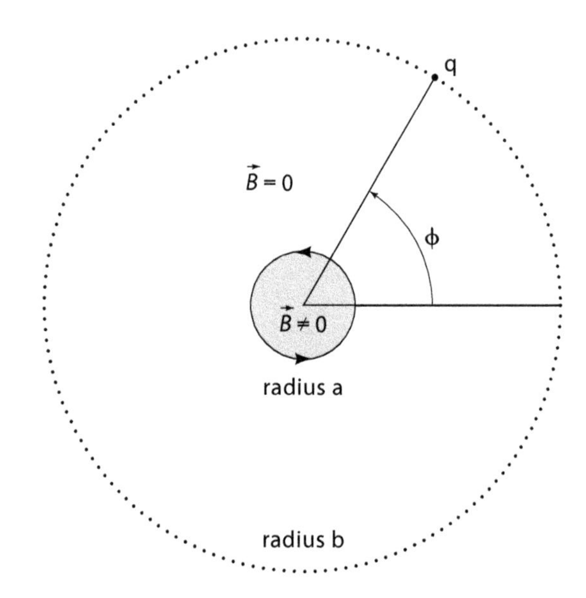

FIGURE 10.6 A view downward from the positive 3-axis on a particle constrained to move on a circle of radius b around a central magnetic field region. The magnetic field is confined to the region $\rho < a$. The circular current shown produces a magnetic field coming out of the page.

Note that if the current is counterclockwise we have $B > 0$. In this case the gauge potential phase factor from Equation 10.121 can be taken to be

$$\frac{q}{\hbar c}\int_0^{\phi} b\,d\phi' A_\phi(\phi') = \frac{qF}{\hbar c}\phi. \tag{10.126}$$

Therefore

$$\psi(\phi,t) = \psi(\phi)e^{iE_\phi t} = \psi_0(\phi)e^{i\frac{qF}{\hbar c}\phi}e^{iE_\phi t} \tag{10.127}$$

$$\Rightarrow \frac{\hbar^2}{2mb^2}\frac{\partial^2}{\partial\phi^2}\psi_0(\phi) = E_\phi\,\psi_0(\phi). \tag{10.128}$$

This gives the general form

$$\psi_0(\phi) = \frac{1}{\sqrt{2\pi}}e^{\pm i\frac{b}{\hbar}\sqrt{2mE_\phi}\,\phi}, \tag{10.129}$$

where the appropriate normalization factor for circular integration has been applied. We now require continuity in $\psi(\phi)$:

$$\psi(0) = \psi(2\pi). \tag{10.130}$$

This gives

$$\frac{qF}{\hbar c} \pm \frac{b}{\hbar}\sqrt{2mE_\phi} = n; \quad n = 0,\pm1,\pm2,\ldots \tag{10.131}$$

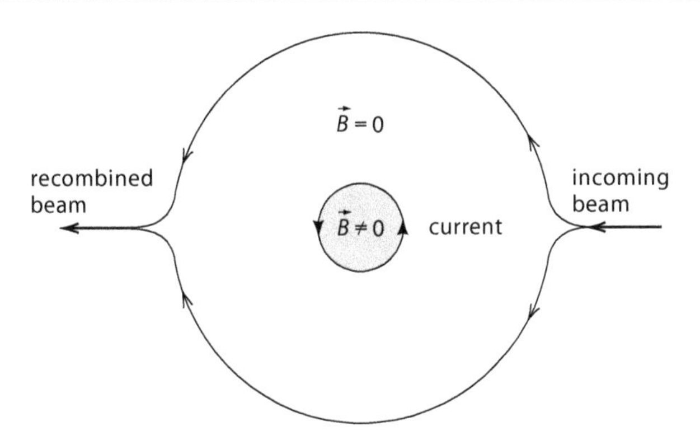

FIGURE 10.7 A view downward from the positive 3-axis on an idealized experiment involving split beams of particles sent around the central magnetic field region.

The plus or minus sign is associated with particle motion in a counterclockwise or clockwise manner, respectively. Solving for the energy now gives

$$E_\phi = \frac{\hbar^2}{2mb^2}\left(n - \frac{qF}{hc}\right)^2. \tag{10.132}$$

Let N_B be the largest magnitude integer (positive or negative) in qF/hc, so that $|N_B| \leq |qF/hc|$. Then from Equation 10.131, for energy levels with $n > N_B$, we deduce that the motion of the particle is counterclockwise; when $n < N_B$ the motion is clockwise. Thus, when the flux is changed, a given energy level may change its motion direction and can increase or decrease in value. Comparing Equation 10.132 to the classical form for a particle rotating with angular momentum L_z in a circle of radius b, $E_\phi = L_z^2/2mb^2$, also suggests that the quantum angular momentum actually is quantized as*

$$L_z = \hbar\left(n - \frac{qF}{2\pi\hbar c}\right). \tag{10.133}$$

The preceding results for energies and angular momentum values are remarkable because $\vec{B} = 0$ on the particle path, so the classical Lorentz force cannot be responsible. Thus, there is a physical effect of the vector potential itself, which is usually not thought of as a measurable field. In the literature, this situation is often referred to as the *solid-state Aharonov–Bohm effect.*

Now consider sending a split beam of particles around this solenoid on the idealized circular path shown in Figure 10.7. It is important to stipulate that the splitting maintains the coherence of the two beams. Then, if the only difference in the wave functions of the two beams is the vector potential $\vec{A}$, according to Equation 10.121 there will be a phase

* F. Wilczek, *Phys. Rev. Lett.* **48**, 1144 (1982).

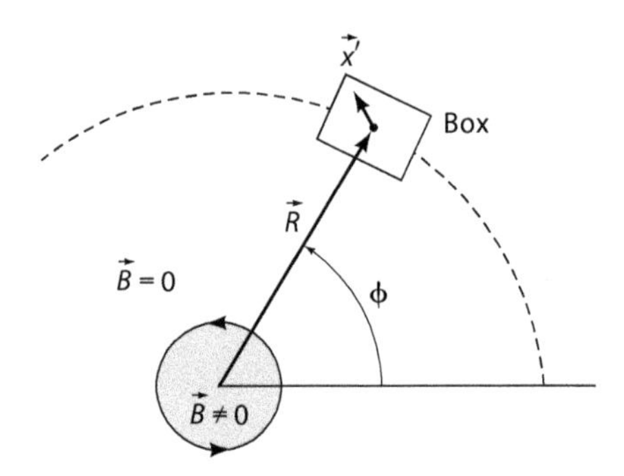

FIGURE 10.8 Close-up of the geometry for the transport of a charged particle in a box around a central magnetic field region.

difference between them as they travel on different sides of the solenoid. These phases are given by Equation 10.126 with $\phi = \pm\pi$:

$$\gamma^{\pm} \equiv \frac{q}{\hbar c} \int_0^{\pm\pi} b\, d\phi' A_\phi(\phi') = \pm \frac{qF}{2\hbar c}. \tag{10.134}$$

For $q > 0$ the phase increases for the particle traveling in the solenoid current direction and decreases for the beam on the other side. If we call the source wave function ψ_0, then when the beams are combined we will have a new wave function ψ such that

$$\psi = \psi_0 \left(e^{i\gamma^+} + e^{i\gamma^-} \right), \tag{10.135}$$

and therefore

$$|\psi|^2 = 2|\psi_0|^2 \left(1 + \cos\left(\frac{qF}{\hbar c} \right) \right). \tag{10.136}$$

This defines the relative intensity for a given flux. There is therefore complete destructive interference for $qF/\hbar c = \pm(2n-1)\pi,\ n = 1, 2, 3, \ldots$. This is a detectable beam interference effect, which has been well verified.* Other similar effects have been proposed and experimentally investigated.†

This beam interference process, the original Aharonov–Bohm effect, is actually another example of a nonholonomic or Berry phase. In order to confirm this, we will have to invent an external parameter in order to use the formulas in Section 10.4. To this end, imagine enclosing a particle in an impenetrable box in order to physically transport a charged particle around the solenoid. The situation is similar to Figure 10.6 but with a box now enclosing the charge q. A close-up picture of the situation is shown in Figure 10.8.

* The best experimental verification is in A. Tonomura et al., *Phys. Rev. Lett.* **56**, 792 (1986).

† For further exploration, see the excellent popular article: H. Batelaan and A. Tonomura, "The Aharonov–Bohm Effects: Variations on a Subtle Theme," *Physics Today* **62**, 38 (2009).

Our wave functions satisfy

$$\psi(\vec{x},t)=\psi_R(\vec{x})e^{-iE_0t};\quad \psi_R(\vec{x})=\psi_0(\vec{x}-\vec{R})e^{i\frac{q}{\hbar c}\int_{\vec{R}}^{\vec{x}}d\vec{x}'\cdot\vec{A}(\vec{x}')}. \tag{10.137}$$

The important point here is that the $\vec{x}-\vec{R}$ variable is *local* to the confined geometry. In addition, it is convenient to imagine $\vec{R}$ directed to a point in the box, although it need not be. Although the particles are confined, we assume the vector potential itself penetrates and is unmodified inside it. The confining condition means that the particles carry only a local phase from the vector potential, not a global phase as in our first example. The associated boundary conditions were applied to ψ in the solid-state version earlier but need to be applied to ψ_0 here. So, in this case

$$\left(-\frac{\hbar^2}{2m}\vec{\nabla}'^2+q\Phi\right)\psi_0(\vec{x}')=E_0\,\psi_0(\vec{x}'), \tag{10.138}$$

and $\vec{x}'\equiv\vec{x}-\vec{R}$. In addition, the boundary conditions on the enclosing box are

$$\psi_0(\vec{x}')\Big|_{\vec{x}_b'}=0, \tag{10.139}$$

where $\vec{x}_b'$ give the coordinates of the box's surface. This determines the energy values, which in this case are independent of the vector potential. We symbolize this by letting $E_0\to E_n$, $\psi_0\to\psi_n$, $\psi_R\to\psi_R^n$ for such an energy level. We also assume that ψ_n is correctly normalized. Inserting spatial completeness gives the form

$$\gamma^C=i\oint_C d\vec{R}\cdot\langle n,\vec{R}|\vec{\nabla}_R|n,\vec{R}\rangle=iR\int_0^{2\pi}d\phi_R\,\hat{e}_{\phi R}\cdot\int_{box}d^3x\,\psi_R^{n^*}(\vec{x}-\vec{R})\vec{\nabla}_R\psi_R^n(\vec{x}-\vec{R}), \tag{10.140}$$

from Equation 10.87, where the particle can be in any of the eigenstates of the box. We now use Equation 10.137, noting that

$$\vec{\nabla}_R\mathrm{e}^{i\frac{q}{\hbar c}\int_{\vec{R}}^{\vec{x}}d\vec{x}'\cdot\vec{A}(\vec{x}')}=-i\frac{q}{\hbar c}\vec{A}(\vec{R})\,\mathrm{e}^{i\frac{q}{\hbar c}\int_{\vec{R}}^{\vec{x}}d\vec{x}'\cdot\vec{A}(\vec{x}')}, \tag{10.141}$$

which, with the form for $\vec{A}(\vec{R})$ from Equation 10.124, gives

$$\gamma^C=R\int_0^{2\pi}d\phi_R\int_{box}d^3x\left[\frac{qF}{2\pi R\hbar c}|\psi_n(\vec{x}-\vec{R})|^2+i\hat{e}_{\phi R}\cdot\psi_n^*(\vec{x}-\vec{R})\vec{\nabla}_R\psi_n(\vec{x}-\vec{R})\right]. \tag{10.142}$$

Now use

$$\int_{\text{box}}d^3x\,|\psi_n(\vec{x}-\vec{R})|^2=1, \tag{10.143}$$

$$\int_{\text{box}} d^3x\, \psi_n^*\left(\vec{x}-\vec{R}\right)\vec{\nabla}_R \psi_n\left(\vec{x}-\vec{R}\right) = -\int_{\text{box}} d^3x\, \psi_n^*\left(\vec{x}-\vec{R}\right)\vec{\nabla}\psi_n\left(\vec{x}-\vec{R}\right) = 0. \tag{10.144}$$

The first equation is just particle normalization, and the second uses the $\vec{x}-\vec{R}$ argument of the wave function as well as the fact that the expectation value of the momentum operator for a bound state (at rest in the lab) is zero (Problem 10.6.2). Thus

$$\gamma^C = \frac{qF}{\hbar c}, \tag{10.145}$$

which is just the difference of the two phases seen in Equation 10.134. Note that transportation of the box itself, if treated quantum mechanically, would give a rotationally boosted wave function, $\psi_n\left(\vec{x}-\vec{R}\right) \rightarrow e^{iL_3^{\text{box}}\phi_R/\hbar}\psi_n\left(\vec{x}-\vec{R}\right)$, with $L_3^{\text{box}} = n\hbar$, but would not affect the overall particle phase factor value, $e^{i\gamma^C}$.

10.7 TIME-DEPENDENT PERTURBATION THEORY AND TRANSITIONS

Let's go all the way back to Equation 10.8, which started all this. I made the point then that these equations can be solved *perturbatively*, which basically means a power series in the potential matrix elements, $V_{nn'}$. Let us set the lowest-order coefficients to $C_n^{(0)}(0) = \delta_{an}$, indicating a system that starts in state a. The superscript tells us the order of the approximation process. We will assume again that $V_{nn}(t) = 0$ for the diagonal elements. Putting $C_n^{(0)}(0)$ on the right-hand side of Equation 10.8 then gives the first nontrivial additional contribution,

$$C_n^{(1)}(t) = -\frac{i}{\hbar}\int_0^t dt'\, V_{na}(t')e^{i\omega_{na}t'}, \tag{10.146}$$

where the integration constant is set to zero. Notice how when $n = a$ we get $C_a^{(1)}(t) = 0$ for $V_{aa}(t) = 0$. Substituting this back again, we obtain

$$C_n^{(2)}(t) = \left(\frac{i}{\hbar}\right)^2 \sum_{n'} \int_0^t dt'\, V_{nn'}(t')e^{i\omega_{nn'}t'} \int_0^{t'} dt''\, V_{n'a}(t'')e^{i\omega_{n'a}t''}. \tag{10.147}$$

This expression is also good for all n states. This process can be continued indefinitely. The whole series is then given as

$$C_n(t) = \sum_{i=0}^{\infty} C_n^{(i)}(t). \tag{10.148}$$

Note how one can trace the time order of the interaction levels by reading the amplitudes from right to left in Equation 10.147, and that the number of matrix element factors is reflected in the approximation order superscript of $C_n^{(i)}(t)$.

Consider two states a and b; a is initial, b is final. Also, consider a matrix element V_{ba}, which is constant in time. From Equation 10.146 we have

$$C_b^{(1)}(t) \approx -\frac{2i}{\hbar\omega_{ba}} V_{ba} e^{\frac{i}{2}\omega_{ba}t'} \sin\left(\frac{1}{2}\omega_{ba}t\right). \tag{10.149}$$

Notice the similarity to Equation 10.49 but with different matrix elements and no adiabatic phases. The probability of transition is given by

$$|C_b^{(1)}(t)|^2 = 4\frac{|V_{ba}|^2}{\hbar^2}\frac{\sin^2\left(\frac{1}{2}\omega_{ba}t\right)}{\omega_{ba}^2}. \tag{10.150}$$

The system oscillates harmonically with a period given by $\frac{2\pi}{|\omega_{ba}|}$ with a response magnitude of $\left|\frac{2V_{ba}}{\hbar\omega_{ba}}\right|^2$. The frequency dependence of the probability distribution is illustrated in Figure 10.9. Note this requires a multitude of closely spaced energy levels E_b close to E_a for this transition probability to make sense. Also notice the fact that most of this probability is concentrated in the region $|\omega_{ba}| < 2\pi/t$. Letting the distribution width be $\Delta\omega_{ba}$, we have

$$\Delta\omega_{ba} \approx \frac{2\pi}{t} \quad \Rightarrow \quad \Delta E_{ba}\, t \approx h. \tag{10.151}$$

This harkens to the energy–time uncertainty relation studied in Chapter 2, Section 2.8, and in fact has the same meaning. t is regarded as the transition time associated with the change in energy states. If one makes this

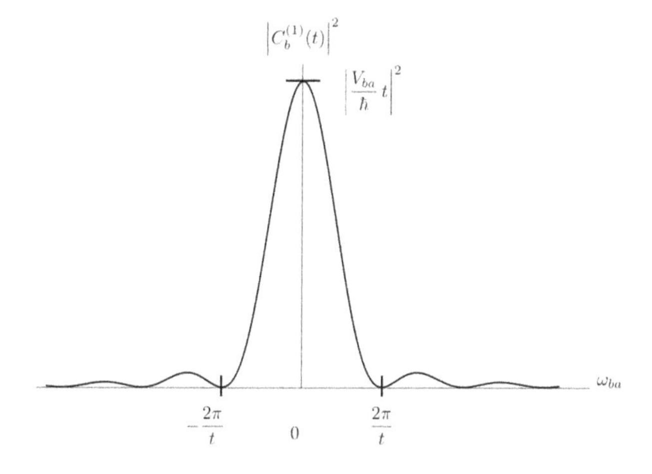

FIGURE 10.9 The probability of occupation of the b state, Equation 10.150, as a function of the difference of angular frequencies ω_{ba}.

larger, then the energy change or uncertainty is smaller and vice versa. In the following, we will consider the limit of very large t values, in which case the probability distribution becomes very narrow.

Let us say then E_a is fixed, but a continuum of E_b values are available; this is the situation for elastic scattering, as we will see in Section 10.8. Let us change notation a bit to symbolize this better; let $b \to [b]$ to represent this new realization. Then, the average rate of transition, $\bar{R}_{a\to[b]}(t)$, is defined to be the total change in the probability sum over final states divided by the total transition time:

$$\bar{R}_{a\to[b]}(t) \equiv \frac{1}{t}\sum_{[b]} \left|C_{[b]}^{(1)}(t)\right|^2. \tag{10.152}$$

Any attempt to define the *probability* in this context is doomed. Only the transition rate, not the probability, can be defined since we have no idea how many states are contributing. Because an untold number of closely spaced states are contributing, we change notation to indicate a continuum integration, rather than a sum,

$$\sum_{[b]} \left|C_{[b]}^{(1)}(t)\right|^2 \to \int_{-\infty}^{\infty} dE' \rho_{[b]}(E') \left|C_b^{(1)}(t)\right|^2, \tag{10.153}$$

where $E' \equiv E_b - E_a$ and $\rho_{[b]}(E')$ is the density of such states for a given E' value. We will come back to this function in Sections 10.8 and 10.9. For now we simply substitute Equation 10.153 into 10.152, which gives

$$\begin{aligned}\bar{R}_{a\to[b]}(t) &= \int_{-\infty}^{\infty} dE' \rho_{[b]}(E') \frac{1}{t} \left|C_b^{(1)}(t)\right|^2, \\ &= 4\int_{-\infty}^{\infty} dE' \rho_{[b]}(E') \,|\, V_{ba}|^2 \, \frac{\sin^2\left(\dfrac{1}{2\hbar}E't\right)}{tE'^2}.\end{aligned} \tag{10.154}$$

The asymptotic value of this rate for large times is a very significant quantity. Let us give it a special designation,

$$R_{a\to[b]} \equiv \lim_{t\to\infty} \bar{R}_{a\to[b]}(t). \tag{10.155}$$

The following scaling form defines a Dirac delta function:*

$$\delta(x) = \lim_{\varepsilon\to 0^+} \frac{1}{\varepsilon} f\left(\frac{x}{\varepsilon}\right); \quad \int_{-\infty}^{\infty} dx\, f(x) = 1. \tag{10.156}$$

Let us choose

$$f(x) = \frac{1}{2\pi} \frac{\sin^2(2x)}{x^2}, \tag{10.157}$$

* W. Wilcox and C. Thron, *Macroscopic Electrodynamics*, Section 2.1.

and make the replacements $\varepsilon \to 4/t$, $x \to E'/\hbar$. Then

$$\lim_{t\to\infty} \frac{\sin^2\left(\frac{1}{2\hbar}E't\right)}{tE'^2} = \frac{\pi}{2\hbar}\delta(E'). \tag{10.158}$$

Thus, for $tE'/\hbar \to \infty$ (E' fixed) the transition time is longer than any period in the system, and the sinusoidal time behavior of the system completely disappears. This gives

$$R_{a\to[b]} = \frac{2\pi}{\hbar}|V_{[b]a}|^2\rho_{[b]}, \tag{10.159}$$

where $\rho_{[b]} \equiv \rho_{[b]}(0)$ indicates the density of states for the system for $E_b = E_a$. Under these circumstances, the theory predicts a *constant* transition rate for $a \to [b]$. Equation 10.159 is called *Fermi's golden rule for time-constant perturbations.*

Now consider the harmonic case. Let us assume the general form

$$V_{ba}\left(=V_{ab}^*\right) = \hat{V}_{ba}e^{i\omega t} + \hat{V}_{ba}^*e^{-i\omega t}, \tag{10.160}$$

where $\hat{V}_{ba}$ and $\hat{V}_{ba}^*$ are expansion coefficients of the matrix element V_{ba} and ω is a positive driving frequency. We now get two terms instead of the one from Equation 10.146:

$$C_b^{(1)}(t) \approx -\frac{2i}{\hbar}\left[\frac{\hat{V}_{ba}}{\omega_{ba}+\omega}e^{\frac{i}{2}(\omega_{ba}+\omega)t'}\sin\left(\frac{1}{2}(\omega_{ba}+\omega)t\right) + \frac{\hat{V}_{ba}^*}{\omega_{ba}-\omega}e^{\frac{i}{2}(\omega_{ba}-\omega)t'}\sin\left(\frac{1}{2}(\omega_{ba}-\omega)t\right)\right]. \tag{10.161}$$

Let us define what we mean by emission and absorption of radiation in this context. Since a is the initial state, for emission we define $E_b = E_a - \hbar\omega$, and for absorption $E_b = E_a + \hbar\omega$, where $\hbar\omega$ is viewed as the energy of the associated photon. Thus ω_{ba} is negative for emission and positive for absorption. Similar to Figure 10.9, we will have that the transitions are limited to a very small probability band around $\omega = -\omega_{ba}$ for emission and $\omega = \omega_{ba}$ for absorption for large time-transition times $t \gg 2\pi/|\omega_{ba}|$. Thus, the first term in Equation (10.161) dominates for emission and the second term dominates for absorption, with very little overlap. This means we may simply take the single probability form,

$$\left|C_b^{(1)}(t)\right|^2 \approx 4\frac{|\hat{V}_{ba}|^2}{\hbar^2}\frac{\sin^2\left(\frac{1}{2}(|\omega_{ba}|-\omega)t\right)}{(|\omega_{ba}|-\omega)^2}, \tag{10.162}$$

for both cases. Mathematically, the situation is essentially the same as the constant potential case, Equation 10.150, but with $\omega_{ba} \to \pm(|\omega_{ba}| - \omega)$. To calculate the transition rate, we may repeat the preceding steps with the redefinition $E' \equiv |E_b - E_a| - \hbar\omega$. Then for asymptotically large times, $t \gg 2\pi/|\omega_{ba}|$, we find the rate

$$R_{a\to[b]} = \frac{2\pi}{\hbar}\left|\hat{V}_{[b]a}\right|^2 \rho_{[b]}^{\text{ems,abs}}, \tag{10.163}$$

where the density of states factor is evaluated either for emission or absorption at $E' = 0$, which means $\omega = |\omega_{ba}|$. Equation 10.163 is called *Fermi's golden rule for harmonic perturbations.* This is very similar to the time-constant case, Equation 10.159, except we are using the expansion coefficients of the matrix element, $\hat{V}_{[b]a}$, and we will need a different density of states factor in the two cases. In applying Equation 10.163 to atomic transitions for emission, the density of states is associated with the different momentums of the final photon; for absorption, the density is associated with the broadening of the final atomic state due to its finite lifetime. We will see explicit examples of these factors in an example in Section 10.8 for emission and in the discussion in Section 10.9 for absorption.

Forms of Fermi's formula are ubiquitous throughout physics. We have already had occasion to use its general form in Chapter 9 when deriving the statistical result for the occupation number n_i for interacting particles in equilibrium. In addition, it will be generalized for scattering in Chapter 11, Section 11.13, and we will again use it in Appendix F on weak interactions. To deepen our understanding of the different ways it can be employed, we will now consider additional examples.

10.8 APPLICATIONS OF FERMI'S GOLDEN RULE

Fermi's golden rule for a time-constant perturbation gives us a quick starting point for deriving one of the most important scattering formulas in physics. The type of scattering in which the incoming and outgoing sets of particles are scattered without loss of energy is called *elastic scattering.* The simplest scenario is the scattering of one particle from a heavy, essentially immovable central potential associated with a second particle, with no loss of kinetic energy. An example would be so-called Coulomb scattering, where a beam of light particles, such as electrons, scatters from the electrostatic field of a heavy particle, such as a proton (see Chapter 11, Section 11.12).

First, we need the definition of the so-called *elastic cross section.* Let us define

$$\frac{d\sigma}{d\Omega} \equiv \frac{R_{\vec{k}\to[\vec{k}']}}{\text{Incident flux}}, \tag{10.164}$$

where $R_{\vec{k}\to[\vec{k}']}$ is the appropriate rate factor, $\vec{k} = \vec{p}/\hbar$ is the incoming vector wave number, and $\vec{k}' = \vec{p}/\hbar$ is the outgoing wave number. For elastic scattering $|\vec{k}| = |\vec{k}'| \equiv k$. The interaction is steady or time independent, so we use the rate factor $R_{\vec{k}\to[\vec{k}']}$ of Equation 10.159 with appropriate initial and final wavenumber states:

$$R_{\vec{k}\to[\vec{k}']} = \frac{2\pi}{\hbar} |\langle \vec{k}' | V(\vec{r}) | \vec{k} \rangle|^2 \rho_{[\vec{k}']}. \tag{10.165}$$

This matrix element is actually an integral. In evaluating this form, we will be using the box normalization ideas from Chapter 9, Section 9.7. This has the nice aspect that an internal check on the calculations is that the length factors, L, of the box length should all cancel. Inserting a complete set of spatial states before and after $V(\vec{r})$ gives (we use $V = L^3$ for the volume of the box)

$$\langle \vec{k}' | V(\vec{r}) | \vec{k} \rangle = \int d^3r \int d^3r' \frac{e^{-i\vec{k}'\cdot\vec{r}'}}{L^{3/2}} \langle \vec{r}' | V(\vec{r}) | \vec{r} \rangle \frac{e^{i\vec{k}\cdot\vec{r}}}{L^{3/2}} = \frac{1}{L^3} \int d^3r\, e^{i(\vec{k}-\vec{k}')\cdot\vec{r}} V(\vec{r}), \tag{10.166}$$

where the second step occurs because $\langle \vec{r}' | V(\vec{r}) | \vec{r} \rangle = V(\vec{r})\delta(\vec{r}' - \vec{r})$. The incident flux of incoming particles is given by the current associated with the incoming and outgoing plane waves:

$$\text{Incident flux} = |\vec{J}| = \left| \frac{\hbar}{m} \text{Im}\left(u^*(\vec{r}) \vec{\nabla} u(\vec{r}) \right) \right|. \tag{10.167}$$

For box normalization, we have

$$u(\vec{r}) = \frac{e^{i\vec{k}\cdot\vec{r}}}{\sqrt{V}} \Rightarrow \text{Incident flux} = \frac{\hbar k}{mV}. \tag{10.168}$$

The necessary density of states factor for the outgoing scattering states has already been calculated in Equation 9.135. For a particle of mass m with $E = p^2/2m$ and $p = \hbar k$ we have ($D = 1$ and we have changed notation slightly)

$$\rho_{[\vec{k}']} = \frac{V}{h^3} p^2 \frac{dp}{dE} = \frac{L^3}{(2\pi)^3} \frac{mk}{\hbar^2}. \tag{10.169}$$

Putting the pieces together gives

$$\frac{d\sigma}{d\Omega} = \left(\frac{m}{2\pi\hbar^2} \right)^2 \left| \int d^3r\, e^{i(\vec{k}-\vec{k}')\cdot\vec{r}} V(\vec{r}) \right|^2. \tag{10.170}$$

This is called the *Born approximation* for elastic scattering and is associated with weakly interacting particles. We will see the analog to this cross section emerge from the scattering considerations in the next chapter, when the amplitude in Equation 11.229 is applied in Equation 11.171. However, the Chapter 11 result is in the context of a dynamic two-body interaction formalism, whereas the potential here, $V(\vec{r})$, is fixed in space for a single scattering particle.

Our second application of this formalism is to electromagnetic transitions; in particular, we will examine the 2p ($n = 2$, $\ell = 1$) to 1s ($n = 1$, $\ell = 0$)

transition in hydrogen. This will also be an opportunity to use the creation and annihilation formalism of Section 9.8 (Chapter 9) for the electromagnetic field. The correct starting point for the calculation is the electromagnetic Hamiltonian, Equation 10.117. In order to use the formalism we have developed, we will need to put the Hamiltonian in the form of Equation 10.1 in order to identify the interaction term to use in Fermi's golden rule. Therefore, let us expand the square in Equation 10.117 and keep only the term linear in the vector potential, $\vec{A}(\vec{r},t)$. (I will comment on the deleted $\vec{A}^2$ term at the end of this section.) Our starting point is thus given by

$$H_0 = \frac{\vec{p}^2}{2m} + q\Phi(\vec{r}); \quad V(t) = -\frac{q}{2mc}\left(\vec{A}(\vec{r},t)\cdot\vec{p} + \vec{p}\cdot\vec{A}(\vec{r},t)\right), \tag{10.171}$$

where the vector potential operator is given by Equations 9.136 and 9.137 in Chapter 9. H_0 is the atomic Hamiltonian and $V(t)$ gives the interaction with the electromagnetic field, which we are treating as a perturbation. Regarding $\vec{p}$ as an operator, we note that a switch in the order of $\vec{A}$ and $\vec{p}$ gives

$$\vec{A}\cdot\vec{p} = \vec{p}\cdot\vec{A} - \frac{\hbar}{i}\vec{\nabla}\cdot\vec{A} \tag{10.172}$$

(cf. Chapter 3, Equation 3.82). However, in our case we have $\hat{\varepsilon}_{\alpha,\vec{k}}\cdot\hat{k} = 0 \ \Rightarrow \vec{\nabla}\cdot\vec{A} = 0$, as we saw before in Equation 9.139, and we need not worry about the order of the $\vec{p}$ and $\vec{A}$ operators.

The initial state of our matrix element consists of an excited atom, symbolized by a_A and no photons. The final state consists of an atom in a lower, or ground state, along with a single emitted photon. In a composite atomic/photonic notation, we take these states to be represented by

$$|a\rangle \to |0\rangle|a_A\rangle; \quad \langle b| \to \langle b_A|\langle 1_{\alpha,\vec{k}}|. \tag{10.173}$$

$|0\rangle$ is the initial vacuum state, or the state of no photons, and $\langle 1_{\alpha,\vec{k}}|$ represents the final single-photon state with polarization α and wave number . It is understood that there is actually a group of states that will contribute to this matrix element. At this point we identify the appropriate matrix element to use in Equation 10.163 from the comparison (see Equation 10.160)

$$\hat{V}_{[b]a}e^{i\omega t} \leftrightarrow -\frac{q}{mc}\langle b_A|\vec{p}\cdot\langle 1_{\alpha,\vec{k}}|\vec{A}(\vec{r},t)|0\rangle|a_A\rangle. \tag{10.174}$$

The matrix element is (see Chapter 9, Problem 9.8.1):

$$\langle 1_{\alpha,\vec{k}}|\vec{A}(\vec{r},t)|0\rangle = c\sqrt{\frac{2\pi}{\omega V}}\hat{\varepsilon}_{\alpha,\vec{k}}e^{-i\left(\vec{k}\cdot\vec{r}-\omega t\right)} \tag{10.175}$$

(We continue using real polarization vectors $\hat{\varepsilon}_{\alpha,\vec{k}}$.) So far, we have made no approximations in this calculation. At this point, however, it behooves us to think about the phase factor $e^{i\vec{k}\cdot\vec{r}}$ in Equation 10.175. Let us suppose the transition frequency energy $\hbar\omega$ is given in an approximate sense by a generalization of the hydrogen atom energy equation,

$$\hbar\omega \sim \frac{Ze^2}{R_A}, \tag{10.176}$$

where Z counts the number of protons (central charge), R_A characterizes the size of the atom, and e is the magnitude of the charge on an electron. Then this implies

$$\frac{\lambdabar}{R_A} \sim \frac{1}{\alpha Z} \approx \frac{137}{Z}, \tag{10.177}$$

where $\lambdabar$ is the reduced transition wavelength and $\alpha = e^2/\hbar c$ is the fine structure constant. Thus, the associated wavelength is much larger than the size of the atom and we may approximate $e^{i\vec{k}\cdot\vec{r}} \approx 1$ as long as $Z \ll 137$. Therefore, to a very good approximation we have

$$\hat{\varepsilon}_{\alpha,\vec{k}} \cdot \langle b_A | \vec{p} e^{i\vec{k}\cdot\vec{r}} | a_A \rangle \approx \hat{\varepsilon}_{\alpha,\vec{k}} \cdot \langle b_A | \vec{p} | a_A \rangle. \tag{10.178}$$

We also modify this matrix element to involve the operator $\vec{r}$ and not $\vec{p}$:

$$[x, H_0] = [x, \frac{\vec{p}^2}{2m}] = \frac{i\hbar}{m} p_x \tag{10.179}$$

$$\Rightarrow \hat{\varepsilon}_{\alpha,\vec{k}} \cdot \langle b_A | \vec{p} | a_A \rangle = -\frac{im}{\hbar} \hat{\varepsilon}_{\alpha,\vec{k}} \cdot \langle b_A | [\vec{r}, H_0] | a_A \rangle = -im\omega\, \hat{\varepsilon}_{\alpha,\vec{k}} \cdot \langle b_A | \vec{r} | a_A \rangle, \tag{10.180}$$

for $\omega = (E_a - E_b)/\hbar$. Note that at this point we have simply

$$\hat{V}_{[b]a} \to -q\hat{\vec{E}} \cdot \langle b_A | \vec{r} | a_A \rangle, \quad \text{where} \quad \vec{E} = -\frac{1}{c}\frac{\partial \vec{A}}{\partial t} \to \hat{\vec{E}} \equiv -i\omega\sqrt{\frac{2\pi}{\omega L^3}}\, \hat{\varepsilon}_{\alpha,\vec{k}}.$$

This reminds one of the electric dipole potential energy from way back in Chapter 1, Equation 1.24. This long wavelength reduction is therefore termed the *electric dipole approximation.*

Almost everything is in place now. The density of states for outgoing photons can be evaluated similar to Equation 10.169. From Equation 9.135 we have ($p = \hbar k$, $k = \omega/c$, $E = pc$ and $D = 1$ since we are not summing over polarizations):

$$\rho_{[b]}^{\text{ems}} \to \rho_{[\vec{k}]} = \frac{L^3}{h^3} p^2 \frac{dp}{dE} = \frac{L^3\omega^2}{(2\pi)^3 \hbar c^3}. \tag{10.181}$$

Fermi's golden rule, Equation 10.163, now gives the differential emission rate into a particular polarization α and final momentum state, $\vec{k}$, located with spherical angles θ and ϕ as

$$R_{a\to[b]} \to \frac{dR_{ba}^{\alpha}(\theta,\phi)}{d\Omega} \equiv \frac{2\pi}{\hbar}(q\omega)^2 \left(\frac{2\pi\hbar}{L^3\omega}\right)\left(\frac{L^3\omega^2}{(2\pi)^3 \hbar c^3}\right) \left|\hat{\varepsilon}_{\alpha,\vec{k}} \cdot \langle b_A | \vec{r} | a_A \rangle\right|^2$$

$$= \frac{q^2\omega^3}{2\pi\hbar c^3} \left|\hat{\varepsilon}_{\alpha,\vec{k}} \cdot \langle b_A | \vec{r} | a_A \rangle\right|^2. \tag{10.182}$$

The final spontaneous emission rate is now given by an integral over all the final states, including momentum and atomic states b_A:

$$R_{ba} \equiv \sum_{\alpha,b_A} \int d\Omega \frac{dR_{ba}^{\alpha}(\theta,\phi)}{d\Omega} = \frac{q^2\omega^3}{2\pi\hbar c^3} \sum_{\alpha,b_A} \int d\Omega \, |\hat{\varepsilon}_{\alpha,\vec{k}} \cdot \langle b_A | \vec{r} \, | a_A \rangle |^2, \quad (10.183)$$

$$\hat{k} = (\sin\theta\cos\phi, \sin\theta\sin\phi, \cos\theta), \quad (10.184)$$

$$\hat{\varepsilon}_1 = (-\sin\phi, \cos\phi, 0); \quad \hat{\varepsilon}_2 = (-\cos\theta\cos\phi, -\cos\theta\sin\phi, \sin\theta). \quad (10.185)$$

This gives

$$\sum_{\alpha} \int d\Omega \, |\hat{\varepsilon}_{\alpha,\vec{k}} \cdot \langle b_A | \vec{r} \, | a_A \rangle |^2 = \frac{8\pi}{3} |\langle b_A | \vec{r} \, | a_A \rangle |^2, \quad (10.186)$$

$$\Rightarrow R_{ba} \equiv \frac{4q^2\omega^3}{3\hbar c^3} \sum_{b_A} |\langle b_A | \vec{r} \, | a_A \rangle|^2. \quad (10.187)$$

In the case we will study, the ground state b_A is degenerate in spin, and we must sum over these final states to get the total emission rate.

As a point of comparison, the classical electrodynamic expression for the instantaneous electric dipole power emitted by a charged particle undergoing accelerated motion, called the *Larmor formula*, has a similar form*

$$P_{\text{class}}(t) = \frac{2q^2\left[\ddot{\vec{r}}(t_0)\right]^2}{3c^3}, \quad (10.188)$$

where $t_0 = t - r/c$ is called the "origin retarded time." So, clearly, one difference between the classical but relativistic theory and nonrelativistic quantum mechanics is the delay time associated with a light signal. If one postulates a harmonic source

$$P_{\text{class}}(t) \equiv \hbar\omega R_{\text{class}}(t); \quad \vec{r}(t_0) = \vec{r}_0 \cos(\omega t_0); \quad t_0 = t - r/c, \quad (10.189)$$

one then has the time-averaged result

$$\langle R_{\text{class}}(t) \rangle = \frac{2q^2\omega^3 \left\langle [\vec{r}(t_0)]^2 \right\rangle}{3\hbar c^3} \quad (10.190)$$

for the classical rate. Although it is tempting to compare Equations 10.187 and 10.190 and speculate that the factor of 2 difference is associated with the two photon polarizations, this would be wrong. These quantities correspond to different processes and cannot be compared directly, although the summed rates can be reconciled in semiclassical approximation.†

Back to our quantum transition considerations! The 2p to 1s transition in hydrogen is an excellent opportunity to employ the spin-angle formalism of Chapter 8, although it is not absolutely necessary at this level of description (see Problem 10.8.1). The eigenstates are

* W. Wilcox and C. Thron, *Macroscopic Electrodynamics*, Equations 11.99 and 11.152.

† See J. Schwinger, *Quantum Mechanics: Symbolism of Atomic Measurements*, edited by B.-G. Englert (Springer, 2001), Section 12.9.

$$u_{n\ell jm}(\vec{r}) \equiv \frac{R_{n\ell}(r)}{r} y_\ell^{j=\ell\pm 1/2,m}(\theta,\phi), \qquad (10.191)$$

where the $y_\ell^{j,m}(\theta,\phi)$ are spin-angle functions of Equation 8.94. (Please don't get the radial dependence $R_{n\ell}(r)$ for the wave function mixed up with the R_{ba} for the rate!) Note that we will ignore the small fine structure splitting between the initial $2p_{3/2}$ and $(2s_{1/2}, 2p_{1/2})$ states (see Chapter 8, Figure 8.1), and will use a value of

$$\omega = \frac{3e^2}{8\hbar a_0} = \frac{3\mu c^2}{8\hbar}\alpha^2 = 1.55037\times 10^{16}\ \text{sec}^{-1}, \qquad (10.192)$$

for the transition angular frequency, where a_0 is the Bohr radius, μ is the reduced mass of the proton/electron system, α is again the fine structure constant, and $q = -e$. Our initial and final states are characterized as follows:

Initial:

$2p_{3/2}\ (m_j = 3/2, 1/2, -1/2, -3/2)$, $2p_{1/2}\ (m_j = 1/2, -1/2)$,
$2s_{1/2}\ (m_j = 1/2, -1/2)$;

Final:

$1s_{1/2}\ (m_j = 1/2, -1/2)$.

There are thus eight possible initial states and two final states. In order to easily do the integrals involved, we need to decompose the position vector into the complex conjugate of spherical harmonics. Using the list in Problem 6.10.3 (see Chapter 6), one can show $(Y_{\ell m}^* \equiv Y_{\ell m}^*(\theta,\phi))$

$$\frac{\vec{x}}{r} = \frac{1}{2}\left(\frac{8\pi}{3}\right)^{1/2}\left(\left(Y_{1-1}^* - Y_{11}^*\right)\hat{e}_1 - i\left(Y_{1-1}^* + Y_{11}^*\right)\hat{e}_2 + \sqrt{2}Y_{10}^*\hat{e}_3\right). \qquad (10.193)$$

For the initial $2p_j$ state, we have $(y_\ell^{j,m} \equiv y_\ell^{j,m}(\theta,\phi))$

$$\langle b_A|\vec{x}|a_A\rangle \to \langle 1s_{1/2}, m_s\,|\,\vec{x}\,|\,2p_j, m_j\rangle = \frac{1}{2}\left(\frac{8\pi}{3}\right)^{1/2}\int_0^\infty dr\, r R_{10}(r) R_{21}(r)$$
$$\times\int d\Omega\left(\left(Y_{1-1}^* - Y_{11}^*\right)\hat{e}_1 - i\left(Y_{1-1}^* + Y_{11}^*\right)\hat{e}_2 + \sqrt{2}Y_{10}^*\hat{e}_3\right) y_0^{\frac{1}{2},m_s\dagger}\cdot y_1^{j,m}. \qquad (10.194)$$

Using orthonormality of the spherical harmonics (see Equation 6.226 in Chapter 6), the reader may verify:

$$\langle 1s_{1/2}, m_s\,|\,\vec{x}\,|\,2p_{3/2}, m_j\rangle = \left(\frac{1}{6}\right)^{1/2}\Im\times\begin{cases} \overbrace{-\hat{e}_1 - i\hat{e}_2}^{m_s=\frac{1}{2}} & \overbrace{0}^{m_s=-\frac{1}{2}} & m_j = \frac{3}{2} \\ \frac{2\hat{e}_3}{\sqrt{3}} & -\frac{1}{\sqrt{3}}\left(\hat{e}_1 + i\hat{e}_2\right) & m_j = \frac{1}{2} \\ \frac{1}{\sqrt{3}}\left(\hat{e}_1 - i\hat{e}_2\right) & \frac{2\hat{e}_3}{\sqrt{3}} & m_j = -\frac{1}{2} \\ 0 & \hat{e}_1 - i\hat{e}_2 & m_j = -\frac{3}{2} \end{cases}$$

(10.195)

$$< 1s_{1/2}, m_s \,|\, \vec{x} \,|\, 2p_{1/2}, m_j >$$

$$= \left(\frac{1}{6}\right)^{1/2} \Im \times \begin{cases} \overbrace{-\frac{2\hat{e}_3}{\sqrt{3}}}^{m_s=\frac{1}{2}} & \overbrace{-\frac{1}{\sqrt{3}}\left(\hat{e}_1 + i\hat{e}_2\right)}^{m_s=-\frac{1}{2}} & m_j = \frac{1}{2} \\ -\frac{1}{\sqrt{3}}\left(\hat{e}_1 - i\hat{e}_2\right) & \frac{2\hat{e}_3}{\sqrt{3}} & m_j = -\frac{1}{2} \end{cases} \quad (10.196)$$

where the common radial integral is

$$\Im \equiv \int_0^\infty dr\, r\, R_{10}(r) R_{21}(r). \quad (10.197)$$

On the other hand, $< 1s_{1/2}, m_s \,|\, \vec{x} \,|\, 2s_{1/2}, m_j >$ is immediately zero from orthogonality of spherical harmonics. The sum over $m_s = \pm 1/2$ in Equation 10.187 for the 2p state is now seen to be independent of the initial $j = 3/2$, $1/2$ and m_j:

$$\sum_{m_s} |< 1s_{1/2}, m_s \,|\, \vec{x} \,|\, 2p_j, m_j >|^2 = \frac{1}{3} \left| \int_0^\infty dr\, r\, R_{10}(r) R_{21}(r) \right|^2, \quad (10.198)$$

where

$$R_{10}(r) = \frac{2r}{\left(a_0\right)^{3/2}} e^{-r/a_0}; \quad R_{21}(r) = \frac{r^2}{\sqrt{24}\left(a_0\right)^{5/2}} e^{-r/2a_0}. \quad (10.199)$$

The rest of the evaluation is left as an exercise (Problem 10.8.2). The result for the rate is

$$R_{1s\,2p} = \frac{\alpha^5 \mu c^2}{\hbar} \cdot \left(\frac{2}{3}\right)^8. \quad (10.200)$$

For quantum transitions, if state a is depopulated because of a transition to another state b at a rate R_{ba}, then the rate of change of occupation probability, $P_a(t)$, is

$$\frac{dP_a(t)}{dt} = -R_{ba} P_a(t) \Rightarrow P_a(t) = P_a(0) e^{-R_{ba} t}. \quad (10.201)$$

The time $\tau = 1/R_{ba}$ associated with a fraction e^{-1} of the original sample remaining is called the *mean life* and corresponds to the expectation value of the decay time. It also corresponds to the peak in the distribution of lifetimes in time. In contrast, the *half-life*, $t_{1/2}$, is associated with decay of half the sample, $e^{-R_{ba} t_{1/2}} = 1/2$, and is related to the mean life by $t_{1/2} = \tau \ln 2$. Finally, using up-to-date parameter values, one finds

$$R_{1s\,2p} = 6.2649 \times 10^8 \,\text{sec}^{-1}; \quad \tau_{1s\,2p} \equiv \frac{1}{R_{1s\,2p}} = 1.5962 \times 10^{-9} \,\text{sec}. \quad (10.202)$$

As pointed out earlier, this result ignores the fine structure effect energy splitting for 2p levels found in Chapter 8. These would be expected to contribute at relative order of $\alpha^2 \sim 5\times10^{-5}$, which is why I am only carrying out the evaluation to a five-place accuracy. One of the published experimental results* for this lifetime is $\tau^{\text{exp}}_{1s,2p} = 1.596211(16)\times10^{-9}$ sec, where the numbers in parentheses give the uncertainty in the last two digits. So, we have done pretty well in the dipole approximation! Note the fact that the lifetime ~10^{-9} sec is much longer than the period $\sim 4\times10^{-16}$ sec of the emitted photon; this is consistent with our large time assumption for calculating the rate through Fermi's golden rule.

We found that the m_j values of the initial state did not matter for the lifetime above. This is because the decay rate does not depend on the spatial orientation of the initial state. It also did not depend on the initial $j = 3/2$, $1/2$ value, which describes an internal property, but it could and would in a more accurate calculation. Because the operator connecting the initial and final states is the position vector, $\vec{r}$, one thing that must change in transitioning from the initial to final state is the angular momentum ℓ value. In this case we have $\Delta\ell = \ell_{\text{final}} - \ell_{\text{initial}} = -1$. In general, for electric dipole transitions we must have the "selection rule" $|\Delta\ell| = 1$. This can be established with a very powerful theoretical tool called the *Wigner–Eckart theorem*, which is the subject of Problem 10.8.3. This rule explains why the $2s_{1/2}$ to $1s_{1/2}$ decay rate was found to be zero in electric dipole approximation. Actually, this decay occurs, but is mediated by the $\vec{A}^2$ term we actually dropped in Equation 10.171 at the very beginning of this calculation. This type of term implies the decay occurs via a *two*-photon emission process with a very long lifetime of about 1/7 of a second; but that is beyond the scope of our considerations here. Note that along with emission processes, the Fermi rule formalism can describe atomic absorption of photons as well. Some aspects of these processes will be examined in Problem 10.8.4.

10.9 EXPONENTIAL TIME DECAY AND DECAY WIDTHS

The 2p state of hydrogen is unstable, and it follows from the energy–time uncertainty relation that it has an energy uncertainty of

$$\Delta E \geq \frac{\hbar}{\tau_{1s\,2p}} \approx 4.1\times10^{-7}\,\text{eV}, \tag{10.203}$$

from Chapter 2, Equation 2.93. This is tiny compared to the actual 1s–2p energy difference of about 10.2 eV. Nevertheless, how does one incorporate this small but finite uncertainty in a description of the energy structure and decay properties of such a state?

Let us try to make contact with Equation 10.201. We will assume that

$$\psi(t) = \psi(0)\,e^{-iEt/\hbar} \tag{10.204}$$

* V.G. Pal'chikov, Yu. L. Sokolov, and V.P. Yakovlev, *Physica Scripta* **55**, 33 (1997).

as usual represents the time wave function of an unstable state. However, let us set

$$E = E_0 - i\Gamma/2; \quad \Gamma \equiv \hbar R = \frac{\hbar}{\tau}, \tag{10.205}$$

where R is the decay rate. The new concept we are considering is that there is a quantum mechanical meaning to assigning an imaginary part to a state's energy. This implies

$$|\psi(t)|^2 = |\psi(0)|^2 e^{-Rt}, \tag{10.206}$$

for probability densities, which gives an exponential decay law as in Equation 10.201. The Fourier transform of this wave function is

$$\psi(\omega) \equiv \frac{1}{\sqrt{2\pi}} \int_0^\infty d\omega\, \psi(t) e^{i\omega t} = \frac{\psi(0)}{\sqrt{2\pi}} \frac{i\hbar}{(\hbar\omega - E_0) + i\Gamma/2}. \tag{10.207}$$

Let us identify $E \equiv \hbar\omega$. Then

$$|\psi(\omega)|^2 = \frac{\hbar^2}{2\pi} \frac{|\psi(0)|^2}{(E - E_0)^2 + (\Gamma/2)^2}, \tag{10.208}$$

which is a probability distribution in energy space. To normalize it, let us now require

$$P_\Gamma(E) = C|\psi(\omega)|^2; \quad \int_{-\infty}^{\infty} dE\, P_\Gamma(E) = 1. \tag{10.209}$$

One can then show the normalized probability distribution is

$$P_\Gamma(E) = \frac{1}{2\pi} \frac{\Gamma}{(E - E_0)^2 + (\Gamma/2)^2}. \tag{10.210}$$

This is called a *Breit–Wigner probability distribution* and is illustrated in Figure 10.10. Notice it is symmetric about the resonance peak, located at $E = E_0$. The quantity Γ is called the *full width*, since it gives the width from side to side of the resonance curve at half of its maximum height.

The form of Equation 10.210 holds if there is a single decay or formation possibility for a resonance. One way to deal with this situation is to introduce a form

$$P_{\Gamma_n}(E) \equiv \frac{1}{2\pi} \frac{\Gamma_n}{(E - E_0)^2 + (\Gamma/2)^2}, \tag{10.211}$$

if there are n Breit-Wigner decay possibilities. Γ_n is called the *partial width* for a particular decay. We then have $\sum \Gamma_n = \Gamma$ ensuring probability conservation, which in turn implies $\sum_n^n \frac{1}{\tau_n} = \frac{1}{\tau}$ for the associated partial lifetimes, τ_n.

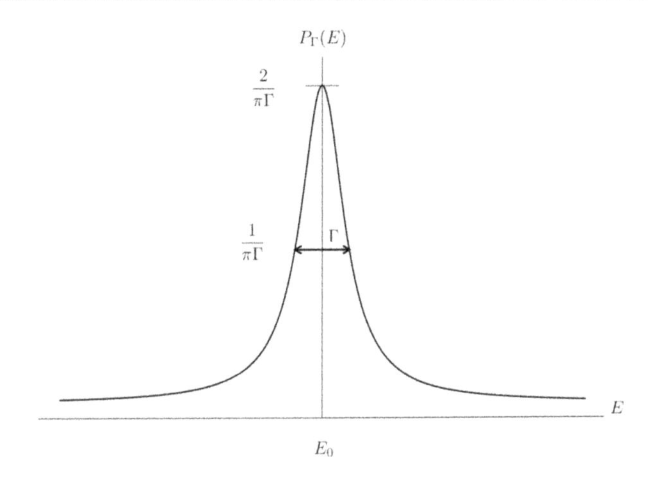

FIGURE 10.10 The shape of the Breit–Wigner probability distribution, Equation 10.210.

Using the scaling form $f(x)=\frac{1}{\pi}\frac{1}{x^2+1}$ in Equation 10.156 gives

$$\lim_{\varepsilon\to 0^+}\frac{1}{\pi}\frac{\varepsilon}{x^2+\varepsilon^2}=\delta(x). \tag{10.212}$$

Therefore with $\varepsilon=\Gamma/2$ and $x=(E-E_0)$ we have

$$\lim_{\Gamma\to 0^+} P_\Gamma(E)=\delta(E-E_0). \tag{10.213}$$

Thus, a stable state is associated with a delta function in energy, as it should. Note that the energy or line width probability can be interpreted in general as the density of states factor associated with absorption processes. That is, the density

$$\rho_{[b]}^{\text{abs}}=\frac{1}{2\pi}\frac{\Gamma_b}{\left(E_a-E_b\right)^2+\left(\Gamma/2\right)^2} \tag{10.214}$$

is what's needed to simulate a broadening of the b state for absorption in Equation 10.163.

The same relation between the imaginary part of the energy and the lifetime of the particle is also found in the scattering formalism of Chapter 11, Sections 11.5 and 11.10 for elastic cross sections, which also display resonance behavior associated with a finite lifetime (see for example Figure 11.11). Such resonance forms are found throughout atomic, molecular, nuclear, and particle physics. An interesting particle physics example is shown in Figure 10.11, where production of a particle called the $K^*(892)^0$ has been examined by the ALICE (A Large Ion Collider Experiment) collaboration at the Large Hadron Collider (LHC) in proton–proton collisions of at a center of mass energy of 7×10^{12} eV (7 TeV). You can try your hand at fitting a particle resonance in Problem 10.9.2!

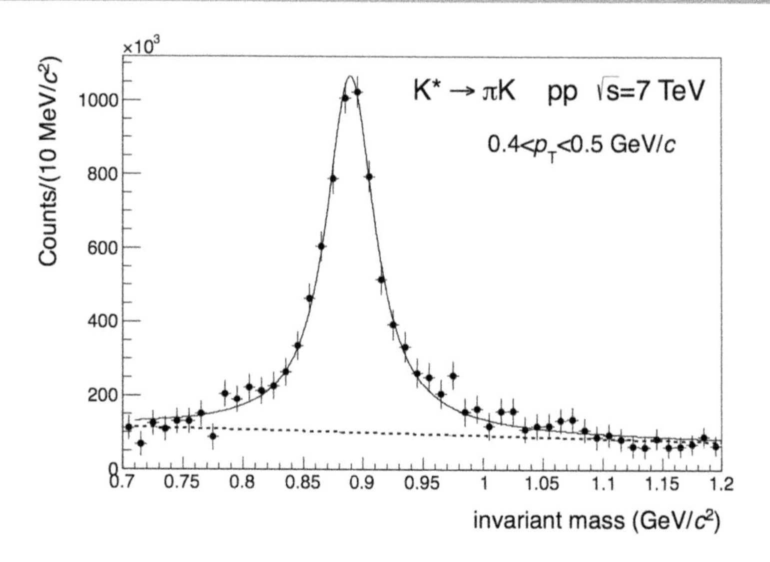

FIGURE 10.11 A Breit–Wigner fit to a resonance in proton-proton collisions at the LHC. The horizontal axis gives the invariant or center-of-mass mass combination from detected kaon and pion particles, and the vertical axis gives the number of such combinations found in 10 MeV/c^2 bins. The resonance produced is a neutral spin 1 meson consisting of down and anti-strange quarks (or its antiparticle) called the $K^*(892)^0$. The dotted line gives the background contribution. (See Figure 3 of the ALICE collaboration journal paper, *Eur. Phys. J. C* **72**, 2183 (2012).)

PROBLEMS

10.1.1 Derive Equation 10.8 from the preceding equations.

10.1.2 Derive Equation 10.14 using the completeness statement, Equation 10.4.

10.1.3 Show that $\omega_n^A(t)$ in Equation 10.33 is a real quantity (cf. Chapter 5, Problem 5.2.2).

10.2.1 (a) As explained in the text, a weak interaction decay can change the atomic number, Z, of a nucleus by one unit, $Z \to Z \pm 1$, by the emission of an electron and an antielectron neutrino or their antiparticles. Now consider a system consisting of a nucleus with atomic number Z with a single bound atomic electron in its 1s ground state. It decays into a $Z+1$ nucleus but keeps its single bound electron. Using the electron 1s wave function, Equation 7.264 (Chapter 7), and considering this a sudden change, show the probability, $P_{Z\to Z+1}$, that the electron will be found in the new 1s ground state after the decay is

$$P_{Z\to Z+1} = \frac{Z^3(Z+1)^3}{(Z+\frac{1}{2})^6}.$$

(b) Show that for $Z \gg 1$ this probability is approximately

$$P_{Z\to Z+1} \approx 1 - \frac{3}{4Z^2}.$$

(c) Find the expectation value of the work done on the electron in this transition and show it grows like Z.

10.2.2 (a) A one-dimensional harmonic oscillator in its ground state has a normalized wave function given by

$$u_0(x) = \left(\frac{m\omega}{\pi\hbar}\right)^{1/4} e^{-\left(\frac{m\omega}{2\hbar}\right)x^2},$$

where m is the mass and ω is the angular frequency. Consider a sudden change: $\omega^2 \to \frac{\omega^2}{F}$ (due to a spring constant change $k \to \frac{k}{F}$) for $F > 0$. Show that the probability that the harmonic oscillator remains in its ground state, P_0, is given by

$$P_0 = \frac{2}{F^{1/4} + F^{-1/4}}.$$

(b) Calculate the expectation value of the final energy of the harmonic oscillator if ω is changed quickly, E^{fast}, or changed slowly (adiabatically), E^{slow}. Show that

$$\frac{E^{\text{fast}}}{E^{\text{slow}}} = \frac{1}{2}\left(F^{1/2} + F^{-1/2}\right) \geq 1.$$

10.2.3 Give a proof of the following statement: The adiabatic phase, γ_n, associated with the slow movement of the origin of a bound state system, is zero. [Hints: Consider a wave function given by $\psi_n(\vec{x} - \vec{x}_0(t)) \equiv \langle \vec{x} \mid n,t\rangle$ and use Equations 10.40 and 10.33. Note also the "Hints" of Problem 10.6.1.]

10.3.1 Consider a simple two-state problem with kets $|n\rangle$, $n = 1, 2$; normalization $\langle n \mid n\rangle = 1$; and a Schrödinger equation

$$i\hbar\frac{\partial}{\partial t}|\psi,t\rangle = H|\psi,t\rangle.$$

The action of H on these states is given as

$$H|1\rangle = K|2\rangle, \quad H|2\rangle = K|1\rangle,$$

where K is a constant.

(a) Show that the normalized eigenstates are such that

$$H\left(\frac{|1\rangle + |2\rangle}{\sqrt{2}}\right) = K\left(\frac{|1\rangle + |2\rangle}{\sqrt{2}}\right), \quad H\left(\frac{|1\rangle - |2\rangle}{\sqrt{2}}\right) = -K\left(\frac{|1\rangle - |2\rangle}{\sqrt{2}}\right),$$

and that one has

$$e^{iHt/\hbar}|1\rangle = \cos\frac{Kt}{\hbar}|1\rangle - i\sin\frac{Kt}{\hbar}|2\rangle,$$

$$e^{iHt/\hbar}|2\rangle = -i\sin\frac{Kt}{\hbar}|1\rangle + \cos\frac{Kt}{\hbar}|2\rangle.$$

(b) Given the $t=0$ initial state

$$|\psi,t=0\rangle = \frac{e^{i\delta}|2\rangle + |1\rangle}{\sqrt{2}},$$

where δ is a real number, show that the probabilities $P_1^{\psi}(t)$ and $P_2^{\psi}(t)$ (that the time-evolved ket $|\psi,t\rangle$ is in the $|1\rangle$ or $|2\rangle$ states at time t) are given by

$$P_1^{\psi}(t) = \frac{1}{2}\left(1+\sin\delta\sin\left(\frac{2Kt}{\hbar}\right)\right),$$

$$P_2^{\psi}(t) = \frac{1}{2}\left(1-\sin\delta\sin\left(\frac{2Kt}{\hbar}\right)\right).$$

10.3.2 Verify Equations 10.58 and 10.59 for $C_1(t)$ and $C_2(t)$.

10.3.3 (a) Using Equation 10.14, find the expectation value of the potential $V(t)$ when $t>0$ for the first two-state problem in Section 10.3, which has initial conditions $C_1(0^+)=1, C_2(0^+)=0$. Show that it gives

$$\langle\psi,t|V(t)|\psi,t\rangle = \frac{4\gamma^2(\omega_{12}-\omega)}{\hbar\,\kappa(\omega)^2}\sin^2\left(\frac{\kappa(\omega)t}{2}\right).$$

(b) Now find $\langle\psi,t|H(t)|\psi,t\rangle$ for the same system, where $H(t)=H_0+V(t)$. Does your solution make sense when $\omega=0$?

10.3.4 We solved the $\bar{C}_n(t)$ problem in Section 10.3 by first solving for the $\tilde{C}_n(t)$ and then using Equation 10.66. However, one can solve for the $\bar{C}_n(t)$ directly using Equation 10.26 alone. This can be shown as follows.

(a) Starting with Equation 10.18, show that ($n'\neq n$ and $t>0$)

$$\langle n',t|\frac{\partial}{\partial t}|n,t\rangle = \frac{\langle n',t|\frac{\partial H(t)}{\partial t}|n,t\rangle}{(\bar{E}_n-\bar{E}_{n'})}.$$

[Hint: See Chapter 5, Section 5.2.]

(b) Use this in the two-state problem to derive the new coupled differential equations

$$\dot{\bar{C}}_1(t) = i\omega_1^A\bar{C}_1(t) + i\frac{\gamma\omega}{\hbar\kappa(0)}e^{i\kappa(0)t}\bar{C}_2(t),$$

$$\dot{\bar{C}}_2(t) = i\omega_2^A\bar{C}_2(t) + i\frac{\gamma\omega}{\hbar\kappa(0)}e^{-i\kappa(0)t}\bar{C}_1(t).$$

Extra: Now show for example that

$$\frac{d^2}{dt^2}\left(e^{-i\gamma_1}\bar{C}_1(t)\right)+i(\omega_1^A-\omega_2^A)\frac{d}{dt}\left(e^{-i\gamma_1}\bar{C}_1(t)\right)$$
$$+\left(\frac{\gamma\omega}{\hbar\kappa(0)}\right)^2\left(e^{-i\gamma_1}\bar{C}_1(t)\right)=0,$$

which demonstrates that these equations can be decoupled and solved independently.

10.3.5 Write a numerical code that calculates and displays $|\tilde{C}_1(t)|^2$ from Equation 10.69 and $|\tilde{C}_2(t)|^2$ from Equation 10.70 for the set of parameters: $\omega_{12}=1.0, \omega=0.1, \gamma/\hbar=0.5$ (units arbitrary) for one period of the driving angular frequency $T=2\pi/\omega$. Also display your verification of $|\tilde{C}_1(t)|^2+|\tilde{C}_2(t)|^2=1$. Show that compared to the adiabatic limits given in Equation 10.78, one actually has $|\tilde{C}_1(t)|^2\geq x_2^2$ and $|\tilde{C}_2(t)|^2\leq y_2^2$.

10.3.6 In the adiabatic limit, Equation 10.71 has the form

$$e^{i\theta_1}\bar{C}_1(t)\approx\frac{2i\omega(\gamma/\hbar)}{\kappa(0)^2}e^{-\frac{i}{2}(\omega+\omega_1+\omega_2)t}\sin\left(\frac{1}{2}\left(\kappa(0)-\frac{\omega\omega_{12}}{\kappa(0)}\right)t\right),$$

where a Taylor series expansion has been made for $\kappa(\omega)$, whereas Equation 10.49 (using Equation 10.22) predicts that

$$e^{i\theta_1}\bar{C}_1(t)\approx-2\frac{\langle 1,t|\frac{\partial}{\partial t}|2,t\rangle}{\bar{\omega}_1-\bar{\omega}_2}e^{\frac{i}{2}(\theta_1+\theta_2+\gamma_1+\gamma_2)}$$
$$\sin\left(\frac{1}{2}(\theta_2-\theta_1+\gamma_2-\gamma_1)t\right),$$

when $n=1$ and $a=2$. In order to establish the equality of these two forms and given that $\bar{\omega}_1-\bar{\omega}_2=\kappa(0)$, show that

(a) $\langle 1,t|\frac{\partial}{\partial t}|2,t\rangle=-\frac{i\omega(\gamma/\hbar)}{\kappa(0)}$

(b) $\theta_2+\theta_1+\gamma_2+\gamma_1=-(\omega+\omega_1+\omega_2)t$

(c) $\theta_2-\theta_1+\gamma_2-\gamma_1=\left(\kappa(0)-\frac{\omega\omega_{12}}{\kappa(0)}\right)t$

10.3.7 Verify Equation 10.74 for ω_2^A and 10.75 for ω_1^A directly from Equation 10.33 and the explicit eigenvectors, Equations 10.63 and 10.64.

10.3.8 Verify Equation 10.81 for ω_1^A in the adiabatic limit with $\bar{C}_1(0)=1$ and $\bar{C}_2(0)=0$ by the following steps:

(a) Show $\tilde{C}_1(t)\approx x_1e^{i\omega_{\bar{1}}t}$.

(b) Show $\tilde{C}_2(t)\approx y_1e^{i(\omega_{\bar{1}}-\omega_{12}+\omega)t}$.

(c) Show that using (a) and (b) in Equation 10.67 gives $\bar{C}_1(t) \approx e^{i\omega_1^A t}$ with ω_1^A given by Equation 10.81.

10.4.1 Look up the paper T.F. Jordan and J. Maps, *Am. J. Phys.* **78**, 1188 (2010), which considers the special path associated with a Foucault pendulum transported along a line of latitude θ_0 (which we measure here relative to the poles) by the Earth's rotation. Based on this, see if you can present an argument that the angle of rotation for an oscillating pendulum transported around a line of latitude on the (nonrotating) Earth is actually $\theta = 2\pi(1-\cos\theta_0)$. Note: From this result, Fig. 10.4 in Section 10.5 indicates how to complete the argument that $\theta = \Omega_{\text{closed path}}$ in general!

10.4.2 (a) Regarding the Hamiltonian as a function of external coordinates, $H = H(\vec{R})$, show that (cf. Problem 10.3.4; $n' \neq n$)

$$\langle n',\vec{R}\,|\,\vec{\nabla}_R\,|\,n,\vec{R}\rangle = \frac{\langle n',t\,|\,\vec{\nabla}_R H(\vec{R})\,|\,n,t\rangle}{(\bar{E}_n - \bar{E}_{n'})}.$$

(b) Employing the vector identity

$$\vec{\nabla}\times\left(\phi\vec{A}\right) = \vec{\nabla}\phi\times\vec{A} + \phi\vec{\nabla}\times\vec{A},$$

and inserting completeness ($\sum_{n'} |n',\vec{R}\rangle\langle n',\vec{R}| = 1$), show that the Berry phase expression Equation 10.90 can be rewritten as

$$\gamma_n^C = i\sum_{n'\neq n}\int_S da\,\hat{n}\cdot\left[\frac{\langle n,\vec{R}\,|\,\vec{\nabla}_R H(\vec{R})\,|\,n',\vec{R}\rangle\times\langle n',\vec{R}\,|\,\vec{\nabla}_R H(\vec{R})\,|\,n,\vec{R}\rangle}{\left(\bar{E}_n - \bar{E}_{n'}\right)^2}\right].$$

10.5.1 Numerically investigate the "first set" of equations

$$i\hbar\dot{C}_1(t) = \frac{\gamma}{2}C_2(t)e^{i\omega_{12}t}\left(e^{i\omega t} + e^{-i\omega t}\right),$$

$$i\hbar\dot{C}_2(t) = \frac{\gamma}{2}C_1(t)e^{-i\omega_{12}t}\left(e^{i\omega t} + e^{-i\omega t}\right).$$

Simultaneously, investigate the "second set"

$$i\hbar\dot{C}_1(t) = \frac{\gamma}{2}C_2(t)e^{i(\omega_{12}-\omega)t},\quad i\hbar\dot{C}_2(t) = \frac{\gamma}{2}C_1(t)e^{i(\omega-\omega_{12})t},$$

using the values $\omega_{12} = 1.0$, $\omega = 0.99$, and $\gamma/\hbar = 0.01$ (units arbitrary). Plot $|C_1(t)|^2$ and $|C_2(t)|^2$ for both cases for one period of the driving angular frequency $T = 2\pi/\kappa(\omega)$. Can you see the additional high frequency component appearing in the first set? Make some quantitative observations regarding this additional component: approximate amplitude and period.

10.5.2 Confirm the result $\gamma_2^{\text{loop}} = -\frac{\Omega_{\text{bottom}}}{2}$ in the Berry phase calculation using Stokes' theorem and the area evaluation.

10.6.1 Try your hand at proving the following statement: The expectation value of the momentum operator $\langle \vec{p} \rangle$ vanishes for bound state wave functions subject to a time-independent potential $V(\vec{x})$. [Hints: One way starts with a proof that bound wave functions can always be chosen as functions of a single complex phase, and the other uses Equation 4.159 in Chapter 4 to calculate $\frac{d\langle \vec{x} \rangle}{dt}$. The results of Problem 6.2.1 in Chapter 6 may also be helpful.]

10.7.1 Go back to the two-state system of Section 10.3 with Hamiltonian given by Equation 10.51. Assuming $C_1(0^+) = 1$, $C_2(0^+) = 0$, calculate $C_2^{(1)}(t)$ in perturbation theory and compare with the exact expression, Equation 10.59, for $C_2(t)$. Make the approximation $2|\gamma/\hbar| \ll |\omega_{12} - \omega|$ and show they are the same.

10.7.2 An important theoretical tool in perturbation theory is called the *interaction picture*, which allows the efficient organization of perturbative expansions. It harkens to the time evolution pictures discussed in Chapter 4, Section 4.11 and is intermediate between the so-called Schrödinger and Heisenberg pictures. I will just give a taste of it here. It starts with a modification of Equation 10.6, which reads

$$|\psi,t\rangle_I \equiv e^{iH_0t/\hbar}|\psi,t\rangle$$
$$= \sum_{n'} C_{n'}(t)|n'\rangle.$$

In addition, the interaction picture evolution operator $U_I(t,t_0)$ is defined implicitly by

$$|\psi,t\rangle_I \equiv U_I(t,t_0)|\psi,t_0\rangle_I,$$

where t_0 is the initial time. Given these definitions, show that

(a) $i\hbar\frac{\partial}{\partial t}|\psi,t\rangle_I = V_I(t)|\psi,t\rangle_I$, where $V_I(t) \equiv e^{iH_0t/\hbar}V(t)e^{-iH_0t/\hbar}$

(b) $C_n(t) = \langle n|U_I(t,t_0)|\psi,t_0\rangle_I$

(c) $i\hbar\frac{dU_I(t,t_0)}{dt} = V_I(t)U_I(t,t_0)$

(d) $U_I(t,t_0) = 1 - \frac{i}{\hbar}\int_{t_0}^{t} dt'\, V_I(t')U_I(t',t_0)$, given $U_I(t_0,t_0) = 1$

10.8.1 Using the alternate m_ℓ, m_s hydrogen wave function basis

$$u_{n\ell m_\ell m_s}(\vec{x}) \equiv \frac{R_{n\ell}(r)}{r} Y_{\ell,m_\ell}(\theta,\phi)\chi_{m_s},$$

where $\chi_{m_s} = \begin{pmatrix} 1 \\ 0 \end{pmatrix}$ for $m_s = 1/2$ and $\chi_{m_s} = \begin{pmatrix} 0 \\ 1 \end{pmatrix}$ for $m_s = -1/2$, verify that

$$\sum_{m_s'} |< 1s, m_s' \,|\, \vec{x} \,|\, 2p, m_\ell, m_s >|^2 = \frac{1}{3}\left|\int_0^\infty dr\, r\, R_{10}(r) R_{21}(r)\right|^2.$$

To keep it simple, just do one case, say $m_\ell = 1$, $m_s = 1/2$. The rate is actually independent of the initial m_ℓ, m_s values.

10.8.2 Using Equations 10.187 and 10.198, finish the evaluation of the 2p to 1s lifetime calculation by doing the radial integral. See if you can get the rate equation in the form

$$R_{1s\,2p} = \frac{\alpha^5 \mu c^2}{\hbar} \cdot \left(\frac{2}{3}\right)^8,$$

where $\alpha = e^2/\hbar c$.

10.8.3 There is a famous and useful theorem in mathematics called the Wigner–Eckart theorem. It is based on the angular momentum considerations of Chapter 8 but can be applied to matrix elements of other operators. It states

$$\langle J', M' \,|\, T_{Kq} \,|\, J, M \rangle = \langle J\, K; M q \,|\, J\, K; J' M' \rangle \frac{\langle J' \,||\, T_K \,||\, J \rangle}{\sqrt{2J+1}},$$

where $\langle J' \,||\, T_K \,||\, J \rangle$ is the "reduced matrix element" (a number independent of M and M'), the T_{Kq} are "spherical tensors,"* and $\langle J\, K; M q \,|\, J\, K; J' M' \rangle$ are just the Clebsch–Gordan coefficients of Chapter 8. The Clebsch–Gordan coefficients on the right-hand side immediately imply the so-called triangle inequalities $|J-K| \le J' \le J+K$ and $|J'-K| \le J \le J'+K$ (first seen for angular momentum addition in Equation 8.44) as well as the magnetic quantum number addition rule, $M' = M + q$ (seen in Equation 8.33).

You are given the electromagnetic radiation Hamiltonian,

$$H^{\text{rad}} = \sum_{L=0,1,2\ldots} \sum_{m=-L}^{L} \sum_{X=(E,M)} A_{Lm}^{(X)} T_{Lm}^{(X)},$$

where the $A_{Lm}^{(E)}$, $A_{Lm}^{(M)}$ (E = electric, M = magnetic) are expansion coefficients (assume $A_{00}^{(M)} = A_{00}^{(E)} = 0$) and the $T_{Lm}^{(E)}$, $T_{Lm}^{(M)}$

* These are basically generalizations of spherical harmonics. See J.J. Sakurai and J. Napolitano, *Modern Quantum Mechanics*, 2nd ed. (Addison-Wesley, 2011), Section 3.11.

are spherical tensors. The tensors $T_{Lm}^{(E,M)}$ have the parity properties

$$\mathbb{P}T_{Lm}^{(E)}\mathbb{P} = (-1)^L T_{Lm}^{(E)} (L \geq 1),$$

$$\mathbb{P}T_{Lm}^{(M)}\mathbb{P} = (-1)^{L+1} T_{Lm}^{(M)} (L \geq 1),$$

where $\mathbb{P}$ is the parity operator, introduced in Chapter 3, Section 3.2. (Note that $\mathbb{P}^2 = 1$ can be inserted in matrix elements where convenient.) Now consider the matrix element in Fermi's golden rule, modeling atomic electromagnetic transitions:

$$\langle J_2, m_{J_2}, \pi_{J_2} \mid H^{\text{rad}} \mid J_1, m_{J_1}, \pi_{J_1} \rangle.$$

The states $| J, m_J, \pi_J \rangle$ have good total angular momentum J, J_z, and parity π_J. (Note the requirement that J_1 and J_2 both must be integer or half-integer.)

(a) Given the magnetic quantum number rule, $m_{J_2} = m_{J_1} + m$, is always satisfied (i.e., no m sum is necessary), for general J_1, J_2 values, how many L terms in the sum in H^{rad} give nonzero contributions? Give, for example, the operators that contribute when $J_1 = 3/2$, $J_2 = 1/2$, and $\pi_{J_1}\pi_{J_2} = -1$, which is the case in the hydrogen 2p to 1s transition. (The lowest L value corresponds to the electric dipole operator.)

(b) Assuming the terms with smallest L give the leading contribution to the transition, give the transition selection rules for $| \Delta J | \equiv | J_2 - J_1 | = 0, 1, 2, 3$ when $\pi_{J_1}\pi_{J_2} = 1$ or $\pi_{J_1}\pi_{J_2} = -1$. That is, determine the allowed L and (E,M) values in each $| \Delta J |$, $\pi_{J_1}\pi_{J_2}$ case.

10.8.4 The photon absorption cross section associated with a transition from an initial energy level $E_i = \hbar\omega_i$ to a final value $E_n = \hbar\omega_n$ for a laboratory material is defined as

$$\sigma_{ni}^{\text{abs}}(\omega) \equiv \frac{R_{i\to n}\, \hbar\omega}{\left\langle \vec{S}\cdot\hat{n} \right\rangle},$$

where $R_{i\to n}$ is the absorption rate given by Fermi's rule Equation 10.163, and $\left\langle \vec{S}\cdot\hat{n} \right\rangle$ is the time average of the normal component of the electromagnetic Poynting vector. Using the Hamiltonian Equation 10.171 and treating the vector potential $\vec{A}$ as a classical field, it can be shown in perturbation theory that this results in

$$\sigma_{ni}^{\text{abs}}(\omega) = \frac{4\pi\hbar\alpha}{m^2\omega} | \langle n | e^{i\vec{k}\cdot\vec{x}}\hat{\varepsilon}\cdot\vec{p} | i \rangle |^2\, \delta(\hbar\omega - \hbar\omega_{ni}),$$

where $\omega_{ni} \equiv \omega_n - \omega_i$. Note that the absorption density of states factor has been replaced with a delta function reflecting the discrete nature of the final state: $\rho^{\text{abs}}(E - E_{ni}) = \delta(\hbar\omega - \hbar\omega_{ni})$.

(a) Using the preceding results, show that in dipole approximation and with $\hat{\varepsilon} = \hat{x}$ one has that

$$\int_0^\infty d\omega \sum_n \sigma_{ni}^{\text{abs}}(\omega) = \frac{2\pi^2\hbar\alpha}{m}\sum_n f_{ni},$$

where

$$f_{ni} \equiv \frac{2m\omega_{ni}}{\hbar}|\langle n|x|i\rangle|^2.$$

The f_{ni} are called atomic "oscillator strengths."

(b) By considering the double commutator, $\langle i|\left[[x,H_0],x\right]|i\rangle$, and inserting a complete set of atomic states $\sum_n |n\rangle\langle n| = 1$ between terms, show that one has the simple result

$$\sum_n f_{ni} = 1 \Rightarrow \int_0^\infty d\omega \sum_n \sigma_{ni}^{\text{abs}}(\omega) = \frac{2\pi^2 e^2}{mc},$$

from which Planck's constant has completely disappeared!

10.9.1 Show that Equation 10.210 satisfies the normalization condition in Equation 10.209. (Analytic means are preferred but not required!)

***10.9.2** (See Chapter 11, Section 11.10 for an introduction to the concept of scattering cross sections.) The following table gives experimental data on center-of-mass (CM) pion–proton scattering cross sections* in the energy range from about 1130 MeV to 1310 MeV (listed without error bars).

***E*(MeV)**	$\frac{\sigma(\pi^+ - p \to \Delta^{++})k^2}{8\pi}$
1138.5	0.03866
1159.5	0.1228
1176.8	0.2800
1178.0	0.2987
1190.4	0.4857
1192.5	0.5237
1205.9	0.7693
1210.1	0.8377
1215.3	0.9050
1226.5	0.9881
1243.8	0.9611
1261.2	0.8333
1280.1	0.6876
1300.9	0.5508

* A.A. Carter, *Nucl. Phys. B* **26**, 445 (1971); J.R. Haskins, *Am. J. Phys.* **53**, 988 (1985).

The quantity $\sigma\left(\pi^+ - p \rightarrow \Delta^{++}\right)$ is the scattering cross section for $\pi^+ - p$ to produce the Δ^{++} resonance, k is the CM wavenumber of the system ($k = p_{CM} / \hbar$) and E is the CM energy. Multiplying the cross section by $k^2 / 8\pi$ gives a dimensionless quantity that should peak at 1 if the Δ^{++} decays only into these same two particles, which it does to an excellent approximation. This data may be fit by a *nonsymmetric* Breit–Wigner resonance formula given by*

$$\frac{\sigma\left(\pi^+ - p \rightarrow \Delta^{++}\right)k^2}{8\pi} = \frac{\left(\frac{\Gamma}{2}\cos\delta + (M_p c^2 - E)\sin\delta\right)^2}{(M_p c^2 - E)^2 + \Gamma^2 / 4},$$

where Γ is the decay width, M_p is called the "pole mass," E is the CM energy, and the angle δ measures the extent of mixing with a resonant background contribution.

(a) Fit the data with the given functional form using a nonlinear fitting package. (I used *Mathematica*'s© "NonLinearModelFit".) Note that the pole mass, M_p, is related to the Breit–Wigner mass, M_{BW}, listed in particle tables by

$$M_{BW}c^2 = M_p c^2 - \frac{\Gamma}{2}\tan\delta.$$

What values for Γ, M_p, δ, and M_{BW} do you get from the fit?

(b) Compare relevant parameters with values from the Particle Data Group review pages for the $\Delta(1232)$ resonance.

* S. Ceci, M. Korolija, and B. Zauner, *Phys. Rev. Lett.* **111**, 112004 (2013).

Quantum Particle Scattering

Synopsis: Students are introduced to the ideas of particle scattering in quantum mechanics in a one-dimensional context. The integral Schrödinger equation is deduced using a Wronskian technique, and reflection and transmission amplitudes are defined. Several simple scattering problems are solved, and the Born series for the energy-wave function is explained. A three-dimensional spherical basis is introduced, which allows the integral Schrödinger equation in this context to be formulated. The relation between plane waves and spherical waves is found from a solution to the unit-source Helmholtz equation, again using the Wronskian technique. Partial waves, phase shifts, and cross sections are defined, and finite range scattering is considered. The chapter concludes with a consideration of identical particle scattering and an application to the proton–proton case, where a dramatic manifestation of the underlying substructure emerges at high energies.

11.1 INTRODUCTION

Particle scattering is the principal tool in the physicist's arsenal of atomic and subatomic discovery. The analysis of scattering can tell us about the strength and range of the interaction as well as shedding light on possible composite structures of the particles involved.

You have probably encountered scattering in a classical mechanics context. There, particles follow deterministic trajectories that result from Newton's laws. One counts the number of particles scattered by the interaction into a particular solid angle. The number of particles scattered into this angle is then divided by the number of incoming particles to obtain a quantity independent of the incoming flux, called the differential cross section. Instead, in the quantum realm, particles have a wave nature, and we must follow the dynamics of wave propagation from a careful study of the Schrödinger equation. The principal object of study is called the scattering amplitude. Instead of counting particles, one takes the absolute square of this amplitude to construct the quantum cross section. In spite of the completely different approaches, in cases where the two descriptions overlap (i.e., the potential is varying slowly enough; see the Chapter 5 WKB discussion), the results from these two approaches are in agreement.

We have already encountered particle scattering in this text. In the simpler context of one dimension, we studied the transmission and reflection coefficients of a step potential in Chapter 3. We will deepen our study of these types of situations in this chapter. We will see that most of the concepts of quantum particle scattering are contained in the one-dimensional case. We will then move on to the more realistic case of three-dimensional scattering for finite-range potentials and some simplified Coulomb scattering considerations.

11.2 THE ONE-DIMENSIONAL INTEGRAL SCHRÖDINGER EQUATION

We start with the abstract eigenvalue/eigenvector statement,

$$H|E+\rangle = E|E+\rangle, \tag{11.1}$$

where $|E+\rangle$ is a scattering state with an initial wave moving in the $+x$-direction, and $E>0$. We will work in a two-body context throughout this chapter (see Chapter 7, Section 7.4). This gives ($H = T + V$)

$$(E-T)|E+\rangle = V|E+\rangle. \tag{11.2}$$

The formal solution of this is

$$|E+\rangle = C(k)|k+\rangle + \frac{1}{E-T}V|E+\rangle. \tag{11.3}$$

With appropriate spatial boundary condition specifications, which have not yet been specified, this is called the *Lippmann–Schwinger equation*. The first term on the right-hand side represents a plane wave, with wave number

$$k = \frac{\sqrt{2\mu E}}{\hbar},$$

moving in the $+x$-direction. It satisfies

$$\left(E-T\right)|k+\rangle = 0. \tag{11.4}$$

The constant, $C(k)$, in Equation 11.3 is arbitrary and arises for the same mathematical reason as the constant, $C(\lambda)$, in Equation 5.4 (Chapter 5). Physically, this constant represents the initial amplitude of the incoming wave. Our scattering results will be independent of its value. We will choose $C(k) = (2\pi)^{1/2}$ for simplicity in the following, which will correspond to a unit incoming amplitude, once Equation 11.3 is projected into coordinate space (see the wave normalization in Equation 11.7.)

The coordinate projection alluded to when coordinate completeness (Chapter 2, Equation 2.141) is supplied gives

$$u_{E+}(x) = e^{ikx} + \int_{-\infty}^{\infty} dx' \langle x|\frac{1}{E-T}|x'\rangle_{+} \langle x'|V|E+\rangle. \tag{11.5}$$

The "+" sign on the coordinate space matrix element of $1/(E - T)$ in the integral is independent of the "+" sign on $u_{E+}(x)$, and will be explained later. We have defined

$$u_{E+}(x) \equiv \langle x|E+\rangle, \tag{11.6}$$

where the new wave-number basis $|k\pm\rangle$ projects as (dimensionless form alluded to in Chapter 6, Section 6.1)

$$\langle x|k\pm\rangle = \frac{e^{\pm ikx}}{(2\pi)^{1/2}}. \tag{11.7}$$

The plane normalization in Equation 11.7 is such that ($k, k' > 0$)

$$\int_{-\infty}^{\infty} dx' \langle k\pm|x'\rangle\langle x'|k'\pm\rangle = \delta(k-k'), \tag{11.8}$$

$$\int_{-\infty}^{\infty} dx' \langle k\pm|x'\rangle\langle x'|k'\mp\rangle = 0. \tag{11.9}$$

Coordinate completeness also gives ($\langle x'|V|x''\rangle = \delta(x'-x'')V(x')$ for a "local" potential):

$$\overset{\downarrow\,\text{"1"}}{\langle x'|V|E+\rangle} = V(x')u_{E+}(x'). \tag{11.10}$$

Defining

$$g_+(x,x') \equiv \left(\frac{\hbar^2}{2\mu}\right)\langle x|\frac{1}{E-T}|x'\rangle_+, \tag{11.11}$$

we obtain finally

$$u_{E+}(x) = e^{ikx} + \left(\frac{2\mu}{\hbar^2}\right)\int_{-\infty}^{\infty} dx' g_+(x,x')V(x')u_{E+}(x'). \tag{11.12}$$

What are the properties of the function $g_+(x, x')$? Consider building a differential equation:

$$\left[\frac{-\hbar^2}{2\mu}\frac{d^2}{dx^2} - E\right]g_+(x,x') = \frac{\hbar^2}{2\mu}\left[\frac{-\hbar^2}{2\mu}\frac{d^2}{dx^2} - E\right]\langle x|\frac{1}{E-T}|x'\rangle_+. \tag{11.13}$$

We may write

$$-\frac{\hbar^2}{2\mu}\frac{d^2}{dx^2}\langle x| = \langle x|T, \tag{11.14}$$

where T is the Hilbert-space kinetic energy operator. I have used

$$T = \frac{p_x^2}{2\mu}, \tag{11.15}$$

and Equation 2.170,

$$\langle x|p_x = \frac{\hbar}{i}\frac{d}{dx}\langle x|, \tag{11.16}$$

in Equation 11.14. Thus,

$$\left[-\frac{\hbar^2}{2\mu}\frac{d^2}{dx^2} - E\right]g_+(x,x') = \frac{\hbar^2}{2\mu}\langle x|(T-E)\frac{1}{E-T}|x'\rangle_+; \tag{11.17}$$

with $E = \hbar^2 k^2 / 2\mu$, this simplifies to $\left(\langle x | x' \rangle = \delta(x - x')\right)$

$$\left(\frac{d^2}{dx^2} + k^2\right) g_+(x, x') = \delta(x - x'). \tag{11.18}$$

The solutions to the second-order differential equation, Equation 11.10, when $x \neq x'$ are just proportional to $e^{\pm ikx}$. Moreover, we assume the coordinate boundary conditions,

$$g_+(x, x') \sim e^{ikx}, \quad x > x', \tag{11.19}$$

$$g_+(x, x') \sim e^{-ikx}, \quad x' > x, \tag{11.20}$$

corresponding to *outgoing* scattered waves. A "backward" or "forward" moving wave is defined relative to the assumed time dependence, $e^{-i\omega t}$, for all amplitudes. Thus, Equations 11.19 and 11.20 describe outgoing waves moving in the $+x$ and $-x$ directions, respectively. A compact way of writing Equations 11.19 and 11.20, which ensures continuity in $g_+(x, x')$ when $x = x'$, is

$$g_+(x, x') = C e^{ikx_>} e^{-ikx_<}, \tag{11.21}$$

where the constant C is undetermined, and we introduce the notation,

$$x_{\gtrless} \equiv \text{the} \begin{pmatrix} \text{greater} \\ \text{lesser} \end{pmatrix} \text{of } x, x'. \tag{11.22}$$

Although $g_+(x, x')$ is continuous everywhere, the delta function on the right-hand side of Equation 11.18 implies a discontinuity in the derivative of $g_+(x, x')$ at $x = x'$. Let us integrate across this point in Equation 11.18:

$$\int_{x'^-}^{x'^+} dx \left(\frac{d^2}{dx^2} + k^2\right) g_+(x, x') = \int_{x'^-}^{x'^+} dx \; \delta(x - x'). \tag{11.23}$$

The notation is

$$x'^{\pm} \equiv x' \pm \varepsilon, \tag{11.24}$$

where ε is a positive, infinitesimal quantity. Given continuity in $g_+(x, x')$ at $x = x'$, the second term on the left-hand side of Equation 11.23 vanishes in the ε limit, and one obtains the statement

$$\frac{d}{dx} g_+(x, x') \Big|_{x'^-}^{x'^+} = 1. \tag{11.25}$$

Using Equation 11.21 in 11.25, we learn that

$$C\left[e^{-ikx'} \frac{d}{dx'} e^{ikx'} - e^{ikx'} \frac{d}{dx'} e^{-ikx'}\right] = 1. \tag{11.26}$$

We have encountered an important mathematical quantity in Equation 11.26 called the "Wronskian." By definition,

$$W_x\left[F(x),G(x)\right] \equiv F\frac{dG}{dx} - G\frac{dF}{dx}, \tag{11.27}$$

where the subscript on W tells us the derived variable. In our case,

$$W_{x'}\left[e^{-ikx'}, e^{ikx'}\right] = 2ik, \tag{11.28}$$

by direct calculation. Thus one may write

$$g_+(x,x') = \frac{1}{2ik}e^{ik(x_> - x_<)}, \tag{11.29}$$

or equivalently,

$$g_+(x,x') = \frac{1}{2ik}e^{ik|x-x'|}. \tag{11.30}$$

Another mathematical solution to Equation 11.18 is

$$g_-(x,x') \equiv -\frac{1}{2ik}e^{-ik|x-x'|}, \tag{11.31}$$

corresponding to incoming "scattered" waves, which is not physically relevant for us.

Putting Equation 11.30 into 11.12 now gives

$$u_{E+}(x) = e^{ikx} + \left(\frac{\mu}{i\hbar^2 k}\right)\int_{\infty}^{\infty} dx' e^{ik|x-x'|}V(x')u_{E+}(x'). \tag{11.32}$$

Please note that Equation 11.32 does not represent a solution for $u_{E+}(x)$ but simply gives an integral equation for this quantity. This integral equation is entirely equivalent to the time-independent differential Schrödinger equation,

$$\left[-\frac{\hbar^2}{2\mu}\frac{d^2}{dx^2} + V(x)\right]u_{E+}(x) = \frac{\hbar^2 k^2}{2\mu}u_{E+}(x), \tag{11.33}$$

given a $+x$-direction initial wave boundary condition. In fact, you can check that substituting Equation 11.32 into 11.33 just gives an identity as it should.

11.3 REFLECTION AND TRANSMISSION AMPLITUDES

In one-dimensional scattering, the objects of prime interest are the reflection and transmission *amplitudes*, which will be denoted by $r(k)$ and $t(k)$, respectively. They are related to the reflection and transmission *coefficients* by

$$R \equiv |r(k)|^2, \tag{11.34}$$

$$T \equiv |t(k)|^2. \tag{11.35}$$

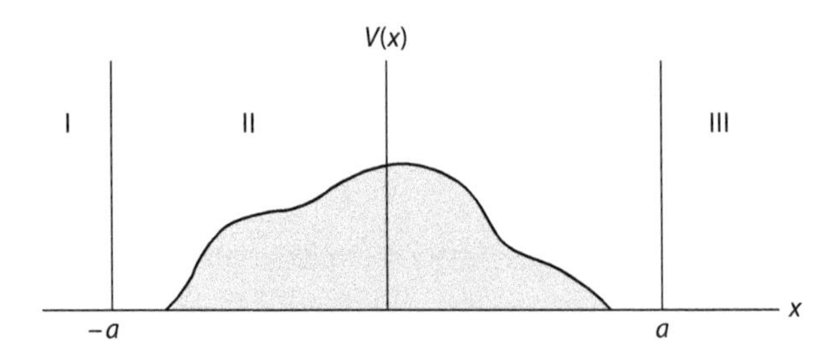

FIGURE 11.1 An arbitrary localized potential in one spatial dimension.

These coefficients were studied for the step-function potential barrier in Chapter 3, Section 3.4.

It is possible to get general expressions for these quantities for localized potentials as follows. Let us assume an arbitrary potential, $V(x)$, which is finite in extent, as in Figure 11.1. Notice that the points $x = \pm a$ are chosen to lie in the zero potential parts of the figure, defining Regions I, II, and III. In Region I, Equation 11.32 gives ($x_< = x$, $x_> = x'$)

$$u_{E+}(x) = e^{ikx} + \left(\frac{\mu}{i\hbar^2 k}\right) e^{-ikx} \int_{-a}^{a} dx' e^{ikx'} V(x') u_{E+}(x'). \tag{11.36}$$

The coefficient of the backward-moving phase factor, e^{-ikx}, defines the reflection amplitude:

$$r(k) \equiv \left(\frac{\mu}{i\hbar^2 k}\right) \int_{-a}^{a} dx' e^{ikx'} V(x') u_{E+}(x'). \tag{11.37}$$

Likewise, in Region III we will have $\left(x_< = x', x_> = x\right)$

$$u_{E+}(x) = e^{ikx} + \left(\frac{\mu}{i\hbar^2 k}\right) e^{ikx} \int_{-a}^{a} dx' e^{-ikx'} V(x') u_{E+}(x'), \tag{11.38}$$

which gives the transmission amplitude,

$$t(k) \equiv 1 + \left(\frac{\mu}{i\hbar^2 k}\right) \int_{-a}^{a} dx' e^{-ikx'} V(x') u_{E+}(x'). \tag{11.39}$$

The absolute squares of these amplitudes yield probabilities. The conservation of probability for such quantities follows from the probability current found in Equation 2.129 (Chapter 2):

$$j(x) = \frac{i\hbar}{2\mu}\left[\frac{\partial \psi}{\partial x}\psi^* - \psi^* \frac{\partial \psi}{\partial x}\right]. \tag{11.40}$$

Choosing $\psi = u_{E+}$, this current can be evaluated to give the position-independent results

$$j^{\mathrm{I}} = \frac{\hbar k}{\mu}\left(1 - |r(k)|^2\right), \tag{11.41}$$

in Region I, and

$$j^{\mathrm{III}} = \frac{\hbar k}{\mu}\left|t(k)\right|^2, \tag{11.42}$$

in Region III. Then, the probability conservation statement,

$$j^{\mathrm{I}} = j^{\mathrm{III}}, \tag{11.43}$$

yields

$$\left|r(k)\right|^2 + \left|t(k)\right|^2 = 1, \tag{11.44}$$

which was previously stated in Chapter 3, Equation 3.115 in terms of T and R.

Although Equations 11.37 and 11.39 are exact, this does us little good since $u_{E+}(x)$ in the integrals are unknown. We will find approximate solution methods for these quantities later in the chapter.

11.4 ONE-DIMENSIONAL DELTA-FUNCTION SCATTERING

Let us apply Equation 11.32 to the simplest possible scattering problem in one dimension: the delta-function potential. We choose

$$V(x) = -\lambda\delta(x-a), \tag{11.45}$$

which represents a sharp, attractive delta function, located at $x=a$, if $\lambda > 0$. Equation 11.32 then gives

$$u_{E+}(x) = \mathrm{e}^{ikx} + \frac{i\mu\lambda}{\hbar^2 k}\mathrm{e}^{ik|x-a|}u_{E+}(a). \tag{11.46}$$

To determine $u_{E+}(a)$, set $x=a$ on both sides of Equation 11.46:

$$u_{E+}(a) = \mathrm{e}^{ika} + \frac{i\mu\lambda}{\hbar^2 k}u_{E+}(a), \tag{11.47}$$

$$\Rightarrow u_{E+}(a) = \frac{\mathrm{e}^{ika}}{\left(1 - \dfrac{i\mu\lambda}{\hbar^2 k}\right)}. \tag{11.48}$$

Then, for $x \geq a$,

$$u_{E+}(x) = \mathrm{e}^{ikx} + \frac{i\mu\lambda}{\hbar^2 k}\frac{\mathrm{e}^{ikx}}{\left(1 - \dfrac{i\mu\lambda}{\hbar^2 k}\right)}, \tag{11.49}$$

and

$$u_{E+}(x) = e^{ikx} + \frac{i\mu\lambda}{\hbar^2 k} \frac{e^{2ika} e^{-ikx}}{\left(1 - \frac{i\mu\lambda}{\hbar^2 k}\right)}, \tag{11.50}$$

for $x \leq a$. This identifies the transmission and reflection amplitudes as

$$t(k) = 1 + \frac{i\mu\lambda}{\hbar^2 k} \frac{1}{\left(1 - \frac{i\mu\lambda}{\hbar^2 k}\right)}, \tag{11.51}$$

or

$$t(k) = \frac{1}{\left(1 - \frac{i\mu\lambda}{\hbar^2 k}\right)}, \tag{11.52}$$

and

$$r(k) = \frac{i\mu\lambda}{\hbar^2 k} \frac{e^{2ika}}{\left(1 - \frac{i\mu\lambda}{\hbar^2 k}\right)}. \tag{11.53}$$

One may check that the probability conservation statement, Equation 11.44, is satisfied by these amplitudes.

Although we started with the integral equation, Equation 11.32, to solve this problem, it would have been just as easy and natural to use the Schrödinger differential equation, Equation 11.33. Given the initial $+x$-directed plane wave, one would simply write the general solutions with unknown ($r(k)$ and $t(k)$) coefficients in Regions I ($x \leq a$) and II ($x \geq a$), and use the matching conditions

$$u^{\mathrm{I}}_{E+}(a) = u^{\mathrm{II}}_{E+}(a) \tag{11.54}$$

and

$$\left.\frac{du^{\mathrm{II}}}{dx}\right|_{x=a} - \left.\frac{du^{\mathrm{I}}}{dx}\right|_{x=a} = -\frac{2\mu\lambda}{\hbar^2} u_{E+}(a) \tag{11.55}$$

to determine them.

The amplitudes $r(k)$ and $t(k)$ for this problem display a simple pole in the complex k-plane. That is, $r(k)$ and $t(k)$ may be written as

$$r(k) = \frac{k_0 e^{2ika}}{(k - k_0)}, \tag{11.56}$$

$$t(k) = \frac{k}{(k - k_0)}, \tag{11.57}$$

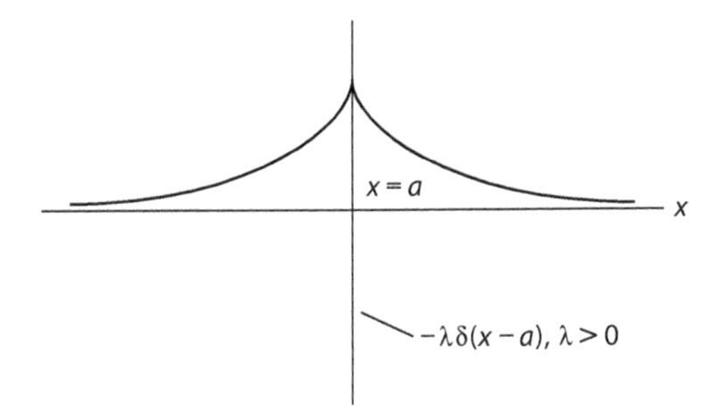

FIGURE 11.2 Qualitative form of the bound-state solution to the delta-function potential, Equation 11.45.

where

$$k_0 \equiv \frac{i\mu\lambda}{\hbar^2}. \tag{11.58}$$

The energy value associated with the momentum, k_0, is

$$E_0 = \frac{\hbar^2 k_0^2}{2\mu} = -\frac{\mu\lambda^2}{2\hbar^2}. \tag{11.59}$$

Amazingly, there is a bound state for the $\lambda > 0$ case with this exact energy! This bound-state solution has the form

$$u_E(x) = C\ \mathrm{e}^{-\kappa|x-a|}, \tag{11.60}$$

where C is a constant. Figure 11.2 shows the solution. This solution must be exponentially decreasing for $x \to \pm\infty$, and is continuous in value but discontinuous in derivative, at $x = a$. Using Equation 11.60 in 11.55 accommodates the discontinuity and determines that

$$\begin{aligned} -2\kappa C &= -\frac{2\mu\lambda}{\hbar^2}C, \\ C \neq 0 \Rightarrow \kappa &= \frac{\mu\lambda}{\hbar^2}, \end{aligned} \tag{11.61}$$

where we encounter the Wronskian evaluated at $x' = a$ but with $k \to i\kappa$. The energy of this state is determined from the time-independent Schrödinger equation,

$$E_0 u_E(x) = \left[-\frac{\hbar^2}{2\mu}\frac{\mathrm{d}^2}{\mathrm{d}x^2} - \lambda\delta(x-a)\right]u_E(x). \tag{11.62}$$

The left-hand side of Equation 11.62 is a continuous function of x, which implies the discontinuous part of the right-hand side must cancel. The energy is thus

$$E_0 = -\frac{\hbar^2\kappa^2}{2\mu} = -\frac{\mu\lambda^2}{2\hbar^2}, \tag{11.63}$$

in agreement with Equation 11.59! (The normalization of this state will be left to a problem.)

The phenomenon we have just seen is completely general. That is, bound states of the Schrödinger equation show up as simple poles in $r(k)$ and $t(k)$ along the positive imaginary axis in the complex k-plane. This is not only true for the transmission amplitudes in one dimension, but turns out to be true for the aforementioned scattering amplitude in three dimensions.

11.5 STEP-FUNCTION POTENTIAL SCATTERING

The reflection and transmission amplitudes for the step function were previously solved for in Chapter 3, Equations 3.116 and 3.119. (We use the same potential depicted in Section 3.4.) Using slightly different notation, we have

$$r(k) = \frac{\frac{i}{2}\left(\frac{k_2}{k} - \frac{k}{k_2}\right)e^{-2ika}\sin(2k_2a)}{\cos(2k_2a) - \frac{i}{2}\left(\frac{k}{k_2} + \frac{k_2}{k}\right)\sin(2k_2a)} \tag{11.64}$$

and

$$t(k) = \frac{e^{-2ika}}{\cos(2k_2a) - \frac{i}{2}\left(\frac{k}{k_2} + \frac{k_2}{k}\right)\sin(2k_2a)}, \tag{11.65}$$

where

$$k = \frac{\sqrt{2\mu E}}{\hbar}, \quad k_2 = \frac{\sqrt{2\mu\left(E - V_0\right)}}{\hbar}, \tag{11.66}$$

where $E - V_0 > 0$ for $V_0 > 0$ and $E > 0$ for $V_0 < 0$.

Based on the ideas of the last section, we anticipate that we can predict the bound states for $V_0 < 0$ from the poles of $r(k)$ and $t(k)$, or the zeros of their denominators, in the complex k-plane. These zeros occur when

$$\cos(2k_2a) = \frac{i}{2}\left(\frac{k}{k_2} + \frac{k_2}{k}\right)\sin(2k_2a). \tag{11.67}$$

There are no solutions of this equation for k, k_2 real. We will analytically continue this equation to search for solutions along the imaginary k-axis by letting $k \to i\kappa$, where $\kappa = \sqrt{-2\mu E}/\hbar$. ($k_2$ remains real if we keep $E - V_0 > 0$.) Then, Equation 11.67 becomes

$$\cos(2k_2a) = \frac{1}{2}\left(\frac{k_2}{\kappa} - \frac{\kappa}{k_2}\right)\sin(2k_2a). \tag{11.68}$$

Defining $R \equiv k_2 / \kappa$, we have

$$\frac{1}{2}\left(e^{2ik_2a} + e^{-2ik_2a}\right) = \frac{1}{4i}\left(R - \frac{1}{R}\right)\left(e^{2ik_2a} - e^{-2ik_2a}\right),$$
$$\Rightarrow e^{2ik_2a}\left(1+iR\right)^2 = e^{-2ik_2a}\left(1-iR\right)^2. \tag{11.69}$$

Taking the square root of both sides gives

$$e^{ik_2a}\left(1+iR\right) = \pm e^{-ik_2a}\left(1-iR\right), \tag{11.70}$$

giving the eigenvalue equation,

$$\kappa \sin(k_2a) = -k_2 \cos(k_2a), \tag{11.71}$$

for the upper sign in Equation 11.70 and

$$\kappa \cos(k_2a) = k_2 \sin(k_2a), \tag{11.72}$$

for the lower sign. You can verify these as the eigenvalue conditions determining the bound-state energies, $E_b = -\hbar^2\kappa^2 / 2\mu$, of a potential well with $V_0 < 0$. Equation 11.71 represents solutions with odd parity wave functions, and Equation 11.72 represents solutions with even parity. These solutions again appear as poles of $r(k)$ and $t(k)$ along the positive, imaginary momentum k-axis.

There is another interesting scattering phenomenon we can investigate with the help of Equation 11.67, which we rewrite as

$$\tan(2k_2a) = -2i\frac{\sqrt{\dfrac{E-V_0}{E}}}{1+\left(\dfrac{E-V_0}{E}\right)}. \tag{11.73}$$

The solutions to this will require complex energies, and we will write

$$E \equiv E_0 - i\frac{\Gamma}{2}, \tag{11.74}$$

where both E_0 and Γ are assumed to be real, positive quantities. Γ is the *decay width* and will be interpreted later. We will investigate solutions of Equation 11.73 in the complex k-plane for two situations:

Case 1: $(E_0 - V_0) \ll E_0 > 0, (E_0 \gg \Gamma/2, V_0 > 0)$

$$\Rightarrow \tan(2k_2a) \approx -2i\sqrt{\frac{E-V_0}{E}}. \tag{11.75}$$

Figure 11.3 shows the associated situation for real scattering energies, E.

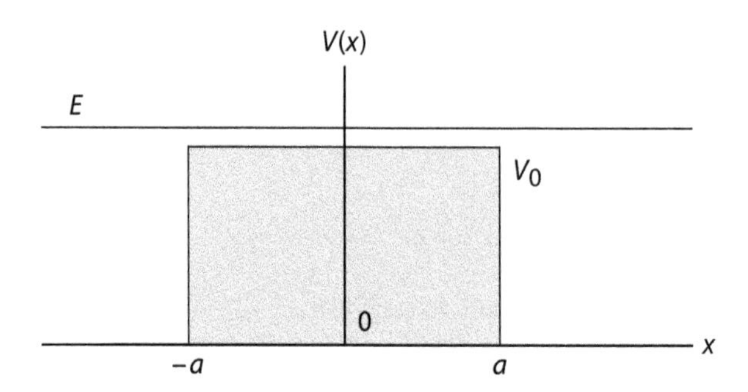

FIGURE 11.3 The Case 1 scattering situation with $E=E_0$ and $V_0>0$.

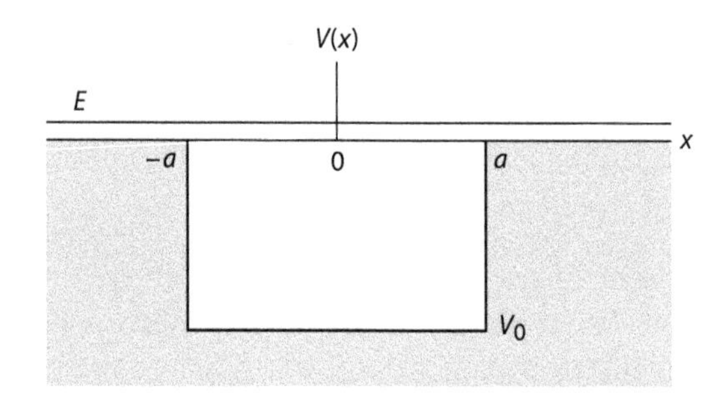

FIGURE 11.4 The Case 2 scattering situation with $E=E_0$ and $V_0<0$.

Case 2: $E_0 \ll -V_0 > 0\,(E_0 \gg \Gamma/2)$

$$\Rightarrow \tan(2k_2a) \approx -2i\sqrt{\frac{-E}{V_0}}. \tag{11.76}$$

Figure 11.4 shows this associated situation, again for real energies, E. Both cases make the right-hand side of Equation 11.73 small, allowing approximate solutions.

The solutions for Case 1 are given approximately by

$$2k_2a \approx \tan^{-1}\left(-2i\sqrt{\frac{E-V_0}{E}}\right) \tag{11.77}$$

or

$$2k_2a \approx n\pi - 2i\sqrt{\frac{E-V_0}{E}}, \tag{11.78}$$

where $n=1, 2, 3, \ldots$. (The $n=0$ case implies $\Gamma=0$, $E_0<V_0$.) It is understood that the first term on the right-hand side dominates in magnitude. Then, given Equation 11.66 for k_2, we may approximately solve for E as

$$E \approx V_0 + \frac{\hbar^2(n\pi)}{8\mu a^2}\left[(n\pi) - 4i\sqrt{\frac{E-V_0}{E}}\right], \tag{11.79}$$

where again, the imaginary term is "small." We have approximately,

$$E_0 \approx V_0 + \frac{(\hbar n\pi)^2}{8\mu a^2}, \tag{11.80}$$

$$\Gamma \approx \frac{\hbar^2 (n\pi)}{\mu a^2}\sqrt{\frac{E_0 - V_0}{E_0}}, \tag{11.81}$$

where the Case 1 condition ($(E_0 - V_0) \ll E_0$) requires

$$\frac{\Gamma}{E_0} \ll \frac{8}{n\pi}, \tag{11.82}$$

which, for $n \geq 3$, becomes more restrictive than the parenthetical Case 1 requirement. The complex solution for E, Equations 11.74, 11.80, and 11.81, is equivalent to the statement that

$$k_0 \approx \frac{\sqrt{2\mu E_0}}{\hbar}\left(1 - i\frac{\Gamma}{4E_0}\right), \tag{11.83}$$

using $E \equiv \hbar^2 k_0^2 / 2\mu$. Mathematically, these approximate solutions correspond to poles in the $r(k)$ and $t(k)$ amplitudes along the real, positive k-axis, but with a small, negative imaginary part.

What does all this mean physically? Let us recall that in Chapter 3, Section 3.4 we found the "resonance" condition (Equation 3.127),

$$2k_2 a = n\pi, n = 1, 2, 3, \ldots,$$

at which the transmission coefficient, T, was 1. (This is equivalent to the requirement that $r(k)$, Equation 11.64, vanishes.) This condition is suspiciously similar to Equation 11.78. The correct conclusion is that each resonance is associated with a pole in the complex k-plane. Let us investigate the form of $t(k)$ in the vicinity of one of these resonances, but on the real k-axis. Starting with Equation 11.65, and expanding the quantity

$$x \equiv 2k_2 a - n\pi, \tag{11.84}$$

where $|x| \ll 1$, we have

$$\cos(2k_2 a) = \cos(n\pi)\cos x - \sin(n\pi)\sin x \approx (-1)^n, \tag{11.85}$$

$$\sin(2k_2 a) = \sin(n\pi)\cos x + \cos(n\pi)\sin x \approx (-1)^n x, \tag{11.86}$$

which gives

$$t(k) \approx \frac{(-1)^n e^{-2ika}}{1 - \frac{i}{2}\sqrt{\frac{E_0}{E_0 - V_0}}\,x}, \tag{11.87}$$

or, restoring the meaning of x,

$$|t(k)|^2 \approx \frac{1}{1+\left(\frac{E_0}{E_0 - V_0}\right)\left(k_2 a - \frac{n\pi}{2}\right)^2}. \tag{11.88}$$

We wish to express this in terms of k, instead of k_2. For very small changes in k and k_2 about a resonance we can write

$$\frac{\Delta k}{\Delta k_2} = \frac{\left(k - \sqrt{2\mu E_0}/\hbar\right)}{\left(k_2 - n\pi/2a\right)}. \tag{11.89}$$

However, we also have

$$k_2^2 = k^2 - \frac{2\mu V_0}{\hbar^2} \Rightarrow \frac{\mathrm{d}k}{\mathrm{d}k_2} = \frac{k_2}{k} \approx \sqrt{\frac{E_0 - V_0}{E_0}}. \tag{11.90}$$

This means that for small changes about resonance we may use

$$\left(k_2 - \frac{n\pi}{2a}\right) \approx \sqrt{\frac{E_0}{E_0 - V_0}}\left(k - \frac{\sqrt{2\mu E_0}}{\hbar}\right). \tag{11.91}$$

However, combining Equations 11.80 and 11.81 gives

$$\Gamma \approx \sqrt{\frac{8}{\mu}}\frac{\hbar}{a}\frac{\left(E_0 - V_0\right)}{\sqrt{E_0}}, \tag{11.92}$$

which, when inserted in Equation 11.88, can be rearranged as

$$|t(k)|^2 = \frac{\left(\frac{\Gamma}{2}\right)^2}{\left(\frac{\Gamma}{2}\right)^2 + 4E_0\left(\sqrt{E} - \sqrt{E_0}\right)^2}, \tag{11.93}$$

where $E = \hbar^2 k^2 / 2\mu$. One final approximation is usually made to get this in a more standard form. Introducing

$$y \equiv \sqrt{E} - \sqrt{E_0}, \tag{11.94}$$

for $|y| \ll \sqrt{E_0}, \sqrt{E}$, we may write

$$\begin{aligned} 4E_0\left(\sqrt{E} - \sqrt{E_0}\right)^2 &= \left(2\sqrt{E_0}\right)^2 y^2 = \left(\sqrt{E} + \sqrt{E_0} - y\right)^2 y^2 \\ &\approx \left(\sqrt{E} + \sqrt{E_0}\right)^2\left(\sqrt{E} - \sqrt{E_0}\right)^2 = \left(E - E_0\right)^2. \end{aligned} \tag{11.95}$$

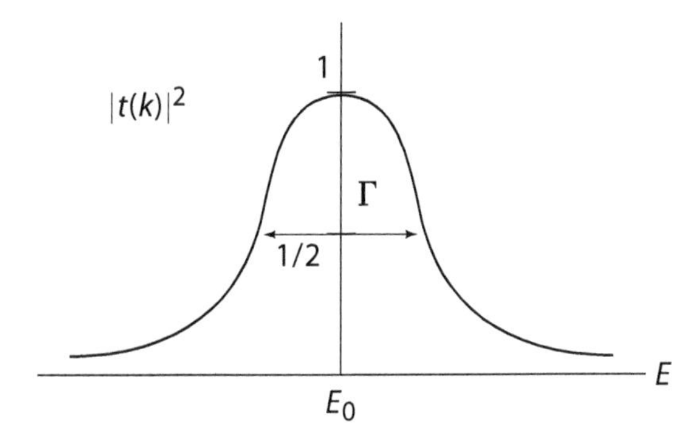

FIGURE 11.5 The one-dimensional Breit–Wigner resonance form, Equation 11.96.

Thus, on the real k-axis near a resonance,

$$|t(k)|^2 = \frac{(\Gamma/2)^2}{(\Gamma/2)^2 + (E - E_0)^2}. \quad (11.96)$$

This is called the *Breit–Wigner resonance form.** It is illustrated in Figure 11.5. It correctly gives $T = |t(\mathrm{Re}(k_0))|^2 = 1$ at resonance. Such a state picks up an exponential time dependence

$$e^{iEt/\hbar} = e^{iE_0 t/\hbar} e^{-\Gamma t/2\hbar}, \quad (11.97)$$

in its amplitude. The decay probability is $|e^{-\Gamma t/2\hbar}|^2$, giving a mean life,† τ, of

$$\tau \equiv \frac{\hbar}{\Gamma}. \quad (11.98)$$

Referring to the figure, we see that Γ has the meaning of the full width of the resonance at half-maximum. Notice that Equation 11.98 is an expression of the time-energy uncertainty relation, Equation 2.93 in Chapter 2, where we associate time uncertainty with τ and the energy uncertainty with Γ.

Figure 11.6 illustrates the structure of the transmission amplitude, $t(k)$, in the complex k-plane. We see the positions of the bound-state poles along the positive imaginary k-axis and the resonance poles arrayed near the real k-axis with negative imaginary part proportional to Γ. The position of these poles "causes" the maximum in $|t(k)|^2$ at $E \equiv E_0$.

Combining Equations 11.80 and 11.81, we can write Γ as

$$\Gamma \approx \frac{(\hbar n\pi)^2}{\mu a^2} \frac{1}{\sqrt{\frac{8\mu a^2 V_0}{\hbar^2} + (n\pi)^2}}. \quad (11.99)$$

* The Breit–Wigner form for the transmission coefficient is slightly different from the form for the partial cross section in three dimensions; see Section 11.10.

† Note the discussion of mean life and half-life at the end of Section 10.8.

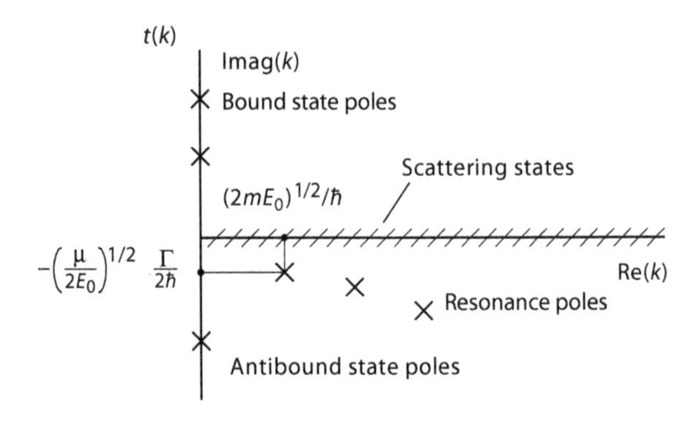

FIGURE 11.6 The analytical pole structure of the transmission amplitude, $t(k)$, in the complex k-plane for a general potential with both bound- and antibound-state solutions as well as scattering resonances.

This formula indicates that the lower-n poles initially move quickly away from the real axis, transitioning over to a slower movement at larger n values. The Case 1 condition

$$\frac{8\mu a^2 V_0}{\hbar^2} \gg (n\pi)^2, \tag{11.100}$$

gives an idea of how many such resonances there will be. We also see in Figure 11.6 a reference to another pole-beast called an "antibound" state, for when we replace $k \to i\kappa$, there may be nonnormalizable solutions of the Schrödinger equation, behaving as $e^{-\kappa|x|}$ for large $|x|$, with $\kappa < 0$. The positions of these poles on the negative imaginary axis are in general independent of the position of the bound-state poles. In the delta-function potential case, such a pole is "bound" for $\lambda > 0$, and "antibound" for $\lambda < 0$.

The Case 2 resonances will be investigated in the problems.*

Although these observations concerning the structure of the transmission amplitude are specific to the one-dimensional problems we have studied, these pole structures, and the general form of the Breit–Wigner resonance equation, Equation 11.96, are quite general.

11.6 THE BORN SERIES

Let us reconsider Equation 11.32, for a potential finite in extent:

$$u_{E+}(x) = e^{ikx} + \left(\frac{\mu}{i\hbar^2 k}\right)\int_{-a}^{a} dx' e^{ik|x-x'|} V(x') u_{E+}(x'). \tag{11.101}$$

Solving this equation (or equivalently, the Schrödinger equation with the same boundary conditions) is usually not possible. However, we need a good expression for $u_{E+}(x)$ to use if we wish to evaluate the reflection and transmission expressions, Equations 11.37 and 11.39.

* See also the journal article by Sprung, Wu, and Martorell, *Am. J. Phys.* **64**, 136 (1996).

One way to proceed is to assume that $V(x')$ in Equation 11.101 is in some sense "small." Then, one has an iterative scheme

$$u_{E+}^{(1)}(x) = e^{ikx}, \tag{11.102}$$

$$\vdots$$

$$u_{E+}^{(n+1)}(x) = e^{ikx} + \left(\frac{\mu}{i\hbar^2 k}\right)\int_{-a}^{a} dx' e^{ik|x-x'|} V(x') u_{E+}^{(n)}(x'). \tag{11.103}$$

For example, for $n = 2, 3$ we have

$$u_{E+}^{(2)}(x) = u_{E+}^{(1)}(x) + \left(\frac{\mu}{i\hbar^2 k}\right)\int_{-a}^{a} dx' e^{ik|x-x'|} V(x') e^{ikx'}, \tag{11.104}$$

$$u_{E+}^{(3)}(x) = u_{E+}^{(2)}(x) + \left(\frac{\mu}{i\hbar^2 k}\right)^2 \int_{-a}^{a} dx' e^{ik|x-x'|} V(x') e^{ikx'} \int_{-a}^{a} dx'' e^{ik|x'-x''|} V(x'') e^{ikx''}. \tag{11.105}$$

This is the *Born series*, named in honor of Max Born. Notice that as one increases the number of terms in the Born series, one acquires a higher number of integrations over the potential, $V(x)$. This is a so-called perturbative expansion for the wave function. We can then approximately evaluate $r(k)$ and $t(k)$ as

$$r^{(n)}(k) \equiv \left(\frac{\mu}{i\hbar^2 k}\right)\int_{-a}^{a} dx' e^{ikx'} V(x') u_{E+}^{(n)}(x'), \tag{11.106}$$

$$t^{(n)}(k) \equiv 1 + \left(\frac{\mu}{i\hbar^2 k}\right)\int_{-a}^{a} dx' e^{-ikx'} V(x') u_{E+}^{(n)}(x'). \tag{11.107}$$

The lowest-order expressions are

$$r^{(1)}(k) = \left(\frac{\mu}{i\hbar^2 k}\right)\int_{-a}^{a} dx' e^{2ikx'} V(x'), \tag{11.108}$$

$$t^{(1)}(k) = 1 + \left(\frac{\mu}{i\hbar^2 k}\right)\int_{-a}^{a} dx' V(x'). \tag{11.109}$$

How do these expressions compare with the exact results for the delta-function potentials, Equations 11.52 and 11.53, and the step-function potential case, Equations 11.64 and 11.65? For the delta-function case, $V(x) = -\lambda\delta(x-a)$, we obtain

$$r^{(1)}(k) = \frac{i\lambda\mu}{\hbar^2 k} e^{2ika}, \tag{11.110}$$

$$t^{(1)}(k) = 1 + \frac{i\lambda\mu}{\hbar^2 k}. \tag{11.111}$$

These are just the first terms in Equations 11.52 and 11.53 with λ considered a small parameter. You will confirm a similar finding for $r^{(1)}(k)$ and $t^{(1)}(k)$ for the step potential in the problems.

There are two important observations to make regarding Born series results. First, the Born series may not converge if the potential parameter is too large. In the delta-function case, it is clear that the exact results, Equations 11.52 and 11.53, will only have their Born series (which are simply arithmetic expansions) converge if (its "radius of convergence")

$$\frac{\mu|\lambda|}{\hbar^2 k} < 1 \quad \text{or} \quad |\lambda| < \hbar\sqrt{\frac{2E}{\mu}}. \tag{11.112}$$

Notice that the requirement is momentum-dependent. In fact, we see that the larger we make $\hbar^2 k/\mu$, the larger is the radius of convergence in $|\lambda|$. The Born series generally converges better at higher energies. This is because the initial ansatz of a plane wave $\sim e^{ikx}$ becomes a better characterization of the actual wave function during the interaction with the potential. Effectively, the larger the momentum, the less we would expect the interaction with the potential to modify the initial plane wave assumption. It is important to realize that the possible failure of the Born series for strong potentials is a mathematical, not a physical failure of the theory. The answers one finds through an exact, nonperturbative approach hold for any strength of the interaction potentials, as we found in the two examples mentioned earlier.

However, there is an additional problem with the Born approximation. The approximate results, Equations 11.110 and 11.111, do not satisfy the probability requirement, Equation 11.44. We have

$$|r^{(1)}(k)|^2 + |t^{(1)}(k)|^2 = 1 + 2\left(\frac{\lambda\mu}{\hbar^2 k}\right)^2. \tag{11.113}$$

Of course, since Equations 11.110 and 11.111 are only correct to order λ, we would expect violations of order $|\lambda|^2$ in such an exact requirement. The best that can be done is to define renormalized quantities,

$$r^{(1R)}(k) \equiv \frac{r^{(1)}(k)}{\sqrt{1+2\left(\frac{\lambda\mu}{\hbar^2 k}\right)^2}}, \tag{11.114}$$

$$t^{(1R)}(k) \equiv \frac{t^{(1)}(k)}{\sqrt{1+2\left(\frac{\lambda\mu}{\hbar^2 k}\right)^2}}, \tag{11.115}$$

which do satisfy Equation 11.44, but clearly this is an ad hoc procedure.

11.7 THE THREE-DIMENSIONAL INTEGRAL SCHRÖDINGER EQUATION

Although the more realistic case of three-dimensional scattering will be significantly more mathematical than the one-dimensional case just examined, we will find that essentially all the major lessons learned there will carry over unchanged. This was a major impetus for covering the one-dimensional case in some detail.

In the three-dimensional case, we will be working in a spherical coordinate basis, as well as a new spherical coordinate wave-number basis,

$$|\vec{k}\rangle = \underset{\substack{\uparrow \\ \text{Magnitude}}}{|k\rangle} \otimes \underset{\substack{\uparrow \\ \text{Angle}}}{|\hat{k}\rangle}, \tag{11.116}$$

where (dimensionless form again)

$$\langle \vec{r} | \vec{k} \rangle \equiv \frac{e^{i\vec{k}\cdot\vec{r}}}{(2\pi)^{3/2}}, \tag{11.117}$$

$$\int d^3k\, |\vec{k}\rangle\langle\vec{k}| = 1, \tag{11.118}$$

$$\langle \vec{k}' | \vec{k} \rangle = \frac{1}{k^2}\delta(k-k')\delta(\cos\theta - \cos\theta')\delta(\phi-\phi'). \tag{11.119}$$

The angular part projects as

$$\langle l,m | \hat{k} \rangle = Y^*_{lm}(\hat{k}), \tag{11.120}$$

where we are using a shortened notation, $Y_{lm}(\hat{k}) \equiv Y_{lm}(\theta,\phi)$. Then Equation 11.3 in three dimensions with $C(k) = (2\pi)^{3/2}$ gives

$$u_{E(\vec{k})}(\vec{r}) = e^{i\vec{k}\cdot\vec{r}} + \left(\frac{2\mu}{\hbar^2}\right)\int d^3r' G_+(\vec{r},\vec{r}')V(r')u_{E(\vec{k})}(\vec{r}'). \tag{11.121}$$

We have assumed a local, spherically symmetric potential, $V(r)$. The "+" on $G_+(\vec{r},\vec{r}')$ in this case corresponds to outgoing scattered spherical waves. (I will have more to say about this as we go along.) We have defined*

$$u_{E(\vec{k})}(\vec{r}) \equiv \langle \vec{r} | E(\vec{k}) \rangle, \tag{11.122}$$

and

$$G_+(\vec{r},\vec{r}') \equiv \left(\frac{\hbar^2}{2\mu}\right)\langle\vec{r}|\frac{1}{E-T}|\vec{r}'\rangle_+, \tag{11.123}$$

which satisfies the analog of Equation 11.18,

$$(\vec{\nabla}^2 + k^2)G_+(\vec{r},\vec{r}') = \delta(\vec{r}-\vec{r}'), \tag{11.124}$$

with $k^2 = 2\mu E/\hbar^2$. The solution of Equation 11.124, which is called the unit-source *Helmholtz equation*, must be a function of $|\vec{r}-\vec{r}'|$ only

* Our state $|E(\vec{k})\rangle/(2\pi)^{3/2}$ is equivalent to the scattering state $|\psi^{(+)}\rangle$ of J.J. Sakurai and J. Napolitano, *Modern Quantum Mechanics*, 2nd ed. (Addison-Wesley, 2011), Section 6.1.

because the right-hand side depends only on $\vec{r}-\vec{r}'$ and $G_+(\vec{r},\vec{r}')$ is a scalar function. It implies that

$$G_+(\vec{r},\vec{r}') = G_+(\vec{r}',\vec{r}), \tag{11.125}$$

in infinite space, which can also be established in other ways. In addition, when $k^2 = 0$, we know the solution to

$$\vec{\nabla}^2 F(\vec{r},\vec{r}') = \delta(\vec{r}-\vec{r}'), \tag{11.126}$$

is just

$$F(\vec{r},\vec{r}') = -\frac{1}{4\pi|\vec{r}-\vec{r}'|}, \tag{11.127}$$

from the Coulomb solution for electrodynamics. We assume that Equation 11.127 gives the approximate solution for $G_+(\vec{r},\vec{r}')$ when $k \ll 1/|\vec{r}-\vec{r}'|$ or $k|\vec{r}-\vec{r}'| \ll 1$.

Far from the source, $r \gg r'$, we would expect any angular dependence based on the position of the source, $\vec{r}'$, to be small. Then, the linearly independent solutions of the radial part of the Helmholtz equation,

$$\frac{1}{r}\frac{\partial^2}{\partial r^2}\big(rG_\pm(\vec{r},0)\big) + k^2 G_\pm(\vec{r},0) = 0, \tag{11.128}$$

are just

$$G_\pm(\vec{r},0) \sim \frac{\mathrm{e}^{\pm ikr}}{r}, \tag{11.129}$$

where we are associating the $\pm$ subscripts with the choice of sign in the exponent.

The natural ansatz for the full $G_\pm(\vec{r},\vec{r}')$ based upon the small $|\vec{r}-\vec{r}'|$ behavior (Equation 11.127), large $r \gg r'$ behavior (Equation 11.129), and the fact that $G_\pm(\vec{r},\vec{r}')$ depends only on $|\vec{r}-\vec{r}'|$ is

$$G_\pm(\vec{r},\vec{r}') = -\frac{\mathrm{e}^{\pm ik|\vec{r}-\vec{r}'|}}{4\pi|\vec{r}-\vec{r}'|}. \tag{11.130}$$

We can confirm this hypothesis by plugging Equation 11.130 into 11.124 and carrying out the required derivatives. Although our function $G_\pm(\vec{r},\vec{r}')$ is in spherical coordinates, we will carry out the derivatives using Cartesian coordinates for ease of calculation. (We will write this function as $G_\pm(\vec{x},\vec{x}')$ to emphasize this.) We have

$$\begin{aligned}\vec{\nabla}^2 G_\pm(\vec{x},\vec{x}') = -\frac{\mathrm{e}^{\pm ik|\vec{x}-\vec{x}'|}}{4\pi}\Bigg[&\big(1 \mp ik|\vec{x}-\vec{x}'|\big)\left(\pm\frac{ik(\vec{x}-\vec{x}')}{|\vec{x}-\vec{x}'|}\cdot\vec{\nabla}\left(\frac{1}{|\vec{x}-\vec{x}'|}\right)\right)\\ &-4\pi\delta(\vec{x}-\vec{x}') \mp \frac{ik(\vec{x}-\vec{x}')}{|\vec{x}-\vec{x}'|}\cdot\vec{\nabla}\left(\frac{1}{|\vec{x}-\vec{x}'|}\right)\Bigg],\end{aligned} \tag{11.131}$$

where Equations 11.126 and 11.127 have been used to produce the $\delta(\vec{x}-\vec{x}')$ term. Canceling terms, this gives

$$\begin{aligned}\vec{\nabla}^2 G_\pm(\vec{x},\vec{x}') &= \delta(\vec{x}-\vec{x}') - \frac{k^2 e^{\pm ik|\vec{x}-\vec{x}'|}}{4\pi}(\vec{x}-\vec{x}')\cdot\vec{\nabla}\left(\frac{1}{|\vec{x}-\vec{x}'|}\right) \\ &= \delta(\vec{x}-\vec{x}') - k^2 G_\pm(\vec{x},\vec{x}'),\end{aligned} \quad (11.132)$$

which reproduces Equation 11.124 when the second term on the right is moved to the left. (A verification of this result in spherical coordinates will be left as a problem.)

The choice of sign in the exponent has a similar physical significance as the choice in one dimension, Equations 11.30 and 11.31. As before, the physical interpretation comes from examining the combination,

$$\lim_{r\gg r'} G_\pm(\vec{r},\vec{r}')e^{-i\omega t} = -\frac{e^{\pm ikr}}{4\pi r}e^{-i\omega t}, \quad (11.133)$$

which identifies $G_+(\vec{r},\vec{r}')$ as being the solution to Equation 11.124 corresponding to outgoing spherical waves, and $G_-(\vec{r},\vec{r}')$ being the incoming spherical wave solution. Our physical boundary condition for scattering demands that $u_{E(\vec{k})}(\vec{r})$ in Equation 11.121 be built up out of outgoing spherical waves, so $G_+(\vec{r},\vec{r}')$ is the relevant solution for scattering events.

Equation 11.121 is again an integral equation, equivalent to the three-dimensional Schrödinger equation (Chapter 7, Equation 7.6) but with an initial incoming plane wave as the source. Either equation may be used to produce solutions for the wave functions, $u_{E(\vec{k})}(\vec{r})$.

11.8 THE HELMHOLTZ EQUATION AND PLANE WAVES

We have two sets of complete solutions to the Helmholtz equation,

$$\left(\vec{\nabla}^2 + k^2\right)u_E(\vec{r}) = 0. \quad (11.134)$$

They are three-dimensional plane waves (Equation 11.117)

$$u_E(\vec{r}) \sim e^{i\vec{k}\cdot\vec{r}}, \quad (11.135)$$

and spherical wave solutions (Equation 7.57)

$$u_E(\vec{r}) \sim j_\ell(kr)Y_{\ell m}(\hat{r}). \quad (11.136)$$

We regard both as expressed in spherical coordinates. It will be crucial in the upcoming formalism to express one in terms of the other. In the usual scattering setup, the incident wave is assumed to be a plane wave ($\hat{k}=\hat{z}$ in Equation 11.121), and the outgoing waves are characterized as spherical waves.

We will now solve Equation 11.124 in a different way using the same Wronskian technique that was used in Section 11.2 in the one-dimensional case. We will assume a solution of Equation 11.124 in the form

$$G_+(\vec{r},\vec{r}') = \sum_{\ell,m} g_\ell(r,r')Y_{\ell m}(\hat{r})Y^*_{\ell m}(\hat{r}'). \tag{11.137}$$

This form is compatible with the symmetry Equation 11.125 if $g_\ell(r,r') = g_\ell(r',r)$. In addition, we will assume a three-dimensional spherical delta function given by

$$\langle \vec{r}|\vec{r}'\rangle = \delta(\vec{r}-\vec{r}') = \frac{1}{r^2}\delta(r-r')\sum_{\ell,m} Y_{\ell m}(\hat{r})Y^*_{\ell m}(\hat{r}'), \tag{11.138}$$

from Equation 6.232 and Equation 7.49. Equation 11.124 now reads

$$\sum_{\ell,m}(\vec{\nabla}^2 + k^2)g_\ell(r,r')Y_{\ell m}(\hat{r})Y^*_{\ell m}(\hat{r}') = \frac{1}{r^2}\delta(r-r')\sum_{\ell,m} Y_{\ell m}(\hat{r})Y^*_{\ell m}(\hat{r}'). \tag{11.139}$$

The spherical coordinate differential operator, $\vec{\nabla}^2$, has been characterized in Equation 6.98 and Equation 7.5 as

$$\vec{\nabla}^2 = \frac{1}{r^2}\frac{\partial}{\partial r}\left(r^2\frac{\partial}{\partial r}\right) - \frac{\vec{L}^2_{op}}{r^2\hbar^2}, \tag{11.140}$$

where the differential operator L^2_{op} satisfies

$$L^2_{op}Y_{\ell m}(\hat{r}) = \hbar^2\ell(\ell+1)Y_{\ell m}(\hat{r}). \tag{11.141}$$

We then obtain

$$\begin{aligned}\sum_{\ell,m}\left(\frac{1}{r^2}\frac{\partial}{\partial r}\left(r^2\frac{\partial}{\partial r}\right) - \frac{\ell(\ell+1)}{r^2} + k^2\right)g_\ell(r,r')Y_{\ell m}(\hat{r})Y^*_{\ell m}(\hat{r}') \\ = \frac{1}{r^2}\delta(r-r')\sum_{\ell,m} Y_{\ell m}(\hat{r})Y^*_{\ell m}(\hat{r}').\end{aligned} \tag{11.142}$$

We may multiply both sides of Equation 11.142 by $Y^*_{\ell' m'}(\hat{r})$ and integrate over θ, ϕ to obtain ($\ell' \to \ell$)

$$\left(\frac{1}{r^2}\frac{\partial}{\partial r}\left(r^2\frac{\partial}{\partial r}\right) - \frac{\ell(\ell+1)}{r^2} + k^2\right)g_\ell(r,r') = \frac{1}{r^2}\delta(r-r'), \tag{11.143}$$

once the common factor of $Y^*_{\ell m}(\hat{r}')$ is removed from both sides. When $r \neq r'$, we recognize Equation 11.143 just as the free-particle radial-wave equation, Equation 7.40. This identification is very handy, because we know the linearly independent solutions to Equation 7.40 are the spherical Bessel functions $j_\ell(kr)$ and $h^{(1)}_\ell(kr)$. In fact, assuming continuity across $r = r'$, we can immediately assume that

$$g_\ell(r,r') = A_\ell(k)j_\ell(kr_<)h^{(1)}_\ell(kr_>), \tag{11.144}$$

where

$$r_{\gtrless} \equiv \text{the} \begin{pmatrix} \text{greater} \\ \text{lesser} \end{pmatrix} \text{of } r, r'.$$

The form of Equation 11.144 is required by our spatial boundary conditions, plus continuity at $r = r'$. We want $g_\ell(r, r')$ to be finite at the origin, which requires the $j_\ell(kr_<)$ factor, and to behave like an outgoing spherical wave for $r > r'$, requiring the $h_\ell^{(1)}(kr_>)$ factor. (The complex conjugate, $h_\ell^{(1)}(kr)^* \equiv h_\ell^{(2)}(kr)$, behaves like an incoming spherical wave.) The discontinuity in Equation 11.143 is now expressed as

$$r'^2 \frac{\partial g_\ell}{\partial r}\bigg|_{r'^-}^{r'^+} = 1, \tag{11.145}$$

when Equation 11.143 is integrated from $r'^- \equiv r' - \varepsilon$ to $r'^+ \equiv r' + \varepsilon$. ($\varepsilon$ is infinitesimal and positive.) Using Equation 11.144 in 11.145 now gives rise to the Wronskian,

$$r'^2 A_\ell(k) W_{r'}\left[j_\ell(kr'), h_\ell^{(1)}(kr') \right] = 1. \tag{11.146}$$

Previously, we were able to compute the Wronskian directly. Now, it is not so easy. Fortunately, the Wronskian may be evaluated in special circumstances where $j_\ell(kr')$, $h_\ell^{(1)}(kr')$ simplify. As far as coordinate dependence is concerned, Equation 11.146 requires that

$$W_{r'}\left[j_\ell(kr'), h_\ell^{(1)}(kr') \right] \sim \frac{1}{r'^2}. \tag{11.147}$$

The unknown coefficient in Equation 11.147, which will give us the $A_\ell(k)$ coefficient in Equation 11.146, can be evaluated using either small-x or large-x expressions for $j_\ell(kr')$, $h_\ell^{(1)}(kr')$. A problem will establish that

$$j_\ell(x) \underset{x \gg \ell}{\rightarrow} \frac{\sin\left(x - \frac{\pi\ell}{2} \right)}{x}, \tag{11.148}$$

$$h_\ell^{(1)}(x) \underset{x \gg \ell}{\rightarrow} \frac{(-i)^{\ell+1} e^{ix}}{x}. \tag{11.149}$$

Now

$$W_x\left[j_\ell(x), h_\ell^{(1)}(x) \right] = j_\ell(x) \frac{dh_\ell^{(1)}(x)}{dx} - h_\ell^{(1)}(x) \frac{dj_\ell(x)}{dx}, \tag{11.150}$$

but

$$\frac{dj_\ell}{dx} = -j_{\ell+1} + \frac{\ell}{x} j_\ell, \tag{11.151}$$

$$\frac{dh_\ell^{(1)}}{dx} = -h_{\ell+1}^{(1)} + \frac{\ell}{x} h_\ell^{(1)}. \tag{11.152}$$

Equation 11.151 comes from Problem 7.2.3 (Chapter 7), which holds for all spherical Bessel functions. We have, therefore,

$$W_x[j_\ell(x), h_\ell^{(1)}(x)] \underset{x \gg \ell}{\rightarrow} -j_\ell(x) h_{\ell+1}^{(1)}(x) + h_\ell^{(1)}(x) j_{\ell+1}(x). \tag{11.153}$$

Plugging Equations 11.148 and 11.149 into this, gives

$$\begin{aligned} W_x[j_\ell(x), h_\ell^{(1)}(x)] &\underset{x \gg \ell}{\rightarrow} \frac{e^{ix}}{x^2}\left[-(i)^{\ell+2}\sin\left(x+\frac{\pi\ell}{2}\right) + (i)^{\ell+1}\sin\left(x+\frac{\pi(\ell+1)}{2}\right)\right] \\ &= \frac{e^{ix}}{x^2}(i)^{\ell+1}\left[\cos\left(x+\frac{\pi\ell}{2}\right) - i\sin\left(x+\frac{\pi\ell}{2}\right)\right] \\ &= \frac{1}{x^2}(i)^{\ell+1} e^{-i\pi\ell/2} = \frac{i}{x^2}. \end{aligned} \tag{11.154}$$

Although this was deduced for large-x arguments of the Bessel functions, the coordinate dependence matches with Equation 11.147, and we have the exact statement

$$W_{r'}\left[j_\ell(kr'), h_\ell^{(1)}(kr')\right] = \frac{ik}{(kr')^2}, \tag{11.155}$$

which means that

$$A_\ell(k) = -ik, \tag{11.156}$$

from Equation 11.146. (No "ℓ" dependence in the coefficient.) This is a complete solution to our problem! Our conclusion from the results in this section and Section 11.7, given the uniqueness of solutions to equations such as Equation 11.124 with specified boundary conditions, is that

$$-\frac{e^{ik|\vec{r}-\vec{r}'|}}{4\pi|\vec{r}-\vec{r}'|} = -ik\sum_{\ell,m} j_\ell(kr_<) h_\ell^{(1)}(kr_>) Y_{\ell m}(\hat{r}) Y_{\ell m}^*(\hat{r}'). \tag{11.157}$$

If we take $r' > r$ on both sides of Equation 11.157, and use the large argument expansion, Equation 11.149, for $h_\ell^{(1)}(kr)$ along with the result

$$\frac{e^{ik|\vec{r}-\vec{r}'|}}{|\vec{r}-\vec{r}'|} \underset{r' \gg r}{\rightarrow} \left(\frac{e^{ikr'}}{r'}\right) e^{-i\vec{k}\cdot\vec{r}}, \tag{11.158}$$

where $\vec{k} \equiv k\hat{r}'$, we finally obtain

$$e^{-i\vec{k}\cdot\vec{r}} = 4\pi\sum_{\ell,m}(-i)^\ell j_\ell(kr) Y_{\ell m}(\hat{r}) Y_{\ell m}^*(\hat{k}). \tag{11.159}$$

Taking the complex conjugation of Equation 11.159 and using Equation 6.243 (see Chapter 6) then gives*

$$e^{i\vec{k}\cdot\vec{r}} = 4\pi\sum_{l,m}(i)^l j_l(kr)Y_{lm}(\hat{r})Y^*_{lm}(\hat{k}). \tag{11.160}$$

Equation 11.160 holds in the entire range $0 \le r \to \infty$ for r since we may maintain the relationship $r' > r$ while taking both quantities to ∞.

The result of Equation 11.160 provides the necessary connection between plane waves and spherical waves to complete the formalism for scattering. In the following, we will need Equation 11.160 specialized to $\hat{k} = \hat{z}$. From Equation 6.233 we have

$$Y_{\ell 0}(\theta,\phi) = \sqrt{\frac{2\ell+1}{4\pi}}P_\ell(\cos\theta). \tag{11.161}$$

Given $P_\ell(1) = 1$,† we may also establish from Equations 6.233, 6.240, and 11.161 that

$$Y^*_{\ell m}(0,\phi) = \delta_{m,0}\sqrt{\frac{2\ell+1}{4\pi}}. \tag{11.162}$$

Finally, this gives

$$e^{ikz} = \sum_\ell (i)^\ell (2\ell+1) j_\ell(kr)P_\ell(\cos\theta). \tag{11.163}$$

Equation 11.163 may also be established more directly from Equation 11.160 by using the so-called addition theorem for spherical harmonics‡

$$P_\ell(\cos\gamma) = \frac{4\pi}{2\ell+1}\sum_{m=-\ell}^{\ell} Y_{\ell m}(\hat{r})Y^*_{\ell m}(\hat{r}'). \tag{11.164}$$

The γ in Equation 11.164 is the angle between $\hat{r}$ and $\hat{r}'$.

$$\hat{r}\cdot\hat{r}' = \cos\gamma. \tag{11.165}$$

This shows that the angle θ in Equation 11.163 has the general meaning of the angle between $\hat{r}$ and $\hat{k}$ in Equation 11.160.

---

* Suggested derivation: Show this relation implies the transformation amplitude connecting Cartesian and spherical momentum descriptions, alluded to in Chapter 7, Section 7.2, is given by

$$\langle \vec{k}|k'\ell,\ell,m\rangle = \frac{(-i)^\ell}{k^2}\delta(k-k')Y_{\ell m}(\theta,\phi).$$

Hint: Use coordinate completeness!

† This may be established by a careful treatment of Equation 6.234 (see Chapter 6).

‡ A proof of Equation 11.164 is contained, for example, in W. Wilcox and C. Thron, *Macroscopic Electrodynamics* (World Scientific, 2016), Section 4.11.

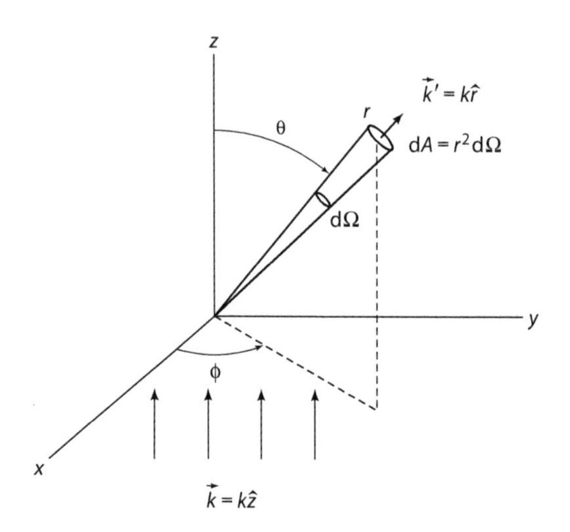

FIGURE 11.7 The cross-section setup, showing the incoming flux in the +z-direction.

11.9 CROSS SECTIONS AND THE SCATTERING AMPLITUDE

We are now ready to define the infinitesimal scattering cross section, dσ (and as illustrated in Figure 11.7):

$$d\sigma \equiv \lim_{r\to\infty}\left[\frac{\text{Particle flux scattered into } dA}{\text{Incident particle flux}}\right].$$

The particle flux is given by the conserved current,

$$\vec{j} = \frac{\hbar}{\mu}\text{Im}(\psi^*\vec{\nabla}\psi). \tag{11.166}$$

The particle fluxes are considered positive quantities, so we write

$$d\sigma = \lim_{r\to\infty}\frac{\left|\hat{r}\cdot\vec{j}^{\,sc}\right|}{\left|j_z^{inc}\right|}r^2 d\Omega. \tag{11.167}$$

The $u_{E(\vec{k})}(\vec{r})$ wave function will have the form $(k_z = k > 0)$

$$u_{E(k_z)}(\vec{r}) \underset{r\to\infty}{\longrightarrow} e^{ikz} + \frac{e^{ikr}}{r} f(\theta,\phi), \tag{11.168}$$

where $f(\theta, \phi)$ is called the "scattering amplitude." Notice that it is defined just as the amplitude associated with an outgoing spherical wave,

$ih_0^{(1)}(kr) = e^{ikr}/r$. The first term in Equation 11.168 is just the incident plane wave. Using $\psi(\vec{r}) = u_{E(k_z)}(\vec{r})$ in Equation 11.166, we have

$$j_z^{inc} = \frac{\hbar k}{\mu} = v, \tag{11.169}$$

$$\hat{r}\cdot\vec{j}^{sc} = \hbar k \frac{|f(\theta,\phi)|^2}{r^2}. \tag{11.170}$$

v is the relative particle speed. With our choice, in a two-body context, of relative particle velocity along $+z$ and outgoing spherical waves, both quantities in Equations 11.169 and 11.170 are positive. Putting these in Equation 11.167 and dividing through by $d\Omega$ defines the *elastic differential cross section:**

$$\frac{d\sigma}{d\Omega} \equiv |f(\theta,\phi)|^2. \tag{11.171}$$

This is the three-dimensional analog to the $|r(k)|^2$ and $|t(k)|^2$ coefficients.

An exact expression for $f(\theta,\phi)$ results from taking the large-r limit of Equation 11.121, with $G_+(\vec{r},\vec{r}\,')$ given by Equation 11.130. In the opposite limit to Equation 11.158, $r \gg r'$, we of course have

$$G_+(\vec{r},\vec{r}\,') \underset{r\gg r'}{\rightarrow} -\frac{1}{4\pi}\left(\frac{e^{ikr}}{r}\right)e^{-i\vec{k}'\cdot\vec{r}\,'}, \tag{11.172}$$

where $\vec{k}' \equiv k\hat{r}$, and we may compare with Equation 11.168 to identify

$$f(k_z,\vec{k}') \equiv f(\theta,\phi) = -\left(\frac{2\mu}{\hbar^2}\right)\frac{1}{4\pi}\int d^3r' e^{-i\vec{k}'\cdot\vec{r}\,'}V(r')u_{E(k_z)}(\vec{r}\,'), \tag{11.173}$$

where I have introduced an alternate notation for the scattering amplitude, which emphasizes its initial (k_z) and final ($\vec{k}'$) wave-number dependence. Equation 11.173 is the scattering analog of Equations 11.37 and 11.39. Like that earlier situation, we cannot solve Equation 11.173 until we can at least approximately solve for $u_{E(k_z)}(\vec{r})$. Note that a spherically symmetric potential, $V(r')$, in Equation 11.173 will yield a scattering amplitude that only depends on the polar angle θ (in $\hat{k}'\cdot\hat{z} = \cos\theta$).

11.10 SCATTERING PHASE SHIFTS

The Schrödinger equation for a spherically symmetric potential is

$$\left[\nabla^2 + k^2 - \frac{2\mu}{\hbar^2}V(r)\right]u_{E(\vec{k})}(\vec{r}) = 0. \tag{11.174}$$

* This cross section is in the center-of-mass frame. For a discussion of the transition between frames of reference, see the online textbook: W. Wilcox, *Modern Introductory Mechanics, Part 2* (Bookboon.com), Chapter 9.

For the region far from the scattering center, where presumably $V(r)$ is small or zero, the general solution for a spherically symmetric potential and initial wave along $+z$ ($\hat{k} = \hat{z}$ in Equation 11.121) satisfies*

$$\lim_{r\to\infty} u_{E(k_z)}(\vec{r}) = \sum_{\ell} a_{\ell}(k)\chi_{\ell}(r)P_{\ell}(\cos\theta). \tag{11.175}$$

The $a_{\ell}(k)$ are wave-number-dependent coefficients, and $P_{\ell}(\cos\theta)$ are the Legendre polynomials. There is no sum on m here because of the spherically symmetric assumption for $V(r)$. (Note that the spherical harmonics, $Y_{\ell m}(\theta,\phi)$, reduce to Legendre polynomials when $m=0$, according to Equation 11.161.)

If we were working with truly free particles even at the origin, the $\chi_{\ell}(r)$ function would just be proportional to the spherical Bessel function $j_{\ell}(kr)$. However, the coordinate origin is not included in the characterization of $u_{E(k_z)}(\vec{r})$ in Equation 11.175 and we must consider the more general case. From Equations 7.126 and 7.127 in Chapter 7 we have

$$j_{\ell}(kr) = -(-k)^{-\ell-1} r^{\ell}\left(\frac{1}{r}\frac{d}{dr}\right)^{\ell}\left(\frac{\sin(kr)}{r}\right), \tag{11.176}$$

$$h_{\ell}^{(1)}(kr) = i(-k)^{-\ell-1} r^{\ell}\left(\frac{1}{r}\frac{d}{dr}\right)^{\ell}\left(\frac{e^{ikr}}{r}\right). \tag{11.177}$$

Two real, linearly independent combinations of $h_{\ell}^{(1)}(kr)$ are

$$\frac{h_{\ell}^{(1)}(kr) + h_{\ell}^{(1)*}(kr)}{2} = j_{\ell}(kr), \tag{11.178}$$

and

$$\frac{h_{\ell}^{(1)}(kr) - h_{\ell}^{(1)*}(kr)}{2i} \equiv n_{\ell}(kr). \tag{11.179}$$

Equation 11.178 may be shown directly from Equations 11.176 and 11.177. Equation 11.179 introduces the *Neumann spherical Bessel function*, $n_{\ell}(kr)$. Since $j_{\ell}(kr)$ and $n_{\ell}(kr)$ are linearly independent, we may take the $\chi_{\ell}(r)$ function most generally as

$$\chi_{\ell}(r) \equiv -\sin(\delta_{\ell}(k))n_{\ell}(kr) + \cos(\delta_{\ell}(k))j_{\ell}(kr), \tag{11.180}$$

outside of an overall amplitude. (This is the reason for $a_{\ell}(k)$ in Equation 11.175.) The real quantity $\delta_{\ell}(k)$ is called the "scattering phase shift." If there was no scattering, $\delta_{\ell} = 0$ and $\chi_{\ell}(r)$ would be proportional to the free-particle solutions involving $j_{\ell}(kr)$ only (Equation 7.57) with $m=0$. These phase shifts are the major new aspect of three-dimensional scattering compared to the one-dimensional case.

---

* The $\lim_{r\to\infty}$ may be replaced by the restriction $r > a$ for a finite range potential, where $V(r > a) = 0$.

We will now compare the large-r forms implied by Equations 11.175 and 11.168, the latter of which is written in the form

$$u_{E(\vec{k})}(\vec{r}) \underset{r\to\infty}{\to} \sum_{\ell} (i)^{\ell} (2\ell+1) \frac{\sin\left(kr - \frac{\pi\ell}{2}\right)}{kr} P_{\ell}(\cos\theta) + \left(\frac{e^{ikr}}{r}\right) \sum_{\ell} (2\ell+1) f_{\ell}(k) P_{\ell}(\cos\theta), \tag{11.181}$$

where we have used Equation 11.163 for e^{ikz}. We have also defined (the $P_{\ell}(\cos\theta)$ are complete in θ-space)

$$f(\theta) \equiv \sum_{\ell} (2\ell+1) f_{\ell}(k) P_{\ell}(\cos\theta), \tag{11.182}$$

where we are denoting the scattering amplitude for spherically symmetric potentials as $f(\theta)$. The $f_{\ell}(k)$ in Equation 11.182 are unknown expansion coefficients we wish to solve for. Equation 11.182 is called the "partial wave expansion" of the scattering amplitude. The large-r form of $\chi_{\ell}(r)$ follows from Equations 11.148 and 11.149, given the form Equation 11.179 for $n_{\ell}(kr)$:

$$\lim_{r\to\infty} \chi_{\ell}(r) = -\frac{\sin\delta_{\ell}}{kr}\left(-\cos\left(kr - \frac{\pi\ell}{2}\right)\right) + \frac{\cos\delta_{\ell}}{kr}\left(\sin\left(kr - \frac{\pi\ell}{2}\right)\right),$$
$$= \frac{\sin\left(kr - \frac{\pi\ell}{2} + \delta_{\ell}\right)}{kr}. \tag{11.183}$$

Putting Equation 11.183 into Equation 11.175 gives the alternate form

$$\lim_{r\to\infty} u_{E(k_z)}(\vec{r}) = \sum_{\ell} a_{\ell}(k) \frac{\sin\left(kr - \frac{\pi\ell}{2} + \delta_{\ell}\right)}{kr} P_{\ell}(\cos\theta). \tag{11.184}$$

One more step is necessary before we can compare the two large-r forms, Equations 11.181 and 11.184. We can expand the sine functions in terms of oscillating exponentials to obtain

$$\lim_{r\to\infty} u_{E(k_z)}(\vec{r}) = \frac{e^{ikr}}{2kr}\left\{\sum_{\ell} (2\ell+1)\left(-(i)^{\ell+1} e^{-\frac{i\pi\ell}{2}} + 2kf_{\ell}(k) P_{\ell}(\cos\theta)\right)\right\} + \frac{e^{-ikr}}{2kr}\left\{\sum_{\ell} (2\ell+1) i^{\ell+1} e^{\frac{i\pi\ell}{2}} P_{\ell}(\cos\theta)\right\}, \tag{11.185}$$

from Equation 11.181, and

$$\lim_{r\to\infty} u_{E(k_z)}(\vec{r}) = \frac{e^{ikr}}{2kr}\left\{\sum_{\ell} a_{\ell}(k)(-i) e^{i\left(\frac{-\pi\ell}{2}+\delta_{\ell}\right)} P_{\ell}(\cos\theta)\right\} + \frac{e^{-ikr}}{2kr}\left\{\sum_{\ell} a_{\ell}(k)\, i e^{-i\left(\frac{-\pi\ell}{2}+\delta_{\ell}\right)} P_{\ell}(\cos\theta)\right\}, \tag{11.186}$$

from Equation 11.184. This is a physical separation in terms of incoming $\sim e^{-ikr}/r$ and outgoing $\sim e^{ikr}/r$ spherical waves. The coefficients of these terms must be the same, which allows us to solve for the coefficients of the complete $P_\ell(\cos\theta)$ functions for each ℓ value. Thus we have

$$(2\ell+1)\left(i^\ell e^{-i\frac{\pi\ell}{2}}+2ikf_\ell(k)\right)=a_\ell(k)e^{i\left(-\frac{\pi\ell}{2}+\delta_\ell\right)}, \tag{11.187}$$

from the outgoing wave coefficients, and

$$(2\ell+1)i^\ell e^{i\frac{\pi\ell}{2}}=a_\ell(k)e^{i\left(\frac{\pi\ell}{2}-\delta_\ell\right)}, \tag{11.188}$$

from the incoming wave coefficients. From Equation 11.188, we have

$$a_\ell(k)=(2\ell+1)i^\ell e^{i\delta_\ell}, \tag{11.189}$$

which implies from Equation 11.187 that

$$f_\ell(k)=\left(\frac{e^{2i\delta_\ell}-1}{2ik}\right) \tag{11.190}$$

or

$$f_\ell(k)=\frac{e^{i\delta_\ell}}{k}\sin\delta_\ell. \tag{11.191}$$

The partial-wave amplitudes, $f_\ell(k)$, satisfy

$$|1+2ikf_\ell(k)|=1, \tag{11.192}$$

which can be shown to be the probability-conserving analog to Equation 11.44 in one dimension. Equation 11.192 is called a statement of "unitarity." Note that $f_\ell(k)$ takes on its maximum magnitude (the "unitarity limit"),

$$f_\ell^{\max}(k)=\frac{i}{k}, \tag{11.193}$$

when the phase shifts satisfy

$$\delta_\ell(k)=\pm\frac{\pi}{2}(2n+1),\quad n=0,1,2,3.... \tag{11.194}$$

The differential cross section has the form

$$\frac{d\sigma}{d\Omega}=\left|\sum_\ell(2\ell+1)f_\ell(k)P_\ell(\cos\theta)\right|^2, \tag{11.195}$$

from Equations 11.171 and 11.182. This may be integrated to give the total cross section, σ_{tot}, defined as (see Equation 11.231 for the necessary angular integral)

$$\sigma_{tot} \equiv 2\pi \int_0^{\pi/2} d\theta \sin\theta \frac{d\sigma}{d\Omega}, \tag{11.196}$$

$$\Rightarrow \sigma_{tot} = 4\pi \sum_{\ell} (2\ell+1) |f_\ell(k)|^2, \tag{11.197}$$

or

$$\sigma_{tot} = \frac{4\pi}{k^2} \sum_{\ell} (2\ell+1) \sin^2(\delta_\ell(k)). \tag{11.198}$$

One can define the so-called lth partial cross section as

$$\sigma_\ell \equiv 4\pi(2\ell+1)|f_\ell(k)|^2. \tag{11.199}$$

Using the same ideas that we encountered in Section 11.5, we expect resonances to again appear as poles in $f_\ell(k)$ in the complex k-plane, below the positive, real k-axis, as seen in Figure 11.6. Introducing the pole value, k_0, as in Equation 11.83, we expect near resonance the partial-wave amplitude $f_\ell(k)$ will be given approximately by

$$f_\ell(k) = \frac{K(k_0)}{k(k-k_0)}, \tag{11.200}$$

where the k^{-1} factor is expected from the general forms, Equations 11.190 or 11.191. The unknown coefficient, $K(k_0)$, can be determined by the unitarity limit at resonance, Equation 11.193.* This gives

$$K(k_0) = \text{Im}(k_0) = -\sqrt{\frac{\mu}{2E_0}} \frac{\Gamma}{2\hbar}. \tag{11.201}$$

Going through similar steps as before, we find that the ℓth partial cross section is given approximately by the Breit–Wigner form,

$$\sigma_l = \frac{4\pi(2l+1)}{k^2} \frac{(\Gamma/2)^2}{(\Gamma/2)^2 + (E-E_0)^2}. \tag{11.202}$$

Although Equations 11.96 and 11.202 represent different physical quantities, these expressions have the same functional dependence on energy and half-width. The factor of k^{-2} will not substantially affect the resonance shape if it is a sufficiently narrow resonance ($\Gamma/2 \ll E_0$).

* A resonance represents a partial wave maximum, but not all partial wave maximums are resonances. We will see an example of this latter situation in the next section.

11.11 FINITE-RANGE POTENTIAL SCATTERING

Once one has ascertained the energy-dependent phase shifts, $\delta_\ell(k)$, the differential cross section, Equation 11.195, is completely determined. However, only in a few cases can this be done analytically. One of these cases is the three-dimensional analog of the step function, the spherical step potential (or spherical well), bound states of which were studied in Chapter 7, Section 7.4. This example will nicely illustrate some general aspects of low-energy scattering and resonances.

The spherical well is an example of a *finite-range* potential. This type of potential can be characterized by a potential-free Region II, for $r > a$, where a is the range of the potential. This is illustrated in Figure 11.8.

If solutions are available in the two regions, it is only necessary to match the value and first radial derivatives of the wave function at $r = a$. That is, we must have

$$u_\ell^{\mathrm{I}}(a) = u_\ell^{\mathrm{II}}(a), \tag{11.203}$$

$$\left.\frac{du_\ell^{\mathrm{I}}}{dr}\right|_{r=a} = \left.\frac{du_\ell^{\mathrm{II}}}{dr}\right|_{r=a}, \tag{11.204}$$

where $u_\ell^{\mathrm{I}}(r)$ and $u_\ell^{\mathrm{II}}(r)$ solve Equation 7.8 with $m \to \mu$, $E = \hbar^2 k^2/2\mu$, for the appropriate spatial boundary conditions. An efficient way of ensuring both conditions hold is to simply require the continuity of the so-called logarithmic derivative (which ensures conservation of the probability current in the radial direction),

$$\beta_\ell(r) \equiv \frac{r}{u_\ell(r)} \frac{du_\ell(r)}{dr}, \tag{11.205}$$

$$\Rightarrow \beta_\ell^{\mathrm{I}}(a) = \beta_\ell^{\mathrm{II}}(a). \tag{11.206}$$

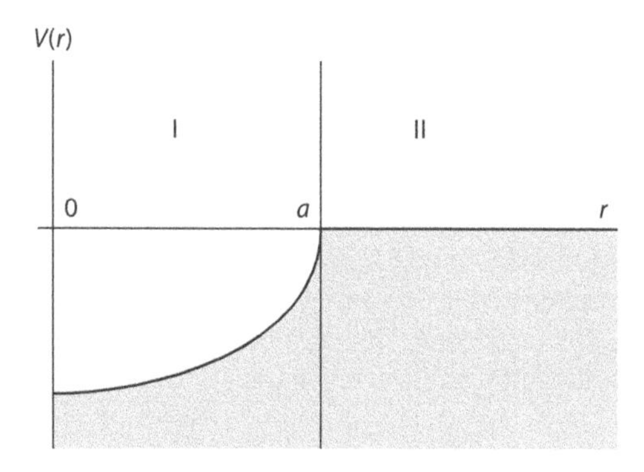

FIGURE 11.8 A generic finite-range spherically symmetric potential.

Using Equations 11.175 and 11.180 to form $\beta_\ell^{II}(r)$, for each ℓ value we have

$$\beta_\ell^{II}(a) = \frac{a\left(\cos\delta_\ell \left.\frac{dj_\ell}{dr}\right|_a - \sin\delta_\ell \left.\frac{dn_\ell}{dr}\right|_a\right)}{\left(\cos\delta_\ell j_\ell(ka) - \sin\delta_\ell n_\ell(ka)\right)}. \tag{11.207}$$

Putting this in Equation 11.206 and solving for $\tan\delta_\ell$ gives

$$\tan\delta_\ell(k) = \frac{\left(a\left.\frac{dj_\ell}{dr}\right|_a - \beta_\ell^{I}(a) j_\ell(ka)\right)}{\left(a\left.\frac{dn_\ell}{dr}\right|_a - \beta_\ell^{I}(a) n_\ell(ka)\right)}. \tag{11.208}$$

If we can form the $\beta_\ell^{I}(a)$, Equation 11.208 in principle completes the solution to the finite-range problem.

An important special case of Equation 11.208 is low-energy scattering, $ka \ll 1$. Then, based upon small argument expansions of the spherical Bessel functions* (for $x \ll \ell$),

$$j_\ell(x) \approx \frac{x^\ell}{(2\ell+1)!!}, \tag{11.209}$$

$$n_\ell(x) \approx -\frac{(2\ell-1)!!}{x^{\ell+1}}, \tag{11.210}$$

where

$$(2\ell+1)!! = (2\ell+1)(2\ell-1)(2\ell-3)\cdots\begin{pmatrix}(1),\ \ell \text{ even}\\(2),\ \ell \text{ odd}\end{pmatrix}, \tag{11.211}$$

which is called the double factorial ($(-1)!! \equiv 1$). Then, using these in Equation 11.208, we have that

$$\tan\delta_\ell(k) \approx \frac{(ka)^{2\ell+1}}{(2\ell+1)\left[(2\ell-1)!!\right]^2}\left(\frac{\ell - \left(\beta_\ell^{I}(a)\right)_0}{\ell+1+\left(\beta_\ell^{I}(a)\right)_0}\right), \tag{11.212}$$

where $(\beta_\ell^{I}(a))_0$ denotes the $k=0$ value of $\beta_\ell^{I}(a)$. Looking back at Equations 11.195 and 11.198, this shows that the $\ell = 0$ partial wave dominates at low energies, and we have

$$\frac{d\sigma}{d\Omega} \approx a^2\left(\frac{\left(\beta_0^{I}(a)\right)_0}{1+\left(\beta_0^{I}(a)\right)_0}\right)^2, \tag{11.213}$$

for $ka \ll 1$, which shows that the scattering is isotropic and approximately energy independent.

* See, for example, Chapter 10 of M. Abramowitz and I. Stegun, eds., *Handbook of Mathematical Functions*, National Bureau of Standards Applied Mathematics Series 55, 1964.

Spherical step-potential scattering can be solved for using these expressions. In Region I, which includes the origin, we must have*

$$\chi_\ell^{\mathrm{I}}(r) \sim j_\ell(\kappa r), \tag{11.214}$$

where

$$\kappa = \frac{\sqrt{2\mu(E - V_0)}}{\hbar}, \tag{11.215}$$

and we assume E and $(E - V_0) > 0$. Then

$$\beta_\ell^{\mathrm{I}}(a) = \frac{a}{j_\ell(\kappa a)} \frac{\mathrm{d}j_\ell}{\mathrm{d}r}\bigg|_{r=a}, \tag{11.216}$$

which, when inserted in Equation 11.208 gives an exact, if not explicit, solution to the scattering expressions. In the $\ell = 0$ case for $ka \ll 1$ and $V_0 < 0$,

$$\left(\beta_0^{\mathrm{I}}(a)\right)_0 = (\kappa_0 a)\cot(\kappa_0 a) - 1, \tag{11.217}$$

and we have

$$\frac{\mathrm{d}\sigma}{\mathrm{d}\Omega} \approx a^2 \frac{\left[(\kappa_0 a)\cot(\kappa_0 a) - 1\right]^2}{(\kappa_0 a)^2 \cot^2(\kappa_0 a)}, \tag{11.218}$$

where

$$\kappa_0 \equiv \frac{\sqrt{-2\mu V_0}}{\hbar}. \tag{11.219}$$

An interesting aspect of this simple finite-range potential is the existence of resonances for $\ell \neq 0$. Figure 11.9 illustrates how such situations can arise for a step potential. One can see that a lip forms on the outside portion of the attractive square well potential, $V_{\mathrm{eff}}(r)$, given by Equation 7.11 (Chapter 7), for $\ell \neq 0$. These are not bound states since they can "tunnel" or leak out through the finite potential barrier. The scattered particle enters such a state, is trapped, and then retunnels out, leading to a time delay compared to the incident particles. We illustrate these remarks with Figure 11.10, which shows the phase shift for $\ell = 3$, $\delta_3(k)$, as a function of dimensionless ka, for $2\mu V_0 a^2/\hbar^2 = -(5.5)^2$.† Notice the extremely fast passage of the phase through $\pi/2$ near $ka = 1.411$. Notice also the slow return of the phase angle to zero as the energy is increased. One can also see that the phase returns to $\pi/2$ near $ka = 8.45$, but this does not correspond to a resonance. That is, there is no scattering amplitude pole associated with

* The connection to the notation of Chapter 7, Section 7.4 is $\kappa \to K$, $\kappa' \to K'$ and $V_0 \to -V_0$.

† This corresponds to the Section 6.7 scattering example in J. J. Sakurai and J. Napolitano, *Modern Quantum Mechanics*, 2nd ed. (Addison-Wesley, 2011).

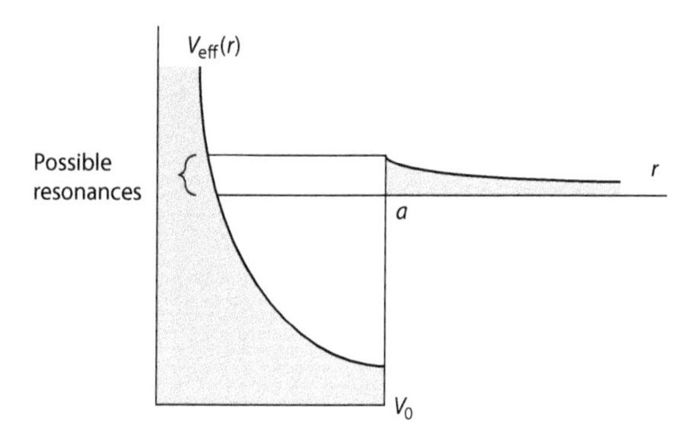

FIGURE 11.9 Illustrating the formation of resonance "lip" in the effective potential, $V_{eff}(r)$, for an attractive spherical step potential with $\ell \neq 0$.

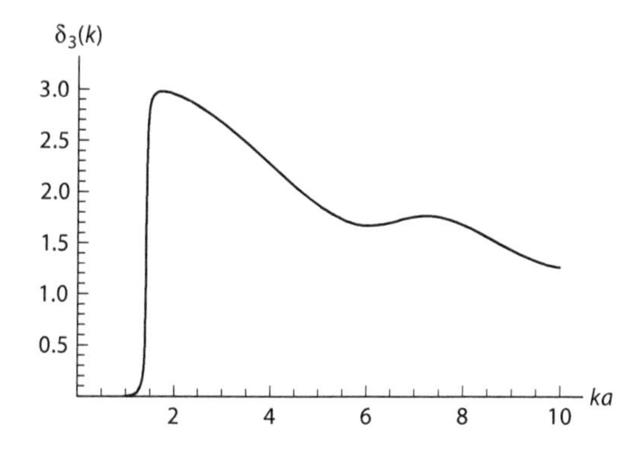

FIGURE 11.10 The $\ell = 3$ phase shift of the attractive spherical step potential, $\delta_3(k)$, as a function of ka for $2\mu V_0 a^2/\hbar^2 = -(5.5)^2$.

this ka value. A way of understanding this behavior is through an important idea called "Levinson's theorem."* It states that

$$\delta_\ell(k=0) - \delta_\ell(k \to \infty) = N_\ell \pi, \tag{11.220}$$

where N_ℓ is the number of bound states of the system for a given ℓ. Defining $\delta_\ell(k \to \infty) = 0$ may always be done. In our case, the potential well is not deep enough to bind any $\ell = 3$ states permanently, so we have $\delta_3(0) = 0$. Thus, although there is a resonance near $ka = 1.411$, the phase shift must ultimately return continuously to zero through $\pi/2$ again, leading to partial-wave maximum, which, as I said, is not a resonance.

The resonance behavior gives rise to a partial-wave cross section, $\sigma_3(k)/a^2$, which is illustrated in Figure 11.11. It has the Breit–Wigner form, Equation 11.202. Its maximum value is given by $4\pi(2\ell+1)/(ka)^2$ which in our case is about 44.2.

Finally, let us notice that the partial-wave amplitude Equation 11.190 may be written as

$$f_\ell(k) = \frac{1}{k}\left(\frac{1}{\cot\delta_\ell(k) - i}\right). \tag{11.221}$$

* See N. Levinson, 1949, https://physics.nyu.edu/LarrySpruch/LevinsonsTheorem.PDF#Levinson_theorem.

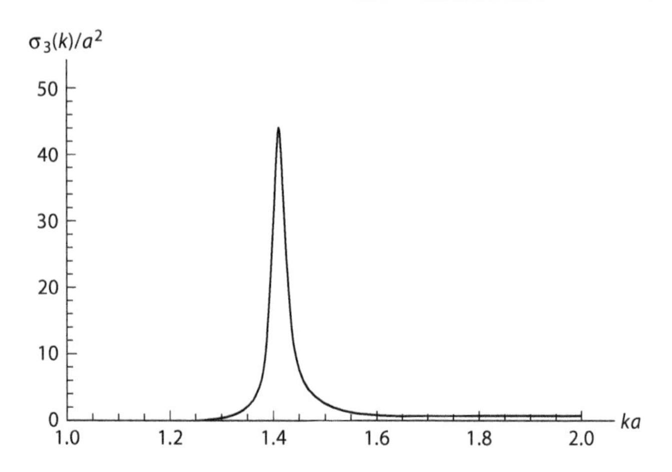

FIGURE 11.11 The dimensionless partial-wave cross section, $\sigma_3(k)/a^2$, with $\sigma_3(k)$ given by Equation 11.199, as a function of ka for the same example as in Figure 11.10.

If simple poles appear in the scattering amplitude, $f_\ell(k)$, when analytically continued into the complex k-plane, we expect this will occur when

$$\cot\delta_\ell(k) = i, \tag{11.222}$$

for some complex value of k. Now, the matching condition, Equation 11.206, for spherical well scattering can be written as

$$\frac{\left.\frac{dj_\ell(\kappa r)}{dr}\right|_{r=a}}{j_\ell(\kappa a)} = \frac{\left.\frac{dj_\ell(kr)}{dr}\right|_{r=a} - \tan\delta_\ell(k)\left.\frac{dn_\ell(kr)}{dr}\right|_{r=a}}{j_\ell(ka) - \tan\delta_\ell(k)n_\ell(ka)}, \tag{11.223}$$

when Equations 11.207 and 11.216 are used. Inserting the condition Equation 11.222 gives

$$\frac{\left.\frac{dj_\ell(\kappa r)}{dr}\right|_{r=a}}{j_\ell(\kappa a)} = \frac{\left.\frac{dh_\ell^{(1)}(kr)}{dr}\right|_{r=a}}{h_\ell^{(1)}(ka)}, \tag{11.224}$$

where the spherical Hankel function, $h_\ell^{(1)}(kr)$, appears on the right-hand side since

$$h_\ell^{(1)}(x) = j_\ell(x) + in_\ell(x), \tag{11.225}$$

from Equations 11.178 and 11.179. When Equation 11.224 is continued along the positive, imaginary k-axis, $k \to i\kappa'$, we now find

$$\frac{\left.\frac{dj_\ell(\kappa r)}{dr}\right|_{r=a}}{j_\ell(\kappa a)} = \frac{\left.\frac{dh_\ell^{(1)}(i\kappa' r)}{dr}\right|_{r=a}}{h_\ell^{(1)}(i\kappa' a)}. \tag{11.226}$$

Equation 11.226 is just the energy eigenvalue condition for spherical well bound states, Equation 7.131. These poles are occurring along the positive, imaginary k-axis at locations given by the energy eigenvalue condition, just as we expect.

11.12 THE THREE-DIMENSIONAL BORN SERIES

Let us return to Equation 11.173, which can give us an alternate, approximate solution to spherically symmetric scattering problems. The integral equation for scattering, Equation 11.121, given an initial $\hat{z}$-directed plane wave is ($k_z = k > 0$)

$$u_{E(k_z)}(\vec{r}) = e^{ikz} + \left(\frac{2\mu}{\hbar^2}\right)\int d^3r' G_+(\vec{r},\vec{r}')V(r')u_{E(k_z)}(\vec{r}'). \qquad (11.227)$$

For small interactions, which we expect will occur at high energies, we can approximate $u_{E(k_z)}(\vec{r})$ to lowest order by the initial plane wave, giving the lowest Born series term,

$$f^{(1)}(k_z,\vec{k}') = f^{(1)}(\theta) \equiv -\left(\frac{2\mu}{\hbar^2}\right)\frac{1}{4\pi}\int d^3r'\, e^{-i\vec{k}'\cdot\vec{r}'}e^{ikz'}V(r'), \qquad (11.228)$$

from Equation 11.173 $(|\vec{k}'| = |\vec{k}| = k)$. For a general initial relative wave number, $\vec{k}$, with the same magnitude, this gives

$$f^{(1)}(\vec{k},\vec{k}') = -\left(\frac{2\mu}{\hbar^2}\right)\frac{1}{4\pi}\int d^3r'\, e^{i\vec{q}\cdot\vec{r}'}V(r'), \qquad (11.229)$$

where $\vec{q} \equiv \vec{k} - \vec{k}'$. Thus, to lowest order, the scattering amplitude is proportional to the three-dimensional Fourier transform of the potential.

What does Equation 11.229 imply for the partial-wave scattering amplitudes, $f_\ell(k)$? We take Equation 11.182, multiply both sides by $P_{\ell'}\cos\theta$, and integrate over $\cos\theta$ using

$$\int d\Omega\, Y_{\ell'0}Y^*_{\ell 0} = \delta_{\ell\ell'}, \qquad (11.230)$$

from Equation 6.226 (Chapter 6) for $m = m' = 0$. Equation 11.230 is equivalent to

$$\int_{-1}^{1} d\cos\theta\, P_\ell(\cos\theta)P_{\ell'}(\cos\theta) = \frac{2}{2\ell+1}\delta_{\ell\ell'}, \qquad (11.231)$$

when Equation 11.161 is used. Thus,

$$f_\ell(k) = \frac{1}{2}\int_{-1}^{1} d\cos\theta\, f(\theta)P_\ell(\cos\theta). \qquad (11.232)$$

We require Equation 11.232 to hold at each level of the Born series:

$$f_\ell^{(1)}(k) = \frac{1}{2}\int_{-1}^{1} d\cos\theta \, f^{(1)}(\theta) P_\ell(\cos\theta). \tag{11.233}$$

We use Equations 11.159 and 11.163, the plane-wave expansions, in this expression, with $f^{(1)}(\theta)$ given by Equation 11.228. This yields the rather cumbersome result,

$$\begin{aligned} f_\ell^{(1)}(k) = -\frac{1}{2}\left(\frac{2\mu}{\hbar^2}\right)\int_{-1}^{1} d\cos\theta \, P_\ell(\cos\theta)\int d^3r' \, V(r') \sum_{\ell',\ell'',m'} (i)^{\ell''}(2\ell''+1) \\ \times j_{\ell''}(kr')P_{\ell''}(\cos\theta')(-i)^{\ell'} j_{\ell'}(kr')Y_{\ell'm'}(\hat{r}')Y^*_{\ell'm'}(\hat{k}'). \end{aligned} \tag{11.234}$$

The $d\Omega'$ solid angle integral requires $m' = 0$. Use of Equations 11.161 and 11.230 now results in

$$\begin{aligned} f_l^{(1)}(k) = -\frac{\mu}{\hbar^2}\sum_{l'}(2l'+1)\sqrt{\frac{4\pi}{2l'+1}}\int_0^\infty dr' r'^2 j_{l'}^2(kr')V(r') \\ \times \int_{-1}^{1} d\cos\theta \, P_l(\cos\theta)Y^*_{l'0}(\theta), \end{aligned} \tag{11.235}$$

where the scattering angle, θ, is given by $\hat{k}'\cdot\hat{z} = \cos\theta$. Utilizing Equation 11.231 gives us finally

$$f_\ell^{(1)}(k) = -\frac{2\mu}{\hbar^2}\int_0^\infty dr' r'^2 j_\ell^2(kr')V(r'). \tag{11.236}$$

We immediately notice that $f_\ell^{(1)}(k)$ is real, and does not satisfy the general form (Equation 11.190) and the unitarity statement (Equation 11.192). This is similar to what we found for the Born series in one spatial dimension. This should not bother us particularly, for Equation 11.236 is just an approximation. We assume the phase shift will come out correctly at higher order, at least if the series is convergent. Assuming $\delta_\ell(k) \ll 1$, Equations 11.191 and 11.236 imply

$$\sin\delta_\ell^{(1)}(k) \approx -\frac{2\mu k}{\hbar^2}\int_0^\infty dr' r'^2 j_\ell^2(kr')V(r'), \tag{11.237}$$

and we can always assume a unitary form,

$$\tilde{f}_\ell^{(1)}(k) = \frac{e^{i\delta_\ell^{(1)}(k)}\sin\delta_\ell^{(1)}(k)}{k}. \tag{11.238}$$

This procedure has the advantage of not changing the magnitude of $\tilde{f}_\ell^{(1)}(k)$ with respect to $f_\ell^{(1)}(k)$, but again is clearly ad hoc.

Substituting the *n*th approximate form, $u^{(n)}_{E(\vec{k})}(\vec{r})$, back into Equation 11.121, we have, for general incoming $\vec{k}$,

$$u^{(n+1)}_{E(\vec{k})}(\vec{r}) = e^{i\vec{k}\cdot\vec{r}} + \left(\frac{2\mu}{\hbar^2}\right)\int d^3r' G_+(\vec{r},\vec{r}')V(r')u^{(n)}_{E(\vec{k})}(\vec{r}'). \qquad (11.239)$$

This is again a perturbation series in powers of the spherically symmetric potential, $V(r')$. Defining $u^{(1)}_{E(\vec{k})} \equiv e^{i\vec{k}\cdot\vec{r}}$, we have, for example,

$$u^{(2)}_{E(\vec{k})}(\vec{r}) = u^{(1)}_{E(\vec{k})}(\vec{r}) + \left(\frac{2\mu}{\hbar^2}\right)\int d^3r' G_+(\vec{r},\vec{r}')V(r')e^{i\vec{k}\cdot\vec{r}'}. \qquad (11.240)$$

If we define (θ is measured from $\vec{k}$ to $\vec{k}'$ in general)

$$f^{(n)}(\theta) \equiv -\left(\frac{2\mu}{\hbar^2}\right)\frac{1}{4\pi}\int d^3r' e^{-i\vec{k}'\cdot\vec{r}'}V(r')u^{(n)}_{E(\vec{k})}(\vec{r}), \qquad (11.241)$$

then the next approximation in the Born scattering series after Equation 11.229 is

$$f^{(2)}(\theta) = f^{(1)}(\theta) - \frac{1}{4\pi}\left(\frac{2\mu}{\hbar^2}\right)^2 \int d^3r' e^{i\vec{q}\cdot\vec{r}'}V(r')\int d^3r'' G_+(\vec{r}',\vec{r}'')V(r'')e^{i\vec{q}\cdot\vec{r}''}.$$

(11.242)

The Born series expansion is illustrated in Figure 11.12, the first term of which is just the Feynman diagram of Figure 9.1. The higher-order terms are called *ladder diagrams*. I have shown the ladder directed in a second spatial direction for clarity since the interaction is still instantaneous in time.

The general form for $f^{(1)}(\theta)$ for spherically symmetric potentials may be simplified to

$$f^{(1)}(\theta) = -\frac{1}{2}\left(\frac{2\mu}{\hbar^2}\right)\int_0^\infty dr' r'^2 V(r')\int_{-1}^{1} d\cos\theta\, e^{iqr'\cos\theta}, \qquad (11.243)$$

or

$$f^{(1)}(\theta) = -\left(\frac{2\mu}{\hbar^2}\right)\frac{1}{q}\int_0^\infty dr'\, r'\, V(r')\sin(qr'), \qquad (11.244)$$

where

$$q = |\vec{k} - \vec{k}'| = 2k\sin(\theta/2). \qquad (11.245)$$

Let us take the specific example,

$$V_Y(r) \equiv -V_0\frac{e^{-r/R}}{r}, \qquad (11.246)$$

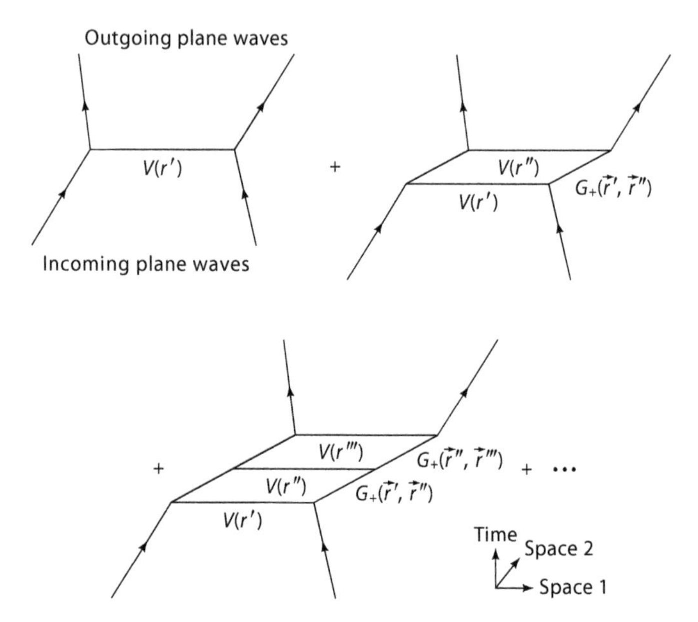

FIGURE 11.12 Illustration of the Born series for the scattering amplitude as a series of "ladder" diagrams. The first two diagrams correspond to the two terms in Equation 11.242.

which is called the *Yukawa potential.* It is the type of potential postulated for the pion exchange model discussed in Chapter 7, Section 7.4, although there we used a simpler finite spherical well. One may show

$$\int_0^\infty \mathrm{d}r' r' V_Y(r') \sin(qr') = \frac{V_0 q}{q^2 + (1/R^2)}. \tag{11.247}$$

This gives

$$f_Y^{(1)}(\theta) = \left(\frac{2\mu}{\hbar^2}\right)\frac{V_0}{q^2 + (1/R^2)}. \tag{11.248}$$

The result is the Born–Yukawa differential cross section,

$$\left(\frac{\mathrm{d}\sigma^{(1)}}{\mathrm{d}\Omega}\right)_Y = \left(\frac{\mu}{2\pi\hbar^2}\right)^2 \frac{(4\pi V_0)^2}{(q^2 + (1/R^2))^2}. \tag{11.249}$$

Sakurai and Napolitano* discuss the validity of the first Born approximation for the Yukawa potential (V_0 here is replaced by $-V_0 R$ there) and concludes that it indeed improves at higher energies. However, the analysis also shows that putting $R \to \infty$ in Equation 11.246 yields a completely invalid Born series for the electrostatic Coulomb potential. Throwing caution to the wind, however, we take the Coulombic limit

$$V_0 = \mathrm{e}^2, \quad R \to \infty,$$

* Sakurai and Napolitano, *Modern Quantum Mechanics.*

in the Born–Yukawa expression, which implies

$$f_C^{(1)}(\theta) = \frac{2\mu e^2}{\hbar^2 q^2}, \tag{11.250}$$

for the Coulomb scattering amplitude, and

$$\left(\frac{d\sigma^{(1)}}{d\Omega}\right)_C = \frac{\mu^2 e^4}{4(\hbar k)^4 \sin^4(\theta/2)} \tag{11.251}$$

for the cross section. This is the famous Rutherford scattering cross section for hydrogen, which was originally derived by him using classical mechanics. That the first Born approximation, recovered in this fashion, and the classical form agree with one another is quite an amazing result. Of course, the classical form does not have reference to hk, but makes the replacement,

$$\hbar^2 k^2 = 2\mu E.$$

The classical/quantum results do not agree in general when identical particles are considered, as we will see in the next section. Related to this, there is another nonclassical quantum effect we should expect. The quasi-classical condition involving the potential we found in Chapter 5, Equation 5.55 was

$$\left|\frac{d\lambda\!\!\!^{-}}{dx}\right| \ll 1,$$

For spherically symmetric potentials, we replace x with r, and I will take the condition here to read

$$\left|\frac{d\lambda\!\!\!^{-}}{dr}\right| \ll 1. \tag{11.252}$$

If we substitute an attractive Coulomb potential, $V(r) = -e^2/r$, use $E = \mu v^2/2$, and take the large-r limit, we obtain the requirement

$$r \gg \left(\frac{\hbar e^2}{\mu^2 v^3}\right)^{1/2}. \tag{11.253}$$

To get a feeling for the energies involved in such a condition, I set this distance equal to a reduced de Broglie wavelength, $\hbar/\mu v$, and solve for v:

$$\frac{v}{c} \gg \alpha. \tag{11.254}$$

($\alpha \equiv e^2/\hbar c$ is the fine structure constant.) Equation 11.254 tells us that classical and quantum scattering results should diverge for very *small* velocities. Indeed, the exact Coulomb scattering amplitude has the same magnitude as Equation 11.250, but begins to differ from it by an overall velocity and θ-dependent phase for velocities smaller than $v/c \sim \alpha$.

Recovering the exact Coulomb scattering amplitude, however, is an advanced topic beyond the level of this text.* In pure nonidentical particle Coulomb scattering, this phase is unobservable. However, its effects can be observed in low-energy proton–proton scattering from the interference of the Coulomb and strong interaction scattering amplitudes, as we will see in Section 11.14.

11.13 IDENTICAL PARTICLE SCATTERING

We begin by realizing that we may write Equation 11.173, using Equations 11.117 and 11.122, as

$$f(\theta) = -\frac{(2\pi)^{1/2}\mu}{\hbar^2}\langle \vec{k}'|V|E(\vec{k})\rangle, \tag{11.255}$$

for a local spherically symmetric potential with general incoming momentum, $\vec{k}$. Let us introduce the so-called transition operator, T, by (remember we choose $C(k) = (2\pi)^{3/2}$ in Equation 11.121)

$$T|\vec{k}\rangle \equiv V\frac{|E(\vec{k})\rangle}{(2\pi)^{3/2}}, \tag{11.256}$$

which allows us to identify

$$f(\theta) = -\left(\frac{4\pi^2\mu}{\hbar^2}\right)\langle \vec{k}'|T|\vec{k}\rangle. \tag{11.257}$$

The operator T is an important concept. (Please do not confuse it with the kinetic energy operator!) It allows us to imagine the scattering event taking place in a symmetrical and physically correct manner between incoming, $|\vec{k}\rangle$, and outgoing, $\langle \vec{k}'|$, plane-wave states. The time ordering in the matrix element is read right to left, as were the earlier Process Diagrams. The operator T is the correct, exact operator to use in Fermi's golden rule, seen in Equation 9.98 (Chapter 9) and Section 10.7 (Chapter 10), and is responsible for the ladder series for the scattering amplitude illustrated in Figure 11.12, if the series converges. To lowest order (first Born approximation), $T = V(\vec{x})$.

Let us now define the two-particle transition operator,

$$T^{(2)} \equiv \frac{1}{2}\sum_{i,j,k,l} a_i^\dagger a_j^\dagger a_k a_l \langle i,j|T|k,l\rangle, \tag{11.258}$$

and generalize Equation 11.257 to a number-basis context. This gives the cross-section expression,

$$\left(\frac{d\sigma}{d\Omega}\right)_{s_i,s_f} \equiv \left(\frac{4\pi^2\mu}{\hbar^2}\right)^2 \left|\langle 1_2,1_{2'},\vec{k}',s_f|T^{(2)}|1_1,1_{1'},\vec{k},s_i\rangle\right|^2, \tag{11.259}$$

* See, for example, the treatment in L. Landau and E. Lifshitz, *Quantum Mechanics*, Vol. 3, 3rd ed. (Pergamon Press, 1977), Section 135.

where I am labeling the occupation numbers of the occupied states as well as the relative wave numbers, and initial (s_i) and final (s_f) spins. We will assume that T is diagonal in spin space. That is, we assume there are no spin interactions between the particles, only a relative spatial interaction. Following the same steps that led to Equation 9.113 (Chapter 9), this gives

$$\begin{aligned}\left(\frac{d\sigma}{d\Omega}\right)_{s_i,s_f} &= \left(\frac{4\pi^2\mu}{\hbar^2}\right)^2 \left|\langle \vec{k}',s_f|T|\vec{k},s_i\rangle \pm \langle \vec{k}',s_f|TP_{12}^s|-\vec{k},s_i\rangle\right|^2 \\ &= \left|f(\theta)\langle s_f|s_i\rangle \pm f(\pi-\theta)\langle s_f|P_{12}^s|s_i\rangle\right|^2.\end{aligned} \tag{11.260}$$

P_{12}^s is the spin particle–label interchange operator, and the top sign is for bosons and the bottom sign for fermions. For example, if $|s_i\rangle = |s\rangle_1 |s'\rangle_2$, where s and s' are individual spin values, then (Figure 11.13)

$$P_{12}^s |s\rangle_1 |s'\rangle_2 = |s'\rangle_1 |s\rangle_2. \tag{11.261}$$

Let us evaluate Equation 11.260 in two cases.

IDENTICAL SPIN 0—SPIN 0 SCATTERING

In this case, we have simply

$$\frac{d\sigma}{d\Omega} = |f(\theta) + f(\pi-\theta)|^2. \tag{11.262}$$

For example, for the Born–Coulomb scattering amplitude, this gives

$$\left(\frac{d\sigma^{(1)}}{d\Omega}\right)_C = \frac{\mu^2 e^4}{4(\hbar k)^4}\left(\frac{1}{\sin^2\frac{\theta}{2}} + \frac{1}{\cos^2\frac{\theta}{2}}\right)^2. \tag{11.263}$$

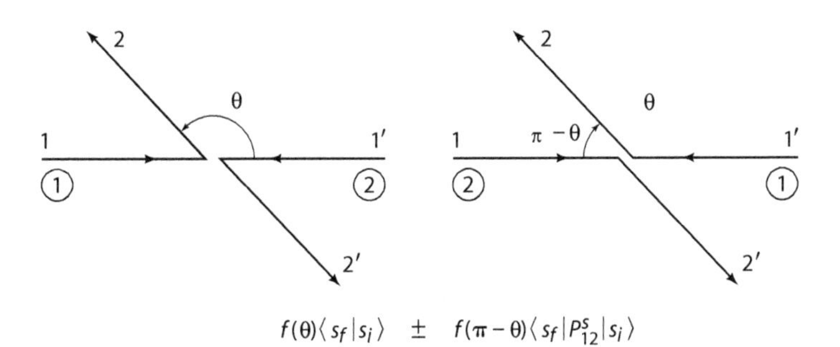

FIGURE 11.13 Symbolic illustration of the two terms contributing in Equation 11.260. These terms have already been illustrated in Section 9.6 (Chapter 9), but we have redrawn these as scattering events in the center-of-mass frame.

This form is symmetric about the scattering angle $\theta = \pi/2$. It is no longer of Rutherford form.

IDENTICAL SPIN 1/2—SPIN 1/2 SCATTERING

We introduce the so-called unpolarized cross section,

$$\left(\frac{d\sigma}{d\Omega}\right)^{\text{unpol}} \equiv \frac{1}{N_i}\sum_{s_i,s_f}\left(\frac{d\sigma}{d\Omega}\right)_{s_i,s_f}. \tag{11.264}$$

The unpolarized cross section corresponds to the measurement of the scattering cross section with an equally mixed incoming spin up/spin down beam, and unmeasured final spin. The phraseology associated with Equation 11.264 states: "average over initial, but sum on final spin." Note that there are $2\times2\times2\times2 = 16$ terms in the sum and that $N_i = 4$ here.

We can evaluate Equation 11.264 (using the lower sign in Equation 11.260) using a coupled or uncoupled spin basis. The most straightforward approach is with an uncoupled basis. Thus, we take the final and initial spin states as

$$|s_i\rangle, |s_f\rangle \rightarrow |m_1, m_{1'}\rangle, |m_2, m_{2'}\rangle, \tag{11.265}$$

with for example

$$P_{12}^s |m_1, m_{1'}\rangle = |m_{1'}, m_1\rangle, \tag{11.266}$$

where all m values are $\pm1/2$. There turns out to be six nonzero contributions ($s_i = {++}$ denotes $m_1 = 1/2$, $m_{1'} = 1/2$ for example):

$$\begin{aligned}\left(\frac{d\sigma}{d\Omega}\right)^{\text{unpol}} &= \frac{1}{4}\Big[\overbrace{|f(\theta) - f(\pi-\theta)|^2}^{(s_i=++,\,s_f=++)} + \overbrace{|f(\theta) - f(\pi-\theta)|^2}^{(s_i=--,\,s_f=--)} + \overbrace{|f(\theta)|^2}^{(s_i=+-,\,s_f=+-)} \\ &\quad + \overbrace{|f(\theta)|^2}^{(s_i=-+,\,s_f=-+)} + \overbrace{|f(\pi-\theta)|^2}^{(s_i=+-,\,s_f=-+)} + \overbrace{|f(\pi-\theta)|^2}^{(s_i=-+,\,s_f=+-)}\Big] \\ &= |f(\theta)|^2 + |f(\pi-\theta)|^2 - \mathrm{Re}\left[f(\theta)f^*(\pi-\theta)\right].\end{aligned} \tag{11.267}$$

The other way to evaluate Equation 11.264 is with a coupled spin basis,

$$|s_i\rangle, |s_f\rangle \rightarrow |s, m_i\rangle, |s, m_f\rangle. \tag{11.268}$$

This actually involves using a different number basis in Equation 11.259, consisting of an appropriate linear combination of spin states. Two such replacements are, for example,

$$\begin{aligned}\left|1_1, 1_{1'}, \vec{k}, s_i\right\rangle &\rightarrow \frac{1}{\sqrt{2}}\left(\left|1_{1+}, 1_{1'-}, \vec{k}\right\rangle + \left|1_{1-}, 1_{1'+}, \vec{k}\right\rangle\right) \\ &\equiv \left|s = 1, m_i = 1\right\rangle,\end{aligned} \tag{11.269}$$

and

$$\left|1_1,1_{1'},\vec{k},s_i\right\rangle \to \frac{1}{\sqrt{2}}\left(\left|1_{1+},1_{1'-},\vec{k}\right\rangle - \left|1_{1-},1_{1'+},\vec{k}\right\rangle\right) \equiv \left|s=0,m_i=0\right\rangle, \tag{11.270}$$

where the 1±, 1′± notation indicates up or down 1 or 1′ states, which are now coupled. The result is still Equation 11.260, but in a different spin basis. The affect of P_{12}^s on all such initial states is

$$\left.\begin{aligned} P_{12}^s\,|s=1,m_i\rangle &= |s=1,m_i\rangle, \\ P_{12}^s\,|s=0,m_i=0\rangle &= -|s=0,m_i=0\rangle. \end{aligned}\right\} \tag{11.271}$$

We then have four nonzero contributions as follows ($s_i = 1, 1$ denotes $s = 1$, $m_i = 1$, for example):

$$\begin{aligned} \left(\frac{d\sigma}{d\Omega}\right)^{\text{unpol}} &= \frac{1}{4}\Big[\overset{(s_i=1,1,\,s_f=1,1)}{\left|f(\theta)-f(\pi-\theta)\right|^2} + \overset{(s_i=1,0,\,s_f=1,0)}{\left|f(\theta)-f(\pi-\theta)\right|^2} \\ &\quad + \overset{(s_i=1,-1,\,s_f=1,-1)}{\left|f(\theta)-f(\pi-\theta)\right|^2} + \overset{(s_i=0,0,\,s_f=0,0)}{\left|f(\theta)+f(\pi-\theta)\right|^2}\Big] \\ &= \left|f(\theta)\right|^2 + \left|f(\pi-\theta)\right|^2 - \text{Re}\left[f(\theta)f^*(\pi-\theta)\right]. \end{aligned} \tag{11.272}$$

This gives the same result as Equation 11.267.

For the Coulomb interaction, using the Born expression from Equation 11.250,

$$\left(\frac{d\sigma^{(1)}}{d\Omega}\right)_C^{\text{unpol}} = \frac{\mu^2 e^4}{4(\hbar k)^4}\left(\frac{1}{\sin^4\frac{\theta}{2}} + \frac{1}{\cos^4\frac{\theta}{2}} - \frac{1}{\sin^2\frac{\theta}{2}\cos^2\frac{\theta}{2}}\right). \tag{11.273}$$

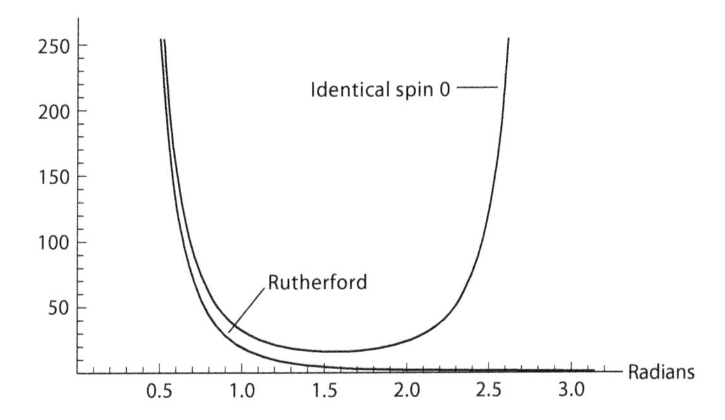

FIGURE 11.14 This shows a comparison of the Rutherford functional form, $1/(\sin^4\theta/2)$ (lower curve), of the cross section versus the identical spin 0 form, $(1/(\sin^2\theta/2) + 1/(\cos^2\theta/2))^2$. Notice the constructive interference for the boson case relative to Rutherford.

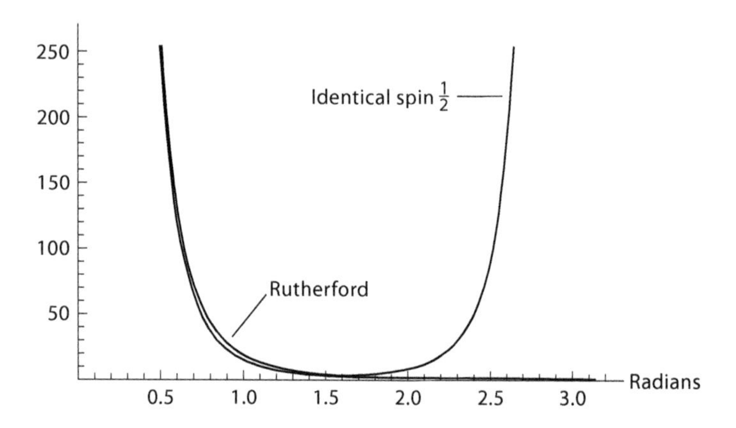

FIGURE 11.15 This is a comparison of the Rutherford functional form, $1/(\sin^4\theta/2)$ with the identical particle unpolarized spin ½ form $(1/(\sin^4\theta/2) + 1/(\cos^4\theta/2) - 1/(\sin^2\theta/2\cos^2\theta/2))$. Notice the destructive interference in this case relative to Rutherford, as well as the agreement between the two forms for $\theta = \pi/2$.

This is again symmetric about $\theta = \pi/2$. Actually, both Equations 11.263 and 11.273 are slightly modified at low energies because the exact $f_C(\theta)$ for Coulomb scattering has a θ-dependent phase angle, as pointed out at the end of the last section. Plots of the relative functions in the Rutherford case, Equation 11.251, as compared to the functions in Equations 11.263 and 11.273, are shown in Figures 11.14 and 11.15.

11.14 PROTON–PROTON SCATTERING

The early expectation after 1932, when James Chadwick discovered the neutral component of the nucleus called the neutron, was that protons and neutrons were fundamental, or elementary, particles. The growing weight of evidence, especially from scattering experiments, showed, however, that the interaction of these supposedly elementary particles was extremely complex. Hideki Yukawa made major theoretical progress in his hypothesis of meson exchange (1935), which helped the interpretation of low-energy nuclear interactions. However, more complications and new particles were soon discovered at higher energies. A hypothesis due to Werner Heisenberg, called "S-matrix theory," but also championed by Geoffrey Chew as the "bootstrap hypothesis" in the 1960s, was that the complicated structure arose out of a self-consistent set of interactions, which gave rise to a complicated particle spectrum. This idea involved a highly mathematical approach to analyzing scattering data in the complex momentum plane, using unitarity as a guiding principle.

We now know through scattering experiments at higher energies that these structure complications are due to actual nonelementary substructure. The quark model, formulated by George Zweig and Murray Gell-Mann independently in 1964, depicts protons and neutrons as bound states of spin-1/2 point particles known now as quarks. Proton–proton scattering is then a matter of bombarding a three-particle bound state against another identical state, leading to a complicated interaction at low energies. Indeed, there are two interactions at play here, the nuclear or strong as well as electromagnetic, since protons are charged.

The scattering of two spin-1/2 protons can take place in a spin $s = 1$ (singlet) or $s = 0$ (triplet) state. In addition, the two protons can be in relative angular momentum quantum states, $\ell = 0, 1, 2, \ldots$. For $\ell = 0$, one has the two possible space/spin states,

$$ {}^1S_0,\ {}^3\cancel{S}_1, $$

where the standard notation gives the $(2s + 1)$ value on the upper left, the angular momentum designation in the middle, and the total j value on the lower right (similar notation as in Chapters 7 and 8). The line through the ${}^3S_1$ designation above indicates that it is forbidden by the Pauli principle. As we learned in Chapter 9, identical fermions are required to be in a totally antisymmetric state. Since s-states are space-symmetric (the exchange operator in space is just the parity operator here), and the triplet spin state is antisymmetric, the ${}^3S_1$ state is totally symmetric and thus forbidden. For $l = 1$, the possible states are

$$ {}^3P_0,\ {}^3P_1,\ {}^3P_2,\ {}^1\cancel{P}_1. $$

Again, the Pauli principle forbids the totally symmetric ${}^1P_1$ state. For $\ell = 2$, we have

$$ {}^1D_2,\ {}^3\cancel{D}_1,\ {}^3\cancel{D}_2,\ {}^3\cancel{D}_3. $$

where the ${}^3D$ states are now all forbidden. The pattern is now clear for higher-ℓ states. Odd-ℓ states will be teamed with a triplet spin state, and even ℓ cases will be combined in a singlet state.

In general, in any two-particle scattering event, we would expect only the total angular momentum quantum number, j, to be conserved instead of ℓ and s individually. However, both the strong and electromagnetic interactions conserve parity; see the comments in Chapter 12, Section 12.4. This "protects" the ${}^1S_0$ and ${}^3P_0$ states from mixing, for example. However, odd-ℓ values differing by two units of angular momentum (and the same j value) can mix since they have the same parity. (The so-called tensor force, mentioned briefly in Chapter 7, Section 7.4 is responsible for this. See Problem 11.14.2)

Let us begin to examine the data for elastic, unpolarized proton–proton scattering. Our investigation into the step potential showed that the s-wave scattering cross section should dominate at low energies. This can also be confirmed from the phase shift expression, Equation 11.237. One can add the known (exact, not Born) Coulomb partial-wave amplitude, $f_C(\theta)$, to an s-wave nuclear amplitude, f_0^{pp}, of the form Equation 11.191 for $\ell = 0$, and fit the resulting expression to the proton–proton differential

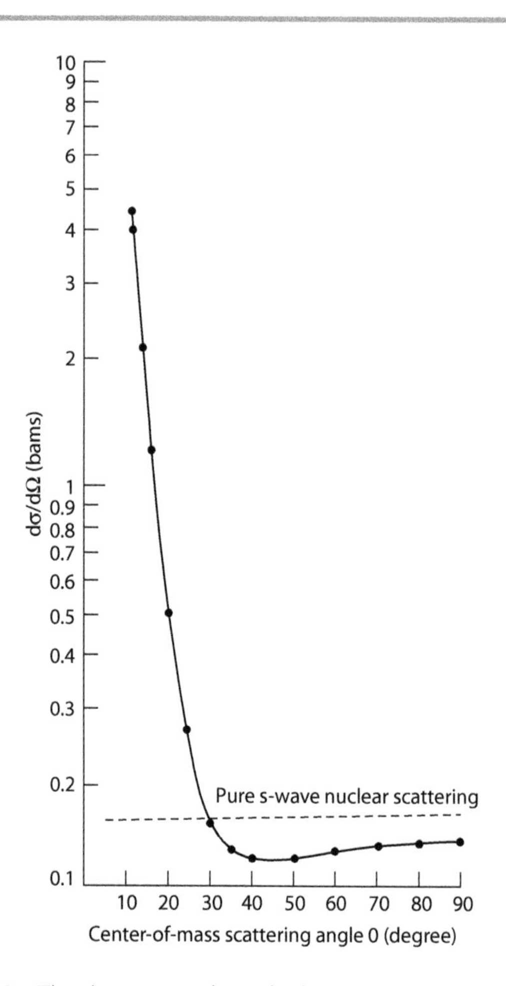

FIGURE 11.16 The data points show the low-energy proton–proton differential cross section as a function of the center-of-mass scattering angle, θ, at a laboratory energy of 3.037 MeV. A fit of the cross-section data using the exact Coulomb scattering amplitude plus a single *s*-wave phase shift is shown.

cross section, as shown for example in Figure 11.16. It displays an excellent fit, using only a single *s*-wave phase shift, $\delta = 50.966°$, at a laboratory energy (one proton at rest in the lab) of 3.037 MeV. This gives the differential cross section as a function of the center-of-mass scattering angle, θ. The cross section for pure nuclear scattering should be isotropic at this energy, and the deduced cross section is also shown in the figure. Interestingly, this shows that destructive interference is taking place between the Coulomb and nuclear amplitudes due to the low-velocity, nonclassical phase in the Coulomb scattering amplitude. Repeating these fits at different energies gives the energy-dependent $^1S_0$ phase shift graph, shown in Figure 11.17.

Phase shift data is where experiment meets theory. The other phase shifts listed earlier become more important at higher energies and have also been measured. A complicating factor at higher energies is the fact that inelastic contributions also occur due to the production of new or

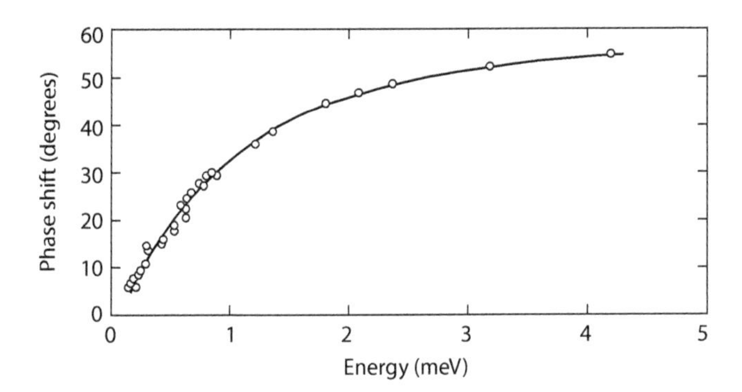

FIGURE 11.17 The energy dependence of the *s*-wave phase shift, $\delta_0(k)$, as a function of laboratory energy. (From K.S. Krane, *Introductory Nuclear Physics*, Figure 4.10, Wiley, 1988. With permission.)

different particles in the final state. The first new "channel" would be pion

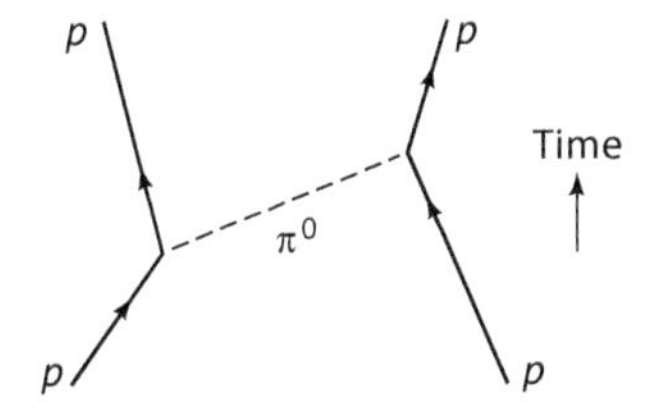

FIGURE 11.18 Feynman diagram for neutral pion exchange between scattering protons.

production starting at a kinetic energy of about 300 MeV/c^2 in the lab frame where one proton is initially at rest. The probability, or unitarity, requirement, Equation 11.192, is then no longer obeyed. This situation may be recharacterized by a description in which the scattering phase shift, originally real, becomes complex (the phases have phases!):

$$\delta_\ell(k) = \zeta_\ell(k) + \frac{i}{2}\ln\left(\eta_\ell^{-1}(k)\right), \tag{11.274}$$

where $\zeta_\ell(k)$ is real and $0 < \eta_\ell(k) < 1$, leading to the form

$$f_\ell^{\text{inel}}(k) = \frac{1}{2ik}\left(\eta_\ell(k)e^{2i\zeta_\ell(k)} - 1\right). \tag{11.275}$$

Elastic data such as in Figure 11.17 demand to be interpreted in terms of nuclear potentials. Extensive analyses* have been done to extract nucleon–nucleon potentials. The explicit forms depend upon the scattering quantum numbers found above. The large distance behavior in the potential for proton–proton scattering is due to neutral one-pion

* M. Lacombe, B. Loiseau, J.M. Richard, R. Vinh Mau, J. Cote, P. Pires, and R. De Tourreil, *Phys. Rev. C* **21**, 861–873 (1980); R. Machleidt, K. Holinde, and C. Elster, *Phys. Rept.* **149**, 1–89 (1987).

exchange. The associated Feynman diagram is shown in Figure 11.18, where the exchanged pion, shown as a dotted line, no longer represents an instantaneous interaction.

At shorter distances (probed at higher energies), an alphabet-soup of other types of spin 0 (scalar) and spin 1 (vector) mesons are exchanged, and one also has double pion exchange diagrams and the excitation of a spin 3/2 state called the Δ. This leads to attraction at intermediate distances but a strong repulsion, or "repulsive core" potential (mentioned briefly in Chapter 7, Section 7.4) at the shortest distances. At higher energies, a multitude of other particle creation channels are opened. Eventually, the meson exchange picture breaks down and the individual quarks themselves begin to interact directly.

A dramatic and explicit manifestation of the substructure of protons emerges at the highest energies. Figure 11.19 shows a proton–proton interaction event at 13 TeV (1 TeV = 10^{12} eV) center-of-mass energy from the Compact Muon Solenoid (CMS) collaboration at the Large Hadron Collider (LHC) near Geneva, Switzerland. The CMS detector is registering two streams of particles called "jets" originating from the collision. The jets are formed from individual high-energy collisions of the quarks and gluons (a force mediating particle) making up the protons. In this case the quark jets come from a bottom (or *b*) quark and a bottom antiquark. Figure 11.20 shows the type of Feynman diagram process responsible for this event. Note that this individual event is just a candidate for such a process as there is a large background of similar events. The jets emerge from the decay of a particle called a Higgs boson, and the electron

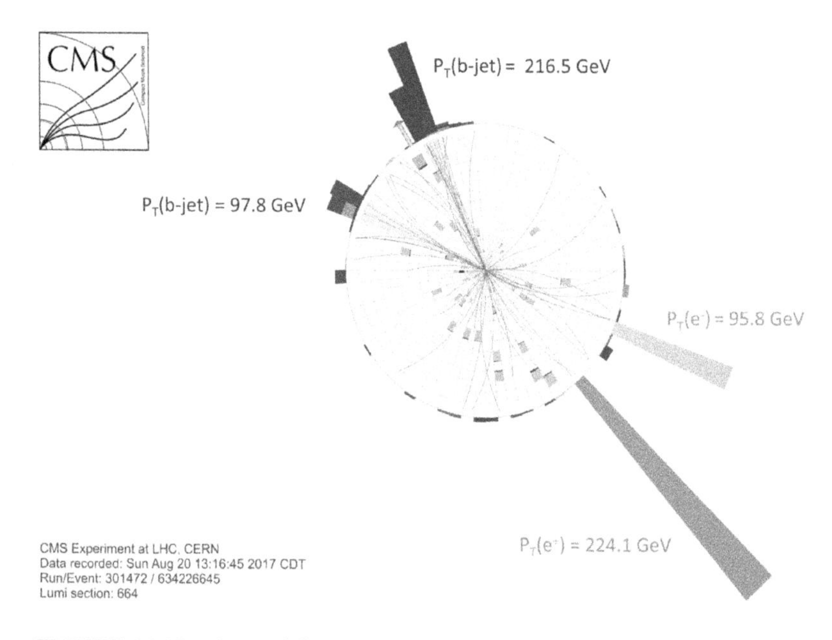

FIGURE 11.19 A candidate event in proton–proton collisions at the LHC from the production and decay of a Higgs boson and a Z boson in the CMS detector. The Higgs boson decays to a bottom quark and a bottom antiquark, and the Z boson decays into an electron and a positron. P_T represents the momentum transverse to the z-axis, which is along the beam axis.

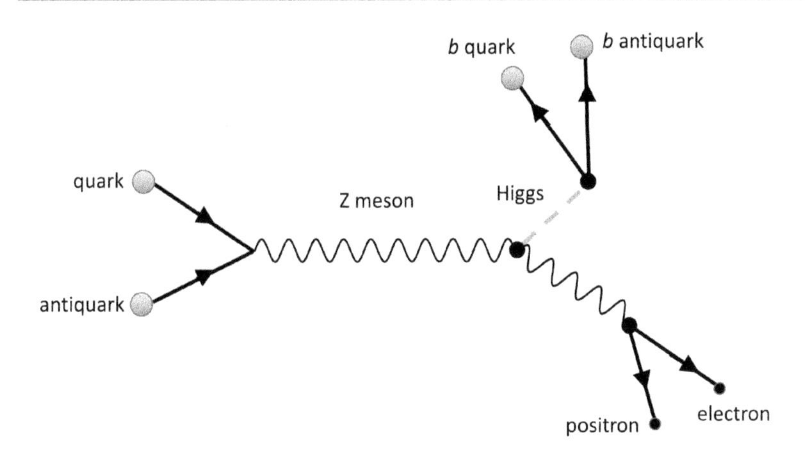

FIGURE 11.20 The Feynman particle interaction diagram responsible for the type of candidate event seen in the previous figure.

and positron come the decay of another particle called a Z boson, a force carrier of the weak interaction. We will learn more about these new particles and interactions as well as the quark substructure of protons in the next chapter.

PROBLEMS

11.1.1 Deduce the Wronskian of sin (x) and cos (x) with respect to x by using Euler's formula, $e^{ix} = \cos(x) + i\sin(x)$, in Equation 11.28. Confirm your answer by direct calculation. [Hint: $W_x[A(x) + B(x), C(x)] = W_x[A(x), C(x)] + W_x[B(x), C(x)]$.]

11.2.2 Check that substituting Equation 11.32 into Equation 11.33 gives an identity, as stated in the text.

11.4.1 Follow the procedure, given in the text, involving Equations 11.54 and 11.55 to determine $r(k)$ and $t(k)$ for the delta-function potential, $V(x) = -\lambda\delta\,(x - a)$.

11.4.2 Check that the $r(k)$ and $t(k)$ in Equations 11.52 and 11.53 (or Equations 11.56 and 11.57) satisfy

$$|r(k)|^2 + |t(k)|^2 = 1.$$

11.4.3 Find the normalization constant, C, in Equation 11.60 (up to an overall phase) by requiring

$$\int_{-\infty}^{\infty} dx\, |u_E(x)|^2 = 1.$$

11.4.4 Consider an initial incoming plane wave moving to the right from Region I and scattering off of two one-dimensional delta-function potentials, one located at $x = -a$ and the other

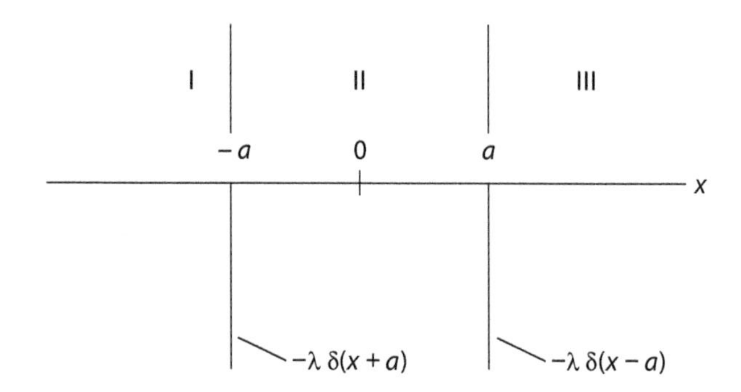

FIGURE 11.21 Two attractive Dirac delta functions, symmetrically located at a and $-a$ along the x-axis. The three scattering regions identified are useful in solving the problem.

at $x = a$ in Figure 11.21. Assume $\lambda > 0$, that is, the delta functions are attractive. Explicitly,

$$V(x) = -\lambda\delta(x+a) - \lambda\delta(x-a).$$

Set up the equations that determine the reflection, $r(k)$, and transmission, $t(k)$, amplitudes, but do not solve them! [Hints: There are two ways to proceed. You may use Equation 11.32 or explicit conditions such as Equations 11.54 and 11.55.]

***11.4.5** (a) The solution to the reflection and transmission coefficients for the last problem is as follows:

$$r(k) = \frac{2i\mu\lambda}{\hbar^2 k}\left(\frac{\cos(2ka) - \left(\dfrac{\mu\lambda}{\hbar^2 k}\right)\sin(2ka)}{\left(1 - \dfrac{i\mu\lambda}{\hbar^2 k}\right)^2 + \left(\dfrac{\mu\lambda}{\hbar^2 k}\right)^2 e^{4ika}}\right),$$

$$t(k) = \frac{1}{\left(1 - \dfrac{i\mu\lambda}{\hbar^2 k}\right)^2 + \left(\dfrac{\mu\lambda}{\hbar^2 k}\right)^2 e^{4ika}}.$$

Show that the implied energies of the associated bound states for large separations, $\mu\lambda a/\hbar^2 \gg 1$, are given approximately by

$$\kappa_{\pm} \approx \frac{\mu\lambda}{\hbar^2}\left(1 \pm e^{-2a\mu\lambda/\hbar^2}\right)$$

where $E_{\pm} \equiv -\hbar^2\kappa_{\pm}^2/2\mu$. [Hint: Set $k = i\kappa$, $\kappa > 0$, and look for approximate pole solutions (zeros of the denominators) of $t(k)$ and $r(k)$.]

(b) Show that the value of k for which resonance occurs, defined to be $r(k_r)=0$, for $\mu\lambda/\hbar^2 k_r \ll 1$, is given approximately by

$$k_r \approx \pm\left(\frac{\pi}{4a}(2n+1) - \frac{2\mu\lambda}{\hbar^2\pi(2n+1)}\right),$$

where $n = 0, 1, 2, \ldots$.

11.4.6 An integral equation for bound states for the Schrödinger equation may be derived as follows. Start with the formal Lippmann–Schwinger equation

$$|\psi\rangle = \frac{1}{E-T}V|\psi\rangle,$$

where E is a *negative* number (set $E = -\hbar^2\kappa^2/2\mu, \kappa > 0$) and V the local potential. To keep things simple, imagine we are in one spatial dimension.

(a) By inserting a complete set of states, show that the equation in (a) leads to the bound-state integral equation,

$$\langle x|\psi\rangle = -\frac{\mu}{\hbar\kappa}\int_{-a}^{a} dx'\, e^{-\kappa|x-x'|}V(x')\langle x'|\psi\rangle,$$

for a potential that is bounded by $-a < x' < a$. (Alternate means of proof: Show by substitution that the above equation exactly solves the one-dimensional Schrödinger equation.)

(b) Using the result of (a), solve for the bound-state energy and wave function of a potential given by a delta function,

$$V(x) = -\frac{\gamma\hbar^2}{2\mu}\delta(x),\ \gamma > 0.$$

11.5.1 Verify that Equation 11.72 represents the eigenvalue equation for bound-state energies of a potential well with $V_0 < 0$, for even parity solutions. [Hints: Assume solutions

$$u(x) = \begin{cases} Ae^{-\kappa x},\ x > a \\ B\cos(k_2 x),\ -a < x < a \\ Ae^{\kappa x},\ x < -a \end{cases}$$

and require continuity of the wave function and its first derivative at $x = a$.]

11.5.2 (a) Show for the step potential that resonances occur at

$$2k_2 a \approx n\pi - 2i\sqrt{-\frac{E_0}{V_0}}$$

for Case 2.

(b) Show for these resonances that ($E = E_0 - i\Gamma/2$)

$$n \geq n_s, \quad \Gamma \approx \frac{\hbar^2 n\pi}{\mu a^2}\sqrt{-\frac{E_0}{V_0}}, \quad E_0 \approx V_0 + \frac{(\hbar n\pi)^2}{8\mu a^2},$$

for $n = n_s, n_s + 1, \ldots,$ where the smallest n value is given by

$$n_s = \left(-\frac{8\mu a^2 V_0}{\pi^2\hbar^2}\right)^{1/2}_{up},$$

and where the right-hand side, which is dimensionless, is rounded up in value. [Note that the second term on the right for E_0 gives the allowed energies for an infinite square well, which means the resonances are appearing where the bound states would otherwise be for $n \geq n_s$.]

11.5.3 Using the results of Problem 11.5.2, show that one has the approximate equality

$$\left(\frac{\mu\Gamma a^2}{2\hbar^2}\right)^2 \approx \frac{2\mu E_0 a^2}{\hbar^2}\left(\frac{n}{n_s}\right)^2,$$

for Case 2 step-potential resonances for $n_s \gg 1$.

***11.5.4** (a) For step-potential scattering, define the phase shifts,

$$\tan(\delta) \equiv \frac{\mathrm{Im}(t(k))}{\mathrm{Re}(t(k))}, \quad \tan(\alpha) \equiv \frac{\mathrm{Im}(r(k))}{\mathrm{Re}(r(k))}.$$

Using Equations 11.64 and 11.65, show that

$$\tan(\delta) = -\cot(\alpha).$$

(b) Show that the expression for the phase shift, δ, may be written as

$$\delta = -2ka + \tan^{-1}\left(\frac{1}{2}\left(\frac{k}{k_2} + \frac{k_2}{k}\right)\tan(2k_2 a)\right),$$

where $k = \sqrt{2\mu E}/\hbar$, $k_2 = \sqrt{2m(E - V_0)}/\hbar$.

11.6.1 (a) Evaluate the first-order Born expressions in Equations 11.108 and 11.109 for the step potential ($V = V_0$, $-a < x < a$; $V = 0$ otherwise), and show that

$$r(k) = -i\frac{\mu V_0}{\hbar^2 k}\sin(2ka), \quad t(k) = 1 - i\frac{2\mu V_0}{\hbar^2 k},$$

to the lowest order in V_0.

(b) Take the full expressions for $r(k)$ and $t(k)$ in Equations 11.64 and 11.65 for the finite-step potential and show that they agree with part (a) for $V_0/E \ll 1$.

11.7.1 The verification of Equation 11.124 in spherical coordinates can be done fairly easily as follows. All quantities here are scalars. Therefore, we can set $\vec{r}\,' = 0$ before the derivatives are taken, then restore $r \to |\vec{r} - \vec{r}\,'|$ afterward. Using such an argument, with

$$\nabla^2(AB) = A\nabla^2 B + B\nabla^2 A + 2\vec{\nabla}A \cdot \vec{\nabla}B,$$

where $A = \mathrm{e}^{\pm ik|\vec{r}-\vec{r}\,'|}$ and $B = 1/|\vec{r} - \vec{r}\,'|$, and given $\nabla^2(1/|\vec{r} - \vec{r}\,'|) = -4\pi\delta(\vec{r} - \vec{r}\,')$, argue that we have

$$\nabla^2 \mathrm{e}^{\pm ikr} = \mathrm{e}^{\pm ikr}\left(-k^2 \pm \frac{2ik}{r}\right),$$

and

$$2\vec{\nabla}\mathrm{e}^{\pm ikr} \cdot \vec{\nabla}\frac{1}{r} = \mp\frac{2ik\mathrm{e}^{\pm ikr}}{r^2},$$

then shift $r \to |\vec{r} - \vec{r}\,'|$ and put the pieces together to show Equation 11.124.

11.7.2 On the basis of the expressions,

$$j_\ell(x) = (-x)^\ell \left(\frac{1}{x}\frac{\mathrm{d}}{\mathrm{d}x}\right)^\ell \left(\frac{\sin(x)}{x}\right), \quad h_\ell^{(1)}(x) = -i(-x)^\ell \left(\frac{1}{x}\frac{\mathrm{d}}{\mathrm{d}x}\right)^\ell \left(\frac{\mathrm{e}^{ix}}{x}\right),$$

establish the large-x expansion of the Bessel functions, Equations 11.148 and 11.149. [Hint: Treat the x in the derivatives as a constant for $x \gg \ell$.]

11.8.1 Given the integral for Legendre polynomials, Equation 11.231,

$$\int_{-1}^{1} \mathrm{d}\cos\theta\, P_\ell(\cos\theta) P_{\ell'}(\cos\theta) = \frac{2}{2\ell + 1}\delta_{\ell\ell'},$$

show that Equation 11.163 gives the Bessel function integral representation,

$$j_\ell(kr) = \frac{1}{2i^\ell}\int_{-1}^{1} d\cos\theta\, e^{ikr\cos\theta} P_\ell(\cos\theta) = \frac{1}{2i^\ell}\int_{-1}^{1} dz\, e^{ikrz} P_\ell(z).$$

Evaluate for $\ell = 0, 1$ and recover known expressions for $j_0(kr)$ and $j_1(kr)$ from Chapter 7.

11.10.1 (a) Show Equation 11.178 from Equations 11.176 and 11.177.

(b) Show that Equations 11.177 and 11.179 give

$$n_\ell(kr) = (-k)^{-\ell-1} r^\ell \left(\frac{1}{r}\frac{d}{dr}\right)^\ell \left(\frac{\cos(kr)}{r}\right).$$

11.10.2 Given (use for example Equations 11.149 and 11.225)

$$n_\ell(kr) \underset{kr \gg \ell}{\rightarrow} -\frac{1}{kr}\cos\left(kr - \frac{\ell\pi}{2}\right),$$

and Equation 11.148, show Equation 11.183.

11.10.3 Show Equation 11.198 from Equation 11.195 using Equations 11.191 and 11.231.

11.10.4 Using Equations 11.182, 11.191, 11.198, and the result for $P_l(1)$ mentioned in Section 11.8, show the "optical theorem,"

$$\sigma_{tot} = \frac{4\pi}{k}\text{Im}(f(0)),$$

where $f(0)$ is the scattering amplitude evaluated in the forward, $\theta = 0$, direction.

11.11.1 Consider scattering off of the "hard sphere" potential,

$$V(r) = \begin{cases} 0, & r > a \\ \infty, & r < a. \end{cases}$$

(a) Find the tangent of the phase shift, $\tan(\delta_\ell(k))$. [Hint: Use Equation 11.180 directly. Answer: $\tan(\delta_\ell(k)) = j_\ell(ka)/n_\ell(ka)$.]

(b) Use Equations 11.209 and 11.210 to show for $ka \ll 1$,

$$\tan(\delta_\ell(k)) \approx -\frac{(ka)^{2\ell+1}}{(2\ell+1)[(2\ell-1)!!]^2}.$$

(c) At low energies, show that the scattering is isotropic and the total cross section is

$$\sigma_{\text{tot}} = 4\pi a^2.$$

[Note the low-energy scattering cross section is four times the geometrical cross section! The object appears larger to a long wavelength probe.]

***11.11.2** (a) Using Equations 11.83, 11.190, 11.200, and 11.201, show that near resonance on the real-k scattering axis we may write

$$e^{2i\delta_l} \approx \frac{k - k_0^*}{k - k_0},$$

where

$$k_0 = \frac{\sqrt{2\mu E_0}}{\hbar} - i\frac{\Gamma}{2\hbar}\sqrt{\frac{\mu}{2E_0}}.$$

(b) At resonance, $k = \text{Re}\,(k_0)$, show that we have

$$e^{2i\delta_l}\Big|_R = -1.$$

(c) Also at resonance, show that

$$k\frac{d\delta_l}{dk}\Big|_R = 4\frac{E_0}{\Gamma},$$

which is very large in the usual scattering limit, $E_0 \gg \Gamma/2$. [Comment: This can help experimentally determine whether true resonance behavior is occurring.]

11.11.3 Show that the exact $\tan(\delta_l)$ for the spherical step potential ($V(r) = V_0$, $r < a$; $V(r) = 0$, $r > a$) is given by

$$\tan(\delta_\ell) = \frac{j_\ell(\kappa a)\dfrac{dj_\ell(kr)}{dr}\Big|_{r=a} - j_\ell(ka)\dfrac{dj_\ell(\kappa a)}{dr}\Big|_{r=a}}{j_\ell(\kappa a)\dfrac{d\eta_\ell(kr)}{dr}\Big|_{r=a} - \eta_\ell(ka)\dfrac{dj_\ell(\kappa a)}{dr}\Big|_{r=a}},$$

where

$$k = \frac{\sqrt{2\mu E}}{\hbar}, \quad \kappa = \frac{\sqrt{2\mu(E - V_0)}}{\hbar}.$$

***11.11.4** Considering V_0 a small quantity ($V_0 \ll E$), use the result of Problem 11.11.3 to show that

$$\tan(\delta_0(k)) \approx -\frac{V_0(ka)}{2E}\left(1 - \frac{\sin^2(ka)}{(ka)^2}\right),$$

to first order in V_0 for $\ell = 0$.

11.11.5 Given the spherical well radius, a, and reduced mass, μ, characterize the shallowest well depth, V_0, for which low-energy square well scattering actually vanishes! [Hint: The observation that cross sections could actually vanish for certain atomic systems, called the *Ramsauer–Townsend effect*, was a mystery before quantum mechanics.]

11.12.1 Evaluate the first Born approximation for the $\ell = 0$ scattering amplitude, $f_0^{(1)}(\theta)$, for the spherical step potential using Equation 11.236. To what extent does your answer agree/disagree with the answer to Problem 11.11.4?

11.12.2 Use Equation 11.229 to show that the full first Born approximation for the scattering amplitude for the three-dimensional spherical step potential is

$$f^{(1)}(\theta) = -\frac{2\mu V_0}{\hbar^2 q}\left[\frac{1}{q^2}\sin(qa) - \frac{a}{q}\cos(qa)\right],$$

where $q \equiv |\vec{k} - \vec{k}'| = 2k\sin(\theta/2)$.

***11.12.3** In Chapter 7, Section 7.7 we learned that the positive energy ($E > 0$) radial-wave function solutions for the Coulomb potential, $V(r) = -\xi/r$ (for atomic physics $\xi = Z_1 Z_2 e^2$ in the attractive case), are

$$R(r) = N(kr)^{\ell+1} e^{-ikr} F\left(\ell + 1 + i\frac{\xi k}{2E}, 2\ell + 2, 2ikr\right),$$

where N is an undetermined normalization constant and $F(a, b, z)$ the confluent hypergeometric function. Show, by the use of the following limiting formula, where x is real and positive*

$$\lim_{x\to\infty} F(a,b,2ix) \to \frac{\Gamma(b)}{\Gamma(a)}(2ix)^{a-b} e^{2ix} + \frac{\Gamma(b)}{\Gamma(b-a)}(-2ix)^{-a},$$

($\Gamma(x)$ denotes the classical gamma function encountered earlier in Section 7.7) that

* J.T. Cushing, *Applied Analytical Mathematics for Physical Scientists* (John Wiley & Sons, 1975), p. 414.

$$\lim_{r\to\infty} R(r) \sim \sin\left(kr - \frac{\pi\ell}{2} + \sigma_\ell - \eta\ln(2kr)\right).$$

Comparing this to Equation 11.183 identifies the Coulomb phase shifts as

$$\tan\sigma_\ell(k) = \frac{\mathrm{Im}(\Gamma(\ell+1+i\eta))}{\mathrm{Re}(\Gamma(\ell+1+i\eta))},$$

where $\eta \equiv \xi\mu/\hbar^2 k = Z_1 Z_2 e^2 \mu/\hbar^2 k$ is called the Sommerfeld parameter. [Hints: Define

$$\Gamma(\ell+1+i\eta) = |\Gamma(\ell+1+i\eta)|\, e^{i\sigma_\ell}.$$

Also the identity

$$(a)^{ib} = e^{ib\ln(a)},$$

where ln denotes the natural log, is extremely useful. *Note*: The slowly varying logarithm in the sine function represents a long-range distortion in the Coulombic wave function; the particles are not "free" even at $r \to \infty$!]

***11.12.4** Using Problem 11.12.3 and a computer software package that computes gamma functions, make a plot of the of the Coulomb phase shift, sin (σ_0), for $\ell = 0$, as a function of the dimensionless Sommerfeld parameter, η, from 0 to 10. Identify the first phase shift maximum. At what value of η does this occur? Is this a resonance?

11.14.1 The deuteron is a bound state of a proton and neutron, nonidentical particles, as we studied in Section 7.4. Its total angular momentum, as measured in the laboratory, is $\hbar$ ($j = 1$), and it has positive parity. Based on these facts, and the fact that the tensor force can mix angular momentum states differing by two units of $\hbar$, as well as the fact that parity is conserved in the interaction, what are the possible ground state configurations, (L,S), of the deuteron? (You can also use the spectroscopic notation of Sect. 11.14.)

11.14.2 In proton–proton scattering, argue that the even ℓ states are unmixed in angular momentum, whereas the odd-ℓ states mix only when j, the total angular momentum quantum number, is even (excluding zero). Assume the tensor force can mix angular momentum states differing by two units of $\hbar$, and that parity is conserved.

Connecting to the Standard Model

Synopsis: The reader is introduced to the diverse ingredients of the Standard Model. Discrete symmetries of parity, time reversal, and charge conjugation are discussed. A description of the 61 particles of the Standard Model is given. Special attention is given to the description of the Standard Model Higgs particle. After a short introduction to the gauge theory concept, the electrodynamic (QED), chromodynamic (QCD), and weak interaction particles and interactions are described. A concluding section describes the ideas of supersymmetry and superstrings, which will likely play important roles in future unified theories incorporating the Standard Model.

12.1 DISCRETE SYMMETRIES

We have come a long way in our understanding and delineation of the quantum aspects of our world. It is now time to begin to apply our quantum description to nature on the smallest scale that can be determined with present technology. We will find a rich, detailed set of interactions and particles that make up the bricks and mortar of our physical world.

We have covered some of the important dynamical quantum principles, but we have not yet met the particles. We will do so momentarily, but first we need a way to distinguish or categorize the particles that exist in our world. We already know about the dynamics associated with some of the properties of particles: mass, momentum, angular momentum, and spin. However, there are also quantum properties that derive from *discrete symmetries*, which characterize a particle and its interactions.

There are two major themes in physics: dynamics and symmetry. "Dynamics" denotes a developmental process, such as the change in a system subject to certain forces and constraints. The word "symmetry" in physics denotes a system balance, regularity, or indistinguishability. Usually, symmetry in a physical system implies a conservation law. Such symmetries can be continuous, as for systems under spatial translation or rotation, which in fact lead to momentum and angular momentum conservation. Other symmetries are discrete. These symmetries also lead to conserved quantities or expected behaviors and are extremely useful in characterizing the properties of particles.

There are three such properties I would like to discuss at this point: parity, time reversal, and charge conjugation. First, we will become reacquainted with the property known as parity, from both a transformation and quantum point of view.

12.2 PARITY

The most general linear transformation of three-dimensional vectors is described by

$$x'_i = \sum_{j=1}^{3} a_{ij} x_j. \tag{12.1}$$

A *matrix* statement of this relationship is

$$\begin{pmatrix} x'_1 \\ x'_2 \\ x'_3 \end{pmatrix} = \begin{pmatrix} a_{11} & a_{12} & a_{13} \\ a_{21} & a_{22} & a_{23} \\ a_{31} & a_{32} & a_{33} \end{pmatrix} \begin{pmatrix} x_1 \\ x_2 \\ x_3 \end{pmatrix}, \tag{12.2}$$

or simply

$$x' = ax, \tag{12.3}$$

where a is understood to be a 3×3 matrix and x and x' are column matrices. We require the preservation of the length of vectors in this formalism. This gives

$$\sum_j (x_i)^2 = \sum_j (x'_i)^2 \Rightarrow \sum_i a_{ij} a_{ik} = \delta_{jk}, \tag{12.4}$$

or in terms of matrices (a^t is the transpose of a),

$$a^t a = aa^t = 1, \tag{12.5}$$

$$\Rightarrow a^t = a^{-1}. \tag{12.6}$$

Equation 12.6 defines *orthogonal transformations.** The aforementioned conditions, Equation 12.5, represent six equations in nine unknowns. This means three a_{ij} values are undetermined and can be used to freely determine rotations about three axes. The determinant of the transformation is defined as

$$\det a \equiv \sum_{i,j,k} \varepsilon_{ijk}\, a_{1i} a_{2j} a_{3k}, \tag{12.7}$$

where the antisymmetric three-index permutation symbol, $\varepsilon_{ijk,}$ has the values $\varepsilon_{123} = \varepsilon_{231} = \varepsilon_{312} = 1$, $\varepsilon_{321} = \varepsilon_{213} = \varepsilon_{132} = -1$, and other values zero. It is easy to show explicitly that $\det a = \det a^t$ for 3×3 matrices. Then, taking the determinant of both sides of Equation 12.5,

$$(\det a^t)\det a) = (\det a)^2 = 1, \tag{12.8}$$

$$\Rightarrow \det a = \pm 1 \text{ only}. \tag{12.9}$$

$\det a = 1$ describes pure rotations. *Passive* rotations rotate the coordinate system leaving the actual vector unchanged. *Active* rotations rotate the actual vector. Of course, one type of rotation is the opposite of the other.

* For a review, see W. Wilcox, *Modern Introductory Mechanics* (Bookboon.com), Chapter 1.

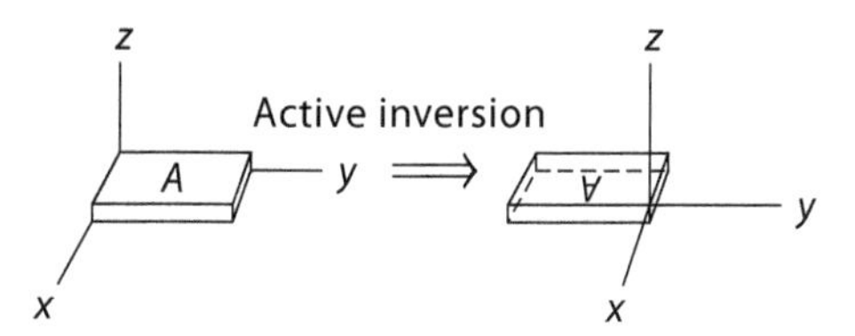

FIGURE 12.1 An active, complete inversion of the object on the left. The transformation makes a new object on the right that cannot be formed by any set of continuous rotations starting from the original.

What does det $a = -1$ describe? Consider the active example in Figure 12.1. This transformation is given by

$$a = \begin{pmatrix} -1 & 0 & 0 \\ 0 & -1 & 0 \\ 0 & 0 & -1 \end{pmatrix}.$$

Such a transformation is a *parity inversion*, which is also orthogonal. Vectors change sign under a complete active inversion, as illustrated in Figure 12.2.

Scalar and *vector* fields, $\phi(\vec{x})$ and $V_i(\vec{x})$, respectively, are defined to transform under orthogonal transformations as

$$\phi'(\vec{x}') = \phi(\vec{x}), \tag{12.10}$$

$$V_i'(\vec{x}') = \sum_j a_{ij} V_j(\vec{x}). \tag{12.11}$$

Equation 12.10 says, for example, that the value of the field at an arbitrary point is the same whether evaluated in the primed or unprimed coordinate system for any orthogonal transformation (active or passive sense). A classical field is just a number-valued quantity, defined throughout space, with specified transformation properties. An example is the electromagnetic scalar potential, $\Phi(\vec{x})$, from which the electric field may be calculated, $\vec{E} = -\nabla\Phi$, for time-independent situations.

An example of a quantity transforming as a scalar is the dot product of two vectors.

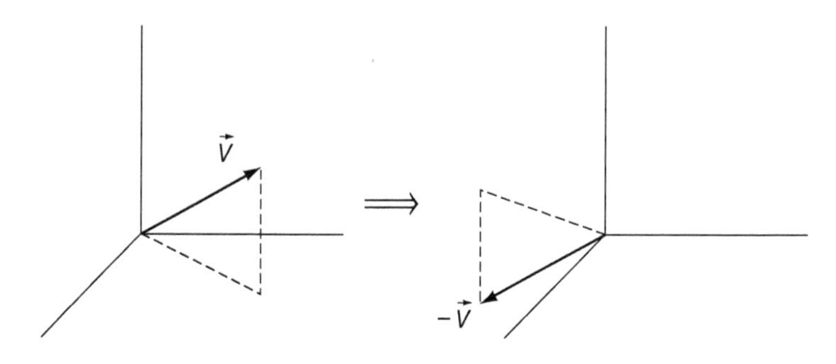

FIGURE 12.2 An active, complete inversion of a vector.

Proof: Given $A_i' = \sum_j a_{ij} A_j;\ B_i' = \sum_k a_{ik} B_k.$

$$\vec{A}' \cdot \vec{B}' = \sum_i A_i' B_i' = \sum_{i,j,k} a_{ij} A_j a_{ik} B_k,$$
$$= \sum_{j,k} \underbrace{\left(\sum_i a_{ij} a_{ik} \right)}_{\delta_{jk}} A_j B_k = \vec{A} \cdot \vec{B}. \tag{12.12}$$

Another example would be $\vec{\nabla} \cdot \vec{A}$, if $\vec{A}$ is a vector and $\vec{\nabla}$ is the gradient operator.

Pseudoscalar and *pseudovector* fields, $\phi(\vec{x})$ and $V_i(\vec{x})$, respectively, transform as

$$\phi'(\vec{x}') = (\det a)\phi(\vec{x}) \tag{12.13}$$

and

$$V_i'(\vec{x}') = (\det a) \sum_j a_{ij} \vec{V}_j(\vec{x}) \tag{12.14}$$

under orthogonal transformations. An example of a pseudovector would be the cross product of two vectors.

Proof: Given $\vec{A}, \vec{B}$ are vectors; then

$$(\vec{A} \times \vec{B})_i = \sum_{j,k} \varepsilon_{ijk} A_j B_k. \tag{12.15}$$

We need the result (Problem 12.2.1)

$$\sum_{j,k} \varepsilon_{ijk}\, a_{j\ell}\, a_{km} = (\det a) \sum_n \varepsilon_{n\ell m}\, a_{in}. \tag{12.16}$$

Thus,

$$\left(\vec{A}' \times \vec{B}'\right)_i = \sum_{j,k} \varepsilon_{ijk} A_j' B_k' = \sum_{j,k,\ell,m} \varepsilon_{ijk}\, a_{j\ell}\, A_\ell\, a_{km} B_m$$
$$= \sum_{j,k,\ell,m} \varepsilon_{ijk}\, a_{j\ell}\, a_{km}\, A_\ell\, B_m = (\det a) \sum_{n,\ell,m} \varepsilon_{n\ell m}\, a_{in}\, A_\ell B_m \tag{12.17}$$
$$= (\det a) \sum_i a_{in} (\vec{A} \times \vec{B})_n.$$

A physical example would be angular momentum, $\vec{L} = \vec{x} \times \vec{p}$. Let us choose

$$a = \begin{pmatrix} 1 & 0 & 0 \\ 0 & 1 & 0 \\ 0 & 0 & -1 \end{pmatrix},$$

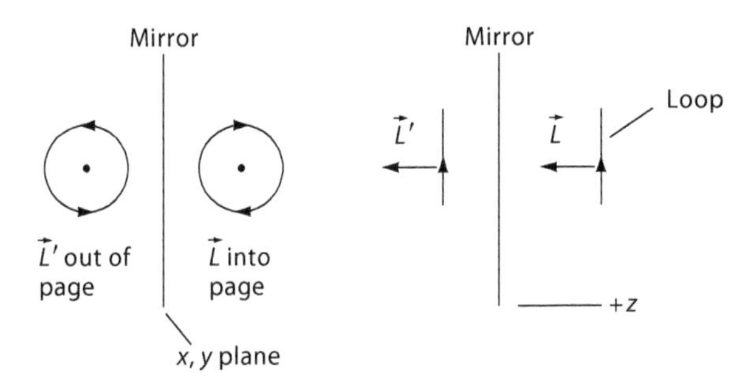

FIGURE 12.3 Behavior of the angular momentum components of a rotating object as reflected in a mirror in the *x*, *y* plane.

which describes a parity inversion in the third (z) axis. Under this a, we have

$$L'_x = -L_x,\ L'_y = -L_y,\ L'_z = +L_z.$$

Notice that since $\vec{L}$ is a pseudovector, the *lack* of change of sign on the z-component under this transformation is what distinguishes it from a vector under the same transformation. This is illustrated in Figure 12.3. Particle spin, $\vec{S}$, also behaves as a pseudovector under coordinate transformations.

The field concept is central to quantum field theories, which generalize quantum mechanics to a relativistic context. Quantum fields, which are dynamical operators in relativistic field theories, can be characterized under the parity operation as well, which constitutes an important particle label. A relativistic set of transformations, carried out at all points in space and time, defines the nature of the fields. These are called *Lorentz transformations*, which preserve only the "proper length," $\Delta s^2 \equiv c^2\Delta t^2 - \Delta\vec{x}^2$, between events, rather than space and time intervals separately.

The preceding paragraphs have been a discussion of the actions of the classical parity operation on fields. However, we know that a *parity operator* also exists in quantum mechanics. We were introduced to some its properties in Chapter 3, Section 3.2:

$$\mathbb{P}|x\rangle \equiv |-x\rangle,\ \langle x|\mathbb{P}^\dagger = \langle -x|,\ \ \mathbb{P}^\dagger = \mathbb{P}. \tag{12.18}$$

We also have $\mathbb{P}^2 = 1$. The crucial point for parity conservation was Equation 3.59:

$$|H,\mathbb{P}| = 0. \tag{12.19}$$

In a spherical basis, we also have Equations 6.244 and 6.249 (see Chapter 6),

$$\langle \hat{n}\,|\,\mathbb{P} = \langle -\hat{n}\,|, \tag{12.20}$$

$$\mathbb{P}\,|\,\ell,m\rangle = (-1)^\ell\,|\,\ell,m\rangle. \tag{12.21}$$

How do operators respond to parity? We have

$$\mathbb{P}\,\vec{x}\,\mathbb{P} = -\vec{x},\ \mathbb{P}\,\vec{p}\,\mathbb{P} = -\vec{p},\ \mathbb{P}\,\vec{L}\,\mathbb{P} = \vec{L},\ \mathbb{P}\,\vec{S}\,\mathbb{P} = \vec{S}. \tag{12.22}$$

This discrete symmetry is also contained in the Lorentz transformations, as are all orthogonal transformations. (Note: The more customary symbol P will be used for this quantum operation in the rest of the chapter.)

We will learn in upcoming sections that the electromagnetic and strong interactions both conserve parity. The strong interaction is responsible for producing bound states from quarks. For example, *mesons* are integer spin-bound states of one quark and one antiquark. These are classified by their parity, just like atomic states. *Baryons*, which are half-integer spin-bound states of three quarks, also have an intrinsic parity. Parity is a *multiplicative* quantum number, that is, the overall parity of a state is determined by the products of the parities of the individual components.

The weak interactions intrinsically do not conserve parity. This is not a small violation for this interaction, but in some sense it is "maximally" violated, and was readily confirmed in the laboratory after physicists figured out where to look. This realization was one of the most revolutionary events in particle physics in the last century.*

12.3 TIME REVERSAL

J.J. Sakurai pointed out in his quantum book that a better name for this discrete symmetry is "reversal of motion," as we will see. All conservative, velocity independent forces in classical mechanics are time-reversal invariant. This can be understood from Newton's law:

$$\left.\begin{aligned} \vec{F} &= m\vec{a}, \\ -\vec{\nabla}V(\vec{x}) &= m\ddot{\vec{x}} \end{aligned}\right\} \text{invariant under } t \to -t. \tag{12.23}$$

This implies if $\vec{x}(t)$ is a possible trajectory (solution), then so is $\vec{x}(-t)$. Thus, if all one had in the world were electric fields any particle trajectory would be invariant under

$$\vec{E} \to \vec{E}, \rho \to \rho, \ t \to -t, \tag{12.24}$$

(ρ represents charge density) since

$$\vec{F} = q\vec{E} = -q\vec{\nabla}\Phi(\vec{x}). \tag{12.25}$$

However, magnetic fields are a little trickier. Imagine tracing an electron trajectory in a magnetic field as in Figure 12.4. Imagine just letting $t \to -t$, $\vec{v} \to -\vec{v}$ at the top of its motion. This leads to Figure 12.5. The electron does *not* retrace its former trajectory. In addition, let us now let

$$\vec{j} \to -\vec{j}, \vec{B} \to -\vec{B}$$

$\uparrow$Makes sense from point of view of *reversal of motion*

* Parity violation was confirmed in 1957 in the β-decay of Cobalt 60: C.S. Wu, E. Ambler, R.W. Hayward, D.D. Hoppes, and R.P. Hudson, *Phys. Rev.* **105**, 1413 (1957). It was first suggested theoretically by C.N. Yang and T.D. Lee in the previous year.

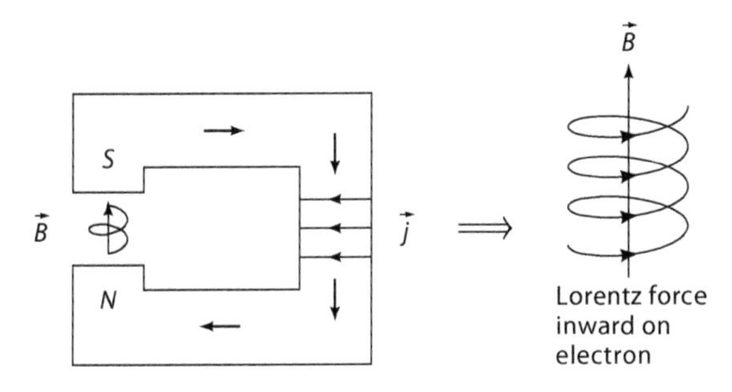

FIGURE 12.4 Physical magnet and associated electron path, following from the Lorentz force law, in the field.

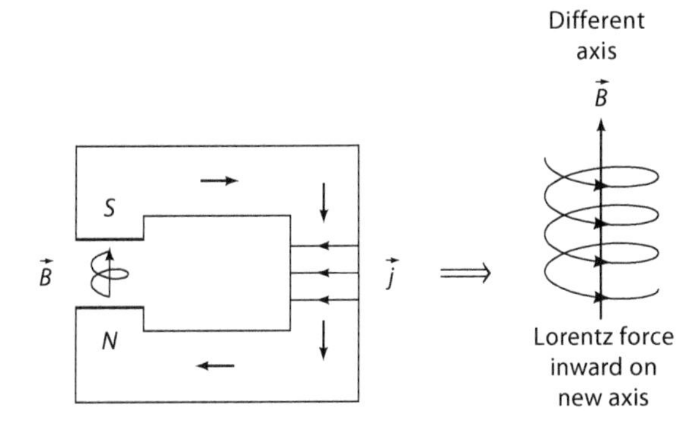

FIGURE 12.5 Original physical magnet and associated electron path, after the velocity has been reversed at the top of its motion in Figure 12.4.

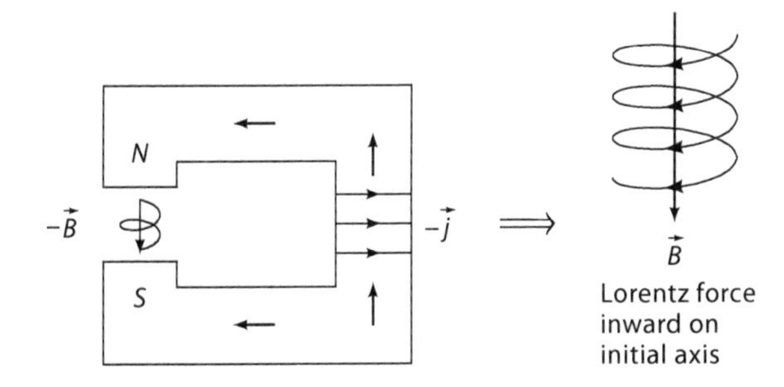

FIGURE 12.6 Physical magnet with reversed magnetic field and associated electron path, after the velocity has been reversed at the top of its motion in Figure 12.4.

Now the trajectory is retraced, as illustrated in Figure 12.6. This makes sense, since Maxwell's equations and the Lorentz force law are invariant under

$$t \to -t,\ \vec{v} \to -\vec{v},\ \vec{E} \to \vec{E},\ \rho \to \rho,\ \vec{B} \to -\vec{B},\ \vec{j} \to -\vec{j}. \quad (12.26)$$

Thus, electromagnetism is time-reversal invariant.

What about the Schrödinger equation in this context? It is a first-order differential equation in time. For a particle in an electric and magnetic field, we have*

$$i\hbar\frac{\partial\psi_\alpha(\vec{x},t)}{\partial t}=\left(-\frac{\hbar^2}{2m}(\vec{\nabla}-i\frac{q}{\hbar c}\vec{A}(\vec{x},t))^2+V(\vec{x},t)\right)\psi_\alpha(\vec{x},t). \quad (12.27)$$

$\psi_\alpha^*(\vec{x},-t)$ (the complex conjugate with negative time parameter) is a solution for a reversed magnetic field:

$$i\hbar\frac{\partial\psi_\alpha^*(\vec{x},-t)}{\partial t}=\left(-\frac{\hbar^2}{2m}(\vec{\nabla}+i\frac{q}{\hbar c}\vec{A}(\vec{x},-t))^2+V(\vec{x},-t)\right)\psi_\alpha^*(\vec{x},-t). \quad (12.28)$$

So we learn that if $\langle\vec{x}\,|\,\alpha\rangle$ is the $t=0$ wave function, the time-reversed (motion-reversed) wave function is given by $\langle\vec{x}\,|\,\alpha\rangle^*=\langle\alpha\,|\,\vec{x}\rangle$.

$$\langle\vec{x}\,|\,\alpha\rangle\underset{\text{Time reversal}}{\Rightarrow}\langle\alpha\,|\,\vec{x}\rangle. \quad (12.29)$$

From this very simple observation, we deduce that if we are to introduce an operator that represents time reversal, it is a very unusual object. The reason is that an operator only takes bras into bras and kets into kets. For example:

$$X=\sum_{i,j}|\,a_i\rangle\langle a_j\,|\,X_{ij}, \quad (12.30)$$

$$\Rightarrow X\,|\,a_k\rangle=\sum_i X_{ik}\,|\,a_i\rangle,\quad \langle a_k\,|\,X=\sum_i X_{kj}\langle a_j\,|. \quad (12.31)$$

What we want to do now is more akin to an *operation*, rather than an operator. For example, we had for Hermitian conjugation (see Chapter 1, Equation 1.193)

$$(|\,\alpha\rangle)^\dagger=\langle\alpha\,|. \quad (12.32)$$

Under "†":

$$(\langle\vec{x}\,|\,\alpha\rangle)^\dagger=\langle\alpha\,|\,\vec{x}\rangle. \quad (12.33)$$

However, now consider the momentum space time-reversed wave function. We have ($t=0$),

$$\tilde{\phi}_\alpha(\vec{p}')=\frac{1}{(2\pi\hbar)^{3/2}}\int d^3x\, e^{-i\vec{p}'\cdot\vec{x}/\hbar}\tilde{\psi}_\alpha(\vec{x}'), \quad (12.34)$$

↑ Time-reversed *momentum* wave function ↑ Time-reversed *position* wave function

$$\tilde{\psi}_\alpha(\vec{x}')=\psi_\alpha^*(\vec{x}'), \quad (12.35)$$

* This form was motivated in Chapter 10 from the classical interaction Hamiltonian (Equation 10.120). See Section 12.6 for a motivation of this form based on gauge invariance.

$$\begin{aligned}\tilde{\phi}_\alpha(\vec{p}\,') &= \frac{1}{(2\pi\hbar)^{3/2}}\int d^3x\, e^{-i\vec{p}\,'\cdot\vec{x}/\hbar}\tilde{\psi}^*_\alpha(\vec{x}\,'),\\ &= \phi^*_\alpha(-\vec{p}\,').\end{aligned} \tag{12.36}$$

We have found that

$$\langle \vec{p}\,'|\alpha\rangle \underset{\text{Time reversal}}{\Rightarrow} \langle\alpha|-\vec{p}\,'\rangle. \tag{12.37}$$

This means that time reversal is not equivalent to "†" (Hermitian conjugation).

We will take the following point of view. Let us define the effect of what I will call the *anti-Hermitian conjugation* (or just anticonjugation) *operation, A,* on basis states:

$$\text{Base kets}\begin{cases}(\langle X_i|)^A \equiv |\tilde{X}_i\rangle = U|X_i\rangle,\\ (|X_i\rangle)^A \equiv \langle\tilde{X}_i| = \langle X_i|U^\dagger.\end{cases} \tag{12.38}$$

If this were all, it would simply be an extended symmetry operation that yields Hermitian conjugation for $U=1$. However, let us also define

$$(C_1\langle X_i|+C_2\langle X_j|)^A \equiv \overset{\downarrow\ \text{No star!}\ \downarrow}{C_1|\tilde{X}_i\rangle + C_2|\tilde{X}_j\rangle}, \tag{12.39}$$

$$(\langle\alpha|\beta\rangle)^A = \langle\alpha|\beta\rangle,\ \ (\langle\alpha|,|\beta\rangle \text{ expandable in base bras, kets}) \tag{12.40}$$

$$(|X_i\rangle\langle X_j|)^A \equiv |\tilde{X}_j\rangle\langle\tilde{X}_i| \tag{12.41}$$

These rules imply, for example (θ is a general operator),

$$(\langle X_i|\theta)^A = \theta^A|\tilde{X}_i\rangle, \tag{12.42}$$

$$(\theta|X_i\rangle)^A = \langle\tilde{X}_i|\theta^A. \tag{12.43}$$

We now define the effect of the time-reversal anticonjugation operation, "T," on $\vec{x}\,', \vec{p}\,'$ base kets (the † and ^{T} operations commute):

$$(\langle\vec{x}\,'|)^T \equiv |\vec{x}\,'\rangle \Rightarrow (|\vec{x}\,'\rangle)^T = \langle\vec{x}\,'|, \tag{12.44}$$

$$(\langle\vec{p}\,'|)^T \equiv |-\vec{p}\,'\rangle \Rightarrow (|\vec{p}\,'\rangle)^T = \langle-\vec{p}\,'|, \tag{12.45}$$

where $U=1$ in Equation 12.38 on $\vec{x}\,'$ eigenstates and one may take $U = P$ (parity operator) on $\vec{p}\,'$ ones. What is the effect on the $\vec{x}, \vec{p}$ operators? Clearly,

$$(\vec{x})^T = \left(\int d^3x'\vec{x}\,'|\vec{x}\,'\rangle\langle\vec{x}\,'|\right)^T = \vec{x}, \tag{12.46}$$

$$(\vec{p})^T = \left(\int d^3p'\,\vec{p}\,'|\vec{p}\,'\rangle\langle\vec{p}\,'|\right)^T = -\vec{p}. \tag{12.47}$$

If the time-reversed state $(|\alpha,\delta t\rangle)^T = (e^{-\delta t H/\hbar}|\alpha\rangle)^T$ (for a time-independent Hamiltonian using energy base kets, say) satisfies

$$(|\alpha,\delta t\rangle)^T = \langle\tilde{\alpha},-\delta t|, \tag{12.48}$$

then the Hamiltonian is time-reversal invariant:

$$(H)^T = H. \tag{12.49}$$

There is an immediate consequence of Equation 12.49, which I will state as a theorem.

Theorem

Given $(H)^T = H$ and the energy eigenkets, $|n\rangle$, are nondegenerate, we may always choose

$$\langle\vec{x}|n\rangle = \langle\vec{x}|n\rangle^*.$$

Proof:

Given: $H|n\rangle = E_n|n\rangle$. Take "$T$": $(|n\rangle)^T = \langle\tilde{n}|$.

$$\Rightarrow \langle\tilde{n}|(H)^T = E_n\langle\tilde{n}| \Rightarrow \langle\tilde{n}| = e^{i\delta}\langle n| \quad (\text{choose } \delta = 0)$$

Expand:

$$|n\rangle = \int d^3x'\,|\vec{x}'\rangle\langle\vec{x}'|n\rangle.$$

Use "T" followed by "$\dagger$":

$$|\tilde{n}\rangle = \int d^3x'\,|\vec{x}'\rangle\langle\vec{x}'|n\rangle^*.$$

Project into $\langle\vec{x}|$:

$$\langle\vec{x}|\tilde{n}\rangle = \langle\vec{x}|n\rangle^*.$$

Using $\langle\tilde{n}| = \langle n|$, we conclude that the wave function can be chosen real, $\langle\vec{x}|n\rangle = \langle\vec{x}|n\rangle^*$.

Summarizing, we have, or may show,

$$(\vec{L})^T = -\vec{L},\ (\vec{S})^T = -\vec{S},\ (\vec{p})^T = -\vec{p},\ (\vec{x})^T = \vec{x}. \tag{12.50}$$

My treatment of time reversal as an *operation*, rather than an *operator*, is not standard. However, much of the Dirac notation formalism developed in earlier chapters must be modified or abandoned if we take the operator point of view. For example, the operator point of view has $\langle\alpha|(\Theta|\beta\rangle) \neq (\langle\alpha|\Theta)|\beta\rangle$, where Θ is the time-reversal *operator*. In addition, even though $(H)^T = H$ characterizes a time-reversal invariant Hamiltonian, unlike parity there is no "conservation of time-reversal quantum number." The operation point of view makes this explicit since there is no operator that commutes with the Hamiltonian!

It turns out that time-reversal invariance is very slightly broken in nature and was detected in 1998 in the neutral kaon* system. This violation is thought to arise from the weak interaction. There is also a potential T-violating term that can arise from the strong interaction (called the θ-term), but it is probably not phenomenologically relevant.

12.4 CHARGE CONJUGATION

Charge conjugation refers to the act of changing particles into antiparticles. We have not yet talked about antiparticles, which are particle states with all *additive* quantum numbers reversed in sign. Additive quantum numbers consist of electric charge, the various quark "flavors," such as strangeness, and the three types of lepton numbers. Calling this operator C, its affect on a state with additive quantum numbers A, B, C, would be

$$C\,|\,A,B,C\rangle = |-A,-B,-C\rangle. \tag{12.51}$$

Given that the electric charge, Q, is such a quantum number, one can immediately see that these operators have the property that they anticommute:

$$\{C,Q\} = 0. \tag{12.52}$$

This last equation implies that a state of nonzero charge cannot be an eigenstate of C. In general, this applies for any additive charge. In addition, applying charge conjugation twice gives back the same state, which says that

$$C^2 = I. \tag{12.53}$$

This equation implies that the eigenvalues of C, called *charge parity*, are ± 1, just like parity, for nondegenerate states with no additive quantum numbers. It also is a multiplicative quantum number. Particles that have no additive quantum numbers can be considered their own antiparticles. They are called *self-conjugate*.

One particle that has no additive charge is the photon. Changing the sign of the charge will change the sign of the photon field, A_μ. We learn in quantum field theory that this field may be used to create or destroy a photon. Thus, each additional photon in a given state changes the sign of the charge conjugation state. In terms of the creation and annihilation operators of Chapter 9, a two-photon state, for example, would be created by $|\,\vec{k},\vec{k}'\rangle = a^\dagger_{\vec{k}} a^\dagger_{\vec{k}'}\,|\,0\rangle$, where $C\,a^\dagger_{\vec{k}} C = -a^\dagger_{\vec{k}}$ and $|0\rangle$ is the vacuum state. Then $C\,|\,\vec{k},\vec{k}'\rangle = |\,\vec{k},\vec{k}'\rangle$, indicating a positive charge conjugation state. This quantum number will be conserved if C commutes with the electromagnetic Hamiltonian causing the decay, and thus the charge conjugation of the initial and finial states are the same. The neutral pion decay,

$$\pi^0 \rightarrow \gamma + \gamma, \tag{12.54}$$

---

* Neutral kaons are mesons involving a strange quark and a down antiquark , or its antiparticle ($\bar{K}^0$). The experimental evidence is in A. Angelopoulos, CPLEAR Collaboration, *Phys. Lett.* B **444**, 43 (1998).

TABLE 12.1 Discrete Transformation Summary

Parity: $\vec{J} \to \vec{J}$ ($\vec{J} = \vec{L}$ or $\vec{S}$), $\vec{p} \to -\vec{p}$, $\vec{x} \to -\vec{x}$.
Time reversal: $\vec{J} \to -\vec{J}$ ($\vec{J} = \vec{L}$ or $\vec{S}$), $\vec{p} \to -\vec{p}$, $\vec{x} \to \vec{x}$.
Charge conjugation: $\vec{J} \to \vec{J}$, $\vec{p} \to \vec{p}$, $\vec{x} \to \vec{x}$. All additive quantum numbers such as electric charge, strangeness, baryon number, and the various lepton numbers change sign.

implies that the neutral pion has a charge parity, $C_{\pi^0} = 1$. This particle is never seen to decay into three photons. The neutral pion state is formed from the strong interaction, and this is a very strong hint that the strong interactions also conserve charge conjugation. Other neutral states in which charge conjugation is a good quantum number are positronium ($e^+ e^-$) and so-called quarkonium states, such as $c\,\bar{c}$ (charm/anticharm). Charge conjugation symmetry is also "maximally" violated in the weak interactions, similar to parity.

Table 12.1 summarizes the transformation properties of states under the aforementioned various discrete operations. The individual discrete symmetries *C*, *P*, and *T*, as well as the bilinear products *CP*, *PT*, and *TC* are all violated in nature. However, as far as is known, the trilinear combination *CPT* (in any order) is conserved. This theoretical constraint means that the *T* violation implies *CP* violation. As far as explicit experimental demonstration is concerned, however, explicit *CP* violation was established in 1964, well before *T* violation.* Both were seen in the neutral kaon system.

12.5 PARTICLE PRIMER

In this section, I will introduce you to the elementary particles† of the so-called Standard Model (a "grotesquely modest name" in the words of F. Wilczek) of particle physics. The theory combining the electromagnetic and weak portions, or *electroweak* interactions, is called the Glashow–Salam–Weinberg (GSW) model, named after Sheldon Glashow, Abdus Salam, and Steven Weinberg, who shared the Nobel Prize in 1979 for its formulation.

Our world is remarkably and intricately made from a collection of 61 elementary particles, some stable from decay into other fundamental particles and some not. All of these particles are now known directly from experiment with one notable exception to be described later.

The particles of the Standard Model can be classified according to their types of interactions: the quarks feel the electromagnetic, "strong," and "weak" interactions, and the leptons participate in electromagnetic and

* J.H. Christenson, J.W. Cronin, V.L. Fitch, and R. Turlay, *Phys. Rev. Lett.* **13**, 138 (1964); A. Abashian et al., *Phys. Rev. Lett.* **13**, 243 (1964).

† What is an elementary particle? I define it colloquially to be an irreducible constituent of matter with a unique mass, lifetime, and set of quantum numbers, except I don't count spin. Massive particles with spin can, in principle, have their spin directions changed by appropriately continuous relativistic Lorentz transformations, which suggests to me that these should not be counted separately. However, I am on shakier ground not to consider the two allowed spin states of massless particles as separate since this argument no longer works, but I will consider and count the photon as a single particle, for example. If you wish, you may add 9 to the particle count of 61 to account for the additional massless photon and 8 gluons to reach 70 particles for the Standard Model.

TABLE 12.2 Quark Additive Quantum Numbers

Flavor	Charge (e)	Baryon Number (A)	C	S	T	B
u (up)	2/3	1/3	0	0	0	0
d (down)	−1/3	1/3	0	0	0	0
c (charm)	2/3	1/3	1	0	0	0
s (strange)	−1/3	1/3	0	−1	0	0
t (top)	2/3	1/3	0	0	1	0
b (bottom)	−1/3	1/3	0	0	0	−1

weak interactions. All three of the fundamental interactions in nature are types of *gauge theories* (see Section 12.6). In addition, there are the particle mediators of these interactions, the so-called *gauge bosons*. For a quick picture of the types of interactions in which these particles can participate, see the pictorial list of allowed particle vertices in Appendix D.

Let us begin first with the quarks. Quarks have spin 1/2 and come in six different strong interaction "flavors" that have come to be called

"up" (***u***),
"down" (***d***),
"charmed" (***c***),
"strange" (***s***),
"top" (***t***), and,
"bottom" (***b***).

These are presented along with their electric charges (in units of the proton's electric charge) and flavor charges in Table 12.2. (There are also antiquarks, $\bar{u}$, $\bar{d}$, $\bar{c}$, $\bar{s}$, $\bar{t}$, and $\bar{b}$ with the opposite color and additive quantum numbers of their particle partners.) The reason for the grouping of two flavors in the table is that each combination (u, d), (c, s), and (t, b) is considered a different "generation" or "family." Note from Table 12.2 that the electric charges in each generation are repeated. Each flavor of quark has a unique mass, but specifying their mass values is difficult because quarks are only seen in bound states; individual quarks, such as u or d, are never detected in the laboratory. As we will see when we get to Sections 12.8 and 12.10, many quantities actually depend on the energies at which they are measured. Thus, quark masses are quoted at agree-upon energy scales.

It turns out that the best way to determine these masses is to compare the theory from lattice QCD (mentioned in Section 7.3; more on this in Section 12.8) and other techniques, with experimental meson and baryon masses. A recent lattice compilation* gives an up-quark mass of 2.16 ± 0.11 MeV/c^2, a down-quark mass of 4.68 ± 0.16 MeV/c^2, and a strange quark mass of 93.8 ± 2.4 MeV/c^2, where the statistical and systematic uncertainties have been added in quadrature and all are quoted at an energy scale of 2 GeV. In addition, the charmed quark has a mass of 1.28 ± 0.025 GeV/c^2 and the bottom quark has a mass of 4.18 ± 0.03 GeV/c^2. The energy scale for the charm and bottom quark masses quoted is the mass itself. The top quark's mass is a whopping 173.21 ± 0.87 GeV/c^2, where again the

* S. Aoki at al., *Eur. Phys. J.* C **74**, 2890 (2014).

statistical and systematic uncertainties have been added. The last three mass results are from the encyclopedic *Review of Particle Physics (RPP)*.*

Each quark also has a *baryon number*, arbitrarily assigned as 1/3, so that three quark combinations, such as the proton or neutron, have a unit value. This quantum number is conserved in all particle interactions as far as is known. There are additional quark additive quantum numbers of C, S, T, and B, standing for the quark properties of charm, strange, top, and bottom, respectively.

Counting up the number of quarks, we have 6 (flavors) × 3 (colors) × 2 (particle/antiparticle) = 36 of the 61 Standard Model particles.

As I mentioned earlier, the group of particles known as leptons feel only the electromagnetic and weak forces. There are six of these particles, just as there are six flavors of quarks, and they also have spin 1/2. Also, just as the (u, d), (c, s), and (t, b) combination of quarks forms a different "generation" or "family," the leptons are similarly grouped, but of course their electric charges are different. There are electron and electron neutrinos, (e^-, ν_e), the muon and muon neutrino, (μ^-, ν_μ), and the tau and tau neutrino, (τ^-, ν_τ). The electron, muon, and tau have identical negative charges, whereas the neutrinos are neutral. And just like the quarks, the masses of each generation increase, $m_e < m_\mu < m_\tau$, with the possible exception of the neutrinos. (See comments in Section 12.9 on their masses.) A recent statement of the charged lepton masses in the *RPP* is $m_e = 0.510998928 \pm 0.000000011$ MeV/c^2, $m_\mu = 105.6583715 \pm 0.0000035$ MeV/c^2, and $m_\tau = 1776.86 \pm 0.12$ MeV/c^2. The neutrinos participate only in weak interactions.

All the leptons in Table 12.3 are fundamental (not decomposable into other particles), but not all of them are stable. For example, the muon and tau leptons decay. The primary decay mode for the muon is $\mu^- \rightarrow e^- \bar{\nu}_e \nu_\mu$. The lifetime associated with the muon is about 2.2×10^{-6} s. Notice that in this decay the lepton numbers defined in Table 12.3 are conserved. The muon has $L_\mu = 1$; this decays into particles having $L_e = 1$ (e^-), $L_e = -1$ $(\bar{\nu}_e)$, and $L_\mu = 1$ (ν_μ). It is now known from recent experiments involving neutrinos that the lepton flavor number is not conserved; this will be explained in more detail in Section 12.9.

The counting of leptons is 6 (types) × 2 (particle/antiparticle) = 12 of the 61.† We are up to 36 + 12 = 48 of 61 particles.

---

* This review is published in even years. See T. Tanabashi et al. (Particle Data Group), *Phys. Rev.* D **98**, 030001 (2018). See the listing reviews "Quark Masses" and "The Top Quark." Also see the PDG website at pdg.lbl.gov for updates.

† A species of spin 1/2 particle that is its own antiparticle (self-conjugate) is called a Majorana neutrino, named after E. Majorana. If the known neutrinos are Majorana, then the counting of leptons would be reduced by three. It is presently not known whether neutrinos are Dirac (particle, antiparticle distinct eigenstates) or Majorana. If neutrinos are Majorana, then one can not assign a lepton number to them, as in Table 12.3, since self-conjugate particles can carry no additive quantum numbers. It would then be the *interaction*, not the *particle*, that effectively conserves lepton number. However, there are expected to be small violations of lepton number for each flavor if neutrinos have a special type of mass called a "Majorana mass." The observation of so-called neutrinoless double beta decay, nucleus (Z,N) –> nucleus $(Z + 2, N - 2) + 2e^-$, (Z = number of protons, N = number of neutrons) would demonstrate that neutrinos are Majorana *and* have a nonzero Majorana mass.

TABLE 12.3 Lepton Additive Quantum Numbers

Lepton	Charge (e)	L_e	L_μ	L_τ
e^- (electron)	−1	1	0	0
ν_e (electron neutrino))	0	1	0	0
μ^- (muon)	−1	0	1	0
ν_μ (muon neutrino)	0	0	1	0
τ^- (tau)	−1	0	0	1
ν_τ (tau neutrino)	0	0	0	1

All forces in nature are mediated by other particles, known as gauge bosons. Bosons have integer spin and gauge bosons all have spin 1. The gauge boson in electrodynamics is the photon, the particle of light; the gauge boson in QCD is called the gluon, mentioned in Chapter 7 and briefly again in Chapter 11. Both these particles are massless. The gluon is confined inside strongly interacting particles called *hadrons*. There are eight gluons. The gauge bosons of the weak interactions are the charged $W^\pm$ and the neutral Z^0 bosons. Both of these particles are massive; in fact, the $W^\pm$ has a mass of about 86 times that of a proton (80.4 GeV/c^2 as opposed to.938 GeV/c^2), and the Z^0 has a mass of about 97 times a proton (91.2 GeV/c^2).

The counting of gauge bosons is 8 (gluons) + 2 ($W^\pm$) + 1 (photon) + 1 (Z^0) = 12. We are up to 60 of 61 particles.

The remaining Standard Model particle is called the *Higgs boson*. It is a spin zero neutral particle named after Peter Higgs of the University of Edinburgh. Early evidence for its existence was provided at a high-energy collider called the Tevatron at Fermilab. However, by the time this machine was decommissioned in September 2011, the evidence accumulated was only suggestive. Conclusive evidence was finally provided at the world's highest energy particle collider, the Large Hadron Collider (LHC) by two detector collaboration groups, ATLAS (*A Toroidal LHC ApparatuS*) and CMS (Compact Muon Solenoid), and was announced with great flourish on July 4, 2012. The LHC is the world's largest machine for any purpose. It is situated in a 27-kilometer long tunnel situated deep underneath the Franco-Swiss border near Geneva, Switzerland. It consists of thousands of superconducting magnets that are operated near absolute zero. It is usually configured to accommodate countercirculating proton beams, although it can also accelerate beams of heavy ions, such as lead or xenon. The ALICE (A Large Ion Collider Experiment) and LHCb (Large Hadron Collider-beauty) are two other detector collaborations that specialize in measurements of quark–gluon plasma properties and properties of beauty and charmed quarks, respectively. Statistical evidence for the Higgs particle was deduced by the ATLAS and CMS collaborations from protons with beam energies of 3.5 or 4.0 TeV (pronounced "teraelectronvolt"; 1 TeV = 10^{12} eV), making a total center of mass beam energy of 7.0 or 8.0 TeV. Accumulation of data at these energies occurred during the time period 2010–2013, and is called "Run 1". After a 2-year upgrade, the LHC was restarted with a total energy of 13 TeV in 2015. It is now (early 2019) undergoing another upgrade to 14 GeV and higher luminosity.

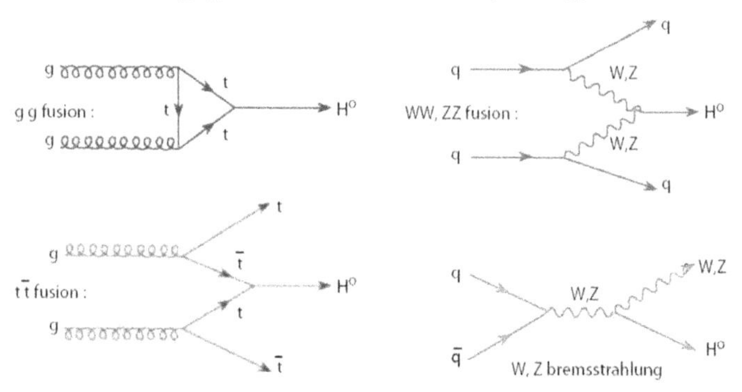

FIGURE 12.7 Production modes for the Higgs boson at the LHC. (From the ATLAS Collaboration.)

Figure 12.7 shows some of the LHC production mechanisms for the Higgs boson, H^0, in proton–proton collisions in Feynman diagram form (time running left to right). Although the gluons are confined in the protons, individual gluons from these protons may interact during the collision. These gluons can produce a single Higgs primarily through their coupling to top/antitop quarks, as shown in the gluon–gluon fusion and top–antitop fusion diagrams. Other quark/antiquark types may also contribute, but the top quark contributes the most because it is the most massive type. As shown at the top right of this figure, there is also a two-quark (or quark/antiquark) mechanism involving the fusion of W^+ and W^- particles or two Z^0 particles for producing a Higgs. Finally, a Higgs particle may be "radiated" in analogy to the way photons are produced in the charged particle process known as bremsstrahlung; this is shown in the bottom right figure. The gluon–gluon fusion mechanism is actually the primary one for Higgs production at the LHC.

Figure 12.8 shows a listing of measurements of the Higgs boson mass at the LHC. There are many decay modes for the Higgs, but they are in general very hard to detect. The two "cleanest" decay modes for the Higgs particle are two-Z^0 decay (each of which decays to two leptons) and two-photon decay. The first decay is denoted as $H \rightarrow ZZ \rightarrow 4l$ in the figure, and the second is listed as $H \rightarrow \gamma\gamma$. Each of these decays is actually quite rare, with the two-photon decays occurring only about 0.25% of the time, and the four-lepton decays even less. However, they can be detected above backgrounds. The results from the ATLAS and CMS collaborations are given separately for these two modes, then the collaboration results are averaged, first for the two modes separately, then for the combined data. The final measured mass of the Higgs is $125.09 \pm 0.24\ \text{GeV}/c^2$, an amazing result with 0.2% precision. Higgs couplings to other states as predicted by the Standard Model have also been verified to be correct within errors. This is crucial in identifying this as the Standard Model Higgs and setting limits on new interactions.

Detection of the Higgs particle has been the crowning achievement of experimental particle physics in the verification of the Standard Model. Many technical hurdles involving superconducting magnets, refrigeration, and data collection were solved, and the total number of scientists

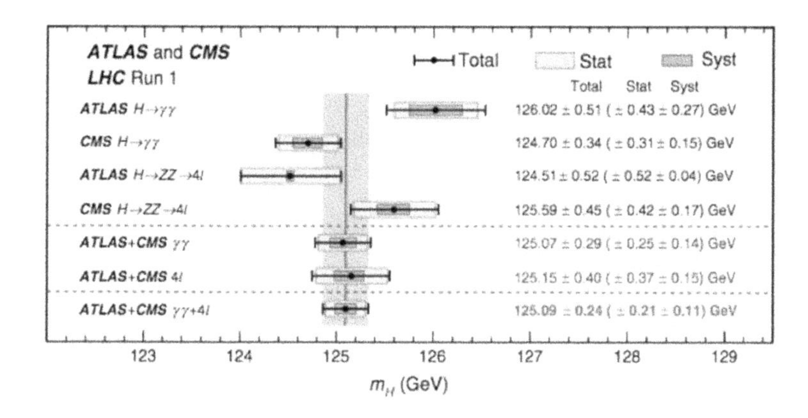

FIGURE 12.8 Combined ATLAS and CMS results on the Higgs mass (in GeV with $c = 1$) from Run 1 at the LHC. Both systematic and statistical error bars are shown for the decay modes. (From G. Gad et al., *Phys. Rev. Lett.* **114**, 191803 (2015). Used with permission.)

and engineers involved is over 10,000. The Nobel Prize in Physics for 2013 was awarded to François Englert and Peter Higgs for their theoretical work. We will have more to say about the Higgs boson in Section 12.9.

These are the 61 particles of the Standard Model of particle physics.

12.6 PARTICLE INTERACTIONS

Three of the four known fundamental forces in nature correspond to *gauge field theories*. The study and development of such theories has been a major focus for particle physics from its inception. The combining of quantum principles and relativity was a major theoretical undertaking, first successfully completed in the theory of quantum electrodynamics by Feynman, Schwinger, and Tomonaga, for which they received the 1965 Nobel Prize.

Classical electrodynamics is centered on the study of Maxwell's equations in various media and boundary conditions. These equations can be derived from the field Euler–Lagrange equations (Problem 12.6.1) using the "Lagrange density" expression,

$$\mathcal{L}_A \equiv -\frac{1}{8\pi}\left\{\begin{aligned}&\sum_i\left(-\frac{1}{c^2}\left(\frac{\partial A_i}{\partial t}\right)^2-(\nabla_i\Phi)^2-\frac{2}{c}\nabla_i\Phi\frac{\partial A_i}{\partial t}\right)\\&+2\sum_{i,j}(\nabla_iA_j\nabla_iA_j-\nabla_iA_j\nabla_jA_i)\end{aligned}\right\}, \tag{12.55}$$

where $\Phi(\vec{x},t)$ is the scalar potential, and $A_i(\vec{x},t)$ is the ith component of the vector potential. The derived Maxwell equations are unchanged under the substitutions,

$$\vec{A} \to \vec{A} + \vec{\nabla}\Lambda, \tag{12.56}$$

$$\Phi \to \Phi - \frac{1}{c}\frac{\partial \Lambda}{\partial t}, \tag{12.57}$$

where $\Lambda(\vec{x},t)$ is arbitrary but continuous. This is called *gauge invariance.* The "action" expression,

$$S_A \equiv c\int d^3x\int dt\,\mathcal{L}_A, \tag{12.58}$$

also shares this invariance.

To describe the electromagnetic interaction of a charged particle, one can add the additional particle action term

$$S_p \equiv \int d^3x\int dt\left[\psi^*(\vec{x},t)\left(-\frac{\hbar^2}{2m}\left(\vec{\nabla}-i\frac{q}{\hbar c}\vec{A}\right)^2 - i\hbar\left(\frac{\partial}{\partial t}+i\frac{q}{\hbar}\Phi\right)\right)\psi(\vec{x},t)\right], \tag{12.59}$$

where q is the particle's charge, to the above. The electromagnetic form of this action expression is dictated by gauge invariance. To get the associated wave equation, recall from Chapter 9 how a minimization procedure of such an expression resulted in the appropriate Schrödinger equation. We do the same here by varying the $\delta\psi^*(\vec{r},t)$ in S_p. In this case, we find

$$\left(-\frac{\hbar^2}{2m}\left(\vec{\nabla}-i\frac{q}{\hbar c}\vec{A}\right)^2 - i\hbar\left(\frac{\partial}{\partial t}+i\frac{q}{\hbar}\Phi\right)\right)\psi(\vec{x},t)=0, \tag{12.60}$$

which is the same as Equation 12.27, with $q\Phi = V(\vec{x},t)$. We now observe that S_p or Equation 12.60 is invariant under the combined substitutions Equations 12.56, 12.57, and

$$\psi(\vec{x},t)\rightarrow \exp\left(\frac{iq}{\hbar c}\Lambda(\vec{x},t)\right)\psi(\vec{x},t). \tag{12.61}$$

Equation 12.61 is called a "local gauge transformation," and the associated gauge group is called $U(1)$.

The three theories known as electromagnetism, weak interactions, and the strong force originate in gauge field theories based on the groups $U(1)$, $SU(2)$ (unitary two-by-two matrices with determinant one), and $SU(3)$ (unitary three-by-three matrices with determinant one), respectively. For the $SU(2)$ and $SU(3)$ cases, the form of the action expression involving the vector field, Equation 12.55, must be modified to maintain gauge invariance. This generalized action expression, postulated in 1954, is called the *Yang–Mills action,* named after C.N. Yang and R. Mills.

In the $SU(2)$ case, the fields Φ and A_i, as well as the arbitrary function $\Lambda(\vec{x},t)$, would all be space- and time-dependent $SU(2)$ matrices. Such matrices have three real, free parameters. The associated parameter space, or "manifold," can be thought of as locating the position of a point on a sphere in four dimensions (a "3-sphere"). Similarly, in the $SU(3)$ case the matrices have eight real parameters and the manifold can be thought of as locating a three-sphere at each point of a five-sphere (like a little flea on big flea). Since these are independent gauge groups, the continuous symmetry of the Standard Model is usually referred to as a $SU(3)\otimes SU(2)\otimes U(1)$, where the "$\otimes$" symbol symbolizes a direct product. However, it turns out the $SU(2)\otimes U(1)$ symmetry is "spontaneously broken" at low energies to

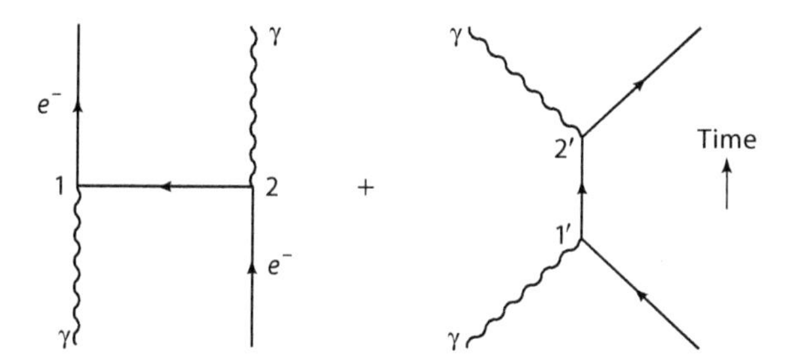

FIGURE 12.9 The lowest-order Feynman diagrams for Compton scattering. Time is flowing upward and the arrows show the flow of charge.

a different $U(1)$ that produces quantum electrodynamics, a concept I will explain in more detail in Section 12.9. Our world is $SU(3)_{\text{strong}} \otimes U(1)_{\text{e\&m}}$ symmetric.

It is now time to examine the three different interactions in the Standard Model.

12.7 QUANTUM ELECTRODYNAMICS

Quantum electrodynamics, or QED, is the best known and most studied aspect of the Standard Model. The basic interaction vertex in electrodynamics is shown in the Appendix D. Using Feynman diagrams, the structure and exact properties of many processes are straightforward to compute. For example, the lowest-order Feynman diagrams for $e^-\gamma \rightarrow e^-\gamma$, Compton scattering, are shown in Figure 12.9.

The vertices in Feynman diagrams can come in any time ordering, and the only thing invariant is the structure or topology of the diagram. For example, in the diagram on the left in Figure 12.9,

If "1" is before "2"; internal line is a positron, e^+.

If "2" is before "1"; internal line is an electron, e^-.

A similar statement can be made about the vertices 1′ and 2′ in the right diagram. The particles that are exchanged in these diagrams, the internal lines in the Feynman diagrams, are called "virtual" particles. We already know a little about the range and lifetime of such particles from the discussion in Chapter 7 regarding the pion. There we learned that the Heisenberg uncertainty principle determined these quantities (see Equations 7.112 and 7.113). It is similar here with the electron; it is exchanged over an approximate range of $\hbar/(m_e c) = 3.9 \times 10^{-11}$ cm, the reduced electron Compton wavelength, with a time uncertainty of $\hbar/(m_e c^2) = 1.3 \times 10^{-21}$ s.

The most fundamental and stringently tested discrete conservation law in nature is the conservation of electric charge. This conservation is intimately connected with the fact that the photon, the gauge boson of QED, is exactly massless, as we have already pointed out, which means its associated range from the uncertainty principle is infinite. Massless particles have an important property related to spin: they have only two degrees of freedom, rather than the expected $2s+1=3$ m_s values of $s=1$. It turns out that relativistic field theory requires the spin of massless particles to only

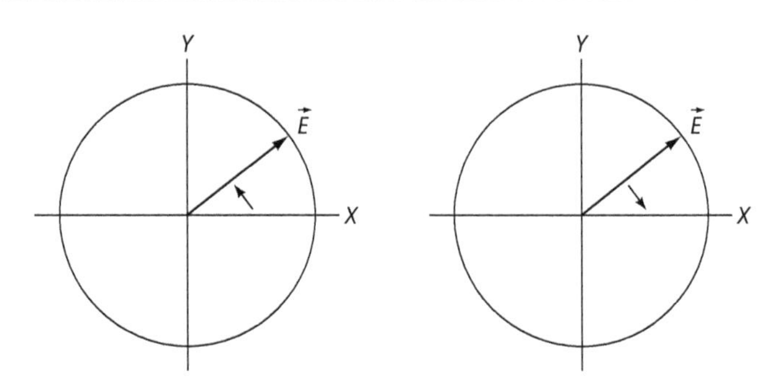

FIGURE 12.10 The two polarizations of a photon. The motion of the photon is out of the page.

point along the direct of motion, $\vec{p}$, or in the opposite direction. That is, the photon's *helicity*, $\vec{S} \cdot \vec{p} / (\hbar |\vec{p}|)$, can take only the values +1 or –1.* This particle property is familiar to us as the two polarizations of light we know about from classical electrodynamics. These two polarizations can be pictured as in Figure 12.10. Let us say that the motion of a photon is out of the page. The left diagram has an $\vec{E}$ field vector that is instantaneously rotating in a circular path in a counterclockwise fashion. This is called left circular polarization and is associated with positive helicity. The diagram on the right represents right circular polarization or negative helicity. Of course, we may use a *linear* polarization basis to describe light beams; these are just linear combinations of these two polarizations. We also learned in the earlier section on charge conjugation that the photon has no additive quantum numbers. This implies that there is no such thing as an antiphoton. The photon can be considered its own antiparticle.

Electrodynamics provides the attraction that makes atoms possible and essentially all the forces that make up the material around us. QED is so well understood that it can be used to test or probe other forces. For example, in the early 1960s a series of experiments, referred to as *deep inelastic scattering*, was done at the Stanford Linear Acceleration Center (SLAC) involving the scattering of electrons off protons and neutrons. ($e^- N \rightarrow e^- X$, where "X" is anything. This is called "inclusive" scattering.) These electromagnetic scatterings involve exchanged "virtual" photons between the incoming electrons and the interior of the nucleons. A careful study of the cross sections revealed that the electrons were hitting point objects within the nucleus and scattering through wide angles, just as Rutherford long ago observed large-angle scattering of β-particles (electrons) from atoms and deduced the existence of the atomic nucleus. These point objects are now identified as quarks. Although the quarks could hide within protons and neutrons, they could not escape detection by the infinite-ranged photon!

QED conserves parity, time-reversal invariance, and charge conjugation. A problem at the end of the chapter illustrates the use of parity in the properties and "selection rules" for electromagnetic transitions.

* The helicity of massless particles is also referred to as their "chirality."

12.8 QUANTUM CHROMODYNAMICS

The theory of the strong interactions is a gauge theory called *quantum chromodynamics*, QCD for short. The force is mediated by the gluons and is extremely strong compared to, for example, electromagnetism. This force is Coulombic $\sim 1/r^2$ at small distances and distance independent (~constant) at large distances. This constant force, or tension, equivalent to about 15 tons, is what is responsible for *quark confinement* (previously mentioned in the confined Coulombic model, Chapter 7, Section 7.7). This tension "string" can break at large enough distances, but instead of getting two pieces of string, another quark–antiquark pair emerges from the vacuum to terminate the string ends, much like what happens to a magnet that is broken in half.

Quarks are confined in hadrons. The energy scales and sizes of hadrons are determined by the strength and range of this force. They are the major building blocks, by mass, of our known physical world. Of these, the up and down quarks dominate; most of hadronic matter is made up of neutrons and protons, which are composed of two down quarks and an up quark, or by two up quarks and a down quark, respectively.* The small mass difference between up and down quarks, down being more massive than up, is responsible for neutrons being slightly more massive than protons. (There is also a smaller electromagnetic effect due to the charge on the proton, which raises its energy. This also affects other particles, like charged and neutral pions.)

Just as electrons carry electric charge, which is absolutely conserved, quarks carry three-color charges, which are also absolutely conserved (call them green, blue, and red). The name "color" is arbitrary but suggestive. Just as white light is composed of a mixture of all the colors in the spectrum, one can make white or colorless combinations of quarks. As far as is known, elementary hadrons can be made in only a few ways: making a quark–antiquark pair (mesons) by combining a color with its opposite, or by combining three different-colored quarks (baryons). There is also the possibility of making a bound state of a mixture of gluons and quark–antiquark pairs. These are called *glueballs*. (Glueballs are expected but not experimentally confirmed. "Hybrid" combinations of quarks and gluons are also possible.) In principle quark matter with any integral multiple of three quarks can also exist. Hadronic physics is the study of the properties and interactions of these composite particles.

As pointed out earlier, in all particle theories forces are mediated by other particles known as gauge bosons. The massless gauge boson in electrodynamics is the photon, the particle of light; the massless gauge boson in QCD is the gluon. Both particles have spin 1 (or helicity ± 1). However, there is one all-important difference between photons and gluons. Photons are electrically neutral, whereas gluons carry color charge. To see why, consider the diagram in Figure 12.11.

---

* There is also an admixture of all other flavors of quarks in protons and neutrons. Quantum field theory tells us that quark–antiquark pairs of all species are continuously appearing and disappearing in the vicinity of hadrons. These are called "sea quarks," and they can have dramatic physical effects, although these flavors are "hidden."

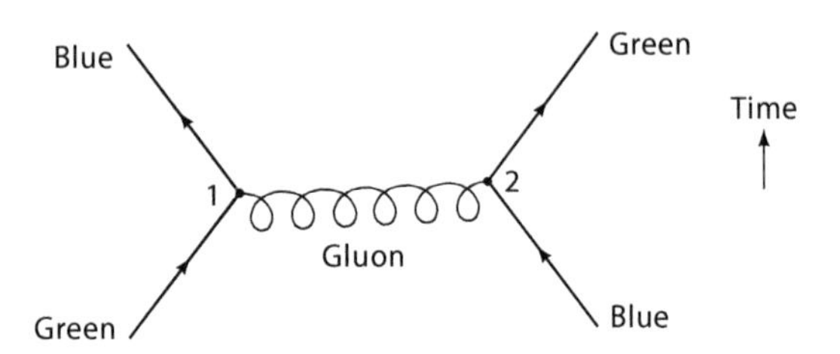

FIGURE 12.11 A hypothetical interaction between a blue and a green quark.

The gluon changes the quark color from one vertex to the other and thus carries color also. This gluon carries the color green/antiblue ($g\,\bar{b}$) if vertex 1 comes before vertex 2, and blue/antigreen ($b\,\bar{g}$) for the opposite ordering. (Remember Feynman diagrams are agnostic on time ordering of vertices, so this specification is really not necessary.) There are nine such combinations of three colors; however, the colorless combination, $g\bar{g} + b\bar{b} + r\bar{r}$, does not correspond to a particle, so there are only eight gluons. This is the same as the number of real parameters of the gauge group, which describes these interactions, $SU(3)$. The basic vertices for quarks and gluons are shown in Appendix D, where we see that unlike photons, gluons can couple to themselves.

The theory of strong interactions cannot be formulated and solved in the usual field theory way of calculating Feynman diagrams, as can QED. QED interactions are characterized by a particle–photon coupling strength given by the fine structure constant, $\alpha = e^2/(\hbar c) \approx 1/137$, which is small compared to one and allows the Feynman diagrams to be summed to very high order. The charge, e, plays the role of the coupling constant, and α the coupling strength. For QCD, the quark–gluon vertex has an analogous coupling constant, g_s, and a strength given by $\alpha_s = g_s^2/(\hbar c)$,* which is not small compared to one, making the series of Feynman diagrams divergent.† As we will see more explicitly in Section 12.10, this coupling strength is a function of the interaction energy and becomes smaller at high energies partly due to sea quark contributions (see a previous footnote) and partly due to the gluon self-coupling referred to earlier. Thus for high-energy interactions, Feynman diagrams are again useful. This property of the QCD coupling strength is known as *asymptotic freedom*. F. Wilczek, D. Gross, and D. Politzer shared the 2004 Nobel Prize in physics for its discovery.

Let us consider an unphysical world where all quarks are "heavy." Then their motions inside hadrons would be nonrelativistic. This is called the "quark model." This model is an incredibly good guide to overall properties, like magnetic moments and mass orderings. QCD does not change any of the flavors into any of the other flavors, so the u, d, etc. quantum numbers

* This is the connection in Gaussian units. Coupling constants and strengths are related by $\alpha_i = g_i^2/(4\pi\hbar c)$ in the Heaviside–Lorentz (Gaussian) units favored by particle physicists.

† Actually, the Feynman series in electrodynamics is also thought to be divergent. However, the QCD series diverges much faster.

are conserved in strong interactions. Since the u, d quarks are considered degenerate in mass in this model, this gives an effective two-valuedness (or, mathematically, the group $SU(2)$; see comments in Chapter 1, Section 1.19) in flavor space, which is called "isospin." The mathematics of isospin is exactly the same as the two-valuedness property of Chapter 1, and different isospins may be added exactly as we learned in Chapter 8. The fact that this quantum number is conserved in the strong interactions allows us to determine the ratio of certain matrix elements involved in particle interactions (see the problems at the end of the chapter).

Tables 12.4 through 12.6 give the flavor-spin wave functions of some of the lowest mass particles expected in the quark model along with their isospin classification. Also shown are the spin, parity (P), and charge parity (C) of these mesons as well as the spin and parity of the ground-state spin 1/2 "octet" baryons. Many more mesonic and baryonic states exist. Note that the various K mesons have a strangeness quantum number, which distinguishes them. If K^+ and K^0 are considered particles, the K^- and $\overline{K^0}$ are their antiparticles. The color wave functions of all mesons are multiplied by a factor of

$$\frac{1}{\sqrt{3}}(r\bar{r}+g\bar{g}+b\bar{b}),$$

whereas baryons have an antisymmetric color wave factor of

$$\frac{1}{\sqrt{6}}(rgb-rbg+gbr-grb+brg-dgr).$$

The technique known as *lattice QCD* (mentioned in Chapter 7, Section 7.3) solves the strong interaction theory directly on a space-time lattice of points and does not depend on the summation of Feynman diagrams. Note that the degrees of freedom in relativistic quantum field

TABLE 12.4 Pseudoscalar Mesons

	Mass (MeV/c²)	Isospin	Spin^{P,C}	Wave Function
π^+ (π^-)	139.6	$I=1, I_3=1\ (-1)$	0^-	$ud(d\bar{u})$
π^0	135.0	$I=1, I_3=0$	0^{-+}	$(u\bar{u}-d\bar{d})/\sqrt{2}$
K^+ (K^-)†	493.7	$I=\frac{1}{2}, I_3=\frac{1}{2}\ (-\frac{1}{2})$	0^-	$u\bar{s}\,(s\bar{u})$
K^0 ($\overline{K^0}$)††	497.6	$I=\frac{1}{2}, I_3=-\frac{1}{2}\ (\frac{1}{2})$	0^-	$d\bar{s}(s\bar{d})$
η	548	$I=0$	0^{-+}	$(2s\bar{s}-u\bar{u}-d\bar{d})/\sqrt{6}$
η'	958	$I=0$	0^{-+}	$(s\bar{s}-u\bar{u}-d\bar{d})/\sqrt{3}$

The nine strong interaction ground state pseudoscalar mesons. The implied antiparticle is given in parentheses. All have in addition singlet (antiparallel) spin wave functions.

† The true isospin particle partners are (K^+, K^0) and ($\overline{K^0}$, K^-), which have $I_3 = 1/2, -1/2$, respectively. Similarly for the vector meson states (K^{*+}, K^{*0}) and ($\overline{K^{*0}}$, K^{*-}).

†† K^0 and $\overline{K^0}$ are mixed by the weak interaction. See the discussion in Section 12.9.

TABLE 12.5 Vector Mesons

	Mass (MeV/c²)	Isospin	Spin^{P,C}	Wave Function
ρ^+ (ρ^-)	775	$I=1, I_3=1\ (-1)$	1^-	$u\bar{d}(d\bar{u})$
ρ^0	775	$I=1, I_3=0$	1^{--}	$(u\bar{u}-d\bar{d})/\sqrt{2}$
K^{*+} (K^{*-})	892	$I=\frac{1}{2}, I_3=\frac{1}{2}(-\frac{1}{2})$	1^-	$u\bar{s}(s\bar{u})$
K^{*0} ($\overline{K^{*0}}$)	896	$I=\frac{1}{2}, I_3=-\frac{1}{2}(\frac{1}{2})$	1^-	$d\bar{s}(s\bar{d})$
ω	783	$I=0$	1^{--}	$(u\bar{u}-d\bar{d})/\sqrt{2}$
ϕ	1019	$I=0$	1^{--}	$s\bar{s}$

The nine strong interaction ground state vector mesons. The implied antiparticle is given in parentheses. All have in addition triplet (parallel) spin wave functions.

TABLE 12.6 Baryon Octet Table

	Mass (MeV/c²)	Isospin	Spin^P	"Spin up" Wave Function (c.p. = cyclic permutations)
p	938.3	$I=\frac{1}{2}, I_3=\frac{1}{2}$	$\frac{1}{2}^+$	$uud(2\lvert{+}{+}{-}\rangle - \lvert{+}{-}{+}\rangle - \lvert{-}{+}{+}\rangle)/(3\sqrt{2}) + \text{c.p.}$
n	939.6	$I=\frac{1}{2}, I_3=-\frac{1}{2},$	$\frac{1}{2}^+$	$-ddu(2\lvert{+}{+}{-}\rangle - \lvert{+}{-}{+}\rangle - \lvert{-}{+}{+}\rangle)/(3\sqrt{2}) + \text{c.p.}$
Λ^0	1115	$I=0, I_3=0$	$\frac{1}{2}^+$	$(uds - dus)\,(\lvert{+}{-}{+}\rangle - \lvert{-}{+}{+}\rangle)/(2\sqrt{3}) + \text{c.p.}$
Σ^+	1189	$I=1, I_3=1$	$\frac{1}{2}^+$	$uus(2\lvert{+}{+}{-}\rangle - \lvert{+}{-}{+}\rangle - \lvert{-}{+}{+}\rangle)/(3\sqrt{2}) + \text{c.p.}$
Σ^0	1192	$I=1, I_3=0$	$\frac{1}{2}^+$	$(uds+dus)(2\lvert{+}{+}{-}\rangle - \lvert{+}{-}{+}\rangle - \lvert{-}{+}{+}\rangle)/6 + \text{c.p.}$
Σ^-	1197	$I=1, I_3=-1$	$\frac{1}{2}^+$	$dds(2\lvert{+}{+}{-}\rangle - \lvert{+}{-}{+}\rangle - \lvert{-}{+}{+}\rangle)/(3\sqrt{2}) + \text{c.p.}$
Ξ^0	1314	$I=\frac{1}{2}, I_3=+\frac{1}{2},$	$\frac{1}{2}^+$	$-ssu(2\lvert{+}{+}{-}\rangle - \lvert{+}{-}{+}\rangle - \lvert{-}{+}{+}\rangle)/(3\sqrt{2}) + \text{c.p.}$
Ξ^-	1321	$I=\frac{1}{2}, I_3=-\frac{1}{2},$	$\frac{1}{2}^+$	$-ssd(2\lvert{+}{+}{-}\rangle - \lvert{+}{-}{+}\rangle - \lvert{-}{+}{+}\rangle)/(3\sqrt{2}) + \text{c.p.}$

Note: For construction of these wave functions, see H. Georgi, *Lie Algebras in Particle Physics* (Benjamin/Cummings, 1982), Chapter XV. The "octet" of baryons has spin 1/2. There is also a ground state decuplet (meaning ten) of baryons, generally higher in mass, which has spin 3/2.

theories are not proportional to the number of particles involved (like in nonrelativistic quantum mechanics), but to the points of space and time themselves. Thus, the simplest field theory already has an infinite number of degrees of freedom! The infinitude of degrees of freedom causes many of the divergences encountered in field theories. To control this situation, imagine restricting the number of points to a finite "lattice" of space-time points—the quark degrees of freedom then reside at these points and interact with one another via the gluon fields, which can be considered connections or "links" between the points. It turns out that the entire theory may be formulated in terms of such site-specific quark fields and gluon links. This lattice theory may be put into computer language and numerical methods used to solve for specific quantities. One important numerical technique used in this program is called *Monte Carlo integration*. The quantities being measured in the lattice simulations can be expressed as an integral over all the degrees of freedom

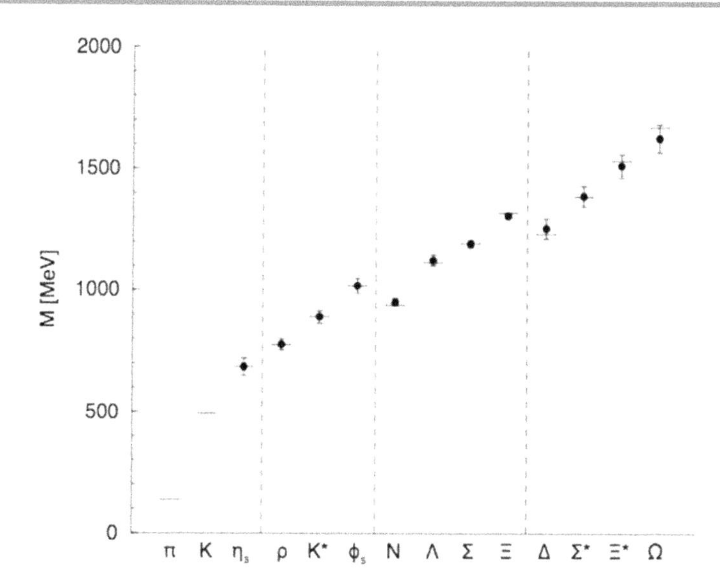

FIGURE 12.12 Masses of the light hadrons (in MeV with $c = 1$) as calculated by the QCDSF-UKQCD Collaboration using lattice QCD methods. Two of the particles, π and K, have no statistical error bars because they are used to fix parameters in the theory. The vertical lines divide the particles into the spin groups 0, 1, 1/2, and 3/2, reading from left to right. (From W. Bietenholz, *Int. J. Mod. Phys. E* **25**, 1642008 (2016). Used with permission.)

of the lattice system. The dimensions of this integral are now very large (billions in current simulations) but finite. The *Metropolis Monte Carlo algorithm* allows an estimation of these integrals by building in a Boltzmann factor and averaging over N likely sets of parameters ("configurations") of the integrand. It also allows an estimation of the likely variation in this value if the simulation was repeated many times. Thus, lattice simulations give values and error bars on physical quantities. The error bars may be reduced of course for a larger, more computer-time-intensive simulation, but they only fall like $1/\sqrt{N}$ in general. Please see Appendix E for a more in-depth explanation of these ideas and techniques in the context of the *Ising model* of spin interactions. Also see Appendix G for a short tutorial on the fascinating and timely subject of quantum computers.

Figure 12.12 shows the results of a recent lattice calculation of masses of 12 light mesons and baryons. The comparison with experiment is impressive.

QCD is thought to conserve parity and charge conjugation symmetries, although a small *CP*-violating (and therefore *T*-violating) term is not excluded, as mentioned earlier.

12.9 WEAK INTERACTIONS

We now come to the most interesting and complicated set of particle interactions, the weak interactions. This theory is also a gauge theory, like electromagnetism (QED) and strong interactions (QCD). In fact, it

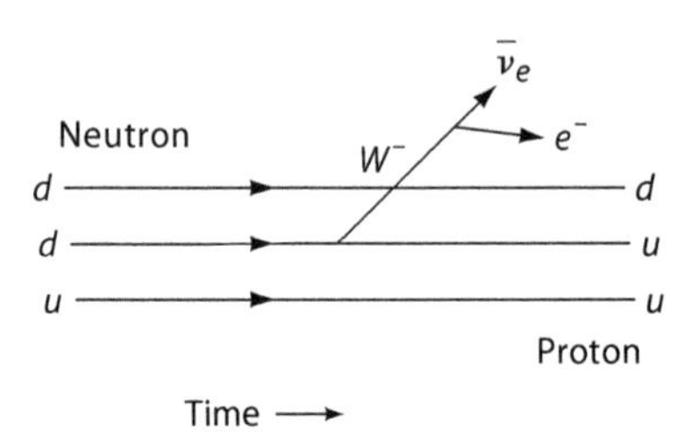

FIGURE 12.13 Stick diagram illustrating the decay of a neutron into a proton, a positron, and an antielectron neutrino. The W^- particle is "virtual." Time is flowing from left to right in this figure.

is considered unified with the electromagnetic interaction* and is often referred to as the weak-electromagnetic gauge theory. However, because the forces are mediated by massive particles, the $W^{\pm}$ and the Z^0, the force is extremely short ranged due to Heisenberg's uncertainty principle. Although the weak interaction is a gauge theory, it has no associated, exactly conserved quantities like electric charge or color charge. The reason for this is quite subtle. We will get back to this point momentarily.

The weak interactions involve both the quarks and leptons, and cause many types of decays of hadronic states. For example, consider the decay $n \to pe^- \bar{\nu}_e$, as depicted in Figure 12.13. Here we see a neutron entering from the left and a proton, an electron, and an antielectron neutrino exiting on the right. This occurs because of the coupling of the W^- particle to both quarks and leptons. This is called a hadronic "charged current" interaction because of the difference in the charge of the *u*, *d* particles at the vertex. The W^- particle in this diagram is virtual, the other particles (proton, neutron, electron, and antielectron neutrino) can be detected in the laboratory. The effects of the Z^0 particle, on the other hand, are much more subtle. We learned earlier that the photon has no additive quantum numbers. The Z^0 is also such a particle–antiparticle combination. This means it is present wherever photons are created or destroyed, but because of its great mass and instability, its range and possible detection are extremely limited.

Although both quarks and leptons participate in the weak interactions, they do so very differently. The quarks are of course all massive, but the neutrinos, for most intents and purposes, are massless. Remember our discussion of particle helicity when discussing photons. For a massless particle, whether it is a fermion or a boson (excluding spin 0), the only allowed physical states have their spin directed either along or antiparallel to the direction of motion of the particle. This is very different from an electron, which also has only two spin degrees of freedom, but these can point "up" or "down" relative to *any* coordinate axis. It turns out that all neutrinos (electron, muon, and tau) participate in interactions as if they were completely left-handed, that is, their spin is pointed antiparallel to their direction of motion. Antineutrinos are right-handed. This association of left-handedness with neutrinos and right-handedness with

* Some would say "mixed" instead of "unified" since there are two different coupling strengths involved, which, like α_s, are functions of energy. More on this in Section 12.11.

antineutrinos is intrinsically and maximally parity-violating. In fact, the weak interactions are known to violate all of the discrete symmetries we learned about at the beginning of this chapter, but strangely, very nearly preserve the combination *CP*.

A new dimensionless coupling strength characterizes the weak interactions analogous to both the electromagnetic, $\alpha = e^2/(\hbar c)$, and strong coupling constant, $\alpha_s = g_s^2/(\hbar c)$. As we will see shortly, its value turns out to be only slightly smaller than the electromagnetic coupling constant in typical low-energy interactions such as nuclear beta decay. The unification or mixing inherent in the GSW electroweak theory explains the similarity of the coupling constants. However, the Heisenberg energy–time uncertainty principle states that for a particle of the mass of a $W^{\pm}$ particle ($M_W = 80.4$ GeV/c^2) interacting as in Figure 12.13, the range will be the associated reduced Compton wavelength, $\hbar/(M_W c) = 0.0025$ fm, a tiny distance on the size of a proton (~1 fm). Thus, it turns out that the weak interactions are weak not because of an intrinsically smaller coupling constant but because the range of the interaction is extremely limited. The much smaller scales require much smaller interaction decay rates and cross sections.

How did the $W^{\pm}$ and Z^0 particles get to be so massive if the weak interactions are just another gauge theory? It is because of an extremely subtle field theory effect known as *spontaneous symmetry breaking*. This effect depends on the existence of the Higgs boson, which is another of the particles without any additive quantum numbers. It is thought that as the temperature of the universe cooled, the Higgs particle scalar field, ϕ, shifted its value: $\phi \rightarrow \phi' + v$, where v is just a number, called the vacuum expectation value (VEV). This is just like the spontaneous formation of little magnetic domains where the direction of the magnetic fields of the atoms is fixed as the temperature of a magnet cools. However, not only does the Higgs field (which is originally a doublet of complex fields having a total of four real components) pick out a direction in its isospin space like a magnet, but also picks up a VEV as it cools. This *Higgs mechanism** is thought to give rise to the masses of the $W^{\pm}$ and Z^0 particles as well as to all the quarks and leptons, via their original couplings (vertices) to the Higgs. In fact, in one of the most colorful phrases in particle physics, the $W^{\pm}$ and Z^0, which were originally massless and therefore had only two degrees of freedom (two helicities), are said to "eat" the three lost degrees of freedom of the Higgs, acquiring the correct three components for massive spin 1 particles. The leftover massive particle is the physical Higgs particle. This mechanism is also thought to give rise to the masses of the quarks and leptons in the Standard Model. Particle physicists would say that the original $SU(2) \otimes U(1)$ symmetry has been spontaneously broken to a shifted $U(1)$ symmetry, which is our old friend, electromagnetism. Thus, the $SU(2)$ symmetry, which would have given rise to conserved quantities, is broken.

Another subtle but spectacular effect in weak interactions is flavor mixing. The weak and strong interactions deal with particle flavor

---

* Also proposed by other physicists, in addition to P. Higgs, including R. Brout, F. Englert, G. Guralnik, C. Hagen, and T. Kibble.

differently. For the quark doublets, we saw earlier that we had the flavor groupings, relevant to the strong interactions:

$$\begin{pmatrix} u \\ d \end{pmatrix}, \begin{pmatrix} c \\ s \end{pmatrix}, \begin{pmatrix} t \\ b \end{pmatrix}. \tag{12.62}$$

For the weak interactions, the flavor groupings are different:

$$\begin{pmatrix} u \\ d' \end{pmatrix}, \begin{pmatrix} c \\ s' \end{pmatrix}, \begin{pmatrix} t \\ b' \end{pmatrix}. \tag{12.63}$$

The matrix that connects these particles is called the Cabibbo–Kobayashi–Maskawa, or *CKM matrix*. It is unitary and three dimensional. That is, one has

$$\begin{pmatrix} d' \\ s' \\ b' \end{pmatrix} = \begin{pmatrix} V_{ud} & V_{us} & V_{ub} \\ V_{cd} & V_{cs} & V_{cb} \\ V_{td} & V_{ts} & V_{tb} \end{pmatrix} \begin{pmatrix} d \\ s \\ b \end{pmatrix}. \tag{12.64}$$

The matrix element V_{ud} in the context of Equation 12.63 for example measures the coupling of u quarks to d quarks, etc. The priming of only the lower components, d, s, and b is just a choice by convention and one could instead have considered the u, c, and t the mixed states.

One effect of this flavor mixing was mentioned in passing in a footnote to Table 12.4. It is the weak interaction mixing of the strong interaction particles K^0 and $\overline{K^0}$, which are particle–antiparticle states with opposite strangeness. These new mixed states, which are not a particle–antiparticle pair, are designated K_L and K_S denoting their decay lifetimes, "L" for long lifetime and "S" for short. The new combinations are almost *CP* eigenstates with values −1 and 1, respectively, but are slightly mixed here also, indicating *CP* violation. They can be thought of as "oscillating" in time between the original K^0 and $\overline{K^0}$. The K_L and K_S, not the K^0 and $\overline{K^0}$, should be considered the "real" particles since they have the unique masses and lifetimes. Another effect of this weak flavor mixing on K-meson particle decays is discussed in Appendix F. The *CP* violation seen in K-mesons is a nonrequired but allowed consequence of having three generations of quarks; a two-generational world would not have this source of *CP* violation. There are also many other consequences of flavor mixing that are too involved to discuss in this short overview. The Standard Model offers no further explanation for the values of the parameters in the CKM matrix.

It has been established that the same sort of flavor mixing and oscillation occurs for neutrinos as well.* This implies that the various lepton numbers listed in Table 12.3 are not conserved across lepton flavor and are, therefore, mixed. The 3×3 neutrino mixing matrix is named after B. Pontecorvo, Z. Maki, M. Nakawaga, and S. Sakata, was introduced in 1962, and is called the *PMNS matrix*. It has the exact same form as Equation 12.64, where

* See K. Nakamura, S. Petcov, and U. Tokyo, "Neutrino Mass, Mixing, and Oscillations" (Particle Data Group) for background and details.

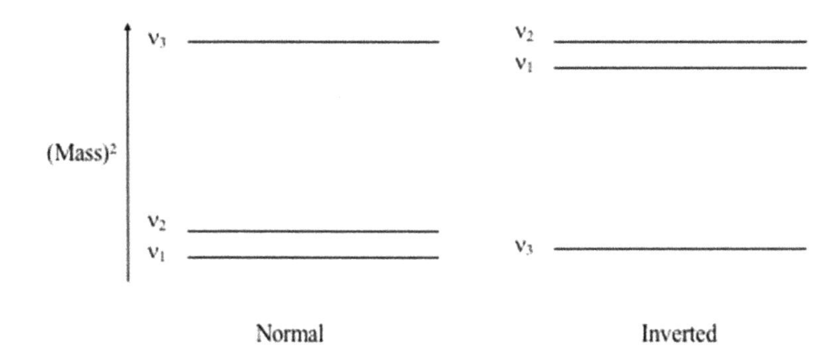

FIGURE 12.14 The two possible mass eigenstate ($\nu_{1,2,3}$) neutrino squared mass spectrums. If the masses of these three types are designated $m_{1,2,3}$, one has $m_2^2 - m_1^2 \simeq 7.4\times10^{-5}\text{eV}^2$ and $|m_3^2 - (m_1^2 + m_2^2)/2| \simeq 2.5\times10^{-3}\text{eV}^2$. The fact that only the magnitude of this last quantity is known allows the inverted scenario.

reference to mixed d, s, and b quarks are replaced with e, μ, and τ left-handed neutrinos; the mass states d', s', and b' are similarly replaced with left-handed ν_1, ν_2, and ν_3. (A naming opportunity!) Unlike the quark flavor mixing via the CKM matrix where the diagonal elements are close to one, the mixing angles for neutrinos are large, which means the generations are strongly mixed. Figure 12.14 shows the possible neutrino mass hierarchy scenarios, but the overall mass scales are uncertain. From astrophysical constraints, one only knows that $\sum_{j=1,2,3} m_j \leq 0.5\text{eV}$. There could even be more than three neutrinos! It is also not known whether the massive neutrinos are Dirac or Majorana (see footnote in Section 12.5) type.

There are many reasons to believe that the Standard Model of particle physics is not the final theory in particle physics, but simply another step toward a more fundamental theory. One potent motivation is the unexplained proliferation of parameters in these matrices and other parameters. Other reasons to believe that a more fundamental theory exists are the wildly different ("unnatural" is a popular physics term) energy scales in particle physics, called the "hierarchy problem," as well as the fact that gravity, the first force to actually be characterized mathematically, is not included in the theory. There is also the question of how the so-called *dark matter* and *dark energy* observed or inferred in astronomy fits in.*

Some of the solutions to these problems may be found in the subjects of supersymmetry and superstrings, which I will now describe.

12.10 SUPERSYMMETRY

Supersymmetry (often personified as SUSY) was developed as a theoretical idea, largely for motivations such as the above, as well as mathematical

* Unfortunately, we do not have the time to go into these astronomical/physical subjects here. An intelligent layman's presentation, also listed in Section 12.13, is given in P. Gagnon, *Who Cares About Particle Physics?* (Oxford University Press, 2016).

beauty, rather than direct experimental input.* Basically, supersymmetry posits that the elementary bosons and fermions are paired in such a way that their degrees of freedom are equal and can be interchanged under a continuous symmetry operation. For example, paired with spin-1/2 quarks are twice as many spin 0 (scalar) "squarks." If the supersymmetry were exact, the quark and squark masses would also match. Clearly, this is not the case and supersymmetry must be a broken symmetry in nature. The supersymmetric partners of the known Standard Model particles, contained in the so-called *Minimal Supersymmetric Standard Model* (MSSM), proposed in 1981 by S. Dimopoulos and H. Georgi, must generally have higher masses, and the search for these is another major motivation in experimental high-energy physics. The lightest supersymmetric particle, possibly a new type of neutrino, is a leading candidate for the universe's dark matter. The other supersymmetric partners to our Standard Model world will hopefully not be too heavy to be detected at the LHC. Indeed, it is only if the supersymmetric partners of the ordinary particles are not too heavy that one can use supersymmetry to convincingly solve the hierarchy problem.

The supersymmetry in the MSSM is called "global," which denotes a symmetry applicable at all space-time points. If the supersymmetry in the theory is made "local," or applicable at individual points in space-time, it contains Einsteinian gravity as a necessary consequence, and the theory is referred to as *supergravity*. The simplest such theory involves a massless spin 2 boson known as the *graviton* and a massless spin 3/2 particle known as a *gravitino*. Many supergravity theories are consistent only in 10 space + 1 time dimensions.

There are many adjustable parameters in the MSSM, which seem to make the arbitrariness found in the Standard Model even worse than before. However, an early circumstantial but convincing piece of evidence that the MSSM is a theory in the right direction came from the associated interaction or coupling strengths.† The interaction strengths for strong, weak, and electric interactions change or "run" logarithmically with the interaction energy (which I will denote as "μ"). This circumstance was already discussed in the context of the strong interaction, and the associated reduction of the coupling at high energies was referred to as "asymptotic freedom."

One of the aspirations of particle physicists is to unify the forces of nature into a simpler, but more comprehensive theory. Figure 12.15 shows that for the Standard Model such a unification of coupling strengths does not occur at high energies for the simplest unifying group, *SU*(5). The scale in this figure is logarithmic, and so many orders of magnitude in energy are involved in the extrapolation. Actually, it is the inverse coupling strengths ($\alpha_{1,2,3}^{-1}(\mu)$), which are shown in Figure 12.15. $\alpha_3(\mu) = \alpha_s(\mu)$ in this figure refers to the strong interaction and has an approximate value

* Supersymmetry was proposed, in the context of hadronic physics, in 1966 by H. Miyazawa. It was independently rediscovered by J. Gervai and B. Sakita (1971), Y. Golfand and E. Likhtman (1971), D. Volkov and V. Akulov (1972), and J. Wess and B. Zumino (1974).

† S. Dimopoulos, S. Raby, and F. Wilczek, *Phys. Rev. D* **24**, 1681 (1981); U. Amaldi, W. de Boer, and H. Fürstenau, *Phys. Lett.* B **260**, 447 (1991).

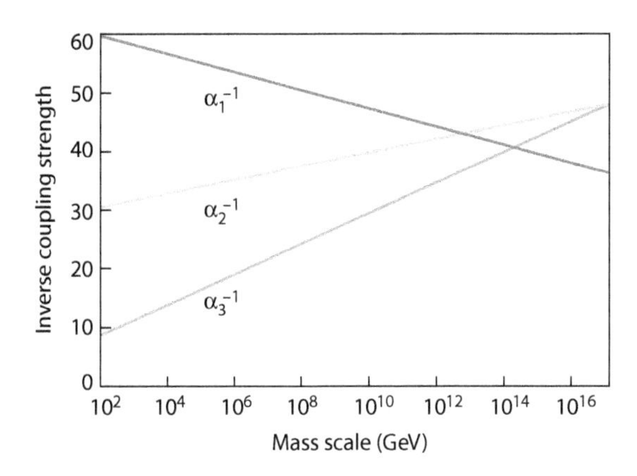

FIGURE 12.15 Energy extrapolation of the Standard Model inverse coupling strengths. The three values clearly fail to meet at a point at high energies. (Based on S. Dimopoulos, S.A. Raby, and F. Wilczek, *Physics Today* **44** (10), 25 (1991). Used with permission.)

of 0.1181 ± 0.0011 at $\mu = M_Z c^2$, where M_Z is the Z^0 boson mass.* $\alpha_1(\mu)$ and $\alpha_2(\mu)$ are the other two independent running coupling constants in the Standard Model for the groups $U(1)$ and $SU(2)$, respectively. At low energies in this graph these are related to the running electromagnetic coupling strength, $\alpha_{EM}(M_z c^2)$ ($\alpha_{EM}(M_Z c^2) \approx 1/128$), by

$$\alpha_1 \equiv \frac{5}{3}\frac{g'^2}{\hbar c} = \frac{5}{3}\frac{\alpha_{EM}}{\cos^2\theta_W}, \tag{12.65}$$

$$\alpha_2 \equiv \frac{g^2}{\hbar c} = \frac{\alpha_{EM}}{\sin^2\theta_W}, \tag{12.66}$$

where the $U(1)$ and $SU(2)$ Gaussian coupling constants, g and g', are related by

$$g' = g \tan\theta_W. \tag{12.67}$$

θ_W is called the weak or Weinberg mixing angle, which has been measured as $\sin^2\theta_W(M_Z C^2) = 0.23122 \pm 0.00004$. (The factor 5/3 in Equation 12.65 is the square of a generalized Clebsch–Gordan coefficient for the unifying group $SU(5)$.) The size of the experimental errors in the extrapolations is reflected in the width of the lines in Figure 12.15.

On the other hand, with SUSY minimal $SU(5)$ unification works! See Figure 12.16. One can even tell roughly where the supersymmetry masses "turn on" by a best-fit analysis assuming unification. This turns out to be about 1 TeV, which is good news for experimentalists, and supports the idea that the SUSY breaking scale is not extremely large compared to the Standard Model mass scales. The unification energy turns out to be about 10^{16} GeV, three orders of magnitude smaller than the scale associated with gravity, 10^{19} GeV, called the Planck scale.† Presumably, a single *grand*

* C. Tanabashi et al. (Particle Data Group), *Phys. Rev.* D **98**, 030001 (2018).

† $E_{\text{Planck}} \equiv \sqrt{\hbar c^5/G} \approx 122 \times 10^{19}\,\text{GeV}$, where G is Newton's gravitational constant.

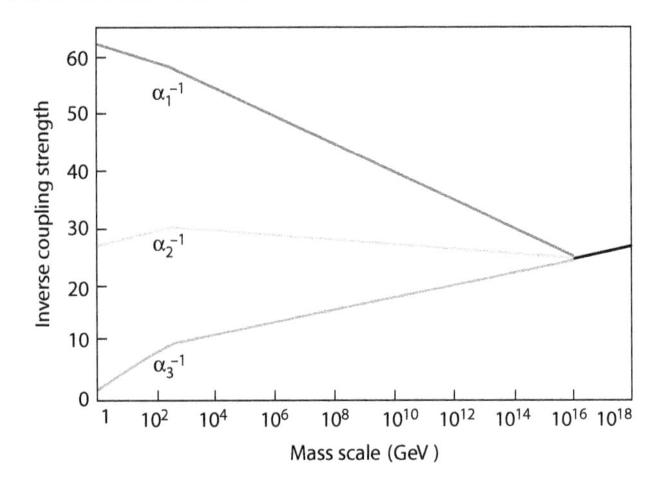

FIGURE 12.16 Energy extrapolation of the Minimal Supersymmetric Standard Model inverse coupling strengths, given the supersymmetric particles appearing at roughly 1 TeV. The three values unify at approximately 10^{16} GeV. (Based on S. Dimopoulos, S.A. Raby, and F. Wilczek, *Physics Today* **44** (10), 25 (1991). Used with permission.)

unified theory (GUT) exists between these energy scales. Unification of the other forces with gravity would then occur at the Planck scale. There turns out to be other nonsupersymmetric ways of arranging the unification of the coupling constants, but they are not nearly as natural as MSSM supersymmetry. Supersymmetry could also help solve the dark matter problem through the existence of a lightest supersymmetric particle, or LSP, which would be absolutely stable and would provide an undetectable mass background to the universe.

There is a price to be paid for supersymmetry: It is a theory with about 100 adjustable parameters. However, now that the Higgs mass is known precisely, it is more constrained than before. In addition, there are constraints from the measurements being done at the LHC in searching for squarks and other supersymmetric particles, as well as measuring Higgs decay channels more accurately. The highest priority mission for the LHC now is to continue this search. The hope is that a 14 TeV proton–proton machine will develop enough energy to discover it; if not, perhaps the discovery will have to wait for a FCC (Future Circular Collider) or an ILC (International Linear Collider).

12.11 SUPERSTRINGS

The frontier of high-energy theoretical physics has for some time been centered on the idea that the fundamental entities in nature that carry mass and energy are not point particles but small relativistic strings. The generic name for the field is "string theory." String theories that contain fermions are automatically supersymmetric, so experimental establishment of supersymmetry would help validate superstring theory. To state that such theories are fascinating and mathematically alluring would be a severe understatement. I can only provide a short, historically based introduction to some of the main string concepts here. Please see the lists of particle and string books that follow for more in-depth discussions.

The history of string theory is fascinating as well. The beginnings of string theory lie in the 1960s attempts to explain the many strong interaction resonances. Quarks were in the mix of theoretical ideas then, but there was resistance to the idea of fractionally charged particles, which had not been observed directly. More popular were the bootstrap models, which grew out of the S-matrix paradigm begun by W. Heisenberg (see Section 11.14). A major advance in this campaign was the theoretical construction of a scattering amplitude leading to the so-called *dual resonance model* by G. Veneziano in 1968. It was soon realized by Y. Nambu, H. Nielson, and L. Susskind that this model, which actually did not fit experimental data very well, really described an interacting theory of *closed strings.* Closed strings are string loops; *open strings* have two ends. The vibrational states of these objects appear to give rise to different "particles" at energies low compared to the string scale. This research took an unexpected turn in 1974 when J. Schwartz and, independently, T. Yoneya suggested that strings were better suited to a description of gravitation and the associated graviton particle than strong interactions. One attractive feature of such stringy gravitational theories is the freedom from infinities that plague traditional point-particle gravitational theories. This is due to the fact that the interactions of fundamental point particles take place at a point, as in Figure 12.17a, resulting in an infinite momentum transfer from the uncertainty principle. On the other hand, a string-like interaction, either open as in Figure 12.17b or closed, Figure 12.17c, does not have a localizable point vertex and avoids the infinities.

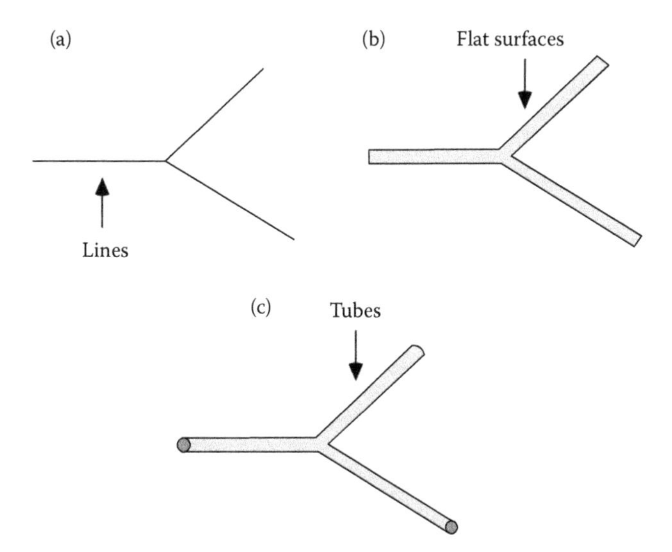

FIGURE 12.17 (a) A point-particle vertex; (b) an open string vertex; (c) a closed string vertex.

String theory was largely ignored during the rest of the 1970s, but its practitioners did not give up. Among these were M. Green and J. Schwartz. It was realized that these new theories involving fermions were only consistent in 10 space-time dimensions; this leaves six "extra" space dimensions. The extra dimensions are assumed to be either large but unobservable, or small and *compactified*. Compactification had been introduced into physics earlier in the *Kaluza–Klein model*, which was a five-dimensional unified theory of gravity and electromagnetism put forward by T. Kaluza and O. Klein in 1921.

The neglect of string theory ended in the 1980s. The "first superstring revolution" began in 1984 when Green and Schwartz showed that field theory *anomalies** canceled out in one type of string theory, resulting in consistent theories in higher dimensions. Other important developments followed in the next several years, and many physicists began to work on string theories. The "second superstring revolution" occurred in the mid-1990s when E. Witten and others found strong evidence that the known string theories were different limits of a grander theory, termed *M-theory*. The "*M*" stands for "mystery" or "membrane" or "matrix," depending on one's point of view. Such a theory is thought to be 11-dimensional, like supergravity, and requires the existence of other objects generically called *p-branes*, and a class of ending objects subject to certain ("Dirichlet") boundary conditions called *D-branes*. (A $p=0$ brane is a point, a $p=1$ brane is a line or string, a $p=2$ brane is a surface or membrane, etc.) Such objects have been useful in the formation and exploration of cosmological models, known as *brane worlds*. In the decade of the 2000s it was generally realized that there are many more discretely different string vacuum states in M-theory than previously thought, giving rise to different theories, on the order of 10^{500} or some such intimidating number. This has given rise to the contemplation of a vast string theory "landscape"† of possible theoretical models.

The initial objective of these theoretical endeavors was the unification of the known forces and, hopefully, an explanation of the parameters found in the Standard Model. A transitory belief developed after the first superstring revolution that there were only five consistent 10-dimensional superstring theories. It is rather ironic and disappointing that this line of thought has instead led to a myriad of possible consistent theories, the opposite of what was hoped for. Nevertheless, this myriad of possibilities may not necessarily be a bad thing. New phenomena and symmetries could arise from experiment that could lead to a winnowing of parameters. Clarifications of theoretical structures will occur. More twists and turns in string theory (pun intended) are certain.

---

* Anomalies are violations of identities involving relativistic space-time currents that occur when perturbative corrections are taken into account. These only vanish if the right particle content is present in the theory. For example, anomaly cancellation implies that equal numbers of generations of quarks and leptons exist.

† A concept introduced to string theory by L. Susskind.

12.12 POSTLUDE

I end this text on a positive note of progress and possibility. Great strides have already been taken in the search for a self-consistent, comprehensive theory of nature. Theoreticians and experimentalists in physics must both participate in this endeavor if we are to make further progress. Many have gone before and revolutionized our understanding of nature, and many are persisting now. It is truly a privilege to be a physicist!*

> Nothing is too wonderful to be true if it be consistent with the laws of nature.
>
> **—Michael Faraday**

12.13 HELPFUL BOOKS ON PARTICLE AND STRING PHYSICS

There are many good books explaining recent ideas and discoveries in particle and string physics. The following list centers on books by experts that are introductory and semipopular in nature, while still maintaining a high degree of accuracy in their discussions.

F. Close, *Neutrino* (Oxford University Press, 2011).

F. Close, *Particle Physics: A Very Short Introduction* (Oxford University Press, 2004).

P. Gagnon, *Who Cares About Particle Physics?* (Oxford University Press, 2016).

B. Greene, The Hidden Reality: Parallel Universes and the Deep Laws of the Cosmos (Vintage, 2011).

N. Mee, *Higgs Force: The Symmetry-Breaking Force that Makes the World an Interesting Place* (Lutterworth Press, 2012).

B.R. Martin, *Particle Physics: A Beginner's Guide* (Oneworld Publications, 2011).

B.A. Schumm, *Deep Down Things: The Breathtaking Beauty of Particle Physics* (Johns Hopkins University Press, 2004).

B. Still, *Particle Physics Brick by Brick: Atomic and Subatomic Physics Explained* (Firefly Books, 2018).

L. Susskind, *The Cosmic Landscape* (Little, Brown and Company, 2006).

M. Veltman, *Facts and Mysteries in Elementary Particle Physics* (World Scientific, 2003).

S. Weinberg, *The First Three Minutes: A Modern View of the Origin of the Universe*, 2nd ed. (Basic Books, 1993).

The following lists some more advanced books on particle and string physics.

I.J.R. Aitchison, *Supersymmetry in Particle Physics: An Elementary Introduction* (Cambridge University Press, 2007).

K. Becker, M. Becker, and J. Schwartz, *String Theory and M-Theory: A Modern Introduction* (Cambridge University Press, 2007).

M. Creutz, *Quarks, Gluons and Lattices* (Cambridge University Press, 1984).

D.J. Griffiths, *Introduction to Elementary Particles*, 2nd ed. (Wiley-VCH, 2008).

E. Henley and A. Garcia, *Subatomic Physics*, 3rd ed. (World Scientific, 2007).

* This phrase echoes the title of the book by Victor Weisskopf, *The Privilege of Being a Physicist* (W.H. Freeman, 1990).

B.R. Martin, *Nuclear and Particle Physics: An Introduction*, 2nd ed. (John Wiley & Sons, 2009).

B.R. Martin and G. Shaw, *Particle Physics (Manchester Physics Series)*, 4th ed. (John Wiley & Sons, 2017).

V.P. Nair, *Concepts in Particle Physics: A Concise Introduction to the Standard Model* (World Scientific, 2018).

M. Thomson, *Modern Particle Physics* (Cambridge University Press, 2013).

V. Schomerus, *A Primer on String Theory* (Cambridge University Press, 2017).

B. Zwiebach, *A First Course in String Theory*, 2nd ed. (Cambridge University Press, 2009).

PROBLEMS

12.2.1 Starting with the definition of the determinant of a 3×3 matrix,

$$(\det a) = \sum_{\ell,m,n} \varepsilon_{\ell mn} a_{1\ell} a_{2m} a_{3n},$$

show that

$$\sum_{j,k} \varepsilon_{ijk} a_{j\ell} a_{km} = \det(a) \sum_{n} \varepsilon_{n\ell m} a_{in},$$

where a_{ij} are elements of a general orthogonal transformation.

12.2.2 Show that the matrix element $\langle s' | \vec{J} \cdot \vec{p} | s' \rangle$ vanishes for good parity states, $|s'\rangle$.

12.3.1 For the time-reversal *operation*, show:

(a) $(AB)^T = B^T A^T$ [For simplicity, assume $A = \sum_{i,j} A_{ij} |X_i\rangle\langle X_j|$, $B = \sum_{k,\ell} B_{k\ell} |X_k\rangle\langle X_\ell|$, where the $|X_i\rangle$ are discrete basis kets with $\langle X_i | X_j\rangle = \delta_{ij}$.]

(b) $(\vec{r} \times \vec{p})^T = -\vec{r} \times \vec{p}$

(c) $\langle \hat{n} | \widetilde{\ell, m} \rangle = \langle \hat{n} | \ell, m \rangle^* \Rightarrow |\widetilde{\ell, m}\rangle = (-1)^m \, | \ell, -m\rangle$

12.3.2 In the usual treatment of time reversal, one defines a time-reversal operator, Θ, such that $\Theta = KU$, where U is a unitary operator and K is the complex-conjugate operator, which changes numbers into their complex conjugates ($KC \equiv C^*$). Given a general ket, expanded in a given basis,

$$|\alpha\rangle = \sum_i \alpha_i \, | X_i \rangle,$$

show that

$$\Theta | \alpha \rangle = ((|\alpha\rangle)^T)^\dagger$$

***12.3.3** Use time-reversal symmetry to show that the opposite direction transmission and reflection coefficients from Chapter 11, defined as $t(-k)$ and $r(-k)$, respectively, for any one-dimensional interaction potential are related to the original direction coefficients, $t(k)$ and $r(k)$, by ($|t(k)|^2 + |r(k)|^2 = 1$)

$$t(-k) = t(k), \; r(-k) = -r(k)^* \frac{t(k)}{t(k)^*}.$$

[Hint: The complex conjugate of the time-independent part of a coordinate space wave function represents the time-reversed solution, according to Equation 12.29.]

12.4.1 Deduce the effect of P, C, and T individually on a helicity operator, $\vec{S}\cdot\vec{p}/(\hbar|\vec{p}|)$ (defined in Section 12.7).

12.6.1 Let us define the electromagnetic Lagrangian,

$$\mathcal{L} \equiv \mathcal{L}_A - \rho(\vec{x},t)\Phi(\vec{x},t) + \vec{J}\cdot\vec{A},$$

where $\mathcal{L}_A$ is defined in Section 12.6, and I have included source terms involving the electric charge density, ρ, and the current, $\vec{J}$. Using the scalar Euler–Lagrange equation,

$$\frac{\delta\mathcal{L}}{\delta\Phi} - \sum_i \nabla_i \frac{\delta\mathcal{L}}{\delta(\nabla_i\Phi)} = 0,$$

show that one recovers the scalar Maxwell equation,

$$\nabla^2\Phi + \frac{1}{c}\frac{\partial}{\partial t}\left(\sum_i \nabla_i A_i\right) = -4\pi\rho.$$

[The other Maxwell equations can be recovered in a similar fashion.]

12.6.2 Show that the action, S_p, is invariant under the combined substitutions Equations 12.56, 12.57, and

$$\psi(\vec{x},t) \to \psi(\vec{x},t)\exp\left(\frac{iq}{\hbar c}\Lambda(\vec{x},t)\right).$$

12.6.3 There exists nine linearly independent 3×3 matrices. If we take the identity matrix as one, the remaining eight, λ_i, can be taken to satisfy ($i,j = 1, 2, \ldots, 8$)

$$Tr\,(\lambda_i\lambda_j) = 2\delta_{ij},\ [\lambda_i\ \lambda_j] = 2i f_{ijk}\,\lambda_k \text{ (understood } k \text{ sum)},$$

("Tr" denotes the trace) where the f_{ijk} are antisymmetric in all three indices and are called the Lie algebra "structure constants." The λ_i are Hermitian, $\lambda_i^\dagger = \lambda_i$, and are called the generators of the fundamental representation of $SU(3)$, the strong "color" group.

(a) Show that the structure constants are given by

$$f_{ijk} = \mathrm{Tr}\big(\big[\lambda_i,\lambda_j\big]\lambda_k\big)/4i.$$

(b) Show the structure constants are real.

(c) Given the structure constants are real, show that

$$[\bar{\lambda}_i,\bar{\lambda}_j] = 2i\,f_{ijk}\,\bar{\lambda}_k,$$

where $\bar{\lambda}_i \equiv -\lambda_i^*$. This is called the complex-conjugate representation of $SU(3)$ and is used to characterize antiquarks.

12.7.1 Problem 10.8.3 introduced the Wigner–Eckart theorem, which is useful in many matrix element evaluations in atomic, nuclear, and particle physics. I restate it here:

$$\langle J',M' | T_{Kq} | J,M\rangle = \langle JK;Mq | JK;J'M'\rangle \frac{\langle J' | T_k | J\rangle}{\sqrt{2J+1}}.$$

In this case, you are given the electromagnetic Hamiltonian density,

$$H^{em} = \sum_{L=0,1,2\ldots} \sum_{m=-L}^{L} \sum_{X=(E,M)} A_{Lm}^{(X)} T_{Lm}^{(X)}$$

where the $A_{Lm}^{(E)}$, $A_{Lm}^{(M)}$ (E = electric, M = magnetic) are expansion coefficients (assume $A_{00}^{(M)} = 0$) and the $T_{Lm}^{(E)}$, $T_{Lm}^{(M)}$ are spherical tensors. The operators $T_{Lm}^{(E,M)}$ have the properties

$$PT_{Lm}^{(E)}P = (-1)^L T_{Lm}^{(E)} (L \geq 1),$$

$$PT_{Lm}^{(M)}P = (-1)^{L+1} T_{Lm}^{(M)} (L \geq 1),$$

where P is the parity operator. Now consider the expectation value (modeling, for example, static *EM* dipole ($L = 1$), quadrupole ($L = 2$), etc. moments):

$$\langle J,m_J,\pi_J | H^{\text{em}} | J,m_J,\pi_J\rangle,$$

where $m_J = J$. (The states $|J, m_J, \pi_J\rangle$ have good total angular momentum, J_z, and parity, π_J.)

(a) For given J ($J = 0$, 1/2, 1, 3/2, 2, ...), what is the largest L value that can contribute? Can a spin-1/2 particle, for example, have a quadrupole ($L = 2$) moment? Is the spin-1/2 dipole ($L = 1$) moment electric or magnetic?

(b) For given J, which L terms in H^{em} give nonzero contributions? Which L values contribute to E operators? Which contribute to M operators?

12.8.1 Particle physicists usually express dimensionful quantities in terms of MeV (megaelectronvolts), $\hbar$, and c. For example,

$$1\,\text{kg} = 5.61 \times 10^{29} \frac{\text{MeV}}{c^2},$$

$$1\,\text{cm}^{-1} = 1.97 \times 10^{-11} \frac{\text{MeV}}{\hbar c},$$

$$1\,\text{s}^{-1} = 6.58 \times 10^{-22} \frac{\text{MeV}}{\hbar}.$$

Using these values, convert the QCD string tension of 15 tons (1 ton = 8.9×10^3 Newtons) into MeV/fm units.

12.8.2 (a) One may use the Wigner–Eckart theorem from Problem 12.7.1 also in an isospin context. Define the isospin states:

$$|\Delta^{++}\rangle = |\frac{3}{2},\frac{3}{2}\rangle,\ |\Delta^{+}\rangle = |\frac{3}{2},\frac{1}{2}\rangle,$$

$$|p\rangle = |\frac{1}{2},\frac{1}{2}\rangle,\ |n\rangle = |\frac{1}{2},-\frac{1}{2}\rangle,$$

$$|\pi^{+}\rangle = |1,1\rangle,\ |\pi^{0}\rangle = |1,0)\rangle.$$

Use Equation 8.47 (Chapter 8) with $I_{\pm}^{\Delta} = I_{\pm}^{\pi} + I_{\pm}^{N}$ and

$$|p\rangle|\pi^{+}\rangle = |\frac{3}{2},\frac{3}{2}\rangle$$

to show that

$$|\frac{3}{2},\frac{1}{2}\rangle = \sqrt{\frac{1}{3}}\,|n\rangle|\pi^{+}\rangle + \sqrt{\frac{2}{3}}\,|p\rangle|\pi^{0}\rangle.$$

(b) Using the golden rule proportionality for the transition rate, w (good for the isospin parts of wave functions in strong interaction decays *only*)

$$w(\text{Initial} \to \text{Final}) \propto |\langle \text{Final} \,|\, \text{Initial}\rangle|^{2},$$

find the decay branching ratio:

$$\frac{w(\Delta^{+} \to p + \pi^{0})}{w(\Delta^{+} \to n + \pi^{+})}.$$

12.8.3 Using the Wigner–Eckart theorem, and Problem 12.8.2b, find the ratio of rates for the decays indicated:

(a) $$\frac{w(K^{*+} \to K^{0} + \pi^{+})}{w(K^{*+} \to K^{+} + \pi^{0})} = ?$$

(b) $$\frac{w(K^{*0} \to K^{+} + \pi^{-})}{w(K^{*0} \to K^{0} + \pi^{0})} = ?$$

The K mesons have $I = 1/2$; you can consider K^{+} the $|1/2, 1/2\rangle$ state and K^{0} the $|1/2, -1/2\rangle$ state. The $K^{*}(892)$ mesons are excited states of the K and also have $I = 1/2$. The pion of course has $I = 1$. The rate ratios are related to the squared ratios of Clebsch–Gordon coefficients in Problem 12.8.2b, and these can be found online or in the *RPP* booklets.

12.8.4 Given the isospin assignments,

$$|p\rangle \equiv |\frac{1}{2},\frac{1}{2}\rangle,\ |n\rangle \equiv |\frac{1}{2},\frac{1}{2}\rangle$$

$$|\pi^{+}\rangle \equiv |1,1\rangle,\ |\pi^{0}\rangle \equiv 1,0\rangle,\ |\pi^{-}\rangle \equiv |1,-1\rangle;$$

along with

$$I_\pm \,|\, I, I_3\rangle = \sqrt{(I \mp I_3)(I \pm I_3 + 1)}\,|\, I, I_3 \pm 1\rangle,$$

fill in the remaining entries (A, B, C, and D) in the isospin decomposition table ($|\pi^+ p\rangle \equiv |\pi^+\rangle\, |p\rangle$, etc.):

	$\left\|\frac{3}{2},\frac{3}{2}\right\rangle$	$\left\|\frac{3}{2},\frac{1}{2}\right\rangle$	$\left\|\frac{3}{2},-\frac{1}{2}\right\rangle$	$\left\|\frac{3}{2},-\frac{3}{2}\right\rangle$	$\left\|\frac{1}{2},\frac{1}{2}\right\rangle$	$\left\|\frac{1}{2},-\frac{1}{2}\right\rangle$
$\|\pi^+ p\rangle$	1	0	0	0	0	0
$\|\pi^- p\rangle$	0	0	$\frac{1}{\sqrt{3}}$	0	0	$-\sqrt{\frac{2}{3}}$
$\|\pi^0 n\rangle$	0	0	$\sqrt{\frac{2}{3}}$	0	0	$\frac{1}{\sqrt{3}}$
$\|\pi^+ n\rangle$	0	A	0	0	C	0
$\|\pi^0 p\rangle$	0	B	0	0	D	0
$\|\pi^- n\rangle$	0	0	1	0	0	0

[Hint: Use the raising and lowering operators, $I_\pm \equiv I_\pm^\pi + I_\pm^N$ on known states.]

12.8.5 Using the "spin up" neutron wave function from Table 12.6, find

(a) The probability of finding an up quark with spin up; similarly for the up quark with spin down.

(b) The probability that the two down quarks are coupled together with total spin 1; similarly, the probability that the two down quarks are coupled with spin 0. Can you explain your answers from symmetry considerations?

12.8.6 Find the magnetic moment of the neutron in the nonrelativistic quark model. Use the normalized "spin up" wave function of the neutron from Table 12.6. Define

$$\mu_z = \langle n,\ up\ |(\mu_z)_{\text{op}}\rangle |\, n,\ up\ \rangle,\ (\mu_z)_{\text{op}} = \frac{e\hbar}{2mc}\sum_{i=1,2,3} e_i\sigma_{iz},$$

where m is a constituent u,d quark mass. The σ_{iz} are the usual z-component $\vec{\sigma}$ matrices working in the ith quark space, and the fractional charges are

$$e_i = \begin{cases} \dfrac{2}{3}, & u \text{ quark} \\[2ex] -\dfrac{1}{3}, & d \text{ quark} \end{cases}.$$

[Hints: One may work out the answer for one cyclic combination as in Table 12.6 and then multiply by 3. The mass, m, needed to get a reasonable answer can be viewed as a phenomenological parameter. Answer: $(\mu_z)_n = e\hbar/(2mc)(-2/3)$. If this calculation is also done for the proton, one obtains $(\mu_z)_p = e\hbar/(2mc)$, giving a ratio $(\mu_z)_p/(\mu_z)_n = -1.5$, compared to the actual value −1.46.]

12.9.1 The following gives a list of particle decays that have not been detected, either because they are strictly forbidden or greatly suppressed. Using Tables 12.3, 12.4, and 12.5, in each case please provide the physical property or reason why the reaction is not seen. (In some cases there may be more than one reason; in this case provide the "strongest" reason.)

(a) $n \rightarrow \pi^+ + \pi^-$
(b) $\mu^+ \rightarrow e^+ + \nu_e + \nu_\mu$
(c) $\pi^0 \rightarrow 3\gamma$
(d) $\Sigma- \rightarrow \overline{p} + \pi^0$
(e) $\rho^0 \rightarrow \pi^0 + \pi^0$
(f) $\eta^0 \rightarrow \pi^+ + \pi^-$

12.11.1 Find a friendly particle physicist and interview her/him on their specialty in any topic in contemporary research. Present this either in class or in written form.

12.11.2 Make a class presentation of one of the particle physics subjects mentioned (or not) in this chapter. Some of the possible topics are dark matter, neutrino masses, neutrino astronomy, observational aspects of supersymmetry, extra dimensions, and sparticles. Investigate the ideas of anomalies, technicolor, or brane worlds.

Appendix A: Notation Comments and Comparisons

Notation is a hard taskmaster. Choosing a good notation is a balancing act. On the one hand, it is important to try to carefully distinguish between different types of mathematical structures. On the other hand, such distinctions can result in a higher learning barrier for new students and be cumbersome for the practicing physicist. Matrices, quantum mechanical operators, and differential operators should technically have their own symbols, even if they represent the same "physical" quantity. Notational devices such as boldface or tildes can be used to distinguish these. However, other than using a prime to distinguish a number from an operator, using explicit indices on matrices, and explicit forms of differential operators when they appear, I incline toward a leaner (pun intended) notation. The contexts, plus textual comments, are almost always sufficient to clarify the meanings of the symbols. However, this point of view calls for some comparisons of different Dirac bra–ket notations to be crystal clear.

Our bra–ket object, $\langle\alpha|\theta|\beta\rangle$ (just a number), has other notations in the literature. Using the same symbols, here are two alternate examples:

$$\text{Notation 1:}^{*}\ \langle\theta^{\dagger}\alpha|\beta\rangle \equiv \langle\alpha|\theta\beta\rangle \tag{A.1}$$

$$\text{Notation 2:}^{\dagger}\ (\theta^{\dagger}\psi_{\alpha},\psi_{\beta}) \equiv (\psi_{\alpha},\theta\psi_{\beta}). \tag{A.2}$$

In both of these notations, quantum mechanical operators are always taken to act to the right, so such definitions are needed. Here, however, the operator θ may act "democratically" to the left or right. However, if one insists on always acting to the right, then in the present notation one would write the following:

$$\text{Democratic Dirac:}^{\ddagger}\ ((\theta^{\dagger}|\alpha\rangle)^{\dagger}|\beta) = \langle\alpha|\theta|\beta\rangle. \tag{A.3}$$

As far as matrices are concerned, assuming a finite matrix representation, Equations A.1 through A.3 are all equivalent to:

$$((\theta^{\dagger})_{ji}\alpha_i)^{*}\beta_j \equiv \alpha_i^{*}(\theta)_{ij}\beta_j, \tag{A.4}$$

where there are understood sums on i and j, and the "†" symbol is interpreted as "complex conjugate + transpose."

---

* Used by D.J. Griffiths, *Introduction to Quantum Mechanics*, 2nd ed. (Prentice Hall, 2004), as well as R.L. Liboff, *Introductory Quantum Mechanics*, 2nd ed. (Addison-Wesley, 1992), except in his Appendix A.

† Used by E. Merzbacher, *Quantum Mechanics*, 3rd ed. (John Wiley, 1998).

‡ Used by J.S. Schwinger, *Quantum Mechanics: Symbolism of Atomic Measurements*, edited by B.-G. Englert (Springer, 2001); J.J. Sakurai and J. Napolitano, 2nd ed., *Modern Quantum Mechanics* (Addison-Wesley, 2011); K. Gottfried, *Quantum Mechanics* (W.A. Benjamin, 1974); and A. Messiah, *Quantum Mechanics* (John Wiley, 1958).

Please do not mistake the above for the alternate identity involving the complex conjugate of the matrix element:

$$\text{Notation 1:} \left\langle \beta \,|\, \theta^{\dagger}\alpha \right\rangle^{*} \equiv \left\langle \alpha \,|\, \theta\beta \right\rangle \tag{A.5}$$

$$\text{Notation 2:} (\psi_{\beta}, \theta^{\dagger}\psi_{\alpha})^{*} \equiv (\psi_{\alpha}, \theta\psi_{\beta}) \tag{A.6}$$

$$\text{Democratic Dirac:} \left\langle \beta \,|\, \theta^{\dagger} \,|\, \alpha \right\rangle^{*} \equiv \left\langle \alpha \,|\, \theta \,|\, \beta \right\rangle \tag{A.7}$$

All of these notations are consistent and appealing. However, by training and inclination I prefer the democratic Dirac notation.

Appendix B: Lattice Models

It is of interest to see to what extent the continuum description of position and momentum measurements can be recovered from the Process Diagram–type measurements of physical observables in Chapter 1. In Chapter 2, we learned that the position and momentum space kets are related as

$$|x'\rangle = \int dp'_x \,|p'_x\rangle\langle p'_x|x'\rangle, \tag{B.1}$$

$$\langle p'_x|x'\rangle = \frac{1}{\sqrt{2\pi\hbar}} e^{-ix'p'_x/\hbar}. \tag{B.2}$$

It is easy to formulate a discrete form of a measurement system on an infinite lattice of points (mathematicians would call it a *noncompact representation*). Based on the symmetrical differential difference found in Equation 2.190, we can postulate (a is the distance between points, we limit ourselves to one dimension, and $p \equiv px$ and $q \equiv x$ for convenience)

$$p|q'_j\rangle = \frac{i\hbar}{2a}\left(|q'_{j+1}\rangle - |q'_{j-1}\rangle\right) \tag{B.3}$$

$$q|q'_j\rangle = q'_j|q'_j\rangle, \tag{B.4}$$

$$q'_j = j\,a. \tag{B.5}$$

($j = -\infty, \ldots, -1, 0, 1, \ldots, \infty$.) Then the associated matrices are as follows:

$$(p)_{jk} \equiv \langle q'_j|p|q'_k\rangle = \frac{i\hbar}{2a}\left(\delta_{j,k+1} - \delta_{j+1,k}\right), \tag{B.6}$$

$$(q)_{jk} \equiv \langle q'_j|q|q'_k\rangle = j\,\delta_{j,k}\,a, \tag{B.7}$$

$$\Rightarrow \left([q,p]\right)_{jk} = \frac{i\hbar}{2}\left(\delta_{j+1,k} + \delta_{j,k+1}\right), \tag{B.8}$$

($\delta_{j,k}$ is a Kronecker symbol.) Equation B.8 is the discrete form of the commutator, Equation 2.189. Notice it is actually nondiagonal in this context.

The problem with such a representation is that there is no way of directly providing a discrete form of Equations B.1 and B.2 for a strictly infinite number of points. We must attempt to recover such statements as a *limit*.

There is not a unique way of proceeding at this point.* However, let us postulate a finite (*compact*) N-dimensional (N even) representation with

$$|q'_k\rangle = \sum_j (U)_{kj} \, | p'_j\rangle, \tag{B.9}$$

$$(U)_{kj} = \frac{1}{\sqrt{N}} \exp\left(-i\frac{2\pi}{N}\left(k-\frac{N}{2}\right)\left(j-\frac{N}{2}\right)\right) \quad \left(U^{-1}=U^{\dagger}\right) \tag{B.10}$$

($\{j, k\} = 1, \ldots, N$) as the discrete forms of Equations B.1 and B.2. This is a "periodic" lattice since formally $|q'_{i+N}\rangle = |q'_i\rangle$. To start off, we insist that the allowed eigenvalues be

$$q'_k = a\left(k-\frac{N}{2}\right), \tag{B.11}$$

$$p'_k = \frac{2\pi\hbar}{aN}\left(k-\frac{N}{2}\right). \tag{B.12}$$

The Process Diagram of a selection of q followed by a separation of p values is illustrated as in Figure B.1 for $N=4$.

One way of introducing the operator, p, in this context is to use *periodic* raising and lowering operators. These are closely related to the spin raising and lowering operators $S_+/\hbar$ and $S_-/\hbar$. I define (N matrix dimensions)

$$(W_+)_{jk} \equiv \delta_{j+1,k} + \delta_{j,N}\delta_{k,1}, \tag{B.13}$$

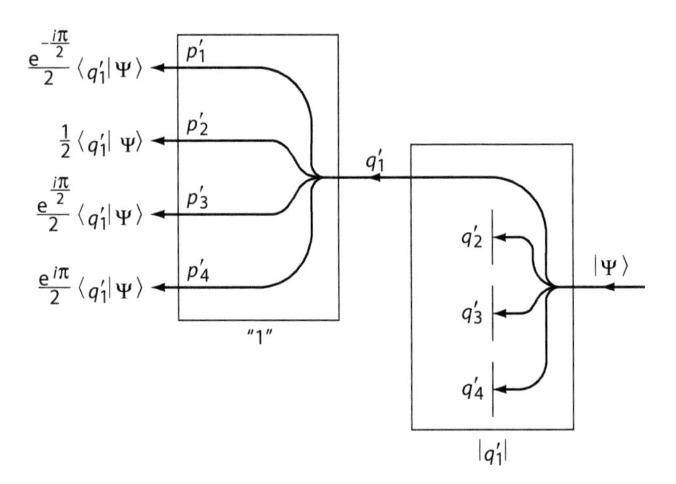

FIGURE B.1 A selection measurement of position followed by a separation of p'_j values using completeness in momentum space for $N=4$. Given an incoming wave function $|\psi\rangle$, the outgoing momentum amplitudes from Equations B.9 and B.10 are shown.

* See an alternate point of view in J. Schwinger, *Quantum Mechanics: Symbolism of Atomic Measurements*, edited by B.-G. Englert (Springer, 2001), Sections 1.14–1.16.

$$(W_-)_{jk} \equiv \delta_{j,k+1} + \delta_{j,1}\delta_{k,N}, \tag{B.14}$$

where, without the extra double Kronecker delta terms on the right one would have $W_+ \to S_+/\hbar$ and $W_- \to S_-/\hbar$, up to an overall scale factor.

For example ($N=4$),

$$W_+ = \begin{pmatrix} 0 & 1 & 0 & 0 \\ 0 & 0 & 1 & 0 \\ 0 & 0 & 0 & 1 \\ 1 & 0 & 0 & 0 \end{pmatrix}, \tag{B.15}$$

$$W_- = \begin{pmatrix} 0 & 0 & 0 & 1 \\ 1 & 0 & 0 & 0 \\ 0 & 1 & 0 & 0 \\ 0 & 0 & 1 & 0 \end{pmatrix}. \tag{B.16}$$

In addition,

$$(W_+)^N = (W_-)^N = 1, \tag{B.17}$$

$$(W_+)^t = W_-, \tag{B.18}$$

$$W_+W_- = W_-W_+ = 1. \tag{B.19}$$

("1" is the unit matrix in Equations B.17 and B.19 and "t" is transpose.) Equation B.17 implies, for example, that the W_+ eigenvalues are given by

$$(W'_+)^N = 1, \tag{B.20}$$

$$\Rightarrow W'_+ = \exp\left(i\frac{2\pi}{N}\left(j - \frac{N}{2}\right)\right). \quad (j = 1,\ldots,N) \tag{B.21}$$

The matrix that diagonalizes W_+ and W_- is just U from earlier:

$$(U\,W_+U^\dagger)_{jk} = \delta_{j,k}\exp\left(i\frac{2\pi}{N}\left(j - \frac{N}{2}\right)\right), \tag{B.22}$$

$$(U\,W_-U^\dagger)_{jk} = \delta_{j,k}\exp\left(-i\frac{2\pi}{N}\left(j - \frac{N}{2}\right)\right), \tag{B.23}$$

A natural postulate for the operator p is, therefore, that

$$\sin\left(\frac{pa}{\hbar}\right) \equiv \frac{i}{2}(W_- - W_+), \tag{B.24}$$

which can be viewed as a periodic version of Equation B.6. The left-hand side of Equation B.24 is defined by the power series expansion for sin(x). We can then define the periodic version of the position operator from

$$\sin\left(\frac{2\pi q}{aN}\right) \equiv U \sin\left(\frac{pa}{\hbar}\right) U^{\dagger}. \tag{B.25}$$

Using Equations B.22, B.23, and B.24, this gives a diagonal representation,

$$\left(\sin\left(\frac{2\pi q}{aN}\right)\right)_{jk} = \delta_{j,k} \sin\left(\frac{2\pi}{N}\left(j - \frac{N}{2}\right)\right), \tag{B.26}$$

analogous to Equation B.7, which shows the eigenvalues q'/a are $i - N/2$, for $i = 1, \ldots, N$, as expected. Since only the periodic momentum function $\sin(pa/\hbar)$ is defined on this lattice, the best that one can do to define the energy of a free particle is to set

$$E_{\text{lat}}(p) \equiv \frac{\hbar}{2ma^2} \sin^2\left(\frac{pa}{\hbar}\right), \tag{B.27}$$

and assume $N \gg 1$. The function in Equation B.27 is illustrated schematically in Figure B.2 for $N = 12$.

Note that the energy, $E_{\text{lat}}(p)$, agrees with the continuum relation only for small momenta. Since the two sides in Figure B.2 can be thought of as being connected by the periodicity of the lattice, we see that there are two positions where $E(p) = 0$: $p = 0$ and $p = \pi\hbar/a$. We are encountering the so-called *species doubling* problem on the lattice. We thought we were describing the momentum and position of a single particle, but it turns out the momentum space description implies there are two different particles involved!

Let us now consider an alternate interpretation. I now insist that the correct energy–momentum relation hold,

$$E_{\text{lat}}(p) \equiv \frac{p^2}{2m}, \tag{B.28}$$

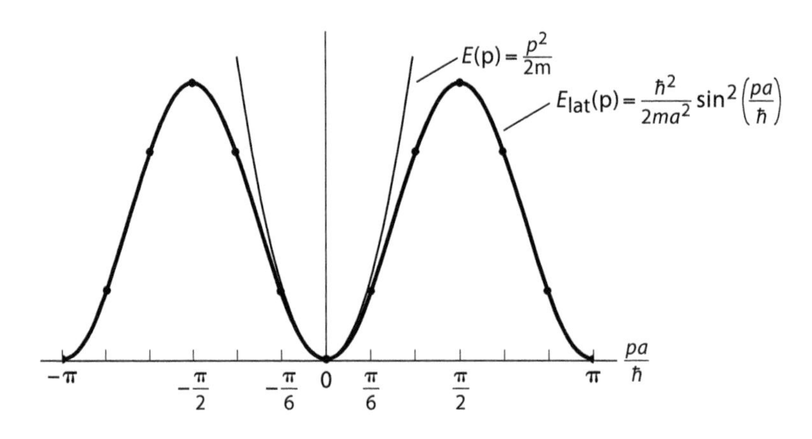

FIGURE B.2 Lattice and continuum energy–momentum relationship on a one-dimensional lattice with $N = 12$.

in an attempt to avoid species doubling. I continue to require that Equations B.9 and B.10 hold between the $|p_i'\rangle$ and $|q_i'\rangle$ bases. In this case, one sets

$$p \equiv \frac{i\hbar}{2a}(W_- - W_+). \tag{B.29}$$

It now turns out that the position and momentum eigenstates are sine functions of the previous set,

$$q_i' = \frac{aN}{2\pi}\sin\left(\frac{2\pi}{N}\left(i - \frac{N}{2}\right)\right), \tag{B.30}$$

$$p_i' = \frac{\hbar}{a}\sin\left(\frac{2\pi}{N}\left(i - \frac{N}{2}\right)\right). \tag{B.31}$$

That is, the values of p_i' and q_i' are proportional to the sine of $2\pi/N$ times the previous set of allowed integer values. Interesting aspects of such a description, which I call a "sine lattice," are brought out in Figure B.3 ($N = 12$ case).

Notice that most positions are doubled, which can be termed "back" and "front" representations, as one moves in discrete steps around the unit circle. The fixed angular step size is

$$\Delta\phi = \frac{2\pi}{N}. \tag{B.32}$$

One can think of the positions other than $\phi = \pm\pi/2$ as accommodating two noninteracting (via nearest neighbors, that is) particles. (If N is even, but not evenly divisible by 4, the top and bottom positions on the lattice are also doubly occupied, and there are no points at $\phi = \pm\pi/2$.) Thus, this interpretation turns out again to have a doubling problem, but this time in coordinate space rather than momentum space! In the limit $N \gg 1$ either description gives a q, p representation consistent with the noncompact forms B.6 through B.8, but only in the vicinity of either the two momentum zeros in Figure B.2 or the two position origins in Figure B.3. It might be argued in the second interpretation that Equations B.9 and B.10 no longer represent a discrete Fourier

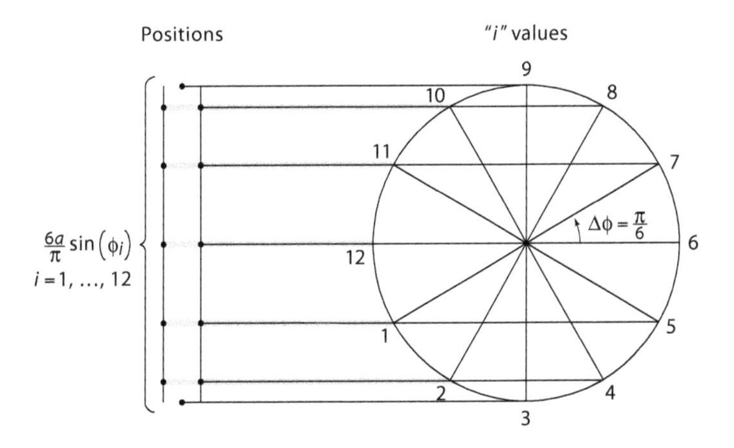

FIGURE B.3 Positions assigned to particles on a "sine lattice" where the momentum operator is given by Equation B.29. A small vertical separation of "back" and "forward" representations is shown for clarity.

transform because U is not of the form $\exp(-ip'q'/\hbar)$, where q' and p' are the eigenvalues; however, the conventional Fourier form is recovered for $N \gg 1$ in the vicinity of the origins.

PROBLEMS

B.1 Show Equation B.8 follows from Equations B.6 and B.7.

B.2 Verify the relations Equations B.17, B.18, and B.19 using Equations B.15 and B.16 for the $N=4$ case.

B.3 Verify Equations B.22 and B.23 from Equations B.10, B.15, and B.16 for the $N=4$ case.

Appendix C: 2-D Harmonic Oscillator Wave Function Normalization

I will derive here the orthonormalization condition on the $u_{n+n-}(\rho, \phi)$, alluded to in Chapter 7, Section 7.6. We start out with the generating function result, Equation 7.195:

$$\sum_{n_+,n_-=0}^{\infty} \frac{\lambda_+^{n_+}\lambda_-^{n_-}}{\sqrt{n_+!n_-!}} u_{n_+n_-}(\rho,\phi) = \frac{1}{\sqrt{\pi}} \exp\left[-\frac{1}{2}\rho^2 + \rho\left(\lambda_+ e^{i\phi} + \lambda_- e^{-i\phi}\right) - \lambda_+\lambda_-\right]. \tag{C.1}$$

I now multiply Equation C.1 by its complex conjugate, except that λ_+ and λ_- are replaced by the independent parameters λ'_+ and λ'_-. I also integrate:

$$\sum_{\substack{n_+,n_-=0\\ n'_+,n'_-=0}}^{\infty} \frac{\lambda_+^{n_+}\lambda_-^{n_-}\lambda_-'^{n_+}\lambda_+'^{n_-}}{\sqrt{n_+!n_-!n'_+ n'_-!}} \int d^2q\; u_{n_+n_-}u_{n'_+n'_-}$$
$$= \frac{1}{\pi}\int d^2q\; \exp\left[-\rho^2 + \rho\left((\lambda_+ + \lambda'_+)e^{i\phi} + (\lambda_- + \lambda'_-)e^{-i\phi}\right) - \lambda_+\lambda_- - \lambda'_+\lambda'_-\right]. \tag{C.2}$$

I now change variables in the integral on the right in Equation C.2 from ρ, ϕ to q_1,q_2 in order to do the integral. I get

(right-hand side of Equation C.2)

$$= \frac{1}{\pi}\int dq_1 dq_2 \exp\left[\begin{array}{c} -\left(q_1^2+q_2^2\right) + \left(\lambda_+ + \lambda'_+\right)\left(q_1 + iq_2\right) \\ +\left(\lambda_- + \lambda'_-\right)\left(q_1 - iq_2\right) - \lambda_+\lambda_- - \lambda'_+\lambda'_- \end{array}\right]. \tag{C.3}$$

I now complete the square in q_1 in order to do the integral:

$$-q_1^2 + q_1\left(\lambda_+ + \lambda_- + \lambda'_+ + \lambda'_-\right) = -\left(q_1 - \frac{1}{2}\left(\lambda_+ + \lambda_- + \lambda'_+ + \lambda'_-\right)\right)^2 + \frac{1}{4}\left(\lambda_+ + \lambda_- + \lambda'_+ + \lambda'_-\right)^2. \tag{C.4}$$

The relevant q_1 part of the integral in Equation C.3 is

$$\int_{-\infty}^{\infty} dq_1 \exp\left[-q_1^2 + q_1\left(\lambda_+ + \lambda_- + \lambda'_+ + \lambda'_-\right)\right]$$
$$= \underbrace{\int_{-\infty}^{\infty} dx\, e^{-x^2}}_{\sqrt{\pi}} \exp\left[\frac{1}{4}\left(\lambda_+ + \lambda_- + \lambda'_+ + \lambda'_-\right)^2\right]. \tag{C.5}$$

Similarly for q_2:

$$\int_{-\infty}^{\infty} \mathrm{d}q_2 \exp\left[-q_2^2 + iq_2\left(\lambda_+ + \lambda'_+ + \lambda_- + \lambda'_-\right)\right]$$

$$= \underbrace{\int_{-\infty}^{\infty} \mathrm{d}x \mathrm{e}^{-x^2}}_{\sqrt{\pi}} \exp\left[-\frac{1}{4}\left(\lambda_+ + \lambda'_+ - \lambda_- + \lambda'_-\right)^2\right]. \tag{C.6}$$

Thus, we have

(right-hand side of Equation C.2)

$$= \exp\left[\frac{1}{4}\left(\lambda_+ + \lambda'_+ + \lambda_- + \lambda'_-\right)^2 - \frac{1}{4}\left(\lambda_+ + \lambda'_+ - \lambda_- - \lambda'_-\right)^2 - \lambda_+\lambda_- - \lambda'_+\lambda'_-\right] \tag{C.7}$$

$$= \exp\left[\lambda_+\lambda'_- + \lambda'_+\lambda_-\right]. \tag{C.8}$$

We may expand Equation C.8 as

$$\exp\left[\lambda_+\lambda'_- + \lambda'_+\lambda_-\right] = \sum_{\substack{n_+=0 \\ n_-=0}}^{\infty} \frac{(\lambda_+\lambda'_-)^{n_+}}{n_+!} \frac{(\lambda'_+\lambda_-)^{n_-}}{n_-!}. \tag{C.9}$$

Our results up to this point can be summarized as

$$\sum_{\substack{n_+n_-=0 \\ n'_+n'_-=0}}^{\infty} \frac{\lambda_+^{n_+}\lambda_-'^{n'_+}\lambda_-^{n_-}\lambda_+'^{n'_-}}{\sqrt{n_+!n_-!n'_+!n'_-!}} \int \mathrm{d}^2q \; u_{n_+n_-} u^*_{n'_+n'_-} = \sum_{n_+n_-=0}^{\infty} \frac{\lambda_+^{n_+}\lambda_-'^{n_+}\lambda_-^{n_-}\lambda_+'^{n_-}}{n_+!n_-!}. \tag{C.10}$$

Comparing equal powers of $\lambda_+, \lambda'_-, \lambda_-, \lambda'_+$ on either side of Equation C.10, we get that

$$\int \mathrm{d}^2q \; u_{n_+n_-} u^*_{n'_+n'_-} = \delta_{n_+n'_+}\delta_{n_-n'_-}. \tag{C.11}$$

This is the desired result. This result is useful when we normalize the Coulomb wave function in Equation 7.252.

Appendix D: Allowed Standard Model Interactions

The generic allowed interaction vertices in Feynman diagrams for the various Standard Model interactions are shown next. All lines can be considered as "real" (external) or virtual (internal). These are just the primitive "vertices"—one must put them together to make complete Feynman diagrams representing physical processes. Of course, individual quark–gluon lines are confined and cannot be external. The Higgs boson couplings are not shown. To be consistent, many "ghost" particle couplings must also be introduced, which are also not shown. I have not attempted to indicate the various interaction "vertex" (or coupling) factors, which would have greatly increased the number of diagrams. One can think of the time axis flowing upward in the following diagrams.

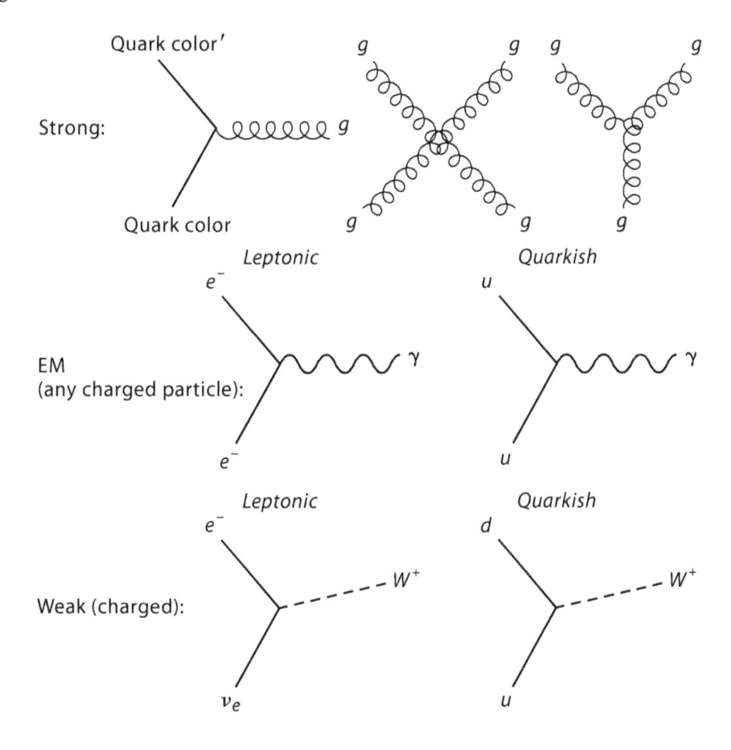

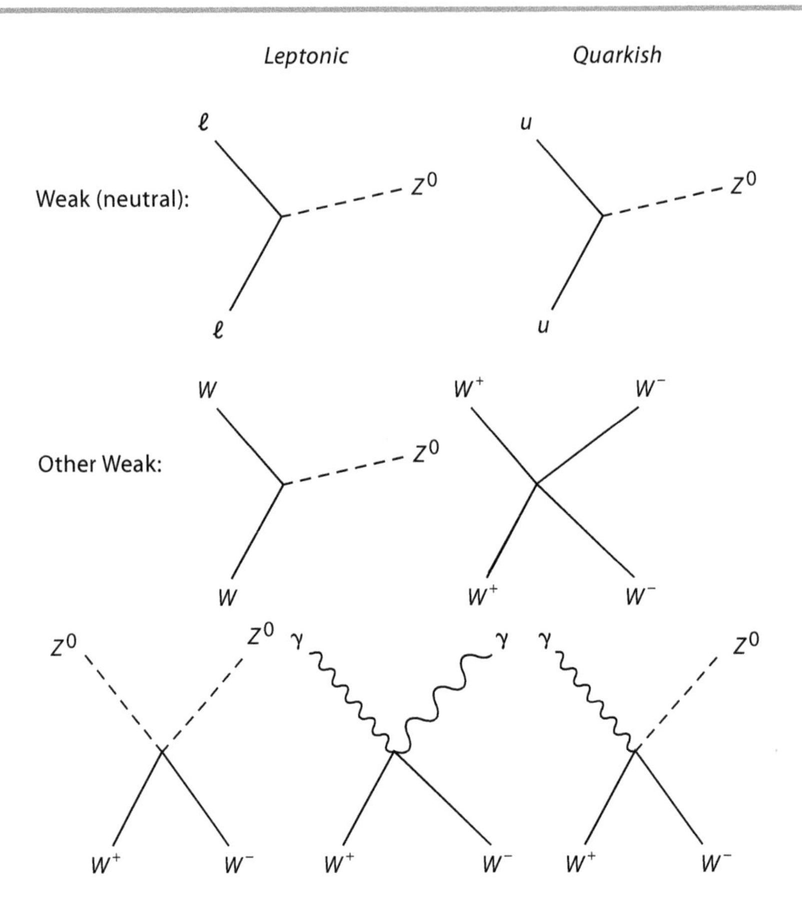

PROBLEM

D.1 Draw the lowest-order (smallest number of vertices) Feynman diagrams for the following processes. You will have to figure out which interaction is responsible for each. Some of them may have more than one topologically distinct lowest-order diagram. For hadrons draw the Feynman diagrams at the quark level. If you have trouble figuring out the nature of a reaction or decay, looking up the rates should tell you which interaction is responsible or dominant. (Make sure you make clear which lines represent which particles.)

(a) $e^+ + \gamma \rightarrow e^+ + \gamma$
(b) $e^+ + e^- \rightarrow \nu_e + \bar{\nu}_e$
(c) $p + \pi^- \rightarrow \Lambda^0 + K^0$
(d) $\nu_\mu + p \rightarrow \mu^- + \Delta^{++}$
(e) $\Sigma^+ \rightarrow n + \pi^+$
(f) $\Delta^+ \rightarrow n + \pi^+$

Appendix E: The Ising Model and More

E.1 THE ONE-DIMENSIONAL ISING MODEL

The Ising model of spin interactions on a background lattice is regarded as a fundamental theoretical paradigm in physics.* Although originally formulated in a statistical mechanics setting, the solution and simulation can also teach us important things in other physics contexts. Here I will use it as a model for some of the concepts and methods that are useful in the formulation of strong interaction Monte Carlo simulations.

I will limit the discussion here to the one-dimensional version, which is exactly solvable. The model consists of a chain of N two-valued spins (spin 1/2) interacting with nearest neighbors along a one-dimensional lattice with fixed lattice spacing. In addition, a magnetic field is introduced. The dimensionless energy of the system is defined to be

$$E(\{S_k\}) \equiv -\varepsilon \sum_{k=1}^{N-1} S_k S_{k+1} - H \sum_{k=1}^{N} S_k, \tag{E.1}$$

where ε is a constant and $S_k = \pm 1$ for all k values. The quantity H will be called the magnetic field. (It is actually missing a magnetic moment factor, μ, if one is picky.) The nonperiodic (the two sides of the lattice are not coupled to one another at this point) partition function† is given by

$$Z_N^{np} \equiv \sum_{S_1=\pm 1} \sum_{S_2=\pm 1} \dots \sum_{S_N=\pm 1} \exp(-\beta E(\{S_k\}), \tag{E.2}$$

where $\beta \equiv 1/kT$, T is the temperature, and k the Boltzmann constant. Using the definition of Equation E.2, one may show that the average energy,

$$\langle E \rangle \equiv \frac{\sum_{S_1=\pm 1} \sum_{S_2=\pm 1} \cdots \sum_{S_N=\pm 1} E(\{S_k\}) \exp(-\beta E\{S_k\})}{\sum_{S_1=\pm 1} \sum_{S_2=\pm 1} \cdots \sum_{S_N=\pm 1} \exp(-\beta E\{S_k\})}, \tag{E.3}$$

and the average magnetization,

$$\langle M \rangle \equiv \frac{\sum_{S_1=\pm 1} \sum_{S_2=\pm 1} \cdots \sum_{S_N=\pm 1} \left(H \sum_{k=1}^{N} S_k \right) \exp(-\beta E\{S_k\})}{\sum_{S_1=\pm 1} \sum_{S_2=\pm 1} \cdots \sum_{S_N=\pm 1} \exp(-\beta E\{S_k\})}, \tag{E.4}$$

* Some of the names associated with the model's formulation and solution are William Lenz, Ernest Ising, Lars Onsager, and C.N. Yang. Lenz originally formulated the model, Ising solved the one-dimensional version in 1925, and Onsager solved the two-dimensional model in the absence of a magnetic field in 1944. Yang considered the nonzero magnetic field case and proved an important spontaneous magnetization result for the two-dimensional model in 1952.

† The partition function is a standard tool in statistical mechanics and is explained in all books on the subject. It has an analog in particle physics simulations, as we will see.

are given by

$$\langle E \rangle = kT^2 \frac{\partial \log Z_N^{np}}{\partial T}, \tag{E.5}$$

and

$$\langle M \rangle = HkT \frac{\partial \log Z_N^{np}}{\partial H}. \tag{E.6}$$

The solution of this system is simplified by the use of the *transfer matrix* concept. As a first step, let us rewrite

$$\sum_{k=1}^{N} S_k = \frac{(S_1 + S_N)}{2} + \frac{1}{2} \sum_{k=1}^{N-1} (S_k + S_{k+1}), \tag{E.7}$$

and define the two by two transfer matrix, T, with elements

$$(T)_{S_1,S_2} = (T)_{S_2,S_1} \equiv \exp(\beta(\varepsilon S_1 S_2 + \frac{1}{2} H(S_1 + S_2)), \tag{E.8}$$

or

$$T = \begin{pmatrix} \exp(\beta(\varepsilon + H)) & \exp(-\beta\varepsilon) \\ \exp(-\beta\varepsilon) & \exp(\beta(\varepsilon - H)) \end{pmatrix}. \tag{E.9}$$

The partition function is thus

$$Z_N^{np} = \sum_{S_1,S_N} \left(\exp\left(\frac{\beta H}{2} (S_1 + S_N) \right) \sum_{S_2,\ldots,S_{N-1}} T_{1,2} T_{2,3} \ldots T_{S_{N-1},S_N} \right), \tag{E.10}$$

where I am using a more symbolic representation of the many sums involved. As I indicated, the superscript *np* stands for nonperiodic since the two ends of the lattice are not coupled. However, if we introduce an additional energy coupling term,

$$\Delta E^p \equiv -\varepsilon S_1 S_N, \tag{E.11}$$

then the periodic partition function becomes

$$Z_N^P \equiv \sum_{S_1,\ldots,S_N} T_{1,2} T_{2,3} \ldots T_{S_{N1} S_1} = \mathrm{Tr}[T^N], \tag{E.12}$$

where Tr denotes the matrix trace. The matrix T may be diagonalized via

$$T_D \equiv S^{-1} T S, \tag{E.13}$$

so that (using the cyclic property of the trace)

$$Z_N^p = \mathrm{Tr}\left[(T_D)^N \right] = \lambda_+^N + \lambda_-^N, \tag{E.14}$$

where

$$\lambda_\pm \equiv e^{\beta\varepsilon}\left(\cosh(\beta H) \pm \sqrt{\sinh^2(\beta H) + e^{-4\beta\varepsilon}}\right), \tag{E.15}$$

are the positive eigenvalues of T. We have $\lambda_+ > \lambda_-$, so that for a long lattice, $N \gg 1$, we can write

$$\lim_{N\gg1} Z_N^p \simeq \lambda_+^N. \tag{E.16}$$

Using Equation E.16 we may then derive

$$\begin{aligned}\frac{\langle E\rangle}{N} \simeq& -H\frac{\sinh(\beta H)}{\sqrt{\sinh^2(\beta H)+e^{-4\beta\varepsilon}}}\\ &-\varepsilon\left(\frac{\sqrt{\sinh^2(\beta H)+e^{-4\beta\varepsilon}}\cosh(\beta H)+\sinh^2(\beta H)-e^{-4\beta\varepsilon}}{\sqrt{\sinh^2(\beta H)+e^{-4\beta\varepsilon}}\cosh(\beta H)+\sinh^2(\beta H)+e^{-4\beta\varepsilon}}\right),\end{aligned} \tag{E.17}$$

$$\frac{\langle M\rangle}{N} \simeq H\frac{\sinh(\beta H)}{\sqrt{\sinh^2(\beta H)+e^{-4\beta\varepsilon}}}, \tag{E.18}$$

from Equations E.5 and E.6. Note from Equation E.18 that $\langle M\rangle = 0$ for $H = 0$, so there is no spontaneous magnetization, as occurs in real magnets for low temperatures. (This does happen for the two-dimensional Ising model.) However, for low temperatures $\sinh^2(\beta H) \gg e^{-4\beta\varepsilon}$, and therefore $\langle M\rangle = NH$, which means all spins are aligned. This implies long-range order, or correlation, of the spins near $T = 0$.

The nonperiodic case lacks the term in Equation E.11, and the spin variables S_1 and S_N are uncoupled. Defining the two-by-two matrix (just T in Equation E.8 with $\varepsilon = 0$),

$$(T_0)_{S_1,S_N} \equiv \exp\left(\frac{\beta H}{2}(S_1 + S_N)\right), \tag{E.19}$$

we now have

$$Z_N^{np} = \mathrm{Tr}\left[T_0 (T)^{N-1}\right]. \tag{E.20}$$

The rest of the derivation in this case will be left as an exercise. One finds,

$$Z_N^{np} = \lambda_+^{N-1}\chi_+ + \lambda_-^{N-1}\chi_-, \tag{E.21}$$

where

$$\chi_\pm \equiv \cosh(\beta H) \pm \frac{\sinh^2(\beta H) + e^{-2\beta\varepsilon}}{\sqrt{\sinh^2(\beta H) + e^{-4\beta\varepsilon}}} > 0. \tag{E.22}$$

In the $N \gg 1$ limit both Equations E.14 and E.21 clearly produce the same physics.

Having an exact solution available makes this model an excellent test bed for numerical simulation methods. Such simulations involve some or all of the following important algorithms or techniques.

Monte Carlo Techniques

The Monte Carlo (named by John von Neumann after a casino in the city/country of Monaco) technique for the numerical evaluation of high-dimensional integrals or summations is ridiculously easy. Given a high-dimensional integral/sum symbolized as (all integrals over finite ranges)

$$F \equiv \sum_{\{s\}} \int \ldots \int \prod_i \mathrm{d}x_i \, f(\{s\};\{x\}), \tag{E.23}$$

$\{s\}$ denotes a set of discrete variables and $\{x\}$ the continuous ones. Given a set of N_s sampling points that are uniformly but randomly distributed over the volume of integration and values of the summation, this quantity is estimated by the Monte Carlo integration/summation method to be

$$F \approx \frac{1}{N_s} \sum_{i=1}^{N} f_i \pm \frac{\sigma_F}{\sqrt{N_s}}, \tag{E.24}$$

where the f_i are the sampled values of the function $f(\{s\};\{x\})$ in Equation E.23. The summations and integrals are simply replaced by an average of the values sampled. This method is only approximate, and there is an associated error in the procedure, given by the second term in Equation E.24. The quantity σ_F is called the standard deviation, and, under the assumption of a Gaussian distribution of values, this implies that there is a 68.3% chance that the true value lies within upper and lower limits; this is called a one-standard deviation range. Similarly, the probability that the true value is within a range given by replacing $\sigma_F \to 2\sigma_F$ in Equation E.24 is 95.4%, and is called a two-standard deviation range. The Monte Carlo technique is very much like performing an actual experiment, and thus one speaks of the Monte Carlo computer "measurement" of a quantity. Notice the error bar in Equation E.24 shrinks only very slowly as one increases the number of measurements, N_s.

The trick in getting the Monte Carlo method to work in statistical mechanics simulations is the implementation of the Boltzmann factor, $e^{-\beta E}$, in the large parameter space. Typically, one is really evaluating a ratio of integrals/summations,

$$\langle \theta \rangle \equiv \frac{\sum_{\{s\}} \int \ldots \int \prod_i \mathrm{d}x_i \theta(\{s\};\{x\}) \exp(-\beta E(\{s\};\{x\}))}{\sum_{\{s\}} \int \ldots \int \prod_i \mathrm{d}x_i \exp(-\beta E(\{s\};\{x\}))}, \tag{E.25}$$

which represents the normalized, thermal expectation value (in a statistical sense) of a physical observable, $\theta(\{s\};\{x\})$. The denominator is just a generalization of the partition function we saw in Equation E.2. A particular set of values of the discrete and continuous variables is termed a *configuration*, and represents a point in the parameter space. If one can correctly assign the Boltzmann weighting associated with a given configuration in the partition function, then the Monte Carlo method gives the result

$$\langle \theta \rangle \approx \frac{1}{N_c} \sum_{i=1}^{N_c} \theta_i \pm \frac{\sigma_\theta}{\sqrt{N_c}}, \tag{E.26}$$

where the sum is over N_c configurations rather than points in the parameter space.

Metropolis Algorithm*

The Metropolis algorithm is a simple way of selecting configurations with a Boltzmann weighting.

The algorithm starts out anywhere in parameter space initially. One speaks of a "cold start" or a "hot start" for the system variables. Typically, a cold start consists of an identical setting of all variables across spatial sites, as if they were frozen in place. In our case, one would set the Ising spins to spin up $S_k = 1$, for example. A hot start consists of a random setting of variables in space, which would translate to a random set of spins in the Ising model. As we saw, the Monte Carlo procedure was very much like doing an actual experiment; the Metropolis algorithm says that the experiment being done is a simulation on a static thermal system (after equilibrium is reached). The algorithm is as follows:

Step 1: From the starting configuration, denoted as $\{C\}$, obtain a new configuration, $\{C'\}$, by changing one of the parameters in $\{C\}$ randomly.
Step 2: Evaluate the change in the energy, $\Delta E = E_{\{C'\}} - E_{\{C\}}$.
Step 3: If $\Delta E < 0$ accept the change.
Step 4: If $\Delta E > 0$, accept the change with the conditional probability, $P(\{C\} \to \{C'\}) = e^{-\beta\Delta E} < 1$.
Step 5: Return to Step 1 with either $\{C\}$ or $\{C'\}$ as the updated configuration.

That's it! This is a minimization procedure for selecting the ground state of the system, but one that allows statistical fluctuations like a real thermal system. Steps 1–5 must be run through many times before the system reaches equilibrium. How does one decide how many iterations are necessary to reach equilibrium? This is empirically determined by monitoring some instantaneous system variable, such as internal energy in the Ising case, until the system equilibrates in a statistical sense. The iterations done may consist of sweeps across all positions in the lattice and must always allow the possibility of changing *any* variable in the system (such systems are called *ergodic*). After the initial equilibration, measurements of system variables may be taken. However, it is critically important that sample configurations be separated by a sufficient number of Monte Carlo sweeps to ensure that the new configuration is statistically independent of the previous one. If this is not done, the average and error bars in Equation E.26 will not be accurate. In particular, if the sweep separation is not sufficient, one can, for example, seriously underestimate the error bars since the system is then limited in fluctuations.

Note that well-separated Monte Carlo configurations may always be saved for later use to measure any physical quantity.

Error Bars

Given a set of N_c independent Monte Carlo measurements of an observable, θ, the "sample variance" is

$$S_\theta^2 \equiv \frac{1}{N_c - 1}\sum_{i=1}^{N_c}\left(\theta_i - \bar{\theta}\right)^2, \tag{E.27}$$

where the $N_c - 1$ in the denominator reflects the constraint that the mean, $\bar{\theta}$, has already been determined from the same data set. In the $N_c \gg 1$ limit, one assumes

$$\lim_{N_c \gg 1} S_\theta^2 = \sigma_\theta^2, \tag{E.28}$$

* N. Metropolis et al., *J. Chem. Phys.* **21**, 1087 (1953).

where σ_θ^2 is the square of the standard deviation for this quantity (see Equation E.26).

It is not always possible to make independent measurements of quantities on individual configurations. For example, it might be necessary to combine such data to fit a curve. Then the question is: How good is this fit? One of the popular and easy methods to implement statistical methods in a more general context is called the "jackknife method," introduced by M. Quenouille in 1956 and developed further by J. Tukey.* In our application, let us say that we are extracting measurements of an observable, θ, from our N_c configurations. The jackknife method estimates the variance in this measurement from the formula

$$S_{J\theta}^2 \equiv (N_c - 1)\sum_{i=1}^{N_c}\left(\theta_i^J - \bar{\theta}_i^J\right)^2, \tag{E.29}$$

where $\theta_i^J (i = 1, ..., N_c)$ represents a measurement of the quantity with the ith configuration *removed*, and

$$\bar{\theta}^J \equiv \frac{1}{N_c}\sum_{i=1}^{N_c}\theta_i^J. \tag{E.30}$$

The $\bar{\theta}^J$ in Equation E.30 is the average of these N_c removed measurements. Note that in general $\bar{\theta}^J \neq \bar{\theta}$, where $\bar{\theta}$ represents the measurement of the quantity using all configurations. However, this procedure gives the same sample variance as in Equation E.27 when estimating the variance of the mean of a set of numbers (see Problem E.5). It has the advantage of being applicable not just to the standard deviation of some quantity but also, for example, to the standard deviation of the standard deviation, which can help establish the robustness of the error estimate. The disadvantage is the extra computations required (about N_c times as much for the mean) as compared to Equation E.27. Usually, though, this is small compared to the simulation time itself.

Many more types of numerical methods are necessary for analyzing Monte Carlo data. For example, internal data correlations and curve fitting must also be considered. Unfortunately, I do not have the time to discuss these here.†

Pseudorandom Numbers

I have referred to random changes in the state variables in Step 1 of the Metropolis algorithm. In addition, Step 4 involves a probabilistic choice. This choice can be implemented by generating a random number, x, uniformly distributed in the interval [0.1). If $e^{-\beta\Delta E} < x$, then we accept the change, if not, we ignore it. Thus, a random number generator is essential for Monte Carlo simulations.

Truly random numbers can be accessed, for example, from hardware-supplied quantum fluctuation phenomena, but such methods are often impractical. In addition, there is the issue of reproducibility, which is important in algorithm debugging. This can only be arranged in such a setting by storing a sufficiently large set of numbers, which must then be quickly available to the algorithm.

The numbers that are generated by computer algorithms cannot possibly be truly random, or unpredictable. They are arranged to be reproducible given an initial starting *seed* and are called *pseudorandom*. Pseudorandom numbers can be generated by a variety of algorithms. The generator should run fast so that little time in a simulation is used for their extraction. Typically, such

* Described in M. Yang and D. Robinson, *Understanding and Learning Statistics by Computers* (World Scientific, 1980).

† See P. Bevington and D. Robinson, *Data Reduction and Error Analysis for the Physical Sciences*, 3rd ed. (McGraw-Hill, 2003), for a discussion of many useful numerical techniques for data analysis.

generators have an iteration cycle period after which the numbers get repeated or are correlated. Obviously, one's simulation must run a smaller number of iterations than this cycle period or risk inaccurate results from the Monte Carlo.

A simple, fast pseudorandom number generator (PRNG) is the linear congruence method (LCM) developed by D. Lehmer in 1951.* It is based on the algorithm

$$y_{i+1} = ay_i + b \pmod{c}. \tag{E.31}$$

That is, one performs the integer multiplication on the right-hand side, then divides by c and assigns the remainder to y_{i+1}. The large integer c is often chosen to be 2^{32} on 32-bit computers. There are important restrictions on the values of the integers a and b to avoid unwanted numerical correlations. The iteration period in this algorithm is at most c. Numbers from Equation E.31 are mapped into the interval [0,1) by using

$$x_i = \frac{y_i}{c}. \tag{E.32}$$

LCM-type algorithms are implemented on many computer compilers. However, because of subsequences that can occur that have a period much smaller than c, they are not considered suitable for large Monte Carlo simulations. Newer PRNGs are available that correct problems found in the LCM but are still fast.†

E.2 MONTE CARLO SIMULATIONS OF GAUGE-FIELD THEORIES

All these concepts and more are necessary in the Monte Carlo simulations of gauge-field theories. However, there are differences in variables, scope, and context compared to the Ising model. Instead of a simple spin variable, one has gauge-field "links," which are matrices, as described in Chapter 12, Section 12.8. More sophisticated Monte Carlo algorithms are used, including multiple-hit Metropolis and heat bath varieties.‡ In addition, there are fermion and antifermion degrees of freedom in the gauge theory. The simulations take place in a four-dimensional lattice world called Euclidean space, where no distinction is made between spatial and time direction variables. In this space, Feynman's "path integral"§ becomes the partition function! In addition, the quantity that is used in the Boltzmann factor is no longer the system energy, but the so-called Euclidean action (a generalization of the classical action used to derive Newton's laws), which includes both gluonic and fermionic variables.

The inclusion of the fermionic part of the action in the simulation causes a vast increase in the computational complexity and computer cycles in the form of the *fermion determinant.* When fermions are included, one must begin to select configurations using the weight ($\det(M)\, e^{-\beta S_E^G}$), where S_E^G is the gluonic Euclidean action and M the *fermion matrix,* which describes fermion propagation and interactions. The fermion matrix is large but sparse; it contains many zeros in each row or column, since, like the Ising model, only nearest-neighbor fermion interactions are usually included. An exact evaluation of the determinant is impractical for lattices that are large enough to be realistic. This limitation forced earlier simulations to drop the determinant

* D. Knuth, *The Art of Computer Programming. Volume 2: Seminumerical Algorithms,* 3rd ed. (Addison-Wesley, 1997).
† J. Gentle, *Random Number Generation and Monte Carlo Methods,* 2nd ed. (Springer, 2003).
‡ M. Creutz, *Quarks, Gluons and Lattices* (Cambridge, 1983).
§ Oh, how I wish there was time and space to cover this concept! This is usually covered in advanced books on quantum mechanics. For an introductory coverage, see J.J. Sakurai and J. Napolitano, *Modern Quantum Mechanics,* 2nd ed. (Addison-Wesley, 2011). Also see Creutz, *Quarks,* Chapter 3.

completely. This is known as the *quenched approximation,* and corresponds to ignoring the effect of the sea of virtual fermion–antifermion pairs that spontaneously emerge from the vacuum at any point in space. These are called *fermion loops,* and the effect is known as *vacuum polarization.* Better implementation methods and faster computers can now avoid this approximation.

A Monte Carlo simulation of a gauge theory produces configurations that represent "snapshots" of the theory's vacuum state. Interestingly, there are no user-supplied values of the interlattice spacing in such simulations. All the numbers from the simulation are considered dimensionless, and the lattice spacing is set by internal consistency checks. The analog of $\beta = 1/kT$ in the gauge-field case is termed the *beta function* (sorry, not very original). I will denote this as $\beta(a)$, where a is the lattice spacing. Indeed, just like the one-dimensional Ising model, one finds that correlation lengths in the gauge simulation increase as $\beta(a)$ increases. If the lattice one is employing has linear dimensions greater than all physical length scales, then the physics should be independent of the spatial boundary conditions (which are usually taken to be periodic), and the lattice is said to be in the *continuum limit.*

Once one has the lattice vacuum state, combinations of fermion creation and annihilation operators can act on it to produce new states. For quarks the important combinations correspond to quark–antiquark pairs (mesons) and three quark states (baryons).

The propagation of fermions on the lattice is dealt with very simply. The solution to the linear equations (understood sum on j),

$$(M)_{ij}\, x_j = b_i, \tag{E.33}$$

where $(M)_{ij}$ are elements of the fermion matrix, and x_i and b_i are column vectors, is formally just

$$x_i = (M^{-1})_{ij}\, b_j. \tag{E.34}$$

It is essential to have numerical techniques that can quickly and efficiently solve the sparse system of linear equations in Equation E.33 since this step occurs many times in a given simulation.* The inverse of the quark matrix, M^{-1}, is called the *fermion propagator.* Note that the whole fermion propagator is much too large to construct for each configuration. (This N-by-N matrix might have N on the order of 10^6.) The "source vector," b, specifies the initial starting properties of the fermions (including starting position or positions), and the solution vector, x, gives the amplitude associated with the propagation of the fermion to all final states (including again positions). The rows and columns of the propagator give the amplitudes for propagation in the lattice:

$$M^{-1}_{\underbrace{\{x',c',D'\}}_{\text{Final particle properties}},\ \underbrace{\{x,c,D\}}_{\text{Initial particle properties}}}$$

$\{x',c',D'\}$ represents a collective row index including position (x), color (c), and "Dirac" (D) index (giving spin, particle/antiparticle indices); $\{x,c,D\}$ is likewise a collective column index. The columns represent initial properties and the rows give final ones. When the source vector b is only nonzero at a particular lattice position, the solution vector gives the amplitudes for fermion propagation to all final positions. These can then be used to make hadronic amplitudes and form many different types of lattice "measurements."

* See, for example, Y. Saad, *Iterative Methods for Sparse Linear Systems,* 2nd ed. (SIAM, 2003).

I have just scratched the surface of one aspect of the vast and developing subject of computational physics here. One of the newest topics, pregnant with possibilities, is the subject of quantum computing; see Appendix G for a short tutorial on this fascinating and timely subject. As far as QCD applications are concerned, the interested reader can explore the reviews listed in the footnote as well as the resource letter prepared by A. Kronfeld and C. Quigg.*

PROBLEMS

E.1 Show Equations E.5 and E.6 are correct expressions.

E.2 Show Equations E.17 and E.18 for the Ising model.

***E.3** Show the results Equations E.21 and E.22 from Equations E.20 for the nonperiodic case. (I had to use computer symbolic manipulation methods.)

***E.4** Write a computer code to simulate an N-site one-dimensional Ising model with periodic boundary conditions. Use the Monte Carlo method with the Metropolis algorithm. Compare the output (values + error bars) to the theoretical results for $\langle E\rangle/N$ and $\langle M\rangle/N$ for a variety of temperatures. (You may set the Boltzmann constant, k, equal to one.)

E.5 (a) In the jackknife method, show that we always have $\bar{\theta} = \bar{\theta}^J$ if the θ_i are simply a set of numbers.

(b) Use (a) to show that $S^2_{J\theta} = S^2_\theta$, that is, the standard deviations from Equations E.30 and E.28 are exactly the same in this case. [Note: In more complicated situations, $\bar{\theta}^J \neq \bar{\theta}$, and the use of the jackknife average of averages, $\bar{\theta}^J$, in Equation E.30 is necessary to correct a "bias" in the $\bar{\theta}_i$ statistic.†]

E.6 Briefly describe the random number generator algorithm on your computer or local machine.

* See S. Hashimoto, J. Laiho, and S. Sharpe, "Lattice Quantum Chromodynamics" (Particle Data Group). Also see Z. Fodor and C. Hoelbling, *Rev. Mod. Phys.* **84**, 449 (2012); and A. Kronfeld and C. Quigg, "Resource Letter: Quantum Chromodynamics," *Am. J. Phys.* **78**, 1081 (2010).

† Yang and Robinson, *Understanding and Learning Statistics.*

Appendix F: Weak Flavor Mixing

In this appendix, I will explain more about the significance of flavor mixing in the weak interactions. I will base this on a very insightful article in *Physics Today* by Howard Georgi, which appeared in April 1988. The interesting point made here is that the structure of flavor interactions can be understood from a straightforward analogy with coupled harmonic oscillators. I will try to keep my explanation mostly in line with the concepts and ideas already introduced in the text, so the emphases and presentation will differ a little from Georgi's article. I am presenting this in an appendix because the text does not require this coverage for the flow of ideas and also because it is not guaranteed that all the concepts encountered here will have had an appropriate pedagogical introduction. Georgi's article is much more refined and detailed than mine, and I highly recommend the original for a more complete and expert point of view.

Our starting Hamiltonian is just

$$H_0 = \Omega\hbar \sum_j A_j^\dagger A_j \quad \text{(Drop constant term; called "normal ordering")}$$

which we recognize as the Hamiltonian for the three-dimensional harmonic oscillator. There are symmetries here. Let

$$A_j \to \sum_k U_{jk} A_k, \tag{F.1}$$

$$\Rightarrow A_j^\dagger \to \sum_k A_k^\dagger U_{jk}^* = \sum_k A_k^\dagger U_{kj}^\dagger. \tag{F.2}$$

$$H_0 \to \Omega\hbar \sum_{j,k,l} A_k^\dagger U_{kj}^\dagger U_{jt} A_\ell = \Omega\hbar \sum_{k,\ell} A_k^\dagger A_\ell \left(\sum_j U_{kj}^\dagger u_{j\ell} \right) \tag{F.3}$$

U is unitary:

$$U^\dagger U = 1, \Rightarrow \sum_j U_{kj}^\dagger u_{j\ell} = \delta_{k\ell}, \tag{F.4}$$

$$\Rightarrow H_0 \to \Omega\hbar \sum_{k,\ell} A_k^\dagger A_\ell \delta_{k\ell} = H_0 : \text{unchanged}. \tag{F.5}$$

U is a unitary, 3×3 matrix $\Rightarrow$ the group known as $SU(3)$. Actually, there is a further symmetry here. Let

$$A_j \to e^{i\alpha} A_j, \quad \Rightarrow A_j^\dagger \to e^{-i\alpha} A_j^\dagger, \tag{F.6}$$

$$H_0 \to \Omega\hbar \sum_j A_j^\dagger A_j e^{-i\alpha} e^{i\alpha} = H_0 : \text{unchanged}. \tag{F.7}$$

Complete symmetry: $SU(3)\otimes U(1)$. The $SU(3)$ symmetry consequence is that the N_k are conserved:

$$\left[H_0, N_k\right] = \Omega\hbar \sum_j \left[A_j^\dagger A_j, A_k^\dagger A_k\right]. \tag{F.8}$$

In Chapter 9 we learned that

$$\left[N_i, N_k\right] = \left\lfloor A_j^\dagger A_j, A_k^\dagger A_k \right\rfloor = 0. \tag{F.9}$$

Problem 6.1.1 (see Chapter 6) shows that the degeneracy factor for the state nth quantum state is $1/2(n+1)(n+2)$. Let us go through it here to refresh our memories:

$$E = \hbar\omega\left(n_1 + n_2 + n_3 + \frac{3}{2}\right) = \hbar\omega\left(n + \frac{3}{2}\right),\ n_1, n_2, n_3 = \{0,1,2\ldots.\}.$$

Set $n_1 = 0$	How many ways?	$n+1$
Set $n_1 = 1$	"	n
.	.	.
.	.	.
.	.	.
Set $n_1 = n$	How many ways?	1

$$\Rightarrow \sum_{i=1}^{n+1} i = \frac{1}{2}(n+1)(n+2). \quad \left(\text{Given} \sum_{i=1}^{n} i = \frac{1}{2}n(n+1)\right)$$

The first few energies are:

$$6\text{ states}\begin{cases} n_1 = 2, n_2 = 2, n_3 = 2\ (\text{others} = 0) \\ n_1 = n_2 = 1, n_3 = 0 \\ n_1 = n_3 = 1, n_2 = 0 \quad E = \frac{7}{2}\hbar\omega \\ n_2 = n_3 = 1, n_1 = 0 \end{cases}$$

$$3\text{ states}\begin{cases} n_1 = 1,\ n_2 = n_3 = 0 \\ n_2 = 1,\ n_1 = n_3 = 0 \quad E = \frac{5}{2}\hbar\omega \\ n_3 = 1,\ n_1 = n_2 = 0 \end{cases}$$

$$1\text{ state}\begin{cases} n_1 = n_2 = n_3 = 0 \quad E = \frac{3}{2}\hbar\omega \end{cases}$$

Now cease to think of these as energy states. Consider n_1, n_2, n_3 to be particle occupation numbers of a Bose–Einstein system. Remember, I showed in Chapter 9 that the algebra of

a_i and $a_i^\dagger$ (creation and annihilation operators),

for the multiparticle bosonic states is identical to

A and A^+ (raising and lowering operators),

for the harmonic oscillator. If instead n_1, n_2, n_3 were the particle occupation numbers of a Fermi–Dirac system or Bose–Einstein, we would have the Cartesian classification and degeneracies:

	Fermi–Dirac degeneracy	Bose–Einstein degeneracy
$E = 9/2\,\hbar\omega$	1	10
$E = 7/2\,\hbar\omega$	3	6
$E = 5/2\,\hbar\omega$	3	3
$E = 3/2\,\hbar\omega$	1	1

Now introduce an interaction (let's not worry about units):

$$H_{\text{int}} = \vec{B} \cdot (\vec{p} \times \vec{r}) = -\vec{B} \cdot \vec{L}. \tag{F.10}$$

Why is such a term reasonable? I argued in Chapter 8 (based on the Chapter 1 discussion) that a magnetic field interacts with a magnetic dipole according to

$$H'_{\text{int}} = -\vec{m} \cdot \vec{B}. \tag{F.11}$$

Given a gyromagnetic ratio such that

$$\vec{m} = \gamma \vec{L}, \tag{F.12}$$

($\gamma = q/(2mc)$ classically), then

$$H'_{\text{int}} = -\gamma \vec{B} \cdot \vec{L}. \tag{F.13}$$

With the harmonic oscillator raising and lowering operators,

$$A_j = \frac{1}{\sqrt{2m\hbar\Omega}} p_j - i\sqrt{\frac{m\Omega}{2\hbar}} r_j, \tag{F.14}$$

$$A_j^{\dagger} = \frac{1}{\sqrt{2m\hbar\Omega}} p_j + i\sqrt{\frac{m\Omega}{2\hbar}} r_j, \tag{F.15}$$

we can evaluate $H_{\text{int}} = \vec{B} \cdot (\vec{p} \times \vec{r})$ as (Exercise F.2):

$$\vec{B} \cdot (\vec{p} \times \vec{r}) = i\hbar \sum_{i,j,k} \varepsilon_{ijk} B_i A_j^{\dagger} A_k. \tag{F.16}$$

Writing this as

$$H_{\text{int}} = \sum_{j,k} A_j M_{jk} A_k, \tag{F.17}$$

we identify

$$M_{jk} = i\hbar \sum_{i} \varepsilon_{ijk} B_i. \tag{F.18}$$

Notice that

$$M^*_{jk} = M_{kj} \Rightarrow \text{Hermitian.} \tag{F.19}$$

Also notice (do the transformation again) that under

$$A_j \to \sum_k U_{jk} A_k, \quad A_j^\dagger \to \sum_k A_k^\dagger U_{kj}^\dagger, \tag{F.20}$$

we have $H_0 \to H_0$, but

$$H_{\text{int}} \to \sum_{j,k,\ell,m} A_\ell^\dagger U_{\ell j}^\dagger M_{jk} U_{km} A_m = \sum_{\ell,m} A_\ell^\dagger \widetilde{M}_{\ell m} A_m, \tag{F.21}$$

where

$$\widetilde{M}_{\ell m} = \sum_{j,k} U_{\ell j}^\dagger M_{jk} U_{km}. \tag{F.22}$$

$$\text{As matrices: } \widetilde{M} = U^\dagger M U. \tag{F.23}$$

Since we do not get $H_{\text{int}} \to H_{\text{int}}$, we have that the $SU(3)$ symmetry is broken by H_{int}. The consequence is that the individual N_k are no longer conserved, just the total $N = \sum_k N_k$ (the $U(1)$ part). That is, we have

$$\begin{aligned} [H_{\text{int}}, N_j] &= [i\hbar \sum_{i,k,\ell} \varepsilon_{ik\ell} B_i A_k^\dagger A_\ell^\dagger, A_j^\dagger A_j] \\ &= i\hbar \sum_{i,k,\ell} \varepsilon_{ik\ell} B_i \left[A_k^\dagger A_\ell^\dagger, A_j^\dagger A_j \right]. \end{aligned} \tag{F.24}$$

But

$$\left[A_i^\dagger A_j, A_k^\dagger A_\ell \right] = \delta_{jk} A_i^\dagger A_\ell - \delta_{i\ell} A_k^\dagger A_j, \tag{F.25}$$

so $(i = j)$

$$\left[A_j^\dagger A_j^\dagger, A_k^\dagger A_\ell \right] = \delta_{jk} A_j^\dagger A_\ell - \delta_{j\ell} A_k^\dagger A_j. \tag{F.26}$$

$$\begin{aligned} \Rightarrow \left[H_{\text{int}}, N_j \right] &= -i\hbar \sum_{i,k,\ell} \varepsilon_{ik\ell} B_i \left(\delta_{jk} A_j^\dagger A_\ell - \delta_{j\ell} A_k^\dagger A_j \right). \\ &= -i\hbar \sum_{i,\ell} \varepsilon_{ij\ell} B_i \left(A_j^\dagger A_\ell + A_\ell^\dagger A_j \right) \neq 0. \end{aligned} \tag{F.27}$$

But notice, however, that

$$\sum_j \left[H_{\text{int}}, N_j \right] = -i\hbar \sum_{i,\ell,j} \varepsilon_{ij\ell} B_i \left(A_j^\dagger A_\ell + A_\ell^\dagger A_j \right) = 0. \tag{F.28}$$

Now let us go back to

$$\widetilde{M} = U^{\dagger} M U. \tag{F.29}$$

We know that we can diagonalize any Hermitian matrix, M, in this fashion. Why do we want to do this? Because if we can find a U such that

$$\widetilde{M}_{ij} = \hbar \omega_i \delta_{ij}, \tag{F.30}$$

then

$$H_{\text{int}} = \sum_{\ell,m} A_\ell^{\dagger}\, \widetilde{M}_{\ell m} A_m \rightarrow \hbar \sum_{\ell} \omega_\ell A_\ell^{\dagger} A_\ell. \tag{F.31}$$

Let us find the ω's for our explicit case.

$$M = i\hbar \begin{pmatrix} 0 & B_3 & -B_2 \\ -B_3 & 0 & B_1 \\ B_2 & -B_1 & 0 \end{pmatrix}. \tag{F.32}$$

Get the ω's from the characteristic equation. (See Chapter 4, Section 4.4.) We form,

$$\det\left(M - \omega_i\right) = 0, \tag{F.33}$$

$$\Rightarrow \det \begin{pmatrix} -\omega & B_3 & -B_2 \\ -B_3 & -\omega & B_1 \\ B_2 & -B_1 & -\omega \end{pmatrix} = 0, \tag{F.34}$$

$$\Rightarrow \omega^3 - \omega\left(B_1^2 + B_2^2 + B_3^2\right) = 0. \tag{F.35}$$

Three solutions:

$$\begin{cases} \omega_1 = |\vec{B}|, \\ \omega_2 = -|\vec{B}|, \\ \omega_3 = 0. \end{cases} \tag{F.36}$$

Therefore,

$$H_{\text{int}} = \hbar \sum_i \omega_i A_i^{\dagger} A_i = \hbar |\vec{B}| \left(A_1^{\dagger} A_1 - A_2^{\dagger} A_2\right). \tag{F.37}$$

Then using first-order perturbation theory,

$$\langle H_{\text{int}} \rangle = \hbar |\vec{B}| \left(n_1 - n_2\right). \tag{F.38}$$

The energy levels now appear as in Figure F.1.

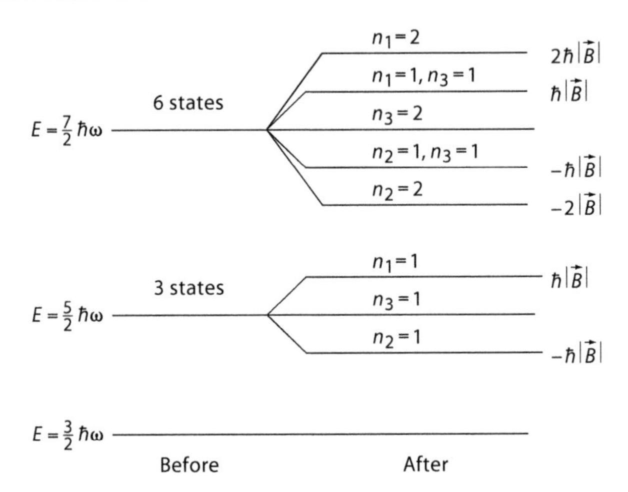

FIGURE F.1 Energy levels of a harmonic oscillator before (left) and after (right) the interaction in Equation F.38.

In general, the splitting is

$$\langle H_{\text{int}} \rangle = \left\langle \sum_i \hbar\omega_i A_i^\dagger A_i \right\rangle = \sum_i \hbar\omega_i n_i = \hbar\left(n_1\omega_1 + n_2\omega_2 + n_3\omega_3\right). \tag{F.39}$$

($n_1 + n_2 + n_3 = n$.) For $n = 1$, the splittings are as follows:

$$\begin{aligned} n_1 &= 1 \rightarrow \omega_{1,} \\ n_2 &= 1 \rightarrow \omega_{2,} \\ n_3 &= 1 \rightarrow \omega_3. \end{aligned}$$

For $n = 2$, the splittings are:

$$\begin{aligned} n_1 &= 2 &&\rightarrow 2\omega_{1,} \\ n_2 &= 2 &&\rightarrow 2\omega_{2,} \\ n_3 &= 2 &&\rightarrow 2\omega_{3,} \\ n_1 &= n_2 = 1 &&\rightarrow (\omega_1 + \omega_2), \\ n_1 &= n_3 = 1 &&\rightarrow (\omega_1 + \omega_3), \\ n_2 &= n_3 = 1 &&\rightarrow (\omega_2 + \omega_3). \end{aligned}$$

Thus the new Hamiltonian is just

$$H_A = \hbar \sum_j \left(\Omega + \omega_j\right) A_j^\dagger A_j, \tag{F.40}$$

which defines H_A.

Now let us add another interaction, H_B, analogous to the first:

$$\left[A_i, B_j\right] = 0, \quad \left[A_i, B_j^\dagger\right] = 0, \tag{F.41}$$

$$H_A + H_B = \hbar \sum_j \left[\left(\Omega_A + \omega_{Aj}\right) A_j^\dagger A_j + \left(\Omega_B + \omega_{Bj}\right) B_j^\dagger B_j\right]. \tag{F.42}$$

We will see that we can view the H_A Hamiltonian as representing the charge 2/3 quarks (*u*, *c*, and *t*), and the H_B Hamiltonian as representing the charge −1/3 quarks (*d*, *s*, and *b*), at least when we are talking about the $n = 1$ states. States are represented by the number ket:

$$|n_{A_1}, n_{A_2}, n_{A_3}, n_{B_1}, n_{B_2}, n_{B_3}\rangle. \tag{F.43}$$

Total N_A and N_B are separately conserved for this form.

Now also add (we're almost there!)

$$H_W = \hbar W \sum_j \left(A_j^\dagger B_j + B_j^\dagger A_j\right), \tag{F.44}$$

which is a stand-in for the weak interaction. Notice that H_W is Hermitian ($H_W^\dagger = H_W$). If we were to drop the ω_{A_j}, ω_{B_j} terms from earlier, we would have a remaining symmetry, as follows:

$$\left.\begin{aligned} A_j &\to \sum_k U_{jk} A_k \\ B_j &\to \sum_k U_{jk} B_k \end{aligned}\right\} \text{same } U_{jk}. \tag{F.45}$$

For example,

$$H_W \to \hbar W \sum_{\ell,j,k} \left[A_k^\dagger \overbrace{U_{kj}^\dagger U_{j\ell}}^{\delta_{k\ell}} B_\ell + B_\ell^\dagger \overbrace{U_{\ell j}^\dagger U_{jk}}^{\delta_{\ell k}} A_k \right] = H_W. \tag{F.46}$$

This is a *combined* $SU_{A+B}(3) \otimes U(1)$ symmetry and has the consequence that $(N_{Ai} + N_{Bi})$ is conserved in this case. That is

$$\left[N_{Ai} + N_{Bi}, H_W\right] = 0. \tag{F.47}$$

However, now let the ω_{Aj} and ω_{Bj} both be nonzero. The $SU_{A+B}(3)$ symmetry is now broken in general. The resultant full Hamiltonian can be written as

$$H \equiv H_A + H_B + H_W = \hbar \sum_j \left(A_j^\dagger B_j^\dagger\right) \begin{pmatrix} \Omega_A + \omega_{A_j} & W \\ W & \Omega_B + \omega_{B_j} \end{pmatrix} \begin{pmatrix} A_j \\ B_j \end{pmatrix}. \tag{F.48}$$

Now, it is only the *total* $N_A + N_B = \sum_i \left(N_{A_i} + N_{B_i}\right)$ that is conserved.

Let

$$\widehat{M} \equiv \begin{pmatrix} \Omega_A + \omega_{A_j} & W \\ W & \Omega_B + \omega_{B_j} \end{pmatrix}. \tag{F.49}$$

Just as before, diagonalize it to get the energies. Define $\Omega_A + \omega_{Ai} \equiv a_i$, $\Omega_B + \omega_{Bi} \equiv b_i$. We have

$$\det \begin{pmatrix} a_i - \lambda^i & W \\ W & b_i - \lambda^i \end{pmatrix} = 0. \tag{F.50}$$

The eigenvalues are

$$\lambda_a^i \equiv \frac{a_i + b_i}{2} + \frac{a_i - b_i}{2}\sqrt{1 + \frac{4W^2}{\left(a_i - b_i\right)^2}}, \tag{F.51}$$

$$\lambda_b^i \equiv \frac{a_i + b_i}{2} + \frac{b_i - a_i}{2}\sqrt{1 + \frac{4W^2}{\left(a_i - b_i\right)^2}}. \tag{F.52}$$

These roots are chosen so that $\lambda_a^i \to a_i$, $\lambda_b^i \to b_i$ for $W \to 0$. What about the eigenvectors? H is now diagonalized as

$$H = \hbar\sum_i \left(\lambda_a^i \alpha_i^\dagger \alpha_i + \lambda_b^i \beta_i^\dagger \beta_i\right), \tag{F.53}$$

where

$$\alpha_i \equiv \cos\Theta_i A_i + \sin\Theta_i B_i, \tag{F.54}$$

$$\beta_i \equiv -\sin\Theta_i A_i + \cos\Theta_i B_i. \tag{F.55}$$

The angle Θ_i is given by the various expressions,

$$\tan\Theta_i = \frac{\lambda_a^i - a_i}{W} = \frac{W}{\lambda_a^i - b_i} = -\frac{\lambda_b^i - b_i}{W} = -\frac{W}{\lambda_b^i - a_i}, \tag{F.56}$$

and one can also show that with appropriate root choices,

$$\cos 2\Theta_i = \frac{1}{\sqrt{1 + \frac{4W^2}{\left(a_i - b_i\right)^2}}}, \tag{F.57}$$

$$\sin 2\Theta_i = \frac{2W\left(a_i - b_i\right)}{\left(a_i - b_i\right)^2\sqrt{1 + \frac{4W^2}{\left(a_i - b_i\right)^2}}}, \tag{F.58}$$

$$\tan 2\Theta_i = \frac{2W}{\left(a_i - b_i\right)}. \tag{F.59}$$

Given the above definitions, I will ask you to work backward in a problem to show Equation F.53 is equal to the earlier Hamiltonian, Equation F.48, which was

$$H = H_A + H_B + H_W = \hbar\sum_i \left(a_i A_i^\dagger A_i + b_i B_i^\dagger B_i + W\left(A_i^\dagger B_i + B_i^\dagger A_i\right)\right), \tag{F.60}$$

to make sure nothing has changed.

So far, we have just set up some interesting static eigenstates, which are supposed to have a similar energy level structure to the quarks. However, if we are to model the weak interactions in action, we will have to introduce another interaction that we can use to produce transitions between the various states.

Remember Fermi's golden rule. From Chapters 9 and 11,

$$\text{rate} \sim |\langle 2|T|1\rangle|^2, \tag{F.61}$$

for a 1 to 2 transition, where T is the transition operator. Georgi suggests we take

$$T = \sum_i \left(A_i^\dagger A_i - B_i^\dagger B_i\right) = N_A - N_B. \tag{F.62}$$

Some caveats are necessary here. In reality, transitions would be mediated by the weak interaction Hamiltonian itself, and the transition operator would incorporate a factor of the weak coupling constant, which we will see is proportional to W. However, in our model, viewing Equation F.62 as part of the weak Hamiltonian would be counterproductive because it would produce mixtures, rather than transitions, between the α and β states. Note that if we had taken a plus sign in Equation F.62, there would be no transitions at all, since $N_A + N_B$ is still a good quantum number. In this model we must understand that we are introducing the weak interactions in a two-step process; Equation F.44 represents the effect on the states, which does use a coupling constant, and Equation F.62 is the transition part, which does not. Eventually, in Equation F.67, we will also mix the quark states via the analog of the *CKM* matrix. Note that weak interaction transitions between real quarks require the emission of a charged W boson, such as is seen in the Appendix D weak (charged) quarkish diagram, which is not part of the model here.

Let us take an example:

$$|1_{\alpha_j}\rangle = \alpha_j^\dagger|0\rangle, \quad |1_{\beta_j}\rangle = \beta_j^\dagger|0\rangle, \tag{F.63}$$

$$\left\langle 1_{\alpha_i}|T|1_{\beta_j}\right\rangle = \langle 0|\left(\cos\Theta A_j + \sin\Theta B_j\right)\sum_k \left(\underbrace{A_k^\dagger A_k}_{N_{A_k}} - \underbrace{B_k^\dagger B_k}_{N_{B_k}}\right)\left(-\sin\Theta A_j^\dagger + \cos\Theta B_j^\dagger\right)|0\rangle \tag{F.64}$$

$\left(\left[N_{A_k}, A_j^\dagger\right] = A_j^\dagger \delta_{kj}, \left[N_{B_k}, B_j^\dagger\right] = B_j^\dagger \delta_{kj}\right)$. Thus

$$\begin{aligned}\left\langle 1_{\alpha_i}|T|1_{\beta_j}\right\rangle &= \langle 0|\left(\cos\Theta_j A_j + \sin\Theta_j B_j\right)\left(-\sin\Theta_j A_j^\dagger - \cos\Theta_j B_j^\dagger\right)|0\rangle, \\ &= -\cos\Theta_j \sin\Theta_j \langle 0|\left(A_j A_j^\dagger + B_j B_j^\dagger\right)|0\rangle, \\ &= -2\cos\Theta_j \sin\Theta_j = -\sin\left(2\Theta_j\right).\end{aligned} \tag{F.65}$$

Likewise, one can show (problem; all $i \neq j$):

$$\left\langle 1_{\alpha_i}|T|1_{\alpha_j}\right\rangle = 0, \left\langle 1_{\beta_i}|T|1_{\beta_j}\right\rangle = 0, \left\langle 1_{\alpha_i}|T|1_{\beta_j}\right\rangle = 0. \tag{F.66}$$

This gives rise to the interactions as pictured in Figure F.2 (take the $n = 1$ case again).

We see that interactions involving the same generation ($i = j$) of quarks are allowed. These are analogous to the same generation charged current quark transitions, involving the $W^\pm$ bosons, in the real world. The transition matrix element, Equation F.65, identifies the amplitude for this process to be approximately $2W/(\Omega_B - \Omega_A)$ $(\text{for}\,|\omega_{A_i}| << \Omega_A, |\omega_{B_i}| << \Omega_B)$ from Equation F.58, which is playing the role of the Fermi constant (assuming $\Omega_B > \Omega_A$). We do not yet have the smaller charged current interactions involving different generations. However, they are not hard to model. We need only make one more change to our Hamiltonian.

Here is the interesting part. Let us simply change the aforementioned H_W to

$$H_W' = \hbar W \sum_{j,k}\left(A_j^\dagger V_{jk} B_k + B_j^\dagger V_{jk}^\dagger A_k\right), \left(H_W'^\dagger = H_W'\right) \tag{F.67}$$

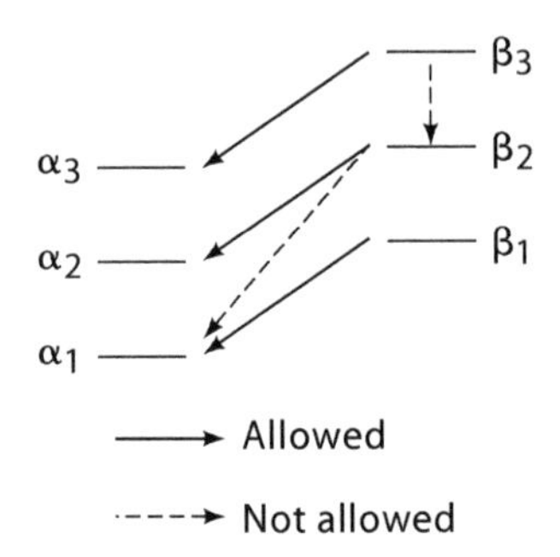

FIGURE F.2 Figure illustrating the allowed and forbidden interactions between the α and β $N = 1$ oscillator levels.

where V_{jk} is a unitary matrix. We now have mixings of generations involving all the charged current interactions! Note that this would be completely equivalent to old H_W if the $\omega_{A_j}, \omega_{B_j}$ terms in H_A and H_B were dropped or were simply all the same. For example, let

$$A_j \to \sum_k V_{jk} A_k, \quad A_j^\dagger \to \sum_k A_k^\dagger V_{kj}^\dagger, \tag{F.68}$$

then

$$\begin{aligned} H_W' &= \hbar W \sum_{\ell,j,k} \left(A_\ell^\dagger \underbrace{V_{\ell j}^\dagger V_{jk}}_{\delta_{\ell k}} B_k + B_j^\dagger \underbrace{V_{jk}^\dagger V_{k\ell}}_{\delta_{i t}} A_\ell \right), \\ &= \hbar W \sum_j \left(A_j^\dagger B_j^\dagger + B_j A_j \right), \text{ old form!} \end{aligned} \tag{F.69}$$

However, if we try to do this redefinition in the full $H = H_A + H_B + H_W$, it would *not* be equivalent to Equation F.48 or F.53. That is, *without* the $SU_A(3)$, $SU_B(3)$ symmetry-breaking terms, this new form is completely equivalent to the old and would not add anything new. However, because of the ω_{Aj}, ω_{Bj} terms, we now have all charged transitions in the previous picture, even those between different generations ($i \neq j$). Getting all the amplitudes would involve rediagonalizing the full H, giving new eigenstates α_i' and β_i', which I will not do here. Georgi, in his article, uses a perturbative approach to display these amplitudes.

As I said, the analog of the transitions from the β_i' to the α_i' states is the various charged current interactions involving the $W^\pm$ particles. The V_{jk} matrix is analogous to the *CKM* mixing matrix in quark weak interactions. There are still no *direct* generation-changing *neutral currents* (those connecting α_i' and α_j' or β_i' and β_j' for $i \neq j$), but now consider the transitions illustrated in Figure F.3.

α_1' is seen to be reached from either β_2' and β_1', which means, from the magic of quantum mechanics, that β_2' and β_1' are actually mixed. This is because a virtual second-order transition could first take β_2' into the state α'_1, then back up to β_1'. However, such transitions are doubly suppressed. First, this transition only occurs in second order due to the presence of the off-diagonal elements of V_{jk}. In addition, these only contribute because of the presence of the ω_{Aj}, ω_{Bj} terms. However, it is not just that the ω_{Aj}, ω_{Bj} terms are there, but that they are all *different* that allows these small transitions. In the real world, these small symmetry-breaking terms arise from the different masses of the quarks. An example of such a process in the Standard

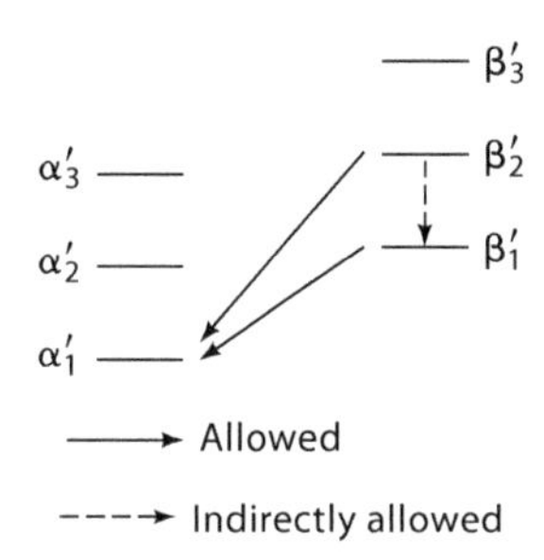

FIGURE F.3 Figure illustrating two of the allowed interactions between the new "α′" and "β′" $n = 1$ oscillator levels with the new $H_w^{'}$ from Equation F.67. Neutral current interaction is also indirectly allowed.

Model is $K^- \rightarrow \pi^- \nu_e \bar{\nu}_e$ called a *strangeness-changing weak neutral current* event, which has been observed, but occurs at a very low rate.

Originally, it was a great mystery why these amplitudes were so small. This was first explained by S. Glashow, J. Iliopoulos, and L. Mianai in their model in 1970, and the process by which such events are suppressed is called the *GIM mechanism*. Only three quarks were known at the time. This scenario required a fourth quark, charm, which was subsequently observed in a mesonic bound system called *charmonium* in 1974. Glashow was awarded the 1979 Nobel Prize in physics partly for these insights.

PROBLEMS

F.1 Let us see if we can also understand the ½(n + 1)(n + 2) degeneracy factor from the *radial* classification of states. (Compare with the Cartesian classification in Problem 6.1.1.) Go back to Chapter 7, Problems 7.5.1 and 7.6.6. We compared (3-D oscillator)

$$\left(\frac{d^2}{d\rho^2} - \frac{\ell(\ell+1)}{\rho^2} - \rho^2 + 2\left(n + \frac{3}{2}\right)\right) R_{n\ell}(r) = 0,$$

to (2-D oscillator)

$$\left[\frac{d^2}{d\rho^2} - \frac{m^2 - \frac{1}{4}}{\rho^2} + 2\left(|m| + 2n_r + 1\right) - \rho^2\right] \sqrt{\rho} P_{n,m}(\rho) = 0.$$

Should have found:

$$\begin{array}{cc} \text{2D} & \text{3D} \\ m^2 - \frac{1}{4} & \rightarrow \ell(\ell+1), \\ 2n_r + |m| + 1 & \rightarrow n + \frac{3}{2}, \end{array}$$

or

$$|m| \rightarrow \ell + \frac{1}{2}, n_r \rightarrow \frac{n-\ell}{2}, \quad (n_r = 0,1,2,3,\ldots).$$

(a) Considering the n even and n odd cases separately, show that the degeneracy is ½ $(n + 1)(n + 2)$.

(b) Write the radial quantum numbers (n, ℓ, m) of the $E = 3/2\hbar\omega, 5/2\hbar\omega, 7/2\hbar\omega$ states explicitly.

F.2 Show

$$\vec{B}\cdot(\vec{p}\times\vec{r}) = i\hbar\sum_{i,j,k}\varepsilon_{ijk}B_iA_j^\dagger A_k.$$

F.3 Show

$$\lfloor N_{A_i} + N_{B_i}, H_W \rfloor = 0.$$

F.4 Given Equation F.56, show that Equation F.59 follows directly. (Do not simply use Equations F.57 and F.58!)

F.5 After the change of variables described in the text, show that the Hamiltonian can still be written as:

$$H = \hbar\sum_i\left(a_iA_i^\dagger A_i + b_iB_i^\dagger B_i + W\left(A_i^\dagger B_i + B_i^\dagger A_i\right)\right).$$

F.6 Explicitly show (before the introduction of the V_{jk} matrix; $(i \neq j)$)

$$\langle 1_{\alpha_i}|T|1_{\alpha_i}\rangle = 0, \langle 1_{\beta_i}|T|1_{\beta_i}\rangle = 0, \langle 1_{\alpha_i}|T|1_{\beta_i}\rangle = 0.$$

F.7 Using Appendix D, show the two almost-canceling second-order Feynman diagrams involving $s \to u \to d$ and $s \to c \to d$ for the weak decay $K^- \to \pi^- \nu_e\bar{\nu}_e$.

Appendix G: Quantum Computing

G.1 INTRODUCTION

It is strongly suspected that computers that function as quantum mechanics simulators, so-called quantum computers, will prove to be inherently more powerful than conventional or classical computers. The main reason for this is that quantum computers, whose functions are based on manipulating coherent quantum amplitudes, will have a huge advantage in algorithmic speed. A number of fundamental algorithms require fewer logical steps to solve on a quantum computer than its conventional counterpart. Another intrinsic advantage is in the expansion of the memory space of computations. The state space of such machines will someday overwhelm conventional machines in terms of the storage and manipulation of large data sets.

Quantum computers are now a reality, but we are only in the initial stages of understanding the things such machines are capable of doing. I cannot in any way do justice to the burgeoning field of quantum computation and information here, but I will attempt to impart some of the excitement of the concepts involved and make connections to helpful references that can paint a more complete picture. In any case, a student in quantum mechanics is actually in a very advantageous position to understand, utilize, and manipulate such new resources, and so a modest introductory tutorial on this topic seems very appropriate.

G.2 COMPUTATIONAL BASICS

What is a quantum computer?* First, let us distinguish it from what we are terming a classical computer. A classical computer works its way serially through a problem, one step at a time. This mode of operation has been greatly improved upon over the past several decades by incorporating the concept of *parallel processing*: a problem can be broken into pieces and many independent calculations can be done simultaneously. However, a quantum computer will do better: it can manipulate all possibilities at once without the need to break the problem into pieces with the attendant overheads. The classical computer works on irreversible steps of logic, pulling items out of an inventory, whereas the quantum computer works on reversible global amplitudes that are immediately accessible. The logic necessary for a quantum computer is a generalization of the ways we learned how to manipulate spin 1/2 amplitudes in Chapter 1. Quantum computation involves all the concepts we learned about in our quantum excursions: indeterminism, interference, uncertainty, and superposition. There will only be a change in terminology here, with no change in the underlying quantum concepts. Mathematically, quantum computation is just an application of linear algebra, but it is important to get used to the language in this new field.

Let us first think about the basic algorithmic elements of a quantum computer. The basic computational component of a classical computer is the bit, which is a logical unit that can take on only two values, conventionally labeled 0 and 1, whereas the basic unit on a quantum computer

* See the early overview article for a more eloquent orientation: L. Grover, "Quantum Computing," *The Sciences*, July/August 1999, 24–30.

is a two-valued *qubit* (*qu*antum *bit*), whose general structure we learned about in Chapter 1. The general state of this qubit $|\psi\rangle$ is the linear superposition

$$|\psi\rangle = \alpha|0\rangle + \beta|1\rangle, \tag{G.1}$$

where bra–ket notation is being used, and the states are orthonormal: $\langle 0|0\rangle = \langle 1|1\rangle = 1$, $\langle 0|1\rangle = \langle 1|0\rangle = 0$. α and β are complex numbers with the normalization condition: $\langle\psi|\psi\rangle = |\alpha|^2 + |\beta|^2 = 1$. Thus $|0\rangle$ is simply "spin up" ($|0\rangle \equiv |+\rangle$ from Chapter 1), and $|1\rangle$ is "spin down" ($|1\rangle \equiv |-\rangle$):

$$|0\rangle \leftrightarrow \begin{pmatrix} 1 \\ 0 \end{pmatrix},\ |1\rangle \leftrightarrow \begin{pmatrix} 0 \\ 1 \end{pmatrix}. \tag{G.2}$$

The states $|0\rangle$ and $|1\rangle$ are called the *computational basis states*, and are the analog of the bits 0 and 1 in conventional computer logic. The column–matrix equivalent form of the state in Equation G.1 is of course

$$\psi = \begin{pmatrix} \alpha \\ \beta \end{pmatrix}. \tag{G.3}$$

After normalization, the qubit parameter space (or manifold) is characterized by two numbers. Going back to bra–ket notation, the most general state may be written

$$|\psi\rangle = \left(\cos\frac{\theta}{2}|0\rangle + e^{i\phi}\sin\frac{\theta}{2}|1\rangle\right) \tag{G.4}$$

(cf. Equations 1.80 and 1.222 in Chapter 1 and Equation 10.100 in Chapter 10) outside of an irrelevant overall phase, for the real parameters θ and ϕ. Based upon our experience with spin 1/2, we know that θ and ϕ may be interpreted as the spherical coordinate direction angles associated with the vector pointing in the geometrical "up" spin direction; θ is the polar angle and ϕ is the azimuthal one. This one-to-one mapping of the state in Equation G.4 to a direction in an internal space is embodied in the so-called *Bloch sphere*, shown in Figure G.1.

Defining a set of labels x_i=(0,1) for qubit i, a direct product of qubits, $|x_1,x_2,\ldots\rangle \equiv |x_1\rangle \otimes |x_2\rangle \otimes \ldots$, may be used as computational basis states for the computer state just as a conventional spin basis for a set of particles, $|m_1,m_2,\ldots\rangle \equiv |m_1\rangle \otimes |m_2\rangle \otimes \ldots$, can be used for the spin state, just as we studied in Chapter 8. So, for a two-qubit system, the basis states can be taken to be $|0,0\rangle$, $|1,0\rangle$, $|0,1\rangle$, and $|1,1\rangle$. The state of such a system can be in a mixture of such basis states, for example

$$|\psi\rangle = \alpha|0,0\rangle + \beta|1,0\rangle + \chi|0,1\rangle + \delta|1,1\rangle, \tag{G.5}$$

where $|\alpha|^2 + |\beta|^2 + |\chi|^2 + |\delta|^2 = 1$. Such a multiqubit state is said to be "entangled."

Now consider a system of n qubits. The number of different states of such a system is 2^n. The computational memory of a conventional one terabyte (10^{12} bytes) machine would be approximately equivalent to a quantum computer with only $n = 40$ qubits. Computer hardware has improved at an exponential rate for well over 50 years. A conventional encapsulation of this

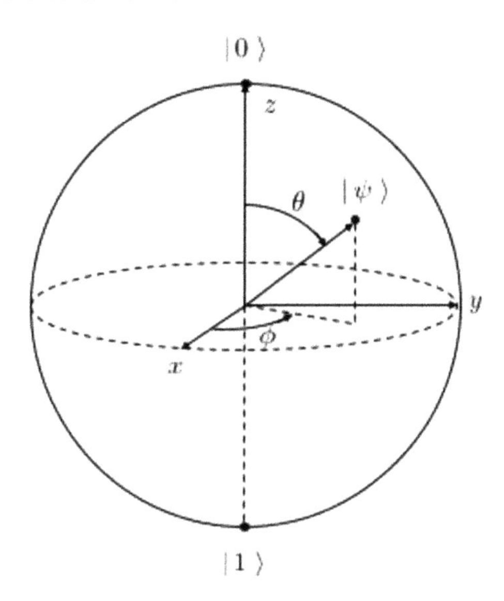

FIGURE G.1 The Bloch sphere representation of the state of a single qubit.

progress is embodied in Moore's law,* which states that market forces drive a doubling in computational power roughly every 2 years. However, this level of advancement in conventional memory would be equivalent to simply adding a single additional qubit to a quantum computer in the same amount of time!

A new computational field has emerged in which conventional computers are used to simulate quantum ones. Such simulations can be used for testing and reliability considerations. Conventional parallel computers with better algorithms have so far kept pace with their quantum cousins. However, a point will come where quantum computers, which are capable of doing 2^n mathematical operations simultaneously, will outpace all conventional ones. This point of separation is termed *quantum supremacy.*

G.3 SOME LOGIC GATE CONCEPTS

In conventional computers, logic is performed by so-called *logic gates* on bits. Four examples are shown in Figure G.2. The NOT gate is the only single-bit logic gate. The three others are called AND, OR, and NOR gates. They have incoming bits labeled a and b; the output bit is determined by the logic of the gate, which is represented by a symbol. Notice that in the last three cases the action is irreversible: 2 bits in, 1 bit out. In each case, an associated *truth table* is also given for the gate. The horizontal lines represent the causal flow from left to right.

Quantum logic gates are fundamentally different. The number of gates on input and output are always the same. In contrast to the conventional computer single-bit gate, there are an unlimited number of single-qubit logic gates. In fact, the operations on single-qubit gates are equivalent to 2×2 unitary matrices U in linear algebra, and the action on the gates is always reversible.

* Named after Gordon Moore, a computer entrepreneur who made his observation/forecast in 1975. More technically, the law involves the number of transistors on a central processing unit (CPU). This rate has held roughly constant until the present (2019), but this trend must end eventually, as microprocessors are reaching the limits of the classical world, where quantum indeterminacy is definitely not an advantage. The latest prognostication is that the law will last until around 2025.

Operation	Input	Symbol	Output	Truth Table
NOT	a		NOT a	a | NOT a 0 | 1 1 | 0
AND	a b		a AND b	a b | a AND b 0 0 | 0 0 1 | 0 1 0 | 0 1 1 | 1
OR	a b		a OR b	a b | a OR b 0 0 | 0 0 1 | 1 1 0 | 1 1 1 | 1
NOR	a b		a NOR b	a b | a NOR b 0 0 | 1 0 1 | 0 1 0 | 0 1 1 | 0

FIGURE G.2 Some standard logic gates for conventional computers.

Some standard single-qubit operations and their associated unitary transformations are given in Figure G.3. Note that such operations are specified by the action on the computational basis $|x_1,x_2\rangle \rightarrow U|x_1,x_2\rangle$, which is analogous to the truth tables for conventional logic gates. Multiple operations may be performed on a qubit, with the right-flowing causal order indicated by a horizontal line. We found it useful in Chapterl to explain the action of operators on quantum ket-states by the use of Process Diagrams, in which there was an understood causal flow of symbols from right to left. Similarly, workers in the quantum information field have found it useful to illustrate multiple qubit operations as such a flow. Qubits are conventionally shown as flowing horizontal lines, with operations taking place on them in causal order from left to right.

In addition to the single-qubit operations, there are also operations involving multiple qubits. One of the two-qubit types of logic gates is called *controlled* operations, in that one qubit state is used to control the output of another qubit. The symbolic form of the *controlled-U* operation is shown next. Note that the gates are labeled from top to bottom.

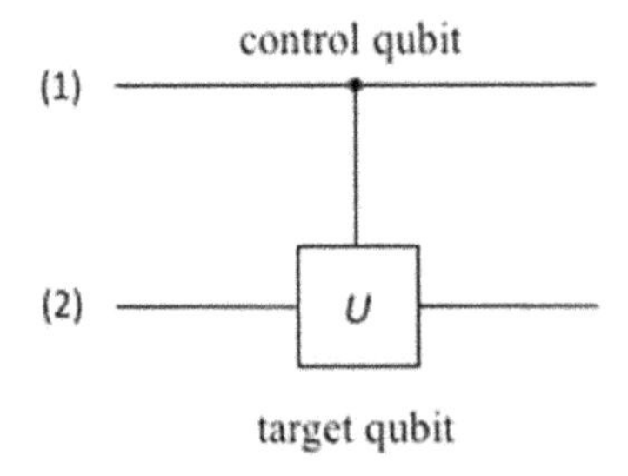

The *control qubit* is used to set a condition, whereas the *target qubit* is conditionally modified with one of the single-qubit logic gates, U. The control qubit is said to be "set" when its computational basis x-value is 1. That is, when qubit 1 is the control and qubit 2 is the target, one has the controlled $U^{(1)}$ unitary operation:

Name	Symbol	U matrix
Hadamard	H	$\frac{1}{\sqrt{2}}\begin{pmatrix} 1 & 1 \\ 1 & -1 \end{pmatrix}$
Pauli X	X	$\begin{pmatrix} 0 & 1 \\ 1 & 0 \end{pmatrix}$
Pauli Z	Z	$\begin{pmatrix} 1 & 0 \\ 0 & -1 \end{pmatrix}$
Phase	S	$\begin{pmatrix} 0 & 1 \\ 0 & i \end{pmatrix}$
π/8	T	$\begin{pmatrix} 0 & 1 \\ 0 & e^{i\pi/4} \end{pmatrix}$

FIGURE G.3 Some single-qubit logic gates and associated matrices.

$$|x_1,x_2\rangle \xrightarrow{U^{(1)}} \begin{cases} |x_1,x_2\rangle,\ x_1 = 0, \\ |x_1\rangle U|x_2\rangle,\ x_1 = 1. \end{cases} \tag{G.6}$$

When control and target are reversed, one has the controlled $U^{(2)}$ operation:

$$|x_1,x_2\rangle \xrightarrow{U^{(2)}} \begin{cases} |x_1,x_2\rangle,\ x_2 = 0, \\ U|x_1\rangle|x_2\rangle,\ x_2 = 1. \end{cases} \tag{G.7}$$

A fundamental controlled two-qubit operation, called CNOT, is defined by

$$|x_1,x_2\rangle \xrightarrow{\text{CNOT}(1)} |x_1,x_1 \oplus x_2\rangle, \tag{G.8}$$

when the first qubit is the control, and where the addition of the labels on the target qubit, $x_1 \oplus x_2$, is done in *mod(2)* arithmetic. It is symbolized by

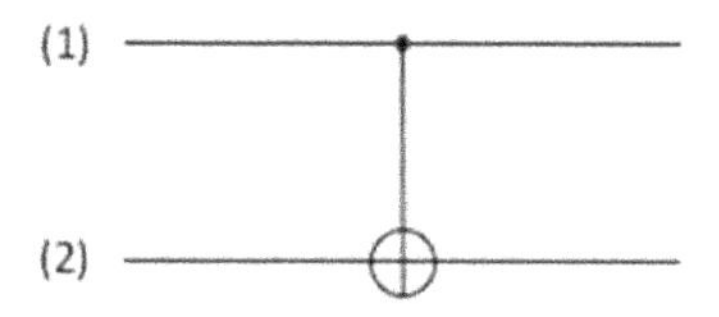

This operation is actually equivalent to a controlled *Pauli X* operation:

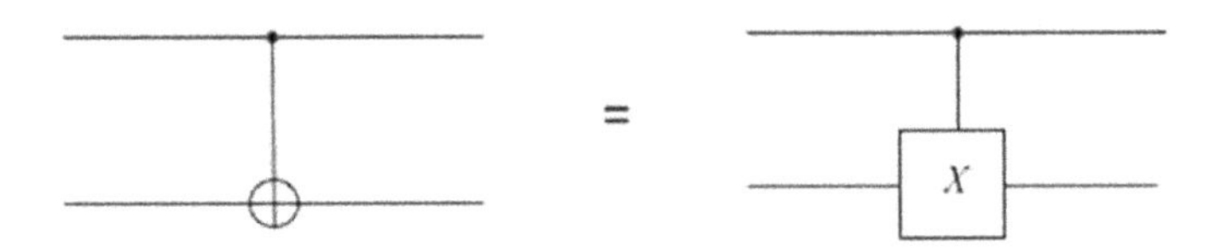

The equivalence is the subject of Problem G.3.

There are many other conditional qubit operations involving three or more qubits. For example, the following gate symbol shows a *Toffoli gate,* which is just a multiqubit generalization of the two-qubit CNOT gate. In this case, both control gates must be "set" for the third gate to act.

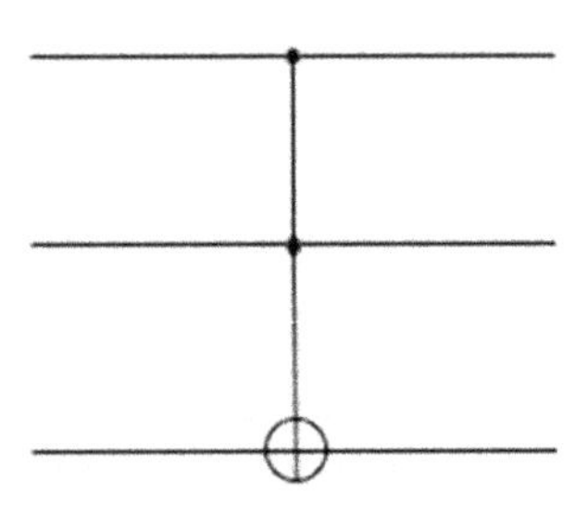

G.4 OUTLINE OF ALGORITHMS

What are some of the known algorithms that quantum computers are expected to excel at?*

- **Quantum Fourier transform**. We have already encountered the quantum version of the traditional Fourier transform in Appendix B, Equations B.9 and B.10. It is a unitary transformation on qubits, and so can be considered a logic gate for quantum information. For a set of 2^n qubits, the number of logic gates necessary is of order n^2. This beats the traditional *Fast Fourier Transform* (FFT) that requires $n2^n$ classical logic gates. The efficient implementation of the quantum Fourier transform is a key ingredient in many quantum algorithms, including the following.

- **Factorization of numbers**. Calculating the prime factors of a very large odd composite number N is considered essentially impossible for a conventional computer, but turns out to be considerably less computationally intensive for a quantum one. A quantum computer uses $\sim (\log N)^2(\log\log N)(\log\log\log N)$ (base 2 logarithms) logic gates, whereas a conventional computer, using a so-called *number field sieve,* uses $\sim \exp(c(\log N)^{1/3}(\log\log N)^{2/3})$ gates, where $c \approx 1.9$. Such factorizations, which are easy to assign but hard to decode, are used every day for commercial encryption purposes, which means that quantum computers could decipher such communications. This is obviously an impetus for the development of study of such machines! The quantum factorization code was developed by Peter Shor in 1994.†

- **Quantum search algorithms**. The problem of doing searches through data is intrinsically more efficient on quantum computers than classical ones. It can be shown that a search through N items can be done with $\sqrt{N}$ logic steps as opposed to the classical

* See the Wiki "Quantum Algorithms" (https://en.wikipedia.org/wiki/Quantum_algorithm) for a more detailed list.

† P. Shor, "Algorithms for Quantum Computation: Discrete Logarithms and Factoring," *Proceedings of the 35th Annual Symposium on Foundations of Computer Science,* IEEE Press (1994).

number of steps N. This algorithm can also be used to speed up cryptographic searches. The algorithm was developed by Lov Grover in 1996.*

- **Quantum simulations.** Is it possible to do an actual quantum simulation† of quantum systems that arise in atomic, nuclear, or particle physics? Classical simulations of quantum systems use the Monte Carlo algorithm described in Appendix E. This is a way of summing over a huge set of possible states. Is it possible to simulate such theories directly on a set of spatially arrayed qubits? This possibility was speculated upon by Feynman in a visionary talk in 1981.‡ His point was that to simulate the responses of quantum systems, the best thing to do is to make them quantum mechanical also. There is an impressive amount of work in this direction already.§

G.5 COMPUTER REALIZATIONS

How does one build a quantum computer? One needs a physical realization of the qubit, an ability to maintain its quantum state, as well as an ability to control their interactions. In addition, a quantum computer has to be completely isolated from its environment if it is to maintain its quantum state. One also needs a way to control input and output. Such machines will in fact be hybrid devices with classical interfaces controlling and interacting with the quantum computing device. Many companies and organizations are working on such devices, and many different ideas are being tested. The race is on to build these machines and demonstrate their usefulness.

It is likely that anything one writes concerning the operational state of these devices will be found to be obsolete within a few years. However, there will always be a need to summarize existing technologies for the incoming student. As of the writing of this appendix (2019), the leading edge of the maturing technologies seems to be based upon the following types of devices.

- **Superconducting Josephson circuits.** Superconducting circuits called Josephson circuits or junctions may be put into a coherent state between various isolated low-lying charge, flux, or phase states. Such devices can be conveniently designed and implemented on integrated chips.

- **Trapped ions.** Individual charged atoms or ions can be trapped using an oscillating electric field in a so-called Paul trap. Tuning with lasers can also place the atoms in a superposition of two of their standard atomic states.

- **Quantum dots.** Small spheres can be fabricated on superconducting silicon chips on which an electron is placed, whose operation is controlled by microwaves. The electronic spin state or electron position state in a quantum potential well can form the qubits.

- **Topological quasiparticles.** Two-dimensional quasiparticle combinations called *anyons* may be formed from electrons on braided integrated circuits. The braiding forms topological structures with conserved quantum numbers.

---

* L. Grover, "A Fast Quantum Mechanical Algorithm for Database Search", *Proceedings of the 28th Annual ACM Symposium on the Theory of Computing*, ACM Press, 212 (1996). L. Grover, "Quantum Mechanics Helps in Searching for a Needle in a Haystack," *Phys. Rev. Lett.* **79**, 325 (1997).

† T. Johnson, S. Clark, and D. Jaksch, "What Is a Quantum Simulator?" *EPJ Quantum Technology* **1**, 10 (2014).

‡ R. Feynman, "Simulating Physics with Computers," *Int. J. Theor. Phys.* **21**, 467 (1982). Such devices were also considered in Yu. Manin, "Computable and Noncomputable" (in Russian), *Sov. Radio.* **13** (1980).

§ See, for example, the articles in A. Trabesinger, ed., "Quantum Simulation," *Nat. Phys.* **8** (2012). A recent imposing simulation is described in J. Zhang et al., "Observation of a Many-Body Dynamical Phase Transition with a 53-Qubit Quantum Simulator," *Nature* **551**, 601 (2017).

This is just a mere sampling of the technologies that are being tested.* The list given now for the leading candidates may be completely different in just a few years!

A major problem with such machines is the tendency for the quantum state created to lose quantum coherence, or "decohere." It turns out the system decoherence time decreases as the number of qubits increases. The decoherence time for n qubits is roughly $1/n$ times shorter than the decoherence time for an individual qubit.† Maintaining a quantum state in the macroscopic world is hard to do! The decoherence time for Josephson circuits is the shortest of the technologies listed, on the order of microseconds, whereas the decoherence time for trapped ions is the greatest, on the order of seconds. Decoherence times for quantum dots falls somewhere in between, and it is thought that the quasiparticle approach could provide a very stable system. Besides the coherence issue, one must also consider the timescale necessary for individual gate processes to occur, which vary greatly over the various technologies. The time necessary for the algorithm to complete is the gate process speed times the number of gates, and this must be much smaller than the system decoherence time for a workable machine.

Part of the cure for the decoherence issue are quantum error-correction and fault tolerance codes. A major theoretical discovery in the 1990s was that quantum computers could reliably compute despite the imperfections and faults introduced by quantum noise. The theory of such quantum error protection is remarkably well developed. The basic idea is to fight entanglement with entanglement and redundancy. Amazingly, it turns out that reliable computation can be performed with faulty logic gates, as long as the probability of error on gates is below a certain threshold. However, the reliability encoding thus introduced means many more qubits must be introduced to control the "working" ones. In order to factor a 300 digit number ($\approx 2^{10^3}$) using Shor's algorithm, it is optimistically estimated‡ to require about 10^4 qubits, but with error correction and encoding, it could take $10^3 \times 10^4 = 10^7$ qubits to actually work. Thus, unless qubits turn out to be easily protectable (as is possible with the quasiparticle approach) or as easy to form as transistors on an integrated circuit, this could be a problem!

G.6 REFERENCES

A short list of some popular and recent introductory textbooks on quantum computing follows.

P. Kaye, R. Laflamme, and M. Mosca, *An Introduction to Quantum Computing* (Oxford University Press, 2007).

N. Mermin, *Quantum Computer Science: An Introduction* (Cambridge University Press, 2007).

M. Nielson and I. Chuang, *Quantum Computation and Quantum Information* (Cambridge University Press, 2010).

E. Rieffel and W. Polak, *Quantum Computing: A Gentle Introduction* (MIT Press, 2011).

N. Yanofsky and M. Mannucci, *Quantum Computing for Computer Scientists* (Cambridge University Press, 2008).

PROBLEMS

G.1 The qubit state

$$|\psi\rangle = \frac{1}{\sqrt{2}}\left(|0\rangle + |1\rangle\right)$$

* See the Wiki "Quantum computing" (https://en.wikipedia.org/wiki/Quantum_computing).

† M. Dyakonov, "Is Fault-Tolerant Quantum Computation Really Possible?" *Future Trends in Microelectronics Workshop*, 4–18 (Wiley, 2007).

‡ Dyakonov, "Fault-Tolerant Quantum Computation."

is represented on the Bloch sphere at the point (1,0,0) in (x, y, z) coordinate space, because comparing to Equation G.4 one has $\theta = \pi/2$ and $\phi = 0$. Given this initial state, find the action of the single-qubit operators H, Z, S, and T on $|\psi\rangle$ and the associated final position on the Bloch sphere in (x, y, z) space.

G.2 Show the single-qubit identity

$$HZH = X.$$

G.3 Show that the CNOT operation is equivalent to the controlled *Pauli X* operation, as discussed and illustrated in Section G.3.

G.4 Show the identity

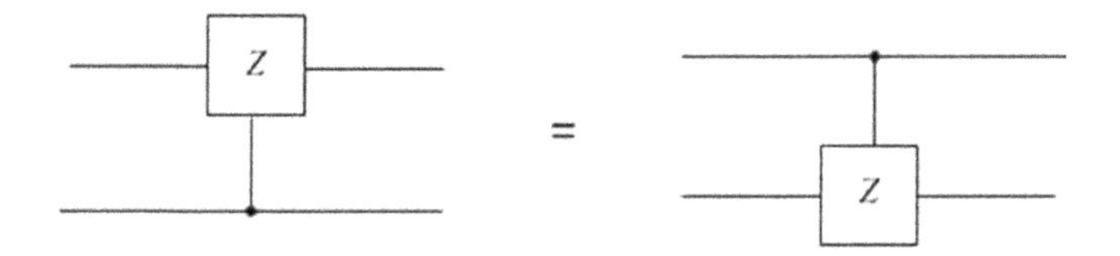

That is, it is possible to interchange control and target qubits for this operation?

G.5 Show the symbolic identity

Notice the CNOT operation has switched sides.

G.6 Prove directly that the quantum Fourier transform defined in Appendix B (Equations B.9 and B.10) is unitary and therefore can be considered a qubit gate operation.

Index

For Product Safety Concerns and Information please contact our EU
representative GPSR@taylorandfrancis.com
Taylor & Francis Verlag GmbH, Kaufingerstraße 24, 80331 München, Germany

www.ingramcontent.com/pod-product-compliance
Ingram Content Group UK Ltd.
Pitfield, Milton Keynes, MK11 3LW, UK
UKHW051834180425
457613UK00022B/1254